Lecture Notes in Artificial Intelligence   4819

Edited by J. G. Carbonell and J. Siekmann

Subseries of Lecture Notes in Computer Science

Takashi Washio   Zhi-Hua Zhou
Joshua Zhexue Huang   Xiaohua Hu
Jinyan Li   Chao Xie   Jieyue He
Deqing Zou   Kuan-Ching Li
Mário M. Freire (Eds.)

Emerging Technologies in Knowledge Discovery and Data Mining

PAKDD 2007 International Workshops
Nanjing, China, May 22, 2007
Revised Selected Papers

 Springer

Series Editors

Jaime G. Carbonell, Carnegie Mellon University, Pittsburgh, PA, USA
Jörg Siekmann, University of Saarland, Saarbrücken, Germany

Volume Editors

Takashi Washio, Osaka University, Japan (washio@ar.sanken.osaka-u.ac.jp)

Zhi-Hua Zhou, Nanjing University, China (zhouzh@nju.edu.cn)

Joshua Zhexue Huang, The University of Hong Kong, China (jhuang@eti.hku.hk)

Xiaohua Hu, Drexel University, USA (thu@cis.drexel.edu)

Jinyan Li, Nanyang Technological University, Singapore (jyli@ntu.edu.sg)

Chao Xie, Georgia State University, USA (cxie@cs.gsu.edu)

Jieyue He, Southeast University, China (jieyuehe@seu.edu.cn)

Deqing Zou, Huazhong University of Science and Technology, China
(deqingzou@hust.edu.cn)

Kuan-Ching Li, Providence University, Taiwan (kuancli@pu.edu.tw)

Mário M. Freire, University of Beira Interior, Portugal (mario@di.ubi.pt)

Library of Congress Control Number: 2007941074

CR Subject Classification (1998): I.2, H.2.7-8, H.3, H.5.1, G.3, J.3

LNCS Sublibrary: SL 7 – Artificial Intelligence

ISSN   0302-9743
ISBN-10   3-540-77016-X Springer Berlin Heidelberg New York
ISBN-13   978-3-540-77016-9 Springer Berlin Heidelberg New York

Springer is a part of Springer Science+Business Media

springer.com

© Springer-Verlag Berlin Heidelberg 2007
Printed in Germany

Typesetting: Camera-ready by author, data conversion by Scientific Publishing Services, Chennai, India
Printed on acid-free paper   SPIN: 12197576   06/3180   5 4 3 2 1 0

Preface

The techniques of knowledge discovery and data mining (KDD) have rapidly developed along the significant progress of the computer and its network technologies in the last two decades. The attention and the number of researchers in this domain continue to grow in both international academia and industry. The Pacific-Asia Conference on Knowledge Discovery and Data Mining (PAKDD) is a worldwide representative international conference on the research areas of KDD. Under the current spread of KDD techniques in our society, PAKDD 2007 invited the organizers to an industrial track on KDD techniques and, moreover, called for enterprising proposals of workshops on novel and emerging topics in KDD studies. The PAKDD industrial track and the PAKDD workshops both attracted more than 350 paper submissions from 22 countries in total, of which only 81 high-quality papers were accepted after strict reviews. The conference provided many informal and vibrant opportunities for researchers and industry practitioners to share their research positions, original research results and practical development experiences on specific new challenges, emerging issues, new technology trends and solutions to real-world problems.

The objective of this volume is to offer the excellent presentations to the public, and to promote the study exchange among researchers worldwide. Sixty-two outstanding papers in the industrial track and the workshops were further selected, among the 81 paper presentations, under even more rigorous reviews by the organizers, and these papers were carefully brushed up and included in this post-proceedings volume.

The first part of this volume contains ten outstanding papers presented in the industrial track. This track was organized to attract papers on new technology trends and real-world solutions in different industry sectors. The succeeding chapters include papers selected from the three workshops of BioDM 2007, HPDMA 2007 and SSDU 2007. The 2nd BioDM Workshop on Data Mining for Biomedical Applications (BioDM 2007) aimed at attracting top researchers, practitioners and students from around the world to discuss data mining applications in the field of bioinformatics. There are many computationally challenging problems in the analysis of the diverse and voluminous data provided in the field, and the readers will find 11 challenging papers on these problems. The 2007 International Workshop on High-Performance Data Mining and Applications (HPMDA 2007) addressed high-performance data mining methods and applications from both algorithmic and system perspectives. It includes 25 papers on how to achieve efficient mining of useful information from the data available, as well as on the topics of parallel hardware platforms, clusters and large-scale distributed computing infrastructures. The 2007 International Workshop on Service, Security and Its Data Management for Ubiquitous Computing (SSDU 2007) was held to foster research in the areas of security and intelligence integration into ubiquitous computing and data management technology. Sixteen papers discussing the many security risks and problems in the ubiquitous computing environment are included in the volume.

We hope this book contributes to the growth of the worldwide community of KDD researchers.

September 2007   Takashi Washio
                                                                 Zhi-Hua Zhou

Organization and Editorial Board

The paper selection of the industrial track and the workshops was made by the Program Committee of each respective organization. After the paper selection, the book was edited and managed by the volume editors.

Volume Editors

Organization and Editorial Board

Table of Contents

PAKDD Industrial Track and Workshops 2007

Industrial Track

Workshop of BioDM 2007

Workshop of HPDMA 2007

Workshop of SSDU 2007

PAKDD 2007 Industrial Track Workshop

Joshua Zhexue Huang[1] and Yunming Ye[2]

[1] E-Business Technology Institute
The University of Hong Kong, Hong Kong, China
`jhuang@eti.hku.hk`
[2] E-Business & Intelligent Enterprise Computing Research Center
Shenzhen Graduate School of Harbin Institute of Technology, Shenzhen 518055, China
`yeyunming@hit.edu.cn`

1 Workshop Overview

The PAKDD 2007 Industrial Track Workshop was held on the 22nd of May 2007 in conjunction with The 11th Pacific-Asia Conference on Knowledge Discovery and Data Mining (PAKDD 2007) at the Mandarin Garden Hotel in Nanjing, China. Following the PAKDD conference tradition to promote industry applications of new data mining techniques, methodologies and systems, the theme of this workshop was defined as "Data Mining for Practitioners". Industrial oriented papers on innovative applications of data mining technology to solve real world problems were solicited with emphasis on the following areas:

1. New data mining methodologies
2. Data mining application systems
3. Data cleansing and transformation in data mining
4. New application problems and data mining solutions
5. Data mining in government applications
6. Innovative data mining case studies

The workshop received 82 submissions from 9 countries and regions in Asia, Europe and North America. The topics of these submissions covered new techniques, algorithms, systems, traditional applications, such as banking and finance, and new applications in manufacturing, facility management, environment and government services. After a rigorous review by the committee members, 13 papers were selected for presentation at the workshop. The acceptance rate was about 16%. For the industrial track papers in PAKDD series conferences, this low acceptance rate was unprecedented.

After reevaluation of the 13 presentations, we selected 10 papers for publication in this proceedings. The first paper by (Xiaojun Chen et al.) gives a survey and evaluation on the current open source data mining systems. The rest 9 papers present real world data mining applications in five areas: (1) Financial Market by (Calum Robertson et al. and Hia-Jong Teoh et al.), (2) new Web Applications by (Flora Tsai et al. and Cemal Köse et al.), (3) Social Services by (Longbin Cao et al. and Huidong Jin), (4) Construction and Power Industry by (Shihai Zhang et al. and Ming Zhou et al.), and (5) Environment by

T. Washio et al. (Eds.): PAKDD 2007 Workshops, LNAI 4819, pp. 1–2, 2007.

(Ke Zhang et al.). These applications demonstrate that innovative use of data mining technology has a lot of potentials in solving various real world problems.

As Co-chairs of this workshop committee, we wish to place on record our sincere thanks to many people who have contributed to the Industrial Track Workshop. Firstly, we wish to thank all the authors of the 82 submissions. Without them, the workshop would not be able to take place. Secondly, we must express our appreciations to the committee members who did an excellent job in completing the quality paper review work in a very short time. Finally, we would like to thank the PAKDD 2007 organizing committee and local organizers for their assistance in organizing this workshop.

2   Program Committee

Workshop Chairs

Workshop Chair	Joshua Zhexue Huang (The University of Hong Kong, China)
Workshop Co-chair	Yunming Ye (Harbin Institute of Technology, China)

Program Committee

Graham Williams	ATO, Australia
Ke Wang	Simon Fraser University, Canada
William Song	University of Durham, UK
Xuelong Li	University of London, UK
Warren Jin	CSIRO, Australia
Simeon J. Simoff	University of Technology, Sydney, Australia
Tao Lin	SAP Research, USA
Xiongfei Li	Jilin University, China
Shuicheng Yan	University of Illinois at Urbana-Champaign, USA
Dacheng Tao	University of London, UK
Wen Jun Yin	IBM China
Yuan Yuan	Aston University, UK
Dong Xu	Columbia University, USA
Yoon-Joon Lee	KAIST, Korea
Longbing Cao	University of Technology in Sydney, Australia
Liping Jing	University of Hong Kong, China
Jun Xu	University of Hong Kong, China
Chunkai Zhang	HIT, China
Shijie Zhou	UESTC, China
Feilong Tang	SJTU, China
Hongbing Wang	SEU, China
Xiaoyuan Jin	NJUPT, China

A Survey of Open Source Data Mining Systems*

Xiaojun Chen[1], Yunming Ye[1,**], Graham Williams[2], and Xiaofei Xu[1]

[1] Shenzhen Graduate School, Harbin Institute of Technology,
Shenzhen 518055, China
yeyunming@hit.edu.cn, xjchen.hitsz@gmail.com
[2] Australian Taxation Office, Australia
graham.williams@togaware.com

Abstract. Open source data mining software represents a new trend in data mining research, education and industrial applications, especially in small and medium enterprises (SMEs). With open source software an enterprise can easily initiate a data mining project using the most current technology. Often the software is available at no cost, allowing the enterprise to instead focus on ensuring their staff can freely learn the data mining techniques and methods. Open source ensures that staff can understand exactly how the algorithms work by examining the source codes, if they so desire, and can also fine tune the algorithms to suit the specific purposes of the enterprise. However, diversity, instability, scalability and poor documentation can be major concerns in using open source data mining systems. In this paper, we survey open source data mining systems currently available on the Internet. We compare 12 open source systems against several aspects such as general characteristics, data source accessibility, data mining functionality, and usability. We discuss advantages and disadvantages of these open source data mining systems.

Keywords: Open source software, data mining, FLOSS.

1 Introduction

Open source software has a solid foundation with, for example, the GNU project [1] dating from 1984. Open source software (also referred to as free, librè, open source software, or FLOSS) is well known through the GNU software suite upon which GNU/Linux is based, but also through the widely used MySQL, Apache, JBoss, and Eclipse software, just to highlight a few. Open source host site sourceforge.net, for example, lists over 100,000 open source projects.

Open source for business intelligence (BI) has also been gathering momentum in recent years. The open source database, MySQL [2], is widely used in building data warehouses and data marts to support BI applications. The open

---

* This paper was supported by the National Natural Science Foundation of China (NSFC) under grants No.60603066.
** Corresponding author.

T. Washio et al. (Eds.): PAKDD 2007 Workshops, LNAI 4819, pp. 3–14, 2007.

source data mining platform, Weka [3], has been a popular platform for sharing algorithms amongst researchers. In the past 6 months, for example, there have been between 16,353 and 29,950 downloads per month[1].

Open source data mining is particularly important and effective for small and medium enterprises (SMEs) wishing to adopt business intelligence solutions for marketing, customer service, e-business, and risk management. Due to the high cost of commercial software and the uncertainty associated with bringing data mining into an enterprise, many SMEs look to adopt a low cost approach to experimenting data mining solutions and in gaining data mining expertise. With open source software an enterprise can easily initiate a data mining project using the most current technology. Often the software is available at no cost, allowing the enterprise to instead focus on ensuring their staff can freely learn the data mining techniques and methods. Open source ensures that staff can understand exactly how the algorithms work by examining the source code, if they so desire, and can also fine tune the algorithms to suit the specific purposes of the enterprise. Therefore, SMEs no longer need be trailing the beneficial data mining technology.

However, issues such as diversity, stability, scalability, usability, documentation and support hinder the wide adoption of open source data mining in business. Such issues present challenges and incur costs for open source data mining projects, whereby effort must be expended by the business user on dealing with these deficiencies rather than solving business problems.

In this paper, we present a survey of 12 popular open source data mining systems available on the Internet. We evaluate the comparative characteristics of these open source systems, their data access functionality, their data mining functionality, and their usability, including user interfaces and extensibility. Based on the analysis, We discuss their advantages and disadvantages.

The rest of this paper is organized as follows. Section 2 provides a brief introduction to data mining and data mining systems. Section 3 surveys and evaluates 12 commonly used open source data mining systems, discussing their advantages and disadvantages.

2   Data Mining and Data Mining Systems

Data mining refers to the process of extracting new and useful knowledge from large amounts of data [4]. Data mining is widely used to solve many business problems, such as customer profiling [5], customer behavior modeling [6], credit scoring, product recommendation, direct marketing [7], cross selling, fraud detection [8,9]. Data mining is adopted in many industries, e.g., retail [10], bank, finance [11], and so on [12,13].

The solution to a data mining problem is carried out in the data mining process, which varies depending on the application domain. In general, a data mining process consists of the following seven steps [14]:

---

[1] http://sourceforge.net/project/stats/?group_id=5091&ugn=weka&type=&mode =year

1. Identify the business problems.
2. Identify and study data sources, and select data.
3. Extract and preprocess data.
4. Mine the data, e.g., discover association rules or build predictive models.
5. Verify the mining results.
6. Deploy models in the business process.
7. Measure the return on investment (ROI).

Each data mining process is composed of a sequence of data mining operations, each implementing a data mining function or algorithm. We can categorize data mining operations into the following groups:

1. **Data Understanding Operation:** Access data from various sources and explore the data to become familiar with it and to "discover" the first insights.
2. **Data Preprocessing Operation:** Generally involves data filtering, cleaning, and transformation, to construct the final dataset for the modeling operations.
3. **Data Modeling Operation:** Implements the data mining algorithms, such as k-means clustering. These operations are used to build data mining models. The common modeling operations include classification, prediction, clustering, association rule, and interactive exploration such as link analysis.
4. **Evaluation Operation:** Used to compare and select data mining models by choosing the best one. Common operations include confusion matrix, lift chart, gain chart, cluster validation and visualization.
5. **Deployment Operation:** Involves deploying a data mining model to make decisions, such as using a predictive model to predict potential customer churn or a campaign model to score customers for a target campaign.

A *data mining system* is a software system that integrates many operations, and provides an easy-to-use (and often graphical) user interface to effectively perform the data mining process. Goebel and Gruenwald [15] previously investigated 43 open source and closed source data mining systems and presented several important features for data mining to accommodate its users effectively. We use some of the same characteristics for our comparison, but also limit our study to just open source software. Also, their study was conducted over 7 years ago, and most of the open sources systems considered here did not exist then.

In the next section, we survey the commonly used open source data mining systems and evaluate their characteristics.

3   Survey of Open Source Data Mining Systems

In this section we survey 12 commonly used open source data mining systems, evaluate them, and discuss their advantages and disadvantages.

3.1 Open Source Systems

Open source software provides users with the freedom to run, copy, distribute, study, change and improve the software (see [16] for a detailed definition). To adopt open source software (or in fact any software) we must understand its license and the limitations that it places on us. Unlike closed source licenses which aim to limit your rights, open source software aims to give you the right to do whatever you please with the software. The common open sources licenses include the GPL, LGPL, BSD, NPL, and MPL. Bruce has discussed each license with suggestions for choosing a proper license [17].

In the past decades, open source products such as GNU/Linux, Apache, BSD, MySQL, and OpenOffice have achieved great success, clearly demonstrating that open source software can be as robust, or even more robust, than commercial and closed source software. Using open source software generally also means saving on the software costs, and allowing an enterprise to instead invest in skilling its people. Wang and Wang [18] have suggested many criteria for adopting open source software.

In the next subsection we discuss some important features to categorize open source data mining systems.

3.2 Important Features of Open Source Data Mining Systems

Open source systems are diverse in design and implementation. Although developed for data mining, they are very different in many aspects. To understand the characteristics of these diverse open source data mining systems and to evaluate them, we look into the following important features:

- **Ability to access various data sources.** Data come from databases, data warehouses, and flat files in different formats. A good system will easily access different data sources.
- **Data preprocessing capability.** Preprocessing occupies a large proportion of the time in a data mining process [19]. Data preparation is often the key to solving the problem. A good system should provide various data preprocessing functions tasks easily and efficiently.
- **Integration of different techniques.** There is no single best technique suitable for all data mining problems. A good data mining system will integrate different techniques (preprocessing functions and modelling algorithms), providing easy access to a wide range of different techniques for different problems.
- **Ability to operate on large datasets.** Commercial data mining system, such as SAS Enterprise, can operate on very large datasets. This is also very important for open source data mining systems, so scalability is a key characteristic.
- **Good data and model visualization.** Experts and novices alike need to investigate the data and understand the models created.
- **Extensibility.** With new techniques and algorithms it is very important for open source data mining systems to provide an architecture that allows

incorporation of new methods with little effort. Good extensibility means easy integration of new methods.

– **Interoperability with other systems.** Open standards means that systems (whether open or closed source) can interoperate. Interoperability includes data and model exchange. A good system will provide support of canonical standards, such as CWM [20] and PMML [21].

– **Active development community.** An active development community will make sure that the system is maintained and updated regularly.

These features categorize open source data mining systems. In the following, we investigate commonly used open source data mining systems with respect to these features.

3.3 Survey of Open Source Data Mining Systems

In this work we investigated 12 open source data mining systems accessible from the Internet. We studied these systems from four aspects: general characteristics, data source accessibility, data mining functionality, and usability. The survey results are summarized in the following four tables.

General Characteristics

We consider some general system features, including Activity, License, Programming Language and Operating Systems. The results are listed in Table 1.

Table 1. General characteristics of the 12 systems

Product	Activity	License	Language	Liunx	Mac	Windows
ADAM [22]	Medium	Unknown	Python	x	x	x
AlphaMiner [23]	High	GPL	Java	x	x	x
Databionic ESOM [24]	High	GPL	Java	x	x	x
Gnome Data Miner [25]	Low	GPL	C++ Python	x	x	x
KNIME [26]	High	Other	Java	x	x	x
Mining Mart [27]	High	Unknown	Java	x	x	x
MLC++[28]	Low	Other	C++	x	.	x
Orange [29]	High	GPL	C++ Python	x	x	x
Rattle [30]	High	GPL	R	x	x	x
TANAGRA [31]	High	Other	C++	.	.	x
Weka [3]	High	GPL	Java	x	x	x
YALE [32]	High	GPL	Java	x	x	x

The activity is measured by the frequency of updates and time of latest update. In terms of the listed operating systems, it is noteworthy that open source data mining systems that run on GNU/Linux will also run on Unix in general, and specific other Unix variances such as Solaris and HPUX.

Data Source Aspect
In real world applications data comes from different sources in different formats. The ability to access different data formats is important in selecting an open source system. In Table 2 we list 10 commonly encountered data sources and identify which systems are able to access which data sources. An indication of the data size that the system is know to be able to deal with is also provided.

Table 2. Data source characteristics of 12 systems

Product	Oracle	Sybase	SQLServ	MySQL	Access	ODBC	JDBC	ARFF	CSV	Excel	Data Size
ADAM [22]	.	.	.	.	.	.	.	x	.	.	Large
AlphaMiner [23]	.	.	.	.	x	x	.	x	x	x	Medium
Databionic ESOM [24]	.	.	.	.	.	.	.	.	.	.	Medium
Gnome Data Miner [25]	.	.	.	.	.	.	.	.	x	.	Medium
KNIME [26]	.	.	.	x	x	x	x	x	x	.	Medium
Mining Mart [27]	x	.	.	.	x	.	.	.	.	.	Unknown
MLC++ [28]	.	.	.	.	.	.	.	.	.	.	Large
Orange [29]	.	.	.	x	.	.	.	.	.	.	Medium
Rattle [30]	.	.	.	x	x	x	.	.	x	x	Large
TANAGRA [31]	.	.	.	.	.	.	.	x	.	x	Medium
Weka [3]	.	.	.	.	.	.	x	x	x	.	Medium
YALE [32]	x	x	x	x	.	.	x	x	x	x	Medium

Functionality Aspect
To be able to solve different data mining problems, the functionality of an open source data mining system is an important feature. Table 3 lists the summary of functionality of the 12 systems. The functions can be divided into six groups: data preprocessing, classification and prediction, clustering, association rules, evaluation, and visualization. For data preprocessing, we score each according to their capability, with 3 as the highest score, and 0 meaning not supported. We include the basic data mining tools that are offered by the commercial closed source systems, decision trees, neural networks, kmeans clustering, and

association rules, but also the more modern techniques of support vector machines (SVM), boosting, and random forests. It is clear that the current open source data mining systems already include the commonly used data mining functions. The difference lies primarily in visualization capabilities.

Table 3. Functionality of the 12 systems

Product	Preprocess	Bayes	DTree	NNet	SVM	Boosting	Forests	KMeans	Associations	Evaluation	Data Vis	Model Vis
ADAM [22]	3	x	x	x	.	.	.	x	x	x	3	3
AlphaMiner [23]	3	x	x	x	x	x	x	x	x	x	3	3
Databionic ESOM [24]	1	.	.	.	.	.	.	.	.	x	3	3
Gnome Data Miner [25]	0	x	x	.	.	.	.	.	.	x	1	1
KNIME [26]	3	x	x	x	x	x	x	x	x	x	3	3
Mining Mart [27]	3	.	.	.	.	.	.	.	.	x	3	3
MLC++ [28]	3	x	x	x	.	.	.	.	.	x	3	3
Orange [29]	3	x	x	.	x	x	x	.	x	x	3	3
Rattle [30]	2	x	x	x	x	x	x	x	x	x	3	3
TANAGRA [31]	3	x	x	x	x	x	.	x	x	x	1	1
Weka [3]	3	x	x	x	x	x	x	x	x	x	3	3
YALE [32]	3	x	x	x	x	x	x	x	x	x	3	3

Usability Aspect

The usability aspect describes how easy an open source data mining system can be used in solving real world business problems in different data and system environments. Here, we consider human interaction, interoperability and extensibility.

Human Interaction indicates how much interaction is required with the discovery process. Autonomous indicates that the system requires no tuning, you simply load the data and it builds the model. Guided indicates that the system provides significant assistance with the process, and Manual indicates very little guidance in the data mining process.

Interoperability essentially indicates whether PMML (the Predictive Modelling Markup Language) is supported, or whether the system is only internally "interoperable".

Clearly, there are many differences in usability in the current open source data mining systems.

Table 4. Usability aspect of 12 systems

Product	Human Interaction	Interoperability	Extensibility
ADAM [22]	Autonomous	Self	Simple
AlphaMiner [23]	Manual	PMML	Excellent
Databionic ESOM [24]	Manual	Self	None
Gnome Data Miner [25]	Guided	Self	Simple
KNIME [26]	Manual	PMML	Excellent
Mining Mart [27]	Manual	Self	Simple
MLC++ [28]	Guided	Self	Simple
Orange [29]	Manual	Self	Excellent
Rattle [30]	Guided	PMML	Simple
TANAGRA [31]	Manual	Self	Simple
Weka [3]	Manual	Self	Excellent
YALE [32]	Manual	Self	Excellent

3.4   Evaluation

From the surveyed characteristics of current open source data mining systems, we are able to evaluate these systems. We have selected 14 characteristics from the 4 aspects and assigned a weight to each characteristic as given in Table 5. We remark that the weight was assigned subjectively according to our own belief and experience with data mining in practise, and specifically from a business point of view.

According to the weight value in each characteristic and the surveyed results of the 12 systems, we calculated a score for each system in each aspect. Figures 1(a), 1(b), 1(c) and 1(d) plot the scores of the 12 systems, and the overall scores of the 12 systems are shown in Figure 1(e).

It must be noted here that we lack some of the background information for Mining Mart and MLC++ which has an impact on figures 1(b) and 1(e).

From the overall scores in figure 1(e), we can see that KNIME, AlphaMiner, Weka, Rattle and YALE are good open source data mining systems. A Google search also confirms that Weka, YALE, AlphaMiner and Rattle are popular (primarily because of their longer history). KNIME is a new entry focused on bioinformatics and has good performance.

Based on the above analysis, we summarize the advantages of open source data mining systems as follows:

1. **Supporting multi-platform.** GNU/Linux, Mac, and MS/Windows are supported by almost all systems.
2. **Good human interaction.** Almost all these systems support some degree of human guided discovery process, and more than half of them offer work-flow style processes. AlphaMiner, KNIME, Mining Mart, Orange, Weka, and YALE provide drag-and-drop style case constructing.

Table 5. The weighted distributions of characteristics for evaluation

Perspective	Characteristic	Weight(%)
General	Activity	9
	Operating Systems	6
Data Source	Data sources	7
	Size of Data	12
Functionality	Preprocessing	12
	Classification/Prediction	9
	Clustering	3
	Associations	3
	Evaluation	4
	Data Visuals	3
	Model Visuals	3
Usability	Human Interaction	9
	Interoperability	10
	Extensibility	10

3. **Plentiful algorithms.** Each system has integrated plentiful algorithms for data preprocessing and modeling. Some contain more than 100 algorithms. However, this often actually works against the usefulness of the tool, as it confuses the business user.
4. **Reuse of cases.** All of them can reuse the cases. AlphaMiner, KNIME, and Rattle can export models in PMML format, which means that they can share models with other PMML compliant applications.
5. **Simple extensibility.** Most systems can integrate new operations in the form of plug-ins, some of which can be called in scripts.

There are also some serious disadvantages in these systems:

1. **Lack of support for diverse data resources.** YALE, KNIME, AlphaMiner, and Rattle support most file formats and relational databases through ODBC and JDBC . Some systems only focus on a smaller number of file formats.
2. **Difficulty with large volumes of data.** Real world applications often deliver very large datasets, with millions of rows. Most open source data mining systems are great for demonstrating and comparing the range of algorithms on smaller datasets but there has not been a focus on dealing with very large datasets. This limitation hinders the use of open source data mining systems in real applications, especially in business.
3. **Poor documentations and services.** During the investigation, we found that most of these systems have poor documentations, of limited use for a novice user wanting to quickly master the systems. Also support services is a serious issue, relying on the good will of a community of users.

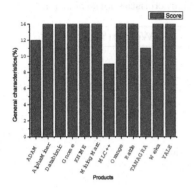

(a) General characteristics

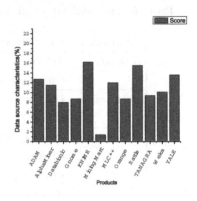

(b) Data source characteristics

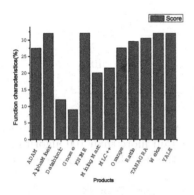

(c) Functionality characteristics

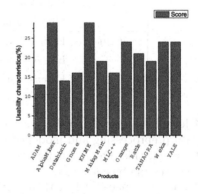

(d) Usability characteristics

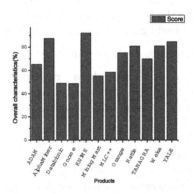

(e) Overall characteristics

Fig. 1. Score of 12 systems

4   Conclusions

In this paper we have presented important features for open source data mining systems, and use these criteria to investigate 12 of the most commonly used open source data mining systems. Based on the analysis, we find that most systems have excellent functionality, and offer powerful tools for users in education and researchers. Bur for commercial use, there is still quite some effort in making the tools accessible.

We present the four following important points for open source data mining systems to gain greater success in deployment:

1. **Supporting various data sources**
   A good open source data mining system should support most commonly used data sources, such as the open source and commercial databases, csv files, and user defined formats.
2. **Providing high performance data mining**
   Most open source data mining systems can't operate on large volumes of data. To offer high performance data mining we need to either rewrite the algorithms (e.g. parallel and distributed algorithms) or more simply to improve the hardware on which the software is running.
3. **Proving more domain-specific techniques**
   Most data mining systems integrate many algorithms at the whim of the researchers, rather than for the benefit of business. We need to better identify algorithms that match the data to be processed. One approach will be to provide domain-specific techniques based on a generic platform.
4. **Better support for business application**
   Real business application are complex, placing many demands on the data mining system. Open source data mining systems need to in improve scalability, reliability, recoverability, and security [33].

References

1. Free Software Foundation: The GNU project, Website (2007), http://www.gnu.org
2. DuBois, P.: MySQL. Sams (2005)
3. University of Waikato, New Zealand: Weka 3.4.9, Website (2006), http://www.cs.waikato.ac.nz/ml/Weka/index.html
4. Han, J., Kamber, M.: Data Mining: Concepts and Techniques. Morgan Kaufmann, San Francisco (2000)
5. Adomavicius, G., Tuzhilin, A.: Using data mining methods to build customer profiles. Computer (2001)
6. Bounsaythip, C., Rinta, E.: Overview of data mining for customer behavior modeling. Technical report, VTT Information Technology (2001)
7. Ling, C.X., Li, C.: Data mining for direct marketing: Problems and solutions. American Association for Artificial Intelligence (1998)
8. Rygielski, C., Wang, J.-C., Yen, D.C.: Data mining techniques for customer relationship management. Technology in Society 24, 483–502 (2002)

9. Apte, C., Liu, B., Pednault, E.P.D., Smyth, P.: Business applications of data mining. Communications of the ACM 45, 49–53 (2002)

10. Ahmed, S.R.: Applications of data mining in retail business. In: Proceedings of the International Conference on Information Technology: Coding and Computing (2004)

11. Kovalerchuk, B., Vityaev, E.: Data Mining in finance: Advances in Relational and Hybrid Methods. Kluwer Academic Publishers, Dordrecht (2000)

12. Han, J., Altman, R.B., Kumar, V., Mannila, H., Pregibon, D.: Emerging scientific applications in data mining. Communications of the ACM 45, 54–58 (2002)

13. Grossman, R., Kamath, C., Kegelmeyer, P., Kumar, V., Namburu, R.: Data Mining for Scientific and Engineering Applications. Kluwer Academic Publishers, Dordrecht (2001)

14. Huang, J.: Data mining overview. Technical report, E-Business Technology Institute (2006)

15. Goebel, M., Gruenwald, L.: A survey of data mining and knowledge discovery software tools. In: SIGKDD Explorations, vol. 1, pp. 20–33. ACM SIGKDD (1999)

16. Open Source Initiative: The open source definition, Website (2007), http://www.opensource.org/docs/definition_plain.html

17. Perens, B.: The open source definition, Website (2007), http://perens.com/Articles/OSD.html

18. Wang, H., Wang, C.: Open source software adoption: A status report. IEEE SOFTWARE (2001)

19. Pyle, D.: Data Preparation for Data Mining. Morgan Kaufman, San Francisco (1999)

20. Object Management Group: Common warehouse metamodel (cwm), Website (2007), http://www.omg.org/cwm/

21. Data Mining Group: Predictive model markup language (pmml) (2005)

22. Information Technology and Systems Center (ITSC) at the University of Alabama in Huntsville: Algorithm development and mining system, Website (2005), http://datamining.itsc.uah.edu/adam/

23. HIT-HKU BI Lab: Alphaminer 2.0 (2006) Website: http://bi.hitsz.edu.cn/AlphaMiner/

24. Data Bionics Research Group, University of Marburg: Databionic esom tools, Website (2006), http://databionic-esom.sourceforge.net/

25. Williams, G.J.: Gnome data mining tools, Website (2006), http://www.togaware.com/datamining/gdatamine/

26. Chair for Bioinformatics and Information Mining, University of Konstanz, Germany: Knime 1.2.0, Website (2007), http://www.knime.org/

27. MiningMartResearch Team: Mining mart 1.1, Website (2006), http://mmart.cs.uni-dortmund.de/

28. Stanford: Mlc++, Website (1997), http://www.sgi.com/tech/mlc/

29. Artificial Intelligence Laboratory, University of Ljubljana, Slovenia: Orange 0.9.64, Website (2007), http://www.ailab.si/orange/

30. Williams, G.J.: Rattle 2.1.116, Website (2006), http://Rattle.togaware.com/

31. Ricco RAKOTOMALALA, University Lyon, France: Tanagra 1.4.12, Website (2006), http://chiroube.univ-lyon2.fr/~ricco/tanagra/en/tanagra.html

32. Artificial Intelligence Unit, University of Dortmund, Germany: Yale 3.4, Website (2006), http://rapid-i.com/

33. Kleissner, C.: Data mining for the enterprise. In: Proceeding of the 31st Annual Hawaii International Conference on System Science, pp. 295–304 (1998)

Predicting the Short-Term Market Reaction to Asset Specific News: Is Time Against Us?

Calum Robertson[1], Shlomo Geva[1], and Rodney Wolff[2]

[1] Information Research Group, Queensland University of Technology
[2] School of Economics and Finance, Queensland University of Technology
2 George Street, Brisbane, Queensland Australia 4000
{cs.robertson,s.geva,r.wolff}@qut.edu.au

Abstract. The efficient market hypothesis states that investors immediately incorporate all available information into the price of an asset to accurately reflect its value at any given time. The sheer volume of information immediately available electronically makes it difficult for a single investor to keep abreast of all information for a single stock, let alone multiple. We aim to determine how quickly investors tend to react to asset specific news by analysing the accuracy of classifiers which take the content of news to predict the short-term market reaction. The faster the market reacts to news the more cost-effective it becomes to employ content analysis techniques to aid the decisions of traders. We find that the best results are achieved by allowing investors in the US 90 minutes to react to news. In the UK and Australia the best results are achieved by allowing investors 5 minutes to react to news.

Keywords: Document Classification, Stock Market, News, SVM, C4.5.

1 Introduction

Not so long ago most traders relied upon newspapers and magazines to supply the information they required to invest. However the last two decades has seen a rapid increase in the availability of both real-time prices and media coverage via electronic sources. It has become difficult for a single person to keep abreast of all available information for a single asset, and impossible for multiple [1].

There is a plethora of research which shows that specific asset markets react to news. This includes the reaction to newspaper, magazine, and other real-time sources such as websites [2-7]. There is also plenty of evidence that markets react to macroeconomic announcements (information released by governments to provide an indication of the state of the local economy) [8-14]. It has also been shown that investors take heed of analyst recommendations [15-17] to aid their investing decisions. Interestingly it has also been shown that the futures prices of oranges are influenced by the release of weather reports [18].

Ederington and Lee found that volatility on Foreign Exchange and Interest Rate Futures markets increases within one minute of a macroeconomic news announcement, and the effect lasts for about 15 minutes [8]. They later determined that the same markets begin to react to news within 10 seconds of macroeconomic news announcements, with weak

T. Washio et al. (Eds.): PAKDD 2007 Workshops, LNAI 4819, pp. 15–26, 2007.

evidence that they tend to overreact to news within the first 40 seconds after news, but settle within 3 minutes [9]. Graham et al. established that the value of stocks on the S&P 500 index is influenced by scheduled macroeconomic news, however, they didn't investigate any intraday effect [12]. Nofsinger and Prucyk concluded that unexpected bad macroeconomic news is responsible for most abnormal intraday volume trading on the S&P 100 Index option [14].

Despite strong evidence that the stock market does react to macroeconomic news, there is far more asset specific news than macroeconomic news. Furthermore, unlike macroeconomic news, most asset specific news isn't scheduled and therefore investors have not formed their own expectation, or adopted analysts' recommendations about the content of the news.

Wutherich et al. analysed the content of news available on a well known website and used it to predict what the given index would do on the next day [19]. They achieved a statistically significant level of accuracy on their forecasts, though it is somewhat easier to predict the direction of an index in a day than it is to predict the reaction of a single asset in less than a day.

Fung et al. examined the effect of all asset specific news on a limited number of stocks and found that they could make money based on the predictions of a system which processed the content of the news [20]. However they didn't report the classification accuracy of their system so it is difficult to determine how good their results are.

Mittermayer investigated the effect of Press Announcements on the New York Stock Exchange and the NASDAQ and determined that the content of news can be used to predict, with reasonable accuracy, when the market will experience high returns within an hour of the announcement [7]. Unfortunately press announcements are only a fraction of asset specific news, so further investigation is required to determine how the stock market reacts, if at all, to this type of news.

It is vitally important to examine how rapidly the market reacts to asset specific news, in order to capitalise on its content. The faster the market reacts to news the more cost-effective it becomes to employ content analysis techniques to aid the decisions of traders. Conversely if the market is slow to react to asset specific news then there is little point in utilising content analysis to highlight interesting articles.

In this paper we formulate a methodology to determine how quickly the stock market reacts to news. We apply this to stocks traded in the US, UK and Australian markets to ascertain how strongly these markets react, if at all, to asset specific news. Firstly we describe the data used in our tests, and the methodology we employed, before presenting our results and conclusions.

2   Data

All data for this research was obtained using the Bloomberg Professional® service. The dataset consists of stocks which were in the S&P 100, FTSE 100, and ASX 100 indices as at the 1st of July 2005 and continued to trade through to the 1st of November 2006, which is a total of 283 stocks. For each stock the Trading Data, and News were collected for the period beginning 1st of May 2005 through to and including the 31st of October 2006.

The set defined in Eq. (1) consists of each distinct minute where trading occurred for the stock (s), within all minutes for the period of data collection (T_A), and the average price for trades during that minute. However we are only interested in the business time scale (minutes which occurred during business hours for the market on which the stock trades). Furthermore we want a homogeneous time series (i.e. an entry for every business trading minute for the stock, regardless of whether any trading occurred). Therefore we produce the date and price time series for all minutes in the business time scale (T_B) with the definitions in Eqs. (2)-(3), where we define the price at time t as the price of the last actual trade for the stock prior to or at the given time. Note that if the stock was suspended from trading for a whole day then the day is excluded from T_B.

$$I_{(s)} = \{I_1, I_2, \ldots, I_m\} \mid I_{(s,z)} = (d_{(s,z)}, p_{(s,z)}) \land z \in T_A \tag{1}$$

$$D_{(s)} = \{D_1, D_2, \ldots, D_n\} \mid D_{(s,t)} > D_{(s,t-1)} \land D_{(s,t)} \in T_B \land T_B \subseteq T_A \tag{2}$$

$$P_{(s)} = \{P_1, P_2, \ldots, P_n\} \mid P_{(s,t)} = (p_{(s,z)} \mid z = \max(z \mid d_{(s,z)} \le D_{(s,t)})) \tag{3}$$

The news search facility within the Bloomberg Professional® service was used to download all relevant articles for each stock within the dataset. These articles include Press Announcements, Annual Reports, Analyst Recommendations and general news which Bloomberg has sourced from over 200 different news providers.

The set defined in Eq. (4) consists of each distinct news article for the stock, which occurred during business hours excluding the first and last Δt minutes of the day, and contains the time and content of the article. The first Δt minutes are excluded as the market behaves differently during this period and therefore this could skew the results of a classifier. The last Δt minutes are excluded as the market doesn't have time to react to the news within the period, and this could also skew the results of a classifier.

All documents are pre-processed to remove numbers, URL's, email addresses, meaningless symbols, and formatting. Each term in the content C of the document is stemmed using the Porter stemmer [21]. The stemmed term index defined in Eq. (5) is created with the stemmed terms which appear in the document, and the number of times they appear.

$$A_{(s)} = \{A_1, A_2, \ldots, A_p\} \mid A_{(s,\lambda)} = (d_{(s,\lambda)}, C_{(s,\lambda)}) \land d_{(s,\lambda)} \in T_B$$
$$\land \min(time(T_B)) + \Delta t \le time(d_{(s,\lambda)}) < \max(time(T_B)) - \Delta t \tag{4}$$

$$C_{(s,\lambda)} = \{T_1, T_2, \ldots, T_q\} \mid T_{(s,\lambda,\varpi)} = \{S_{(s,\lambda,\varpi)}, SC_{(s,\lambda,\varpi)}\} \land SC_{(s,\lambda,\varpi)} = \#\{\forall S_{(s,\lambda,\varpi)} \in C_{(s,\lambda)}\} \tag{5}$$

3 Methodology

The methodology section is split into sections covering the Classification of the documents and the Training procedure.

3.1 Classification

In order to determine the accuracy of a classifier it is necessary to have specific measures of how the market reacts to news. To do so it is necessary to perform time series analysis on the data and classify each document according to how the market reacted shortly after its arrival. The return time series defined in Eq. (6) investigates the log returns over the period Δt for the stock. The return time series is one of the most interesting to investors as it demonstrates the amount of money which can be made. Abnormal returns should correlate more highly to the arrival of news because the market suddenly has more information to process.

The variable M in Eq. (7) defines the average number of trading minutes per month by using the average number of trading minutes per business day for the relevant country, and multiplying by the average number of trading days per month (20).

In Eq. (8) we define the mean for each trading minute in the return time series, by taking the mean value for the M trading minutes which preceded the start of the current trading day. In Eq. (9) we define the standard deviation for each trading minute in the return time series, by again using the M trading minutes which preceded the start of the current trading day. Note that if a stock was suspended from trading during the last 20 trading days for the stock exchange, we use the last 20 days which the stock traded on.

$$R_{(s,\Delta t)} = \{R_1, \ldots, R_m\} \mid R_{(s,t,\Delta t)} = \log(P_{(s,\ t)}) - \log(P_{(s,\ t-\Delta t)}) \tag{6}$$

$$M = 20 \times m \mid \{m_{US} = 390,\ m_{UK} = 510,\ m_{AU} = 360\} \tag{7}$$

$$\mu R_{(s,t,\Delta t)} = \frac{\displaystyle\sum_{j=t_0-M}^{t_0-1} R_{(s,j,\Delta t)}}{M} \mid t_0 = \min\left(\begin{cases} \forall T_{(B,i)} \mid time(T_{(B,i)}) = \\ \min(time(T_B)) \wedge T_{(B,i)} \le t \end{cases}\right) \tag{8}$$

$$\sigma R_{(s,t,\Delta t)} = \sqrt{\frac{\displaystyle\sum_{j=t_0-M}^{t_0-1} \left(R_{(s,j,\Delta t)} - \mu R_{(s,t,\Delta t)}\right)^2}{M}} \mid t_0 = \min\left(\begin{cases} \forall T_{(B,i)} \mid time(T_{(B,i)}) \\ = \min(time(T_B)) \wedge T_{(B,i)} \le t \end{cases}\right) \tag{9}$$

We classify the outcome O of each article in Eq. (4) using the definition in Eq. (10) in which we require that the return within τ minutes is equal or exceeds δ standard deviations from the mean function value.

$$O_{(s,\Delta t,\tau,\delta)} = \{O_1, O_2, \ldots, O_p\} \mid O_{(s,\Delta t,\tau,\delta,\lambda)} =$$

$$\left(\exists t \mid d_{(s,\lambda)} < t \le d_\lambda + \tau \wedge \left(\begin{matrix} R_{(s,t,\Delta t)} \ge \mu R_{(s,t,\Delta t)} + \delta \times \sigma R_{(s,t,\Delta t)} \\ \vee R_{(s,t,\Delta t)} \le \mu R_{(s,t,\Delta t)} - \delta \times \sigma R_{(s,t,\Delta t)} \end{matrix}\right) ? 1 : 0\right) \tag{10}$$

3.2 Training

The stocks for each country c are grouped together using Eq. (11) to form a large dataset of related stocks. Each document for each stock within each country is then classified using the return time series with the chosen parameters. For each country and value of δ we create 10 training and tests sets for the sake of robustness.

We create training sets by taking Φ documents at random, of which Ψ were classified as interesting, and the rest were not. The remaining documents are used as the test set.

A dictionary is created using Eq. (12) for each term which appears in at least one document for a stock in the training set. We store the term count TC, document count DC, interesting term count ITC, and interesting document count IDC for each term. The TC is the total number of times the given term appears in all documents in the training set. The DC is the total number of documents which contain the given term. The ITC is the total number of times the given term appears in all documents which are classified as interesting in the training set. The IDC is the total number of documents which are classified as interesting in the training set which contain the given term.

$$G_{(c)} = \{G_1, G_2, \ldots, G_v\} \tag{11}$$

$$
\begin{aligned}
X_{(c,\Delta t,\tau,\delta)} &= \{X_1, X_2, \ldots, X_w\} \\
\mid\ X_{(c,\Delta t,\tau,\delta,\varpi)} &= \{S_{(c,\Delta t,\tau,\delta,\varpi)}, TC_{(c,\Delta t,\tau,\delta,\varpi)}, DC_{(c,\Delta t,\tau,\delta,\varpi)}, ITC_{(c,\Delta t,\tau,\delta,\varpi)}, IDC_{(c,\Delta t,\tau,\delta,\varpi)}\} \\
\wedge TC_{(c,\Delta t,\tau,\delta,\varpi)} &= \sum SC_{(s,\lambda,\varpi)} \mid s \in G_{(c)} \\
\wedge DC_{(c,\Delta t,\tau,\delta,\varpi)} &= \#\{\forall C_{(s,\lambda)} \mid SC_{(s,\lambda,\varpi)} > 0 \ \wedge \ s \in G_{(c)}\} \\
\wedge ITC_{(c,\Delta t,\tau,\delta,\varpi)} &= \sum SC_{(s,\lambda,\varpi)} \mid O_{(s,\Delta t,\tau,\delta,\lambda)} = 1 \ \wedge \ s \in G_{(c)} \\
\wedge IDC_{(c,\Delta t,\tau,\delta,\varpi)} &= \#\{\forall C_{(s,\lambda)} \mid SC_{(s,\lambda,\varpi)} > 0 \ \wedge \ O_{(s,\Delta t,\tau,\delta,\lambda)} = 1 \ \wedge \ s \in G_{(c)}\}
\end{aligned} \tag{12}
$$

A sub-dictionary is formed by taking the top ψ terms based on a given term ranking algorithm. Firstly we chose the term frequency inverse document frequency (TFIDF) algorithm defined in Eq. (13). It is calculated by combining the product of the term frequency (first part of equation) with the inverse document frequency (log part of equation). Note the Φ in Eq. (13) is the number of documents in the training set. We sort the values in descending order such that terms that occur most frequently are chosen, as this is the order which generally gives the best results when querying documents.

$$TFIDF_{(c,\Delta t,\tau,\delta,\varpi)} = TC_{(c,\Delta t,\tau,\delta,\varpi)} \times \log_{10}\left(\frac{\Phi}{DC_{(c,\Delta t,\tau,\delta,\varpi)}}\right) \tag{13}$$

Secondly we chose the binary version of Quinlan's Gain ratio [22], as defined in Eq. (15). This algorithm selects terms which provide the most information, i.e. split the data between the classes most effectively. In Eq. (15) the E(Ψ, Φ) part of the equation is the entropy value for the ratio of interesting documents (Ψ) to documents (Φ) in the training set. The next part of the equation calculates the entropy value for the ratio of interesting documents to documents which contain the term, scaled by the ratio of documents which contain the term. The last part of the equation calculates the entropy value for the ratio of uninteresting documents to documents which contain the term, scaled by the ratio of documents which don't contain the term.

$$E(n,N) = -\left(\frac{n}{N}\log_2\left(\frac{n}{N}\right) + \left(1-\frac{n}{N}\right)\log_2\left(1-\frac{n}{N}\right)\right) \mid n \le N \tag{14}$$

$$GAIN_{(c,\Delta t,\tau,\delta,\varpi)} = E(\Psi,\Phi) - \frac{DC_{(c,\Delta t,\tau,\delta,\varpi)}}{\Phi} \times E\left(IDC_{(c,\Delta t,\tau,\delta,\varpi)}, DC_{(c,\Delta t,\tau,\delta,\varpi)}\right) -$$
$$\frac{\Phi - DC_{(c,\Delta t,\tau,\delta,\varpi)}}{\Phi} \times E\left(DC_{(c,\Delta t,\tau,\delta,\varpi)} - IDC_{(c,\Delta t,\tau,\delta,\varpi)}, DC_{(c,\Delta t,\tau,\delta,\varpi)}\right) \tag{15}$$

Finally we adapted Robertson and Spärk Jones's BM25 algorithm (Best Match) [23] to get the Average Document BM25 value (ADBM25) defined in Eq. (16), where k_1 and b are constants. The ADBM25 algorithm is the same as the BM25 algorithm if Φ were equal to 1, or in other words if there was only one document. The first part of the equation normalises the term frequency by taking into account the length of the document which contains the term and the average document length. This ensures that if a term occurs frequently in a very long document, it isn't given unwarranted significance. The log part of the equation normalises results by factoring in the number of interesting documents which contain the term (IDC), the number of documents which contain the term (DC) and the total number of interesting documents (Ψ) and documents (Φ). This favours terms which provide more information, i.e. split the two classes most efficiently.

$$ADBM\,25_{(c,\Delta t,\tau,\delta,\varpi)} = \frac{1}{\Phi}\sum_{d=1}^{\Phi} \frac{(k_1+1)\times TC_{(c,\Delta t,\tau,\delta,\varpi)}}{\left(k_1\times\left((1-b)+b\times\frac{dl_{(d)}}{avdl}\right)\right)+TC_{(c,\Delta t,\tau,\delta,\varpi)}} \times$$
$$\log\frac{\left(IDC_{(c,\Delta t,\tau,\delta,\varpi)}+0.5\right)\times\left(\Phi - DC_{(c,\Delta t,\tau,\delta,\varpi)} - \Psi + IDC_{(c,\Delta t,\tau,\delta,\varpi)}+0.5\right)}{\left(DC_{(c,\Delta t,\tau,\delta,\varpi)} - IDC_{(c,\Delta t,\tau,\delta,\varpi)}+0.5\right)\times\left(\Psi - IDC_{(c,\Delta t,\tau,\delta,\varpi)}+0.5\right)} \tag{16}$$
$$\mid dl_{(\lambda)} = \sum \forall SC_{(s,\lambda,\varpi)} \,\wedge\, avdl = \frac{1}{\Phi}\sum_{\rho=1}^{\Phi} dl_{(\rho)}$$

A binary vector is created for each document in the training and test set where each entry specifies whether the given term (which is a member of the sub-dictionary) occurred in the document. These vectors are used to train and test Quinlan's C4.5 decision tree [22], and Vapnik's support vector machine (SVM) [24] using the SVM Light Classifier [25].

The C4.5 decision tree [22] classifies documents by building a tree where the root node is the term which produces the highest Gain value (Eq. (15)). Each leaf node which branches from the root, or from subsequent branches, chooses the combination of terms which produces the highest Gain value. This value is calculated by combining the terms and their values (i.e. contains the term or not) of every node above the current node and adding one extra term. Only documents which have the given term values (i.e. contain or don't contain the specified terms as required) are included in the Gain equation. This ensures that the extra terms are appended based on their ability to separate the remaining documents into the two classes.

Vapnik's support vector machine (SVM) [24] projects the terms and their values into higher dimensional space (e.g. one dimension per term). It produces a classifier by identifying the hyperplane which most effectively separates the two classes.

To measure the performance of the classifiers we use the overall accuracy of the classifier defined in Eq. (17). It is calculated by dividing the total number of correctly classified documents (True Positives and True Negatives) by the total number of documents in the test set (Φ).

$$Accuracy = \frac{\#TP + \#TN}{\Phi}.$$ (17)

4 Results

Each document is classified using the return time series with $\Delta t = \tau$ equal to the specified number of minutes, and $\delta = 4$. For all tests there are 1,000 documents in the training set (Φ), of which 500 are interesting (Ψ). We chose an equal split so as not to skew the results of the classifier. Furthermore we use $k_1 = 1$ and $b = 0.5$ for the ADBM25 term ranking algorithm in Eq. (16). Finally we run tests with varying φ values (100,200,500,1000,2000,5000).

The characteristics of the datasets are shown in Table 1 where Docs are the total number of documents which were released to the given country during the times specified in Eq. (4). The Int. Docs are those which were categorised as causing an abnormal return. The ratio is the percentage of documents which are interesting.

Table 1. Characteristics of different datasets, with $\delta = 4$, $\Delta t = \tau$ having the listed values, G consisting of all stocks for the given country (US, UK, Australia)

$\Delta t, \tau$ (minutes)	US			UK			Australia		
	Docs	Int. Docs	Ratio	Docs	Int. Docs	Ratio	Docs	Int. Docs	Ratio
5	133,019	2,414	1.81%	81,528	2,046	2.51%	33,165	933	2.81%
10	129,370	2,760	2.13%	80,245	2,487	3.10%	31,728	1,259	3.97%
15	124,616	2,539	2.04%	78,871	2,756	3.49%	30,455	1,388	4.56%
30	112,907	2,205	1.95%	74,664	2,851	3.82%	27,054	1,232	4.55%
45	100,245	1,710	1.71%	70,054	2,698	3.85%	23,910	1,030	4.31%
60	89,159	1,452	1.63%	65,230	2,503	3.84%	20,835	913	4.38%
90	68,056	973	1.43%	54,230	1,899	3.50%	14,588	560	3.84%

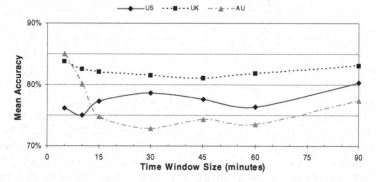

Fig. 1. The mean accuracy for the best classifier for each time window ($\Delta t = \tau$) and country

The ratio actually increases as the time window size is increased before reducing again. This is because τ allows more time for the market to react to the news and Δt isn't large enough to limit the number of documents which occur in the time period.

The results in Fig. 1 show the results of the best classifier for each time window size ($\Delta t = \tau$) and country. The details of the term ranking algorithm, the classifier, the number of terms (φ) which produced these results are included in Table 2. This also includes the standard deviation of the classifier as well as the maximum achieve with the given parameters, where the best results are bolded.

Table 2. The characteristics of the best classifier for each time window ($\Delta t = \tau$) and country

Country	$\Delta t, \tau$	Term Ranking	Classifier	Terms (φ)	Accuracy	Maximum Accuracy
US	5	ADBM25	SVM	5,000	76.12±02.39%	78.95%
US	10	ADBM25	SVM	100	74.97±02.30%	79.73%
US	15	ADBM25	SVM	5,000	77.28±01.13%	79.02%
US	30	ADBM25	SVM	5,000	78.67±02.05%	83.32%
US	45	ADBM25	SVM	5,000	77.58±01.86%	80.30%
US	60	ADBM25	SVM	5,000	76.39±01.19%	78.75%
US	**90**	**GAIN**	**SVM**	**200**	**80.36±01.20%**	**82.55%**
UK	**5**	**GAIN**	**C4.5**	**100**	**83.72±01.33%**	**86.46%**
UK	10	GAIN	C4.5	100	82.46±01.08%	85.24%
UK	15	GAIN	C4.5	100	82.09±01.57%	84.52%
UK	30	GAIN	C4.5	100	81.47±00.93%	83.04%
UK	45	GAIN	C4.5	100	81.06±01.61%	84.92%
UK	60	GAIN	C4.5	100	81.84±00.92%	84.22%
UK	90	GAIN	C4.5	100	83.10±00.73%	84.80%
AU	**5**	**ADBM25**	**SVM**	**5,000**	**85.00±00.64%**	**86.60%**
AU	10	ADBM25	SVM	2,000	80.13±01.30%	82.75%
AU	15	ADBM25	C4.5	100	74.74±02.71%	78.98%
AU	30	ADBM25	C4.5	100	72.85±01.74%	76.56%
AU	45	ADBM25	C4.5	100	74.27±01.94%	77.73%
AU	60	ADBM25	C4.5	100	73.54±01.35%	75.40%
AU	90	ADBM25	C4.5	100	77.37±01.24%	80.81%

In the US it appears that investors react to some news within 5 minutes and tend to consistently react within 30 minutes of news. However the best results are achieved by allowing 90 minutes for the market to react to news. This indicates that investors in the US tend to pay more attention to market movement than to the release of news. This is because the US market is the largest stock market in the world and therefore has the most frequent trading. Therefore investors need to pay close attention to how other market participants are behaving in order to make their decisions. This leaves them less time to read news and therefore probably only read news once they have noticed abnormal returns which they may capitalise on. Furthermore there is significantly more news released to the US market than the other two markets.

Therefore investors are less likely to read many articles as they are likely to see most news as irrelevant unless it has a catchy headline.

Investors in the UK tend to react quickly and decisively to news within 5 minutes and continue to react in a similar fashion for long after. This indicates that investors in the UK not only pay close attention to news but they also consistently react in a similar fashion to news with similar content. The slight reduction in accuracy over time is an indication that market noise has introduced articles which themselves where not responsible for any change.

It appears that investors in the Australian market are less rational than those in the other two markets. This is because they react quickly and decisively to news within 5 minutes but the accuracy dramatically reduces when the time window is increased. This should be expected as the Australian market is significantly smaller than the other two markets and therefore has considerably less trading than the others. Therefore there can be long periods where there has been little or no trading which leads to a lower standard deviation of the return. However some investors, most likely large institutional investors, must pay close attention to news in order to react to the news consistently and quickly. The rest of the market however probably either has less or delayed access to public information and therefore don't tend to react in a similar fashion. Alternatively it could mean that investors in Australia are somewhat irrational as they don't consistently react in the same way to the same news.

Table 3. The characteristics of the best classifier with φ limited to 100 and 200 for each country and term ranking algorithm

Country	$\Delta t, \tau$	φ	TFIDF		GAIN		ADBM25	
			Accuracy	Class	Accuracy	Class	Accuracy	Class
US	90	100	73.45±01.15%	SVM	**78.24±10.07%**	SVM	75.56±01.68%	SVM
US	90	200	71.77±01.61%	SVM	**80.36±01.20%**	SVM	76.40±02.02%	SVM
UK	5	100	69.52±04.29%	SVM	**83.72±01.33%**	C4.5	75.11±01.94%	SVM
UK	5	200	68.50±04.05%	SVM	**79.67±02.03%**	C4.5	75.80±01.82%	SVM
AU	5	100	**83.74±01.06%**	SVM	81.50±11.32%	C4.5	82.28±00.71%	SVM
AU	5	200	**83.46±00.92%**	SVM	74.75±15.72%	SVM	82.70±00.81%	SVM

The results in Table 3 show the most accurate classifier for each country and term ranking algorithm with the best time window and a limit of 100 and 200 terms. The most accurate results are bolded. Combining the results with those in Table 2 it is clear that the SVM is the best classifier for the US and Australian markets, whilst the C4.5 classifier is more useful in the UK market. This suggests that longer rules are required to classify documents in the UK market as the SVM tends to ignore terms which individually have little impact, whilst the C4.5 classifier is more comprehensive. However when the number of terms (φ) is increased beyond 200 the SVM is better than the C4.5 classifier in the UK though produces worse results than for the 100 and 200 terms tests.

In the US and UK the best results are obtained using the Gain term ranking algorithm. This implies that the training sets are indicative of the entire dataset. However with time windows ($\Delta t = \tau$) less than 90 minutes it is necessary to include more terms in the US and in that case the ADBM25 term ranking algorithm is better. This is because annual reports released in the US are very long whilst some analyst recommendations and similar news are very short. Therefore it is necessary to account for the length of the document when ranking the effect of the term.

In the Australian market the TFIDF term ranking algorithm yields the best results when the number of terms (φ) is limited. This indicates that the relevance of the term is less important in Australia. We can assume this because both the Gain and ADBM25 term ranking algorithms account for the relevance of the term, whilst TFIDF does not. This could imply that investors in Australia are less rational because they don't react consistently to similar news. Alternatively it could mean that news providers for the Australian market don't use the same terminology. Therefore there are many different ways to say the same thing which leads to difficulty in seeing the similarity between documents.

The Receiver Operating Characteristic (ROC) curves in Fig. 2 are for the classifier, term ranking algorithm, and time window combination which produced the most accurate classifier for each country. All results are clearly better than the line of no-discrimination though the number of true positives found is quite low. This is because it is necessary to choose a classifier with the least number of false positives, as there are considerably more uninteresting documents. However the results show that it is possible to predict whether news will cause abnormal returns.

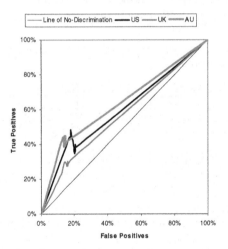

Fig. 2. The ROC curves of the best classifier for each country

5 Conclusion

We have classified news based to abnormally large returns and then employed the SVM and C4.5 classifiers to forecast the short-term market reaction to news. We have also utilised the TFIDF, Gain, and a modified version of the BM25 term ranking algorithms to aid the decisions of the classifiers.

We have found that the Gain term ranking algorithm is superior in the US and UK markets. This implies that the training sets are representative of the entire dataset as the Gain term ranking algorithm chooses terms which have high information value within the training set. This means that investors in the US and UK appear to be more rational as they react the same way to similar documents.

The SVM classifier was discovered to have the best performance in the US and Australian markets whilst the C4.5 classifier was better for the UK market. This suggests that longer rules are required to classify documents in the UK market as the SVM tends to ignore terms which individually have little impact, whilst the C4.5 classifier is more comprehensive.

The fact that the Gain and ADBM25 term ranking algorithms are less effective in the Australian market when the number of terms (φ) is limited, indicates that Australian investors are somewhat irrational (i.e. make decisions without any new information). This is probably because the Australian market is substantially smaller than the other two markets and therefore there are less large institutional investors, who tend to pay close attention to the market. Alternatively it could mean that news providers in the Australian market don't use consistent terminology. This means there are many ways to say the same thing, and therefore the relevance of the document is less important.

We have found that the 90 minute time window yields the most accurate classifier in the US. This suggests that investors in the US tend to pay more attention to actual market behaviour than to the release of news. This is because the US stock market is the largest in the world and therefore has more frequent trading. Therefore investors must pay attention to trading and therefore they have less time to read news. Furthermore there is far more news released in the US than the other two markets. Therefore investors probably tend to ignore articles unless they have compelling headlines or they notice the market behaving differently.

The UK and Australian markets react quickly and decisively to some news within 5 minutes. This is most likely annual reports, or earnings announcements which investors were anxiously awaiting and therefore the documents have similar content. However the classification accuracy drops as the time window is increased before eventually rising again. This effect is significantly more noticeable in the Australian market than the UK. This implies that investors in the UK pay close attention to news and are rational as they consistently react in the same way to news with similar content. In Australia however it appears that the large institutional investors are responsible for the decisive reaction within 5 minutes. This is because they have undelayed access to most news and the staff to read it. However other investors either have delayed access or are slightly irrational as they don't consistently react to news with similar content.

These results are promising and it would appear that it is cost-effective to develop a system which highlights interesting news for investors based on its content.

References

1. Oberlechner, T., Hocking, S.: Information Sources, News, and Rumours in Financial Markets: Insights into the Foreign Exchange Market. Journal of Economic Psychology 25, 407–424 (2004)
2. Cutler, D.M., Poterba, J.M., Summers, L.H.: What Moves Stock Prices? Journal of Portfolio Management 15, 4–12 (1989)
3. Goodhart, C.A.E.: News and the foreign exchange market. In: Proceedings of Manchester Statistical Society, pp. 1–79 (1989)

4. Goodhart, C.A.E., Hall, S.G., Henry, S.G.B., Pesaran, B.: News Effects in a High-Frequency Model of the Sterling-Dollar Exchange Rate. Journal of Applied Econometrics 8, 1–13 (1993)
5. Mitchell, M.L., Mulherin, J.H.: The Impact of Public Information on the Stock Market. Journal of Finance 49, 923–950 (1994)
6. Melvin, M., Yin, X.: Public Information Arrival, Exchange Rate Volatility, and Quote Frequency. Economic Journal 110, 644–661 (2000)
7. Mittermayer, M.-A.: Forecasting Intraday Stock Price Trends with Text Mining Techniques. In: Proceedings of 37th Annual Hawaii International Conference on System Sciences (HICSS 2004), Big Island, Hawaii, p. 30064b (2004)
8. Ederington, L.H., Lee, J.H.: How markets process information: News releases and volatility. Journal of Finance 48, 1161–1191 (1993)
9. Ederington, L.H., Lee, J.H.: The short-run dynamics of the price adjustment to new information. Journal of Financial & Quantitative Analysis 30, 117–134 (1995)
10. Ederington, L.H., Lee, J.H.: Intraday Volatility in Interest-Rate and Foreign-Exchange Markets: ARCH, Announcement, and Seasonality Effects. Journal of Futures Markets 21, 517–552 (2001)
11. Almeida, A., Goodhart, C.A.E., Payne, R.: The Effects of Macroeconomic News on High Frequency Exchange Rate Behavior. Journal of Financial & Quantitative Analysis 33, 383–408 (1998)
12. Graham, M., Nikkinen, J., Sahlstrom, P.: Relative Importance of Scheduled Macroeconomic News for Stock Market Investors. Journal of Economics and Finance 27, 153–165 (2003)
13. Kim, S.-J., McKenzie, M.D., Faff, R.W.: Macroeconomic News Announcements and the Role of Expectations: Evidence for US Bond, Stock and Foreign Exchange Markets. Journal of Multinational Financial Management 14, 217–232 (2004)
14. Nofsinger, J.R.: Option volume and volatility response to scheduled economic news releases. Journal of Futures Markets 23, 315–345 (2003)
15. Hong, H., Lim, T., Stein, J.C.: Bad news travels slowly: Size, analyst coverage, and the profitability of momentum strategies. Journal of Finance 55, 265–295 (2000)
16. Womack, K.L.: Do Brokerage Analysts' Recommendations Have Investment Value? Journal of Finance 51, 137–167 (1996)
17. Michaely, R., Womack, K.L.: Conflict of Interest and the Credibility of Underwriter Analyst Recommendations. Review of Financial Studies 12, 653–686 (1999)
18. Roll, R.: Orange Juice and Weather. American Economic Review 74, 861–880 (1984)
19. Wuthrich, B., Permunetilleke, D., Leung, S., Cho, V., Zhang, J., Lam, W.: Daily Stock Market Forecast from Textual Web Data. In: Proceedings of IEEE International Conference on Systems, Man, and Cybernetics, pp. 2720–2725 (1998)
20. Fung, G.P.C., Yu, J.X., Wai, L.: Stock Prediction: Integrating Text Mining Approach using Real-Time News. In: Proceedings of IEEE International Conference on Computational Intelligence for Financial Engineering, Hong Kong, pp. 395–402 (2003)
21. Porter, M.F.: An Algorithm for Suffix Striping. Automated Library and Information Systems 14, 130–137 (1980)
22. Quinlan, J.R.: C4.5: Programs for Machine Learning. Morgan Kaufmann, San Francisco (1993)
23. Robertson, S., Jones, K.S.: Simple, Proven Approaches to Text Retrieval. University of Cambridge Computer Laboratory Technical Report no. 356 (2006)
24. Vapnik, V.: The Nature of Statistical Learning Theory. Springer, Heidelberg (1999)
25. Joachims, T.: SVM Light Classifier (2007), Available, http://svmlight.joachims.org/

Frequency-Weighted Fuzzy Time-Series Based on Fibonacci Sequence for TAIEX Forecasting

Hia Jong Teoh[1,2], Tai-Liang Chen[1], and Ching-Hsue Cheng[1]

[1] Department of Information Management,
National Yunlin University of Science and Technology,
123, section 3, University Road, Touliu, Yunlin 640, Taiwan, R.O.C.
[2] Department of Accounting Information, Ling Tung University
1, Ling Tung Road, Nantun, Taichung 408, Taiwan, R.O.C.
{g9320824,g9320817,chcheng}@yuntech.edu.tw

Abstract. This paper proposes a new fuzzy time-series model for promoting the stock price forecasting, which provides two refined approaches, a frequency-weighted method, and the concept of Fibonacci sequence in forecasting processes. In empirical analysis, two different types of financial datasets, TAIEX (Taiwan Stock Exchange Capitalization Weighted Stock Index) stock index and HSI (Hong Kong Heng Seng Index) stock index are used as model verification. By comparing the forecasting results with those derived from Chen's, Yu's, and Hurang's models, the authors conclude that the research goal has been reached.

Keywords: Fuzzy Time-series, Stock Price Forecasting, Fibonacci Sequence, Fuzzy Linguistic Variable.

1 Introduction

Time-series models have utilized the fuzzy theory to solve various domain forecasting problems, such as temperature forecasting, university enrollments forecasting [1-6] and financial forecasting [7-12]. Especially in the area of stock price forecasting, fuzzy time-series models are often employed [10-12]. As Dourra (2002) notes, it is common practice to "deploy fuzzy logic engineering tools in the finance arena, specifically in the technical analysis field, since technical analysis theory consists of indicators used by experts to evaluate stock price."[7]

In stock technical analysis fields, Elliott (1938) proposed the Elliott Wave Principle which has been playing an important role in stock analysis for more than six decades [13-15]. The theory is closely related to time-series because it applies the Fibonacci sequence to predict the *timing* of stock price fluctuation. Therefore, this paper applies the sequence in time-series models to forecast stock prices.

Lastly, based on the drawbacks of previous models, we propose a new fuzzy time-series model and recommend two refined processes in the forecasting processes. By employing a ten-period of TAIEX(Taiwan Stock Exchange Capitalization Weighted Stock Index) and HSI (Hong Kong Heng Seng Index) stock index as experiment datasets, the verifications show that our model outperforms conventional and advanced fuzzy time-series models.

T. Washio et al. (Eds.): PAKDD 2007 Workshops, LNAI 4819, pp. 27–34, 2007.
© Springer-Verlag Berlin Heidelberg 2007

The remaining content of this paper is organized as follows: Section 2 introduces the related literature of fuzzy time-series models; section 3, propose the new model and algorithm; section 4 evaluates the performance of the proposed model; and section 5 concludes this paper.

2 Related Works

Fuzzy theory was originally developed to deal with the problems involving human linguistic terms [16-18]. Time-series models had failed to consider the application of this theory until fuzzy time-series was defined by Song and Chissom (1993) who proposed the definitions of fuzzy time-series and methods to model fuzzy relationships among observations [1]. In following research, Song and Chissom (1994) continued to discuss the difference between time-invariant and time-variant models [2]. Besides these researchers, Chen (1996) proposed another method to apply simplified arithmetic operations in forecasting algorithm rather than the complicated max-min composition operations used in Song and Chissom's model [3].

Huarng (2001) pointed out that the length of intervals affects forecast accuracy in fuzzy time-series and proposed a method with distribution-based length and average-based length to reconcile this issue [19-20]. The method applied two different lengths of intervals to Chen's model and the conclusions showed that distribution-based and average-based lengths could improve the accuracy of forecast. Although this method has excellent performance, it creates too many linguistic values to be identified by analysts. According to Miller (1956), establishing linguistic values and dividing intervals would be a trade off between human recognition and forecasting accuracy [21].

Time-series models will generate terrible inaccurate forecasts whenever unexpected events have occurred before forecasting. To deal with the problem, a group decision-making method was utilized to integrate the subjective forecast values of all decision makers. Fuzzy weighted method was then combined with subjective forecast values to produce an aggregated forecast value. Yu (2004) proposed a weighted model to tackle two issues, recurrence and weighting, in fuzzy time-series forecasting [10]. The researcher argued that recurrent fuzzy relationships should be considered in forecasting and recommended that different weights be assigned to various fuzzy relationships. In Yu' research, it was concluded that the weighted model outperforms one of the conventional fuzzy time series models [10, 22].

Among recent research, Genetic Algorithms and Neural Networks Algorithms are frequenters and efficiently improve the forecasting accuracy of fuzzy time-series models [6, 23]. However, the authors argue that there are other reasonable approaches such as stock analysis theory or techniques to apply in fuzzy time-series whenever the forecasted targets are stock price materials [12]. From the reviewed literature, we also have found that most of conventional fuzzy time-series models lack the considerations in discovering recurrence information. To reconcile these problems above, we propose a new fuzzy time-series model.

3 Proposed Model

In this section, the researchers propose a new fuzzy time-series model based on the Fibonacci sequence and introduce the algorithm of the proposed model.

3.1 Research Model

Based on the thoughtlessness of the reviewed models, we propose a new fuzzy time-series model. There are two refined processes factored into our model as follows: (1) Use a frequency-weighted method to represent the patterns of stock price fluctuations in history. Traditionally, the weight assigned for each FLR (Fuzzy logical relationship) is determined either based on chronological order or experts' knowledge. Since most of FLRs will occur again within a specific time period, the occurrence frequency in history can be used as the estimated probability of reoccurrence in near future. Therefore, classifying FLRs and weighting their occurrence is a reasonable approach for making predictions [10]. and (2) Apply the Fibonacci sequence ($F_n = F_{n-1} + F_{n-2}$) in forecasting process. The Elliott Wave Principle utilizes the Fibonacci sequence to predict the *timing* of stock price fluctuation [13-15] and has become one of major stock technical analysis theories. For the reason mentioned above, applying this sequence in fuzzy time-series models is a reasonable and theorizing approach. In next subsection, we provide the proposed algorithm.

3.2 The Algorithm

In this subsection, the algorithm of the proposed model is detailed and several numerical examples are employed for explaining each step.

Step 1: Reasonably define the universe of discourse, $U = [Low, Up]$, and decide that how many intervals the universe will be partitioned into. For example, the minimum and maximum of TAIEX stock prices, from 1999/1/1~1999/10/31, is 5474 and 8608, respectively. Then the universe of discourse can be defined as $U = [5474, 8608]$. By using fuzzy method, low bound can be roughly expanded by 100 smaller from 5474 to make *Low* become 5400 and up bound can be roughly expanded by 100 larger from 8608 to make *Up* become 8700. As a result, the defined universe of discourse, $U = [5400, 8700]$, can cover every occurred stock price in the 1999-year period. In this study, we initially partition the universe of discourse into "*seven*" linguistic [21]. Take TAIEX stock trading data as example, the initial seven linguistic values and intervals for the universe of TAIEX stock price are demonstrated in Table 1.

Table 1. Seven linguistic intervals of TAIEX stock price

Linguistic Interval	Range for Interval	Midpoint of Interval
I_1	[5400, 5870]	5636
I_2	[5871, 6341]	6107
I_3	[6342, 6812]	6578
I_4	[6813, 7283]	7049
I_5	[7284, 7754]	7520
I_6	[7755, 8225]	7991
I_7	[8226, 8700]	8462

Step 2: Establish related fuzzy set for each observation in training dataset. Define the fuzzy set, $L_1 L_2 ... L_k$, on the universe of discourse by equation (1). The value of a_{ij} indicates the grade of membership of u_j in fuzzy set L_i, where $a_{ij} \in [0,1]$,

$1 \leq i \leq k$ and $1 \leq j \leq m$. Find out the degree of each stock price belonging to each L_i ($i=1,...,m$). If the maximum membership of the stock price is under L_k, then the fuzzified stock price is labeled as L_k. The fuzzy logical relationships are generated based on the fuzzified stock price. In this paper, the seven fuzzy linguistic vales, L_1 = (very low price), L_2 = (low price), L_3 = (little low price), L_4 = (normal price), L_5 = (little high price), L_6 = (high price) and L_7 = (very high price), are applied [3].

$$L_1 = a_{11}/u_1 + a_{12}/u_2 + ... + a_{1m}/u_m$$
$$L_2 = a_{21}/u_1 + a_{22}/u_2 + ... + a_{2m}/u_m$$
$$\vdots$$
$$L_k = a_{k1}/u_1 + a_{k2}/u_2 + ... + a_{km}/u_m$$

(1)

Each observation in training dataset can be labeled a specific linguistic value by equation (1). Table 2 demonstrates the assignments of linguistic value for eight periods of stock prices based on the intervals from Table 1.

Table 2. Assign related linguistic value to stock price

Time	Stock Price	Linguistic Value
$t=1$	6152	L_2
$t=2$	6199	L_2
$t=3$	6404	L_3
$t=4$	6421	L_3
$t=5$	6406	L_3
$t=6$	6363	L_3
$t=7$	6319	L_2
$t=8$	6241	L_2

Step 3: Establish fuzzy logic relationships for linguistic value time-series. One FLR (Fuzzy Logical Relationship) is composed of two consecutive linguistic values. For example, the FLR ($L_i \rightarrow L_j$) is established by $L_i(t-1)$ and $L_j(t)$. Table 3 demonstrates the FLR establishing processes of the linguistic value time-series based on Table 2.

Table 3. FLR Table

$L_2(t=1) \rightarrow L_2(t=2)$
$L_2(t=2) \rightarrow L_3(t=3)$
$L_3(t=3) \rightarrow L_3(t=4)$
$L_3(t=4) \rightarrow L_3(t=5)$
$L_3(t=5) \rightarrow L_3(t=6)$
$L_3(t=6) \rightarrow L_2(t=7)$
$L_2(t=7) \rightarrow L_2(t=8)$

Step 4: Establish FLR groups and produce fluctuation-type matrix. The FLRs with the same LHS (Left Hand Side) linguistic value can be grouped into one FLR group. All FLR groups will construct a fluctuation-type matrix. Table 4

shows the fluctuation-type matrix produced by the FLRs from Table 3. Each row of the matrix represents one FLR group and each cell represents the occurrence frequency of each FLR.

Table 4. An example of occurrence frequency of each FLR

P(t+1) / P(t)	L_1	L_2	L_3	L_4	L_5	L_6	L_7	total
L_1	0	0	0	0	0	0	0	0
L_2	0	2	1	0	0	0	0	3
L_3	0	1	3	0	0	0	0	4
L_4	0	0	0	0	0	0	0	0
L_5	0	0	0	0	0	0	0	0
L_6	0	0	0	0	0	0	0	0
L_7	0	0	0	0	0	0	0	0

Step 5: Assign fluctuation weight for each FLR group. Each FLR within the same FLR group should be assigned a weight. For example, in Table 4, the total occurrences of FLR group of L_2.is $L_2 \rightarrow L_1, L_2, L_3, L_4, L_5, L_6, L_7$ is 3. The FLR of $L_2 \rightarrow L_2$ occurs twice and the weight is assigned 0.667(2/3), and the FLR of $L_2 \rightarrow L_3$ occurs 1, then, the weight is assigned 0.333(1/3) In the weighted method, the FLR weight is determined by its occurs divide by total occurrences of FLR group.

Table 5. An example of frequency-weighted matrix

P(t+1) / P(t)	L_1	L_2	L_3	L_4	L_5	L_6	L_7	$\sum\limits_{k=1}^{i} W_k$
L_1	0	0	0	0	0	0	0	0
L_2	0	0.667	0.333	0	0	0	0	1
L_3	0	0.25	0.75	0	0	0	0	1
L_4	0	0	0	0	0	0	0	0
L_5	0	0	0	0	0	0	0	0
L_6	0	0	0	0	0	0	0	0
L_7	0	0	0	0	0	0	0	0

Table 5 demonstrates the frequency-weighted matrix converted from the data in Table 4. The sum of the weight of each FLR should be standardized to obtain the frequency-weighted matrix, $W(t)$. The standardized weight matrix equation is defined in equation (2). For example, by equation (2), the standardized weights for the FLR group of L_2 are specified as follows: $W_1 = 0$, $W_2 = 2/3$, $W_3 = 1/3$, $W_4 = 0$, $W_5 = 0$, $W_6 = 0$ and $W_7 = 0$.

$$W(t) = [W_1', W_2', \ldots, W_i'] = \left[\frac{W_1}{\sum\limits_{k=1}^{i} W_k}, \frac{W_2}{\sum\limits_{k=1}^{i} W_k}, \ldots, \frac{W_i}{\sum\limits_{k=1}^{i} W_k} \right] \quad (2)$$

Step 6: Compute forecasting values. From step 5, the standardized weight matrix can be both obtained. In this step, we use the midpoints of the linguistic intervals to generate initial forecasts. This process is call "defuzzification" (defined in equation (3)), where $L_{df}(t)$ is the deffuzified matrix composed of the midpoint of each interval, and $W_n(t)$ is the weight matrix obtained from Step 5.

$$Defuzzify(t) = L_{df}(t) \cdot W_n(t) \tag{3}$$

Step 7: Use the Fibonacci forecasting equation (defined in equation (4)) to produce conclusive forecasting results. By using the two linear parameters, α and β, as adjustments, the proposed model can produce the adopted results with forecasting errors as small as possible.

$$
\begin{aligned}
Fibonacci\_forecast(t+1) = P(t) + &\alpha * (defuzzify(t) - p(t)) \\
+ &\beta * (defuzzify(t-1) - P(t-1))
\end{aligned}
\tag{4}
$$

Step 8: Evaluate forecasting performance. The RMSE (Root Mean Squared Error, defined in equation (5)) is employed as a performance indicator to evaluate the proposed model.

$$RMSE = \sqrt{\frac{\sum_{i=1}^{n}(actual_t - final(t))^2}{n}} \tag{5}$$

4 Model Verifications

In empirical analysis, two different types of financial datasets, TAIEX (Taiwan Stock Exchange Capitalization Weighted Stock Index) stock index and HSI (Hong Kong Heng Seng Index) stock index, are used as experiment datasets. To verify the forecasting performance of the proposed model, we employ two other fuzzy time-series model, Chen's (1996) [3] and Yu's (2004) models [10], as comparison models.

4.1 Forecasting for TAIEX

In this experiment, a ten-year period of TAIEX data, from 1990 to 1999, is selected as experimental datasets. Previous ten-month of each year, from January to October, is used for training and the rest, from November to December, for testing [10]. To validate the improvements in stock price forecasting, a recent advanced fuzzy time-series models, Huarng's model (2005) [23], is added in comparison models. From the literatures [10, 23], we produce a performance comparison table (listed in Table 6.) to illustrate forecasting performances for four different models. Table 6 shows that the proposed model wins 27 out of 29 RMSE among these models in ten testing periods.

Table 6. Forecasting performance for testing periods (TAIEX)

Models	1990	1991	1992	1993	1994	1995	1996	1997	1998	1999
Chen's model [3]	227	80	60	110	112	79	54	148	167	149
Yu's model [10]	220	61	67	**105	135	70	54	133	151	142
Huarng model [23]	No data	54	54	**107	79	74	73	141	121	109
Proposed model	175*	38*	41*	108	75*	55*	50*	132*	113*	108*

* Win.

** Lose comparisons on minimum RMSE value.

4.2 Forecasting for Hong Kong HSI

In this experiment, a ten-year period of HSI data, from 1990 to 1999, is selected as experimental datasets. Previous ten-month of each year, from January to October, is used for training and the rest, from November to December, for testing [10]. To validate the efficiency forecasting in stock price, from the literatures [3, 10], we produce a performance comparison table (listed in Table 10.) to illustrate forecasting performances for three different models. Table 7 shows that the proposed model wins 19 out of 20 RMSE among these models in ten testing periods.

Table 7. Forecasting performance for testing periods (HSI)

Models	1990	1991	1992	1993	1994	1995	1996	1997	1998	1999
Chen's model [3]	45	105	197	336	313	214	169	512	282	306
Yu's model [10]	40	40	183	**203	198	105	141	375	258	273
Proposed model	26*	38*	131*	208	143*	79*	140*	257*	205*	231*

* Win.
** Lose comparisons on minimum RMSE value.

5   Conclusions and Future Research

In this paper, we propose a new fuzzy time series based on frequency-weight method and the Fibonacci sequence to improve forecasting accuracy. From verification section, the research goal has been reached and two further conclusions are made as follow: (1) Assigning proper fluctuation weights to fuzzy relationships makes more reasonable descriptions for the past patterns of stock price fluctuations; and (2) the forecasting equation originated from the Fibonacci sequence makes the conclusive forecasts match late fluctuation patterns more closely.

However, there is still room for testing and improving the hypothesis of this model as follows: (1) Employing other stocks and financial materials as testing datasets to evaluate the performance; (2) simulating the model to trade in stock market, and sum up the profits of these trades to evaluate the profit making; and (3) reconsidering the factors affecting the behavior of the stock markets, such as trading volume, news and financial reports which might impact stock price in the future.

References

1. Song, Q., Chissom, B.S.: Forecasting enrollments with fuzzy time-series - Part I. Fuzzy Sets and Systems 54, 1–10 (1993)
2. Song, Q., Chissom, B.S.: Forecasting enrollments with fuzzy time-series - Part II. Fuzzy Sets and Systems 62, 1–8 (1994)
3. Chen, S.M.: Forecasting enrollments based on fuzzy time-series. Fuzzy Sets and Systems 81, 311–319 (1996)
4. Chen, S.M.: Forecasting Enrollments Based on High-Order Fuzzy Time Series. Cybernetics and Systems 33, 1–16 (2002)

5. Chen, S.M., Hsu, C.C.: A New Method to Forecast Enrollments Using Fuzzy Time Series. Applied Science and Eng. 2, 234–244 (2004)
6. Chen, S.M., Chung, N.Y.: Forecasting Enrollments Using High-Order Fuzzy Time Series and Genetic Algorithms. International Journal of Intelligent Systems 21, 485–501 (2006)
7. Dourra, H., Pepe, S.: Investment using technical analysis and fuzzy logic. Fuzzy Sets and Systems 127, 221–240 (2002)
8. Faff, R.W., Brooks, R.D., Ho, Y.K.: New evidence on the impact of financial leverage on beta risk: A time-series approach. North American Journal of Economics and Finance 13, 1–20 (2002)
9. Wang, Y.F.: Predicting stock price using fuzzy grey prediction system. Experts Systems with Applications 22, 33–39 (2002)
10. Yu, H.K.: Weighted fuzzy time-series models for TAIEX forecasting. Physica A 349, 609–624 (2004)
11. Huarng, K.H., Yu, H.K.: A Type 2 fuzzy time-series model for stock index forecasting. Physica A 353, 445–462 (2005)
12. Leon Lee, C.H., Liu, Alan, Chen, W.S.: Pattern Discovery of Fuzzy Time Series for Financial Prediction. IEEE Transactions on Knowledge and Data Engineering 18, 613–625 (2006)
13. Fischer, R.: The New Fibonacci Trader: Tools and Strategies for Trading Success. John Wiley & Sons, New York (2001)
14. Collins, C.J., Frost, A.J., Robert Jr., R.: Prechter: Elliott Wave Principle: Key to Market Behavior. John Wiley & Sons, New York (2003)
15. Jordan, K.: An Introduction to the Elliott Wave Principle. Alchemist 40, 12–14 (2004)
16. Zadeh, L.A.: The concept of a linguistic variable and its application to approximate reasoning I. Information Science 8, 199–249 (1975)
17. Zadeh, L.A.: The concept of a linguistic variable and its application to approximate reasoning II. Information Science 8, 301–357 (1975)
18. Zadeh, L.A.: The concept of a linguistic variable and its application to approximate reasoning III. Information Science 9, 43–80 (1976)
19. Huarng, K.H.: Effective lengths of intervals to improve forecasting in fuzzy time-series. Fuzzy Sets and Systems 123, 387–394 (2001)
20. Huarng, K.H.: Heuristic models of fuzzy time-series for forecasting. Fuzzy Sets and Systems 123, 137–154 (2001)
21. Miller, G.A.: The magical number seven, plus or minus two: some limits on our capacity of processing information. The Psychological Review 63, 81–97 (1956)
22. Yu, H.K.: A refined fuzzy time-series model for forecasting. Physica A 346, 657–681 (2005)
23. Huarng, K.H., Yu, T.H.-K.: The application of neural networks to forecast fuzzy time series. Physica A 336, 481–491 (2006)

Probabilistic Techniques for
Corporate Blog Mining

Flora S. Tsai, Yun Chen, and Kap Luk Chan

School of Electrical & Electronic Engineering,
Nanyang Technological University, Singapore, 639798
fst1@columbia.edu

Abstract. With the proliferation of blogs, or weblogs, in the recent years, information in the blogosphere is becoming increasingly difficult to access and retrieve. Previous studies have focused on analyzing personal blogs, but few have looked at corporate blogs, the numbers of which are dramatically rising. In this paper, we use probabilistic techniques to detect keywords from corporate blogs with respect to certain topics. We then demonstrate how this method can present the blogosphere in terms of topics with measurable keywords, hence tracking popular conversations and topics in the blogosphere. By applying a probabilistic approach, we can improve information retrieval in blog search and keywords detection, and provide an analytical foundation for the future of corporate blog search and mining.

Keywords: Weblog search, blog mining, probabilistic latent semantic analysis, corporate blog, business blog, web mining.

1 Introduction

After a slow start in the early 1990s, blogging rapidly gained in popularity in the mid 1990s. A blog, or weblog, is a type of website with content displayed in reverse chronological order. Blogs often provide commentary or news on a particular subject, but many function more as personal online diaries. Blogosphere is the collective term encompassing all blogs as a community or social network. According to David Sifry of Technorati, over 57 million blogs were being tracked by Technorati in October 2006; on average, a new weblog is created every second [19]. Because of the huge volume of existing blog posts and their free format nature, the information in the blogosphere is rather random and chaotic; effective access and retrieval techniques are needed to improve the searching quality.

The business world has experienced significant influence by the blogosphere. A hot topic in the blogosphere may affect a product's life period. An exposure of an inside story in the blogosphere may influence a company's reputation. Studies on the blogosphere include measuring the influence of the blogosphere [7], analyzing the blog threads for discovering the important bloggers [14], determining the spatiotemporal theme pattern on blogs [12], focusing the topic-centric view of the blogosphere [2], detecting the blogs growing trends [8], tracking the propagation

T. Washio et al. (Eds.): PAKDD 2007 Workshops, LNAI 4819, pp. 35–44, 2007.

of discussion topics in the blogosphere [9], and detecting cyber security threats in blogs [20]. Although some complex algorithms have been developed to track the trends and influences of the blogosphere, other machine learning techniques can be applied in the tracking of blogs.

Many studies have focused on analyzing personal blogs, but few have looked at corporate blogs, which are published and used by organizations. Existing studies on corporate blogs have focused on blogging strategies and case studies [3,10], not on blog search and mining. In this paper, we focus on analyzing business blogs, which are blogs providing commentary or analysis of companies, and external company blogs, which can be an important link to customers and potential clients.

Although not as common as personal blogs, corporate blogs are not new. More than 8% of the Fortune 500 companies blog [1] externally, and market research shows that 35% of large companies plan to institute corporate blogs in 2006 [5]. According to the research, nearly 70% of all corporate website operators will have implemented corporate blogs by the end of 2006 [5]. Blogs can be more efficient than any other corporate communications medium because of their low cost and ability to reach millions of people with very little investment. Companies can use blogs to listen to what people are saying about the product, company, or category, and give them a chance to respond.

Some key differences between corporate blogging and other communications channel are [18]:

1. **Publishable:** Anyone can publish a blog, and each post can be immediately accessible to others worldwide.
2. **Searchable:** Blogs can be searched by subject, author, or both.
3. **Social:** Through blogs, people with shared interests build friendships unrestricted by geographic boundaries.
4. **Viral:** Information can spread faster through blogs than via a news service.
5. **Syndicated:** By creating a subscription, RSS-enabled blogs can be delivered to users, saving search time.
6. **Linkable:** Because each blog can link to others, every blogger has access to other bloggers.

Although any one of these elements can be found elsewhere, combined together they form one of the most powerful interactive Internet communication tools ever developed. Organizations who do not include blogging as an part of their overall marketing strategy may miss out on significant opportunities and be at a disadvantage when competitors adopt blogs. We hope to take advantage of the corporate blogging trend by pioneering work in the growing area of corporate blog mining.

In our work, we first built a dual-functional blog search engine, which differ from existing blog-specific search engines in that it searches the full text of the blog entry and ranks the results based on similarity measures. Moreover, to broaden the usefulness of the blog search engine, an additional function was created to detect the keywords of various topics of the blog entries, hence tracking the trends and topics of conversations in the blogosphere. Probabilistic Latent

Semantic Analysis (PLSA) was used to detect the keywords from various corporate blog entries with respect to certain topics. By using PLSA, we can present the blogosphere in terms of topics with measurable keywords.

The paper is organized as follows. Section 2 reviews the related work on blog search and mining. Section 3 describes an overview of the Probabilistic Latent Semantic Analysis model for mining of blog-related topics. Section 4 presents experimental results, and Section 5 concludes the paper.

2   Review of Related Work

This section reviews related work in developing blog-specific search engines and extraction of useful information from blogs.

2.1   Blog-Specific Search Engines

Search engines have been widely used on the Internet for decades, with Google and Yahoo currently the most popular. As blogging becomes a hit in the air, blog-specific search engines are created suit the demand. Past studies [7,15] have examined the characteristics, effectiveness and distinctiveness of blog-specific search engines. Currently, blog search engines are still in their infancy [13], and many blog-specific search engines index only XML (Extensible Markup Language) feeds, which usually consist of the summary or the first few sentences of the blog entries [15]. Moreover, a study on blog search [13] concluded that blog searches have different interests than normal web searches, suggesting that blog searches tend to track references to known entities and focuses on specific themes or topics.

As of October 2006, Technorati[1] has indexed over 57 millions blogs [19]. It is believed to be tracking the largest number of links in real-time. RSS (Really Simple Syndication) or XML feeds automatically send notification from blogs to Technorati quickly, and the thousands of updates per hour that occur in the blogosphere are tracked by Technorati. Technically, Technorati supports open microformat standards, and indexes and searches posts tagged with `rel-tag`, using the full boolean search technique on the RSS/XML files. This full boolean search functionality provides `AND`, `OR` and `NOT` functionality, which narrows down the searching results, and has made tagging more valuable. In order to provide the users with more options, search results of synonymous queries are listed. For example, when "car" is the search query, there is an option of refining the results on "auto", "automotive", "automobile", "autos", "life", "vehicles", "work".

Similarly, many other blog-specific search engines, such as Bloglines[2], Feedster[3], and BlogPulse[4], index and search the RSS/XML feeds of blogs, using the

---

[1] http://www.technorati.com
[2] http://www.bloglines.com
[3] http://www.feedster.com
[4] http://www.blogpulse.com

boolean search. However, the ranking of the results are not in the order of relevance, but rather in the time of posting, the "popularity" (number of links to the blog), or the update frequency. More importantly, the existing blog search engines do not necessarily use complex models to achieving better results, due to the constraints of the real-time nature of blogs and its extremely rapid speed of update in the blogosphere. Therefore, to increase the quality and accuracy of the search results, more complex models and information retrieval techniques are required for mining the blogs.

2.2 Extraction of Useful Information from Blogs

Current blog text analysis focuses on extracting useful information from blog entry collections, and determining certain trends in the blogosphere. NLP (Natural Language Processing) algorithms have been used to determine the most important keywords and proper names within a certain time period from thousands of active blogs, which can automatically discover trends across blogs, as well as detect key persons, phrases and paragraphs [8]. A study on the propagation of discussion topics through the social network in the blogosphere developed algorithms to detect the long-term and short-term topics and keywords, which were then validated with real blog entry collections [9]. On evaluating the suitable methods of ranking term significance in an evolving RSS feed corpus, three statistical feature selection methods were implemented: χ^2, Mutual Information (MI) and Information Gain (I), and the conclusion was that χ^2 method seems to be the best among all, but full human classification exercise would be required to further evaluate such method [17]. Probabilistic approaches based on PLSA was proposed in [12] for viewing spatiotemporal life cycle patterns of blogs and in [20] to analyze cyber security threats in blogs. These studies illustrate that the PLSA-based approach using probabilistic mixture models can be effectively blog-related analysis.

Our work differs from existing studies in two respects: (1) We focus on corporate blog entries which may contain less extraneous and more useful information than personal blogs, potentially resulting in higher quality search results (2) We have combined a blog search engine with topic and keyword extraction, and use probabilistic models to extract popular keywords for each topic.

3 Probabilistic Latent Semantic Analysis Model for Blog Mining

Probabilistic Latent Semantic Analysis (PLSA) [11] is based on a generative probabilistic model that stems from a statistical approach to LSA (Latent Semantic Analysis) [4]. PLSA is able to capture the polysemy and synonymy in text for applications in the information retrieval domain. Similar to LSA, PLSA uses a term-document matrix which describes patterns of term (word) distribution across a set of documents (blog entries). By implementing PLSA, topics are generated from the blog entries, where each topic produces a list of word usage, using the maximum likelihood estimation method, the expectation maximization (EM) algorithm.

The starting point for PLSA is the *aspect model* [11]. The aspect model is a latent variable model for co-occurrence data associating an unobserved class variable $z_k \in \{z_1, \ldots, z_k\}$ with each observation, an observation being the occurrence of a keyword in a particular blog entry. There are three probabilities used in PLSA:

1. $P(b_i)$ denotes the probability that a keyword occurrence will be observed in a particular blog entry b_i,
2. $P(w_j|z_k)$ denotes the class-conditional probability of a specific keyword conditioned on the unobserved class variable z_k,
3. $P(z_k|d_i)$ denotes a blog-specific probability distribution over the latent variable space.

In the collection, the probability of each blog and the probability of each keyword are known, while the probability of an aspect given a blog and the probability of a keyword given an aspect are unknown. By using the above three probabilities and conditions, three fundamental schemes are implemented:

1. select a blog entry b_i with probability $P(b_i)$,
2. pick a latent class z_k with probability $P(z_k|b_i)$,
3. generate a keyword w_j with probability $P(w_j|z_k)$.

As a result, a joint probability model is obtained in asymmetric parameterization. After the aspect model is generated, the model is fitted using the EM algorithm. The EM algorithm involves two steps, namely the expectation (E) step and the maximization (M) step. The E-step computes the posterior probability for the latent variable by implying Bayes' formula. The M-step is to update the parameters based on the expected complete data log-likelihood depending on the posterior probability resulted from the E-step.

The EM iteration is continued to increase the likelihood function until the specific conditions are met and the program is terminated. These conditions can be a convergence condition, or a cut-off stopping, which is specified for the reaching the local maximum, rather than a global maximum.

In short, the PLSA model selects the model parameter values that maximize the probability of the observed data, and returns the relevant probability distributions, by applying the EM algorithm. Word usage analysis with the aspect model is a common application of the aspect model. Based on the pre-processed term-document matrix, the blogs are classified onto different topics. For each topic, the keyword usage, such as the probable words in the class-conditional distribution $P(w_j|z_k)$, is determined. By applying probabilistic techniques, the relationship among blogs documents, keywords, and topics can be measured quantitatively. Thus, the PLSA model for blog mining can discover the most important keywords for a group of blog documents, which can potentially lead to automatic classification of new blog documents based on important keywords.

Empirical results indicate the advantages of PLSA in reducing perplexity, and high performance of precision and recall in information retrieval [11]. Furthermore, LSA can be used to better initialize the parameters of a corresponding

PLSA model, with the resultant of the combination of the advantages of both techniques [6].

4   Experiments and Results

We have created a corporate blog data set, built a blog search system using the latent semantic analysis model, and applied the probabilistic model for blog mining on our data set of corporate blogs. The blog search system demonstrates the ranking of corporate blog entries by similarity measures, and allows for full-text searching in different categories. We extract the most relevant categories and show the topics extracted for each category. Experiments show that the probabilistic model can reveal interesting patterns in the underlying topics for our data set of corporate blogs.

4.1   Data Set

For our experiments, we created a corporate blog data corpus (CBlogs06) that focuses on blogs created by companies or about companies. During the period from April to September 2006, we extracted a set of corporate blogs through the following methods:

1. Search corporate blog entries from various CEOs' blog sites[5].
2. Search corporate blog entries from various companies' blog sites [1].
3. Search particular corporate blog entries using the existing blog search engines, such as Bloglines, Technorati, and Google Advanced Blog Search [6].

Meaningful blog entries from these blog sites were extracted and stored into our database. There are a total of 86 companies represented in the blog entries, and Table 1 summarizes the top companies in the CBlogs06 data corpus, and Table 2 lists the fields in the database.

We then categorize the 1269 blog entries into four categories based on the contents or the main description of the blog: Company, Finance, Marketing, and Product. The Company category deals with news or other information specific to corporations, organizations, or businesses. The Finance category relates to financing, loans, credit information. The Marketing category deals with marketing, sales, and advertising strategies for companies. Finally, the Product category describes the blog entries on specific company products, such as reviews, descriptions, and other product-related news. Table 3 summarizes the number of blog entries in each category.

Each blog entry is saved as a text file in its corresponding category, for further text preprocessing. For the preprocessing of the blog data, we performed lexical analysis by removing stopwords and stemming using the Porter stemmer [16].

---

[5] http://blogwrite.blogs.com   Note: only a few CEOs' blogs write about purely business matters.
[6] http://blogsearch.google.com/blogsearch/advanced_blog_search   RSS news subscription feeds were eliminated from the search results.

Table 1. List of Top Companies

Company
Microsoft
eBay
Samsung
Dell
Amazon
Sony
Google
Apple
Palm
Yahoo

Table 2. Database schema for blog entry

Field	Type
ID	int(11)
Title	text
Author	text
Publish_Date	date
URL	text
Content	text
Category_ID	int(11)
Type_ID	int(11)
Company_ID1	int(11)
Company_ID2	int(11)

Table 3. Categories for the CBlogs06 Data Corpus

Categories	Number of entries
Company	265
Finance	348
Marketing	269
Product	387

The text files are then used as the input for the Text to Matrix Generator (TMG) [21] to generate the term-document matrix for input to the blog search and mining system.

4.2 Blog Search System

We implemented a blog search system for our corporate blog data. We used the LSA model [4] for constructing the search system, as LSA is able to consider blog entries with similar words which are semantically close, and calculate a similarity measure based on documents that are semantically similar to the query terms. The similarity measure used is the cosine similarity measure:

$$sim(q, b_j) = \frac{p_{qj}}{|x_q||b_j|} \tag{1}$$

where q is the query, b_j the selected blog document in the jth column of the term-document matrix, p_{qj} is the inner product query with blog document, and $|x_q|$ is the norm of the query vector.

In this system, a user can choose select the type and the category, and enter a word or a phrase as the query to search for blog entries matching the query. The results are ranked in the order of similarity. The title of each searched blog entry is then displayed in order of similarity. Clicking on the title results in a full text of the blog entry, as well as the original hyperlink. Figure 1 shows the blog entry results for a search on the company eBay.

1. Interview with Mike Effle of Vendio (Similarity = 0.788489)
2. eBay Express in Germany launches next week (Similarity = 0.768930)
3. Developers: Town Hall with Bill Cobb - (Similarity = 0.764883)
4. eBay Express Supports Auction Listings with BIN (Similarity = 0.746711)
5. Enhancements coming to eBay Express (Similarity = 0.741025)
6. eBay: The OS for E-Commerce? (Similarity = 0.735754)
7. We're hiring (Similarity = 0.714972)
8. Going to the Source Workshop (Similarity = 0.679372)
9. A Sad Day at eBay (Similarity = 0.679187)
10. Wise Words podcast with Greg Isaacs (Similarity = 0.641112)

Fig. 1. Blog search results for query on "eBay", ranked in order of similarity

Table 4. List of keywords for Topic 1 (Company)

Keyword	Probability
blog	0.01099289
ebay	0.00924013
amazon	0.00734261
google	0.00546131
web	0.00521787
develop	0.00512831
api	0.00501918
site	0.00478636
search	0.00449269
product	0.00425156

Table 5. List of keywords for Topic 2 (Finance)

Keyword	Probability
save	0.01575988
money	0.01410636
debt	0.00743223
year	0.00680512
financ	0.00625563
financi	0.00621918
credit	0.00593993
card	0.00591865
college	0.00591530
invest	0.00570176

Table 6. List of keywords for Topic 3 (Marketing)

Keyword	Probability
market	0.01182965
company	0.00715767
custom	0.00650473
busi	0.00578797
firm	0.00473713
advertise	0.00384597
brand	0.00370062
product	0.00358449
corpor	0.00328202
client	0.00324660

Table 7. List of keywords for Topic 4 (Product)

Keyword	Probability
mobile	0.01320409
battery	0.00845365
device	0.00826390
phone	0.00825595
window	0.00673951
tablet	0.00650617
umpc	0.00603427
samsung	0.00555756
keyboard	0.00475929
apple	0.00474569

4.3   Results for Blog Mining of Topics

We conducted some experiments using PLSA for the blog entries. Tables 4-7 summarizes the keywords found for each of the four topics (Company, Finance, Marketing, and Product).

By looking at the various topics listed, we are able to see that the probabilistic approach is able to list important keywords of each topic in a quantitative fashion. The keywords listed can relate back to the original topics. For example, the keywords detected in the Product topic features items such as mobile products, batteries, phones, and umpc (ultra mobile PC). In this way, it is possible to list popular keywords and track the hot topics in the blogosphere.

5   Conclusions

This paper presents results using probabilistic and latent semantic models for search and analysis of corporate blogs. To our knowledge, this is the first such study focusing on corporate blogs. We have created a corporate blog data corpus for this study, and categorized the data set into four classes. We have also developed a corporate blog search system that is based on latent semantic analysis, which is able to rank the results in terms of blog document similarity to the query. Our experiments on our data set of corporate blogs demonstrate how our probabilistic blog model can present the blogosphere in terms of topics with measurable keywords, hence tracking popular conversations and topics in the blogosphere. We hope that this work will contribute to the growing need and importance for search and mining of corporate blogs.

There are some aspects that can be further improved in the study for better accuracy and quality, such as larger dataset collection and automatic categorization mechanism. In addition, experiments have also been done to compare the qualities of search of alternative PLSA schemes with LSA, and the results show significant noise reduction superiority to the singular vector decomposition method used in LSA. Therefore, alternative methods can be further evaluated and implemented for corporate blog mining in the future.

References

1. Anderson, C., Mayfield, R.: Fortune 500 Business Blogging Wiki (2006), available at: http://socialtext.net/bizblogs
2. Avesani, P., Cova, M., Hayes, C., Massa, P.: Learning Contextualised Weblog Topics. WWW 2005 Workshop on the Weblogging Ecosystem: Aggregation, Analysis and Dynamics (2005)
3. Cass, J., Munroe, K., Turcotter, S.: Corporate blogging: is it worth the hype? (2005), available at: http://www.backbonemedia.com/blogsurvey/blogsurvey2005.pdf
4. Deerwester, S., Dumais, S., Landauer, T., Furnas, G., Harshman, R.: Indexing by latent semantic analysis. Journal of the American Society of Information Science 41(6), 391–407 (1990)

5. Dowling, W.G., Daniels, D.: Corporate Weblogs: Deployment, Promotion, and Measurement. The JupiterResearch Concept Report (2006)
6. Farahat, A., Chen, F.: Improving Probabilistic Latent Semantic Analysis with Principal Component Analysis. In: EACL 2006 (2006)
7. Gill, K.E.: How Can We Measure the Influence of the Blogosphere? WWW 2004 Workshop on the Weblogging Ecosystem: Aggregation, Analysis and Dynamics (2004)
8. Glance, N.S., Hurst, M., Tomokiyo, T.: BlogPulse: Automated Trend Discovery for Weblogs. WWW 2004 Workshop on the Weblogging Ecosystem: Aggregation, Analysis and Dynamics (2004)
9. Gruhl, D., Guha, R., Liben-Nowell, D., Tomkins, A.: Information Diffusion Through Blogspace. In: WWW 2004 (2004)
10. Lee, S., Hwang, T., Lee, H.-H.: Corporate blogging strategies of the Fortune 500 companies. Management Decision 44(3) (2006)
11. Hofmann, T.: Probabilistic Latent Semantic Indexing. In: SIGIR 1999 (1999)
12. Mei, Q., Liu, C., Su, H., Zhai, C.: A Probabilistic Approach to Spatiotemporal Theme Pattern Mining on Weblogs. In: WWW 2006 (2006)
13. Mishne, G., de Rijke, M.: A Study of Blog Search. In: Lalmas, M., MacFarlane, A., Rüger, S., Tombros, A., Tsikrika, T., Yavlinsky, A. (eds.) ECIR 2006. LNCS, vol. 3936, pp. 289–301. Springer, Heidelberg (2006)
14. Nakajima, S., Tatemura, J., Hino, Y., Hara, Y., Tanaka, K.: Discovering Important Bloggers based on Analyzing Blog Threads. In: WWW 2005 Workshop on the Weblogging Ecosystem: Aggregation, Analysis and Dynamics (2005)
15. Pikas, C.K.: Blog Searching for Competitive Intelligence, Brand Image, and Reputation Management. Online 29(4), 16–21 (2005)
16. Porter, M.F.: An algorithm for suffix stripping. Program 14(3), 130–137 (1980)
17. Prabowo, R., Thelwall, M.: A Comparison of Feature Selection Methods for an Evolving RSS Feed Corpus. Information Processing and Management 42, 1491–1512 (2006)
18. Scoble, R., Israel, S.: Naked Conversations: How Blogs Are Changing the Way Businesses Talk with Customers. John Wiley & Sons, Chichester (2006)
19. Sifry, D.: Sifry's Alerts: State of the Blogosphere (2006), Available at: http://www.sifry.com/alerts/archives/000443.html
20. Tsai, F.S., Chan, K.L.: Detecting Cyber Security Threats in Weblogs Using Probabilistic Models. In: Yang, C.C., et al. (eds.) LNCS, vol. 4430, pp. 46–57. Springer, Heidelberg (2007)
21. Zeimpekis, D., Gallopoulos, E.: TMG: A MATLAB Toolbox for generating term-document matrices from text collections. In: Grouping Multidimensional Data: Recent Advances in Clustering, pp. 187–210. Springer, Heidelberg (2005)

Mining Chat Conversations for Sex Identification

Cemal Köse, Özcan Özyurt, and Guychmyrat Amanmyradov

Department of Computer Engineering
Faculty of Engineering,
Karadeniz Technical University,
61080 Trabzon, Turkey
{ckose,oozyurt,guychmyrat}@ktu.edu.tr

Abstract. Chat mediums are becoming an important part of human life in societies and provide quite useful information about people such as their current interests, habits, social behaviors and tendencies. In this study, we have presented an identification system to identify the sex of a person in a Turkish chat medium. Here, the sex identification is taken as a base study in the information mining in chat mediums. This system acquires data from a chat medium, and then automatically detects the chatter's sex from the information exchanged between chatters and compares them with the known identities of the chatters. To do this task, a simple discrimination function is used to determine the sex of the chatters. A semantic analysis method is also proposed to enhance the performance of the system. The system with the semantic analyzer has achieved accuracy over 90% in the sex identification in the real chat medium.

Keywords: Mining Chat Conversations, Sex Identification, Textual Data Knowledge Extraction, Mining Machine Learning.

1 Introduction

Internet mediums such as chat medium contain a vast amount of information, which is potentially relevant to a society's current interests, habits, social behaviors, crime tendency and other tendencies [1], [2], [3]. In other words, everyday millions of people chat and generate large amount of data. Mining this large amount of data in acceptable times and detecting people aiming some certain bad activities such as committing crime is becoming more important in these mediums [4], [5]. In order to do this, users may spend a large portion of their time in these mediums. An intelligent system may help the users find the interested information in these mediums [1], [4], [6], [7]. One of our major targets is to develop a system that automatically determines persons with criminal tendencies in chat mediums. In a conversation, chatters consider the corresponding chatter's sex, and the course and contents of the conversation may be formed according to the corresponding persons' sexual identity [8], [9]. Therefore, a sample identification system is implemented to determine chatter's sex identity in chat mediums. To do this, many conversations are acquired from a specially designed chat medium, and then statistical results are derived from the conversations [1], [6], [10], [11]. Topics of these conversations are not predetermined and

T. Washio et al. (Eds.): PAKDD 2007 Workshops, LNAI 4819, pp. 45–55, 2007.
© Springer-Verlag Berlin Heidelberg 2007

may be about any subjects. These results are used to determine weighting coefficients of the proposed discrimination function. The proposed function includes some important parameters representing a group of words and signs such as abbreviations, interjections, shouting, and sex and interest related words. Each weighting coefficient of the proposed function is determined in terms of usage frequency of words in a group and determinative characteristic of each word group. Words are also grouped according to their conceptual and relational meaning and each word group is also weighed with respect to determinative importance of the group. A semantic analysis method is also employed to enhance accuracy of the system.

In this paper, we presented a mining system to collect and evaluate information obtained from chat mediums for sex identification. The system with the semantic analysis method is evaluated on the data acquired from a specially designed chat system. Performance of the system is also measured in real chat mediums. The rest of this paper is organized as follows. Some notable characteristics of Turkish language are given in Section 2. Morphological analysis methods for Turkish are also explained in the same section. The proposed discrimination function and semantic analysis methods for sex identification are presented in Section 3. A detailed description of methods used in the system is given in the same section. The implementation and results are discussed in Section 4. The conclusion and future work are given in Section 5.

2 Turkish Language

World Languages may be classified according to their structures and origins. Ural-Altaic languages are one of the most commonly spoken languages in the world. Turkic languages that belong to the group of Ural-Altaic languages form a subfamily of the Altaic languages, and Turkish is an Oghuz oriented member of Turkic languages family. Turkish language has several noticeable characteristics. One of them is that Turkish is an agglutinative language with respect to word structures formed by productive affixations of derivational and inflectional suffixes to root words [12], [13], [14]. Hence, many different words can be constructed from roots and phonemes to describe objects and concepts. Inflectional suffixes determine state, condition, case, number, person and tense of the words such as *ev+de (at home), okul+dan (from school), resim+im (my picture)* and *gel+ecek (he will come)*. Adding a derivational suffix to a noun or a verb changes the meaning of the word and generates a new word used for a different concept such as *iş (work) > iş+çi (worker), söz (word) > söz+lük (dictionary)* and *göz (eye) > göz+lük (glasses)*.

Another characteristic of the Turkish language is the vowel and consonant harmony. Turkish languages have two kinds of vowel harmony. The vowels in Turkish language are grouped as front and back vowels. The front vowels are produced at the front of the mouth and back vowels are produced at the back of the mouth. Turkish words can contain only one kind of vowel and all suffixes added to the word must conform to the vowel of the syllable preceding them [12]. Turkish language has also consonant harmony as well. If a morpheme follows a word ending in consonants k, p, t or ç, the last consonant of the word changes to constants g-ğ, b, d or c respectively.

Most of the world languages have three grammatical genders (masculine, feminine, and neuter) but Turkish language has only one (neuter). Here, some words in English may expose the sex of the chatter but they don't expose in Turkish. For example, "Ben onun kardeşiyim" in Turkish may be translated into English as "I am his brother/sister". Therefore, the identification of chatters' sex from a conversation in Turkish may be more difficult than in English.

2.1 Morphological Analysis

Turkish is an agglutinative language with respect to word structures formed by productive affixations of derivational and inflectional suffixes to root words [12], [13]. Therefore, many different words can be constructed from roots and phonemes to describe objects and concepts. Basically, the identification system takes a sentence from a chat session and analyzes all the words morphologically. A two-level morphological model is used to analyze morphological structures in the identification system [13], [14], [15]. These levels are surface and lexical levels. The surface level is the input as represented in original language. The lexical level is decomposed form of the input into morphemes. The sequence of morphemes, appearing in a word, is determined by morphotactics of a language. Our morphological analyzer recognizes eight types of phonemes; punctuation, possessives, proper names, short forms, quoted items, roots, words and suffixes. Punctuations are signs such as comma, full stop, question mark, and semi-colon. The roots and words may be a noun, verb, pronoun, adjective, adverb, preposition, conjunction, numeral, and interjection. The words are grouped as noun, verbs, adjectives, simple numbers, pronouns and connectives. Suffixes may be derivational and inflectional. Almost all of most commonly used Turkish words and suffixes are included in the database of the system.

The morphologic analysis in Turkish is realized in three steps; determination of the root of an input word, morphological tests, and determinations of the morphemes of the input word. The system tries to locate the root and possible following morphemes by checking possible constructions. For example, an internal representation of an input word like "kedisi" (his/her cat) can be explained as "kedi + sH" which is created with the help of vowel and consonant harmony rules.

3   The Identification System

To evaluate the identification system, real information is collected and extracted from chat mediums. Also some statistical data collected from the specially designed medium (SDCM) is used to evaluate the discrimination function and semantic analysis method [8], [9]. Here, male or female dominant words and some sentence structures are used to design the semantic analysis method. On the other hand, weighting coefficients of each word of the function are determined by considering the statistical information. In practice, the usage frequencies of the words may increase dramatically if the number of chatters and duration of the conversation increase. Therefore, a normalization process is applied and the normalized values are used in the identification function. The most frequently used signs obtained from the SDCM, and the mIRC or real Internet medium (RIM) are also used to evaluate the system.

Table 1. Word groups and some of the most frequently used words

Abbreviation and signs	Slang and jargon words	Politeness delicacy words	Interjections Shouting words
Hi (Slm)	My son! (Oglum)	Nice (Güzel)	Hey!/Man! (Yaw)
Answer (Cvp)	Man! (Lan)	Thanks (Tşk)	Hmm (Hımm)
What is the news (Nbr)	Uncle! (Dayı!)	Well done (Aferin)	And, soo (Ee)
You! (u)	Go away! (Defol)	Yes! (Efendim)	Oh! (Aa)
Thank you (tşk)	Repentance! (Töv-be!)	You (Siz)	Well (İi)
Messages (Msj)	Father! (Baba!)	My dear (Canım)	Hello! (alo)
One? (Bi)	Older brother (Abi)	Please (Lütfen)	O (Ooo)
Telephone (tlf, tel)	Swear! (a.q)	Pardon! (Pardon)	Hey!, Hey u (Hişt)
Okay (Ok)	Maniac (Manyak)	My lamp! (Kuzum)	What (be)
Age and sexuality related words	**Question words**	**Particle and conjunction words**	**Other words**
Age (Yaş)	What (for)? (Niye?)	Such/so/that (Öyle)	You (Sen)
Sexuality (Cinsi-yet)	Why? (Neden?)	If not/otherwise (Yoksa)	I/me (Ben)
My love (Aşkım)	Which? (Hangi?)	In order to (Diye)	If only.. (Olsun)
My lady (Baya-nım)	Where? (Nerde?)	Another /Other/ (Başka)	You (Seni)
My man/gent. (Erkeğim)	Where are you? (nerdesin?)	Thus/so/such (Böyle)	Look (Bak)
Girlfriend (Manita)	Who are you? (Kim-sin?)	Now (Şimdi)	Wrong (Yanlış)
My girl! (Kızım!)	What is it? (Neyi)	Thing/stuf (Şey)	Name (İsim)
Handsome (Yakı-şıklı!)	Why (Niçin)	Like this (Şöyle)	Answer (Cevap)
My son! (Oğlum!)	Who (Kim)	No (Yok)	Really (Gerçekten)

3.1 Classifying Words and Word Groups

For sex identification, words used in chat conversations may be grouped to identify chatter's sex easily. Hence, eighth word groups are defined to cover as many sex related concepts and subjects as possible in a chat medium. These groups and some words in the groups are listed in Table 1. These word groups are manually built by considering the conceptual relations and usage frequencies of the words used by female or male chatters. These groups are *abbreviations and signs*, slang *and jargon* words, *politeness and delicacy words, interjections and shouting, sex and age related words, question words, particle and conjunction words,* and the *other word* groups. The weighting coefficients of each word group are determined experimentally and by considering their usage frequencies.

3.2 The Discrimination Function

In this application, a simple discrimination function is used to identify the sex of a person in a chat medium. This function evaluates each word in conversations separately and collectively. Here, statistical information related to each word is obtained from the SDCM and Internet chat mediums. By considering the statistical information, a weighting coefficient is assigned for each word in each word group. Practically, a normalized (into the 0.0-1.0 interval) weighting coefficient of each word is calculated. Equation (1) is used to calculate a sexual identity value for each conceptually related word group.

$$g_i = (\alpha_{i1}w_1 + \alpha_{i2}w_2 +,...,+\alpha_{ik}w_k)/\beta_i. \tag{1}$$

$$\beta_i = \alpha_{i1} + \alpha_{i2} +,.. + \alpha_{ik}. \tag{2}$$

where, g_i varies from 0.0 to 1.0 and determines the chatters' sexual identities as female or male for ith word group, α_{ij} is the weighting coefficient of jth word in ith word group and varies related to the number of words in the interested text, w_j represent the existing jth words in the interested text (if a word exists in the text, then $w_j = 1.0$ else $w_j = 0.0$), k is the number of word in ith word group and β_i is normalization divider for the current number of existing words in the ith word group and calculated by equation (2). If a word is female dominant, α varies from 0 to 0.5, but if the word is male dominant, α varies from 0.5 to 1.0.

In the application, some conceptually related words can be emphasized collectively. Then, words are classified in several groups according to their conceptual relations. So, several word groups are defined by considering words acquired from the conversations in the chat mediums. Hence, a weighting coefficient is determined for each word group. Finally, the discrimination function as Equation (3) is formed for the sex identification of any chatter in a conversation.

$$\gamma = (\lambda_{g1} * g_1 + \lambda_{g2} * g_2 +,...,+\lambda_{gn} * g_n)/\theta. \tag{3}$$

where, γ varies from 0 to 1 and determines the chatters sexual identity as female or male, λ_{gi} is the weighting coefficient for ith female or male word group and θ is normalization divider for the current number of existing groups in a conversation and calculated by equation (4).

$$\theta = \lambda_{g1} + \lambda_{g2} +,... + \lambda_{gn}. \tag{4}$$

where, λ_{gi} is the weighting coefficient of i*th* groups. Hence, the weighting coefficients of each group are determined according to dominant sexual identity of the group. Then, the sex of the chatters may be identified as female when γ is determined to be between 0.0 and 0.5. On the other hand, chatters may be identified as male when γ is determined to be between 0.5 and 1.0.

3.3 Semantic Analysis and Sex Identification

Generally, we may ask whether it is possible to further improve the accuracy of the identification system by adding a morphological and semantic analyzer to the system. A morphologic and a semantic analyzer are employed to produce the semantic network of conversation [16], [17]. Semantically, some sentence structures in a conversation such as questions, answers and addressed sentences may expose the sex of the chatter [1], [3], [4]. For example *"How are you John?"*, *"I am fine Alice"*, *"This is David"*, *"What is your name? (ismin neydi?)* → John", "Name? *(ismin?)*→ Alan", "Who are you? *(sen kimsin?)* → David", "U *(U)* → Buket" and etc. In some conversations, many addressed sentences such as "Hi John *(Merhaba John)*", "How are you David *(Nasılsın? David)*", "What is the news Ahmet *(Nbr Ahmnet?)*", "I am fine Ali *(İyidir Ali)*" and etc. may also be used.

Table 2. An example of semantic analysis used for sex identification

	Chatter	Sentence	English
(1)	GencPrens	:)	Smile.
(2)	GencPrens	yoksa kafana göre birini bulamadın mı?	Couldn't you find someone who is like-minded with you?
(3)	merix	kafama göre birini bulamadım.	I couldn't find someone like-mined with me.
(4)	GencPrens	O zaman sen evde kalırsın bu gidişle.	At this rate you are not to be able to get married.
(5)	merix	Zaten olmasında.	Anyway, it should not become too.

The analysis of the relation between subjects and personal suffixes may also conclude important clues about the chatting persons in a conversation as presented in Table 2. For example, "It seem you are not going to be able to get married *(O zaman sen evde kal-ır-sın bu gidişle)*". In the sentence, the suffix "-sın" expresses that other chatter (merix) is female because the sentence, "you are not to be able to get married", is used for female persons.

Table 3. Personal suffixes used in semantic analysis

Suffixes	Persons
-m	Singular first person
-(y)Im (im, ım, um, üm, yim, yım, yum, yüm)	Singular first person
-n	Singular second person
-sIn (sin, sın, sun, sün)	Singular second person

Sometimes many implication sentences may also be accounted in these conversations. Here, sex of the chatters is identified indirectly through the implying sentences. For example, "I am not Murat, I am his older brother *(Ben murat değilim abisiyim)*", "No, I am a house girl *(yok ev kızıyım)*" and etc. In the semantic analysis, idiomatic expressions, and the relations between subjects and suffixes are specially analyzed for the identification. Here, the personal suffixes used in the semantic analysis are listed in Table 3.

Table 4. Personal suffixes used in semantic analysis

Verb	Example sentences	Morphological analysis
gitmek (to go)	Gittim. (**I** went)	git-ti-m >> git: verb; -ti: simple past tense suffix; -m: singular first person suffix.
	Gittin. (**You** went)	git-ti-n >> git: verb; -ti: simple past tense suffix; -n: singular second person suffix.

Some sentences, phrases and expressions can also be used to determine the persons' sexual identities. Here, the personal suffixes used in the key phrases and expressions are determined to define the sexual identity of the subject. For example, "Yakışıklıyım (I am handsome)" can morphologically be analyzed as yakışıklı-(y)ım. Here, "handsome" determines the dominant sex and the suffix "-(y)ım" determines the singular first person. Then, the chatter can be identified as male.

The following algorithm gives a simple semantic analysis approach for sex identification. Here, the SFP and SSP represent the singular first and second person respectively.

For affirmative sentences or phrases this algorithm gives the formal representation of the semantic analyzer.

```
Semantic Analysis (S)
   if S ≠ Male dominant and S ≠ Female then
      return No Result
   if S = Male dominant then
      if S.Suffix = Personal then
         if Personal suffix = SFP then
            return Chatter is male
         if Personal suffix = SSP then
            return Other chatter is possible male
      else
         return No Result
   else
      if S.Suffix = Personal then
         if Personal suffix = SFP then
            return Chatter is female
         if Personal suffix = SSP then
            return Other chatter is possible female
      else
         return No Result
```

Table 5. A generalized semantic analysis

Input sentences	Dominant	Suffixes	Result
Affirmative Sentences	Male	SFP	Chatter is Male
		SSP	Other Chatter is possibly Male (Check the next response for more precise decision)
	Female	SFP	Chatter is Female
		SSP	Other Chatter is possibly Female (Check the next response for more precise decision)
Negative Sentences	Male	SFP	No Decision (Check the next response for more precise decision)
		SSP	Other Chatter is possibly Male (Check the next response for more precise decision)
	Female	SFP	No Decision (Check the next response for more precise decision)
		SSP	Other Chatter is possibly Female (Check the next response for more precise decision)
Questions	Male	SFP	Chatter is possibly Male
		SSP	Other Chatter is possibly Male (Check the next response for more precise decision)
	Female	SFP	Chatter is possibly Female
		SSP	Other Chatter is possible Female (Check the next response for more precise decision)
Negative Questions	Male	SFP	Chatter is Male
		SSP	No Decision (Check the next response for more precise decision)
	Female	SFP	Chatter is Female
		SSP	No Decision (Check the next response for more precise decision)

A more complex semantic analysis method analyzes input with dominant male/female phrases or words. The first simple approach analyzes only the affirmative sentences but a more generalized method should analyze negative, question and negative question sentences as well. This generalized method is summarized in Table 5. Here, some input may generate more precise and some others less precise outputs. As can be seen from the table, the reply or replies must confirm the previous sentence. For example, if the firs sentence is "Güzelmiyim? (Am I beautiful?) > Güzel+mi+(y)im > {ADJ N+Question Suffix+(y)Im}", the reply may be "Evet (Yes you are)" and it confirms the first sentence.

Hence, semantic relations may contribute to the final decision and strengthen the accuracy of the identification system. Equation (5) and (6) combines the statistical and semantic identification outputs and produces a single identification output.

$$\lambda = (\lambda_{sta} * \gamma_{sta} + \lambda_{sem} * \gamma_{sem}) / \eta. \qquad (5)$$

$$\eta = \lambda_{sta} + \lambda_{sem}. \qquad (6)$$

where λ is the final result that identifies the sex of the chatter, λ_{sta} and λ_{sem} are the statistical and semantic weigh coefficients respectively, γ_{sta} and γ_{sem} are statistical and semantic identifications respectively, and η is the normalization divider. Here, the accuracy of the results increases that it shows female or male gender when λ approaches to 0.0 and 1.0 respectfully.

4 Results

In this paper, we have presented a full-scale implementation of a chat system to collect information from conversations and developed a method identifying chatters profiles. This method describes how to use a discrimination function for sex identification in the medium. About two hundreds conversations have been collected from SDCM and RIM. Forty-nine of the conversations made among ninety-eight chatters (forty-four female and fifty-four male) are chosen as the training set for testing. Experimental results indicate that the proposed discrimination function has sufficient discriminative power for the sex identification in the chat mediums. We also found that the system can quite accurately predict the chatter's sex in the mediums.

About 1.27 MB of text data was obtained from more than two hundred conversations. Duration of the conversations varies from few minutes to few hours. The identification system was run on PC with P4-3.2 GHz CPU and 512 MB RAM. For sex identification, these conversations are processed in 4.34 minutes on the system. These results prove that our system is quite promising for large-scale mining applications.

Table 6. The general result of sex identification for the specially designed medium

	Male Chatters	Female Chatters	General
Number of chatters	54	44	98
Number of correct decision	47	40	87
Number of wrong decisions	4	4	8
Number of undecided results	3	0	3
Percentage of correct decision considering NUR	92.2%	90.9%	91.6%
Percentage of correct decision	87.0%	90.9%	88.7%
Percentage of wrong decisions	7.4%	9.1%	8.2%
Percentage of undecided results	5.5%	0.0%	3.1%

Table 6 presents the sex classification results for the conversations between chatters in the medium. Two correct decision percentages are presented in the table. These reflect the percentages of the correct decisions with and without considering the

Number of Undecided Results (NUR). The accuracy of decision of the system reaches to 92.2%.

More than one hundred conversations are collected from mIRC (mIRC is a shareware Internet Relay Chat) and thirty of them are chosen randomly. The test results for the chat medium are given in Table 7. Here, five persons are chosen to determine the sex of the chatters in these conversations. These people read the collected conversations in detail and decide about the sex of the chatters. Their decisions are compared with the decision of the identification system. The decision results of the identification system are listed in the table. Here, the three chatters are not taken into consideration because the chosen peoples do not make a decision about the sex of these chatters. These results show that general accuracy of system reaches to 90%.

Table 7. The performance of identification system in the mIRC medium

	Male Chatters	Female Chatters	General
Number of chatters	19	8	27
Number of correct decisions	17	8	25
Number of wrong decisions	2	-	2
Number of undecided decisions	-	-	-
Percentage of correct decisions	89.5%	100%	92.6%
Percentage of wrong decisions	10.5%	0%	7.4%

5 Conclusions and Future Work

Nowadays chat mediums are becoming an important part of human life and provide quite useful information about people in a society. In this paper, sex identification as an information-mining problem in a chat medium is taken as a base study, and design of an identification system is addressed. A simple discrimination function with semantic analysis method is proposed for the sex identification. This identification system with the discrimination function achieves accuracy over 90% in the sex identification in the mediums.

In the literature, several other methods are also employed for text classification. In text classifications, these methods yields accuracies vary from 75% to 90% related to the categorization subjects [3], [4]. The proposed simple identification method for binary classifications (Female, Male) has a quite good performance but a comparison between other machine learning methods and this method is needed. So, comparing this proposed method with other machine learning methods such as Support Vector Machine (SVM) and Naive Bayes (NB) on the same data set may be considered as another important future work [18], [19].

In this application, misleading questions and answers are not taken into account. In the future implementation of the system, a chat engine would be employed to minimize or eliminate these misleading sentences. Another future work, a Neuro-Fuzzy method considering the intersection of the word groups, can be employed to determine the weighting coefficients of the proposed discrimination function. Then, the weighting coefficients of the proposed discrimination function would be calculated more precisely and accuracy of the identification system could be improved.

References

1. Khan, F.M., Fisher, T.A., Shuler, L.A., Tianhao, W., Pottenger, W.M.: Mining Chat-room Conversations for Social and Semantic Interactions. Lehigh University Technical Report LU-CSE-02-011 (2002)
2. Elnahrawy, E.: Log-Based Chat Room Monitoring Using Text Categorization: A Comparative Study. In: The International Conference on Information and Knowledge Sharing, US Virgin Islands (2002)
3. Baumgartner, R., Eiter, T., Gottlob, G., Herzog, M., Koch, C.: Information extraction for the semantic. In: Eisinger, N., Małuszyński, J. (eds.) Reasoning Web. LNCS, vol. 3564, pp. 275–289. Springer, Heidelberg (2005)
4. Haichao, D., Siu, C.H., Yulan, H.: Structural analysis of chat messages for topic detection, Online Information Review. 30(5), 496–516 (2006)
5. Ville, H.T., Henry, T.: Combining Topic Models and Social Networks for Chat Data Mining. In: International Conference on Web Intelligence (WI 2004), Beijing, China, pp. 206–213 (2004)
6. Harksoo, K., Choong-Nyoung, S., Jungyun, S.: A dialogue-based information retrieval assistant using shallow NLP techniques in online domains. IEICE Trans. Inf. & Syst. 5, 801–808 (2005)
7. Tianhao, W., Khan, F.M., Fisher, T.A., Shuler, L.A., Pottenger, W.M.: Error-Driven Boolean-Logic-Rule-Based Learning for Mining Chat-room Conversations. Lehigh University Technical Report LU-CSE-02-008 (2002)
8. Kose, C., Nabiyev, V., Özyurt, O.: A statistical approach for sex identification in chat mediums, The international scientific conference on Problems of Cybernetic and Informatics (PCI), 17–20 (2006)
9. Ozyurt, O., Kose, C.: Information extraction in the chat mediums: statistical and semantic approaches for sex identification, ELECO 2006, Electrical- Electronics-Computer Engineering Workshop (2006)
10. Tianhao, W., Pottenger, W.M.: A Semi-supervised Algorithm for Pattern Discovery in Information Extraction from Textual Data. In: Whang, K.-Y., Jeon, J., Shim, K., Srivastava, J. (eds.) PAKDD 2003. LNCS (LNAI), vol. 2637, pp. 117–123. Springer, Heidelberg (2003)
11. Bing, L., Xiaoli, L., Wee, S.L., Philip, S.Y.: Text Classification by Labeling Words. Nineteenth National Conference on Artificial Intelligence, 425–430 (2004)
12. Hengirmen, M.: Türkçe Dilbilgisi. Engin Yayınevi, Ankara (2002)
13. Oflazer, K.: Two-level Description of Turkish Morphology. Literary and Linguistic Computing 9, 137–148 (1994)
14. Eryigit, G., Adali, E.: An Affix Stripping Morphological Analyzer For Turkish. In: Proceedings of the IASTED International Conference on Artificial Intelligence and Applications, Innsbruck, Austria, pp. 299–304 (2004)
15. Külekci Oğuzhan, M., Özkan, M.: Turkish word segmentation by using morphological analyzer. In: Proceedings of Eurospeech, pp. 1053–1056 (2001)
16. Kanagaluru, C.S., Janaki, R.D.: The dynamics of language understanding. Language Engineering Conference, Hyderabad, India, 197–199 (2002)
17. Gao, X., Zhang, M.: Learning knowledge bases for information extraction from multiple text based Web sites. In: IEEE/WIC International Conference on Intelligent Agent Technology, pp. 119–125 (2003)
18. Joachims, T.: Text categorization with support vector machines: learning with many relevant features. In: Nédellec, C., Rouveirol, C. (eds.) Machine Learning: ECML-98. LNCS, vol. 1398, pp. 137–142. Springer, Heidelberg (1998)
19. George, H.J., Pat, L.: Estimating Continuous Distributions in Bayesian Classifiers. In: Proceedings of the Eleventh Conference on Uncertainty in Artificial Intelligence, pp. 338–345. Morgan Kaufmann, San Mateo (1995)

Mining High Impact Exceptional Behavior Patterns*

Longbing Cao[1], Yanchang Zhao[1], Fernando Figueiredo[2], Yuming Ou[1], and Dan Luo[1]

[1] Faculty of Information Technology, University of Technology, Sydney, Australia
[2] Centrelink, Australia
{lbcao,yczhao,yuming,dluo}@it.uts.edu.au,
fernando.figueiredo@centrelink.gov.au

Abstract. In the real world, exceptional behavior can be seen in many situations such as security-oriented fields. Such behavior is rare and dispersed, while some of them may be associated with significant impact on the society. A typical example is the event September 11. The key feature of the above rare but significant behavior is its high potential to be linked with some significant impact. Identifying such particular behavior before generating impact on the world is very important. In this paper, we develop several types of high impact exceptional behavior patterns. The patterns include frequent behavior patterns which are associated with either positive or negative impact, and frequent behavior patterns that lead to both positive and negative impact. Our experiments in mining debt-associated customer behavior in social-security areas show the above approaches are useful in identifying exceptional behavior to deeply understand customer behavior and streamline business process.

1 Introduction

High impact exceptional behavior refers to customers' behavior, for instance, actions taken by them, aiming or leading to specific impact on certain business or societies. The *impact* can take form of an event, disaster, government-customer debt or other interesting entities. For instance, in social security, a large volume of isolated fraudulent and criminal customer activities can result in a large amount of government customer debt. Similar problems may be widely seen from other emerging areas such as distributed criminal activities, well-organized separated activities or events threatening national and homeland security, and self-organized computer network crime [5]. Activities or events in traditional fields such as taxation, insurance services, telecommunication network, drug-disease associations, customer contact center and health care services may also result in impact on related organization or business objectives [8]. Therefore, it is important to specifically discover such impact-oriented behavior to find knowledge about what types of behavior is exceptionally associated with target impact of high interest to management.

---

* This work is sponsored by Australian Research Council Discovery Grant (DP0773412, DP0667060) and Linkage Grant (LP0775041).

T. Washio et al. (Eds.): PAKDD 2007, LNAI 4819, pp. 56–63, 2007.

There are the following characteristics of impact-targeted exceptional behavior. First, impact-targeted exceptional behavior specifically refers to those behavior itself, rather than behavior outcomes such as events, which has resulted or will result in big impact on the running of a business. Second, impact-targeted exceptional behavior is normally rare and dispersed in large customer populations and their behavior. They present unbalanced class and itemset distributions.

In this paper, we present lessons learnt in discovering low frequent and sequential exceptional behavior but associated with high impact in the social security domain. First, a strategy involving domain knowledge is discussed to partition and re-organize unbalanced data into *target set*, *non-target set* and *balanced set*, and construct impact-targeted activity baskets or sequences individually. We then mine exceptional behavior frequently leading to either *positive* (say {*P-->T*}, T refers to impact) or *negative* [6] (e.g., {*P--> $\bar{T}$* }) impact in unbalanced data. *Impact-contrasted* exceptional behavior *patterns* identify significant difference existing in two frequent patterns discovered on the same behavior basket or sequence in target set and non-target set, respectively.

We illustrate our approaches through analyzing exceptional behavior patterns leading to debt and non-debt in debt-related social-security activity data in Centrelink [1]. The outcomes of this research are of interest to Centrelink for understanding, monitoring and optimizing government-customer contacts, to prevent fraudulent activities leading to debt, and to optimize social security processes, therefore improving government payment security and policy objectives.

2   Preparing Exceptional Behavior Data

In practice, high impact exceptional behavior is a very small fraction of the whole relevant behavior records. It presents *unbalanced* [7] *class distribution* and *unbalanced itemset distribution*. Such unbalanced data makes it difficult to find useful behavior patterns due to many reasons.

To deal with the imbalance of exceptional behavior classes and itemsets, as shown in Table 1, unbalanced exceptional behavior data is organized into four data sets: original *unbalanced set*, *balanced set*, *target set*, and *non-target set*. For instance, all exceptional behavior instances related to debt in social security area are extracted into debt activity set, while those unlikely linked to debt go to non-debt set. A balanced activity set is to extract the same number of non-debt activity baskets/sequences as that of debt-related ones. The partition and re-organization of unbalanced activity data can deduce the imbalance effect, boost impact-oriented exceptional behavior, and distinguish target and non-target associated instances. In this way, impact-targeted exceptional behavior patterns easily stand out of overwhelming non-impact itemsets.

In social security government-customer contacts, the method to build activity basket/sequence is as follows. For each debt, those activities within a time window immediately before the occurrence of a debt are put in a basket/sequence. The time window is then moved forwards targeting the second debt for building another basket/sequence starting at the new occurrence of the debt.

Table 1. Partitioning unbalanced exceptional behavior data into separated sets

Set	Description
Unbalanced set	The original data set including both target and non-target exceptional behavior with unbalanced class distribution
Balanced set	A boosted data set including both target and non-target exceptional behavior with balanced class distribution
Target set	A data set solely including data of target-oriented class
Non-target set	A data set solely including data of non-target-oriented class

Furthermore, it is a strategic issue to determine the size of sliding time window. Domain knowledge, descriptive statistics and domain experts [2] are used to determine the window size. There may be varying methods to build such sliding window separating behavior instances in terms of the occurrence of an impact (1) with *changing window size*, namely the window covers all behavior instances between two impacts, (2) with *fixed window size*, namely the left hand side of the window always covers the impact, (3) with *tilt window*, namely the size can be either fixed or changing, the left hand side of the window always stop at a target impact, while the window may cover a long time period with coarser granularity for earlier behavior and finest for the latest.

For instance, Figure 1 shows two strategies, where a_i ($i=1,\ldots, m$) denotes a normal behavior and d_j ($j=1,\ldots, n$) is a debt closely associated with a series of behavior instances. In Changing Window mode, all behavior instances between the occurrences of two debts are packed into one window. This mode is more suitable for those applications with a frequent targeted impact. For instance, debt d_3 window includes behavior instance sequence $\{a_{14}, a_{15}, a_{16}, a_{17}, a_{18}, d_3\}$. Fixed Window mode fixes the length of the sliding time window, and packs all behavior instances in the window exactly before the occurrence of an impact into one window, say $\{a_8, a_9, a_{10}, a_{11}, a_{12}, a_{13}, d_2\}$. In Tilt Window mode, behavior instances happened in early time are considered but with low weight. This can be through sampling. For instance, for the scenario in Figure, we build sequence $\{a_6, a_9, a_{11}, a_{13}, a_{14}, a_{15}, a_{16}, a_{17}, a_{18}, d_3\}$.

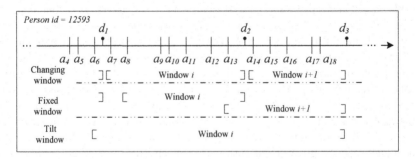

Fig. 1. Modes for constructing exceptional behavior window

On the other hand, *negative* impact targeted baskets/sequences are useful for contrast analysis. The strategy we use for building non-debt related customer behavior in social security area is to match with the positive impact scenario. Using the Fixed

Window mode, we can insert a non-debt impact into the behavior sequence if there is no debt happened then.

3 Mining High Impact Exceptional Behavior Patterns

3.1 Positive/Negative Impact-Oriented Exceptional Behavior Patterns

An impact-oriented exceptional behavior pattern is in the form of $\{P\text{-->}T\}$, where the left hand P is a set or sequence of exceptional behavior and the right hand of the rule is always the target impact T. Based on the impact type, both *positive* and *negative* impact oriented exceptional behavior patterns may be discovered.

Definition 1. *Frequent positive impact-oriented exceptional behavior patterns* (P --> T, or $\bar{P}$ --> T) refer to those exceptional behavior more likely leading to positive impact, resulting from either the appearance (P) or disappearance ($\bar{P}$) of a pattern.

Definition 2. *Frequent negative impact-oriented exceptional behavior patterns* (P --> $\bar{T}$, or $\bar{P}$ --> $\bar{T}$) indicate the occurrence of negative impact ($\bar{T}$), no matter whether an activity itemset happens or not.

In unbalanced set, a frequent impact-oriented exceptional behavior sequence leads to positive impact T: $\{P$ --> $T\}$ if P satisfies the following conditions:

- P is frequent in the whole set,
- P is far more frequent in target data set than in non-target set, and
- P is far more frequent in target set than in the whole data set. To this end, we define the following interestingness measures.

Given an activity data set A, based on exceptional behavior sequence construction methods, a subset D of A consists of all exceptional behavior baskets/sequences which are associated with positive impact, while the subset $\bar{D}$ includes all exceptional behavior baskets/sequences related to negative impact. For instance, in social security network, an exceptional behavior itemset P ($P = \{a_i, a_{i+1}, ...\}$, $a_i \in A$, $i=0, 1,....$) is associated with debt T: $\{P$ --> $T\}$. The count of debts (namely the count of sequences enclosing P) resulting from P in D is $|P, D|$, the number of debts resulting from P in A is $|P, A|$, $|P, \bar{D}|$ is the count of non-debts resulting from P in non-debt subset $\bar{D}$, $|A|$ is the count of debts in set A, and $|\bar{D}|$ is the count of non-debts in set $\bar{D}$. We define the following interestingness measures.

Definition 3. The *global support* of a pattern $\{P$ --> $T\}$ in activity set A is defined as $Supp_A(P,T) = |P, A|/|A|$.

$Supp_A(P,T)$ reflects the global statistical significance of the rule $\{P$ --> $T\}$ in unbalanced set A. If $Supp_A(P,T)$ is larger than a given threshold, then P is a frequent exceptional behavior sequence in A leading to debt.

Definition 4. The *local support* of a rule $\{P$ --> $T\}$ in target set D is defined as $Supp_D(P,T) = |P, D|/|D|$. On the other hand, the local support of rule $\{P$ --> $\bar{T}\}$ in exceptional behavior set $\bar{D}$ (i.e., non-debt exceptional behavior set) is defined as $Supp_{\bar{D}}(P,\bar{T}) = |P, \bar{D}|/|\bar{D}|$.

Definition 5. The *class difference rate* $Cdr(P, |_{\overline{D}}^{D})$ of P in two independent classes D and $\overline{D}$ is defined as

$$Cdr(P, |_{\overline{D}}^{D}) = Supp_D(P,T) / Supp_{\overline{D}}(P,\overline{T}).$$

This measure indicates the difference between target and non-target sets. An obvious difference between them is expected for positive frequent impact-oriented exceptional behavior patterns. If $Cdr(P, |_{\overline{D}}^{D})$ is larger than a given threshold, then P far more frequently leads to positive than negative impact.

Definition 6. The *relative risk ratio* $Rrr(P, |_{\overline{T}}^{T})$ of P leading to target exceptional behavior classes D and non-target class $\overline{D}$ is defined as

$$Rrr(P, |_{\overline{T}}^{T}) = Prob(T|P) / Prob(\overline{T}|P) = Prob(P,T) / Prob(P,\overline{T})$$

$$= Supp_A(P,T) / Supp_A(P,\overline{T}).$$

This measure indicates the statistical difference of a sequence P leading to positive or negative impact in a global manner. An obvious difference between them is expected for positive frequent impact-targeted exceptional behavior patterns. In addition, if the statistical significance of P leading to T and $\overline{T}$ are compared in terms of local classes, then relative risk ratio $Rrr(P, |_{\overline{T}}^{T})$ indicates the difference of a pattern's significance between target set and non-target set. If $Rrr(P, |_{\overline{T}}^{T})$ is larger than a given threshold, then P far more frequently leads to debt than results in non-debt.

Based on the above and other existing metrics such as *confidence*, *lift* and *Z-Score*, frequent impact-oriented exceptional behavior patterns can be studied to identify positive impact-oriented exceptional behavior patterns and negative impact-oriented exceptional behavior patterns.

3.2 Impact-Contrasted Exceptional Behavior Patterns

Difference between target activity set D and non-target set $\overline{D}$ may present useful contrast information in finding impact-targeted exceptional behavior patterns. For instance, exceptional behavior itemset P may satisfy one of the following scenarios:

- $Supp_D(P,T)$ is high but $Supp_{\overline{D}}(P,\overline{T})$ is low,
- $Supp_D(P,T)$ is low but $Supp_{\overline{D}}(P,\overline{T})$ is high.

In each of the above two cases, if there is a big contrast between two supports, say if $Supp_D(P,T)$ is much greater than $Supp_{\overline{D}}(P,\overline{T})$, it indicates that P is more or less associated with positive rather than negative impact, or vice versa.

In practice, those frequent itemsets P in D ($\{P \text{-->} T\}$) but not in $\overline{D}$ ($\{P \to \overline{T}\}$) are interesting because they tell us which exceptional behavior or exceptional behavior sequences lead to positive impact. In other cases, those frequent items in $\overline{D}$ ($\{P \to \overline{T}\}$) but not in D $\{P \to \overline{T}\}$ may help understand which activity sequences could prevent positive impact. Therefore, we define *impact-contrasted patterns* $P_{T\overline{T}}$ and $P_{\overline{T}T}$ as follows.

Definition 7. Given local frequent exceptional behavior itemset P, a *positive impact-contrasted pattern* $P_{T\backslash\overline{T}}$ exists if P is frequent in set D but not in set $\overline{D}$.

$$P_{T\backslash\overline{T}}: \{ P \rightarrow T , \; P \rightarrow \overline{T} \}$$

Definition 8. Given local frequent activity itemset P, a *negative impact-contrasted pattern* $P_{\overline{T}\backslash T}$ exists if P is frequent in $\overline{D}$ but not in D.

$$P_{\overline{T}\backslash T}: \{ P \rightarrow \overline{T}, \; P \rightarrow T \}.$$

After mining $P_{T\backslash\overline{T}}$, those itemsets with negative impact can be checked to see whether they trigger patterns $\{P \text{-->} T\}$ or not. It is useful in applications where an exceptional behavior or exceptional behavior sequence leads to non-target impact. If yes, then they more likely lead to positive impact. $P_{\overline{T}\backslash T}$ represents frequent itemsets that are potentially interesting for non-target exceptional behavior.

Further, to measure the interestingness of frequent impact-contrasted exceptional behavior patterns, we define contrast supports and contrast lifts for $P_{T\backslash\overline{T}}$ and $P_{\overline{T}\backslash T}$, respectively.

Definition 9. Given a positive impact-contrasted exceptional behavior pattern $P_{T\backslash\overline{T}}$, the *positive contrast support* $CSupp_D(P_{T\backslash\overline{T}})$ and *positive contrast lift* $CLift_D(P_{T\backslash\overline{T}})$ are defined as follows. They tell us how much the lift of $P_{T\backslash\overline{T}}$ is.

$$CSupp_D(P_{T\backslash\overline{T}}) = Supp_D(P,T) - Supp_{\overline{D}}(P,\overline{T})$$

$$CLift_D(P_{T\backslash\overline{T}}) = Supp_D(P,T) / Supp_{\overline{D}}(P,\overline{T}) = Cdr(P, \tfrac{|D}{\overline{D}})$$

Definition 10. Given a negative impact-contrasted pattern $P_{\overline{T}\backslash T}$, the *negative contrast support* $CSupp_{\overline{D}}(P_{\overline{T}\backslash T})$ and *negative contrast lift* $CLift_{\overline{D}}(P_{\overline{T}\backslash T})$ are defined as follows. They tell us how much the lift of $P_{\overline{T}\backslash T}$ is:

$$CSupp_{\overline{D}}(P_{\overline{T}\backslash T}) = Supp_{\overline{D}}(P,\overline{T}) - Supp_D(P,T)$$

$$CLift_{\overline{D}}(P_{\overline{T}\backslash T}) = Supp_{\overline{D}}(P,\overline{T}) / Supp_D(P,T) = Cdr^{-1}(P, \tfrac{|D}{\overline{D}}).$$

4 Experiments

We tested [1] the above-discussed patterns on Centrelink debt-related activity data [3]. We used four data sources, *activity files* recording activity details, *debt files* containing debt details, *customer files* containing customer profiles, and *earnings files* storing earnings details. Our experiments analyzed activities related to both income and non-income related debts. To analyze the relationship between activity and debt, data from activity files and debt files was extracted. The timeline used in the activity data was between the 1st Jan and the 31st Mar 2006. We extracted 15,932,832 activity transactions recording government-customer contacts for 495,891 customers, leading to 30,546 debts in the first three months of 2006.

Based on the proposed activity construction strategy, we construct 454,934 sequences: 16,540 (3.6%) activity sequences associated with debts and 438,394 (96.4%) sequences with non-debts. These sequences contain 16,540 debts and 5,625,309 activities.

Table 2 shows frequent impact-oriented activity patterns discovered from the above unbalanced activity dataset. In the table, "LSUP" and "RSUP" denote the supports of a pattern's antecedent and consequent respectively. "CONF", "LIFT" and "ZSCORE" stand for the confidence, lift and z-score of the rule. From the table, we can see that the rule ("a_1, $a_2 \rightarrow$ DET") has high confidence and lift. Its supports are very low, which are actually caused by the unbalanced class size (only 3.6% are activity sequences of debts). The second rule ("$a_1 \rightarrow$ DET") is also of high lift (6.5), but the appearance of "a_2" triples the lift of the first rule.

Table 2. Frequent debt-oriented activity patterns discovered in unbalanced set

Frequent patterns	LSUP	RSUP	SUPP	CONF	LIFT	Z-SCORE
$a_1 \rightarrow$ DET	0.0626	0.0364	0.0147	0.2347	6.5	175.7
$a4 \rightarrow$ DET	0.1490	0.0364	0.0162	0.1089	3.0	99.3
$a1$, $a4 \rightarrow$ DET	0.0200	0.0364	0.0125	0.6229	17.1	293.7
a_1, $a_2 \rightarrow$ DET	0.0015	0.0364	0.0011	0.7040	19.4	92.1

Table 3 presents a sequential impact-contrasted pattern discovered in target and non-target data sets. The pattern pair, "$a_4 \rightarrow$ DET and $a_4 \rightarrow$ NDT", has $CLift_D$ =3.24, shows that a_4 is 2.24 times more likely to lead to debt than non-debt.

Table 3. Impact-contrasted activity pattern identified in target and non-target sets

Patterns (-->DET/NDT)	$Supp_D$	$Supp_{\bar{D}}$	$CSup_D$	$CLift_D$	$CSup_{\bar{D}}$	$CLift_{\bar{D}}$
a_4	0.446	0.138	0.309	3.24	-0.309	0.31
a_5	0.169	0.117	0.053	1.45	-0.053	0.69
a_4, a_5	0.335	0.107	0.227	3.12	-0.227	0.32
a_5, a_4, a_6	0.241	0.077	0.164	3.13	-0.164	0.32

In this section, we illustrate some examples of high impact exceptional government-customer contact patterns identified in the Australian social-security activity data. The work was produced in close cooperation and on-spot assessment and iterative refinement by senior business analysts and managers in Centrelink. The Summary Report [4] delivered to the Centrelink Executives, the findings are deemed as very interesting to understand customer behavior, streamlining business processes, and preventing customer government debt.

5 Conclusions

High impact exceptional behavior is hard to be identified in massive data in which only rare and dispersed high impact behavior is of interest. The high impact exceptional behavior data presents special structural complexities, in particular, *unbalanced class* and *itemset distribution*. Mining rare exceptional behavior leading to significant impact to business is worthwhile data mining research.

In this paper, we have identified the following types of interesting impact-oriented exceptional behavior patterns in unbalanced activity data: (1) impact-oriented exceptional behavior patterns leading to either positive or negative impact, (2) impact-contrasted exceptional behavior patterns differentiating the significance of the same exceptional behavior resulting in contrast impact in target and non-target sets. New technical interestingness metrics have been developed for evaluating the above impact-targeted exceptional behavior patterns.

We have demonstrated the proposed impact-targeted activity patterns in analyzing Australian social-security activity data. The findings are of interest to Centrelink. The identified approach is also useful for analyzing exceptional behavior in many other applications, say national security and homeland security for counter-terrorism, distributed crimes and frauds, financial security, social security, intellectual property security etc.

Acknowledgement

Thanks are given to Ms Yvonne Morrow, Mr Rick Schurmann, Mr Peter Newbigin, Ms. Carol Ey, Ms. Michelle Holden at Centrelink Australia for their domain knowledge and data.

References

[1] Zhao, Y., Cao, L.: Full report: Improving income reporting (May 31, 2006)
[2] Cao, L., Zhang, C.: Domain-driven data mining: a practical methodology. Int. J. of Data Warehousing and Mining (2006)
[3] Centrelink. Integrated activity management developer guide, Technical Report, 30 (September 1999)
[4] Centrelink: Summary report: Improving income reporting (June 2006)
[5] Chen, H., Wang, F., Zeng, D.: Intelligence and security informatics for homeland security: information, communication, and transportation. IEEE Transactions on Intelligent Transportation Systems 5(4), 329–341 (2004)
[6] Wu, X., Zhang, C., Zhang, S.: Efficient Mining of Both Positive and Negative Association Rules. ACM Transactions on Information Systems 22(3), 381–405 (2004)
[7] Zhang, J., Bloedorn, E., Rosen, L., Venese, D.: [7] Zhang, J. In: Perner, P. (ed.) ICDM 2004. LNCS (LNAI), vol. 3275, pp. 571–574. Springer, Heidelberg (2004)
[8] Cao, L., Zhao, Y., Zhang, C., Zhang, H.: Activity Mining: from Activities to Actions. International Journal of Information Technology & Decision Making 7(2) (2008)
[9] Cao, L., Zhao, Y., Zhang, C.: Mining Impact-Targeted Activity Patterns in Imbalanced Data. IEEE Trans. on Knowledge and Data Engineering (to appear)

Practical Issues on Privacy-Preserving Health Data Mining

Huidong (Warren) Jin[1,2]

[1] NICTA, Locked Bag 8001, Canberra ACT, 2601, Australia
`Huidong.Jin@nicta.com.au`
[2] RSISE, the Australian National University, Canberra ACT, 2601, Australia

Abstract. Privacy-preserving data mining techniques could encourage health data custodians to provide accurate information for mining by ensuring that the data mining procedures and results cannot, with any reasonable degree of certainty, violate data privacy. We outline privacy-preserving data mining techniques/systems in the literature and in industry. They range from privacy-preserving data publishing, privacy-preserving (distributed) computation to privacy-preserving data mining result release. We discuss their strength and weaknesses respectively, and indicate there is no perfect technical solution yet. We also provide and discuss a possible development framework for privacy-preserving health data mining systems.

Keywords: Data anonymisation, secure multiparty computation, encryption, privacy inference, health data privacy.

1 Introduction

Health information, according to the Australian Commonwealth Privacy Act [1], is defined to be

1. *information or an opinion about:*
   (a) *the health or a disability (at any time) of an individual; or*
   (b) *an individual's expressed wishes about the future provision of health services to him or her; or*
   (c) *a health service provided, or to be provided, to an individual;*
   that is also personal information; or
2. *other personal information collected to provide, or in providing, a health service; or*
3. *other personal information about an individual collected in connection with the donation, or intended donation, by the individual of his or her body parts, organs or body substances.*

As important *personal information*, health information is classified as being one type of *sensitive information* [1].

With the development of powerful data mining tools/systems, we are facing the dilemma that a health data mining system should satisfy user requests for

T. Washio et al. (Eds.): PAKDD 2007 Workshops, LNAI 4819, pp. 64–75, 2007.

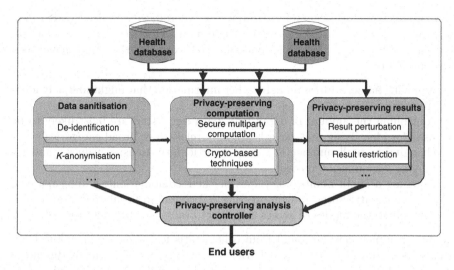

Fig. 1. Illustration of privacy-preserving health data mining system

discovering valuable knowledge from databases [2,3,4,5,6], while guarding against the ability to infer any privacy about individuals. The identification of an individual person or organisation (by the third party) should not be able to be made from mining procedures or results that we release. Furthermore, information attributable to an individual person or organisation should not be disclosed. We should develop policy, procedures as well as new techniques for privacy confidentiality with the aim of meeting legislative obligations. We only concentrate on technical issues in this paper.

Unfortunately, *privacy* faces changes over time, and a common understanding of what is meant by "privacy" is still missing [7]. For example, it is not directly defined in the Australian Commonwealth Privacy Act [1]. This fact has led to the proliferation of a wide variety of different techniques for privacy-preserving data mining. They range from privacy-preserving data publishing, privacy-preserving (distributed) computation, and privacy-preserving result release, as illustrated in the middle of Fig. 1. These techniques are crucial to develop privacy-preserving health data mining systems. We review their typical techniques and discuss their pros and cons in Sections 2, 3 and 4, respectively. We try to answer the question whether they are sufficient for a privacy-preserving health data mining system in practice. In last section, we suggest a system development framework in order to make use of their strength to protect health data privacy thoroughly.

2   Privacy-Preserving Data Publishing

The first category of privacy-preserving data mining techniques are for privacy-preserving data publishing. These techniques mainly de-identify, perturb, swap, re-code, anonymise, or even simulate raw data before conducting data mining or publishing [8,9].

A typical real-world example is Public Use Microdata Sample (PUMS) files, provided by U.S. Census Bureau [10]. These data files have been edited to protect the confidentiality of all individuals. For example, they may have been de-identified by

- removing name, address and any other information that might uniquely identify any individual;
- controlling the amount of detail (say, data items that are most likely to be used for identifying unit records are only released in broad categories);
- changing a small number of values - particularly unusual values - and removing very unusual records;
- changing a small number of values or their combinations, adding noise to continuous attributes and suppressing cell counts in tabulated data;
- controlling the modes of access to restrict access to more detailed data.

Similar examples may be found from other statistical agencies including Confidentialised Unit Record Files (CURF) provided by the Australian Bureau of Statistics (ABS) [11] and Statistics New Zealand [12]. The CURF microdata are used widely by universities, government and private sector researchers and analysts.

Latanya Sweeney [13] argued whether such kind of de-identification is enough for data privacy protection. She showed that such kind of protection of individual sources does not guarantee protection when sources are cross-examined: a sensitive medical record, for instance, can be uniquely linked to a named voter record in a publicly available voter list through some shared attributes. To eliminate such opportunities of inferring private information through link threats, Latanya Sweeney [13] introduced a very important model for protecting individual privacy, k-anonymity, a notion that establishes that the cardinality of the answer to a quasi-identifier will be at least k. The larger the k, the more difficult it is to identify an individual using the quasi-identifier. There are several techniques to randomise data to satisfy the k-anonymity. For example, starting from a very generalised data set, Bayardo and Agrawal [14] proposed an efficient technique to specialise the data set to satisfy the k-anonymity. This top-down specialisation is natural and efficient for handling both categorical and continuous attributes. Fung et al. [15] presented an efficient algorithm for determining a generalised version of data that masks sensitive information and remains useful for modelling classification. Compared to [14], this approach is greedy, thus does not guarantee optimality. However, they shew that the optimality is not needed for some problems such as classification. Actually the optimal data is the raw data without any generalisation, but such data is overfitting. The greedy algorithm is significantly faster than the one in [14]. However, optimal k-anonymisation is computationally expensive in general [14]. Data mining researchers have also established other randomisation techniques for specific data mining functionality, such as association analysis [16].

Along the direction of k-anonymity [17], data mining researchers have proposed k-anonymity model for continuous attributes [18], templates to limit sensitive inferences [19,20], l-diversity model [21] and (α, k)-anonymous model [22].

The latter two models mainly aim to handle the uniformity for individuals with the same quasi-identifier value, which can bring privacy threats. Template-based privacy-preservation [19,20] used 'confidence' to handicap the dominance of a sensitive value in an equivalence class (with the same quasi-identifier value). These models can increase privacy protection level.

2.1 Strength

Privacy-preserving data publishing techniques are found to be efficient, useful and appropriate for a data custodian (such as statistical agencies) who may make the same dataset available to many thousands of different users. The data custodian only needs to prepare one anonymised version of raw data, and doesn't have to consider which analyses will be conducted. There is no fear of litigation. For example, in Australia, under the *Census and Statistics Act 1905*, the ABS is authorised to release unit record data provided that it is done in a manner that is not likely to enable identification of a particular person or organisation to which it relates.

2.2 Weaknesses

For data privacy protection, there are several weak points if we would like to anonymise raw health data for publishing.

1. The quasi-identifiers required in the k-anonymity, the l-diversity and the (α, k)-anonymity models are quite difficult to specify beforehand, especially when we don't know what kind of information adversaries may have. This renders privacy-preserving data publishing techniques difficult for verification in practice.
2. Data mining results based on these published data can be rather different from the true, and we may have to develop special or complicated data mining techniques to compensate for anonymised or perturbed data.
3. In general, optimal anonymisation, say, k-anonymisation, is computationally expensive [14] in order to maintain health data accuracy as high as possible.

3 Privacy-Preserving (Distributed) Computation

In this paper, our definition of privacy-preserving (distributed) computation implies that nothing other than the final computation result is revealed during the whole computation procedure. In other words, intermediate computation results in a data mining procedure don't disclose privacy of the data.

This definition is equivalent to the "security" definition used in the Secure Multiparty Computation (SMC) literature. Developing privacy-preserving (distributed) computation techniques is the mainstream of privacy-preserving data mining in the literature. Vaidya and Clifton [23] gave a summary of these SMC-based techniques for vertically partitioned or horizontally partitioned data. For these SMC-based techniques, there exists a general solution based on circuit evaluation: Take a Boolean circuit representing the given functionality and produce a protocol for evaluating this circuit [24]. Circuit evaluation protocol scans the circuit from input wires to output wires, processing a single gate in each basic step. When entering each basic step, the parties hold shares of the values of the input wires, and when the step is completed they hold shares of the output wire. Thus evaluating the circuit "reduces" to evaluating single gates on values shared by both parties. This general solution is elegant in its simplicity and generality and proves the existence of a solution, but, is highly inefficient! The reason for this inefficiency is the all-to-all communication operations required during the protocol. Such operations are very costly in large-scale distributed networks [25]. Typically, specific solutions for specific problems can be much more efficient [23]. A key insight is to trade off computation and communication cost for accuracy, i.e., improve efficiency over the generic SMC method. Extending the SMC protocol, Gilburd *et al.* [25] have recently proposed the *k-privacy* definition and *k*-TTP concepts in order to scale-up SMC to hundreds of parties, where *k-privacy* is defined as the privacy attained when no party learns statistics of a group of less than *k* parties.

Some privacy-preserving (distributed) computation approaches lies on canonical encryption techniques. Comparing with other techniques, a stronger encryption scheme can be more effective and acceptable in protecting data privacy. For example, homomorphic encryption is a powerful cryptographic tool where certain computation operators are allowed to performed on encrypted data without prior decryption. Wright and Yang [26] used a homomorphic encryption technique to construct Bayesian networks from two parties but without revealing anything about their data to each other. O'Keefe *et al.* [27] used it to establish several protocols for privacy-preserving data linkage and extraction across databases of sensitive information about individuals, in an environment of constraints on organisation's ability to share data and a need to protect individuals' privacy and confidentiality.

A third possible strategy is to only release aggregate data from a party to the others rather than unit records. For example, only data summaries of at least *k* unit records will be released for final clustering [18]. There are a couple of techniques available to generate clusters based on summary statistics of groups of unit records, such as [28,4].

3.1   Strength

This category of techniques are easy for the data mining community to accept since they lie on well-developed secure computation and encryption techniques.

3.2   Weaknesses

There are several issues should bear in mind when we deploy privacy-preserving (distributed) computation as a solution.

1. Most of the privacy-preserving computation protocols lie on the semi-honest model, which assumes that participating parties follow the prescribed protocol but try to infer private information using the messages they receive during the protocol. Although the semi-honest model is realistic in many settings, there are cases where it may be better to use the "malicious model" in which we try to prevent any malicious behaviour by using more expensive cryptographic techniques.
2. Encryption/decryption operations are usually very costly in large-scale distributed networks [25], especially for the malicious model.
3. Existing efficient SMC protocols are mainly suitable for a proportion of data mining functionalities. A complete but concise set of privacy-preserving computation primitives are still open for further research [23].
4. Last but not the least, these privacy-preserving (distributed) computation techniques alone cannot provide a solution for protecting data privacy even for distributed databases. For example, as illustrated in Example 1, what if data mining results themselves disclose data privacy? More examples about frequent sequential patterns containing identification and sensitive attribute values can be found in [29]. We turn to this problem in the next section.

4   Privacy-Preserving Results Release

This category of techniques attempt to disseminate data mining results without undue risk of disclosure of individual information. Put it in other words, the disclosure of discovered knowledge do not open up the risk of privacy breaches, especially *indirect privacy divulgence*. Indirect privacy divulgence can take place by performing *inference* on pieces of superficially "insensitive" information (say, rules or patterns). In other words, inference is the process of deducing sensitive/private information from the legitimate responses received to user queries or data mining results [30]. A *privacy inference channel* indicates a series of released information from which it is probable to infer sensitive information such as identification or sensitive attribute values. Huang *et al.* discussed how to infer private information from randomised data by considering data correlation. In this section, we only discuss the privacy inference from data mining results.

A real-world privacy-preserving result release scenario is to protect the patients' privacy, such as identification and health status, in the healthcare sector. In Australia, for example, the government agency Medicare Australia holds data on drug prescriptions, while each state government holds local hospitalisation data including diagnoses [31,6]. To enhance healthcare quality, government agencies could analyse the health events and release knowledge discovered, e.g., frequent itemsets.

Example 1. Bob gets 2 itemsets from the above healthcare databases:

1. $\{a, b, c\}$ with support 1000, i.e., 1000 patients having a, b and c. a and b indicate, say, drugs while c one condition;
2. $\{a, b\}$ with support 1001;

These frequent itemsets represent a number of individuals as required by the minimum support threshold, and seemingly do not compromise privacy. However, these released frequent itemsets alone can indirectly divulge privacy. For example, Bob can easily infer that if a patient took Drugs a and b, he/she most likely suffered Condition c, which can be sensitive like HIV. In addition, Bob knows one and only one patient in the databases suffering c but not taking Drugs a and b. Through linkage with other data sources, this patient can be re-identified. This results in privacy leakage via linking attacks [22].

There are several strategies we may use in order to present results in a privacy-preserving way. One is **result perturbation** or **sanitisation**, where results are perturbed or sanitised to remove possible inference channels for privacy disclosure before releasing. They can be rounded up, recoded, dropped, suppressed, or perturbed. The second one is **result restriction**, where certain sensitive analyses are restricted if the disclosure of their results casting high risk on privacy. There are a volume of literature in statistical databases about how to restrict statistical analyses or queries [32]. Normally, a privacy-preserving health data mining system needs to use both strategies.

The Remote Data Access Laboratory (RADL) [33] provided by the ABS gives a good industrial example. It provides secure online access to a range of basic CURFs that have previously been released on CD-ROM, as well as new expanded datasets that contain more CURF data than those can be made available on CD-ROM. Using the RADL, at any time from their desktops, researchers are able to interrogate/query CURF data that they are approved to access. The ABS imposes a number of privacy and confidentiality preserving restrictions on the nature of queries and the nature and size of the outputs returned. For example, outputs will only be returned automatically if they pass *print/table limit checks* and *the total output size check* [33]. A similar system, Privacy-Preserving Analytics[TM] [34,35], was developed by CSIRO. Different from query support in the RADL, Privacy-Preserving Analytics[TM] provides a set of data analytic tools (e.g., exploratory analysis and regression) with the objective of ensuring that no unit record is released and also ensuring that information released about unit records is insufficient to make identification of any individual likely. The system also contains some data mining functionalities like sequential pattern mining [29].

In the data mining community, there are mainly two kinds of research efforts to release data mining results without compromising privacy, based on whether or not system designers have *a priori* knowledge of what is private (or sensitive). (1) If private information is given before hand, new technologies are developed to perturb original data in order to protect these sensitive information [36,37]. Unfortunately, these techniques may inevitably introduce some fake knowledge, e.g., infrequent itemsets may become frequent. (2) The other efforts focus on individual's *privacy*, which is mainly concerned with the *anonymity* of individuals. That is, without any background knowledge of what is sensitive, we have to protect the anonymity of individuals in the analysed data [13]. This is a concrete objective. For example, Atzori *et al.* [38] proposed the k-anonymous patterns concept, where a pattern P is said to be k-anonymous if its support is 0 or larger than an anonymity threshold k for a given binary database. A pattern, such as $(a \vee b) \wedge \neg c$, is defined as a logical sentence built by AND ($\wedge$), OR($\vee$), and NOT ($\neg$) logical connectives on binary attributes in the database. For instance, $supp((a \wedge b) \wedge \neg c) = 1$ in Example 1. Atzori et al. [38] studied privacy inference channels that involve only frequent itemsets. They further developed two methods for blocking these privacy inference channels by distorting released frequent itemsets [39,38]. However, there exists more complicated inference channels. For example, Fienberg and Slavkovic [40] discussed possible inferential disclosure following the release of information on one or more association rules. Jin et al. [29] proposed k-anonymous sequential patterns and α-dissociative sequential patterns as two concrete objectives for privacy protection in order to release frequent sequential patterns.

4.1 Strength

Privacy-preserving result release is an essential component of a privacy-preserving health data mining system. As we discussed above, privacy-preserving (distributed) computation alone cannot construct a privacy-preserving data mining system.

Compared with privacy-preserving data publishing techniques in Section 2, there is no need for special and complex analysis techniques to compensate for anonymised data. Standard statistical analysis or data mining tools are generally applicable, therefore the analysis and the interpretation of the results are made easier. For example, there is no need to adjust data mining techniques just because noise has been added to one or more attributes.

The released results can be generated from original raw data, and they can be quite accurate even after sanitisation [29].

Sometimes it is easier to sanitise or restrict data mining results rather than anonymise data for privacy protection.

4.2 Weaknesses

Some possible weaknesses of privacy-preserving result release techniques are listed as follows.

1. The result sanitisation procedures can be computationally expensive, though there exist some efficient sanitisation techniques for specific data mining functionalities, say, privacy-preserving sequential pattern release [29].
2. The sanitised results may still disclose some sensitive information even after deliberate computation. For example, there exist privacy inference channels for frequent itemsets besides those discussed and blocked by Atzori et al. [39,38].
3. At this stage, the sanitisation procedures are mostly developed for a specific form of knowledge, for example, privacy-preserving sequential pattern

release [29]. There is little investigation for privacy-preserving result release when data mining results from different functionalities are released.

4. Sometimes, a denial of service in a privacy-preserving system can be informative for privacy divulgence too.

5   Discussions and Possible Solutions

There are a number of research and development efforts on privacy-preserving data mining techniques/systems, ranging from privacy-preserving data publishing, privacy-preserving (distributed) computation, and privacy-preserving result release. We have discussed that they are suitable for different scenarios, e.g., most of privacy-preserving result release techniques have been developed for specific patterns or rules. They need to be incorporated with different legislation and policy. Usually such a single technique alone cannot provide an ideal technical solution [41] for privacy-preserving health data mining systems.

One possible privacy-preserving health data mining system development framework is illustrated in Fig. 1. The main motivation is to integrate various existing techniques together in order to take use of their strength at the appropriate time. The basic idea is to add a privacy-preserving data mining controller to determine which kinds of techniques should be used for a coming analysis query. An original data set may be first de-identified and/or k-anonymised, then the analysis is computed in a privacy-preserving way if data are distributed, and final data mining results are examined and sanitised before releasing. It must assess risk, especially taking into account interactions between different forms of releases (data or data mining results). It also can assess utility, accounting for data or analyses that become unreleasable. Such a health data mining system may as well be developed from a single data mining functionality first, and then add more functionalities. It stops when it is becomes impossible to include more data mining functionalities. For a set of users with different data mining requirements, a privacy-preserving health data mining system may provide modes of access appropriate to the level of data mining available.

Besides developing privacy-preserving data mining techniques, we still need to placing restrictions on how the data mining results or released data are used. For example, every organisation and individual user may sign a legal undertaking with the data mining service provider. A user should

- not attempt to identify particular persons or organisations implied in the released data or data mining results or both;
- not disclose, directly or indirectly, the private information to individuals or organisations who have not signed an appropriate legal undertakings with the service provider;
- not attempt to match, with or without using identifiers, the information with any other list of persons or organisations;
- comply with any other direction and requirements specified by the data mining service provider.

Acknowledgements

The author thanks the useful discussions and suggestions from previous colleagues Christine M. O'Keefe, Ross Sparks, Jie Chen, Hongxing He and Damien McAullay in CSIRO; He also thank Dr. Jiuyong Li at the University of Southern Australia for discussions on privacy-preserving data publishing issues. Partial financial support for this project from the Australian Research Council is gratefully acknowledged. NICTA is funded by the Australian Governments Department of Communications, Information Technology, and the Arts and the Australian Research Council through Backing Australia's Ability and the ICT Research Centre of Excellence programs.

References

1. The Office of Legislative Drafting: Privacy Act (Cth) (1988), Attorney-general's Department, Canberra, Australia (2004), http://www.privacy.gov.au/act/privacyact
2. Han, J., Kamber, M.: Data Mining: Concepts and Techniques, 2nd edn. Morgan Kaufmann, San Francisco (2006)
3. Jin, H.D., Shum, W., Leung, K.S., Wong, M.L.: Expanding self-organizing map for data visualization and cluster analysis. Information Sciences 163, 157–173 (2004)
4. Jin, H., Wong, M.L., Leung, K.S.: Scalable model-based clustering for large databases based on data summarization. IEEE Transactions on Pattern Analysis and Machine Intelligence 27(11), 1710–1719 (2005)
5. Jin, H., Chen, J., He, H., Williams, G.J., Kelman, C., O'Keefe, C.M.: Mining unexpected temporal associations: Applications in detecting adverse drug reactions. IEEE Transactions on Information Technology in Biomedicine (2007)
6. Jin, H., Chen, J., Kelman, C., He, H., McAullay, D., O'Keefe, C.M.: Mining unexpected associations for signalling potential adverse drug reactions from administrative health databases. In: Ng, W.-K., Kitsuregawa, M., Li, J., Chang, K. (eds.) PAKDD 2006. LNCS (LNAI), vol. 3918, pp. 867–876. Springer, Heidelberg (2006)
7. Crompton, M.: What is privacy? In: Privacy and Security in the Information Age Conference, Melbourne (2001), http://www.privacy.gov.au/news/speeches/sp51note1.html
8. Oliveira, S.R.M., Zaane, O.R.: Protecting sensitive knowledge by data sanitization. In: ICDM 2003. Proceedings of the Third IEEE International Conference on Data Mining, pp. 613–616. IEEE Computer Society Press, Los Alamitos (2003)
9. Kargupta, H., Datta, S., Wang, Q., Sivakumar, K.: On the privacy preserving properties of random data perturbation techniques. In: Proceedings of the Third IEEE International Conference on Data Mining, pp. 1–9. IEEE Computer Society Press, Los Alamitos (2003)
10. U.S.Census Bureau: Public-use microdata samples (PUMS) (2007) (Accessed on 21 January 2007), http://www.census.gov/main/www/pums.html
11. Australian Bureau of Statistics: Confidentialised unit record file (CURF) (2007) (Accessed on 20 January 2007), http://www.abs.gov.au
12. Statistics New Zealand: Confidentialised unit record file (CURF). (2007) (Accessed on 21 January 2007), http://www.stats.govt.nz/curf-programme
13. Sweeney, L.: k-anonymity: a model for protecting privacy. Fuzziness and Knowledge-based Systems 10(5), 557–570 (2002)

14. Bayardo, R.J., Agrawal, R.: Data privacy through optimal k-anonymization. In: ICDE 2005, pp. 217–228 (2005)
15. Fung, B., Wang, K., Yu, P.: Top-down specialization for information and privacy preservation. In: Proceedings of 21st International Conference on Data Engineering (ICDE 2005), pp. 205–216 (2005)
16. Agrawal, R., Srikant, R.: Privacy-preserving data mining. In: Proceedings of SIG-MOD 2000, pp. 439–450. ACM Press, New York (2000)
17. Li, J., Wang, H., Jin, H., Yong, J.: Current developments of k-anonymous data releasing. In: Proceedings of ehPASS 2006, Brisbane, Australia, pp. 109–121 (2006)
18. Jin, W., Ge, R., Qian, W.: On robust and effective k-anonymity in large databases. In: Ng, W.-K., Kitsuregawa, M., Li, J., Chang, K. (eds.) PAKDD 2006. LNCS (LNAI), vol. 3918, pp. 621–636. Springer, Heidelberg (2006)
19. Wang, K., Fung, B.C., Yu, P.S.: Template-based privacy preservation in classification problems. In: ICDM 2005: Proceedings of the Fifth IEEE International Conference on Data Mining, pp. 466–473. IEEE Computer Society Press, Los Alamitos (2005)
20. Wang, K., Fung, B.C.M., Yu, P.S.: Handicapping attacker's confidence: An alternative to k-anonymization. Knowledge and Information Systems: An International Journal (2006)
21. Machanavajjhala, A., Gehrke, J., Kifer, D., Venkitasubramaniam, M.: ℓ-diversity: Privacy beyond κ-anonymity. In: Proceedings of the 22nd IEEE International Conference on Data Engineering (ICDE 2006), vol. 11(3), pp. 345–368 (2007)
22. Wong, R., Li, J., Fu, A., Wang, K. (alpha,k)-anonymity: An enhanced k-anonymity model for privacy-preserving data publishing. In: KDD 2006, pp. 754–759 (2006)
23. Vaidya, J., Clifton, C.: Privacy-preserving data mining: Why, how, and when. IEEE Security & Privacy 2(6), 19–27 (2004)
24. Yao, A.: Protocols for secure computations. In: Proceedings of the twenty-third annual IEEE Symposium on Foundations of Computer Science, pp. 160–164. IEEE Computer Society Press, Los Alamitos (1982)
25. Gilburd, B., Schuster, A., Wolff, R.: k-TTP: a new privacy model for large-scale distributed environments. In: Proceedings of the 2004 ACM SIGKDD international conference on Knowledge discovery and data mining, pp. 563–568. ACM Press, New York (2004)
26. Wright, R., Yang, Z.: Privacy-preserving Bayesian network structure computation on distributed heterogeneous data. In: Proceedings of the 2004 ACM SIGKDD international conference on Knowledge discovery and data mining, pp. 713–718. ACM Press, New York (2004)
27. O'Keefe, C.M., Yung, M., Gu, L., Baxter, R.: Privacy-preserving data linkage protocols. In: WPES 2004, pp. 94–102 (2004)
28. Jin, H., Leung, K.S., Wong, M.L., Xu, Z.B.: Scalable model-based cluster analysis using clustering features. Pattern Recognition 38(5), 637–649 (2005)
29. Jin, H., Chen, J., He, H., O'Keefe, C.M.: Privacy-preserving sequential pattern release. In: PAKDD 2007. LNCS, vol. 4426, pp. 547–554. Springer, Heidelberg (2007)
30. Woodruff, D., Staddon, J.: Private inference control. In: CCS 2004: Proceedings of the 11th ACM conference on Computer and communications security, pp. 188–197. ACM Press, New York (2004)
31. Li, J., Fu, A.W.C., He, H., Chen, J., Jin, H., McAullay, D., Williams, G., Sparks, R., Kelman, C.: Mining risk patterns in medical data. In: KDD 2005, pp. 770–775 (2005)

32. Adam, N.R., Worthmann, J.C.: Security-control methods for statistical databases: a comparative study. ACM Comput. Surv. 21(4), 515–556 (1989)
33. Australian Bureau of Statistics: Remote access data laboratory (RADL) – user guide (2006) (Accessed on 20 January 2007), http://www.abs.gov.au
34. Sparks, R., Carter, C., Donnelly, J., Duncan, J., O'Keefe, C., Ryan, L.: A framework for performing statistical analyses of unit record health data without violating either privacy or confidentiality of individuals. In: Proceedings of the 55th Session of the International Statistical Institute, Sydney (2005)
35. Sparks, R., Carter, C., Donnelly, J., O'Keefe, C., Duncan, J., Keighley, T., McAullay, D., Ryan, L.: Privacy-preserving analytics: remote access methods for exploratory data analysis and statistical modelling. Under review, CSIRO (2006)
36. Fule, P., Roddick, J.F.: Detecting privacy and ethical sensitivity in data mining results. In: Proceedings of ACS 2004, pp. 159–166 (2004)
37. Oliveira, S.R.M., Zaïane, O.R., Saygin, Y.: Secure association rule sharing. In: Dai, H., Srikant, R., Zhang, C. (eds.) PAKDD 2004. LNCS (LNAI), vol. 3056, pp. 74–85. Springer, Heidelberg (2004)
38. Atzori, M., Bonchi, F., Giannotti, F., Pedreschi, D.: k-anonymous patterns. In: Jorge, A.M., Torgo, L., Brazdil, P.B., Camacho, R., Gama, J. (eds.) PKDD 2005. LNCS (LNAI), vol. 3721, pp. 10–21. Springer, Heidelberg (2005)
39. Atzori, M., Bonchi, F., Giannotti, F., Pedreschi, D.: Blocking anonymity threats raised by frequent itemset mining. In: ICDM 2005, pp. 561–564 (2005)
40. Fienberg, S.E., Slavkovic, A.B.: Preserving the confidentiality of categorical statistical data bases when releasing information for association rules. Data Mining and Knowledge Discovery 11(2), 155–180 (2005)
41. Bayardo, R.J., Srikant, R.: Technological solutions for protecting privacy. IEEE Computer 36(9), 115–118 (2003)

Data Mining for Intelligent Structure Form Selection Based on Association Rules from a High Rise Case Base

Shihai Zhang[1,2], Shujun Liu[1], and Jinping Ou[1]

[1] School of Civil Engineering, Harbin Institute of Technology, Harbin 150090, China
[2] Department of Civil Engineering, Nanyang Institute of Technology,
Nanyang 473004, China
zshhit@vip.sina.com, lsj.6290@163.com

Abstract. This paper presents a uniform model for high-rise structure design information and a case base containing 1008 high-rise buildings around the world. A case management system has been implemented with functions of data recording, modification, deletion, inquiry, statistical analysis and knowledge discovery. We propose a data-mining process of mining quantitative association rules for structure form selection from the case base and a method for mining fuzzy association rules. In the fuzzy association rule mining, we present a method for fuzzy interval division and fuzzification of quantitative attributes of the real cases. We demonstrate the application of the Apriori algorithm to generate association rules that can be used in building design. This data mining approach provides a new technical support for design efficiency, quality and intelligence.

Keywords: High-rise building structure, intelligent form selection, case base, data mining, association rules, Apriori algorithm.

1 Introduction

Selection of structure forms is a key problem in design of high-rise buildings because of the gap between the structural design theories and real practices. Selection of an optimal form for a high rise building is very complicated and with high risks, and a lot of knowledge and expertise is required. It has become a'bottle-neck'problem in intelligent structure design. Currently, different technologies, including engineering modeling, engineering resource knowledge base, data mining and knowledge discovery, etc., are used to solve this problem.

Intelligent solutions to the structure form selection problem have been widely investigated. Maher and Fenves (1985) [1] developed the expert system, HI-RISE, for high-rise primary structural design. Bailey and Smith (1994) [2] implemented the cases-based primary structural design system CADRE. Soibelman and Pena-Mora (1998, 2000)[3],[4] proposed the distributive multi-inferential method system M-RAM

T. Washio et al. (Eds.): PAKDD 2007 Workshops, LNAI 4819, pp. 76–86, 2007.

for high-rise structure concept design. Other researchers, such as Weimin Lu(1991) [5], Guangqian He(1991) [6], Ming Lai (1996)[7], Jinping OU (1997)[8], Xila Liu (1999)[9], Guangyuan Wang(2000) [10]–[13] have introduced the structural form-selection or scheme design methods based on fuzzy synthesis evaluation, fuzzy reasoning and neural networks, respectively.

In the past decades, a large number of high-rise buildings have been constructed around the world. Those construction cases accumulated plenty of data and expertise knowledge, such as strategy, experience and relative criterion, etc., which are undoubtedly important references for designing new high rise buildings [4, 14]. However, these past cases are very difficult to use effectively because of their large data volume, multi-factors, high noise, and complexity of parameters. Hence, acquiring cases of high rise structural scheme design from various sources, developing a case-based system, adopting appropriate processing method, and discovering and reusing the useful knowledge in these existing cases to guide the new design have become a new research direction in modern construction industry. In this paper, we present the establishment of a case base of high-rise structures, and discuss its application to intelligent structural form selection.

2 Modeling High-Rise Structure Cases

2.1 Information Representation of High-Rise Structural Cases

In order to acquire, store and use existing structure information from the aspect of structure design, the case information of high rise structures is divided into four types: general situation of (construction) works, surrounding condition, parameterized construction scheme and parameterized structure scheme.

The general situation of works includes the name of the work, the construction site, design time, completion time, construction time, construction units, design units, the total architecture area, the total cost etc. The surrounding condition concerns many items, such as site classification, soil-layer space distribution, liquefacient exponent, groundwork-soil bearing-weight standard, ground soil water measurement and compressing-pattern measurement, water table, site-soil concretion type, basic intensity, basic wind pressure, restriction of the near buildings, etc. The parameterized construction scheme includes the function, number of floors, the main building height, plane form and scale of the first floor, plane form and scale of the standard-floor, the three-dimensional form, the length-to-width ratio, the high-to-width ratio, the floor-height and distribution, the typical space characteristic, etc. The parameterized structure scheme includes the parameterized upper structure system and the nether structure system. The parameterized nether structure system includes the groundwork basic structure system, the basement structure system, etc. The information structure of the parameterized upper structure system is shown in Fig. 1.

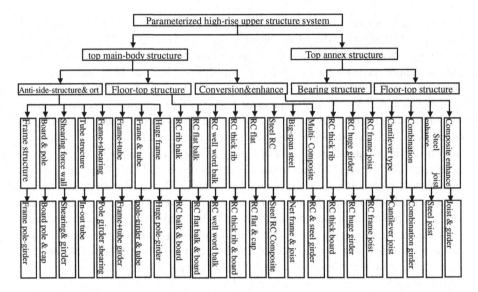

Fig. 1. High-rise building structure system

2.2 The Sources of High-Rise Structure Cases

Based on the above information structure, the cases of 1008 high-rise buildings are represented in 43 tables in a relational database. The cases contain, among others, 100 highest buildings in the world[15], high-rise buildings in the West Pacific countries[16], 100 high-rise buildings in China built before 1998[17], high-rise buildings built in 1980s in China[18], high-rise buildings over 200m high in China[19], high-rise buildings over 60m high in Japan[19,20], 40 high-rise official business buildings in China[21], 25 other high-rise buildings in China[22]. Among those buildings in China, 372 are in Shanghai, 59 in Beijing, 46 in Shenzhen, 35 in Guangzhou, and others in Wuhan, Changsha, Hangzhou, Foshan, Zhongshan, Fuzhou, Hefei, Chongqing, Chengdu, Dalian, respectively.

The high-rise buildings in the world include 205 in Japan, 26 in New York. Table 1 lists some characteristics of these cases.

3   Establishment of the High-Rise Case-Base System

3.1   Structure of the High-Rise Case-Base

By analyzing the case data, we have designed the structure of the high-rise case base as follows. It mainly includes the table of the general situation of the works, the surrounding condition table, the architecture scheme characteristic table, the group the table of main body structure forms, the group table of annex structure forms, the group table of building-covers, and the group table of basic structure forms. Fig. 2 shows the tree-structure that link these tables.

Table 1. Some characteristics of the cases in the case base

Frame	Shearing-force	Tube	Frame+Shear	Frame+Tube	Tube-in-tube	Integration
17.66%	15.34%	2.03%	35.74%	19.10%	6.08%	4.05%
Height(24-64)	(64-104)	(104-144)	(144-184)	(184-224)	(224-244)	(>244)
36.89%	23.75	11.32%	10.52%	8.04%	5.41%	4.07%
Floors(8-22)	(23-34)	(35-46)	(47-58)	(59-70)	(71-82)	(>82)
45.65%	21.44%	13.56%	10.46%	4.99%	2.40%	1.50%
Length-to-width(1-2)	(2-3)	(3-4)	(4-5)	(5-6)	(6-7)	(7-8)
73.04%	15.65%	6.96%	1.74%	0.87%	1.74%	0
Height-to-width(1-2)	(2-3)	(3-4)	(4-5)	(5-6)	(6-7)	(7-8)
16.07%	22.32%	31.25%	18.75%	7.14%	3.58%	0.89%
Official	Residence	Hotel	Integration			
32.09%	30.12%	15.55%	22.24%			

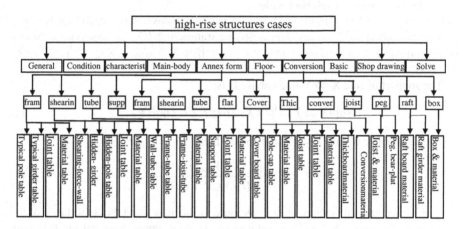

Fig. 2. The relational structure of the high-rise structure case base

3.2   Implementation of the High-Rise Case Base System

We implemented the case base (the interface of case-base inquiry is omitted.) in Microsoft SQL Server2000 that stores the relational structure of the high-rise case base and a case management system in Delphi 6.0. We have also implemented the functions of data recording, modification, deletion, inquiry, statistical analysis and data mining. The interface of the case-base management system is omitted here due to the space limitation.

4   Association Rule Mining for Form Selection from the Case Base

4.1   General Process of Association Rules Mining

The general process of association rules mining from the case base is shown in Fig. 3.

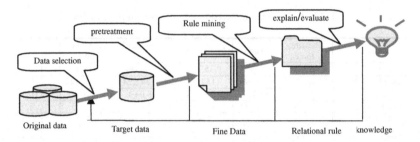

Fig. 3. The process of association rule mining

4.2 Association Rule Mining for High-Rise Structure Form Selection

4.2.1 Overview of Association Rules

Let $I=\{i_1,i_2,\ldots,i_m\}$ be the itemset, D the transaction database, where each transaction T is an item subset $(T \subseteq I)$, and has a unique transaction ID. An association rule is expressed as $X \Rightarrow Y$, meaning that "IF X, THEN Y." where $X \subset T$, $Y \subset T$, and $X \cap Y=\varphi$.

The support of rule $X \Rightarrow Y$ in database D is:

$$sup(X \Rightarrow Y)=\left|\{T: X \cup Y \subseteq T, T \in D\}\right|/\left|D\right| \tag{1}$$

The confidence of rule $X \Rightarrow Y$ in database D is:

$$conf(X \Rightarrow Y)=\frac{\left|\{T: X \cup Y \subseteq T, T \in D\}\right|}{\left|\{T: X \subseteq T, T \in D\}\right|} \tag{2}$$

Confidence and support are used to measure the accuracy and the importance of the association rule, respectively. The support shows how representational the rule is in all transactions. Obviously, the bigger the support is, the greater the association rule is covered by the data. The confidence indicates how accurate the association rule can predict the item Y given X. It should be noted that some association rules may have higher confidence but lower support which means that such rules have little data supporting. This may imply that the collected data are not enough or the data are not belonging to this class.

The rule that meets the min-support and min-confidence simultaneously is known as the strong rule.

4.2.2 Quantitative Association Rules

Mining association rules can be seen as seeking associations among the Boolean values in a relational table. In a record of the relational table, if the value of an attribute is "1(T)", it represents that an item is contained in the record (transaction) otherwise, if the attribute value is "0(F)", the corresponding item is not included. The discovery of association rules under this situation is known as the Boolean association rules problem. But the attributes in civil engineering are not always Boolean type. For example, the main-building height, over-ground floors and many other are represented as quantitative attributes, while the main-body structure form, the function and the fortifying intensity are represented as discrete attributes (enumerative attributes). For the sake of simplification, both the quantitative and discrete attributes are treated as the quantitative attributes in this paper. The quantitative association rules problem can be converted to the Boolean relational rules problem by means of threshold based classification with respect to the characteristics of the case attributes to transform the quantitative attributes to the Boolean attributes.

4.2.3 Mining Steps

Mining quantitative association rules for high-rise structure form selection can be carried out in three steps: 1) Selecting data under the instruction of the domain experts to create the target data for structure form selection, determining the interval division of every quantitative attribute range and performing Boolean conversion according to the division thresholds; 2) Discovering all the frequent item-sets; 3) Constructing and evaluating the frequent item-sets to get the association rules useful in structure form selection. The details are given as follows.

4.2.4 Boolean Representation of Target Cases

4.2.4.1 Data Selection

The main-body structure form and its relationship with other attributes as well as its relative degree are more concerned by users when they use the high-rise case-base to mine the structure form selection rules. Therefore, the data used by mining algorithms should be restricted. In line with the instruction of the experts, we choose nine attributes for structure form-selection. They are the main-building height, over-ground floors, the length-to-width ratio, the height-to-width ratio, the maximum wind pressure, function, the fortifying intensity, site classification and the main-body structure form. From the original high-rise structure case base, we selected nine records to establish the target database D and mine the quantitative relational rules. The target database D is shown in Table 2.

4.2.4.2 Attribute Conversion

In the light of the characteristics of their values, the attributes of the target database are divided into three groups. The first group includes the main-building height, over-ground floors, the length-to-width ratio, the height-to-width ratio and the maximum wind pressure. The second group includes the main-body structure form and function. The third group contains the fortifying intensity and site classification.

Table 2. Target database D

No.	Main-body structure form	Main-building height(m)	Overground floors	Length-to-width	Height-to-width	Maximum wind pressure (N/m^2)	Function	Fortifying intensity	Site classification
1	frame+shearing	66.74	23	1	2.5	981	official business	7	II
2	shearing force wall	93.51	29	1.31	1.46	1200	official business	8	II
3	tube in tube	100.60	24	1.929	2.661	1500	residence	9	II
4	frame+shearing	101	30	2.02	3.771	1600	official business	8	II
...	...	...	...	...	...	...	...	...	...

The attributes in the first group are material figures, which are divided equidistantly by experts through iterative argumentations. The defaults are provided systematically as: main-building height—10 (m), over-ground floors—3, the length-to-width ratio—0.05, the height-to-width ratio—0.05, the maximum wind pressure—500 (N/m^2). The attributes in the second group are the main-body structure form and function. Since these two attributes take the synonymous and approximate values, we can adopt the value standardization to get their typical terms in thesaurus as: frame, shearing force wall, tube, frame+shearing, frame+tube, tube in tube, colligation, official business, residence, hotel and integration. These values can be transformed into Boolean ones directly according to their corresponding typical values. The attributes in the third group are simple. Their values can be transformed into Boolean ones directly. Table 3 shows some interval divisions and classifications of the attributes being transformed into Boolean values.

Table 3. Some attributes interval division and classification

overground floors 8-11	overground floors 12-14	overground floors 15-17	overground floors 18-20	overground floors 21-23	overground floors 24-26	overground floors 27-29	...
main-building height m(20,30)	main-building height m(30,40)	main-building height m(70,80)	main-building height m(70,80)	main-building height m(70,80)	main-building height m(70,80)	main-building height m(70,80)	...
site classification I	site classification II	site classificationIII	site classificationIV	...			

4.2.4.3 Data Preprocessing

The attributes in Table 2 can be transformed into Boolean attributes in accordance with the determined classifications and the intervals of the values. The result is shown in Table 4. We can see that all the quantitative attributes are converted into Boolean attributes and one original attribute is represented in multi-attributes. Thus, the quantitative association rules' mining is transformed into the Boolean association rules mining.

Table 4. Boolean database D

No.	Main-body structure form Frame+she-aring	...	Main building height 70~80m	...	Over groun d floors 20~30 0	...	Lengt h-to-width 1.5~2	...	Height -to-width 1.5~2	...	Maximu m wind pressure 100~150 0N/m^2	...	Functi on Offici al bussin ess	...	Fortif ying intensi ty 8	...	Site classif ic-ation	...
1	T	...	F	...	T	...	F	...	F	...	F	...	F	...	F	...	T	...
2	F	...	F	...	T	...	F	...	F	...	F	...	T	...	T	...	T	...
3	F	...	F	...	T	...	T	...	F	...	T	...	F	...	F	...	T	...
4	T	...	F	...	T	...	F	...	F	...	F	...	T	...	T	...	T	...
...	...	...	...	...	...	...	...	...	...	...	...	...	...	...	...	...	...	...

4.2.4.4 Strong Association Rules Generated by the Apriori

The Apriori algorithm was used to generate association rules from the database shown in Table 4. We first found the k-item frequent item-set L_k. Then, we obtained the strong rules of structure form from L_k.

The Apriori process is as follows. (1) Scanning database D and computing the support of every single item in each record i in D. The frequent 1-itemset L_1 is made of items satisfying the support threshold; (2) Using L_1 to mine L_2; (3) Repeating this process until no new frequent itemset is discovered. At this point, all association rules are generated. Whether the discovered association rules are proper or not needs evaluation and explanation.

Based on the domain experts' evaluations on the discovered strong rules, some knowledge about generation of the structural forms can be obtained. The following are some examples of the strong rules we got:

Rule 1: function- residence=> main structural form _shearing wall (support 7.64%, confidence 50.33%).

Experts' explanation: The works with functions as residence and the main building structural forms as shearing wall takes up 7.64% in all the works; 50.33% of the works whose functions are residence have shearing walls as their main building form.

Rule 2: main building height _20.00~40.00, function_complex=>main structural form_frame (support 3.47%, confidence 79.55%).

The above structural-form-generating rules we got obviously have important referential values for guiding the structural form-selection design.

5 Mining Fuzzy Type Structural Form-Selection Rules

5.1 Fuzzification of the Quantitative Attributes Intervals

During the mining process of quantitative relational rules, a tiny change of the attributes on each interval may lead to a big change of the Boolean transformation result. Meanwhile, considering that some quantitative attributes have comparatively large and varied ranges, it is difficult to work out accurate associations between different value cutoff points or value ranges of every attribute. In this case, association rules have limitations in effectively expressing the associations between data. Therefore, we developed a mining method for fuzzy association rules through fuzzification of the first group attributes. For an attribute x whose values distribute in Interval [a, b], if the attribute is equally divided into n-1 cutoff points, we may ascertain n+1 fuzzy intervals whose centers are at the n-1 dividing points, and the left and right edges are at two ends. Every interval corresponds to a fuzzy value, the ith (i=[2, n]) fuzzy interval is [(i-2)(a-b)/n, (a-b)i/n]. The membership function of interval i=[1, n+1] is given as formula (3) below:

$$\mu_i(x) = \begin{cases} 1 - \dfrac{|(i-1)(a-b) + n(a-x)|}{(a-b)} \\ (x \in (a + \dfrac{(i-2)(a-b)}{n}, a + \dfrac{i(a-b)}{n})) \\ 0 \qquad (others\ ) \\ where\ ,\ i \in [1, n+1], x \in [a,b] \end{cases} \qquad (3)$$

From the divisions of fuzzy intervals and their corresponding membership functions introduced above, we can get the membership degrees of the quantitative attributes in each fuzzy interval. The qualitative attributes of Group 2 and Group 3 can be regarded as the fuzzy numerals 1 and 0 respectively. Table 5 gives four fuzzy intervals of the main building height and the membership degree values in every fuzzy interval.

Table 5. Fuzzy intervals and their membership degree of main building height

...	main building height (m) (40-60)	main building height (m) (50-70)	main building height(m) (60-80)	main building height (m) (70-90)	...
...	0	0	0.19	0.81	...
...	0.53	0.47	0	0	...
...	0	0	0.11	0.89	...

5.2 Mining Fuzzy Association Rules

By setting fuzzy threshold, we can transform the fuzzy values in Table 5 into Boolean attribute values of "0（F）and 1（T）". Accordingly, we can use the above quantitative association rules mining method to mine fuzzy association rules for structural form-selection.

5.3 Identification of Strong Fuzzy Association Rules

Based on domain experts' evaluations on the discovered strong fuzzy rules, some knowledge about generation of the structural forms can be obtained. The following are two strong fuzzy rules identified.

Rule 1: Main building height_20~40 and fortifying intensity_6=>main structural form _frame (support 2.58%, confidence 81.25%).

Rule 2: Main building height_40.00~60.00m and function–residence and fortifying intensity_7=>main structural form _shearing wall (support 4.56%, confidence 61.33%).

6 Application of Association Rules to Intelligent Structure Design

The association rules we discovered from the case base can not only guide the intelligent form-selection design, but also reveal the key influential factors related to different structural forms. Such information is surely of high value to the designer. Besides, the association rules can also be applied to case-reasoning. In the process of ascertaining the characteristic parameters, the distorted reasoning result is likely to occur due to the noise interference. In this case, through analysis of association rules between the characteristic attributes, we can eliminate noises and obtain influential factors closely related to the results. At the same time, in the process of distance-based case-reasoning, structural form-selection evaluation and optimal selection decision, the designer usually has to ascertain the weight coefficients expressing the relative importance between different factors. In this case, the size of appearance frequency of attributes in the association rules can be used as a reference.

7 Conclusions

In this paper, we have presented a case base system that contains 1008 high-rise structure cases for knowledge discovery in structure form selection applications and discussed mining quantitative and fuzzy association rules from the case base for structure form-selection. The following conclusions are drawn from this work.

1) On the basis of analyzing the characteristics of high-rise structure cases, we proposed a uniform method to represent case information. This method offers a uniform module structure for systematically expressing various characteristics of cases. Based on this method, we designed 43 cases information questionnaires and obtained 1008 high-rise structural cases through literature survey and consultation with experts. All of these lay a sound foundation for fully utilizing the information from existing cases.

2) We established a relationship structure about the high-rise structural case-base. We built a case base with Microsoft SQL Server 2000 and implemented a management system in Delphi 6.0. The system provides a means for uniformly storing and managing case information, and implementing the functions of information recording, modification, deletion, inquiry and statistical analysis. This system makes a useful attempt on intelligently and automatically mining knowledge for guiding the structural scheme design.

3) Considering high-rise structure intelligent form selection, we developed a process and method for mining quantitative and fuzzy association rules from the case base. Furthermore, we investigated the methods for fuzzy interval division and fuzzification of quantitative attributes. Finally, we discussed possible applications of the association rules in intelligent engineering design. This work shows a new approach to fully utilizing the information in high-rise works and improving efficiency, quality and intelligence of structural design.

References

1. Maher, M.L., Fenves, S.J.: HI-RISE: An expert system for the preliminary structure design of high rise buildings. In: Gero, S. (ed.) Knowledge engineering in computer-aided design[M], pp. 125–146. Elsevier Science Publishers, North-Holland, Amsterdam (1985)
2. Bailey, S.F., Smith, I.F.C.: Case-based preliminary building design. J. Computing in Civ.Engrg[J], ASCE 8(4), 454–468 (1994)
3. Soibelman, L.: The exploration of an integrated representation for the conceptual design phase of structural design for tall buildings through distributed multi-reasoning algorithms[D]. PhD thesis, Massachusetts Institute of Technology, Cambridge, Mass (1998)
4. Soibelman, L., Pena-Mora, F.: Distributed Multi-Reasoning Mechanism to Support Conceptual Structural Design. Journal of Structural Engineering [J] 126(6), 733–742 (2000)
5. Lu, W., Yang, J.: Anti-seismic Expert System Primary Designing of Multi High-rise RC House [J]. Engineering Mechanics 8(4), 80–87 (1991)

6. He, G., Lin, S., et al.: High-rise structure primary design experts system. [A], Key project of NSFC (The research and applications of intelligent aided decision system in engineering construction)[R] 11 (1991)
7. Lai, M., Yang, B.: Anti-seismic Form Selection and Fuzzy Synthesis Evaluation of High-rise RC Structure [J]. Journal of Building Structures 17(8), 44–51 (1996)
8. Ou, J., Zhang, S., et al.: Anti-seismic Form Selection and Fuzzy Expert System of High-rise RC Structure[J]. Earthquake Engineering and Engineering Vibration 17(2), 82–91 (1997)
9. Liu, X., Li, C.: System Selection Based on NN [J]. Journal of Building Structures 20(5), 36–41 (1999)
10. Zhang, S., Wang, G., Ou, J.: Intelligent evaluation of high-rise structure aseismatic form selection. World seism engineering[J] 18(3), 161–167 (2002)
11. Zhang, S., Liu, X., Ou, J., Wang, G.: Intelligent optimal selection of high-rise basic aseismatic form selection. Seism engineering and engineering vibration[J] 22(3), 28–35 (2002)
12. Zhang, S., Ou, J., Wang, G., Liu, S.: Data Mining and Its Applications for High-rise Structure Intelligent Form-optimization Based on Genetic Algorithm. In: IEEE The 6th World Congress on Intelligent Control and Automation(WCICA 2006) Conference Proceedings, vol. 11(12), pp. 8779–8784 (2006)
13. Wang, G., Lv, D., Zhang, S.: Some key problems of structure form selection. Journal of Haerbin architecture university [J] 33(1), 1–7 (2000)
14. Fayyad, U.M., Shapiro, G.P.: Advances in Knowledge Discovery and Data Mining [M]. MA: AAAI/MIT Press (1996)
15. Zhao, X.: Modern high-rise building structure design, vol. 1[M]. Science press (2000)
16. High-rise building structure utility design method (third version) [M]. pp. 10–17, 35-42. Tongji university press (1998)
17. Yuan, C.: High-rise building structure design [M], pp. 83–91. Heilongjiang science and technology press (2000)
18. Shen, G., Chen, C., Liu, B., et al.: Shanghai high-rise building structure design in 1980s [M]. Shanghai science popularization press (1994)
19. Chen, Y.: building structure design omnibus (China, Japan) [M]. China architecture industry press (2001)
20. Cui, H., Zhou, W.: High-rise building structure design examples [M]. China architecture industry press (1989)
21. Wu, J.: High-rise building design [M]. pp. 39–40, 49–50. China architecture industry press (1987)
22. Building structure excellent design drawing collection edit committee. Building structure excellent design drawing collection 1,2 [M]. China architecture industry press (1999)

CommonKADS Methodology for Developing Power Grid Switching Orders Systems

Ming Zhou[1,2], Jianwen Ren[2], Jianxun Qi[1], Dongxiao Niu[1], and Gengyin Li[2]

[1] School of Business Management, North China Electric Power University
[2] The Key Laboratory of Power System Protection and Dynamic Security Monitoring and Control under Ministry of Education, North China Electric Power University, Zhuxinzhuang Dewai 102206, Beijing, China
zhouming@ncepu.edu.cn

Abstract. Modeling knowledge and expertise is necessary but difficult in formation of switching operation orders in power systems. Switching operations of electrical components are very frequent in normal changes of operation modes and in system restoration. The switching operations have to accord with the requirements from the system security level as well as the component's switching security level. In order to assure correct switching operations, the switching orders sheet for switching operations is a compulsory mechanism. The formation of a switching-orders sheet involves a lot of expertise and knowledge that has to be represented with an effective modeling methodology. This paper presents the use of the CommonKADS methodology, a structured development approach for knowledge and expertise representation in designing and developing a power grid switching orders generation system. The practical applications have shown that the CommonKADS methodology has apparent merits in knowledge modeling and maintainability of the system.

Keywords: Power grid switching operation, expertise knowledge modeling, CommonKADS methodology, automatic formation, task template, ontology.

1 Introduction

The modern power system is one of the most complicated and vital systems in modern human life. A power system is composed of a large number of electrical components, such as generators, transformers, lines, circuit breakers, etc., and the new types of components keep coming rapidly. All components are subject to switching operations, including normal switching as well as restoring switching due to faults or emergent situations. Any switching operation must comply with the correct sequence and accord with the grid security requirements and the equipment's switching demand. Otherwise, any improper switching may cause a serious accident. In order to assure the power grid security, there are strict rules and management procedures to make sure correct switching operations on the power system. In practice, the system for generating the switching orders sheet is strongly required in the Chinese power grid operation.

A switching orders sheet consists of a switching task and corresponding switching steps that are formulated according to the power grid security requirements and the

T. Washio et al. (Eds.): PAKDD 2007 Workshops, LNAI 4819, pp. 87–98, 2007.

current operating state. Before switching, the order sheet is strictly examined by several experienced staff on duty. Preparation of a correct switching orders sheet requires a lot of professional knowledge such as the power grid topological structure, the power flow knowledge, security operating rules, the grid operating state, etc. The mix of different kinds of knowledge is difficult to model in a deterministic way and is often expressed as so-called expert rules in conventional applications.

Traditionally, a switching orders sheet is mainly drafted by one domain expert, and then examined for proof by several other experts due to the complexity of the involved knowledge. It is a heavy and tense daily task for the operators to draft switching-order sheets frequently. The automatic generation of switching-order sheets by aid of computer is becoming important in running power systems. The procedure for generating a switching orders sheet can be viewed as a process of "extracting" knowledge from human experts and transferring it to the machine usable formats. This is essentially a process of knowledge engineering. Therefore, the methodologies for building knowledge-based systems are used in developing the power grid switching orders systems (PGSOS).

Since 1980s, many efforts have been put on PGSOS [1], [2], [3], [4], [5]. The rule based expert system is the most common and direct approach to developing the PGSOS [1], [1], [3], in which relevant switching knowledge is extracted from human experts and transferred to the form of "if/then" production rules stored in the rule base. These rules are processed in the specific inference engine to automatically generate the switching orders sheet. Different power systems use different switching rules. When a PGSOS system that was developed for one power grid is used in another power grid, the rule base has to be changed. Often, the inference engine that processes the rules has to be modified as well. Communications with other automatic applications in power systems, such as SCADA (Supervisory Control And Data Acquisition) and power flow computation, are difficult in rule-based systems. Another method is to simulate switching by human experts on the single-line diagram of a power grid on screen or a whiteboard to produce the corresponding orders sheet [3]. The automatic error-checking function through basic logic reasoning is equipped with this approach. The effectiveness of this approach is limited by the logic rules stored in the system. The support and reuse of this system are difficult. The multi-agent system is a novel approach to the power system switching orders system. It has ability to solve complicated power system operations[4] [5]. Whatever a method is adopted, the biggest difficulty in developing PGSOS is to model the large amount of expertise knowledge. The rule based system has a common drawback, i.e., the size of rule base increasing rapidly as the size of the power system expands.

The CommonKADS methodology is a structured development approach to building knowledge-based systems (KBS) that provide support for high-level analysis and design, including knowledge acquisition. This paper presents the use of the CommonKADS methodology for knowledge and expertise representation in designing and developing a power grid switching orders generation system. The practical applications have shown that the CommonKADS methodology has apparent merits in knowledge modeling and maintainability of the system.

2   Survey of the CommonKADS Methodology

The CommonKADS methodology was developed in the European Esprit program in 1983. After the project ended early in 1990, the methodology has been actively in use throughout Europe and is gaining popularity in US. In the CommonKADS methodology, development of a knowledge based system (KBS) entails construction of a set of engineering modes of problem solving behaviors in its concrete organization and application context. This modeling concerns not only expert knowledge, but also various characteristics of how that knowledge is embedded and used in the organizational environment. The different models are the means of capturing the different sources and types of requirements that play a role in real applications. A KBS, then, is a computational realization of a collection of these models.

A CommonKADS project is composed of a suite of models, including the organization model, the task model, the expertise model, the agent model, the communication model and the design model. Fig.1 summarizes the suite of models involved in a CommonKADS project [6]. The expertise model is the central model in the CommonKADS methodology and it models the problem solving behavior of an agent in terms of knowledge that is applied to perform a certain task. Other models capture relevant aspects of the reality, such as the task supported by an application; the organizational context; the distribution of tasks over different agents; the agents' capabilities and communication; and the computational system design of the KBS. These are engineering-type models and serve engineering purposes. The models are considered not as "steps along the way," but as independent products in their own right that play an important role during the life cycle of the KBS. In the CommonKADS methodology, expertise modeling is an important aspect of the KBS development that distinguishes it from other types of software development.

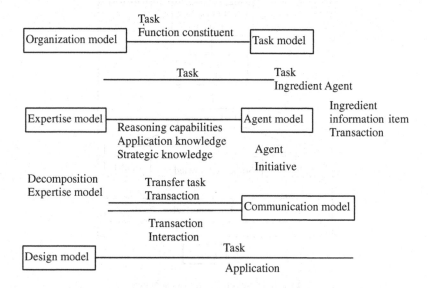

Fig. 1. The CommonKADS suite of models

3 CommonKADS Methodology Based PGSOS Knowledge Modeling

We adopt the CommonKADS methodology in PGSOS knowledge modeling and development. Fig. 2 shows the overall function scheme. The PGSOS part is contained in the dashed rectangle and interfaced to SCADA and DMIS (Dispatching Management Information System), two existing applications in the power grid, which provide real-time measurement and state data of the grid for the PGSOS. Furthermore, the generated switching orders sheet is carried out through these two systems. Therefore, the communication model is an important bridge between these two systems and the PGSOS, and a necessary link among the internal models of the PGSOS. The organization model analyzes the major features of an application to discover problems and opportunities for the PGSOS development. The templates that define object and relation types are used in each model. In the following discussions of several main models, we will give examples to explain related templates. The task model is tightly associated with the switching task in a dominant way, since a switching orders sheet is composed of a task and relevant switching steps which are specific for this task. The task model is presented as a hierarchy of tasks in which the composite tasks are general and need to be further decomposed into several subtasks. In the task model, inputs and outputs of tasks, the task features and task requirements are also modeled. The task model also specifies the distribution of tasks over agents. An agent is an executor of a task or a subtask. In the agent model, the

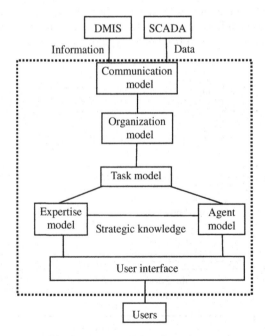

Fig. 2. The overall function scheme

capabilities of each agent are described, including relevant constraints, the rules, the needed computation, etc. There is more than one agent involved in the PGSOS. The user interface is an important and friendly man-machine interface. In our system, the hierarchical graphical interface showing real-time states is provided for users to select the object to be switched, browse the detailed switching steps generated by the PGSOS and perform relevant operations.

In the CommonKADS Methodology based PGSOS, the expertise model is the kernel model for reasoning required by tasks. A CommonKADS-based switching orders expertise model consists of three categories of knowledge: *domain*, *inference*, and *task*, each consisting of a group of associated knowledge structures. The first category contains a description of the domain knowledge of a PGSOS application. This description has to be as independent as possible from the role that the domain knowledge plays in the reasoning process. In the PGSOS, such usage-independent descriptions of domain knowledge are formalized as templates in a topological, component based structure.

The *inference knowledge* of the PGSOS expertise model describes the reasoning steps (inference actions) that can be performed with respect to the domain knowledge in a way that the domain knowledge is used in these inference steps. The PGSOS formalizes the inference knowledge as a metarule of the domain knowledge. Moreover, the PGSOS represents each inference action with a logic that is defined by a system of axioms in an order-sorted, component-based logic. An inference action's inputs and outputs (called knowledge roles) correspond to these logic arguments. Knowledge roles are described in the domain-independent terminology. The PGSOS represents the relations between the inference steps, through their shared input/output roles, with a dependency graph among inference steps and knowledge roles. Such a graph, called an inference structure, specifies only the data dependencies among the inferences, not the order in which they should execute.

The execution order among the inference steps is specified as *task knowledge* (*task model*). For this purpose, the PGSOS uses a simple procedural language with primitive procedures to execute inference steps and logic rules to test the contents of knowledge roles. The PGSOS can combine these procedures via sequences, conditions, or iterations. The PGSOS formalizes the procedural language through the quantified dynamic logic.

3.1 Modeling Domain Knowledge

The domain knowledge model describes the object/knowledge structure of static information and knowledge of an application. It can be divided into two parts: domain type and domain knowledge base. In the PGSOS, domain knowledge describes the topology of the power grid, which consists of primary equipments and secondary equipments (protection). Nowadays, the connection modes of power substations are standardized. In general, typical primary equipments and secondary equipments are integrated. By summarizing these standardized connections and the large amount of historical switching orders sheets, we can define the domain knowledge type.

Domain type. Domain type presents schematic descriptions to specific domain knowledge and information, which describes static domain knowledge/knowledge structure. The dominant feature of power system domain knowledge is hierarchical.

The high level is composed of tie lines, generating plants and power distribution substations; the middle level consists of buses and compartments; and the bottom level consists of primary equipments, e.g., generators, transformers, feeders, breakers, PT, etc., as well as the secondary protection. So domain type is designed in a three-level mode: grid knowledge type, compartment type, and equipment type.

Fig.3 shows the domain type of generating plants or power distribution substations in the grid level, since generating plants have similar structure with distribution substations except that they have generators, while the substations have source feeders.

Domain type of substation and plant
Member :
Name
Identification number
Type (Plant or Substation)
Voltage level

Fig. 3. The domain type of plant/substation

Tie lines are the links between plants and substations, and their domain type is shown in Fig.4.

Domain type of tie line
Member :
Name
ID
Voltage level
One substation linked
Another substation linked
Status
Main protection 1
Main protection 2
Backup protection 1
Backup protection 2
…….…

Fig. 4. The domain type of tie lines

The compartment is specifically defined as a group of relevant equipments in the power system operation. For example, Fig. 5 shows a line compartment, which consists of a breaker and several relevant switches (including feeder switches and grounding switches), and the equipped protection relays (now explicitly shown). This compartment is often called Compartment 2212 in practice. In most cases, the switching operations are focused on the devices in a compartment, having little relation with other devices, especially for simple tasks. For those composite switching tasks, by task decomposition to be introduced below, they are finally divided into several simple tasks whose operations are limited to several particular compartments. The coordination between compartments is inducted by the task control.

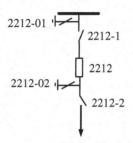

Fig. 5. A feeder compartment

The secondary protection knowledge. Each primary component in power systems is equipped with a suite of advanced secondary protections. The related protection operations are generally accompanied with the protected primary equipments and subject to coordinated operation rules. Even if the operation task is indeed to take on/off certain protection, or adjust the protection setting value, one has firstly to operate the protected primary equipment and then proceed with protection operations. Hence, the specific protection domain knowledge is absorbed into one of the protected primary equipments, including characteristics of the protection name, type, protecting zone, action conditions, and operation mode, etc. When establishing a domain knowledge base for an application, a proper protection is selected from the protection knowledge base to equip the relevant primary equipments. Accordingly, task models also define the involved protection operation attached to the protected primary equipment if needed. That is called "the secondary following the primary".

3.2 Modeling Task Knowledge

The task model describes the tasks that are performed or will be performed in the application where the expert system will be installed. In the CommonKADS based PGSOS, the task model is represented in a layered structure from the grid-level tasks to component-level tasks. On the bottom of the task model, the inference and transformation functions are directly related to tasks. These tasks are called basic tasks and others called composite tasks.

A specification of a CommonKADS task is divided into two parts: task goal and task control. The task definition is a declarative specification of the goal of the task, describing what must be achieved. The task control specifies a procedure, and prescribes the activities to accomplish the task. The task control describes how the goal can be achieved.

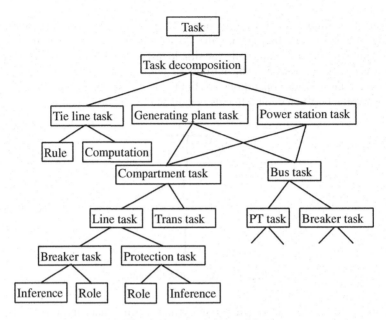

Fig. 6. The task decomposition

Fig.6 shows the task decomposition procedure in the PGSOS application. In the PGSOS, a task is in general very complicated and needs to be decomposed into many subtasks in the form of hierarchy in terms of the status of the involved task in the power grid. For example, the task is firstly decomposed into tie line tasks and generating plant tasks or power substation tasks. If it is a tie line task, then the corresponding rules and computation are invoked to complete the associated inference and produce the detailed switching steps with the aid of domain knowledge and inference knowledge to be introduced below. If it is a generating plant task or a distribution substation task, the further decomposition is in process. Fig. 7 shows a bus task is decomposed into basic subtasks.

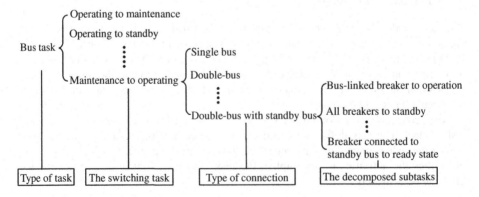

Fig. 7. A bus task decomposition

Task Presentation. The task in the PGSOS is presented in the form of templates that consists of two parts as required in CommonKADS. The task definition is a declarative specification of the goal of task, describing what is to be achieved. The task body specifies a procedure and prescribes the activities to accomplish that task. The task body also describes how to realize the task.

Fig.8 shows a template about a line task of a T-connected tie line from maintenance to operation. The selection box on the top specifies the task that is integrated by three parts: switching object (line in this example), connection type (T-connection) and goal state to be achieved (from maintenance to operation). The lower edit boxes describe the procedure of task decomposition and the corresponding task control. There are several subtasks involved in this switching task, each subtask being expressed in involved components, initial state, goal state and descriptive task control. In the specifications of the task body, the mark "*" refers to the specific object that will be replaced by a specific component, and the mark "?" indicates to be determined according to the practical condition by the input role. The switching objects include lines, buses, transformers, breakers, protection relay and their combination. There are five goal states: operating, hot standby, cold backup, maintenance and take-off.

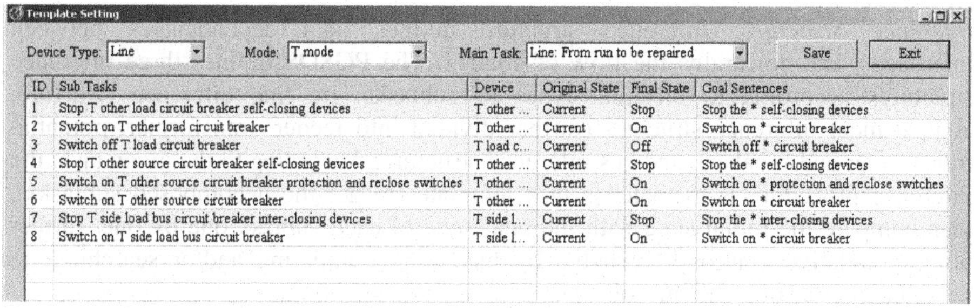

Fig. 8. Task configuration template

3.3 Modeling Inference Knowledge

In CommonKADS, inference knowledge describes how to apply domain knowledge to derive new information, which consists of inferences and roles. An inference is implemented by several operations, including search, match, conflict eliminating and triggering. A role is a label for some class of domain knowledge elements that are used in a particular inference operation, which is similar to that of inference rules in the classical logic. A role is often viewed as an input/output specification derived from an inference. There are two types of roles: static roles and dynamic roles. Static roles indicate underlying domain knowledge that is used by the inference, while dynamic roles are derived from the inference process.

Taking a switching inference as an example, the goal states of each subtask, formed by task decomposition and the present states, can be viewed as dynamic roles or inferred outputs. Static roles include domain knowledge, relevant operating rules and control knowledge. The PGSOS summarizes a large amount of logic combination describing switching rules, and expresses them in the inference structure. The inference is to realize the state transformation from the initial states to the goal states through operations such as search, match, conflict eliminating and triggering. Fig.9 presents an inference structure in the PGSOS.

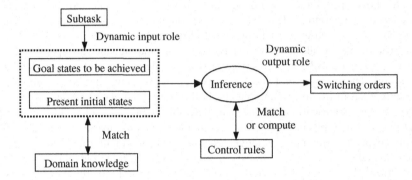

Fig. 9. An inference structure

Inference Structure. Inference structure defines data dependencies between inferences, and constrains the flow of control. The PGSOS defines three inference structures according to different types of subtasks: tie line inference structure, compartment inference structure (further divided into feeder, transformer, potential transformer and breaker), and bus inference structure. According to the given task, the inference procedure is to match a task template (to get the decomposed subtasks), then compare the initial state with the goal state of each subtask one by one, further utilizing relative control knowledge by matching, computing and/or searching, to generate a series of switching steps.

Fig.10 shows the relationship of the above three knowledge models.

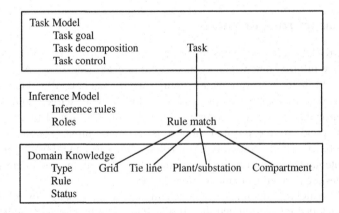

Fig. 10. The relationship of task model, inference model and domain knowledge model

4 Reuse and Maintainability of the System

The switching orders system makes full use of expertise knowledge with little computation in comparison with other analysis applications in power systems. Reuse and maintainability of the system is very important. Since the PGSOS is used to support daily operators, one system that is specially developed for one application may be not suitable for another. Since the design of the PGSOS complies with the CommonKADS methodology, the structural design of CommonKADS, not rules database design, provides more beneficial to reuse of the PGSOS.

The structural design of CommonKADS realizes the reuse and maintainability of the PGSOS from two aspects. The domain knowledge is modeled in the form of ontology and the task knowledge is expressed in templates (or task structures). Except for these structural designs, in our development, the relevant configuration platforms with friendly man-machine interfaces are provided for the users to maintain these ontologies and templates.

The domain knowledge and control-related knowledge are integrated into expertise knowledge models. The domain knowledge is explicitly described in a number of ontologies. These ontologies are metamodels describing the model structure of (part of) the domain knowledge. The ontologies are organized in a multilevel structure, where each level corresponds to a particular type of topological connection. Since the topological structure of a power system has a hierarchical feature with a series of standard substation connections in the infrastructure, such as single-bus connection, double-bus connection, etc. We model these standard connection forms with different ontologies. More importantly, these ontologies can be modified in content through the specially provided configuration platform.

The task knowledge is modeled in the form of templates. Fig.8 gives one example where the star mark and question mark will be instantiated in a practical application. Furthermore, the task body can be modified and extended by a special authorization. The task knowledge is the kernel of the PGSOS, which is generally regarded as the difficult point in realizing the universality of the PGSOS. The structural design of task knowledge provides a good solution to this problem. Furthermore, the task template in a certain extent makes the inference easy to realize since the template tells what the goal is and how to achieve this goal. The inference is in fact a procedure of searching and matching in the domain knowledge base and expertise knowledge base.

5 The Practical Applications

Based on the above knowledge models, we chose C++ to develop the system on top of an Oracle database. The developed PGSOS has been applied to three real power grids. Due to adopting the CommonKADS methodology to design the expertise knowledge, the developed system is easy to apply to a certain power grid. The users just require short time training to draw a system connection diagram with the provided graphical tool (a visualized expression of ontology). The system automatically constructs the grid's topological relations and form relevant domain knowledge bases. The users need to check the expertise knowledge base, especially the task templates, to confirm that the present templates are enough and correct for their own system. If not, they can modify these templates through the configuration platform in advance, and then continue to install the system. The field applications have shown the CommonKADS methodology based PGSOS has high efficiency in modeling, development and maintainability, and a database smaller than the expertise rule-based systems.

6 Conclusions

Switching operation is an important and frequent daily task in power system operation and requires a large amount of expertise knowledge. How to model expertise knowledge is the core of developing the switching orders automation system. The CommonKADS methodology is a structured development approach to building knowledge-based systems that provide support for high-level analysis and design, including knowledge acquisition. This paper adopts the CommonKADS methodology to design the expertise knowledge related to switching operations and develop the power system switching orders generating system. The practical applications have shown the CommonKADS methodology based PGSOS is much more efficient in development, reuse and maintainability of the system.

References

1. Mohinder, S., Sachdev, P.D., Tarlochan, S.S.: A Computer-aided Technique for Generating Substation Interlocking Schemes. IEEE Trans. on Power Delivery 2, 538–544 (2000)
2. Zhou, M., Ren, J.W., Yang, Y.H.: Application of the Switch Control Logic to the Automatically Generated Operation Order System. Automation of Electric Power Systems 21(8), 31–34 (1997)
3. Tang, L., Zhang, B.M, Sun, H.B.: General Cognitive Models of Power Network in Operating Command Expert System. Automation of Electric Power Systems 25(22), 6–8 (2001)
4. Nagata, T., Nakayama, H., Utatani, M.: A Multi-agent Approach to Power System Normal State Operations. In: IEEE PES Summer Meeting Proceedings, pp. 1582–1586. IEEE Computer Society Press, Los Alamitos (2002)
5. Zhou, M., Ren, J.W., Li, G.Y., Xu, X.H.: A Multi-Agent Based Dispatching Operation Instructing System in Electric Power Systems. IEEE PES General Meeting, Toronto, Canada, 13–18, 436–440 (2003)
6. Schreiber, G., Wielinga, B., de Hoog, R., Akkermans, H., Van de Velde, W.: CommonKADS: a comprehensive methodology for KBS development. IEEE Expert 6, 28–37 (1994)
7. van Harmelen, F., Aben, M., Ruiz, F., van de Plassche, J.: Evaluating a formal KBS specification language. IEEE Expert 1, 56–61 (1996)

Discovering Prediction Model for Environmental Distribution Maps

Ke Zhang[1,2], Huidong Jin[1,2], Nianjun Liu[1,2], Rob Lesslie[3], Lei Wang[1,2],
Zhouyu Fu[1,2], and Terry Caelli[1,2]

[1] Research School of Information Sciences and Engineering (RSISE)
Australian National University
[2] National ICT Australia (NICTA), Canberra Lab, ACT, Australia
[3] Bureau of Rural Sciences (BRS), Canberra, Australia
{ke.zhang,huidong.jin,nianjun.liu,lei.wang,
zhouyu.fu,terry.caelli}@rsise.anu.edu.au,
rob.lesslie@brs.gov.au

Abstract. Currently environmental distribution maps, such as for soil fertility, rainfall and foliage, are widely used in the natural resource management and policy making. One typical example is to predict the grazing capacity in particular geographical regions. This paper uses a discovering approach to choose a prediction model for real-world environmental data. The approach consists of two steps: (1) model selection which determines the type of prediction model, such as linear or non-linear; (2) model optimisation which aims at using less environmental data for prediction but without any loss on accuracy. The latter step is achieved by automatically selecting non-redundant features without using specific models. Various experimental results on real-world data illustrate that using specific linear model can work pretty well and fewer environment distribution maps can quickly make better/comparable prediction with the benefit of lower cost of data collection and computation.

Keywords: Environmental distribution map, prediction model, model selection, feature selection.

1 Introduction

Technologies of analysing spatial data such as Environmental Distribution Maps (EDM) have raised great expectations for coping with the natural resource management and policy making. As social and ecological development are becoming more intensively linked through time, it would be very beneficial in the socio–environmental policy making if the human effects on ecosystem are evaluated/predicted with high accuracy. For example, in the planning of land use, we should carefully consider that which activities may generate negative effects on the regional and the local environmental issues. To make an appropriate land plan, the potential human influence to ecosystem need to be well predicted. As shown in Fig 1, the socio-environmental land use planning model [1] is dynamic,

T. Washio et al. (Eds.): PAKDD 2007 Workshops, LNAI 4819, pp. 99–109, 2007.

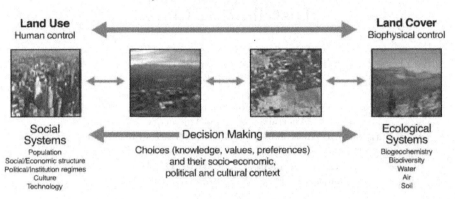

Fig. 1. The framework of socio-environmental land system [1]

where the land planning decisions can be adjusted according to the prediction of environmental issues. In those prediction tasks, all decisive factors represented by large–scale spatial data, which may impact the target assessment and formulation, should be selected and considered carefully.

As many geographic and data mining researchers have been working on the spatial data analysis, several prediction models/tools for crops production, plants in a certain ecosystem and other environmental specific fields have been proposed. In 1998, Priya et al. [2] proposed a multi-criteria prediction model for crop production, which linearly combines several decisive factors together. A GIS-based plant prediction model was proposed by Horssen et al. [3], which uses a geostatistical interpolation method to construct spatial patterns of relevant ecological factors. Besides those academic papers, prediction models based on spatial data analysis are also included in several GIS tools, such as IDRISI and MCAS-S [4], which are often customised for particular problems. However, there are two issues associated with these prediction models: (1) Do the domain experts need to specify a proper type of models for prediction tasks? (2) Are all inputting decisive factors they suggested necessary? The answers are normally no.

In this paper, based on a case study on the EDM data provided by Australian Bureau of Rural Sciences (BRS), we study the common problem existing in the current prediction models from two angles: model selection and model optimisation. Because of the similarity of prediction problems, the proposed methodology can also be applied on other prediction tasks.

The remainder of this paper is organized as follows. In Section 2, the BRS prediction problems and our abstraction framework for prediction problems are introduced. The method of model selection and the procedure of model optimisation are discussed in Section 3 and Section 4 respectively for a specific case of prediction problem. The experimental results are shown in Section 5 followed by concluding comments in Section 6.

2 Problems and Framework

Australian Bureau of Rural Sciences (BRS), which developed the GIS-based decision making tool MCAS–S, provides the scientific advice for government environmental policy making by analysing the prediction of environmental issues. Generally, domain experts in BRS set up the prediction model manually. To illustrate the prediction model clearly, we take a specific case for the prediction of graze total stock in Australian BRS as an example. As shown in the top half of Fig. 2, the decisive EDMs are suggested by domain experts. They include 9 decisive factors: soil fertility/carbon/nitrogen/phosphor, annual/winter/spring rainfall amount, forage productivity, mean annual normalised difference vegetation index (NDVI mean) and mean annual net primary production (NPP mean). The decisive EDMs are assumed to be combined linearly, and their corresponding interactive weights were manually selected based on domain knowledge.

However, we expected that the prediction model could input the decisive maps and output a predicted target map with out restriction on prediction model type. The purpose of the prediction modeling is to minimise the difference between the real–world target map (ground truth) and the output of the prediction model (predicted target map). The framework of our prediction model is illustrated in Fig. 2. In order to minimise the individual impacts on the prediction performance, we use a part of statistics data of graze total stock (ground truth) as training data to learn the parameters in our prediction model (see Fig. 3).

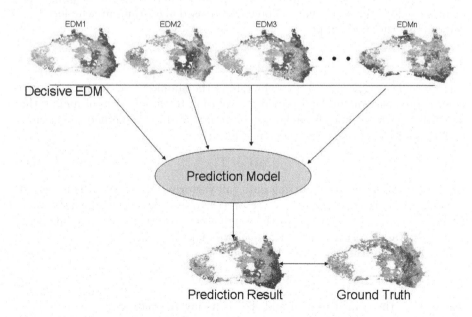

Fig. 2. The framework of our prediction model

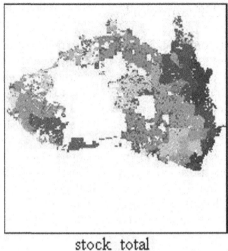

stock_total

Fig. 3. The statistics of graze total stock as ground truth

3   Prediction Model Selection

In the EDM prediction tasks, for the various kinds of decisive factors (inputting data), the accuracy of existing prediction methods may show significant differences. For the efficiency of the discovering procedure, the type of prediction models would be predetermined. Three widely used algorithms are employed for determining the model type.

3.1   Least Square Fitting (LSF)

The linear least squares fitting technique is the simplest and most commonly applied regression model and provides a solution to the problem of finding the best fitting straight line through a set of points [5]. In this method, we assume that those decisive EDMs are combined linearly:

$$AW = T \tag{1}$$

where A is the matrix of inputting data (all vectorised data of EDMs is ranged in column), T is the target data vector (vectorised from the target/training EDM), $W = [w_1, w_2, \cdots, w_N]^T$ is the weight vector of their corresponding decisive EDMs, and N is the number of decisive maps.

The LSF algorithm estimates W as follows:

$$W = (A^T A)^{-1} A^T T \tag{2}$$

where A^T is the transpose of A and A^{-1} is its inverse matrix.

Thus, given the inputting decisive data and the training target, the interactive weights of decisive EDMs can be estimated by Eq 2.

3.2   Support Vector Machine (SVM) Regression

The SVM algorithm is a generalisation of the *Generalised Portrait* algorithm. It was first developed at AT&T Bell Laboratories by Vapnik and his co-workers [6] [7]. Suppose we are given training data $\{(x_1, y_1), \cdots, (x_l, y_l)\} \subset X \times R$, where X denotes the space of the inputting patterns (e.g., $X = R^d$). The purpose of SVM is to find a function $f(x)$ that has at most ε deviation from the actually obtained targets y_i for all the training data, and at the same time is as flat as possible. Suppose function $f(x)$ is linear then it takes the form:

$$f(x) = \langle w, x \rangle + b \ \text{ with } \ w \in X, b \in R \tag{3}$$

where $\langle a, b \rangle$ denotes the dot product in X. The following optimisation problem is solved to obtain the weight vector w.

$$\text{Minimise} \qquad \tfrac{1}{2}\langle w, w \rangle + C \times \sum_{i=1}^{l}(\xi_i + \xi_i^*), \tag{4}$$

$$\text{Subject to:} \qquad y_i - \langle w, x_i \rangle - b \leq \varepsilon + \xi_i \tag{5}$$

$$\langle w, x_i \rangle + b - y_i \leq \varepsilon + \xi_i^*$$

$$\xi_i, \xi_i^* \leq 0.$$

The non-linear SVM algorithm is similar with the linear one except that every inner product is replaced by a non-linear kernel function. In this paper we use Gaussian kernel function. This allows the algorithm to fit the maximum-margin hyper plane in the transformed feature space [8]. For the prediction task, the inputting data in Eq 3 is a matrix containing all data of decisive EDMs in the vector form. Let us say that $x = [x^1, x^2, \cdots, x^N]$, where x^n is the data vector of the nth EDM. The expression of function $f(x)$ can be regarded as the prediction model that combines the inputting EDM data non-linearly to approximate the target.

3.3   Neural Networks

The neural network is a powerful data modeling tool that is able to capture and represent complex input/output relationships. In the EDM prediction case, we chose feed-forward neural network, which functions as follows: each neuron receives a signal from the neurons in the previous layer, and each of those signals is multiplied by a separate weight value. The weighted inputs are summed, and passed through a limiting function that scales the output to a fixed range of values. The output of the limiter is then broadcast to all of the neurons in the next layer [9]. When the decisive factors in a prediction task have a very complicated correlation, the predicting performance of Neural Networks may be better than the linear method and comparable to SVM. For our prediction tasks, each inputting data $x_1, x_2, \cdots, x_N$ can be regarded as a set of data extracted from decisive EDMs and they will be mapped non-linearly in the hidden nodes.

4   Model Optimisation

Since the data collection for the decisive factors takes the main proportion of the project cost, minimising the amount of inputting data will be significantly beneficial. Let F be a full set of decisive factors (can be regarded as features) and T is the target we want to predict. In general, the goal of feature selection can be formalised as selecting a minimum subset F^* such that $P(T|F^*)$ is equal or as close as possible to $P(T|F)$, where $P(T|F^*)$ is the posterior probability distribution of the target given the feature values in F [10]. We call such a minimum subset F^* an optimal subset. The feature selection method used in this paper is Redundancy Based Filter (RBF). The basic idea is using the concept of redundant cover to determine which features should be removed. The correlation between features and the target values are used to determine the features which form a redundant cover for others. There exist broadly two types of measures for their correlations: linear and non-linear [11]. Since linear correlation measures may not be able to capture correlations that are not linear in nature, in the approach we adopt a non-linear correlation measure based on the information-theoretical concept of entropy, a measure of the uncertainty of a random variable [10]. The entropy of a variable X is defined as

$$H(X) = -\sum_i P(x_i) \log_2 P(x_i) \tag{6}$$

and the entropy of X conditioned on variable Y is defined as

$$H(X|Y) = -\sum_j P(y_j) \sum_i P(x_i|y_j) \log_2 P(x_i|y_j) \tag{7}$$

where $P(x_i)$ is the prior probability for all values of X, and $P(x_i|y_j)$ is the conditional probability of X given the values of Y. The amount by which the entropy decrease of X after conditioning reflects additional information about X provided by Y and is called information gain, given by

$$IG(X|Y) = H(X) - H(X|Y). \tag{8}$$

The information gain tends to favour variables with greater differences and can be normalised by their corresponding entropy. We use symmetrical uncertainty (SU) to measure information gains of features and it is defined as:

$$SU(X,Y) = 2 \left[\frac{IG(X|Y)}{H(X) + H(Y)} \right]. \tag{9}$$

The value of SU is ranged within $[0, 1]$. A value of 1 indicates that knowing the values of either feature completely predicts the values of the other; a value of 0 indicates that variables X and Y are independent.

Since the huge differences of the number of discrete values among EDM data, varying from 70 to 2,000, the EDM data is required to discretise into a same scale before calculating their entropy. For the convenience of computation, we

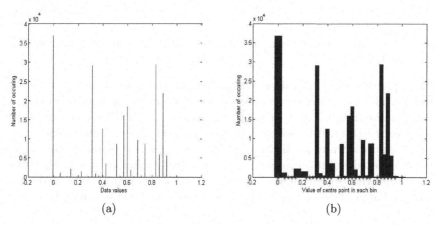

Fig. 4. (a) is the histogram of the Soil Fertility Environmental Distribution (SFED) expression data. (b) is the histogram of SFED after equal frequency discretisation into 30 intervals.

discretise each feature into 30 bins with equal occurring frequency. A result of discretisation for one EDM is illustrated in Fig. 4.

In order to select the non-redundant features explicitly, we differentiate two types of correlation between the features and the target [10]:

Individual T-correlation Symmetrical Uncertainty (ISU): The correlation (represented by SU) between any feature F_i and the target T is denoted by ISU_i.

Combined T-correlation Symmetrical Uncertainty (CSU): The correlation (represented by SU) between any pair of features F_i and F_j $(i \neq j)$ and the target T is denoted by CSU_{ij}. In the computation of CSU, we treat the pair of features F_i and F_j as one single feature $F_{i,j}$.

We assume that a feature with a larger individual T-correlation value contains by itself more information about the target than a feature with a smaller individual T-correlation value. For two features F_i and F_j $(i \neq j)$ with $ISU_i > ISU_j$, we choose to evaluate whether feature F_i can form an approximate redundant cover for feature F_j in order to maintain more information about the target. In addition, if combining F_j with F_i does not provide more predictive power in determining the target than F_i alone, we heuristically decide that F_i forms an approximate redundant cover for F_j. Therefore, an approximate redundant cover can be defined as: For two features F_i and F_j, if $ISU_i \geq ISU_j$ and $ISU_i \geq CSU_{i,j}$, F_i forms an approximate redundant cover for F_j.

The RBF feature selection algorithm can be expressed as follows [10]:

1. Order features based on decreasing ISU values.
2. Initialise F_i with the first feature in the list.

3. Find and remove all features for which F_i forms an approximate redundant cover.
4. Set F_i as the next remaining feature in the list and repeat step 3 until the end of the list.

The algorithm described above can determine the redundant features automatically and it can select non-redundant features independent of the prediction model. For the independency of prediction models, the RBF shows an obvious advantage for EDM feature selection. Since the uncertainty of combination models in the environmental prediction problems, we may not guarantee the performance of selected features by only using a specified prediction model. Therefore, based on the RBF algorithm, the comparably reliable features could be selected among the batch of decisive EDM factors, as substantiated in Table 2.

5   Experimental Results

5.1   Data Description

As mentioned in Section 2, 9 decisive EDMs were recommended to predict the potential production of graze, and we had an EDM of statistics of graze total stock as the training/testing data. All of those EDM data were provided by Australian BRS, with the size of 700×880, float format. Those EDM data had been normalised into the range of $[0, 1]$ as a pre-processing step. The algorithms mentioned in this paper were implemented by MATLAB, and performed on a computer with 3.2GHz CPU.

5.2   Prediction Model Selection

In our experiments, we randomly selected 80% of the original data (graze stock statistics) as training data and take the rest as testing data, and independently repeated this procedure for 30 times with random partitions of training and testing data for each run. The performances of evaluation method were measured by mean square error and their standard deviation (Std.) as well as running time. Table 1 shows the results of training and testing errors for 3 different models.

As shown in Table 1, the difference of testing errors between LSF and SVM is quite significant: $SVM_{TestingError} - LSF_{TestingError} = 2.11\text{e-}5$, which is much larger than the standard deviation of LSF's testing error. Similarly, we can say

Table 1. The prediction accuracy for each method based on 30 independent runs. The best one within a row is indicated in bold.

	LSF	SVM	NN
Training error(mean)	**3.38e-6**	6.67e-5	9.38e-6
Training error(Std.)	2.96e-6	1.89e-5	7.81e-6
Testing error(mean)	**9.53e-6**	3.07e-5	2.63e-5
Testing error(Std.)	1.02e-5	1.13e-4	8.11e-6

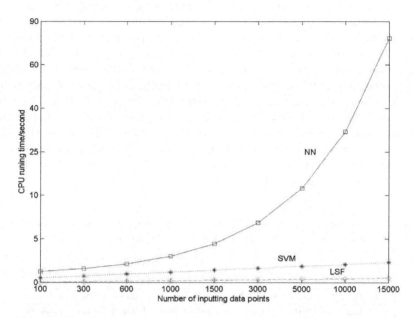

Fig. 5. Curves of computing time to training prediction methods with different training sample sizes

that the accuracy of LSF is significantly higher than that of Neural Networks. Thus, LSF shows obvious advantages compared with the other two methods.

The curve of computing time for each prediction method is illustrated in Fig. 5.

As shown in Table 1 and Fig. 5, it is clear that the linear method (LSF) has higher efficiency in evaluating the prediction model in this project, which has the highest prediction accuracy with least computing cost among those experimented methods. Thus, we can suggest that it would be beneficial if the LSF prediction method can be used to set up the prediction model rather than using a more complicated non-linear prediction methods. This is also confirmed by domain experts.

However, we cannot guarantee that the linear method is suitable to each GIS-based prediction case because of different kinds of decisive factors and large variations in their correlations. From this viewpoint, we may suggest that, for every GIS-based prediction case, model selection should be pre-performed in order to find the prediction method which suits the current case best.

5.3 Model Optimisation

As mentioned in the previous section, a model-independent feature selection method was used in the model optimisation. In order to make the entropy of features comparable, the histograms of all EDMs and the target were discretised into 30 bins with an approximately equal occurring frequency. And then, the conditional probability tables of each feature/combined features given target could be calculated by counting their corresponding discretised data. In this prediction case, we calculated 9 conditional probability tables $P(F_i|T)$, and 81 combined conditional probability tables $P(F_{i,j}|T)$. In the experiments, considering acceptable computational load, we randomly cropped 100,000 data points from the 9 decisive EDMs and the target map. Based on the algorithm described in Section 4, the feature selector removed the decisive EDM of "rainfall amount in winter/spring" as a redundant feature. To check the reliability of the selection result, we had performed a performance comparison of all methods (LSF, SVM and NN) for using all the features and only the 8 selected features. The comparison results are listed in Table 2.

Table 2. Performance comparison by using the 9 and the 8 features based on 30 independent runs. The best one within a row indicated in bold.

	LSF		SVM		NN	
Number of features	9	8	9	8	9	8
Training error(mean)	**3.37e-6**	1.12e-5	6.67e-5	3.03e-5	9.38e-5	1.33e-5
Training error(Std.)	2.95e-6	3.65e-6	1.89e-5	1.03e-4	7.81e-6	1.33e-6
Testing error(mean)	9.53e-6	**2.69e-6**	3.07e-5	1.23e-5	2.63e-5	7.28e-6
Testing error(Std.)	1.01e-5	1.09e-5	1.13e-4	3.32e-4	8.11e-6	9.47e-6
CPU running time/s	0.0381	**0.0349**	0.1536	0.1390	32.6086	27.0955

As shown in Table 2, the feature removed in the RBF algorithm is redundant, and the rest of features can make a better/comparable prediction with lower computation load.

In order to further confirm the effect of the RBF method, we employed "wrapper" feature selector to validate the result of RBF. Our experiments showed that reducing the decisive EDM of "rainfall amount in winter/spring" was the only positive action comparing with the other 8 features. To verify this conclusion, we performed additional experiments that randomly reduced a pair of features and calculated the influence of the rest of 7 features. And also the experiments showed that none of set of 7 features has a comparable prediction accuracy with that of using 9 features or 8 features.

6   Conclusions

We have presented a discovering method to choose a prediction model for EDM data for Australian Bureau of Rural Sciences. The discovering procedure consists of two procedures: model selection by comparing the performance of 3 prediction models which can be learned from ground truth data; and model optimisation which aims to use less environmental data for prediction but without any loss on accuracy. Various experimental results have shown that using a specific linear model can work pretty well and fewer EDMs can quickly make better/comparable prediction with lower cost of data collection and computation. It means that this model may help Australian BRS save a lot of resources in the real-world application. In the future work, we will incorporate spatial information into the prediction model to enhance the accuracy.

Acknowledgement. Financial support for this project from the Australian Research Council is gratefully acknowledged. NICTA is funded by the Australian Government's Department of Communications, Information Technology, and the Arts and the Australian Research Council through Backing Australia's Ability and the ICT Research Centre of Excellence programs.

References

1. Hill, M.: The global land project: An international context for australian analysis of human transformation of ecosystems and landscapes. In: Australian Bureau of Rural Sciences Seminar Series Presents (2005) (accessed on Jan 30, 2007), http://www.affashop.gov.au/PdfFiles/brs_seminar_25nov05.pdf.
2. Priya, S., Shibasaki, R., Ochi, S.: Soil erosion and crop production: A modeling approach. In: Proceedings of Global Environmental Symposium organized by Japanese Society of Civil Engineers, pp. 175–180 (1998)
3. Horssen, P., Schot, P., Barendregt, A.: A gis-based plant prediction model for wetland ecosystems. Landscape Ecology 14, 253–265 (1999)
4. Hill, M., Lesslie, R., Barry, A., Barry, S.: A simple, portable, spatial multi–criteria analysis shell–MCAS–S. In: International Symposium on Modelling and Simulation, University of Melbourne (2005)
5. Chatterjee, S., Hadi, A., Price, B.: Simple linear regression. In: Regression Analysis by Example, 3rd edn., pp. 21–50. Wiley, New York (2000)
6. Vapnik, V., Lerner, A.: Pattern recognition using generalized portrait method. Automation and Remote Control. 24, 774–780 (1963)
7. Vapnik, V., Golowich, S., Smola, A.: Support vector method for function approximation, regression estimation and signal processing. Advances in Neural Information Processing Systems. 9, 281–287 (1997)
8. Cristinaini, N., Shawe-Taylor, J.: An introduction to support vector machines and other kernel–based learning methods. Cambridge University Press, Cambridge (2000)
9. McCollum, P.: An introduction to back-propagation neural networks. encoder. The Newsletter of Seattle Robotics Society (1997)
10. Yu, L., Liu, H.: Redundancy based feature selection for microarray data. In: KDD, pp. 737–742 (2004)
11. He, H., Jin, H., Chen, J.: Automatic feature selection for classification of health data. In: Australian Conference on Artificial Intelligence, pp. 910–913 (2005)

Workshop BioDM'07—An Overview

Jinyan Li[1] and Xiaohua Hu[2]

[1] School of Computer Engineering,
Nanyang Technological University, 639798 Singapore
`jyli@ntu.edu.sg`
[2] College of Information Science and Technology,
Drexel University, 19104 Philadelphia, PA
`thu@cis.drexel.edu`

Abstract. This edited volume contains the papers selected for presentation at the Second Workshop on Data Mining for Biomedical Applications (BioDM'07) held in Nanjing, China on 22nd of May 2007. The workshop was held in conjunction with the 11th Pacific-Asia Conference on Knowledge Discovery and Data Mining (PAKDD 2007), a leading international conference in the areas of data mining and knowledge discovery. The aim of this workshop was to provide a forum for discussing research topics related to biomedical applications where data mining techniques were found to be necessary and/or useful.

1 Introduction

Biomedical research and applications often involve large volume of data, e.g. clinical data, genomic sequence data, protein structure data, gene expression profiles, mass spectra, protein interaction networks, pathway networks, and Medline abstracts. There are many computationally challenging problems in the analysis of these diverse and voluminous data. To address the computational problems and their relevance to biology, we encouraged authors to submit papers that propose novel data mining techniques or that effectively use existing computational algorithms to solve challenging problems in the following topics (but not limited to):

- Rule induction from traditional clinical data
- Biomedical text mining
- In-silico diagnosis, prognosis, and treatment of diseases
- Genomics, proteomics, and metabolomics
- Systems biology
- Protein structure and function
- Topological properties of interaction networks
- Biomedical data integration
- Ontology-driven biomedical systems
- Biomedical data privacy and security.

T. Washio et al. (Eds.): PAKDD 2007 Workshops, LNAI 4819, pp. 110–111, 2007.

The first BioDM workshop was successfully held in April, 2006 in Singapore which attracted 35 full submissions, of which 14 papers were presented in the workshop and published as a volume of LNBI by Springer-Verlag. Again, BioDM 2007 was aimed at attracting top researchers, practitioners, and students from around the world to discuss data mining applications in the field of bioinformatics.

2   Workshop Overview

BioDM'07 received a total of 30 full-length paper submissions from 5 countries. Most of the submitted papers were rigorously reviewed by at least three Program Committee members. Although many papers were worthy of publication, only 13 regular papers can be accepted into the workshop for presentation and publication in this volume. The accepted papers were organized into three sessions according to their topics, with 4 papers on research topics related to protein such as protein subcellular location identification, 3D off-lattice protein modeling, protein unfolding pathways, and protein name entity recognition, 4 papers on gene expression profiling data analysis, and 5 papers on some emerging bioinformatics topics. The distribution of the paper topics indicated that database query, 3D similarity measurement, feature selection, and supervised learning remained the current computational approaches in the field. In addition to the contributed presentation, the BioDM'07 workshop featured a keynote talk delivered by Professor Francis Chin (Hong Kong University), who shared his insightful vision on the bioinformatics research problems related to motif discovery.

After the workshop, we had a careful evaluation for every paper presented in the workshop based on the audience's feedback and the comments from some Program Committee members. Finally, we decided to accept 11 of them to be published in this volume.

This workshop would not be possible without the help of many colleagues. We would like to thank the Program Committee members for their invaluable review and comments. Given the extremely tight review schedule, their effort for completing the review reports before the deadline was greatly appreciated. These comments were very helpful for us in selecting the papers.

Very importantly, we would like to acknowledge the PAKDD'07 Conference Workshop Chair Professor Takashi Washio for coordinating between this workshop and the conference. We also thank other PAKDD'07 officers for their effort and time in workshop registration and other logistics.

3   Workshop Organizing Committee Chair

Shuigeng Zhou   Fudan University, China

4   Program Co-chairs

Jinyan Li   Nanyang Technological University, Singapore
Xiaohua Hu   Drexel University, USA

Extracting Features from Gene Ontology for the Identification of Protein Subcellular Location by Semantic Similarity Measurement

Guoqi Li and Huanye Sheng

Department of Computer Science and Engineering, Shanghai Jiao Tong University,
Shanghai 200240, China
{liguoqi,hysheng}@sjtu.edu.cn, gqli.cn@gmail.com

Abstract. It is necessary to find a computational method for prediction of protein subcellular location (SCL). Many researches have focused on the topic. Among them, methods incorporated Gene Ontology (GO) achieved higher prediction accuracy. However the former method of extracting features from GO have some disadvantages. In this paper, to increase the accuracy of the prediction, we present a novel method to extract features from GO by semantic similarity measurement, which is hopeful to overcome the disadvantages of former method. Testing on a public available dataset shows satisfied results. And this method can also be used in similar scenarios in other bioinformatics researches or data mining process.

Keywords: Protein subcellular location, Gene Ontology, semantic similarity.

1 Introduction

Identification of subcellular localization (SCL) of proteins provides valuable clues for molecular biology research. For example, surface-exposed or secreted proteins are of primary interest due to their potential as vaccine candidates, diagnostic agents (environmental or medical) and the ease with which they may be accessible to drugs[1]. The high throughput genome sequencing projects are producing an enormous amount of raw sequence data, which need to be cataloged and synthesized. Since experimental determination of subcellular location is time consuming and costly, it is necessary to find a computational, fully automatic prediction system for protein's SCL.

There have been many papers on the topic [2-8]. Early methods extracted features from sequence information of proteins. For example, based on the use of N-terminal sorting signals and amino acid composition [6] for feature extraction in conjunction with machine learning approaches such as neural network (NN) [7] and support vector machines (SVM) [3]. By incorporating functional background knowledge of proteins into the prediction of SCL, higher accuracy was achieved. In 2003, Cai and Chou [4] presented an approach to extract features by combining functional domain composition and pseudo-amino acid composition. Each vector in the approach is composed of two parts. One is extracted from pseudo-amino acid composition and the other is from functional domain composition. The combined vectors then were

T. Washio et al. (Eds.): PAKDD 2007 Workshops, LNAI 4819, pp. 112–118, 2007.

inputted to nearest neighbor algorithm for training classifier or for prediction. With the development of Gene Ontology (GO), background knowledge of proteins can be retrieved from GO and its annotation databases. Then a new hybrid approach was given using GO terms instead of functional domain composition [8]. And in 2005, a successor method was used to predict multi-labeled protein subcellular localization for prokaryotic bacteria. The method extracts feature vector of protein from amino acid composition and GO molecular function terms in conjunction with SVM as classifier. Summarily, these methods, combined with GO, achieved higher prediction accuracy than earlier methods. From a computational point of view, GO contains background knowledge of proteins. By combining GO, background knowledge was integrated into the input of classifier. Owing to the fact that the localization of a protein in a cell is closely correlated with its biological function [4], it is reasonable to perform better.

However the former method of extracting vector from GO have some disadvantages. Firstly, the dimension of the vector is too long, typically having about 2,000 elements [8]. Long vectors usually accompanied with high computation cost. And more, since most members of the vectors are zero, there begs information compression. Actually, the vectors extracted from GO molecular function terms are essentially semantic description of proteins' function in a numerical array fashion. From a computational point of view, the former method is not suitable to measure semantic similarity between proteins. This will be described in the next section in detail.

In this paper, we present a novel method, which is hopeful to overcome the disadvantages mentioned above. We firstly select some, about ten, stonemarks in GO and calculate the semantic similarity between GO molecular function terms of protein and every stonemarks. Then we look the numerical array of semantic similarity as the feature vector extracted from GO. Finally, combine them with vector extracted from amino acid composition before classification.

The rest of the paper is organized as follows: In the following section, describe the former method and our new method in detail. Then, presents a testing on a public available dataset to test the method and list the results. At last, we draw a number of conclusions from this study and address our future works.

2 Method

2.1 Former Method and Its Disadvantages

In this subsection, we will briefly describe the former method of extracting feature vectors from GO for SCL. Firstly, search InterPro [9] database for a given protein and find out the corresponding InterPro entries. InterPro is an integrated documentation resource for protein families, domains and sites. Every InterPro entry has an accession number of the form IPRXXXXXXX, where X is a digit. By mapping of InterPro entries to GO, one can get a list of data called "InterProt2GO" (ftp://ftp.ebi.ac.uk/pub/databases/interpro/interpro2go/), where each InterPro entrance corresponds to a GO number. The relationships between IPRXXXXXXX and GO may be one-to-many [8]. Since the GO numbers in InerProt2GO are not increasing successively and orderly, some reorganization and compression procedure was taken to renumber them. For example, after such a procedure, the original GO numbers GO:0000012, GO:0000015, GO:0000030, ..., GO:0046413 would become GO_compress:0000001, GO_compress: 0000002, GO_compress:0000003, ..., GO_compress: 0001930, respectively. The GO database thus obtained is called GO_compress database. Each of the entities in the GO_compress database will serve as a base to define a protein.

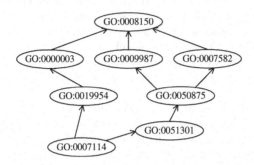

Fig. 1. A sub graph of Gene Ontology to show the semantic similarity between terms

For a given protein, If there is a hit corresponding to the ith number of the GO_compress database, then the ith component of the protein in the GO_compress space is assigned 1; otherwise, 0. Thus, the protein can be formulated as vector P:

$$P = \begin{bmatrix} a_1 & a_2 & \cdots & a_i & \cdots & a_n \end{bmatrix},$$

where $a_i = 1$, if hit found in GO_compress, else $a_i = 0$. n being the dimension of GO_compress.

Actually, P is semantic description of proteins' function in a numerical array fashion. However, it is lack of representation ability. For example, if there are three proteins p1, p2 and p3 and the corresponding vectors extracted with method mentioned above are $P1$, $P2$, $P3$ respectively. If p1 hits GO:0019954, p2 hits GO:0050875 and p3 hits GO:0007582. The distance between $P2$ to $P1$ and $P2$ to $P3$ are same. However, see figure 1, from semantic point of view, it is obvious that p2 are more similar with p3 than p1. The method cannot sufficiently describe the semantic similarity between molecular functions of proteins for the structure information in GO is missed out.

2.2 The New Method

To improve the former method, a logic step is to encode the semantic similarity into vectors. Firstly set some GO terms as references, call them markstones. Suppose that there are n markstones. The relationships between protein and GO term may be one-to-many. For example, if a protein have m GO terms. Then calculate the semantic similarity between the m GO terms and every markstone. Now, we get the new P: (Note that we only consider the molecular function part of GO. In the paper of Cai, C. Z. et al. [8], all the GO terms were considered, however in the later researches [10, 11] only the molecular function part of GO terms were used.)

$$P = \begin{bmatrix} a_1 & a_2 & \cdots & a_i & \cdots & a_n \end{bmatrix}$$

$$a_i = \min_{j=1}^{m} \left(d\left(p_j, m_i \right) \right)$$

where p_j means the jth GO term of the protein and $d\left(p_j, m_i \right)$ being the semantic similarity between the GO term and the markstone. Note that the dimension of the P is equal to the number of markstones, typically about 10 to 20. GO terms are more close in their semantic meanings, the $d\left(p_j, m_i \right)$ between them is more smaller.

Many researches investigated into the semantic similarity of GO[12, 13]. We will choose the distance-based measurement [13, 14] for the following demonstration. Other kind of measurement will be tried in our future research. The distance-based method is deduced from the hypothesis that the distance between the concepts on the upper layer is larger than that between the concepts on the lower layer. It supports the idea that the distance between brother nodes is larger than that between the parent node and the child node, and evaluates the semantic similarity according to the distance between the layers where the concepts exist. Each node on some layer has a value called milestone:

$$milestone(n) = \frac{1/2}{k^{l(n)}}$$

where k is a predefined parameter larger than 1 that denotes the rate for milestone going down along the layers and often defined as 2 [13]. Practical test in the demonstration show that $k = 2$ is best for the dataset we used. $l(n)$ is the depth of the node in the hierarchy (in a DAG, it is the length of the longest path from the node to the root).

$$\begin{cases} d_c\left(c_1, c_2 \right) = d_c\left(c_1, ccp \right) + d_c\left(c_2, ccp \right) \\ d_c\left(c, ccp \right) = milestone(ccp) - milestone(c) \end{cases}$$

where ccp is the closest common ancestor node of c_1 and c_2, $d_c\left(c_1, c_2 \right)$ is the distance between c_1 and c_2.

Another problem is how to select markstones. Here we used the result of a former research. Z. Lu and L. Hunter [11] described their study in the paper "GO Molecular Function Terms Are Predictive of Subcellular Localization". Owing the fact that use of Gene Ontology molecular function terms to extend sequence-based subcellular localization prediction has been previously shown to improve predictive performance. They explore directly the relationship between GO function annotations and localization information, identifying both highly predictive single terms, and terms with large information gain with respect to location. The results identify a number of predictive and informative GO terms with respect to subcellular location, particularly nucleus, extracellular space, membrane, mitochondrion, endoplasmic reticulum and Golgi. There are several clear examples illustrating why the addition of function information provides additional predictive power over sequence alone [11]. In the following demonstration for prediction of SCL, we use the highly discriminating terms GO terms selected by Z. Lu and L. Hunter as markstones. Table 1 shows the highly discriminating terms and the corresponding subcellular localizations.

Table 1. Selected highly discriminating terms (including children) for the six subcellular localization [11]

Location	Predictive GO Molecular Function terms
nucleus	GO:0003676, GO:0008134, GO:0030528
membrane	GO:0004872, GO:0015267, GO:0008528
extracellular	GO:0005125, GO:0030414, GO:0005201
mitochondria	GO:0015078, GO:0004738, GO:0003995, GO:0015290
E.R.	GO:0004497, GO:0016747
Golgi	GO:0016757, GO:0015923, GO:0005384

3 Testing

The dataset we used in the demonstration was gotten from PSORTdb. It is a database of SCL for bacteria that contains both information determined through laboratory experimentation (ePSORTdb dataset) and computational predictions(cPSORTdb dataset) [15]. The set of proteins from Gram-negative bacteria used in ePSORTdb is considered in this work. It is consisted of 1597 proteins with experimentally determined localizations(until December, 2005), in which 1448 proteins are of single localization site. Just single localization samples were selected. The prediction of multi localization proteins is another research topic [10]. And in fact, transmembrane proteins were excluded for they could be quite reliably predicted by some known methods Now. We ignored the proteins, which have no annotation in GO molecular function terms. Consequently, the final dataset are composed of three categories: 194 cytoplasmic, 96 extracellular and 193 periplasmic.

In order to extract feature from amino acid composition, scan the sequence of a protein from the first amino acid to the end. We count the number of every symbol in the set of 20 symbols of amino acids, which may appear in protein sequences. Therefore, got a final numerical vector of 20 dimensions.

To extract features from GO molecular function terms, firstly used the former method. Since the dataset we used is small, it only covers 227 GO molecular function terms, so the dimension of our GO_compress is 227. Then used our new method. Because cytoplasmic, extracellular and periplasmic are not related to nucleus, we only choose 12 GO terms in table 1 as markstones.

Finally, we got three sets of feature vectors for testing, ever set have 483 vectors. "AAC" whose dimension is 20 were extracted from amino acid composition. "AAC and GO former method" whose dimension is 20+227=247 were extracted from amino acid composition and GO with the former method. "AAC and GO new method" whose dimension is 20+12=32 were extracted from amino acid composition and GO with our new method.

SVM (Support Vector machine) is selected as classifier in the research, for it has been proven with high performance in the prediction of PCL [3]. Practically, we used LIBSVM [16], which is integrated software for support vector classification, regression and distribution estimation. Before training the SVM, we must decide which kernel should be select and then the penalty parameter C and kernel parameters are chosen. It has been proven the RBF kernel has many advantages and is suit to the perdition of PCL [3]. After selection of kernel, five-fold cross-validation was used to find the best C and γ.

Every set of features of the three was divided randomly into two parts averagely, one for test and the other for prediction. We repeated the test for six times and list the result in Table 2.

Table 2. Summery of the prediction performance

	Accuracy (%)		
	AAC	AAC and method	GO former	AAC and GO new method
Test 1	82.6446	86.3636		88.0165
Test 2	78.8382	80.0830		81.7427
Test 3	76.4463	79.7521		80.5785
Test 4	81.3278	80.9129		81.3278
Test 5	80.5785	75.2066		83.8843
Test 6	83.4025	86.7220		87.5519
Average	80.5397	81.5067		83.8503

4 Conclusion and Future Work

The testing result shows that the new method achieved higher accuracy. Although the testing dataset is small, it is very promising that the method can work better for large datasets. In fact the method of combining GO with information of sequences for data mining is not limited in the prediction of SCL in bioinformatics research. For example, the prediction of protein secondary structure [17]. And further more, the method can be used in other field with similar scenarios in which background knowledge is begged for input of data mining process. In the following research, we will try other kinds of semantic similarity measurement and test on large datasets.

Any discussion on the topic is welcome and the data and source code of the testing is available by email.

Acknowledgements

The research is supported by the doctoral fund of Education Ministry of China. Project No. 20040248001.

References

1. Rey, S., Acab, M., Gardy, J.L., Laird, M.R., deFays, K., Lambert, C., Brinkman, F.S.L.: PSORTdb: a protein subcellular localization database for bacteria. Nucleic Acids Research 33 (2005)
2. Yu, C.S., Chen, Y.C., Lu, C.H., Hwang, J.K.: Prediction of protein subcellular localization. Proteins-Structure Function and Bioinformatics 64, 643–651 (2006)

3. Hua, S.J., Sun, Z.R.: Support vector machine approach for protein subcellular localization prediction. Bioinformatics 17, 721–728 (2001)
4. Cai, Y.D., Chou, K.C.: Nearest neighbour algorithm for predicting protein subcellular location by combining functional domain composition and pseudo-amino acid composition. Biochemical and Biophysical Research Communications 305, 407–411 (2003)
5. Gardy, J.L., Spencer, C., Wang, K., Ester, M., Tusnady, G.E., Simon, I., Hua, S., deFays, K., Lambert, C., Nakai, K., Brinkman, F.S.L.: PSORT-B: improving protein subcellular localization prediction for Gram-negative bacteria. Nucleic Acids Research 31, 3613–3617 (2003)
6. Nakai, K.: Protein sorting signals and prediction of subcellular localization. Advances in Protein Chemistry 5454, 277–344 (2000)
7. Reinhardt, A., Hubbard, T.: Using neural networks for prediction of the subcellular location of proteins. Nucleic Acids Research 26, 2230–2236 (1998)
8. Chou, K.C., Cai, Y.D.: A new hybrid approach to predict subcellular localization of proteins by incorporating gene ontology. Biochemical and Biophysical Research Communications 311, 743–747 (2003)
9. Mulder, N.J., Apweiler, R., Attwood, T.K., Bairoch, A., Bateman, A., Binns, D., Bradley, P., Bork, P., Bucher, P., Cerutti, L., Copley, R., Courcelle, E., Das, U., Durbin, R., Fleischmann, W., Gough, J., Haft, D., Harte, N., Hulo, N., Kahn, D., Kanapin, A., Krestyaninova, M., Lonsdale, D., Lopez, R., Letunic, I., Madera, M., Maslen, J., McDowall, J., Mitchell, A., Nikolskaya, A.N., Orchard, S., Pagni, M., Pointing, C.P., Quevillon, E., Selengut, J., Sigrist, C.J.A., Silventoinen, V., Studholme, D.J., Vaughan, R., Wu, C.H.: InterPro, progress and status in 2005. Nucleic Acids Research 33, 201–205 (2005)
10. Su, C.-Y., Lo, A., Lin, C.-C., Chang, F., Hsu, W.-L.: A Novel Approach for Prediction of Multi-Labeled Protein Subcellular Localization for Prokaryotic Bacteria. IEEE The Computational Systems Bioinformatics Conference, Stanford (2005)
11. Lu, Z., Hunter, L.: GO Molecular Function Terms Are Predictive of Subcellular Localization. In: Pacific Symposium on Biocomputing, vol. 4-8, World Scientific, Hawaii, USA (2005)
12. Lord, P.W., Stevens, R.D., Brass, A., Goble, C.A.: Investigating semantic similarity measures across the Gene Ontology: the relationship between sequence and annotation. Bioinformatics 19, 1275–1283 (2003)
13. Li, R., Cao, S.L., Li, Y.Y., Tan, H., Zhu, Y.Y., Zhong, Y., Li, Y.X.: A measure of semantic similarity between gene ontology terms based on semantic pathway covering. Progress in Natural Science 16, 721–726 (2006)
14. Zhong, J.W., Zhu, H.P., Li, J.M., Yu, Y.: Conceptual graph matching for semantic search. In: Priss, U., Corbett, D.R., Angelova, G. (eds.) ICCS 2002. LNCS (LNAI), vol. 2393, pp. 92–106. Springer, Heidelberg (2002)
15. Rey, S., Acab, M., Gardy, J.L., Laird, M.R., DeFays, K., Lambert, C., Brinkman, F.S.L.: PSORTdb: a protein subcellular localization database for bacteria. Nucleic Acids Research 33, D164–D168 (2005)
16. Chang, C.-C., Lin, C.-J.: LIBSVM: a library for support vector machines. Software (2001), available at: http://www.csie.ntu.edu.tw/čjlin/libsvm
17. Hua, S.J., Sun, Z.R.: A novel method of protein secondary structure prediction with high segment overlap measure: Support vector machine approach. Journal of Molecular Biology 308, 397–407 (2001)

Detecting Community Structure in Complex Networks by Optimal Rearrangement Clustering

Rui-Sheng Wang[1,2], Yong Wang[3,4], Xiang-Sun Zhang[4,*], and Luonan Chen[2,5,*]

[1] School of Information, Renmin University of China, Beijing 100872, China
[2] Osaka Sangyo University, Osaka 574-8530, Japan
[3] State Information Center, Beijing 100045, China
[4] Academy of Mathematics and Systems Science, CAS, Beijing 100080, China
zxs@amt.ac.cn
[5] Institute of Systems Biology, Shanghai University, Shanghai 200444, China
chen@eic.osaka-sandai.ac.jp

Abstract. Detecting community structure in biological and social networks recently attracts increasing attention in various fields including mathematics, physics and biology. Identifying communities in complex networks can help us to understand and exploit the networks more clearly and efficiently. In this paper, we introduced a method based on a combinatorial optimization problem — traveling salesman problem (TSP) as optimal rearrangement clustering for finding community structure in complex networks. This method can explore the global topology of a network and thus is effective in detecting modularity structure. Unlike most other algorithms for community identification, an advantage of this method is that it does not need to spend much time in finding a proper k, the number of communities in a network. We applied this method to several widely well-studied networks including a protein-protein interaction network, which demonstrates that this method is effective in detecting meaningful communities or functional modules.

Keywords: Complex networks, protein interaction network, community structure, functional modules, optimal rearrangement clustering.

1 Introduction

Many systems in real world can be expressed by a network in which the objects or individuals are represented by nodes and the edges correspond to some relations among the objects, such as social networks [1], technological networks [2] and biological networks [3]. Many of these real world networks have a modularity/community structure which are characterized by groups of individual nodes within which nodes are densely linked and between which the connection is sparser. These modules can represent functional units in their networks, for example, a community in scientific collaboration networks corresponds to a research field or a discipline. Communities in social networks represent social

---

* Corresponding author.

T. Washio et al. (Eds.): PAKDD 2007 Workshops, LNAI 4819, pp. 119–130, 2007.

groups with similar interest or background. The modules in protein-protein interaction networks may be protein complexes or functional categories of proteins.

Community structure is an important property of complex networks. In recent years, many methods have been developed to determine communities in complex networks, such as betweenness-based methods [1,4], spectral clustering methods [5,6,7]. Another class of methods are heuristic search algorithms aiming to maximize a measure Q developed by Newman [8] to evaluate the quality of community structure. Though the modularity Q has some weakness when applied in large and sparse networks [9], it has been broadly used as a modularity measure. Specifically, given an undirected graph/network $G(V, E)$ consisting of node set V and edge set E, its symmetric adjacency matrix is denoted as $A = [a_{ij}]_{n \times n}$, where $a_{ij} = 1$, if nodes i and j are connected and otherwise $a_{ij} = 0$. Let n be the size of the node set. Then, the modularity function Q is defined as:

$$Q(P_k) = \sum_{i=1}^{k} \left[e_{ii} - a_i^2 \right] \tag{1}$$

where P_k is a partition of the nodes into k groups, e_{ij} is the fraction of all edges in the network that link vertices in community i to vertices in community j and $a_i = \sum_j e_{ij}$. This quantity measures the fraction of the edges in the network that connect vertices of the same type (i.e., within-community edges) minus the expected value of the same quantity in a network with the same community divisions but random connections between the vertices [8] and it provides a way to evaluate the quality of community structure in networks. Generally, bigger Q corresponds to more obvious community structure. Therefore, a large class of methods directly based on maximizing modularity have been proposed [10,11]. In contrast to these methods, Gustafsson et al. [12] use some standard ways of clustering data such as k-means, hierarchical clustering to cluster networks. More algorithms for detecting community structure in complex networks can be referred to the review papers [9,13].

An important problem in detecting communities in complex networks is how to determine k, the number of communities in a network. Some of existing algorithms [1,10] are hierarchical or divisive and does not require the value of k, but they need a threshold for hierarchical trees to determine final clusters. Most others spend much time to search a proper k before clustering networks [5,6,12]. In this paper, we proposed a combinatorial optimization method — a traveling salesman approach as optimal rearrangement clustering for finding community structure in complex networks. This method can explore the global topology of a network and thus is effective in detecting modularity structure. An advantage of this method is that it dose not need to spend much time in finding a proper k. So it can deal with networks with ten thousands of nodes. We applied this method to several widely well-studied networks including a protein-protein interaction network, which demonstrates that this method is effective in detecting meaningful communities or function modules.

2   Method

In this section, we proposed a method based on traveling salesman problem as optimal rearrangement clustering for detecting communities in complex networks. The Traveling Salesman Problem (TSP) is a classic problem in combinatorial optimization theory [14] and can be described as: Given n cities with the distances between any two cities, how to visit all n cities in a tour so that each city is visited exactly once and the total distance of traveling is minimized [14]. One of many applications of the TSP is data array clustering [15]. In the traveling salesman-based rearrangement clustering [16], the rows and columns of the data array are viewed as the intermediate cities visited. By selecting an optimal tour to visit the cities, traveling salesman clustering rearranges the order of the rows of data array such that the sum of the distances between two neighboring rows is minimized which of course can be achieved by globally optimizing an objective function. Traveling salesman-based rearrangement clustering has previously been used in other fields including clustering gene expression data [17] and predicting protein function prediction [18].

When traveling salesman approach is used in clustering data, a measure of effectiveness (ME) is defined to represent the overall similarity and acts as the objective function to be maximized. Generally, this measure is based on the similarities among data points. In our paper, ME can be directly based on the adjacency matrix $A = [a_{ij}]_{n \times n}$ or other feature matrices of a network such as shortest path-based feature matrices $P = [p_{ij}]_{n \times n}$, where p_{ij} is the length of the shortest path between nodes i and j. Assume ρ is a permutation of both the rows and the columns since the adjacency matrix or the feature matrix of a network is a symmetric one. It can be considered as a feasible tour visiting all the 'cities'. ME is calculated as the following function for a ρ to characterize its effectiveness (overall similarity)[14]:

$$ME(\rho) = \sum_{i=1}^{n} \sum_{j=1}^{n} s_{\rho(i)\rho(j)} \left(s_{\rho(i+1)\rho(j)} + s_{\rho(i)\rho(j+1)} \right) \qquad (2)$$

which is positively proportional to the total similarity (characterized by the dot products) on the tour determined by ρ. With the symmetry of the feature matrix $S = [s_{ij}]_{n \times n}$ (S can be adjacency matrix, shortest path-based feature matrix or any other symmetric feature matrix), the function mentioned above can be reduced to the following one:

$$ME(\rho) = 2 \sum_{i=1}^{n} \sum_{j=1}^{n} s_{\rho(i)j} s_{\rho(i+1)j}. \qquad (3)$$

With such a measure, detecting community structure (network clustering) becomes a combinatorial optimization problem where the optimal clustering corresponds to the optimal permutation ρ maximizing $ME(\rho)$. It is equivalent to a Traveling Salesman Problem (TSP) [14] looking for an optimal permutation or tour ρ with the distance matrix $D = [d_{ij}]_{n \times n}$, where

$$d_{ij} = -\sum_{k=1}^{n} s_{ik}s_{jk}.$$

To make sure that the matrix entries in D are all non-negative, we can add a big positive constant number C into each entry:

$$d_{ij} = C - \sum_{k=1}^{n} s_{ik}s_{jk}. \qquad (4)$$

Now clustering complex networks can be achieved through rearranging the rows and columns of the network by the traveling salesman problem with the distance matrix D so as to make the corresponding tour optimal. If we are directly given a similarity matrix S instead of adjacent matrix or feature matrix, ME can be simply defined for a tour ρ in such a way: $ME(\rho) = \sum_{i=1}^{n} s_{\rho(i)\rho(i+1)}$ and the distance matrix for TSP is $D = [d_{ij}]_{n \times n}$ and $d_{ij} = C - s_{ij}$.

According to the solution (the optimal tour) to this traveling salesman problem, we can obtain a clustering of this network and find the community structure in it. Specifically, each edge on the optimal tour has a distance value characterizing the difference score between each two adjacent rows (after rearranged). Generally, the neighboring nodes on the tour which belong to one community in a network have high similarities and thus the corresponding edges has lower difference scores, while the neighboring nodes on the tour which are in different communities tend to have high difference scores. So we can plot the difference scores of the edges on the optimal tour and use the 'peaks' as the boundaries of different communities. Note that the peaks we select as the boundaries can not be too near (in our paper, there are at leat two edges between two peaks). We first select a peak with highest difference score, and then the second one (excluding those too near to selected peaks), and so on. We can use Q to decide whether the process should continue or not. In fact, if a network has a distinct modularity structure, the peaks are also obvious. After determining the boundaries

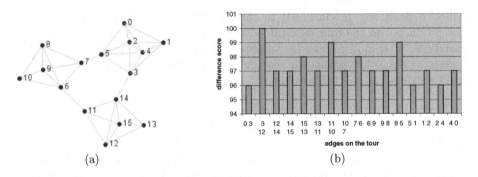

(a)   (b)

Fig. 1. An example. A tour with node order (0, 3 | 12, 14, 15, 13, 11 | 10, 7, 6, 9, 8 | 5, 1, 2, 4, 0) is obtained.

of communities, we can know the number of communities k and can also detect the modularity structure in this network.

In Figure 1, we give an example to illustrate the method. The network in this example has 16 nodes with three communities. These 16 nodes are viewed as 'cities' which will be visited by a traveling salesman. Given a distance matrix constructed by the formula (4) (C=100) based on the adjacency matrix, the algorithm returns a tour by solving the corresponding traveling salesman problem. This tour is (0, 3, 12, 14, 15, 13, 11, 10, 7, 6, 9, 8, 5, 1, 2, 4, 0) and the difference scores of the edges on the tour are plotted in Figure 1(b), from which we can find three distinct peaks: the edges (3, 12), (11, 10) and (8, 5). When these three edges act as the boundaries of communities, we can find three communities $\{0, 1, 2, 3, 4, 5\}, \{6, 7, 8, 9, 10\}, \{11, 12, 13, 14, 15\}$ with modularity $Q = 0.5698$ which are exactly consistent with what we see from the network. If we continue to select (15, 13) and (7, 6) as the boundaries, the modularity Q of the five-cluster partition is 0.2950 and much lower than that of the three-cluster partition, and furthermore, these too peaks are two near to the selected ones.

3   Experimental Results

In this section, we applied TSP-based rearrangement clustering method to several well studied real-world networks and a large protein-protein interaction network to test its ability of detecting meaningful communities. The algorithm is coded and implemented by MatLab 6.5. For the Traveling Salesman Problem, we used the Concorde program which is an award winning TSP solver publicly available at [19] and can solve problems with ten thousands of nodes owing to its linear programming core.

3.1   The Karate Club Network

The famous karate club network analyzed by Zachary [20] is widely used as a test example for community detection methods in complex networks [1,12,13,8]. The network consists of 34 members of a karate club as nodes and 78 edges representing friendship between members of the club which was observed over a period of two years. Due to a disagreement between the club's administrator and the club's instructor, the club split into two smaller ones (See Figure 2, squares denote the supporters of the instructor (node 1) and circles represent the supporters of the administrator (node 34)). Here what we concern is whether we can uncover the potential behavior of the network or not, detect two or more communities, and particularly identify which community each node belongs to.

In this example, we use shortest path matrix to construct distance matrix by the formula (4). The tour returned by the algorithm is (1, 7, 17, 6, 11, 5, 12, 13, 18, 22, 8, 4, 14, 3, 9, 31, 34, 28, 25, 26, 24, 30, 27, 15, 21, 23, 19, 16, 10, 29, 32, 20, 2, 1). We plot the difference scores between two neighboring nodes on the tour (the rearrangement of rows) in Figure 3, in which the scales in the x axis are not node pairs but the edge labels. According to the method for selecting the

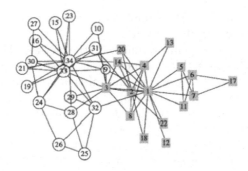

Fig. 2. Zachary's friendship network of the karate club

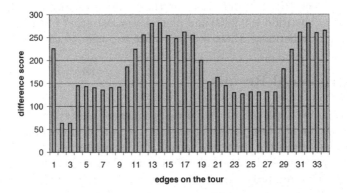

Fig. 3. Difference scores between neighboring nodes on the tour for the karate club network

boundaries of communities in Section 2, one peak is at the edge (3, 9) labeled by 14 and the other is (32, 20) labeled by 32. These two boundaries determine a partition of the network exactly consistent with known communities. At this time the modularity $Q = 0.3715$. Newman [7] obtained a result of $Q = 0.381$ for this network but with node 10 misclassified. Again if the network is further divided by other peaks, the modularity Q will decrease. The result on this example shows the effectiveness of the method.

3.2 The Football Team Network

The second network that we use as a test example is the college football network provided by Girvan and Newman [1]. The network represents the game schedule of the 2000 season of Division I of the U.S. college football league. The nodes in the network represent the 115 teams, while the links represent 613 games played in the course of the year. The teams are divided into conferences of 8-12 teams each and generally games are more frequent between members of the same conference than between teams of different conferences. The natural community structure

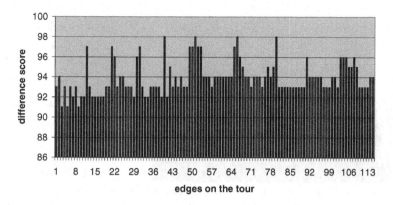

Fig. 4. Difference scores between adjacent nodes on the tour for the football team network

in the network makes it a commonly used benchmark testing example for various community-detecting algorithms [12]. All the teams are divided into 12 conferences, but there is a conference in which teams have more games with teams in other conferences than that with their own. In this example, we construct distance matrix directly based on adjacency matrix. According to the rearrangement of the rows and columns, the difference scores between neighboring rows are plotted in Figure 4. From the peaks in this figure, the football team network is divided into 10 clusters. The number of communities detected by our method is consistent with others (the methods in [12] which are based on shortest paths and sum of paths generate 8, 11, 10 clusters respectively for this network). Since the community structure in this network is known, we introduce a normalized mutual information index as a measure of similarity between two partitions P and P' [9]:

$$I_{NMI}(P', P) = \frac{-2 \sum_{i=1}^{|P'|} \sum_{j=1}^{|P|} n_{ij}^{p'_i p_j} \log(n_{ij}^{p'_i p_j} n / (n_i^{p'_i} n_j^{p_j}))}{\sum_{i=1}^{|P'|} n_i^{p'_i} \log(n_i^{p'_i}/n) + \sum_{j=1}^{|P|} n_j^{p_j} \log(n_j^{p_j}/n)}$$

where $n_i^{p'_i}$ represents the number of nodes in cluster p'_i and $n_{ij}^{p'_i p_j}$ denotes the number of shared elements between clusters p'_i and p_j . Obviously, $0 \le I_{NMI} \le 1$ with $I_{NMI}(P, P) = 1$. We compare the partition obtained by the proposed method with the known partition and the value of I_{NMI} is 0.8574 which indicates that the detected community structure is very near to the original one. In addition, the modularity Q of the original partition is 0.5371 and that of the detected partition is 0.5864, which means the detected community structure is more reasonable than the original one from a view of modularity. In Newman [7], 6 clusters with a modularity $Q = 0.556$ for this networks are obtained.

3.3 The Scientific Collaboration Network

Now we test another network — the scientific collaboration network collected by Girvan and Newman [1] and also used in many algorithms. The community

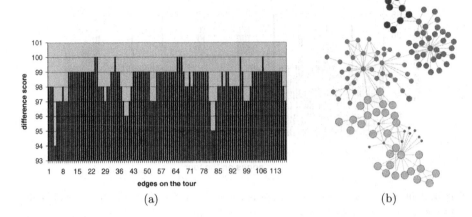

Fig. 5. The scientific collaboration network. (a) Difference scores between adjacent nodes on the tour for the scientific collaboration network. (b) The detected community structure.

structure in this network is not so obvious as football team network, and the deterministic classification membership of each node is also unclear. This network consists of 118 nodes (scientists) and the edges represent the collaboration relationships among scientists. In this example, the distance matrix is constructed directly based on adjacency matrix. According to the rearrangement of the rows and columns, the difference scores between neighboring rows for the scientific collaboration network are plotted in Figure 5(a). From the peaks in this figure, the network is divided into 5 clusters. We can not use I_{NMI} to evaluate the classification accuracy since the true partition is unknown. The primary divisions detected by our algorithm are indicated by different vertex colors and sizes in Figure 5(b). Except some nodes with sparse links (the tiny nodes), other four communities are visually reasonable. Although the deterministic partition of this network is unknown, the detected divisions with $Q = 0.5811$ are consistent with those in [1] on the whole. The community with nodes of largest size mainly contains scientists whose research interest is structure of RNA. The community consisting of nodes with second largest size mainly contains scientists whose research interest is agent-based models. The nodes of the third largest size constitute a community in which the scientists mainly research mathematical ecology. And the scientists in the last community have a research interest in statistical physics.

3.4   A Protein-Protein Interaction Network

Finally, we consider a larger network — a protein-protein interaction network of *S. cerevisiae*. This network has 1257 proteins as nodes and 6835 interactions as edges. The original data set [3] has 1298 proteins involved in 54406 interactions. All of these protein interactions have a confidence score. Here we only

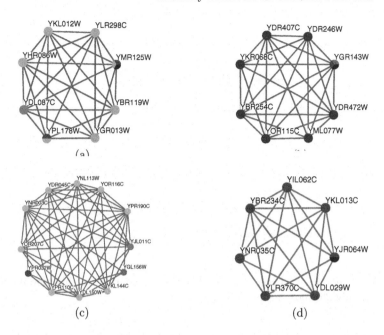

Fig. 6. Functional modules of the protein-protein interaction network. (a) All the proteins in this module have a function of RNA processing except the protein YDL087C without GO function annotation. (b) All the proteins in this module have a function of transport. (c) The module with a main function of transcription except two proteins YPR032W and YGL156W with inconsistent functions and one protein YJL011C without GO function annotation. (d) A cell organization and biogenesis module.

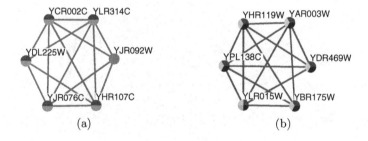

Fig. 7. Modules with multiple main functions. (a) Each protein in this module has two functions: cell organization and biogenesis, cell cycle except the protein YJR092W without GO function annotation. (b) A module with three main functions: DNA metabolism, metabolism, cell organization and biogenesis.

use the reliable interactions (the scores are not lower than 0.3). By applying the proposed method, we obtain a tour indicating the rearrangement of the rows of this network. The difference scores are not plotted here since the nodes are too many. According to the difference scores, we totally find 47 functional modules.

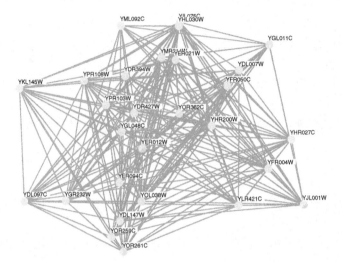

Fig. 8. A big module with one function: protein degradation. In this module, all the proteins (totally 28) have a same function.

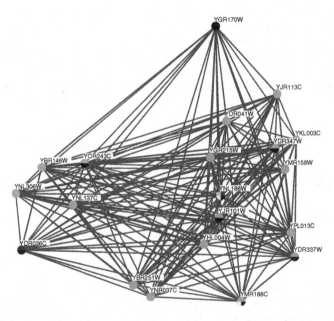

Fig. 9. A big module with two main functions: protein biosynthesis, metabolism

Among these, the largest one has 49 proteins and the smallest one has 4 proteins. Although protein functional modules are formed not only by the topology of the network, protein-protein interactions are believed to be very informative for protein function. We find that most of the communities (functional modules) detected by the method consist of one main function or two main functions based on analysis of function annotation terms of GO, which indicates the effectiveness of the method. Some of these modules are listed in Figures. 6, 7, 8, 9 where the captions of the figures describe their functions. In these figures, nodes with the same color imply that proteins on them have the identical function. On the other hand, proteins on those nodes with multiple colors have multiple functions. Based on these functional modules, the functions of some proteins without GO annotation can be predicted according to the main functions in their modules, such as Figures 6(a), 6(c), 7(a).

4   Conclusion

Detecting community structure in biological and social networks recently attracts increasing attention in various fields. In this paper, we introduce a combinatorial optimization method — traveling salesman problem based optimal rearrangement clustering to detect modularity structure in complex networks. Extensive experiments show the effectiveness of this method. An advantage of this method is that it does not need to spend much time for finding the number of communities in a network. Since the TSP solver used in this paper is based on linear programming and can solve problems with ten thousands of nodes, the method we proposed can be applied for protein-protein interaction networks and other biological networks from most organisms.

Acknowledgement

This research work is supported by the National Nature Science Foundation of China (NSFC) under grant No.10701080, the Research Foundation of Renmin University of China "Several fundamental problems in complex networks" (No.06XNB054) and the Ministry of Science and Technology, China, under grant No.2006CB503905.

References

1. Girvan, M., Newman, M.E.J.: Community structure in social and biological networks. Proc. Natl. Acad. Sci. USA 99(12), 7821–7826 (2002)
2. Gibson, D., Kleinberg, J., Raghavan, P.: Inferring Web communities from link topology. In: Proceedings of the 9th ACM Conference on Hypertext and Hypermedia, ACM Press, New York (1998)
3. Gavin, A.C., Aloyand, P., Grandi, P., et al.: Proteome survey reveals modularity of the yeast cell machinery. Nature 440, 631–636 (2006)
4. Wasserman, S., Faust, K.: Social Network Analysis: Methods and Applications. Cambridge University Press, Cambridge (1994)
5. White, S., Smyth, P.: A spectral clustering approach to finding communities in graphs. In: SIAM International Conference on Data Mining, SIAM Press, USA (2005)

6. Josep, M.P., Béjar, J., Delgado, J.: Clustering algorithm for determining community structure in large networks. Phys. Rev. E 74, 16107 (2006)
7. Newman, M.E.J.: Fast algorithm for detecting community structure in networks. Phys. Rev. E 69, 66133 (2004)
8. Newman, M.E.J., Girvan, M.: Finding and evaluating community structure in networks. Phys. Rev. E 69, 026113 (2004)
9. Danon, L., Daz-Guilera, A., Duch, J., Arenas, A.: Comparing community structure identification. J. Statist. Mech. Theory and Experiment 9, P09008 (2005)
10. Duch, J., Arenas, A.: Community identification using extremal optimization. Phys. Rev. E 72, 027104 (2005)
11. Newman, M.E.J.: Modularity and community structure in networks. Proc. Natl Acad. Sci. USA 103(23), 8577–8582 (2006)
12. Gustafsson, M., Hörnquista, M., Lombardi, A.: Comparison and validation of community structures in complex networks. Physica A 367, 559–576 (2006)
13. Newman, M.E.J.: Detecting community structure in networks. Eur. Phys. J. B 38, 321–330 (2004)
14. Gutin, G., Punnen, A.P.: The Traveling Salesman Problem and Its Variations. Kluwer Academic Publishers, Dordrecht (2002)
15. Lenstra, J.K.: Clustering a data array and the traveling-salesman problem. Operations Research 22(2), 413–414 (1974)
16. Climer, S., Zhang, W.: Rearrangement clustering: Pitfalls, remedies, and applications. Journal of Machine Learning Research 7, 919–943 (2006)
17. Climer, S., Zhang, W.: A travelling salesman's approach to clustering gene expression data. Technical Report, WUSEAS-2005-5, Washington University in St. Louis (2005)
18. Johnson, O., Liu, J.: A travelling salesman approach for predicting protein functions. Source Code for Biology and Medicine 1 (2006) doi:10.1186/1751-0473-1-3
19. http://www.tsp.gatech.edu//concorde/index.html
20. Zachary, W.W.: An information flow model for conflict and fission in small groups. J. Anthropol. Res. 33, 452–473 (1977)

The HIV Data Mining Tool for Government Decision-Making Support

Huijun Liu[1,2], Qunying Xiao[1,2], and Zhengwei Zhu[2]

[1] School of Management, Xi'an JiaoTong University, Xian Ning West Road 28,
710049 Xi'an, P.R. China
[2] School of Public Policy and Administration, Xi'an JiaoTong University, Xian Ning West
Road 28, 710049 Xi'an, P.R. China
liuhui.j@163.com, qy987@sina.com, zwzhu@mail.xjtu.edu.cn

Abstract. To explore more effective policies on HIV epidemic control, this paper designs a new HIV data mining tool for decision-making support at the basis of SIR epidemic dynamical model, and then make it operate in the context of Shenzhen China from 1992 to 2005. Its predicting function proves power and effective by comparing with GM(1,1) model and general dynamical model, and its identifying function and prioritizing function also pass the test in Shenzhen context.

Keywords: HIV epidemic, Decision support, Dynamical model, DM tool.

1 Introduction

It is known that each HIV virus keeper may be epidemic resource, if most of the infectors are still in early process of disease, and the surveillance cannot cover them, they will be still unconscious and become more dangerous. That is the situation faced by Chinese government. By the end of 2005 there are 144,067 HIV cases reported, which is only 21.74% of the estimation (650,000) [1], smaller than 75% as requested. And these cases are most in early process of disease, as only 0.8 HIV/AIDS deaths happened in per 100 deaths on Earth [1]. Furthermore, China is a migrating country. In early 70s there were only 2 million migrations, in 1984 the scale turned to be 20 million, and according to Chinese No.5 Population Survey it has reached 121 million in 2000. Migration is regarded as high-risk factor causing HIV transmission [2], thus giant migrating flow might make the situation worse.

To deal with the above problem, high effective decision-making support system for HIV prevention is badly needed for the government. And such functions should be included in the system, as the predicting function for HIV epidemic tendency forecast, the identifying function for effective policy choice, and the prioritizing function for public resource deployment. Here the first function is the basic one.

Many researchers has developed lots of mathematical models to describe and forecast HIV epidemic tendency. But it is hard to integrate migrating influence into HIV forecast model. Some researchers had contributed to this problem, but seldom satisfied the necessity of real world. The GM(1,1) HIV forecast model in Sec. 1.1, and the general dynamical model in Sec. 1.2, are both models for HIV tendency forecast,

T. Washio et al. (Eds.): PAKDD 2007 Workshops, LNAI 4819, pp. 131–141, 2007.

having been tested in Chinese context and proven effective [3][4]. While both models still don't consider migrating impacts on HIV epidemic, and can't provide other decision support elements except the forecast function.

In this paper a new HIV data mining tool will be developed in Sec. 2, which might integrates both predicting model and decision modules. And the precise and usage of the new tool will be tested in Shenzhen City context.

1.1 The GM(1,1) HIV Forecast Model [3]

With the time series HIV data $X^{(0)}(k) = \{X^{(0)}(1), X^{(0)}(2), X^{(0)}(3), \cdots, X^{(0)}(n)\}$, firstly generate $X^{(1)}(k)$ by accumulative operator like (1); then generate $Z^{(1)}(t)$ by consecutive neighbor operator like (2).

$$X^{(1)}(k) = \sum_{i=1}^{k} X^{(0)}(i); k = 1, 2, \cdots, n \tag{1}$$

$$Z^{(1)}(k) = 0.5X^{(1)}(k) + 0.5X^{(1)}(k-1); k = 2, 3, \cdots, n \tag{2}$$

Let:

$$Y = \begin{bmatrix} X^{(0)}(2) \\ X^{(0)}(3) \\ \vdots \\ X^{(0)}(n) \end{bmatrix}, B = \begin{bmatrix} -Z^{(1)}(2) & 1 \\ -Z^{(1)}(3) & 1 \\ \vdots & \vdots \\ -Z^{(1)}(n) & 1 \end{bmatrix} \tag{3}$$

$$[a, b]^T = (B^T B)^{-1} B^T Y \tag{4}$$

We can get the gray forecast function as follows:

$$X^{(1)}(k+1) = (X^{(0)}(1) - \frac{b}{a})e^{-ak} + \frac{b}{a} \tag{5}$$

$$X^{(0)}(k+1) = X^{(1)}(k+1) - X^{(1)}(k) \tag{6}$$

1.2 The General HIV Dynamical Model

It is a two stages model including two sub-equations for HIV and AIDS separately [4]. We merge them into one and call it general HIV dynamical model. The new model is:

$$X(t+1) = \lambda(t)(N(t) - X(t)) + (1 - \mu)X(t) \tag{7}$$

$$\lambda(t) = \frac{\beta X(t)}{N(t)} \tag{8}$$

Here $N(t)$ is the size of target population, $X(t)$ is the number of HIV cases, μ is the death rate dieing for HIV/AIDS, and β is a coefficient describing the contact depth between sensitive group and infective group. In fact this is a dynamical model with SI mode.

It is very popular to use dynamical model of infectious disease to simulate epidemic diffusion, such as SIR, SI, SIS, SIRS, SIRE and so on, among which SIR is the most basis model. According to the principle the target species (N) can be divided into three basic cabins: sensitive group (NS), infectious group (NI), and recovered group (NR). NI transmits virus to NS by cross infection; after being checked out as HIV keeper, some infectors might be conscious to adopt self-isolated measures or accept a medication to reduce the virulence, thus they are excluded out of infecting process and enter into NR. Here $S + I + R = 1$, $S_0 > 0$, $I_0 > 0$, $R_0 = 0$. The size and the structure of target species and its three subgroups will keep changing, with the influences of population factors (natural birth or death), behavior factors (migration, cross contagions, self-isolation or medication), biological and pathology factors (mother-to-child vertical transmission, dead for disease) etc.

Different from other infectious disease having immune medicine, HIV infectors should obey special excluding mechanism to be NR. Three solutions exist in the literature: (1) self-isolation. Isham thought that once HIV infectors were checked out or turned to AIDS stage, they would like to be self-isolated consciously [5]; (2) virulence reduction. Gumel assumed that HIV infectors should generate immunity after treatment [6]; (3) zero assumption. Hyman and Stanley directly set HIV recovering rate as zero[7]. For its convenience the third method is usually used to model HIV epidemic, for example the general HIV model in [4].

2 The HIV Data Mining Tool

The new model is developed at the basis of SIR epidemic model.

2.1 New Forecast Model

Using λ as the effective contact rate of the cross infection between NI and NS, the cross infection size $\triangle_1$ is:

$$\Delta_1 = \lambda \text{NSI} \qquad (9)$$

Taking μ_2 and μ_3 as natural death rate and HIV/AIDS death rate, the total death size $\triangle_2$ is:

$$\Delta_2 = -(\mu_2 + \mu_3)\text{NI} \qquad (10)$$

Here another method different from [5][6][7] is induced to describe the recovered. As some infectors might be against ethics, the self-isolated assumption was unable to turn into realities completely. Let ϑ be the self-isolated proportion of checked-out infectors, ν be the recovering rate of infectors, γ be the ratio of the confirmed to the estimation, then the recovered size $\triangle_3$ is:

$$\Delta_3 = -NR$$
$$= -(v + \gamma\vartheta)\,NI \tag{11}$$

Thus $\triangle_1$ can be express in other way:

$$S = 1 - I - R$$
$$= 1 - I - (v + \gamma\vartheta)I \tag{12}$$

$$\Delta_1 = (1 - I - v - \gamma\vartheta)\lambda NI \tag{13}$$

Furthermore, take μ_1 as natural birth rate, β as infectious rate of infants born by the pregnancy living with HIV, and ϕ as policy covering probability to interrupt vertical transmission from mother to infant, then new infants size living with HIV $\triangle_4$ is:

$$\Delta_4 = \mu_1 NI\beta(1 - \phi) \tag{14}$$

Here another method is induced to express migrating impacts. Take M_i and M_o as the immigrants and emigrants, N_m and α^1 as the size and the resorting rate of total city migration. Assume the HIV infectious level of immigration to be ρ times than the average infectious level in China (I_i), then HIV cases influenced by migration $\triangle_5$ is:

$$\Delta_5(t) = M_i(t)\rho I(t) - M_o(t)I(t)$$
$$= [N_m(t+1) - aN_m(t)]\rho I(t) - (1-\alpha)N_m(t)I(t) \tag{15}$$

Let $Y = NI$, then:

$$\dot{Y}(t) = \sum_{k=1}^{5} \Delta_k(t) \tag{16}$$

$$Y(t+1) = Y(t) + \dot{Y}(t) \tag{17}$$

$$Y(0) = Y_0 \tag{18}$$

From the above formulas, we can select several parameters to form a multi policy vector $V = [\lambda, \phi, \gamma, \rho, \alpha]$, which mean:

- To reduce cross infection by behavior intervention (the effective contact rate, λ);
- To interrupt mother-to-child vertical transmission (the covering probability of interruption, ϕ);
- To increase self-isolated or medicated cases by expanding surveillance network (the ratio of the checked to the estimation, γ);
- To regulate migration mode by adjusting the average infectious level of immigrants (the times of national infectious level, ρ) and the resorting rate (α).

---

[1] The annual resorting rate means the immigrants having arrived at t year or before, will detain at the same place at t+1 year.

2.2 Policy Modules

Three policy modules are listed in table 1: (1) the reference module: simulate current policy environment; (2) special view module: only regulating one single policy parameter in three points of view; (3) continuous regulating module: keep other parameter unchanged and compute policy utility distribution one by one.

Table 1. Multi policies simulating schemes

Category	Parameter	Methodology
Reference	V	To simulate HIV epidemics of a certain city, compute the outputs of different policy sections and serve as reference for other schemes
Special view modulation	ρ	Three situations: higher or lower than the current 10%, or almost no infectors immigrate into the city
	α	Three situations: higher or lower than the current 10%, or almost no migration will stay in the city
	λ	Three situations: higher or lower than the current 10%, or almost no new cross infection happened
	γ	Three situations: higher or lower than the current 10%, or reach the level of developed country (75%)
	φ	Three situations: the widest scope, the least scope or just wider than the current 10%
Continuous tiny-step modulation	V	Supposing all the environment variables and other policy parameters are kept stable, try to regulate single policy parameter k by the tiny step interval of +0.001 at the scope of [0,1]

2.3 Observation Functions

To identify the simulating outputs three functions are designed for policy quality and priority identified, where $\overline{\dot{Y}}$ means the average of $\dot{Y}$.

1. $J_1(\ . \ )$ provides a simple observation to the contributions of each policies in the first scheme.

$$J_1 = 1 : \frac{\overline{\Delta}_1}{\overline{\dot{Y}}} : \frac{\overline{\Delta}_2}{\overline{\dot{Y}}} : \frac{\overline{\Delta}_3}{\overline{\dot{Y}}} : \frac{\overline{\Delta}_4}{\overline{\dot{Y}}} : \frac{\overline{\Delta}_5}{\overline{\dot{Y}}} \qquad (19)$$

2. Regulate each policy parameters, compute each $\overline{\dot{Y}}$, and then have a comparison in the second scheme.

$$J_2 = \overline{\dot{Y}}_s : \overline{\dot{Y}}_{k1} : \overline{\dot{Y}}_{k2} : \overline{\dot{Y}}_{k3} \qquad (20)$$

3. $J_3 (\ . \ )$ is designed to compute the distribution of the changing rate $\overline{\dot{Y}}$ between i and i+1 step, where diff (.) is a difference function. This function can provide the trace of policy utility. The special structure of (21) also aims at getting ride of the scale difference of different parameters, which is helpful for comparison.

$$J_3 = \text{diff } \frac{(\overline{\dot{Y}}_{k, i+1} - \overline{\dot{Y}}_{k, i})}{\overline{\dot{Y}}_{k, i}} \tag{21}$$

2.4 Parameters Estimation

The following parameters are collected by literature study:

1. λ and μ_3. referring to [4] λ =0.53. At the end of 2005 national average death rate of HIV disease is 5.84%[8]. If it is regarded as city infectious death rate, then μ_3=5.84%.

2. γ, v and ϑ. According to [9] only 20% people living with HIV have been examined out, it is acceptable to set γ =0.20. If the average HIV infectious period equals to 12 years as [7], then $v = 0.08^2$. We estimate ϑ =0.70.

3. β and φ. HIV vertical transmitting probability is 13%-48%, it is reasonable to set β =0.30. According to the data of Shenzhen Health Bureau, we set φ =0.75 [9].

4. ρ and α. Referring to [2], the infectious rate of rural residence with migrating history is 1.77 times as those without migrating history. According to our investigation on Shenzhen migration in 2005 (N=1739), about 79.9% migrants come to the city for the first time, 20.1% visitors have been in another city. It is easy to know ρ =1.16[3]. Further referring to Shanghai immigration survey data [10], let α =0.23.

3 The Application in Shenzhen Context

3.1 Data

As Table 2 we operate the new model by the demography data and HIV monitoring data of Shenzhen and China in 1992-2005. Because China has made a large-scale check to people living with HIV in 2004, so has Shenzhen to migration in 2005, the relative two data points appear strange. We let both re-estimated by curve fitting method. On the other hand, the Statistical Yearbook of Shenzhen only includes one year or more staying migrants, but what we need is all the migration size. Thus the annual migrating data is adjusted with the original data of Chinese No.4 and No.5 demography survey in 1990 and 2000. The number of national and city HIV cases is also adjusted to the estimation level with γ value in Sec.1 and Sec. 2.4. On the other hand, the simulating parameters are inputted in different policy schemes:

- To the reference scheme: V=(0.53,0.75,0.2,1.16,0.23)
- To the scheme under special points of view: ρ =(1.16, 1.28, 1.04, 0.01); α=(0.23, 0.33, 0.13, 0.01); λ=(0.53, 0.63, 0.43, 0.01); γ=(0.20, 0.30, 0.10, 0.75); φ=(0.75, 1.00, 0.00, 0.85)
- To the scheme of continuous tiny-regulation: each policy parameter increases by the step +0.001 in the scope of [0,1]

---

[2] $\Gamma = 1/v$, where Γ is the length of HIV infectious period.

[3] ρ =1.77×0.201+0.799≈1.16.

Table 2. The original data for model simulation

Year	Shenzhen Data						China Data	
	Resorting migrants (10,000)	Estimating migrants (10,000)	Registering population (10,000)	Cumulati ve HIV cases	Natural birth rate (‰)	Natural death rate (‰)	Total population (10,000)	Cumulati ve HIV cases
1992	180.68	249.94	80.22	1	14.07	2.4	117171	969
1993	207.3	318.95	87.69	1	14.36	1.93	118517	1243
1994	241.54	402.41	93.97	2	13.77	1.75	119850	1774
1995	245.54	435.24	99.16	4	13.23	1.78	121121	3341
1996	255.1	475.24	103.38	12	13.00	2.09	122389	5990
1997	270.18	524.3	109.46	23	12.64	2.13	123626	9333
1998	280.36	562.97	114.60	48	11.65	2.09	124761	12639
1999	285.29	589,78	119.85	76	12.58	2.97	125786	17316
2000	308.02	652.85	124.92	103	14.68	2.55	126743	22517
2001	336.72	729.32	132.04	177	14.06	1.73	127627	30736
2002	364.8	805.33	139.45	349	16.60	1.46	128453	40468
2003	406.48	912.67	150.93	540	10.63	1.53	129227	62159
2004	432.42	985.6	165.13	887	11.58	1.37	129988	106990
2005	645.82	1022.4	181.93	1394	12.64	1.41	130756	144089
2006		1081.82						

Note: The data collected from [12][13], the Statistical Yearbook of Shenzhen and China in 1993-2006.

3.2 Results

3.2.1 The Scheme of Reference

Operate the model from 1992 to 2005, and get the results as Fig.1. It seems that the model has nice estimating capacity. Carmines & McIver (1981) said that if card square freedom (X^2/df) less than 2 [14], the estimating model could have an ideal fitting. And $X^2/df = 1.08$ in our model.

We also make GM(1,1) and general model operate in the same context, then compare their forecast results. Firstly using 1992-2000 data to fit GM(1,1) we get (22). By (22) and (6) predict the HIV tendency in 2001-2005. Secondly set $\beta = 0.5256$ and $\mu = 0.0143$ like [4] in the general model to get (23), and operate it from 1992 to 2005. At last analyze the estimating results by D^4, which is the standardized deviation indicator in average. To GM(1,1) model D=0.24; to general model D=0.52; and to our model D=0.06. Apparently the new model is closer to the real world. It might contribute to those environmental variables, and to the integrity of migrating factors.

---

[4] $$D = \frac{\sum_{i=t_0}^{t} \sqrt{((X_t - E(X_t))/X_t)^2}}{t - t_0}$$, where X_t is the real data, and $E(X_t)$ is the estimation.

$$X^{(1)}(k+1) = 55.24 + 60.24e^{0.33k} \tag{22}$$

$$X(t+1) = \frac{0.5256X(t)(N(t) - X(t))}{N(t)} + 0.9857X(t) \tag{23}$$

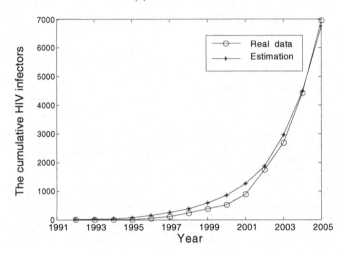

Fig. 1. The estimation to Shenzhen HIV epidemics in 1993-2005

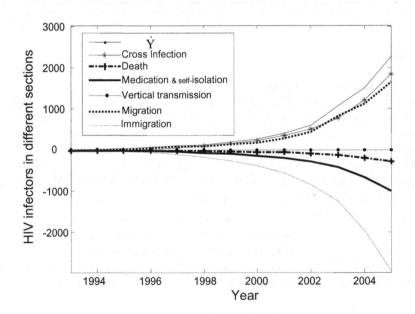

Fig. 2. The different outputs caused by multi policies

The output of different policy sections (as Fig.2) can be ranked as follows: (1) cross infection (0.7959); (2) migrating impacts (0.7414); (3) medication and self-isolation (-0.4237); (4) vertical transmission (-0.1155); (5) death impacts (0.0018). In conclusion: cross infection and migration play key role in HIV epidemic, Surveillance network also takes important position, the death impacts are very low, and HIV vertical transmission have the weakest influence. The government should take care of the immigrating consequence, which is lager than that of emigration today.

$$J_1 = 1: 0.7969: -0.1155: -0.4237: 0.0018: 0.7414 \qquad (24)$$

3.2.2 The Scheme Under Special View
Keep other variables or parameters unchanged,

- when $\rho = (1.16, 1.28, 1.04, 0.01)$, $J_2 = 519:573:466:4$;
- when $\alpha = (0.23, 0.33, 0.13, 0.01)$, $J_2 = 519:498:538:556$;
- when $\lambda = (0.53, 0.63, 0.43, 0.01)$, $J_2 = 519:562:483:372$;
- when $\gamma = (0.20, 0.30, 0.10, 0.75)$, $J_2 = 519:470:579:329$;
- when $\varphi = (0.75, 1.00, 0.00, 0.85)$, $J_2 = 519:519:521:519$.

In conclusion: high-risk behaviors prevention, monitoring & diagnosing network expansion, and migrating mode regulation have great influence on HIV epidemic control, among which the regulation to ρ and γ show more sensitive than the others. While the interruption to HIV vertical transmission has small effects even though it is extended to the extreme.

3.2.3 The Scheme of Continuous Regulation
It is supposed that the increasing of γ and φ value might cause positive policy output, which is contrary to adding the value of λ, ρ and α. The policy parameter regulation is just like the adding or reducing investment of public policy. If we deploy policy resource in one control point continuously, then by (21) we can explore its policy utility distribution. The results show as Fig.3, where X axis represents for 1000 sets of policy schemes with +0.001 step interval and Y axis points to the difference value of $\overline{Y}$'s changing rate.

- To λ and γ. As subplot1 the increasing investments of λ and γ might provide desirable outputs. The policy utilities of λ and γ show exponent distribution, and the next output only depends on current input, thus adding policy resource might expand the utility. It shows the best policy output for the government to increase the investment of HIV medication and surveillance continuously.
- To φ. The utility of φ appears uniform distribution as subplot 2. The utility of φ regulation is upward in a certain level, but it is lower than γ and λ regulation.
- To ρ and α. Migrating mode regulation has complex appearance like subplot3 and subplot4. ρ curve shows fluctuation primarily, then keeps unchanged in zero

for long term; while α curve appear shrinking tendency firstly and then show accelerating reduction afterward. Because both ρ and α regulating result relay on the total migrating size, they cannot be regulated separately. Thus further evidences are required for exact explanation.

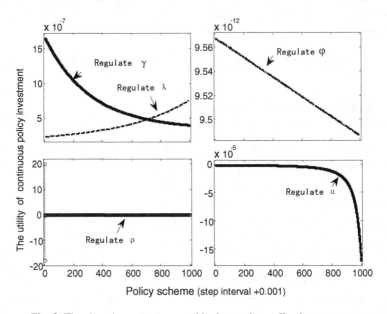

Fig. 3. The changing outputs caused by increasing policy investment

4 Conclusion

This paper contributes to provide a new HIV data mining tool to support city HIV epidemic control. Its predicting function proves power and effective by comparing with GM(1,1) model and general dynamical model, and its identifying function and prioritizing function also pass the test in Shenzhen context.

Most epidemic models seldom care about environmental variables, which might make against public policy output. At this point of views we should try the best to assure model integrity and its support for decision-making. Just as we know, high-risk behaviors result in HIV transmission by many channels and modes, such as injecting drug use, commercial blood and plasma donation, homosexual behavior, commercial sexual behavior and so on. The above tool should be improved by considering high-risk behaviors in details, as well to be applied and tested in other culture context in the future.

Acknowledgements. This work was funded in part by: Chinese National Science Foundation under grants 06CRK004, Shaanxi Province Soft Science Foundation under grants 2006KR86, and Chinese National Education Department "985-2" Project under grants 07200701.

References

1. China Ministry of Health, Joint United Nations Programme on HIV/AIDS, World Health Organization: 2005 Update on the HIV/AIDS Epidemic and Response in China. Beijing, pp. 1–2 (2006)
2. Xu, C., Wu, Z., Zhang, Y.: A Study on HIV Prevalence among Rural to Urban Migrants. Symposium on AIDS Prevention and Control in China, Beijing, pp. 23–24 (1998)
3. Lai, Y., Hong, F., Zeng, X., et al.: Grey Forecast to Epidemic Tendency of HIV/AIDS in Shenzhen. Modern Preventive Medicine 3, 446–447 (2003)
4. Liu, M., Luan, Y., Han, L., Zhou, Y.: Dynamic Models to Predict Future HIV/AIDS Prevalence in China. Journal for China AIDS/STD 6, 335–337 (2003)
5. Isham, V.: Stochastic Models for Epidemics with Special Reference to AIDS. The Annals of Applied Probability 1, 1–27 (1993)
6. Gumel, A.B., Moghadas, S.M., Mickens, R.E.: Effect of Preventive Vaccine on the Dynamics of HIV Transmission. Communications in Nonlinear Science and Numerical Simulation 9, 649–659 (2004)
7. Hyman, J.M., Li, J., Stanley, E.A.: Modeling the Impact of Random Screening and Contact Tracing in Reducing the Spread of HIV. Mathematical Biosciences 181, 17–54 (2003)
8. China Center Government: Strengthen HIV Epidemic Surveillance (March 3, 2006), http://www.gov.cn/node_11140/2006-03/03/content_214916.htm
9. Shenzhen Bureau of Health: The HIV/AIDS Epidemic Style Reprot in Shenzhen (2004)
10. Zhu, B.: The Resorting State and Trend of Shanghai Immigration. Population Science of China 3, 38–45 (1999)
11. Zhang, B.: Economic Forecasting and Decision, p. 96. Economic Science Press, Beijing (2004)
12. Chinese State Council AIDS Working Committee Office, UN Theme Group on HIV/AIDS in China: A Joint Assessment of HIV/AIDS Prevention, Treatment and Care in China, Beijing, pp. 2–6 (2004)
13. Peng, C.: Shenzhen HIV Epidemic Status Analysis in 1992-2003. South China Journal of Prevention Medicine 5, 20–21 (2004)
14. Carmines, E.G., McIver, S.P.: Analyzing Models with Unobserved Variables: Analysis of Covariance Structures. In: Bohrnstedt, G.W., Borgatta, E.F. (eds.) Social Measurement: Current Issues, pp. 65–115. Sage, Beverly Hills, CA (1981)

Negative Localized Relationship Among p70S6 with Smad1, 2, 3 and p38 in Three Treated Human Cancer Cell Lines

Lin Wang[1,3], Minghu Jiang[2,3], Stefan Wolfl[4], and Yinghua Lu[1]

[1] Center for Biomedical Engineering, School of Electronics Engineering,
Beijing University of Posts and Telecommunications, Beijing, 100876, China
[2] Lab of Computational Linguistics, School of Humanities and Social Sciences,
Tsinghua University, Beijing, 100084, China
[3] Center for Psychology and Cognitive Science, Tsinghua Univ., Beijing, 100084, China
[4] Institute of Pharmacia & Molecular Biotech., Dept. Biology – Bioanalytic
University Heidelberg, 69120 Heidelberg, Germany
jiang.mh@tsinghua.edu.cn

Abstract. P70S6 with Smad1, 2, 3 and p38 as very hot and important multi-function proteins are widely studied now. Protein localization is fundamentally important to eukaryotic protein function and cell regulation. In this paper, the relationship among p70S6 with Smad1, 2, 3 and p38 in Three Treated Human Cancer Cell Lines was studied using western blot, FACS and nano-gold phospho-antibody microarray by reaction with our purification of nuclear and cytoplasmic protein as follows: P-p70S6 nuclear decrease whereas p-p38, p-smad1 and p-smad2/3 nuclear expression increase in BMP2–induced apoptosis U937 cells (2000ng/ml BMP2 for 3 days) and in STI571-inhibited cell growth and-induced proliferation of K562 cells (0.2 uM STI571 for 24h). P-p70S6 only cytoplasmic increases whereas p-p38, p-smad1 and p-smad2/3 cytoplasmic inhibition by nuclear decrease and cytoplasmic increase under total protein unchanged in BMP2-induced cell survival and differentiated MCF7 cells (100ng/ml BMP2 for 4h). This result implies the negative localized relationship among p70S6 with Smad1, 2, 3 and p38 in three treated human cancer cell lines.

1 Introduction

With the post-genome era proteomics study for direct analysis of a group of protein become more and more important [1]. Identification of the type of modification and its location often provide crucial information for understanding the function or regulation of a given protein in biological pathways [2]. It is widely acknowledged that proteins rarely act as single isolated species when performing their functions in vivo. The analysis of proteins with known functions indicates that proteins involved in the same cellular processes often interact with each other. Following this observation, one valuable approach for elucidating the function of an unknown protein is to identify other proteins with which it interacts, some of which may have known activities. On a large scale, mapping protein-protein interactions has not only

T. Washio et al. (Eds.): PAKDD 2007 Workshops, LNAI 4819, pp. 142–152, 2007.

provided insight into protein function but facilitated the modeling of functional pathways to elucidate the molecular mechanisms of cellular processes [2].

P70S6 with Smad1, 2, 3 and p38 as very hot and important multifunction proteins are widely studied now. Protein localization data provide valuable information in elucidating eukaryotic protein function [2]. Protein trafficking between nucleus and cytoplasm fundamentally important to cell regulation. As such, the nuclear import and export are pivotal in orchestrating the activities of the key regulators of the cell cycle [2]. Such as: Smad nuclear translocation is a required component of the activin A-induced cell death process in liver cells; ERK-1 and −2 nuclear translocation triggers cell proliferation in vitro models from [3]. ERK activation plays an active role in mediating cisplatin-induced apoptosis of HeLa cells and functions upstream of caspase activation (cyto) to initiate the apoptotic signal [4]. Lorenzini et al verified in senescent cells that no ERK are able to phosphorylate efficiently their nuclear targets [5]; JNK cytoplasmic localization as inhibitor of proliferation from [6] verifying that a murine cytoplasmic protein that binds specifically to JNK [the JNK interacting protein-1 (JIP-1)] and caused cytoplasmic retention of JNK and inhibition of JNK-regulated gene expression; MKK7-JNK/SAPK and MKK6-p38 pathways to cytoplasmic apoptotic activation induced by Fas [7]. p38 MAPK induced cytoplasmic domain-dependent cellular migration (differentiation) of alpha 2 integrin subunit [8]; Phospho-Tyrosine Rak (54 kDa) expressed nuclear and perinuclear regions of the cell in 2 different breast cancer cell lines inhibits growth and causes G(1) arrest of the cell cycle [9]; STAT3 cytoplasmic localization was detected in the pathogenesis of mantle cell lymphoma (MCL) tumours. STAT3 nuclear localization of STAT3 was shown in node-negative breast cancer associated with a better prognosis (differentiation and apoptosis) by tissue microarray analysis [10]; Akt/PKB intranuclear translocation is an important step in signalling pathways that mediate cell proliferation [11]. No Akt is able to phosphorylate efficiently its nuclear targets in senescent cells [3]. p70S6K is localized both in the cytosol and, after cytokine stimulation, also in nucleus in factor-dependent hematopoietic M-07e cells [12].

In this paper, the relationship among p70S6 with Smad1, 2, 3 and p38 in Three Treated Human Cancer Cell Lines was studied using western blot, FACS and nano-gold phospho-antibody microarray by reaction with our purification of nuclear and cytoplasmic protein as follows: P-p70S6 nuclear decrease whereas p-p38, p-smad1 and p-smad2/3 nuclear expression increase appeared in BMP2–induced apoptosis U937 cells (2000ng/ml BMP2 for 3 days) and in STI571-inhibited cell growth and-induced proliferation of K562 cells (0.2 uM STI571 for 24h). P-p70S6 cytoplasmic increases whereas p-p38, p-smad1 and p-smad2/3 cytoplasmic inhibition by nuclear decrease and cytoplasmic increase appeared in BMP2-induced cell survival and differentiated MCF7 cells (100ng/ml BMP2 for 4h). This result implies Negative co-localized Relationship among p70S6 with Smad1, 2, 3 and p38 in Three Treated Human Cancer Cell Lines.

2   Method

2.1   Cell Lines, Chemicals, Antibodies, Apparatus

Leukemia cell line K562 and breast cell line MCF-7 were kindly provided by Dr Joachim Clement. Leukemia cell line U937 provided by Frau Dagmar Haase.

STI571 was kindly provided by Prof. Pachmann and prepared as a 10 mM stock solution in sterile DMSO (Merck Darmstadt, Germany). BMP2 was kindly provided by Dr. Clement stored at -20°C, and dissolved in sterile water as a 1mg/ml stock solution before use. Stock solutions were then diluted in RPMI medium to achieve the desired final concentration. In all of the cases, final concentrations of DMSO were < 0.1% and did not modify responses of cells to STI571. Biotinamidocaproate N-hydroxysuccinimide ester was purchased from SIGMA D2643; Streptavidin-Gold EM.STP5 and LM/EM Silver Enhancement Kit SEKL15 were purchased from British BioCell; dist.PLANO, Germany. BSA from SIGMA; Milk Powder from Roth GmbH, Germany. Cy 3 and Cy 5 mono-reactive dyes were purchased from Amsham, Germany. phospho-p70S6 kinase (Thr389), phospho-Akt (Ser473), phospho-STAT3 (Ser727), phospho-Tyrosine (Tyr100), phospho-SAPK/JNK (Thr183/Tyr185), phospho-p42/44 MAPK (Thr202/Tyr204), phospho-p38 (Thr180/Tyr182) and p38 MAPK purchased from Cell Signalling Technology; c-Myc, phospho-Smad1 (Ser463/465) and phospho-Smad2/3 (Ser433/435), β-actin, secondary antibodies antigoat, mouse, rabbit purchased from Santa Cruz Biotechnology.

2.2 Cell Culture

K562, MCF-7 and U937 human cancer cells were cultured in RPMI 1640 supplemented with 10% heat-inactivated FCS (Hyclone, Logan, UT). They were maintained in a 37°C, 5% CO_2, fully humidified incubator, passed twice weekly, and prepared for experimental procedures when in log-phase growth (4 x 10^5 cells/ml).

2.3 Cytoplasmic and Nuclear Protein Extracts

K562, MCF-7 and U937 human cancer cellular proteins were obtained from cell culture cells as cytoplasmic or nuclear fractions using a buffer system that allows lysis of cells in two steps. In the first step only the plasma membrane is lysed, leaving nucleus intact. The nuclei are pelleted by centrifugation. The supernatant contains the cytoplasmic protein lysate. To obtain nuclear proteins, the nuclei are washed repeatedly and then lysed using the nuclear lysis buffer B which described in details as follows:

K562, MCF-7 and U937 human cancer cells (10^5 to 10^6) were collected from culture cells and washed with 10 ml PBS by centrifugation with 1500 x g for 5 min. The cell pellet was resuspended in 1 ml PBS and transfered to 1.5 ml tube by centrifugation for 15 sec, buffer removed. The cell pellet was resuspended in 400 μl ice cold buffer A (cytoplasmic lysis buffer) and left on ice for 15 min (cells should swell). 25 μl of Np-40 (10%solution) was added and vortexed for 10 sec by centrifugation for 30 sec in 9000 rpm. Supernatant in 1.5 ml tube for cytoplasmic proteins was added 0.11 volume of ice cold buffer C and mixed thoroughly by centrifugation for 15 min at maximum speed.

Nuclear pellet was washed in 500 μl of ice cold buffer A and 20 μl Np-40 and votexed for 10 sec followed by a centrifugation with 9000 rpm for 30 sec. The pellet was resuspended in 50 μl of buffer B and rotated or shaken for 15-20 min at 4°C. Samples were centrifuged for 5 min and supernatants collected and frozen in aliquots of 10 μl (-70 C). Buffer A (lysis buffer) including 10 mM Hepes pH 7.9; 10 mM KCl;

0.1 mM EDTA; 0.1 mM EGTA; 1 mM DTT; 0.5 mM PMSF. Buffer B (nuclear extract buffer) including 20 mM Hepes pH 7.9; 0.4 mM NaCl; 1 mM EDTA; 1 mM EGTA; 1 mM DTT; 1 mM PMSF. Buffer C (cytoplasmic extract buffer) including 0.3 M Hepes pH7.9 ; 1.4 M KCL ; 0.03 M MgCl2.

2.4 Determination of Protein Concentration

Protein concentration was determined according to Bradford (1976). Several dilutions of protein standards (BSA) containing from 1 to 100 µg/ml were prepared. 0.1 ml of standard samples and appropriately diluted samples were placed in dry test tubes. 0.1 ml sample buffer was used as a negative control. 1.0 ml diluted dye reagent was added to each tube and mixed several times by gentle inversion. After 15 min, OD595 values versus reagent negative control were measured. OD595 versus concentration of standard was plotted. The protein of interest was calculated from the standard curve using the Microsoft Excel5 software.

2.5 Western Analysis

Equal amounts of protein (20 µg) were boiled for 10 min, separated by SDS-PAGE (5% stacker and 10% resolving), and electroblotted to nitrocellulose. After blocking in PBS-T (0.05%) and 5% milk for 1h, the blots were incubated in fresh blocking solution with an appropriate dilution of primary antibody for 4h. The source and dilution of antibodies were as follows: phospho-Akt 1:200, phospho-STAT3 (1:100), phospho-Tyrosine (1:100), phospho-p42/44 MAPK (1:200), phospho-p38 (1:200), phospho-Smad1 (1:100) and β-actin (1:1000). Blots were washed 3 x 5 min in PBS-T and then incubated with a 1:2000 dilution of horseradish peroxidase-conjugated secondary antibody for 1h. Blots were again washed 3 x 5 min in PBS-T and then developed by ECL chemiluminescence.

2.6 Silver Staining in Arraytube

50 µg of cellular proteins are diluted to a final volume of 25 µl in buffer (extraction buffer or PBS). If other buffers than indicated below are used for protein extraction, a change of buffers may be necessary before biotinylation (Tris-based buffers can not be used for biotinylation!). For biotinylation 1 µl of NHS-succinimid-Biotin (SIGMA) (100 µg/µl in ultra pure DMSO; water free) was added and left at room temperature for 1h. Reaction was stopped adding 2%BSA. Protein preparations were left at room temperature for another 15 minutes to ensure complete consumption of the biotinylation reagent. Finally, the volume was adjusted to 100 µl with PBS (2% milk powder or 2% BSA). Biotinylated proteins are then ready to add to blocked ArrayTubes™ for binding.

Before incubation with protein extracts, antibody arrays must be blocked. After spotting most of the activated surface is still freely accessible and has a high capacity for protein binding. For this reason, our arrays were blocked with 5% milk powder in PBS for at least 5 minutes at 30° C shaking at 750 rpm (Eppendorf Thermomixer™). Not all milk powder can be used, so I recommend tests with various milk powders

(Sigma). In this case best results were obtained with a milk powder that fully dissolved into a clump free white colored solution at 5% in PBS. Blocking solution is then replaced by the biotinylated protein extract in PBS (supplemented with 2% milk powder or BSA). Arrays are incubated with the protein extracts for 2h at room temperature. Alternatively, I also obtain good results with an overnight incubation at 4°C. After incubation arrays are washed 3 times for 5 min with 500 µl PBS.

The bound biotinylated proteins are detected in a two step detection process. In the first step, streptavidin gold nanoparticles (British BioCell, Plano) are bound to the biotin groups. In the second step, a silver precipitate is formed around the gold particles. This step is monitored online in the ArrayTube™ Reader, which allows to detect the onset of sliver precipitation. In this way errors resulting from saturation effects from final point measurements can be avoided.

Before incubation with the streptavidin gold particles tubes are again blocked with PBS (5% milk powder) for 15 min. Blocking solution is replaced by 100 µl of streptavidin gold particles in PBS and tubes are incubated for 30 min at 30° C shaking at 350 rpm. Excess streptavidin gold is removed in 3 wash steps. with 200 µl of PBS-tween (0.1%) for 10 min at 20° C with 750 rpm three times. For silver staining 100 µl silver developing solution is added. The silver developing solutions contains equal amounts of silver enhancer and developer, which are combined directly before use (here I used the reagents from British BioCel, distributed in Germany by Plano. Other reagents may work equally well). Tubes are placed into the ArrayTube™ Reader and recording of pictures is started. Pictures are taken every minute for 40 minutes to 1 h.

50 µl of solution A and 50 µl of solution B from silver enhancement kit were combined immediately before use and added to ArrayTube to start silver development, collect images for 40 min to 1h at 1 min interval and analysis of images with appropriate software e.g. Partisian IconoClust from Clondiag.

For each experiment, 40 exposures are obtained and all are evaluated by the IconoClust software. This software can automatically produce data including mean, background, Sigma etc. To compare results between each experiment, all values are normalized by the median method. The pictures obtained were analyzed using the PARTISAN IconoClust image analysis software from Clondiag. This software automatically recognises the arrays and overlays a grid to measure the intensities for each spot. All pictures were combined to generate time curves for the increasing signals of all samples (spots). This time course was then used to assign differences in signal intensities. In comparison with pictures taken after a given period of staining, this procedure eliminates the assignment of the wrong intensity due to saturation effects.

3 Results

Our results showed that in BMP2 treated U937 cells, p-p70S6 (thr389) displayed the increase in cytoplasm and the decrease in nucleus in BMP2 treated U937 cells. Total protein (sum of cytoplasm and nucleus) levels of phospho-p70S6 was not altered compared with control; p-p38 displayed the increase in cytoplasm and nucleus of BMP2 treated U937 cells. Total protein (sum of cytoplasm and nucleus) level of

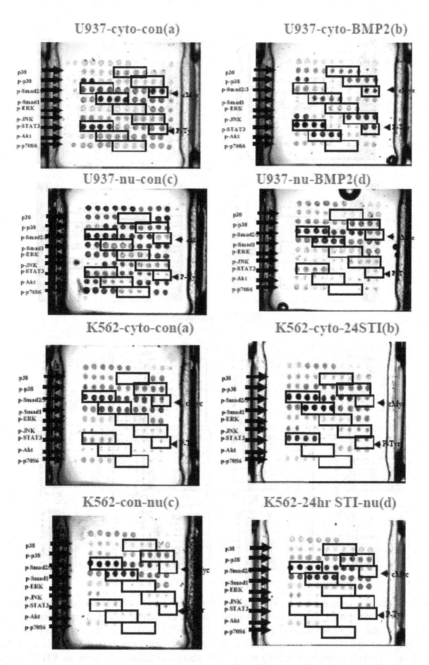

Fig. 1. Antibody micro-array analysis of phosphorylation and localization in control and BMP2 treated U937 cells. Untreated (a, c) and treated (b, d) with 2000ng/ml BMP2 for 3 days. In control and STI571 treated K562 cells. Untreated (a, c) and treated (b, d) with 0.2 μM STI571 for 24h. In control and BMP2 treated MCF7 cells. Untreated (a, c) and treated (b, d) with 100ng/ml BMP2 for 4h. (a, b) Cytoplasm; (c, d) Nucleus. Positions of phospho-p70S6 and phospho-Akt, phospho-ERK antibodies are marked with frame and arrow separately.

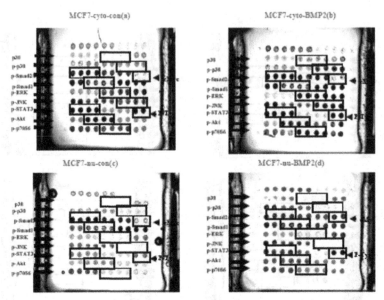

Fig. 1. (*continued*)

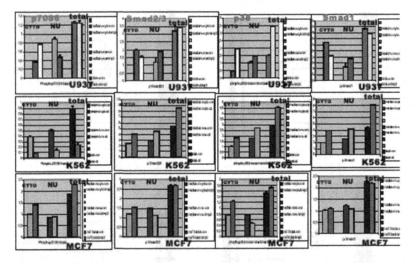

Fig. 2. Graphic numerical value analysis of phosphorylation and localization modulation. In BMP2 treated U937 cells, total phospho-p38 increased with cytoplasmic and nuclear phosphorylation increasing; total phospho-Smad1, phospho-Smad2/3 unchanged with nuclear phosphorylation increasing and cytoplasmic phosphorylation decreasing; Total phospho-p70S6 unchanged with cytoplasmic phosphorylation increasing and nuclear phosphorylation decreasing. In STI571 treated K562 cells, phospho-p70S6 decreased with cytoplasm and nucleus decreasing, total phospho-p38, phospho-Smad1 and phospho-Smad2/3 increased with cytoplasm and nucleus increasing. In BMP2 treated MCF7 cells, total p-Smad1, p-Smad2/3, p-p38 unchanged with cytoplasm increasing and nucleus decreasing, p-P70S6 increased with cytoplasm increasing and nucleus unchanged. Each column indicates a mean of four measurements with standard deviations. Different antibodies have different affinities, arrow indicating each antibody quantity value 1 position.

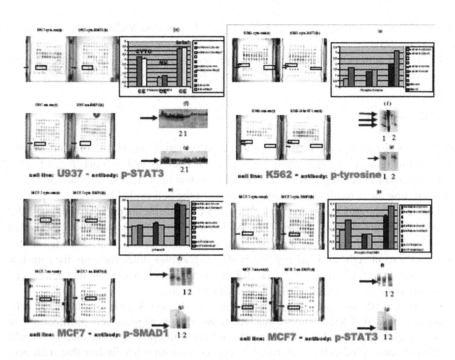

Fig. 3. Western blotting comparison with antibody microarray. (a-d) Antibody micro-array analysis, untreated (a, c) and treated (b, d) with 100ng/ml BMP2 for 4h. (a, b) Cytoplasm; (c, d) Nucleus. Position of antibody is marked with frame and arrow. (e) Graphic display of numerical value analysis. Each column indicates a mean of four measurements with standand deviations. (f) Western blotting analysis is consistent with antibody microarray.

p-p38 is higher in BMP2 treated U937 cells than in control U937 cells. p-Smad1 and p-Smad2/3 displayed the increase in cytoplasm and the decrease in nucleus in BMP2 treated U937 cells. Total protein (sum of cytoplasm and nucleus) level of p-Smad1, p-Smad2/3 were not altered between control and BMP2 treated cells.

In STI571 treated K562 cells, p-p70S6 (thr389) displayed the decrease in cytoplasm and nucleus. Total protein (sum of cytoplasm and nucleus) level of p-P70S6 is lower than in control cells; p-ERK (thr202/tyr204) displayed the decrease in cytoplasm and the increase in nucleus of STI571 treated K562 cells. Total protein (sum of cytoplasm and nucleus) level of p-ERK is higher in STI571 treated K562 cells than in control cells. p-p38, p-Smad1 and p-Smad2/3 displayed the increase in cytoplasm and nucleus of STI571 treated K562 cells. Total protein (sum of cytoplasm and nucleus) level of p-p38, p-Smad1 and p-Smad2/3 are higher in STI571 treated cells than in control.

In BMP2 treated MCF7 cells, p-P70S6 displayed the increase in cytoplasm and unchanged in nucleus. Total protein (sum of cytoplasm and nucleus) level of p-P70S6 was increased compared with control; phospho-p38, p-Smad1, p-Smad2/3 displayed the increase in cytoplasm and the decrease in nucleus of BMP2 treated MCF7 cells. Total protein (sum of cytoplasm and nucleus) level of phospho-p38, p-Smad1, p-Smad2/3 were not significantly changed in BMP2 treated MCF-7 cells and control cells. (Fig. 2-3).

Compared with cell number, FACS and western blotting, antibody array results are consistent with the experiments of cell number, FACS and western blotting (Fig.3).

4 Discussion

Protein localization data provide valuable information in elucidating eukaryotic protein function [2]. Protein trafficking between nucleus and cytoplasm fundamentally important to cell regulation. As such, the nuclear import and export are pivotal in orchestrating the activities of the key regulators of the cell cycle [2].

Our result showed that p-Akt and p-ERK co-cytoplasmic increase appear in BMP2 treated U937 cells (data not shown). Our FACS result also showed apoptosis in 2000ng/ml BMP2 treated U937 cells compared with control (Fig.3). Lorenzini, et al. [5] demonstrated that in senescent cells not ERK nor Akt is able to phosphorylate efficiently their nuclear targets, however, total protein Akt and ERK no significant change. From this point it can be reflected the apoptosis state in BMP2 treated U937 cells.

Our finding showed p-ERK nuclear increase in 0.2 uM STI571 treated K562 cells (data not shown). ERK-1 and –2 nuclear translocation triggers cell proliferation in vitro models [3]. From this point it can be reflected the proliferation state in 0.2 uM STI571 treated K562 cells. This result is consistent with other finding that treatment of Bcr-Abl-expressing cells with STI571 elicits a cytoprotective MAPK activation response [13]. Our study showed p-Akt and p-p70S6 nuclear and cytoplasmic co-decrease in 0.2 uM STI571 treated K562 cells (data not shown). Akt controls cell growth through its effects on the mTOR and p70S6 (in the cytoplasm) kinase pathways. Therefore, Akt and p70S6 nuclear and cytoplasmic co-decrease reflect the growth inhibition in 0.2 uM STI571 treated K562 cells. In sum, the net effect from ERK nuclear increase and Akt and p70S6 nuclear and cytoplasmic co-decrease reflect the growth inhibition and proliferation state in 0.2 uM STI571 treated K562 cells.

Our result showed that P-p70S6 nuclear decrease whereas p-p38, p-smad1 and p-smad2/3 nuclear increase appeared in BMP2–induced apoptosis U937 cells (2000ng/ml BMP2 for 3 days) and in STI571-inhibited cell growth and-induced proliferation of K562 cells (0.2 uM STI571 for 24h) (Fig. 1-2), implying the common mechanism of negative localized relationship among p70S6 with Smad1, 2, 3 and p38. The present study shows association and interactions of p38 MAPK with smad in nucleus which function as apoptosis seen in many papers, such as: Transforming growth factor-beta (TGF-beta)-dependent apoptosis used GADD45b as an effector of TGF-beta-induced apoptosis that the proximal Gadd45b promoter is activated by TGF-beta through the action of Smad2, Smad3, and Smad4.that ectopic expression of GADD45b in AML12 murine hepatocytes is sufficient to activate p38 and to trigger apoptotic cell death, whereas antisense inhibition of Gadd45b expression blocks TGF-beta-dependent p38 activation and apoptosis [14].

Our study demonstrated p-ERK and p-p70S6 co-cytoplasmic increase in BMP2 treated MCF7 cells (data not shown). Edelmann, et al [15] using synchronized Swiss mouse 3T3 fibroblasts that p70 S6 kinase (p70S6k) and mitogen-activated protein kinases (p42MAPK/p44MAPK) are not only activated at the G0/G1 boundary, but also in cells progressing from M into G1. p70S6k activity increases 20-fold in G1 cells released from G0. Throughout G1, S, and G2 it decreases constantly, so that during M phase low kinase activity is measured. The kinase is reactivated 10-fold when cells released from a nocodazole-induced metaphase block enter G1 of the next cell cycle. From this point it can be reflected the G1 state in BMP2 treated MCF7 cells. Our study showed the p-Akt unchanged in BMP2 treated MCF7 cells. Akt controls cell growth through its effects on the mTOR and p70S6 (in the cytoplasm) kinase pathways. p70S6 kinase (p70S6K) belong to multifunctional kinases downstream of phosphatidylinositol 3 kinase (PI3K), such as the 70-kDa ribosomal protein S6 kinase (p70S6K) is itself a dual pathway kinase, signalling cell survival as well as growth through differential substrates which include mitochondrial BAD and the ribosomal subunit S6, respectively [16]. From this point it can be reflected the growth inhibition and cell survival state in BMP2 treated MCF7 cells. From all the above it can be concluded that the net effect of p-Akt unchanged whereas p-ERK and p-p70S6 co-cytoplasmic increase induced cell survival and differentiation in 100ng/ml BMP2 treated MCF7 cells for 4h. Our study showed that p-p70S6 only cytoplasmic increase whereas p-p38, p-smad1 and p-smad2/3 cytoplasmic inhibition by nuclear decrease and cytoplasmic increase under total protein unchanged in BMP2-induced cell survival and differentiated MCF7 cells (100ng/ml BMP2 for 4h), further implying the negative localized relationship among p70S6 with Smad1, 2, 3 and p38. Here, we need to emphasize that p-p38, p-smad1 and p-smad2/3 under total protein (sum of cytoplasm and nucleus) level of p-p38, p-smad1 and p-smad2/3 were not altered in control and BMP2 treated cells displayed an increase of cytoplasmic phosphorylation and decrease of nucleus (Fig 1-2). From this result, it can be deduced that spatial control of cell cycle is through the retention of p-p38, p-smad1 and p-smad2/3 in the cytoplasm, thereby preventing them from physical contact with their substrates or partner by p-p70S6 cytoplasmic inhibition through nuclear decrease and cytoplasmic increase whereas total protein un-changed.

5 Conclusions

In this paper, the relationship among p70S6 with Smad1, 2, 3 and p38 in three treated human cancer cell Lines was studied using western blot, FACS and nano-gold phospho-antibody microarray by reaction with our purification of nuclear and cytoplasmic protein as follows: P-p70S6 nuclear decrease whereas p-p38, p-smad1 and p-smad2/3 nuclear expression increase in BMP2–induced apoptosis U937 cells (2000ng/ml BMP2 for 3 days) and in STI571-inhibited cell growth and-induced proliferation of K562 cells (0.2 uM STI571 for 24h). P-p70S6 only cytoplasmic increase whereas p-p38, p-smad1 and p-smad2/3 cytoplasmic inhibition by nuclear decrease and cytoplasmic increase under total protein unchanged in BMP2-induced cell survival and differentiated MCF7 cells (100ng/ml BMP2 for 4h). This result implies the negative localized relationship among p70S6 with Smad1,2,3 and p38 in three treated human cancer cell lines.

Acknowledgement

This work was supported by National Natural Science Foundation (No. 60673109), National Natural Science Key Foundation of China (No. 60331010) and State Key Lab of Pattern Recognition open fund, Chinese Academy of Sciences.

References

1. James, P.: Protein Identification in the Post-Genome Era: the Rapid Rise of Proteomics. Q. Rev. Biophys 30, 279–331 (1997)
2. Zhu, H., Bilgin, M., Snyder, M.: Proteomics. Annu. Rev. Biochem. 72, 783–812 (2003)
3. Tarnawski, A.S., Pai, R., Wang, H., et al.: Translocation of MAP (Erk-1 and -2) Kinases to Cell Nuclei and Activation of c-fos Gene during Healing of Experimental Gastric Ulcers. J. Physiol Pharmacol 49, 479–488 (1998)
4. Wang, X., Martindale, J.L., Holbrook, N.J.: Requirement for ERK Activation in Cisplatin-Induced Apoptosis. J. Biol. Chem. 275, 39435–39443 (2000)
5. Lorenzini, A., Tresini, M., Mawal-Dewan, M., et al.: Role of the Raf/MEK/ERK and the PI3K/Akt (PKB) Pathways in Fibroblast Senescence. Exp Gerontol 37, 1149–1156 (2002)
6. Dickens, M., Rogers, J., Cavangh, J., et al.: A Cytoplasmic Inkibitto of the JNK Tranatuotlen Pathway. Science 277, 693–696 (1997)
7. Moriguchi, T., Toyoshima, F., Masuyama, N., Hanafusa, H., Gotoh, Y., Nishida, E.: A Novel SAPK/JNK Kinase, MKK7, Stimulated by TNF and Cellular Stresses. EMBO J. 16, 7045–7053 (1997)
8. Klekotka, P.A., Santoro, S.A., Zutter, M.M.: $\alpha2$ Integrin Subunit Cytoplasmic Domain-dependent Cellular Migration Requires p38 MAPK. J. Biol. Chem. 276(12), 9503–9511 (2001)
9. Meyer, T., Xu, L., et al.: Breast Cancer Cell Line Proliferation Blocked by the Src-related Rak Tyrosine Kinase. Int. J. Cancer 104(2), 139–146 (2003)
10. Dolled-Filhart, M., Camp, R.L., et al.: Tissue Microarray Analysis of Signal Transducers and Activators of Transcription 3 (Stat3) and Phospho-Stat3 (Tyr705) in Node-negative Breast Cancer Shows Nuclear Localization is Associated with a Better Prognosis. Clin. Cancer Res. 9(2), 594–600 (2003)
11. Borgatti, P., Martelli, A.M., Bellacosa, A., et al.: Translocation of Akt/PKB to the Nucleus of Osteoblast-Like MC3T3-E1 Cells Exposed to Proliferative Growth Factors. FEBS Lett. 477, 27–32 (2000)
12. Fleckenstein, D.S., Dirks, W.G., Drexler, H.G., et al.: Tumor Necrosis Factor Receptor-Associated Factor (TRAF) 4 is a New Binding Partner for the p70S6 Serine/Threonine Kinase. Leuk Res. 27, 687–694 (2003)
13. Yu, C., Krystal, G., Varticovksi, L., et al.: Pharmacologic Mitogen-Activated Protein/Extracellular Signal-Regulated Kinase Kinase/Mitogen-Activated Protein Kinase Inhibitors Interact Synergistically with STI571 to Induce Apoptosis in Bcr/Abl-Expressing Human Leukemia Cells. Cancer Res. 62, 188–199 (2002)
14. Hata, K., Nishimura, R., et al.: Differential Roles of Smad1 and p38 Kinase in Regulation of Peroxisome Proliferator-activating Receptor Gamma during Bone Morphogenetic Protein 2-induced Adipogenesis. Mol. Biol. Cell 14(2), 545–555 (2003)
15. Edelmann, H.M., Kuhne, C., Petritsch, C., et al.: Cell Cycle Regulation of p70 S6 Kinase and p42/p44 Mitogen-activated Protein Kinases in Swiss Mouse 3T3 Fibroblasts. J. Biol. Chem. 271, 963–971 (1996)
16. Harada, H., Andersen, J.S., Mann, M., et al.: p70S6 Kinase Signals Cell Survival as Well as Growth, Inactivating the Pro-Apoptotic Molecule BAD. Proc. Natl. Acad. Sci. USA 98, 9666–9670 (2001)

Cancer Identification Based on DNA Microarray Data

Yihui Liu

School of Computer Science and Information Technology,
Shandong Institute of Light Industry, Jinan, Shandong, China, 250353
Yihui_liu_2005@yahoo.co.uk

Abstract. In this study we perform wavelet transform on the analysis of DNA microarray data. A set of wavelet features is used to measure the change of gene expression profile. Then wavelet features are input to support vector machine (SVM) to classify DNA microarray data into different diagnostic classes. Experiments are carried out on six datasets of microarray data. On a wide range of data sets, our method displays a highly competitive accuracy in comparison to the best performance of other kinds of classification models.

1 Introduction

Recently, huge advances in DNA microarray have allowed the scientist to test thousands of genes in normal or tumor tissues on a single array and check whether those genes are active, hyperactive or silent. Therefore, there is an increasing interest in changing the criterion of tumor classification from morphologic to molecular. In this perspective, the problem can be regarded as a classification problem in machine learning. Generally, microarray expression experiments allow the recording of expression levels of thousands of genes simultaneously. These experiments primarily consist of either monitoring each gene multiple times under various conditions [1], or alternately evaluating each gene in a single environment but in different types of tissues, especially for cancerous tissues [2]. Those of the first type have allowed for the identification of functionally related genes due to common expression patterns, while the experiments for the latter have shown a promise in classifying tissue types.

Generally speaking, approaches usually use a criterion relating to the correlation degree to rank and select key genes, such as signal-to-noise ratio (SNR) method [3], Pearson correlation coefficient method [4] and t -test statistic method [5]. Independent component analysis [6] also is used in the analysis of DNA microarray data. Li et al. use approximation coefficients, together with some useful features from the high-frequency coefficients selected by the maximum modulus method as features [7]. But approximation coefficients only compress the gene profiles, detail coefficients characterize the changes of gene profiles based on derivative of different order.

In this research we propose a novel feature extraction algorithm based on wavelet analysis for DNA microarray data. A set of orthogonal wavelet basis at different scale is used to measure the changes of gene expression profiles. The most discriminating features projected wavelet basis is used to feed into support vector machine for classifying DNA microarray data. Experiments are carried out on six data sets and experimental results show that our method is efficient and feasible.

T. Washio et al. (Eds.): PAKDD 2007 Workshops, LNAI 4819, pp. 153–161, 2007.

2 Feature Extraction Based on Wavelets

Wavelet technology is applied widely in many research areas. The wavelet-transform method, proposed by Grossmann and Morlet [8], analyzes a signal by transforming its input time domain into a time-frequency domain. For wavelet analysis for gene expression data, a gene expression profile can be represented as a sum of wavelets at different time shifts and scales using discrete wavelet analysis (DWT). The DWT is capable of extracting the local features by separating the components of gene expression profiles in both time and scale. According to DWT, a time-varying function $f(t) \in L^2(R)$ can be expressed in terms of $\phi(t)$ and $\psi(t)$ as follows:

$$f(t) = \sum_k c_0(k)\phi(t-k) + \sum_k \sum_{j=1} d_j(k) 2^{\frac{-j}{2}} \psi(2^{-j}t - k)$$

$$= \sum_k c_{j0}(k) 2^{\frac{-j0}{2}} \phi(2^{-j0}t - k) + \sum_k \sum_{j=j0} d_j(k) 2^{\frac{-j}{2}} \psi(2^{-j}t - k)$$

where $\phi(t), \psi(t), c_0$, and d_j represent the scaling function, wavelet function, scaling coefficients (approximation coefficients) at scale 0, and detail coefficients at scale j, respectively. The variable k is the translation coefficient for the localization of gene

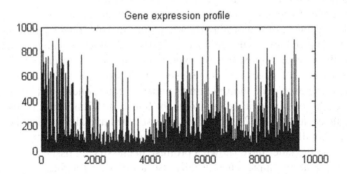

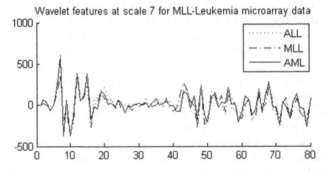

Fig. 1. Gene expression profile of MLL-Leukemia data and wavelet features

expression data. The scales denote the different (low to high) scale bands. The variable symbol j_0 is scale (level) number selected.

For wavelet analysis, the maximum number of scale for gene expression profile is decided by the below formula:

$$(l_{filter} - 1)2^{lev} < l_{gene}$$

l_{filter} is the length of wavelet filter. Here we use Daubechies wavelet of order 7 (db7) [9] for wavelet analysis of DNA microarray data. l_{gene} is the length of gene expression profile. lev is the maximum number of scale.

We use detail wavelet coefficients to measure the changes of gene expression profile. The higher level decomposition correspond to the more "stretched" wavelets, which is compared with the longer portion of the gene profile, thus the coarser the change of gene expression profiles being measured by the wavelet coefficients. Figure 1 shows gene expression profile of MLL-Leukemia data set and wavelet features at 7th level wavelet decomposition.

3   Support Vector Machine (SVM)

The SVM originated from the idea of the structural risk minimization developed by Vapnik [10]. SVMs are an effective algorithm to find the maximal margin hyperplane to separate two classes of patterns. A transform to map nonlinearly the data into a higher-dimensional space allows a linear separation of classes, which could not be linearly separated in the original space. The objects that are located on these two hyperplanes are the so-called support vectors. The maximal margin hyperplane, which is uniquely defined by the support vectors, gives the best separation between the classes. The support vectors can be regarded as selected representatives out of the training data, and are most critical for the separation of the two classes. As usually only few support vectors are used, there are only some parameters adjustable by the algorithm and thus overfitting is unlikely to occur.

We have the training data sets $\{x_k, y_k\}_{k=1}^{l}$, $x_k \in R^q$ are the training data points, $y_k \in \{-1,+1\}$ are the class labels, l is the number of samples and q is the dimension number of gene profile in each sample. SVM maps input vector $x \in R^q$ into a high dimensional feature space $z_k \in R^n, n > q$ using a mapping function $\psi : x \rightarrow z$ in order to find the best liner hyperplane. The support vector machine classifier is based on the idea of margin maximization and it can be found by solving the following optimization problem:

$$\min \frac{1}{2} w^T w + C \sum_{k=1}^{l} \xi_k$$

$s.t. y_k (w^T \psi(x_k) + b) \geq 1 - \xi_k, k = 1,2,...,l, \xi_k \geq 0$, and C is regularization constant.

The decision function of the SVM is linear in the feature space and can be written as:

$$f(x) = sign(\sum_{i \in SV} a_i^0 y_i K(x, x_i) + a_0)$$

SV means set of support vector, a_i^0 can be obtained by solving the following optimization formula:

$$W(a) = \sum_{i=1}^{l} a_i - \frac{1}{2} \sum_{i,j=1}^{l} a_i a_j y_i y_j K(x_i, x_j)$$

Under constraints $\sum_{i=1}^{l} a_i y_i = 0, 0 \le a_i \le C, i = 1,...,l$. Where $K(x_i, x_j)$ is a symmetric kernel function in the input space which satisfies Mercer's theorem: $K(x_i, x_j) = < \psi(x_i), \psi(x_j) >$. Those input vectors x_i with corresponding non-zero a_k are called support vectors. They are located in the boundary margin and contribute to the construction of sepration hyperplane.

An upper bound function on the error is given as bellow:

$$T = \frac{1}{l} \frac{R^2}{r^2} = \frac{1}{l} R^2 \|w\|^2$$

Where r is the distance between the support vectors and the separating hyperplane, and R is the radius of the smallest sphere including all points $\psi(x_k)$ of the training data in the feature space.

In this study radial basis functions (RBF) $K(x_i, x_j) = e^{-\|x_i - x_j\|^2 / r1}$, where $r1$ is a strictly positive constant, is used. Apparently the linear kernel is less complex than the polynomial and the RBF kernels. The RBF kernel usually has better boundary response as it allows for extrapolation, and most high-dimensional data sets can be approximated by Gaussian-like distributions similar to those used by RBF networks [11].

4   Experiments and Results

We perform our feature selection method on six microarray data sets. Firstly we filter genes with small profile variance. Normally we select 25% threshold, which means that genes with a variance less than the threshold are removed from gene expression profiles. Secondly a set of features in wavelet subspace are selected to measure the change of gene expression profiles. Finally the wavelet features for microarray data are fed into SVM classifier to distinguish the diagnostic classes. The experimental results are show in Table 1.

Leukemia (ALL v.s. AML)
Training dataset consists of 38 bone marrow samples (27 ALL and 11 AML) with 7129 attributes from 6817 human genes, and 34 testing samples including 20 ALL and 14 AML [3].

Table 1. The prediction accuracy of SVM classifier based on wavelet features

Dataset	Training	Testing	classes	Error number	Correct rate
Leukemia	38	34	2	1	97.06
MLL-Leukemia	57	15	3	0	100
Breast Cancer	78	12	2	1	91.67
Prostate Cancer	102	34	2	2	94.12
Lung Cancer	32	149	2	2	98.66
Subtypes-Leukemia	163	85	6	0	100

113 features in wavelet subspace are selected. The recognition rate is 97.06%, which is the same as the Bayesian variable method [12] and the partial least squares method [13]. It is also better than 82.3% of the PCA disjoint models [14] and 88.2% of the between-group analysis [15].

MLL-Leukemia (ALL v.s. MLL v.s. AML)
Leukemia data [16] contains 57 training leukemia samples (20 ALL, 17 MLL and 20 AML). Testing data contains 4 ALL, 3 MLL and 8 AML samples. The number of attributes is 12582.

160 wavelet features are selected to describe the gene expression profile. 100% recognition rate is achieved. It is better than 95% of k-nearest-neighbors method [16].

Breast Cancer
The breast cancer dataset [17] contains the training data of 78 patient samples, 34 of which are from patients who had developed distance metastases within 5 years (labelled as "relapse"), the rest 44 samples are from patients who remained healthy from the disease after their initial diagnosis for interval of at least 5 years (labelled as "non-relapse"). Correspondingly, there are 12 relapse and 7 non-relapse samples in the testing data set. The number of genes is 24481.

86 features in wavelet subspace are selected. The recognition rate is 91.67%. It is super than the 89% of error-weight, uncorrelated shrunken centroid algorithms (EWUSC) [18] and 89% obtained by van't Veer et al. [17].

Prostate Cancer
Prostate cancer data [19] contains training set of 52 prostate tumor samples and 50 non-tumor (labelled as "Normal") prostate samples with 12600 genes. An independent set of testing samples is also prepared, which is from a different experiment. The testing set has 25 tumor and 9 normal samples.

86 features in wavelet subspace are obtained and the prediction accuracy is 94.12%. The accuracy is better than 92.47% of bagboosting method [20] and 86% obtained by Singh et al. [19].

Lung Cancer
Lung cancer data [21] contains two kinds of tissue including malignant pleural mesothelioma (MPM) and adenocarcinoma (ADCA) of the lung. There are 181 tissue samples (31 MPM and 150 ADCA) including 32 training samples (16 MPM and 16 ADCA) and 149 testing samples (15 MPM and 134 ADCA). The number of genes of each sample is 12533. They examined (in the test set) the accuracy of multiple ratios

combined to form a simple diagnostic tool. Using two and three expression ratios, they found that the differential diagnoses of MPM and ADCA were 95% and 99% accurate rate, respectively. Any single ratio of the 15 examined was at least 90% accurate in predicting diagnosis.

In our research 105 features in wavelet subspace are obtained and the prediction accuracy is 98.66%. Compared with ratio-based diagnosis [21], our method is more simple and efficient.

Subtypes of Acute Lymphoblastic Leukemia (Subtypes-Leukemia)

This acute lymphoblastic leukemia data [22] has six diagnostic classes (BCR-ABL, E2A-PBX1, Hyperdiploid>50, MLL, T-ALL and TEL-AML1). Each sample has 12558 genes. The number of training and testing samples in each group and the prediction accuracy of each class is listed in the Table 2.

In our research 1080 features in wavelet subspace are obtained and the prediction accuracy is 100% for each class. It is super than the performance in Yeoh's paper [22]. In order to see the SVM classifiers clearly, we perform principal components

Table 2. Subtypes-Leukemia data set and performance for each class

Class	Training (number)	Testing (number)	Accuracy Yeoh et al. [12]	Accuracy Wavelet feature
BCR-ABL	9	6	100%	100%
E2A-PBX1	18	9	100%	100%
Hyperdiploid>50	42	22	99%	100%
MLL	14	6	97%	100%
T-ALL	28	15	100%	100%
TEL-AML1	52	27	96%	100%

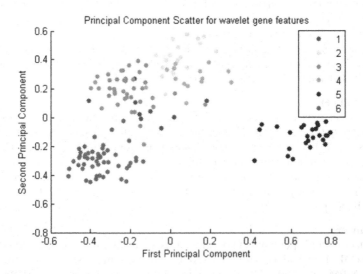

Fig. 2. The distribution of first and second principal components of Subtypes-Leukemia dataset. The different number represents the different subtypes (1 - BCR-ABL, 2 - E2A-PBX1, 3 - Hyperdiploid>50, 4 - MLL, 5 - T-ALL, 6 - TEL-AML1).

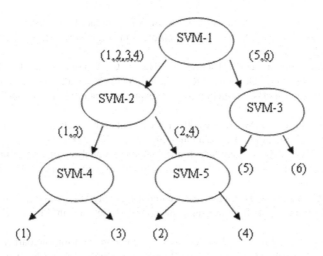

Fig. 3. Hierarchy of SVM classifier. SVM-1 separates Subtypes-Leukemia dataset between (1,2,3,4) and (5,6).

analysis (PCA) [23] on dataset to obtain principal components, which are the representation of microarray data in the new principal component space. Figure 2 illustrates the distribution of first and second principal components of Subtypes-Leukemia dataset. Support vector machines are primarily two class classifiers. We build the hierarchical SVM classifier for the classification of multi-class microarray data. The corresponding SVM classifier is designed for Subtypes-Leukemia dataset in Figure 3.

5 Conclusions

In this paper a novel method is proposed to extract the essential features of high dimensional microarray data. A set of wavelet basis is used to characterize the change of gene expression profile. It is shown that this new method achieves highly competitive accuracy compared to the best performance of other kinds of classification models.

Acknowledgements

This study is supported by research funds of Shandong Institute of Light Industry (12041653).

References

1. Roberts, C.J., Nelson, B., Marton, M.J., Stoughton, R., Meyer, M.R., Bennett, H.A., He, Y.D., Dai, H., Walker, W.L., Hughes, T.R., Tyers, M., Boone, C., Friend, S.H.: Signaling and circuitry of multiple mapk pathways revealed by a matrix of global gene expression profiles. Science 287, 873–880 (2000)

2. Zhang, L., Zhou, W., Velculescu, V.E., Kern, S.E., Hruban, R.H., Hamilton, S.R., Vogel-stein, B., Kinzler, K.W.: Gene expression profiles in normal and cancer cells. Science 276, 1268–1272 (1997)
3. Golub, T.R., Slonim, D.K., Tamayo, P., Huard, C., Gaasenbeeck, M., Mesirov, J.P., Coller, H., Loh, M.L., Downing, J.R., Caligiuri, M.A.: Molecular classification of cancer: class discovery and class prediction by gene expression monitoring. Science 286, 531–537 (1999)
4. Xiong, M., Jin, L., Li, W., Boerwinkle, E.: Computational methods for gene expression-based tumor classification. Biotechniques 29, 1264–1270 (2000)
5. Baldi, P., Long, A.D.: A Bayesian framework for the analysis of microarray expression data: regularized t -test and statistical inferences of gene changes. Bionformatics 17, 509–519 (2001)
6. Huang, D.S., Zheng, C.H.: Independent component analysis-based penalized discriminant method for tumor classification using gene expression data. Bioinformatics 22, 1855–1862 (2006)
7. Li, S., Liao, C., Kwok, J.T.: Wavelet-based feature extraction for microarray data classifi-cation. In: Proceedings of International Joint Conference on Neural Networks, pp. 5028–5033 (2006)
8. Grossmann, A., Morlet, J.: Decomposition of Hardy functions into square integrable wave-lets of constant shape SIAM. Journal of Mathematical Analysis 15, 723–736 (1984)
9. Daubechies, I.: Orthonormal bases of compactly supported wavelets. Communications on Pure and Applied Mathematics 41, 909–996 (1988)
10. Vapnik, V.N.: Statistical learning theory. Wiley, New York (1998)
11. Burges, C.: A Tutorial on Support Vector Machines for Pattern Recognition. Kluwer Aca-demic Publishers, Dordrecht (1998)
12. Kyeong, E.L., Sha, N., Dougherty, E.R., Vannucci, M., Mallick, B.K.: Gene selection: a bayesian variable selection approach. Bioinformatics 19, 90–97 (2003)
13. Danh, V.N., David, M.R.: Tumor classification by partial least squares using microarray gene expression data. Bioinformatics 18, 39–50 (2002)
14. Bicciato, S., Luchini, A., Bello, C.D.: PCA disjoint models for multiclass cancer analysis using gene expression data. Bioinformatics 19, 571–578 (2003)
15. Aedin, C.C., Guy, P., Elizabeth, C.C., Thomas, G.C., Desmond, G.H.: Between group analysis of microarray data. Bioinformatics 18, 1600–1608 (2002)
16. Armstrong, S.A., Staunton, J.E., Silverman, L.B., Pieters, R., den Boer, M.L., Minden, M.D., Sallan, S.E., Lander, E.S., Golub, T.R., Korsmeyer, S.J.: MLL translocations spec-ify a distinct gene expression profile that distinguishes a unique leukemia. Nat. Genet. 30, 41–47 (2002)
17. van 't Veer, L.J., Dai, H., van de Vijver, M.J., He, Y.D., Hart, A.A., Mao, M., Peterse, H.L., van der Kooy, K., Marton, M.J., Witteveen, A.T., Schreiber, G.J., Kerkhoven, R.M., Roberts, C., Linsley, P.S., Bernards, R., Friend, S.H.: Gene expression profiling predicts clinical outcome of breast cancer. Nature 415, 530–536 (2002)
18. Yeung, K.Y., Bumgarner, R.E.: Multiclass classification of microarray data with repeated measurements: application to cancer. Genome. Biol 4, R83 (2003)
19. Singh, D., Febbol, P.G., Ross, K., Jackson, D.G., Manola, J., Ladd, C., Tamayo, P., Renshaw, A.A., D'Amico, A.V., Richie, J.P., Lander, E.S., Loda, M., Kantoff, P.W., Golub, T.R., Sellers, W.R.: Gene expression correlates of clinical prostate cancer behavior. Cancer Cell 1, 203–209 (2002)
20. Dettling, M.: BagBoosting for tumor classification with gene expression data. Bioinfor-matics 20, 3583–3593 (2004)

21. Gordon, G.J., Jensen, R.V., Hsiao, L.L., Gullans, S.R., Blumenstock, J.E., Ramaswamy, S., Richards, W.G., Sugarbaker, D.J., Bueno, R.: Translation of microarray data into clinically relevant cancer diagnostic tests using gene expression ratios in lung cancer and mesothelioma. Cancer research 62, 4963–4967 (2002)
22. Yeoh, E.-J., Ross, M.E., Shurtleff, S.A., Williams, W.K., Patel, D., Mahfouz, R., Behm, F.G., Raimondi, S.C., Relling, M.V., Patel, A., et al.: Classification, subtype discovery, and prediction of outcome in pediatric acute lymphoblastic leukemia by gene expression profiling. Cancer Cell 1, 133–143 (2002)
23. Jolliffe, I.T.: Principal Component Analysis, 2nd edn. Springer, Heidelberg (2002)

Incorporating Dictionary Features into Conditional Random Fields for Gene/Protein Named Entity Recognition

Hongfei Lin, Yanpeng Li, and Zhihao Yang

Department of Computer Science and Engineering,
Dalian University of Technology,
Dalian,China 116024
{hflin,yangzh}@dlut.edu.cn, lyp_8218@163.com

Abstract. Biomedical Named Entity Recognition (BioNER) is an important preliminary step for biomedical text mining. Previous researchers built dictionaries of gene/protein names from online databases and incorporated them into machine learning models as features, but the effects were very limited. This paper gives a quality assessment of four dictionaries derived form online resources, and investigate the impacts of two factors (i.e., dictionary coverage and noisy terms) that may lead to the poor performance of dictionary features. Experiments are performed by comparing performances of the external dictionaries and a dictionary derived from GENETAG corpus, using Conditional Random Fields (CRFs) with dictionary features. We also make observations of the impacts regarding long names and short names. The results show that low coverage of long names and noises of short names are the main problems of current online resources and a high quality dictionary could substantially improve the accuracy of BioNER.

Keywords: BioNER, dictionary feature, CRF.

1 Introduction

Biomedicine literatures are expanding at an exponential rate. Biomedical text mining [1] can be an aid to information seekers who aim at finding knowledge from terabyte-scale texts. Biomedical Named Entity Recognition (BioNER) is the preliminary step of biomedical text miming, but its performance is far below that in the general domain. The best Named Entity Recognition (NER) systems on newswire articles can achieve an F-score over 95% [2][3], while the state-of-the-art performances of BioNER are only between 75%-85% [1] varying with different datasets and evaluation measures. In JNLPBA 2004 task [4], five classes of named entities are required to recognize and the evaluation follows an exact matching criteria, where the top system [5] obtained an F-score of 72.6% which used a

T. Washio et al. (Eds.): PAKDD 2007 Workshops, LNAI 4819, pp. 162–173, 2007.

combination of HMM and SVM models plus post-processing with manual rules. BioCreative 2004 Task 1A [6] is to identify entities of one type which is tagged as "NEWGENE" or "NEWGENE1" and its evaluation measure is an F-score of relax matching where a name can have several representations. Finkel et al. [7] used a MEMM [8] model with carefully designed features plus post-processing of the abbreviations and mismatching brackets. Their system obtained an F-score of 83.2%, which was the best in BioCreative 2004 Task 1A.

Why is this task so difficult? Liu et al. [9] built a large gene/protein database BioThesaurus from online resources. It was reported to have over 2.6 million names or 2.1 million normalized names covering more than 1.8 million UniPro-tKB[1] entries. But the total amount of gene/protein names in the BioCreative 2004 training corpus is less than 10,000 before normalization. Thus the number of unknown names is at least hundreds of times more than names in a closed dictionary, and many long range names have extremely complex structures and many variant names. According to the report of JNLPBA 2004 task [4] most errors occurred at boundaries of long names. In addition, there are also many errors in single-token names. Some of the words have the shapes like common English words that are difficult to identify by orthographic features and many words are acronyms of gene/protein names that may be confused with other chemicals. So not given dictionary knowledge, it is extremely difficult to distinguish them from common English words and other named entities only using shallow parsing information.

Dictionary-based systems suffer from the problem of low coverage and noisy terms. Currently there is no dictionary that contains all gene/protein names mentioned in literatures and most long names can not be found in dictionary. Since these dictionaries are automatically generated, they tend to bring large quantities of noises that will lead to low precision. Rule-based method is an effective way to recognize unknown words, but it is difficult to list all rules to model the structure of biomedical named entities, so it is always used in the post-processing stage. Machine learning method is more robust and can give better answers according to the context. Discriminative models with the structure of Markov Networks, such as CRFs [10] and MEMMs [8], are able to achieve state-of-the-art performances in this task. But there is a puzzling problem in many systems using these models that the performance improved little or decreased when external dictionary information was incorporated as features. In JNLPBA 2004 task, Zhou et al. [5] made use of SwissProt as external dictionary features, which improved the performance by 1.2 percent in F-score, while in BioCreative 2004 task, their performance reduced by 4 percent. Finkel et al. [7] used a lexicon of 1,731,581 entries built from LocusLink, Gene Ontology and internal resources, but the improvement in F-score was less than 1 percent. Settles [11] used a dictionary of five classes of entity, but the overall F-score decreased.

In the following sections, we attempt to find the reasons for this phenomenon. Section 2 describes the implement of our baseline tagger. Section 3 presents what

---

[1] http://www.pir.uniprot.org/database/knowledgebase.shtml

dictionaries are used and how they are incorporated as features. Section 4 and Section 5 present the experiments and result discussion.

2 Baseline Tagger

2.1 Conditional Random Fields

A Conditional Random Field (CRF) [10] is a discriminative probabilistic model with the structure of Markov Network. Our experiment uses the linear chain CRFs, where given an input sequence $\mathbf{o}$ and state sequence $\mathbf{s}$, the conditional probability $P(\mathbf{s}|\mathbf{o})$ is defined as follow:

$$P\left(\mathbf{s}|\mathbf{o}\right) = \frac{1}{Z_o} \exp\left(\sum_i \sum_k \lambda_k f_k(s_{i-1}, s_i, \mathbf{o}, i)\right) \tag{1}$$

where Z_o is a normalization factor of all state sequences. $f_k(s_{i-1}, s_i, o, i)$ is the feature function, and λ_k is the feature's weight. s_i and s_{i-1} refer to the current state and the previous state respectively. The training process is to find the weights that maximize the log likelihood of all instances in training data:

$$LL = \sum_j \log P\left(\mathbf{s_j}|\mathbf{o_j}\right) - \sum_k \frac{\lambda_k^2}{2\sigma^2} \tag{2}$$

where the second term in Formula (2) is a spherical Gaussian prior over feature weights. Once these settings are found, the labeling for a new unlabeled sequence can be done using a modified Viterbi algorithm [10].

CRFs have several advantages for labeling sequence data. Discriminative training make it possible to incorporate rich features that may be overlapped with each other and Markov Network is a powerful model to capture the information of contexts by computing the probability of state transition. Also automatic feature conjunction can be used to enhance the performance.

2.2 Implementation

Our baseline BioNER tagger is derived from Settles' system [11], which is a CRF-based tagger with varieties of orthographic features, and features conjunctions with the window of [-1, 1]. Performance of this system is close to the best score in JNLPBA task, and it uses no external resources or manual rules for post processing. Detail of feature selection and other configuration can be found in Settles' paper [11]. Our system do a little modification by adding POS tags and chunking tags as features, because these features are very important in BioNER and was chosen by many systems in the task. In this way, it can produce comparative results with other systems. The tagger is trained on the GENETAG corpus [12], which is the training and test data of BioCreative 2004 Task 1A, and achieves an F-score of 79.8% on the test set using relax match and 71.5% using exact match.

3   Dictionaries Features

3.1   Dictionaries

There are many online databases (e.g., LocusLink, EntrezGene) that are built to help biomedicine researchers. Many previous researchers [5][7][11][13] built gene/protein lists by extracting names from these databases. In this work, we investigated the qualities of four external dictionaries derived from the following four databases: LocusLink[2], EntrezGene[3], BioThesaurus[4] and ABGene lexicon [14][5].

LocusLink. It presents information on official nomenclature, aliases, sequence accessions, phenotypes, EC numbers, MIM numbers, UniGene clusters, homology, map locations, and related web sites. It organizes information around genes to generate a central hub for accessing gene-specific information for fruit fly, human, mouse, rat and zebra fish.

EntrezGene. It has been implemented to supply key connections in the nexus of map, sequence, expression, structure, function, citation, and homology data. Unique identifiers are assigned to genes with defining sequences, genes with known map positions, and genes inferred from phenotypic information.

BioThesaurus. It combines all UniProtKB protein entries and multiple resources based on database cross-references in iProClass[6]. It was reported to have a coverage of 94% based on the gene names of BioCreative 2004 Task 1B.

ABGene lexicon. Tanabe et al. [14] have used ABGene [15] to generate a large gene and protein lexicon of names found in MEDLINE database. Their approach yielded a final set of 1,145,913 gene names. In their experiments, assessment of a random sample determined the precision to be approximately 82%, and comparison with a gold standard gave an estimated coverage of 61% for exact matches and 88% for partial matches.

GENETAG (Internal dictionary). It includes all the named entities labeled in training and test sets of GENETAG corpus except the alternative answers (the "Correct" files). Note that this dictionary contains all names in gold standard, so models trained on these features will lead to low generalization ability in real data, and it is only used to estimate the upper bound of dictionary features.

For LocusLink and EntrezGene, the fields we used are the same as Cohen [13]. For BioThesaurus we extract names from the second column. We note that there are a lot of single characters and common English words in these dictionaries

---

[2] ftp://ftp.ncbi.nlm.nih.gov/refseq/LocusLink/
[3] ftp://ftp.ncbi.nih.gov/gene/
[4] ftp://ftp.pir.georgetown.edu/databases/iprolink/
[5] ftp://ftp.ncbi.nlm.nih.gov/pub/tanabe/
[6] http://pir.georgetown.edu/iproclass/

such as "a", "the", and "is". Then we use a stopword list of 500 common English words as a filter to remove these obvious noisy terms. In addition, we obtain a dictionary of 6,914,651 entries by combining the four external dictionaries. To our best knowledge, this is the largest gene/protein name dictionary that has been used in BioNER. Table 1 shows the total number of terms in each dictionary.

Table 1. Total number of terms in each dictionary

Dictionary	LocusLink	EntrezGene	BioThesaurus	ABGene	GENETAG
Number of terms	437,219	2,503,038	4,958,804	1,145,913	13,139

3.2 Matching Schemes

Once the dictionary is obtained, the next step is to choose the proper matching scheme. We present three matching schemes to estimate the quality of dictionaries and generate dictionary features.

Method 1. Uppercase letters are converted into lowercase and hyphens are replaced by white spaces.

Method 2. Every name is converted into its normalized form as the following steps.

1. Words are broken into token units. Four classes are defined for each character: uppercase letters, lowercase letters, digits and others. If adjacent characters are in different classes, the word is split into two token units from this point. For example, (e.g., "Abeta42 gene" to "A beta 42 gene")
2. Non-alphabet and non-digit characters are replaced by white spaces. Every digit unit is converted into a single label "[D]". All the Greek letters and Roman letters are replaced by "[G]" and "[R]". (e.g., "A beta 42 gene" to "A [G] [D] gene").
3. Uppercase letters are converted into lowercase. (e.g., "a [g] [d] gene")
4. White spaces are removed. (e.g., "a[g][d]gene")

Method 3. Names are broken into token units as the step 1 of Method 2. Then the last unit is removed, and the rest part is converted into the normalized form as described above (e.g., "Abeta42 gene" to "a[g][d]").

3.3 Incorporating Dictionary Features

Dictionary features used in the experiments are generated by a sliding window based on the maximum matching scheme, where the size of the window is up to the length of a complete sentence. If a name is found in dictionary, the beginning of window moves to the next token. Three tags are used to label the features of each token: "B-GENE", "I-GENE" and "E-GENE" respectively referring to a token that is in the beginning, inner and end of a name in dictionary.

4 Experiments

4.1 Dictionary Coverage

In our experiment, the dictionary coverage is defined as follow:

$$Coverage = \frac{N_{TP}}{N_{AP}} * 100\% \tag{3}$$

where N_{TP} is the number of names found in both the external dictionary and GENETAG dictionary, and N_{AP} is the total number of names in GENETAG dictionary. We used the matching schemes introduced in Section 3.2 to evaluate this figure. We also investigated the dictionary coverage on single-token names, double-token names and multi-token names. The experiment is denoted as **Experiment 1**, and the results are shown in Table 2.

Table 2. Dictionary coverage estimated by different matching schemes

Dictionary	Matching scheme	Coverage (single-token)	Coverage (double-token)	Coverage (multi-token)	Coverage (all)
LocusLink	Method1	49.5	10.5	7.1	23.6
LocusLink	Method2	60.7	14.3	9.9	29.8
LocusLink	Method3	79.8	64.4	24.1	56.1
EntrezGene	Method1	66.1	13.1	8.0	30.8
EntrezGene	Method2	74.0	17.0	10.9	35.8
EntrezGene	Method3	86.9	72.9	25.9	61.7
BioThesaurus	Method1	75.9	20.8	11.9	39.7
BioThesaurus	Method2	83.6	26.1	16.5	43.9
BioThesaurus	Method3	90.1	77.7	31.6	66.3
ABGene	Method1	38.5	48.5	38.7	41.4
ABGene	Method2	54.2	54.2	42.7	50.2
ABGene	Method3	91.2	88.6	57.4	78.7
All	Method1	82.3	52.6	41.9	59.8
All	Method2	91.3	64.4	49.3	69
All	Method3	95	91.7	61	82.3

4.2 Impact of Coverage and Noises

We investigated the impact of coverage and noises of these dictionaries by incorporating dictionary information into the baseline tagger as features (Section 3.3) and then compared the performances on the test set of GENETAG corpus regarding single-token names, double-token names and multi-token names. This experiment is divided into four parts:

Experiment 2. We compared the performances of the external dictionaries and the internal dictionary using the same matching scheme Mehod1 (Section 3.2). The evaluation measures are precision, recall, relax matching F-score and exact matching F-score. Precision, recall and relax matching F-score are the same as BioCreative 2004 Task 1A, and exact matching F-score is the same as the JNLPBA task. The results are shown in Table 3.

Table 3. Performances of dictionary features using Method1

Dictionary	Precision	Recall	F-score (relax)	F-score (exact)
Baseline (no dictionary features)	80.7	79.0	79.8	71.5
LocusLink	83.2	77	80	71.6
EntrezGene	82.1	77.5	79.7	71.3
BioThesaurus	84.3	77.9	81	72.4
ABGene	82.3	77.6	79.9	71.9
All external	84.4	78.7	81.5	73.1
GENETAG	96	98.1	97	96.5

Experiment 3. In this experiment, performances of different matching schemes were compared using features from the combined external dictionary. The results are shown in Table 4.

Table 4. Performance of different matching schemes

Matching scheme	Precision	Recall	F-score (relax)	F-score (exact)
Method1	84.4	78.7	81.5	73.1
Method2	83.7	78.6	81.1	72.6
Method3	81.3	78.9	80.1	72.3

Experiment 4. We assume there is little noise in the internal dictionary and noises are caused mainly by the introduction of external resources. Two dictionaries were used to evaluate the impact. One is the internal dictionary and the other is a dictionary which combines the internal dictionary and all external dictionaries. The coverages of the two dictionaries are both 100%, so the difference on performance reflects the impact of noises. We also investigated the different

Table 5. Impact of noises on F-score

Dictionary	F-score(Fs)	F-score(Fd+Fm)	F-score(Fs+Fd+Fm)
GENETAG	86	89.7	97
GENETAG+External	82.9 (-3.6%)	87.3 (-2.7%)	90.1 (-7.1%)

impact on short names and long names by selecting dictionary features based on single-token (Fs), double-token (Fd) and multi-token (Fm) names. The results are shown in Table 5.

Experiment 5. New dictionaries were generated by mixing various proportions of the internal dictionary and external dictionaries. In this way we were able to "control" the coverages and noises of these dictionaries, thus obtaining more meaningful observation data for investigating the relationship between these factors and for prediction. In this experiment we depict two curves: one reflects the performance of the internal dictionary which we assume has little noise and the other reflects the performance of the mixed dictionary described above, where the impact of noises is the most serious of our available resources. The results are shown in Fig. 1.

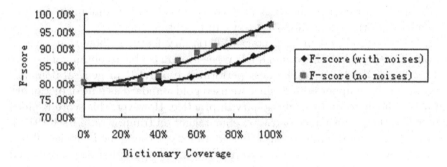

Fig. 1. Relationships between dictionary coverage, noises and BioNER performances

5   Results and Discussion

5.1   Experiment 1

From Table 2, it can be seen that the overall coverages of the four dictionaries are substantially low. No single dictionary is able to obtain a score over 50% in near exact matching scheme (Method1). LocusLink, which was used by many previous researchers, only has a coverage of 23.6%. EntrezGene has a coverage about 7 percent more than LocusLink and 9 percent lower than BioThesaurus. Also multi-token names matched in the former three dictionaries are obviously lower than single-token names. In all the four dictionaries, Method2 and Method3 both improve the coverage greatly, especially Method3 which improves the overall scores by over 20 percent on average. We examined the temple results and found that most of these cases were due to removing head nouns such as "gene", "protein" and "receptor". For example, "fibroblast growth factor receptor 2 gene" was not found in dictionary but "fibroblast growth factor receptor 2" could be found by removing the last token. However this method is certain to introduce many false positive instances. For example, term "gene E75" was converted to "gene".

ABGene lexicon is somewhat different, which has much higher coverage on multi-token names and over all score than any other dictionaries. One explanation is that names in this dictionary were directly extracted from MEDLINE records and our test data, GENETAG corpus, is part of MEDLINE abstracts. However, the generation of it is an automatic procedure and there are also a number of names beyond the dictionary in MEDLINE records.

5.2   Experiment 2

Table 3 shows the impact on F-score of dictionaries with different coverages. It can be seen that the performances of all the external dictionaries differed not much from baseline, and dictionaries with higher coverage generally have slightly better F-score, which is very similar to the experiment results of previous researchers [5][7][11]. It indicates that dictionary features generated by exact matching help little when the dictionary coverage is low. By combining the four dictionaries another slight improvement can be obtained.

The results produced by the internal dictionary features show that given a dictionary with substantially high quality, the system is able to obtain an F-score of 97%, which is similar to the best performance of NER in general domains. However this is an upper bound, since we used gold standard as part of dictionary and it is difficult to find this dictionary in practice. However, this score is much higher than the performance of the baseline tagger on training data itself (91%), and we did not use the full corpus of test data except the named entities. When analyzing the errors, we found most errors occurred at single-token words and some of them can only be inferred by semantic information. This is a good indicator that "enough" context information has been learnt from training data, and further improvement can be done only by enlarging the dictionary.

5.3   Experiment 3

Table 4 shows the performance of different matching schemes. It can be seen that Method1 is better than Method2 and Method3, although it has the lowest dictionary coverage. It indicates that fuzzy matching like Method2 and Method3 introduces many false positive instances that have negative effects on precision and training procedure.

5.4   Experiment 4

It can be seen from Table 5 that the overall performances decreased seriously when adding noises by mixing the internal and the external dictionaries. When using only single-token features, the F-score decreased by 3.6%, which indicated that using a dictionary with a coverage of 100% the score improved not much from the baseline (79.8% to 82.9%). For double-token and multi-token features it decreased by 2.7%. It indicates that the impact of noises on short names is more serious than long names. This is because the probability of matching long names is small and surface word features in a long name can help to reduce the impact of noises. When analyzing the result errors, we find that some of the "noises" are common English words (e.g., "pigs", "damage"), which are real noises. Some are acronyms (e.g., "MG" "IMM") which may be gene/protein names according to the context, so not given information of the context they will mislead the training procedure where CRFs will reduce the weights of the dictionary features, which will lead to low recall. Some are general biomedical terms such as "lung", "cell" and "cDNA". Others are named entities of other substance (e.g., "Grf10" "xth-11") that could both affect the recall and precision. Besides, there are a lot of cases of boundary alternative names. For example, in the training corpus, "goat alpha s1-casein transcription unit" is labeled as a gene name, while only "alpha s1-casein" can be found in the dictionary. In another case, "type 1C-terminal peroxisomal targeting signal (PTS1)" is in the dictionary and "type 1C-terminal peroxisomal targeting signal" is the gold standard. As a result, the weights of the dictionary features (e.g., "I-GENE", "E-GENE") to predict a gene name will be reduced.

5.5 Experiment 5

Fig. 1. shows the relationships between dictionary coverage, noises and BioNER performances. The curve above reflects the performances of the features derived from various proportions of the internal dictionary that we assume has no noises. The other curve reflects the performances of features derived from the mixed dictionary that combines various proportions of the internal dictionary and the large external dictionaries with a total number of 6,914,651 entries. Generally, the F-score is increasing with increment of dictionary coverage and is also affected seriously by noisy terms introduced by external resources. From this graph, we can make prediction that by improving the quality of dictionaries, the performance of a BioNER system will vary between the two curves using the current method. This conclusion is encouraging, since the space for improvement is large. We also compared the best result of our experiment with the top system in BioCreative 2004 Task 1A (Table 6). Note that in our experiments no manual rules for post processing were used.

Table 6. Comparison with the top system of BioCreative 2004

System	Precision	Recall	F-score
Top in BioCreative	82.8	83.5	83.2
Method1	84.4	78.7	81.5
Baseline	80.7	79.0	79.8

6 Conclusions and Future Work

In our experiment, we built several gene/protein dictionaries using online resources and investigated the impact of dictionary coverage and noises on the performance of a CRF-based BioNER tagger with dictionary features. The results show that low coverage of long names and noises of short names are the main problems of these external dictionaries derived from online resources. The features based on exact matching and maximum matching is an effective way to reduce the impact of noises on long names. In addition, fuzzy matching like Method2 and Method3 can substantially improve the dictionary coverage, but with little help to the overall performance of the tagger. The possible reason is that names of different length were treated equally and we should develop more variable features in the next step. Also an internal dictionary derived form the training and test data was built to estimate the upper bound of performance of dictionary features and its relationship with dictionary coverage and noises. Experiment results also show that a high quality dictionary can substantially improve the performance of a BioNER system, but the qualities of current online resources have a distance from that.

We assume there is still a large space for BioNER to improve using machine learning models with dictionary features. For single-token names, the problem is to reduce the noises, and for multi-token names the most important thing is to increase the dictionary coverage by building high quality dictionaries automatically or developing proper fuzzy matching schemes. Also, this strategy is very efficient, because building dictionary is once laborious work, and time-consuming methods (e.g., SVM [5] and web [7]) used in building dictionaries will be much more efficient than that used in the tagging procedure.

Furthermore, the current evaluation of BioNER performance is a relatively ambiguous problem. In the task of BioCreative 2004, both in training data and test data there are alternative gold standards for the same gene/protein name, and in GENIA corpus and JNLPBA 2004 shared task, the answer is unique, so that a large number of errors occurred at boundaries. It indicates that the tasks need agreement in both development and evaluation stages, and the evaluation metric need to improve. Named entity recognition is the preliminary step for advanced text mining or information retrieval, so evaluation binding with next step application will be more practical. For example, the evaluation of NER and NE normalizations should be combined since these two procedures are often joined together, and the more valuable result is the normalized gene ID in database, such as SwissProt ID.

Acknowledgments. This work is supported by grant from the Natural Science Foundation of China (No.60373095 and 60673039) and the National High Tech Research and Development Plan of China (2006AA01Z151).

References

1. Cohen, A.M, Hersh, W.R.: A survey of current work in biomedical text mining. Briefings in Bioinformatics 6(1), 57–71 (2005)
2. Bikel, D., Schwartz, R., Weischedel, R.: An algorithm that learns what's in a name. Machine Learning 34, 211–231 (1997)

3. Tjong, E.F., Sang, K., De Meulder, F.: Introduction to the CoNLL-2003 shared task: Language-independent named entity recognition. In: Proceedings of the Seventh Conference on Natural Language Learning (CoNLL-2003), pp. 142–147 (2003)
4. Kim, J.D, Tomoko, O., Yoshimasa, T., et al.: Introduction to the Bio-Entity Recognition Task at JNLPBA. In: Proceedings of the International Workshop on Natural Language Processing in Biomedicine and its Applications (JNLPBA-04), pp. 70–75 (2004)
5. Zhou, G., Su, J.: Exploring Deep Knowledge Resources in Biomedical Name Recognition. In: Proceedings of the Joint Workshop on Natural Language Processing in Biomedicine and its Applications (JNLPBA-2004), pp. 96–99 (2004)
6. Hirschman, L., Yeh, A., Blaschke, C., Valencia, A.: Overview of BioCreAtIvE: critical assessment of information extraction for biology. BMC Bioinformatics 6(1), S1 (2005)
7. Finkel, J., Dingare, S., Manning, C.D.: Exploring the boundaries: gene and protein identification in biomedical text. BMC Bioinformatics 6(1), S5 (2005)
8. McCallum, A., Freitag, D., Pereira, F.: Maximum entropy Markov models for information extraction and segmentation. In: Proceedings of The Seventeenth International Conference on Machine Learning, pp. 591–598. Morgan Kaufmann, San Francisco (2000)
9. Liu, H., Hu, Z., Torii, M., Wu, C., Friedman, C.: Quantitative Assessment of Dictionary-based Protein Named Entity Tagging. Journal of the American Medical Informatics Association 13(5), 497–507 (2006)
10. Lafferty, J., McCallum, A., Pereira, F.: Conditional random fields: Probabilistic models for segmenting and labeling sequence data. In: Proceedings of the International Conference on Machine Learning, pp. 282–289. Morgan Kaufmann, San Francisco, CA (2001)
11. Settles, B.: Biomedical Named Entity Recognition Using Conditional Random Fields and Novel Feature Sets. In: Proceedings of the Joint Workshop on Natural Language Processing in Biomedicine and its Applications (JNLPBA-2004), pp. 104–107 (2004)
12. Tanabe, L., Xie, N., Thom, L.H., Matten, W., Wilbur, W.J.: GENETAG: a tagged corpus for gene/protein named entity recognition. BMC Bioinformatics 6(1) (2005)
13. Cohen, A.M.: Unsupervised gene/protein entity normalization using automatically extracted dictionaries. In: Linking Biological Literature, Ontologies and Databases: Mining Biological Semantics, Proceedings of the BioLINK2005 Workshop; Detroit, MI: Association for Computational Linguistics, pp. 17–24 (2005)
14. Tanabe, L., Wilbur, W.J.: Generation of a Large Gene/Protein Lexicon by Morphological Pattern Analysis. Journal of Bioinformatics and Computational Biology 1(4), 611–626 (2004)
15. Tanabe, L., Wilbur, W.J.: Tagging gene and protein names in biomedical text. Bioinformatics 18(8), 1124–1132 (2002)

Translation and Rotation Invariant Mining of Frequent Trajectories: Application to Protein Unfolding Pathways

Alexander Andreopoulos[1], Bill Andreopoulos[1,2], Aijun An[1], and Xiaogang Wang[1]

[1] York University, Dept. of Computer Science, Toronto Ontario, M3J 1P3, Canada
[2] Biotechnological Centre, TU Dresden, Germany
{alekos,billa,aan}@cs.yorku.ca, stevenw@mathstat.yorku.ca

Abstract. We present a framework for mining frequent trajectories, which are translated and/or rotated with respect to one another. We then discuss a multiresolution methodology, based on the wavelet transformation, for speeding up the discovery of frequent trajectories. We present experimental results using noisy protein unfolding trajectories and synthetic datasets. Our results demonstrate the effectiveness of the proposed approaches for finding frequent trajectories. A multiresolution mining strategy provides significant mining speed improvements.

1 Introduction

There exist many situations where we are confronted with trajectories describing the movement of various objects. We are often interested in mining the *frequent trajectories* that groups of such objects go through. Trajectory datasets arise in many real world situations, such as discovering biological patterns, mobility experiments, and surveillance [9,7,4]. Of special interest are trajectories representing protein unfolding pathways, which have been derived from high-throughput single molecule force spectroscopy experiments [8]. Such trajectories are represented on a two-dimensional $force \times distance$ grid. The y axis corresponds to the force (pN) involved in pulling the protein out of the cellular membrane via the tip of a mechanical cantilever; the x axis corresponds to the force-induced distance (nm) on the unfolding pathway of the protein. Such trajectories are often very noisy, which makes it difficult to distinguish the frequent subtrajectories from the deluge of irrelevant trajectories. Moreover, frequent subtrajectories may be translated or rotated with respect to one another. Our aim is to find such frequent subtrajectories in datasets resulting from high-throughput experiments. This is useful for identifying different protein unfolding pathways and, therefore, classifying proteins based on their structure. The contributions of this paper are as follows: (i) We present a framework for finding frequent trajectories whose sampling interval is small enough to estimate their first and second order derivatives. (ii) We propose a robust framework for mining frequent translated trajectories and frequent trajectories that are both rotated and translated with respect to each other. (iii) We apply our method

T. Washio et al. (Eds.): PAKDD 2007 Workshops, LNAI 4819, pp. 174–185, 2007.

to find frequent trajectories in protein unfolding pathways. *(iv)* We present a multiresolution framework to speed up the mining process.

This paper is organized as follows. Section 2 presents some related work. Section 3 introduces the general framework we use for mining trajectories. Section 4 describes a method for mining translated and rotated trajectories. Section 5 offers an approach for optimizing the mining speed of frequent trajectories and dealing with noisy trajectories. Section 6 presents experiments testing the proposed approaches. Section 7 concludes the paper.

2   Related Work

In sequential pattern mining we are typically given a database containing sequences of transactions and we are interested in extracting the *frequent sequences*, where a sequence is frequent if the number of times it occurs in the database satisfies a minimum support threshold. Popular methods for mining such datasets include the GSP algorithm [2] - which is an Apriori [1] based algorithm - and the PrefixSpan [10] algorithm. GSP can suffer from a high number of generated candidates and multiple database scans. Pattern growth methods such as PrefixSpan are more recent approaches for dealing with sequential pattern mining problems. They avoid the candidate generation step, and focus the search on a restricted portion of the initial database making them more efficient than GSP [2,10]. The problem that is most related to frequent trajectory mining is sometimes referred to as *frequent spatio-temporal sequential pattern mining* in the literature. The main difference between our work and previous work [9,7,4,5] is that our method assumes that we are dealing with *densely sampled* trajectories - trajectories whose sampling interval is small enough to allow us to extract from a trajectory its first and second derivative. This allows us to define a neighborhood relation between the cells making up our trajectories, allowing us to perform various optimizations. There has been a significant amount of research on defining similarity measures for detecting whether two trajectories are similar [13,3]. However, this research has not focused on mining frequent trajectories. The previous work closest to our approach is given in the innovative work described in [5] where the authors match two candidate subgraphs by comparing the set of angles made by the graph edges. This measure is similar to the curvature measure that we use later on to detect rotation and translation invariant trajectories. However, [5] is not suited for detecting trajectories that are translated but not rotated with respect to each other and does not address various robustness and speed improvements that are introduced in this paper.

3   Apriori Based Mining of Frequent Trajectories

We define a trajectory c as a continuous function $c(s) = [x(s), y(s)]$ in the 2D case and as $c(s) = [x(s), y(s), z(s)]$ in the 3D case. Similar extensions follow for higher dimensional trajectories. The function $c(s)$ is an arc-length parameterization of a curve/trajectory. In other words, the parameter s denotes the length along the trajectory and $c(s)$ denotes the position of the trajectory after traversing distance s. In other words our trajectories do not depend on time, or the speed with which the object/person traverses the trajectory. We assume independence from time and speed for mining the frequent trajectories and subtrajectories.

A trajectory c is *frequent*, if the number of the trajectories $\{c_1, c_2, ..., c_o\}$ that pass through the path described by c satisfy a minimum support count (*minsup*). This definition requires only that there exist *minsup* subtrajectories of all trajectories in $\{c_1, c_2, ..., c_o\}$ that are identical to c; but it does *not* require that $c = c_i$ for a sufficient number of c_i's. More formally, we say that trajectory c over interval $[0, \tau]$ is frequent with respect to a dataset of trajectories if there exist a *minsup* number of compact intervals $[\alpha_1, \alpha_1 + \tau], \cdots, [\alpha_{minsup}, \alpha_{minsup} + \tau]$ such that for all $i \in \{1, \cdots, minsup\}$ and for all $0 \leq s \leq \tau$ we have $c(s) = c_{\pi(i)}(\alpha_i + s)$ (where π is a permutation function of $\{1, \cdots, o\}$).

Fig. 1. A cell sequence representation of a dense trajectory. The cell sequence representation of the dense trajectory consists of the gray cells (in order) that are intersected by the dense trajectory.

The frequent trajectory mining problem for 2D trajectories can be formulated as a sequential pattern mining problem in the following way. The 3D case is similar to the 2D case. Assume that we are observing a square *region* of size $N \times N$ over which all the trajectories occur. By splitting the region into a *grid* of square *cells*, as shown in Figure 1, we denote by (x_i, y_j) the cell located at the ith column and j^{th} row. A potential way of discretizing a region into a grid is by uniformly sampling along the two dimensions. In this paper we create the grid by uniform sampling, even though square cells are not necessary for our approach to work. Then we define:

(i) A trajectory $c(s)$ is referred to as a *dense trajectory* if it is represented by a densely sampled set of points. The sampling interval depends on the problem at hand and should be small enough to obtain accurate first and second derivatives.

(ii) A dense trajectory's *cell sequence* refers to the sequence of cells $((x_{\pi_x(1)}, y_{\pi_y(1)}), \cdots, (x_{\pi_x(n)}, y_{\pi_y(n)}))$ intersected by the dense trajectory (where π_x is a permutation function). The following conditions must hold: a. $\pi_x(i) \neq \pi_x(i+1)$ or $\pi_y(i) \neq \pi_y(i+1)$, and b. $|\pi_x(i) - \pi_x(i+1)| \leq 1$ and $|\pi_y(i) - \pi_y(i+1)| \leq 1$. Thus, we encode the order in which the dense trajectory intersects the cells. As we discuss below, in some situations it is preferable to also associate with each cell (x_i, y_j) from the sequence the arclength/distance over which the trajectory falls in this cell.

(iii) The number of cells in a trajectory's cell sequence is its *length*. For example, the cell sequence $((x_4, y_3), (x_3, y_2), (x_3, y_3), (x_3, y_4), (x_2, y_5))$ has length 5.

(iv) A *continuous subsequence* ω of a trajectory c's cell sequence $((x_{\pi_x(1)}, y_{\pi_y(1)}), \ldots, (x_{\pi_x(n)}, y_{\pi_y(n)}))$ must satisfy $\omega = ((x_{\pi_x(i)}, y_{\pi_y(i)}), (x_{\pi_x(i+1)}, y_{\pi_y(i+1)}), \ldots, (x_{\pi_x(j)}, y_{\pi_y(j)}))$ where $1 \le i \le j \le n$.

Sometimes a trajectory $c(s)$ might be represented by a small number of sample points. We can interpolate those points and subsample the interpolated trajectory, to obtain the dense representation of those trajectories.

Using the cell representation method to represent trajectories, the problem of mining frequent trajectories is defined as finding all the contiguous subsequences of the cell sequences in a database that satisfy a support threshold. We first point out that frequent trajectories satisfy the Apriori property: Any continuous subsequence of a frequent trajectory's cell sequence is frequent. We exploit this property to implement efficient algorithms for mining frequent cell sequences.

If (x_i, y_j) is our current cell position, the next allowable cell position (x_k, y_l) *must* be one of its 8 neighboring cells, such that $|i - k| \le 1$ and $|j - l| \le 1$. We use this constraint to modify the GSP algorithm and generate a much lower number of candidates than the GSP algorithm would generate without this constraint.

Figure 2 shows the pseudocode for the Apriori based mining of frequent trajectories where L_k is the set of frequent length-k cell sequences found in the grid and C_k is the set of candidate length-k cell sequences. The main difference between this algorithm and GSP lies in the `trajectory()` function for generating candidates of length k from frequent cell sequences of length $k - 1$. (Figure 2b). When finding the candidate length-2 cell sequences, it suffices to only join two

```
Inputs: minsup: Minimum support count.
        D: Data set of cell sequences of trajectories.
Output: All the frequent cell sequences.
(1) for each cell sequence t ∈ D
(2)     for each cell g ∈ t
(3)         g.count + +.
(4) L₁ = {cell g|g.count ≥ minsup}.
    //L₁ is the frequent length-1 cell sequences
    //(consisting of a single cell)
(5) for (k=2;Lₖ₋₁ ≠ ∅;k++) {
(6)     Cₖ =trajectory(Lₖ₋₁). //candidate
                             //length-k cell sequences
(7)     for each cell sequence representation t ∈ D {
(8)         Cₜ= the set of contiguous subsequences of t
            that are contained in Cₖ
(9)         for each candidate cell sequence c ∈ Cₜ
(10)            c.count++. //increment the support
                //count of this candidate
(11)    }
(12)    Lₖ = {c ∈ Cₖ|c.count ≥ minsup}.
(13) }
(14) return L = ∪ₖLₖ.
```
(a)

```
Input: Lₖ: The length-k frequent cell sequences.
Output: Cₖ₊₁: The candidate length-(k + 1) cell
        sequences.
(1) if (k = 1) {
(2)     for all pairs of single cells (a₁) ∈ L₁,(b₁) ∈ L₁
        such that (a₁) ≠ (b₁)
(3)         if (a₁) and (b₁) are neighbors, then
(4)             c₂ = (a₁,b₁). //join a₁ and b₁
(5)             C₂ = C₂ ∪ {c₂}.
(6) } else {
(7)     for all pairs of length-k cell sequences
        (a₁,⋯,aₖ) ∈ Lₖ, (b₁,⋯,bₖ) ∈ Lₖ
        such that (a₁,⋯,aₖ) ≠ (b₁,⋯,bₖ)
(8)         if (a₂,⋯,aₖ) = (b₁,⋯,bₖ₋₁), then
(9)             cₖ₊₁ = (a₁,⋯,aₖ,bₖ).
(10)            Cₖ₊₁ = Cₖ₊₁ ∪ {cₖ₊₁}.
(11) }
(12) return Cₖ₊₁.
```
(b)

Fig. 2. (a) Apriori based mining of frequent trajectories. (b) The `trajectory()` function.

length-1 cell sequences i.e., single cells, if the cells are neighboring/adjacent to each other, resulting in a much smaller number of candidates than if we had used GSP to accomplish this without using this neighborhood constraint. For $k \geq 2$, when joining length-k cell sequences to find length-$(k + 1)$ candidate cell sequences, it suffices to only join cell sequences a with b if the last $k - 1$ cells of a and first $k - 1$ cells of b are identical. We notice that by joining two continuous paths, the resulting path is also continuous. Also, notice that there is no pruning step in the candidate generation process of our algorithm. This is because pruning may cause the loss of good candidates since we are mining for frequent *contiguous* subsequences. This is another major difference between the stardard GSP algorithm and this algorithm.

Assume there are r cells into which our $N \times N$ region has been split and there are b neighboring cells for each non-boundary cell. In our case $b = 8$, since each cell is surrounded by at most 8 other cells. Then, the upper bound on the number of length-$(k + 1)$ candidate cell sequences generated is $|L_k| \times b$ since every sequence in L_k can only be extended by its b neighboring cells on one end of the sequence. This is lower than the upper bound of $|L_k| \times r$ that GSP might generate if we were dealing with a sequential pattern mining problem where we could not apply this neighborhood constraint, since typically $b << r$. Because of our definitions, a cell sequence of length n has $O(n^2)$ subsequences, implying that a brute force approach for finding the support count of each candidate would run in polynomial time. However, the number of database scans would be greater than the number of database scans needed by the Apriori based method.

4   Translational and Rotational Invariant Mining

In this section we present two trajectory mining techniques. The first is a method for mining frequent trajectories that are translated with respect to each other. The second method is for mining frequent trajectories that are both translated and rotated with respect to each other. Figures $3(e)$ and $3(f)$ show examples of translated and rotated trajectories, respectively. Such algorithms are useful in situations where we are interested in detecting more complex motion patterns. For example in surveillance situations, the camera which extracts the motion patterns might be rotated and translated by an unknown amount over the course of acquiring the motion patterns. In such situations the best we can hope to accomplish, in terms of frequent trajectory mining, is to make the frequent trajectory extraction invariant to the unknown amount by which the camera and subsequently the trajectories were translated and rotated.

Assume $c_1(s) = [x(s), y(s)]$ and $c_2(s) = [x(s) + 5, y(s) + 3]$. In other words c_2 is a translated version of path c_1. If we take the derivatives $c_1'(s)$, $c_2'(s)$ of these two paths then we notice that $c_1'(s) = c_2'(s)$ for all values of s. We use this fact to mine for frequent trajectories that are translated with respect to each other. An issue to keep in mind is that derivatives tend to magnify noise. In other words, two trajectories that are slightly different due to noise would have an even more different derivative. Below we will discuss methods for dealing with

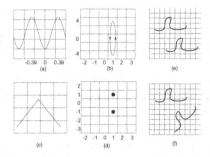

Fig. 3. (a) A trajectory given by function $c(s) = (s, sin(4s))$. (b) The derivative space $c'(s) = (1, 4cos(4s))$. We mine for such patterns to find translated patterns. (c) A trajectory that is not differentiable everywhere. (d) The derivative space of this function consists of two isolated and non-neighboring cells. We either have to smooth the function in (c) to have a continuous derivative space, or we are forced to use standard Apriori to do the mining, which is much slower. (e) Trajectories that are translated with respect to each other. (f) Trajectories that are both translated and rotated.

this problem. We mine for frequent translated trajectories in the following way. For every dense trajectory $c_i(s) = [x_i(s), y_i(s)]$ in our database of trajectories we use finite differences to find its derivative $c'_i(s) = [x'_i(s), y'_i(s)]$. We then use equiwidth binning to discretize the space of derivatives for the x-coordinates and y-coordinates into a number of bins (cells). We then represent every $c'_i(s)$ by a sequence of tuples (dx_i, dy_i), each of which denotes the current *derivative cell* in which the trajectory is located. A new tuple (dx'_i, dy'_i) is added to the sequence of tuples whenever the trajectory's derivative changes significantly enough to be part of a new derivative cell (Figure 3). For example, a linear trajectory is encoded by a sequence of length 1 (a single derivative cell) since its slope is constant. With each such tuple we could also associate a number denoting the arclength/distance over which the cell occurs. We use this measure, as described below, to detect translated trajectories. We can apply on this new trajectory representation the trajectory mining algorithms to find frequent trajectories that are translated with respect to each other.

Note that in our test cases we make the assumption that we are dealing with differentiable functions that do not change abruptly. Figures 3(c) and 3(d) show problems that might arise otherwise. If we wish to mine such trajectories, we could either apply some sort of smoothing such as the wavelet transformation described in the next section to make the function better behaved, or we could apply some sort of standard Apriori/GSP/PrefixSpan mining which does not make the neighborhood assumption for adjacent cells in our cell sequence representation. This would likely be detrimental to our trajectory mining speed.

The *curvature* of an arclength parameterized path $(f(s), g(s))$ at s is given by the derivative with respect to s of the angle θ the path makes with the x axis.

$$\kappa = \frac{d\theta}{ds} = \begin{vmatrix} f' & g' \\ f'' & g'' \end{vmatrix} = f'g'' - f''g' \tag{1}$$

Intuitively, the curvature gives us a measure of the rate with which a curve is changing direction. It is straightforward to show that any rotation and translation of $(f(s), g(s))$ results in the same curvature measure. In other words *curvature for 2D trajectories is rotation and translation invariant*. We can, therefore, use this measure to detect rotationally and translationally invariant patterns in a similar way as we did with translated trajectories. To encode each trajectory $c_i(s)$ we follow the same procedure that we followed for the translationally invariant mining. We can use equiwidth or equidepth binning to discretize the space of curvature measures and encode each trajectory using a sequence of *curvature cells*. We could also associate a number with each curvature cell, denoting the arclength/distance over which the trajectory belongs in this cell before it changes significantly to warrant using another cell in the sequence to encode it.

We wish to point out a potential problem which we have not discussed so far. The problem arises in the above cell sequence representation of derivatives and curvatures if we do not associate the distance over which each dense trajectory belongs to a particular cell. It is possible, for example, to have two trajectories whose cell derivative representation consists of the same two cells. If the distance over which each dense trajectory belongs to each cell is very different, the two trajectories might be very different and should not lead to a match. A solution to this problem is to associate with each cell the distance over which the dense trajectory belongs to the cell and mine this data as an extra dimension in our trajectory, or to simply use a brute force approach to refine the mined frequent trajectories. From our experiments we notice that a brute force approach is feasible in most cases due to the significantly decreased number of trajectories that need to be processed after the initial mining.

It should also be pointed out that because we are dealing with arclength parameterized curves $c(s) = [x(s), y(s)]$ we have $\int_0^s \sqrt{(x'(t))^2 + (y'(t))^2} dt = s$ which implies $\sqrt{(x'(s))^2 + (y'(s))^2} = 1$ which in turn implies $|x'(s)| \leq 1$ and $|y'(s)| \leq 1$. In other words, arclength parameterized curves do not change abruptly, implying that this parameterization makes it feasible for us to discretize the space of derivatives, since all derivative values will be in the range of -1 to 1. If we did not have such a bound on the space of derivatives this approach would be problematic in our opinion, since it would be too difficult to appropriately discretize the real line using a finite number of cells.

5   Wavelet Based Optimization of Mining Speed

Multiresolution techniques are well known in the signal processing community and are of great use for solving difficult problems such as image denoising and image compression[6]. More recently, the applicability of such methods has been demonstrated for various data mining problems. For example WaveCluster is a multiresolution clustering algorithm that uses the Wavelet transform to transform the original data and find dense regions in the transformed space [12]. We now propose a method for speeding up the frequent trajectory mining phase by mining for trajectories on multiple resolutions. Assume **dwt** is a function denoting the 1-D discrete wavelet transform. For example, given as input a vector x, $\mathbf{dwt}(x)$ returns a vector x' denoting the lower resolution version of vector x. At the pre-processing stage when the original *dense* trajectories are processed in order to convert them to their cell sequence representation, we apply $\mathbf{dwt}$ to each dense trajectory in order to get its lower resolution version. Then, we convert the lower resolution trajectory to a lower resolution cell sequence representation, where the cells now have *twice the width and height* they previously had. The effect of this is that the lower resolution cell sequence representation has approximately *half the length* of the original cell sequence representation. We refer to these new cell sequence representations as *scaled down* and we refer to their larger cells as *scaled down cells*. We can use these scaled down cell sequence representations to mine the frequent trajectories in our database at a coarser scale. This significantly decreases the length of our trajectories and the number of cells used to represent our region. As we discuss below, this can lead to significant improvements in the mining speed. A potential objection to this procedure involves the need to apply the wavelet transform. Some might argue, that in order to get a smaller cell representation, it suffices to simply double the cell size used in our equiwidth/equidepth discretization of the trajectories. The reason is to remove noise and high frequency components. This makes the derivative estimates more accurate. It also diminishes the risk that for various signals whose dense representation has a localized high frequency component passing near a cell border, we would needlessly add cells to a trajectory's cell sequence representation. Furthermore, by 'smoothing' a function we decrease the risk of dealing with functions which are not differentiable everywhere (Figures 3(c), 3(d)). A drawback of this method is that we lose precision on the localization of the trajectory since the scaled down cell sequence representations consist of larger cells. We now propose a method for obtaining a better localization of the coordinates through which frequent trajectories pass. When mining for frequent trajectories we are often only interested in finding trajectories that have a minimum non-scaled down length of m. Let S denote the set of all scaled down cells through which a frequent scaled down trajectory of length at least $\lfloor \frac{m}{2} \rfloor$ passes. Then, we are guaranteed that the frequent trajectories with a non-scaled down cell sequence representation of length over m pass through the scaled down cells in S. This means that to refine the accuracy of mining frequent scaled down cell sequence representations, it suffices to consider only the non-scaled down cell sequence representations that pass through some cells in S. For datasets that have very spread out trajectories with a few paths through which frequent trajectories pass, this can also result in significant improvements in mining speed.

6   Experiments

We experimentally evaluated the above mentioned algorithms using a real world dataset containing 139 protein unfolding trajectories, acquired using single molecule force spectroscopy [8]. We also use a dataset of 4,000 synthetic trajectories. In both cases, we are interested in mining translation invariant trajectories. We generated the synthetic trajectories with the same trajectory simulator that was used by the inventors of the TPR-tree algorithm [11] for indexing moving objects. We used MATLAB 7.01 running on an Intel Xeon 3Ghz with 3GB RAM to run our experiments. We utilized progressively larger subsets of the above mentioned trajectories to investigate the accuracy of our algorithms, their scaling properties, and their robustness to noise.

The distinguishing characteristic of the protein unfolding dataset was its extremely noisy trajectories. Moreover, in this dataset it is desirable to find subtrajectories that are translation invariant. The dataset consisted of 139 trajectories which we denoised, before applying the translation invariant methodology to detect the frequent translation invariant subtrajectories. We discretized the derivative space into 8x8 cells, and each trajectory's derivative representation consisted of around 200 cells on average. This served to demonstrate the practicality of our approach on a real world problem and to obtain quantitative results of our translation invariant method's performance on a very noisy dataset.

Fig. 4(a) shows a noisy protein unfolding trajectory. Fig. 4(b) shows the denoised trajectory from (a) using wavelet analysis. As shown, wavelet analysis was successful in identifying peaks, associated with single potential barriers stabilising segments within membrane proteins. Figures 4(c)-(f) show matched subtrajectories between the red arrows. Matched subtrajectories may potentially be manually annotated peaks, corresponding to three-dimensional protein structures [8]. On close inspection of the results we notice that the algorithm is

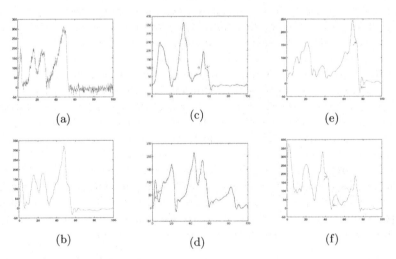

Fig. 4. (a):Noisy protein unfolding trajectory. (b):Denoised trajectory. Trajectories where a translation invariant match was found, with support threshold 3% : (c,d): the twin peaks with x-coordinates approximately in intervals [52, 55] in (c) and [2, 4.7] in (d) (interval denoted with red arrows), are matched as frequent translated subtrajectories. (e,f): In (e) x-coordinate interval [74, 78] matches in (f) interval [43, 47]. The y axes correspond to the force (pN) and the x axes correspond to the force-induced distance (nm) on the unfolding pathway of the protein.

Table 1. Results using the 139 protein unfolding trajectories (translation invariant with standard apriori/not translation invariant with standard apriori)

Number of		Support %		
Trajectories		1	2	3
	T	22693/16054	7411/7579	3206/3792
D=139	L	39/61	28/43	25/35
	F	45677/44937	15537/21507	6617/10105

Table 2. Results using the translation invariant/rotation invariant mining on a synthetic dataset

Number of		Support %		
Trajectories		1	2	3
	T	33351/13	12843/11	6565/11
D=1000	L	43/21	32/21	25/21
	F	35117/338	11267/312	3205/288
	T	53071/24	24984/22	12026/18
D =2000	L	33/21	30/21	12/19
	F	27007/338	10705/312	2515/200
	T	110824/44	47301/41	23934/35
D=4000	L	32/21	30/21	11/19
	F	20128/312	9613/288	1832/200

successful at detecting translated subtrajectories. Table 1 presents some quantitative results with and without the translation invariant mining algorithm. It shows the running time in seconds (T), the length of the longest frequent trajectory discovered (L) and the total number of frequent trajectories discovered (F). The "Support" columns show the minimum support used, as a percentage of the number of trajectories (D).

The next set of experiments was designed to test the translation and rotation invariant mining of trajectories (Section 4). For the translated trajectories, the derivative values of the trajectories were discretized into a 32×32 grid using equidepth bining and the curvature values were split into 32 bins using equiwidth bining. We first applied our mining method to mine these derivative cell representations. See Table 2 for the results using the translation invariant algorithm. The algorithm accurately detected the frequent translated trajectories. In the next section we show how the mining speed could be improved. To test the rotation invariant algorithm, we added to each of the above datasets 100 rotated sinusoidal trajectories having various frequency values in order to obtain a better understanding of how the algorithm performs for longer trajectories. The rotation invariant algorithm is much faster than the other approaches. This is due to the fact that the curvature at any point along a 2D curve is described by a single number while the translation invariant approach and Standard Apriori mining approach using 2D trajectories required two real numbers to describe each point along the curve. Furthermore, on average the curvature cell trajectories were

Table 3. Results using the wavelet based approach on a synthetic dataset. Shows the running time with/without the optimization.

Number of		Support %		
Trajectories		1	2	3
	T	224/630	131/344	95/240
D=1000	L	20/20	20/20	20/20
	F	1057/1133	742/638	634/534
	T	436/1270	262/671	192/474
D =2000	L	20/20	20/20	20/20
	F	1008/1088	738/619	631/526
	T	887/2563	542/1327	388/951
D=4000	L	20/20	20/20	20/20
	F	998/1099	743/614	630/532

much shorter in length resulting in faster mining speeds. We also observe that curvature is very sensitive to noisy trajectories, as expected. This indicates that a preprocessing stage for denoising/smoothing the trajectories is a necessity.

We then proceeded to test the wavelet based method for optimizing the mining speed (Section 4). Table 3 presents our results. The results demonstrate that the wavelet based method significantly improves the mining speed if there are clusters through which frequent trajectories of the desired length m pass, since it provides a quick method of finding which trajectories should be eliminated from further processing. However, if our trajectories are evenly spread out around the region we do not expect to gain a lot in terms of mining speed in general.

To test the approach we used the Standard Apriori based algorithm to compare the mining performance with and without the wavelet based method described in Section 4. In order to demonstrate the effectiveness of the approach we slightly modified the datasets we had used to test the Standard Apriori algorithm. We translated one-fourth of the trajectories in each of the datasets so that they are close to each other and form a cluster of trajectories that pass close by each other and added to this cluster a sufficient number of duplicate trajectories of length at least 15, so that they would end up being discovered as frequent trajectories during the mining process. Note that these were also the longest trajectories in our dataset. The other three-fourths of the trajectories were positioned in locations away from this cluster and were spread out so that it was unlikely to have many frequent trajectories in this group. We instructed our algorithm to search for frequent trajectories of length at least 15. We used a grid spacing of 32×32 cells to represent the scaled down cells and used a 64×64 grid to represent the non-scaled down cells. The algorithm located the trajectories of length 15 while significantly improving the mining speed.

7   Conclusions

We presented various methods for mining frequent trajectories that are translated and/or rotated with respect to each other. We also presented approaches for optimizing the mining speed of such trajectories. The methods were successful in finding peaks in protein unfolding pathways, which may correspond to three-dimensional protein structures. More research needs to be done in using intelligent methods for discretizing the continuous range of values that the trajectories can assume, as this could potentially decrease the number of cells used to encode trajectories.

Acknowledgements

The authors are grateful for the financial support of the Natural Sciences and Engineering Research Council of Canada (NSERC) through the PGS-D program and the Ontario Graduate Scholarships program (OGS). We would also like to thank Annalisa Marsico for kindly providing the protein unfolding dataset.

References

1. Agrawal, R., Srikant, R.: Fast algorithms for mining association rules. Proc. 20th Int. Conf. Very Large Data Bases (1994)
2. Agrawal, R., Srikant, R.: Mining sequential patterns. In: ICDE (1995)
3. Cai, Y., Ng, R.: Indexing spatio temporal trajectories with chebyshev polynomials. In: SIGMOD (2004)
4. Cao, H., Mamoulis, N., Cheung, D.: Mining frequent spatio-temporal sequential patterns. In: Proceedings of the ICDM (2005)
5. Kuramochi, M., Karypis, G.: Discovering frequent geometric subgraphs. In: 2nd IEEE Conference on Data Mining (ICDM) (2002)
6. Mallat, S.: A Wavelet Tour of Signal Processing. Academic Press, London (1999)
7. Mamoulis, N., Cao, H., Kollios, G., Hadjieleftheriou, M., Tao, Y., Cheung, D.W.L.: Mining, indexing, and querying historical spatiotemporal data. In: Conference on Knowledge Discovery in Data (2004)
8. Marsico, A., Labudde, D., Sapra, T., Muller, D.J., Schroeder, M.: A novel pattern recognition algorithm to classify membrane protein unfolding pathways with high-throughput single molecule force spectroscopy. Bioinformatics (2006)
9. Morimoto, Y.: Mining frequent neighboring class sets in spatial databases. In: Conference on Knowledge Discovery in Data (2001)
10. Pei, J., Han, J., Mortazavi-Asl, B., Pinto, H., Chen, Q., Dayal, U., Hsu, M.: Prefixspan: Mining sequential patterns efficiently by prefix-projected pattern growth. In: Proc. Int. Conf. on Data Engineering (ICDE) (2001)
11. Saltenis, S., Jensen, C., Leutenegger, S., Lopez, M.: Indexing the positions of continuously moving objects. In: SIGMOD, pp. 331–342 (2000)
12. Sheikholeslami, G., Chatterjee, S., Zhang, A.: Wavecluster: A multi-resolution clustering approach for very large spatial databases. In: Proceedings of the 24th International Conference on Very Large Databases (1998)
13. Vlachos, M., Hadjieleftheriou, M., Gunopulos, D., Keogh, E.: Indexing multi-dimensional time-series with support for multiple distance measures. In: SIGKDD 2003 (2003)

Genetic-Annealing Algorithm for 3D Off-lattice Protein Folding Model

Xiaolong Zhang, Xiaoli Lin, Chengpeng Wan, and Tingting Li

School of Computer Science and Technology,
Wuhan University of Science and Technology, Wuhan 430081, P.R. China
{xiaolong.zhang,xiaoli.lin,chengpeng.wan,
tingting.li}@mail.wust.edu.cn

Abstract. This paper presents a model used to deal with three-dimensional off-lattice *AB* protein folding. The model is extended from genetic-annealing algorithm that is for the two-dimensional off-lattice *AB* protein model. The three-dimensional model also has only two types of residues, hydrophobic and hydrophilic. Based on a physical model, the problem is converted from a nonlinear constraint-satisfied problem to an unconstrained optimization problem. Also, in contrast to earlier studies using off-lattice *AB* models, our results demonstrate that the proposed methods are very promising for searching the ground states of protein folding in three dimensions.

Keywords: Protein folding, 3D Off-Lattice Model, Genetic-Annealing Algorithm.

1 Introduction

The protein molecule is a very complex system, and understanding the folding of natural protein remains one of the most challenging objectives in bioinformatics research. In recent years there has been an increasing interest in introducing highly simplified models based on general principles that are believed to primarily control the structure formation of proteins. Various optimization methods have been applied to formulations of the ab-initio folding problem. These methods are based on simplified models of protein structure. Based on the minimum free-energy theory [1], many researchers have developed various theoretical computing methods to search for the structures of proteins, such as conformational space annealing (CSA) [2], energy landscape paving minimizer (ELP) [3], pruned-enriched-Rosenbluth method (PERM) [4] etc.

Due to the complexity of the protein folding problem, simplified models such as Dill's hydrophobic-polar (*HP*) model [5] have become one of the major tools for studying protein structure. Stillinger [6] studied a similar off-lattice *AB* protein model in two dimensions which uses only two types of residues, hydrophobic (*A*) and hydrophilic (*B*). Recently, Bachmann, et al. [3] studied the off-lattice *AB* model using the energy landscape paving minimizer (ELP). It is found that their results for the ground state energies are lower than the best values reported in the earlier literature. Kim, et al. [2] also studied the same model by the conformational space annealing, and obtained even better results.

T. Washio et al. (Eds.): PAKDD 2007 Workshops, LNAI 4819, pp. 186–193, 2007.

In this paper we study an extension to three dimensions (3D) of a two dimensions (2D) off-lattice model [6] that was successfully used in Refs. [7, 8]. The 3D model we used has been introduced in Ref. [9], which took account of the torsional energy implicitly. In order to accurately search for the ground-state conformations of the protein, an improved hybrid algorithm that combines genetic algorithm and simulated annealing was proposed.

The rest of the paper is arranged as following. Section 2 briefly introduces the 3D off-lattice *AB* model. Section 3 presents the improved genetic-annealing algorithm (GAA) which is used to search the conformation of the final state of a given protein sequence. Section 4 introduces the experiment results and gives the comparison with previous algorithm. Section 5 summarizes related work. Section 6 concludes this paper and gives the possible future research directions.

2   The Off-lattice AB Model

The off-lattice *AB* model consists of hydrophobic *A* monomers ($\sigma_i = +1$) and hydrophilic *B* monomers ($\sigma_i = -1$). The energy function is given by [9]

$$E = -k_1 \sum_{i=1}^{N-2} \hat{b}_i \cdot \hat{b}_{i+1} - k_2 \sum_{i=1}^{N-3} \hat{b}_i \cdot \hat{b}_{i+2} + \sum_{i=1}^{N-2} \sum_{j=i+2}^{N} E_{LJ}(r_{ij}; \sigma_i, \sigma_j) \quad (1)$$

where $\hat{b}_i$ is the bond vector between the monomers i and $i+1$ with unit length. Since $\hat{b}_i \cdot \hat{b}_{i+1} = \cos\theta_i$, $\hat{b}_i \cdot \hat{b}_{i+2} = \cos\alpha_i$, the N-mer can be specified by the $N-1$ bond vectors $\hat{b}_i$ or by $N-2$ bond angles θ_i and $N-3$ torsional angles α_i. The r_{ij} is just the usual Eucledian distance between sites i and j, and r_{ij} depends on the bond angles and torsional angles. These two angles are the degrees of freedom of the model. The species-dependent global interactions are given by the Lennard-Jones potential.

$$E_{LJ}(r_{ij}; \sigma_i, \sigma_j) = 4C(\sigma_i, \sigma_j)(\frac{1}{r_{ij}^{12}} - \frac{1}{r_{ij}^{6}}) \quad (2)$$

The depth of the minimum of this potential $C(\sigma_i, \sigma_j)$ is chosen to favor the formation of a core of *A* residues and $\sigma_1, \cdots \sigma_N$ is a binary string that specifies the primary sequence.

$$C(\sigma_i, \sigma_j) = \begin{cases} 1 & AA \\ 1/2 & BB , AB \end{cases} \quad (3)$$

The two parameters of the energy function, k_1 and k_2, denote the strength of species-independent local interactions. In Ref. [9], different values for the parameter set (k_1, k_2) were tested and finally set to (−1, 0.5) as a suitable choice for structural stability.

In particular, for any N -residue molecule, the $N-2$ bond angles θ_i and $N-3$ torsional angles α_i should be found when the potential-energy function (1) obtains the energy minimum, which is based on the minimum free-energy hypothesis. Thereby, the objective function of the optimization algorithm is the potential-energy function of the 3D off-lattice model. So the protein folding problem can be described as

$$\min_{\theta_i, \alpha_i \in (-\pi, \pi)} E\left(\theta_2, \cdots, \theta_{n-1}; \alpha_3, \cdots, \alpha_{n-1}\right) \tag{4}$$

3 Genetic-Annealing Algorithm

3.1 Genetic Algorithm and Simulated Annealing Algorithm

Genetic algorithm (GA) [10] is a learning method motivated by an analogy to biological evolution. It disposes directly a population, of which individuals are the possibility solutions of the given problem. Every individual has a fitness which correlates with the capability of solving the problem. The most fitness individual has the larger opportunity which generates the successor by using a crossover and mutation operation than other individuals. Genetic algorithm has been successfully applied to many domains. It is also used in simulation of protein evolution, folding analysis and prediction of the three dimension structure of proteins [11], and finding of low-energy conformations of organic molecules [12]. GA is of high parallel processing, strong robust and global searching ability. However, GA suffers from the premature convergence and poor local searching ability.

Simulated annealing (SA) [13] is a method for obtaining good solutions to difficult optimization problems. The motivation of SA algorithm comes from an analogy for the physical annealing of solids. An important characteristic of SA is that it uses a probability mechanism to control the process of jumping out trap in search. In the period of iterative computing, it accepts the new solution according to the Metropolis criterion. That is, the bad solution also can be accepted by the certain probability besides the superior solution. SA is a type of local search algorithm, which attempts to avoid becoming trapped in a local optimum by the suitable mechanism. Although SA is simple and quick to execute, the disadvantage of the algorithm is that the local minimum found may be far from the global minimum.

To keep the advantages and avoid the disadvantages of both search algorithms, this paper proposes a combined algorithm (GAA: Genetic-Annealing Algorithm) with novel genetic operators and local adjustment mechanism in Ref. [7]. These genetic operators and local adjustment mechanism are involved in the 3D off-lattice model. This paper also uses an initial conformation mechanism to speed up search process.

3.2 Initial Conformation Mechanism

Since the GAA method generates the initial conformation randomly, the results are very unstable. According to the phenomenon that hydrophobic amino acids fold into a hydrophobic core surrounded by hydrophilic amino acids in protein molecule, this paper proposes a heuristic initialization mechanism to speed up the lowest-energy state search.

The idea is as follows: Pick out all *A*-monomers and place them in certain spots in 3D space. All *B*-monomers wrap the hydrophobic (*A*) core, as Fig.1 illustrates.

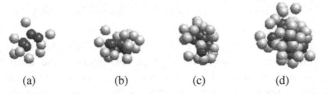

(a)	(b)	(c)	(d)

Fig. 1. The initial conformation generated by the heuristic initialization mechanism. (a) $n=13$; (b) $n=21$; (c) $n=34$; (d) $n=55$. The red balls (or black balls in white-black printing) represent hydrophobic *A* monomers, grey balls represent hydrophilic *B* monomers.

3.3 Genetic-Annealing Algorithm

For the main genetic evolution, GAA is the same as standard genetic algorithm in the reproduction process, but they are different in the genetic operator. The reproduction process is based on the Darwinian principle of survival of the fittest. GAA operates by iteratively updating a pool of hypotheses, called the population. On the each iteration, all members of the population are evaluated according to the fitness function. The hypotheses with higher fitness have a higher survival probability into the next generation.

1) Crossover mechanism

In the crossover process of GAA, suppose that the current hypothesis with minimum energy could be described as $h_{\min} = \{\theta_2^{\min}, \theta_3^{\min}, \cdots, \theta_{n-1}^{\min}, \alpha_3^{\min}, \alpha_4^{\min}, \cdots \alpha_{n-1}^{\min}\}$, where θ_i is the angle of bend among the nonterminal residues and α_i is the torsional angles. The $r * scale$ members are probabilistically selected from a pool of hypotheses, where r is the crossover rate and $scale$ is the population scale. The probability of selecting hypothesis is given by $Random(0...1) > Fitness(i)$. That is, the individuals with higher fitness are of a higher selected probability. Moreover, the individuals with lower fitness also have the certain chance to be selected as the individual in the next population. If each individual selected is defined as $h_i = \{\theta_2^i, \theta_3^i, \cdots, \theta_{n-1}^i, \alpha_3^i, \alpha_4^i, \cdots \alpha_{n-1}^i\}$, which is to do the crossover process with $h_{\min}$, then the new individual is

$$h_i' = \{\theta_2^{i\prime}, \theta_3^{i\prime}, \cdots, \theta_{n-1}^{i\prime}, \alpha_3^{i\prime}, \alpha_4^{i\prime} \cdots \alpha_{n-1}^{i\prime}\} \tag{5}$$

where

$$\theta_m^{i\prime} = \begin{cases} r\theta_m^i + (1-r)\theta_m^{\min}, & \Delta < 0.8 \\ (\theta_m^i + \theta_m^{\min})/2, & \Delta \geq 0.8 \end{cases}, \quad \alpha_m^{i\prime} = \begin{cases} r\alpha_m^i + (1-r)\alpha_m^{\min}, & \Delta < 0.8 \\ (\alpha_m^i + \alpha_m^{\min})/2, & \Delta \geq 0.8 \end{cases} \tag{6}$$

Here, Δ is the percentage of evolution. $\theta_m^{i\prime} = r\theta_m^i + (1-r)\theta_m^{\min}$ and $\alpha_m^{i\prime} = r\alpha_m^i + (1-r)\alpha_m^{\min}$ are the stochastic linearity combination with each parameter of h_i and $h_{\min}$, which can enhance the searching ability in the initial phase. $\theta_m^{i\prime} = (\theta_m^i + \theta_m^{\min})/2$ and $\alpha_m^{i\prime} = (\alpha_m^i + \alpha_m^{\min})/2$ have the strong astringency[14], which can improve the quality of solution or hypotheses, but also easily induce to the premature convergence, so it is used in the later of the evolution. The novel crossover mode is proposed here for improving the performance of the crossover process. Each parameter in the hypothesis selection is to do the crossover with the parent individuals in order to bring the better characters of the parent individuals into the next generation.

2) Annealing mechanism

In an annealing process, a melt, initially at high temperature and disordered, is slowly cooled so that the system at any time is approximately in thermodynamic equilibrium. As cooling proceeds, the system becomes more ordered and approaches a frozen ground state at $T = 0$. Hence the process can be considered as an adiabatic approach to the lowest energy state. If the initial temperature of the system is too low or cooling is done insufficiently slowly, the system may become trapped in a local minimum energy state. In this paper, the cooling schedule that provides necessary and sufficient conditions for convergence is $T_{i+1} = \sigma T_i (0 \le \sigma \le 1)$. When σ inclines to 1, the cooling speed of temperature becomes slow.

The Metropolis criterion is that a state of a thermodynamic system was chosen at energy E and temperature T, holding T constant the change in energy $\Delta E = E - E_{\min}$ is calculated. If $\Delta E < 0$, then the configuration is accepted. If $\Delta E > 0$, then the state is accepted with a probability given by the Boltzmann factor $e^{-(\Delta E / T)}$. This process is repeated sufficient times to give good sampling statistics for the current temperature, and then the temperature is dropped until a frozen state is achieved at $T = 0$.

The general idea of GAA based on 3D off-lattice AB model is follow. The energy of each individual in population is calculated based on off-lattice AB model. Then population is rearranged from minimal energy to maximal energy. At the same time, the minimum energy value and the individual with minimal energy are stored. In the process of annealing, the stochastic linearity combination crossover, mutation operation and local adjustment are employed continuously. Consequently, the ultimate energy value and individual is the global optimum. The steps in detail are described as follows.

Step 1: Set parameters and initialize the algorithm.
Step 2: Generate the population by heuristic initialization mechanism.
Step 3: Calculate the objective function value of each hypothesis in the population, and rearrange population by the energy value.
Step 4: Repeat the following steps throughout the process of the simulate annealing.
    a) Apply the crossover strategy.
    b) Apply the mutation strategy introduced in Ref. [8].
    c) Apply the local adjustment strategy introduced in Ref. [7].
Step 5: Return the global minimum value and conformation of the protein.

4   Experimental Results

GAA has been implemented by C++ in windows XP. In our experiments, the parameters are set as following: the initial temperature 100, the terminal temperature 10^{-7}, the drop rate of temperature 0.96, the loop times under same temperature 100, the mutation rate 0.8, the crossover rate 0.6, and the population scale 1000.

The purpose of these experiments is to see whether GAA algorithm is feasible and efficient for searching the low energy conformations in the 3D off-lattice AB model. Table 1 lists the same Fibonacci sequences that have already been studied in Ref. [7, 8], where A and B behave respectively as hydrophobic and hydrophilic residues. Fibonacci sequences are defined recursively by

$$S_0 = A, \quad S_1 = B, \quad S_{i+1} = S_{i-1} * S_i$$

Where " $*$ " is a concatenation operator. The first few sequences are $S_2 = AB$, $S_3 = BAB$, $S_4 = ABBAB$, etc. Hydrophobic residue A occurs isolated along the chain, while B occurs either isolated or in pairs and the molecules have a hierarchical string structure

$$S_i \equiv S_{i-2} * S_{i-3} * S_{i-2}$$

$$\equiv \left(S_{i-4} * S_{i-5} * S_{i-4}\right) * \left(S_{i-5} * S_{i-6} * S_{i-5}\right) * \left(S_{i-4} * S_{i-5} * S_{i-4}\right)$$

$$\equiv \cdots$$

Table 1 lists the lowest energies obtained about the 3D off-lattice AB model for all Fibonacci sequences with $13 \le N \le 55$, along with the values obtained by CSA [2], ELP [3], ACMC [16] methods for comparison. It can be seen that our results are better than those of the CSA for all the four sequences. For sequence with length 21, our result is slightly better than that of ACMC, and was equal to that of ELP. For other cases, our results are better than the results obtained by other methods except for length 13.

Fig. 2 depicts the lowest energy conformations in the 3D off-lattice model obtained by our GAA algorithm, corresponding to the energies shown in Table 1, where red balls (or black balls in white-black printing) represent hydrophobic A monomers, grey balls represent hydrophilic B monomers. In all cases, the conformations form a single hydrophobic core. This indicates that the AB model in three dimensions with Fibonacci sequences displays the important feature when it is used to simulate the real proteins.

Table 1. Test sequences and the minimal energies obtained by different algorithms

N	SEQUENCE	ELP	ACMC	CSA	GAA
13	ABBABBABABBAB	-26.498	-26.507	-26.471	-26.498
21	BABABBABABBABBABABBAB	-52.917	-51.757	-52.787	-52.917
34	ABBABBABABBABBABABBABABBABBABABBAB	-92.746	-94.043	-97.732	-98.692
55	BABABBABABBABBABABBAB	-172.696	-154.505	-173.980	-174.928
	ABBABBABABBABBABABBABABBABBABABBAB				

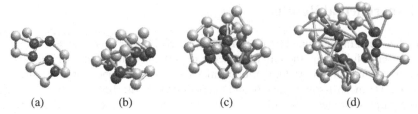

Fig. 2. The lowest energy conformations for the four Fibonacci sequences obtained by GAA algorithm. (a) $n = 13$; (b) $n = 21$; (c) $n = 34$; (d) $n = 55$. The red balls (or black balls in white-black printing) represent hydrophobic A monomers, grey balls represent hydrophilic B monomers.

5 Related Work

The native conformation of a protein is strongly correlated with the sequence of amino acid residues. Therefore, predicting the structure of a protein from its amino acid sequence is one of the most central problems in the field of bioinformatics. Recently many mathematic models have been proposed to solve the protein structure prediction problem. One of the widely used models is the *HP* lattice-model which was proposed by Dill, et al. [5]. However, this model only deals with the hydrophobic of protein and ignores other characters about protein. To reflect the characters of protein in more native way, Frank H. Stillinger [6] proposes off-lattice *AB* model in two dimensions. Similar 3D off-lattice models were also studied in Refs. [2,3,4,9]. The methods used to find low energy conformation of the off-lattice model include steepest-descent minimizations [6], neural networks [15], conformational space annealing (CSA) [2], energy landscape paving minimizer (ELP) [3] and pruned-enriched-Rosenbluth method (PERM) [4] etc. All these approaches have successfully studied in 2D and 3D off-lattice model, but the accuracy and efficiency still needs to be improved. This paper has extended the GAA algorithm to 3D off-lattice model. In comparison with related algorithms, GAA performs the best among them.

6 Conclusion

This paper describes an improved genetic-annealing algorithm. It is according to a 3D off-lattice *AB* model protein consisting of two types of amino acids residues, hydrophobic (*A*) and hydrophilic (*B*) residues. The experimental results show GAA has the advantage over the previous algorithms. The novel strategies proposed in this paper help to improve the searching lowest energy conformation of the protein. As one of the further works, we will study how to apply this algorithm to more complex real protein folding.

Acknowledgements

We thank the members of Machine Learning and Artificial Intelligence Laboratory, School of Computer Science and Technology, Wuhan University of Science and

Technology, for their helpful discussion within seminars. This work was supported in part by the Scientific Research Foundation for the Returned Overseas Chinese Scholars, State Education Ministry, and the Project (No.2004D006) from Hubei Provincial Department of Education, P. R. China, as well as and National Natural Science Foundation of P. R. China (60674115).

References

[1] Anfinsen, C.B.: Principles that govern the folding of protein chains. Science 181, 223 (1973)

[2] Kim, S.-Y., Lee, S.B., Lee, J.: Structure optimization by conformational space annealing in an off-lattice protein model. Phys. Rev. E 72, 011916 (2005)

[3] Bachmann, M., Arkin, H., Janke, W.: Multicanonical study of coarse-grained off-lattice models for folding heteropolymers. Phys. Rev. E 71, 031906 (2005)

[4] Hsu, H.-P., Mehra, V., Grassberger, P.: Structure optimization in an off-lattice protein model. Phys. Rev. E 68, 037703 (2003)

[5] Lau, K.F., Dill, K.A.: A lattice statistical mechanics model of the conformational and sequence space of proteins. Macromolecules 22, 3986 (1989)

[6] Stillinger, F.H., Head-Gordon, T., Hirshfel, C.L.: Toy model for protein folding. Physical review, E 48, 1469–1477 (1993)

[7] Zhang, X., Lin, X.: Protein folding prediction using an improved genetic-annealing algorithm. Proceedings of the 19th Australian Joint Conference on Artificial Intelligence (2006)

[8] Zhang, X., Lin, X.: Effective protein folding prediction based on genetic- annealing algorithm in Toy model. In: 2006 Workshop on Intelligent Computing & Bioinformatics of CAS (2006)

[9] Irback, A., Peterson, C., Potthast, F., Sommelius, O.: Local interactions and protein folding: A 3D off-lattice approach. J. Chem. Phys. 107, 273 (1997)

[10] Forrest, S.: Genetic algorithms: Principles of natural selection applied to computation. Science 261 (1993)

[11] Dandekar, T., Argos, P.: Potential of genetic algorithms in protein folding and protein engineering simulations. Protein Engineering 5(7), 637 (1992)

[12] Meza, J.C., Judson, R.S., Faulkner, T.R., Treasurywala, A.M.: A comparison of a direct search method and a genetic algorithm for conformational searching. Journal of Computational Chemistry 17(9), 1142 (1996)

[13] Eglese, R.W.: Simulated annealing: A tool for operational research. European Journal of Operational Research 46, 271 (1990)

[14] Liu, X., Yu, S.: A genetic algorithm with fast local adjustment. Chinese Journal of Computers 29, 100 (2006)

[15] Stillinger, F.H.: Collective aspects of protein folding illustrated by a toy model. Physical Review, E 52, 2872 (1995)

[16] Liang, F.: Annealing contour Monte Carlo algorithm for structure optimization in an off-lattice protein model. J.Chem. Phys. 120, 6756 (2004)

Biclustering of Microarray Data Based on Singular Value Decomposition

Wen-Hui Yang[1], Dao-Qing Dai[1,2,*], and Hong Yan[2,3]

[1] Center for Computer Vision and Department of Mathematics, Sun Yat-Sen (Zhongshan)
University, Guangzhou 510275, China
Tel.: (86)(20)8411 0141; Fax: (86)(20)8403 7978
yangwenhui204@126.com, stsddq@mail.sysu.edu.cn
[2] Department of Electric Engineering, City University of Hong Kong, 83 Tat Chee Avenue,
Kowloon, Hong Kong
h.yan@cityu.edu.hk
[3] School of Electrical and Information Engineering, University of Sydney, NSW 2006, Australia

Abstract. Biclustering is an important approach in microarray data analysis. Using biclustering algorithms, one can identify sets of genes sharing compatible expression patterns across subsets of samples. These patterns may provide clues about the main biological processes associated to different physiological states. In this study, we present a new biclustering algorithm to identify local structures from gene expression data set. Our method uses singular value decomposition (SVD) as its framework. Based on the singular value decomposition, identifying bicluster problem from gene expression matrix is transformed into two global clustering problems. After biclustering, our algorithm forms blocks of up-regulated or down-regulated in gene expression matrix, so as to infer that which genes are co-regulated and which genes possibly are functionally related. The experimental results on three benchmark datasets (Human Tissues, Lymphoma, Leukemia) demonstrate good visualization and interpretation ability.

1 Introduction

DNA microarrays technology enables the simultaneous monitoring of the expression level of a large number of genes for a given experimental condition [18]. Extracting biologically relevant information from this kind of data is a challenging and a very important task.

Various clustering methods have been proposed for the analysis of gene expression data [9,22], such as K-means clustering, hierarchical clustering, self-organizing maps (SOM) or normalized cuts algorithm. But when analyze the large and heterogeneous collections of gene expression data, conventional clustering algorithms often cannot produce a satisfactory solution, and appear several critical limitations [8,21]. Firstly, as experimental capacity improves in high dimensional expression datasets, similarity may not exist over all conditions. Secondly, related genes may naturally correlate over some conditions and not others. Lastly, genes may have more than one function, and a

---

* Corresponding author.

T. Washio et al. (Eds.): PAKDD 2007 Workshops, LNAI 4819, pp. 194–205, 2007.

gene may be involved in more than one biological process and may exhibit an expression profile that is a result of the regulatory effect of each process. To overcome these limitations, researches have turned to biclustering.

Biclustering refers to the "simultaneous clustering" of both rows and columns of a data matrix [17,19]. Hartigan [10] pioneered this type of analysis in the seventies using two-way analysis of variance to locate constant valued submatrices within datasets. More recently, inspired by Hartigan's so called "direct clustering" approaches [10], the concept was introduced to the area of gene expression analysis by Cheng and Church [8]. They identified the problem of finding significant biclusters as being NP-Hard. Since then, several alternative biclustering approaches have been proposed within gene expression analysis. Tanay et al. [21] mapped expression data onto bipartite graphs and used a probability model to find statistically significant subgraphs. Abdullah and Hussain [1] proposed a new graph-drawing-based biclustering technique based on the crossing minimization technique to work for asymmetric overlapping biclusters. Lazzeroni and Owen [14] developed what they termed a plaid model in which the dataset is represented by a linear function of variables or layers, which correspond to biclusters. Ben-Dor et al. [5] looked for order-preserving submatrices (OPSMs), in which the expression levels of all genes induce the same linear ordering of the experiments. For a review of the proceeding approaches, the reader is directed to [17].

Recently, a number of works had explored Singular Value Decomposition (SVD) for visualization of gene expression data, representation of the data using a smaller number of variables, and detection of patterns in noisy gene expression data, etc. Alter et al. [3,4] used SVD to analyze and model the gene expression data. Kluger et al. [13] put forward a spectral biclustering method based on SVD that simultaneously clusters genes and conditions, finding distinctive 'checkerboard' patterns in matrices of gene expression data. Liu et al. [16] presented a robust SVD analysis method to address the outliers recognition and the missing values imputation for microarray data. Ihmels et al. [11] proposed an iterative signature algorithm (ISA). The signature algorithm was designed to identify groups of co-regulated genes together with the experimental conditions over which the co-regulation is observed ('transcription modules'). Carmona-Saez et al. [6] used the non-smooth non-negative matrix factorization (nsNMF) technique to find biclusters. The main difference between nsNMF and SVD relies in the sparse non-negativity constraints imposed on both basis experiments and basis genes which can produce more localized feature representations of both genes and experiments. Differing from these methods, our method uses the property of SVD to find some special biclusters.

In this paper, we will answer under certain experimental conditions, which genes are up-regulated, and which genes are down-regulated. These information can deliver much valuable regulation information of gene network. We introduce a novel idea, the problem of discovering these local patterns (biclusters) in gene expression matrix is transformed into two global clustering problems. This technique uses SVD as its framework. In the global clustering problem, applying statistical theory, we estimate the threshold in hierarchical clustering to cut the dendrogram at a particular threshold level, and then sorting the clusters by relevant coefficient. Meanwhile, merging two adjacent biclusters with similar expression, so as to form blocks of up-regulated or down-regulated in gene expression matrix. Using this approach, we are able to identify the important co-regulated genes and group the samples efficiently at the same time.

The rest of this paper is organized as follows. In Section 2, we introduce the detailed algorithm. In Section 3, we discuss the experimental results on three gene data sets. Finally, conclusions will be given in Section 4.

2   Proposed Algorithm for Biclustering

Our proposed algorithm consists of three steps: a preprocessing step, a biclustering algorithm step followed by parameters estimation step.

2.1   Preprocessing

We will represent DNA microarray data as a $d \times N$ matrix X whose rows represent the genes, columns represent the experimental conditions. The real-number entries x_{ij} represents the expression level of gene i under condition j. The elements of the i-th row of X form the N-dimensional vector $\mathbf{g}_i$, which we refer to as the transcriptional response of the i-th gene. Alternatively, the elements of the j-th column of X form the d-dimensional vector $\mathbf{s}_j$, which we refer to as the expression profile of the j-th condition.

In the gene expression data, the block structure can be confounded by a number of effects. In particular, different overall expression levels of genes across all experimental conditions or of samples across all genes in multiple tumor datasets can obscure the block structure. Consequently, selecting an appropriate pre-processing method is critical.

In the first step, the following operations are prescribed: Carry out a base 10 logarithmic transformation to the gene expression data, standardize each sample to zero mean and unit variance across the genes, and normalize each gene by subtracting the median of the gene. With the data preprocessing in this step in this fashion, the up-regulated genes or down-regulated genes under certain conditions can be discovered by detecting constant biclusters.

The second step of preprocessing is to eliminate the genes which do not contribute to sample clustering. The main reason is that only small part of genes are effective to the special disease and a large number of genes exhibit near constant expression levels across samples. Firstly, the genes whose values do not vary much across the samples are eliminated. Secondly, we select the genes which are the most relevant to the basis genes which can be obtained by Principal Component Analysis (PCA). For each basis gene b_k, we obtain a set of genes with largest values of the Cosine Measure [7,15]:

$$R_{ik} = \frac{\mathbf{g}_i^T b_k}{\sqrt{\mathbf{g}_i^T \mathbf{g}_i} \sqrt{b_k^T b_k}}, \quad \text{for each gene } \mathbf{g}_i.$$

By this Cosine Measure, some of the irrelevant genes will be eliminated in this stage.

2.2   Biclustering with Singular Value Decomposition

In this section, we will explore how to locate special biclusters on gene expression data by using SVD method.

Singular Value Decomposition and Its Property. After preprocessing, without loss of generality, the gene expression matrix is still denoted by X with size $d \times N$. The SVD of X is the decomposition of X into the product of three matrices as follows:

$$X = U\Lambda V^T = \sum_{i=1}^{r} \lambda_i u_i v_i^T \tag{1}$$

where r is the rank of the matrix X, $U = [u_1, u_2, \cdots, u_r]$ is a $d \times r$ matrix with orthonormal columns, $V = [v_1, v_2, \cdots, v_r]$ is an $N \times r$ matrix with orthonormal columns, and Λ is an $r \times r$ matrix with elements λ_i ($\lambda_i > 0, 1 \leq i \leq r, \lambda_1 \geq \lambda_2 \cdots \geq \lambda_r$) along the diagonal. From the Eq. (1), we can see that the diagonal elements of the matrix Λ actually give a weight to the column vectors of the left singular matrix U and the row vectors of the right singular matrix V. Let $R = U\Lambda^{\frac{1}{2}}$ and $C = V\Lambda^{\frac{1}{2}}$, then the SVD of X can be rewritten as:

$$X = RC^T.$$

Gene expression data are currently rather noisy, and SVD can detect and extract some signals from noisy data. One important result of the SVD of X is that

$$X^{(l)} = \sum_{i=1}^{l} \lambda_i u_i v_i^T \qquad (l \leq r)$$

is the rank-l matrix which minimizes the sum of the squares of the difference of the elements of X and $X^{(l)}$. In order to extract some effective signals from gene expression data and decrease the noise influence, only the first l singular values can be selected, usually, $l < r$. At the same time, the matrix R and C also select only the first l columns, denoted by $R^{(l)}$ and $C^{(l)}$, so $X^{(l)} = R^{(l)}(C^{(l)})^T$.

There are many well-known properties of the SVD that are useful for a variety of applications. We first present an interesting property related to the SVD of a matrix.

Theorem 1. *Let X be a $d \times N$ matrix with rank r and $X^{(l)}(l \leq r)$ be the approximate matrix of X. Let $X^{(l)} = R^{(l)}(C^{(l)})^T$ and $(C^{(l)})^T = [\phi_1, \phi_2, \cdots, \phi_N]$. If two column vectors of X are the same, $s_i = s_j$, then the i-th and j-th columns of $(C^{(l)})^T$ are equal, i.e., $\phi_i = \phi_j$.*

Proof. From the definition of $X^{(l)}$, let $R = [R^{(l)}, R^{(r-l)}]$, $C = [C^{(l)}, C^{(r-l)}]$ and $(C^{(r-l)})^T = [\phi_1', \phi_2', \cdots, \phi_N']$, then $X = [R^{(l)}, R^{(r-l)}][C^{(l)}, C^{(r-l)}]^T$. So, we have

$$s_i - s_j = R \begin{pmatrix} \phi_i - \phi_j \\ \phi_i' - \phi_j' \end{pmatrix}.$$

Since R is the column full rank matrix, then $s_i - s_j$ is zero vector if and only if both $\phi_i - \phi_j$ and $\phi_i' - \phi_j'$ are zero vectors. $\square$

At the same way, let $R^{(l)} = [\varphi_1, \varphi_2, \cdots, \varphi_d]^T$, if $g_i = g_j$, then $\varphi_i = \varphi_j$. Therefore, we can analyze the gene expression data by exploring the matrices $R^{(l)}$ and $C^{(l)}$.

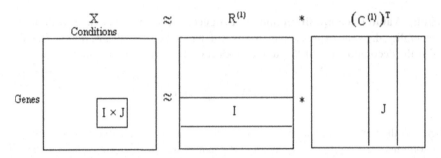

Fig. 1. The approximate SVD of X

Biclustering Technique. The gene expression datasets usually satisfy two assumptions: A set of co-regulated genes are expected to have correlated expression levels. The expression profiles for every two conditions of the same type are expected to be correlated. These correlation relationships can be better observed when averaged over sets of genes of similar expression profiles and different conditions of same type.

These assumptions are supported by Pomeroy et al. [20] on a brain tumors dataset, and are also mentioned in [13]. Under these assumptions, it is very important to find out the submatrices which is composed of subsets of genes and conditions with similar expression. Next, we attempt to automatically identify these submatrices on gene data sets.

Suppose that the block $x_{IJ} = (I,J)$ is a submatrix of X where I is the subset of rows, and J is the subset of columns, from the singular value decomposition $X = RC^T$, we obtain:

$$x_{ij} = \sum_{m=1}^{r} R_{im}C_{jm}, x_{Ij} = \sum_{m=1}^{r} R_{Im}C_{jm}, x_{IJ} = \sum_{m=1}^{r} R_{Im}C_{Jm}.$$

Hence the submatrix (I,J) of X only relies on the I rows of R and on the J columns of C^T.

Considering gene expression data with some noisy, only part of columns of R (that is, $R^{(l)}$) and part of rows of C (that is, $C^{(l)}$) are selected. The approximate SVD of X is shown in Fig. 1.

From Theorem 1, if gene g_i and gene g_j have similar transcriptional profiles, corresponding to the rows of $R^{(l)}$ have also similar profiles. On the sample s_i and the sample s_j, there is the same result corresponding to the columns of $(C^{(l)})^T$. If the I rows of the matrix $R^{(l)}$ are constant rows and the J columns of $(C^{(l)})^T$ are constant columns, then each element of submatrix (I,J) of X is approximately constant. Thus, identifying local structure problem from gene expression matrix transforms into two global clustering problems, that is, the similar expression submatrices can be uncovered by clustering on the rows of $R^{(l)}$ and on the columns of $(C^{(l)})^T$, respectively. These can be implemented by generic clustering methods, such as K-Means, hierarchical clustering. In this article, we use the hierarchical clustering method.

After hierarchical clustering to the rows of matrix $R^{(l)}$, $R^{(l)}$ can be divided into m groups:

$$R_1, R_2, \cdots, R_m.$$

After hierarchical clustering to the columns of matrix $C^{(l)}$, $C^{(l)}$ can be divided into n groups:

$$C_1, C_2, \cdots, C_n.$$

Because the arrangements of the clusters $R_1, R_2, \cdots, R_m$ and the clusters $C_1, C_2, \cdots, C_n$ are random, in order to see the patterns clearly, we adjust the sequence of these clusters in terms of the correlation coefficients of every pair of clusters.

These clusters form $m \times n$ original biclusters. But for any bicluster $(R_i, C_j)(1 \leq i \leq m, 1 \leq j \leq n)$, it may not be effective. In the next subsection, we will further validate these biclusters.

Biclusters Validation and Visualization. Our focus is on finding subsets of genes that are up-regulated or down-regulated across a subset of conditions. After special preprocessing, these patterns can be found by detecting constant biclusters.

A *perfect* constant bicluster is a submatrix (I, J) where all values are equal, for all $i \in I$ and $j \in J$:

$$x_{ij} = \mu.$$

The merit function used to compute and evaluate constant biclusters is the variance [10]:

$$VAR(I, J) = \frac{1}{|I||J|} \sum_{i \in I, j \in J} (x_{ij} - \bar{x}_{IJ})^2.$$

where $\bar{x}_{IJ}$ is the mean of the whole bicluster.

A user-defined threshold δ represents the maximum allowable dissimilarity within the bicluster. For a good bicluster, we have $VAR(I, J) < \delta$ for some $\delta > 0$.

The parameter δ is difficult to determinate owing to the noise. The lower the noise, the less the value of δ. Generally, δ is empirically determined. In our method, the range of the parameter δ is small owing to the special preprocessing, usually, $0.5 \leq \delta \leq 1$. Hence, the appropriate value of δ is easily found.

In addition, we add a parameter $\mu_0 > 0$ to determine the bicluster that is up-regulated, down-regulated, or non-significant. In one δ-bicluster, if $\bar{x}_{IJ} > \mu_0$, we call this bicluster is up-regulated; if $\bar{x}_{IJ} < -\mu_0$, we call this bicluster is down-regulated; or else, we call this bicluster is non-significant. In our experiments, the parameter μ_0 is determined by the half of mean variance of all genes.

If two δ-biclusters are adjacent and they have the same size of row or column, we merge two biclusters when the variance of combining two biclusters is less than δ.

After merging the biclusters, in order to highlight the visualization and interpretation, the values of the significant biclusters (up-regulated or down-regulated) are replaced by the mean value of this bicluster, and the non-significant biclusters are masked, then form the new expression matrix. From this expression matrix, we can investigate the biological significance of these biclusters, including which genes are down-regulated, up-regulated or non-significant in the special condition. Further more, we can perform gene function prediction, gene classification, etc.

2.3 Parameters Estimation

In the previous steps in our algorithm, there are two important parameters: l, the number of principal component, and the threshold level of hierarchical clustering. Different selections of two parameters have great influence to our results. Hence selecting two appropriate parameters are very important.

The Estimation of Number of Principal Component. Presently, parameter l is empirically determined in microarray analysis in most cases. Liu et al. [16] use only one eigenvector on the robust SVD analysis method. Kluger et al. [13] also only utilize the first two or three eigenvectors on the spectral biclustering method.

Gene expression data are currently rather noisy, thus the selection of parameter l is very important. If the parameter l is too small, some hidden patterns can be lost, on the contrary, if the parameter l is too large, some "fake" patterns can appear.

It is very difficult to determine the "right" number of components. The most common approach is the percentage of variance method, retaining eigenvalues that account for approximately 70% or 90% of the variance. Another simple approach is the broken stick method that is based on eigenvalues from random data. Observed eigenvectors are considered interpretable if their eigenvalues exceed a threshold based on random data: the k-th eigenvector is retained if its eigenvalue λ_k exceeds $\sum_{i=k}^{p}(1/i)$. An alternative approach is to examine the eigenvalue spectrum and see if there is a point where the values fall sharply before levelling off at small values (the scree plot test). Jackson [12] provided comparative studies. In this paper, the 'scree plot' approach is used to select the important components. The experiment results will demonstrate that the 'scree plot' approach is effective on gene expression data.

The Threshold Level Estimation of Hierarchical Clustering. In our algorithm, one important step is hierarchical clustering to the rows of $R^{(l)}$ and to the columns of $(C^{(l)})^T$, respectively. Hierarchical clustering generates a hierarchical series of nested clusters that can be graphically represented by a tree, i.e., a dendrogram. The branches of the dendrogram not only record the formation of the clusters but also indicate the similarity of the clusters. By cutting the dendrogram at a particular threshold level, one can obtain a specified number of clusters. Defining the threshold level plays a crucial role in detecting the number of clusters. Next, we will consider how to estimate a suitable threshold level.

From the definition of SVD, $\|u_i\|_2 = 1$ and $\|v_i\|_2 = 1, (i = 1, 2, \cdots, l)$, so $\|R^{(l)}\|_F = \|C^{(l)}\|_F = \sqrt{\sum_{1}^{l} \lambda_i}$, where $\|.\|_F$ denotes the Frobenius-norm. Suppose that the elements u_{ij} of U obey the normal distribution with the mean zero and independent each other, denoted $u_{ij} \sim N(0, \sigma^2)$, then $\xi = \sum_{j=1}^{d}(u_{ij}/\sigma)^2 \sim \chi^2(d)$, from the Expectation value $E(\xi) = d$ and $\sum_{j=1}^{d}(u_{ij}/\sigma)^2 = \|u_i\|_2^2/\sigma^2 = 1/\sigma^2$, so we can estimate the parameter $\sigma^2 = 1/d$. The Expectation of the square distance between the i-th row and the j-th row of $R^{(l)}$ is

$$E(\|r_i^l - r_j^l\|_2^2) = E(\sum_{k=1}^{l} \lambda_k (u_{ik} - u_{jk})^2) = \sum_{k=1}^{l} \lambda_k E(u_{ik} - u_{jk})^2 = 2/d * \sum_{k=1}^{l} \lambda_k. \quad (2)$$

Similarly, the Expectation of the square distance between the i-th row and j-th row of $C^{(l)}$ is

$$E(\|c_i^l - c_j^l\|^2) = 2/N * \sum_{k=1}^{l} \lambda_k. \quad (3)$$

From the Eq. (2) and the Eq. (3), we can estimate the mean square distance of any two rows of $R^{(l)}$ and $C^{(l)}$. Because the distance of any two points in one cluster should be closer than the mean distance in total points, based on this idea, we can set an appropriate threshold in hierarchical clustering of $R^{(l)}$ and $C^{(l)}$. In our experiments, the square distance in one cluster of $R^{(l)}$ is not greater than $1/2d * \sum_{k=1}^{l} \lambda_k$ (just one-fourth of estimated mean square distance) and the square distance in one cluster of $C^{(l)}$ is not greater than $1/2N * \sum_{k=1}^{l} \lambda_k$ (just one-fourth of estimated mean square distance).

3   Experiments and Analysis

We analyze the performance of our algorithm on three data sets: Human Tissues, Lymphoma and Leukemia. Since the classes are known a priori for the data sets, the clustering results are used to find the number of misclassification.

Human Tissues Dataset

Human tissues dataset contain 224 genes and 504 normal and malignant tissues (covering eight tissue types: adipose tissue(6, 6), breast(37, 45), colon(67, 67), kidney(37, 32), liver(19, 20), lung(49, 52), ovary(5, 5), and prostate(29, 29)). Because the number of the samples of adipose tissue and prostate is rather small, these samples do not be analyzed here. The log-transformed and row and column means subtracted data are available at www.samsi.info/200304/dmml/web-internal/bio/data/ data_rsvd.xls. The detail information about this dataset can be found in [16].

Fig. 2(a) shows the image of the unordered log transformed data matrix. In this image, we cannot see any clear patterns. After preprocessing, 177 significant genes were selected. Fig. 2(b) plots the value of each successive eigenvalue against the rank order. The smaller eigenvalues, representing random variation, tend to lie along a straight line. The point where the ninth eigenvalue depart from the line distinguished the "interpretable" and trivial components. Hence, we can estimate $l = 9$. The biclustering results with $l = 9$ are shown in Fig. 3(A).

From Fig. 3(A), we can see that there are some small clusters in column clustering. We merge each of these clusters with the cluster which is the most correlated with itself, and then we partition the sample as the merge results. We label the sample name for each cluster as the vote principle. The partition and label results are also shown in

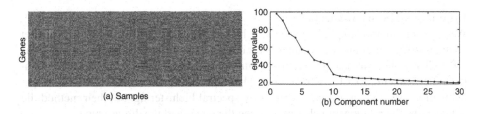

(a) Samples

(b) Component number

Fig. 2. (a) The unordered gene expression data matrix of Human Tissues data set. The rows correspond to the cell lines, and the columns correspond to the genes. In this image, we cannot see any clear patterns. (b) the eigenvalue of the first 30 principal components as descending order.

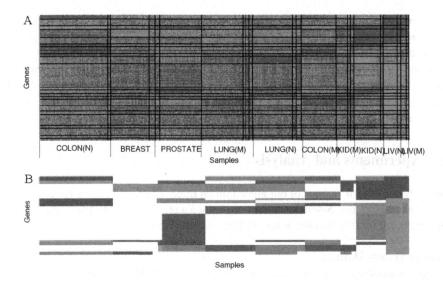

Fig. 3. (A) The biclustering result on Human Tissues data set. ('KID' denotes kidney samples;'LIV' denotes liver samples; 'N' denotes normal tissues; 'M' denotes malignant tissues.); (B) Marking significant biclusters ($\delta = 0.5$) and merging biclusters on Human Tissues dataset.

Fig. 3(A). Except for breast and prostate tissues, the other tissues all have the normal and malignant clusters. The number of wrong classification is 41 in 482 samples, and the total wrong rate is only 8.5%. To this hard problem, this is a satisfied results.

We set the score of each bicluster $\delta = 0.5$, and merge the adjacent biclusters with similar expression. The significant biclusters are shown in Fig. 3(B). From these biclusters, we can get the significant gene group information. Each group has similar gene expression information in special conditions. We can easily observe the expression state to the different tissues and the different gene group. There are different expression information for different tissues, so that we can recognize the unknown sample according to its expression profile. In our experiment, 'UGT1A', 'UGT2B', and 'HSD' are grouped together that are involved in steroid hormone metabolism, at the same time, the duplicate probes of 'UGT1A', 'UGT2B' also correctly clusters together. These results are consistent with [16].

Lymphoma Microarray Dataset

We apply our method to publicly available lymphoma microarray data [2]. This dataset contains 4026 genes and 62 samples: 42 samples of Diffuse Large B-Cell Lymphoma (DLBCL), 9 observations of Follicular Lymphoma (FL) and 11 cases of Chronic Lymphocytic Leukemia (CLL). Kluger et al. [13] obtained the desired partitioning of patients in the second largest eigenvectors by spectral biclustering, but their method did not describe the relationship clearly between the genes and the disease type.

In this dataset, 400 significant genes are selected at the preprocessing stage. From Fig. 4(c), we can estimate $l = 2$ as the 'scree plot' approach. From Fig. 4(b), our algorithm partitions all samples into 5 clusters. The first cluster denotes the FL sample; the

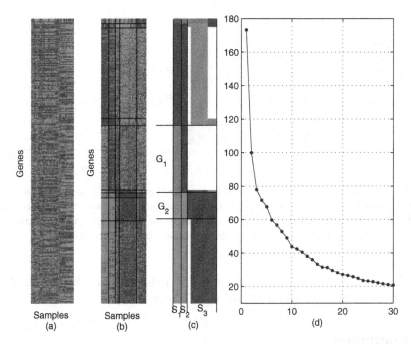

Fig. 4. (a) The unordered gene expression data matrix of Lymphoma Microarray dataset. In this image, we cannot see any clear patterns. (b) Biclustering results of the proposed algorithm for the Lymphoma dataset with $l = 2$. (c) Biclusters validation and merging biclusters with $\delta = 1.2$. S_1 denotes the FL sample; S_2 denotes the CLL sample, S_3 denotes denote the DLBCL sample. (d) The eigenvalues of the first 30 principal components as descending order.

second cluster denotes the CLL sample, the 3-rd to 5-th clusters denote the DLBCL sample. Our algorithm correctly clusters the same disease type together without wrong classification. From Fig. 4(c), the CLL sample and the FL sample have significant difference with DLBCL sample in the 'G_2' gene group; and the FL sample have significant difference with the CLL sample in the 'G_1' gene group.

Leukemia Dataset
Finally, we analyze the public leukemia microarray data set [9]. This data set contains 3571 genes and 72 leukemia samples where these samples are divided into 47 samples of (B- and T-cell) Acute Lymphocytic Leukemia (ALL) and 25 samples of Acute Myelogenous Leukemia (AML).

In this dataset, 166 significant genes are selected at the preprocessing stage. From Fig. 5(c), obviously, the second principal component is the 'scree plot' point, hence we can estimate $l = 2$. The biclustering results with $l = 2$ are shown in Fig. 5(b). Our algorithm partitions the samples into 2 clusters. The first cluster denotes the ALL sample; the second cluster denotes the AML sample. In this experiment, the number of wrong cluster is 2 in 72 samples. In each cluster, there is only one wrong classification. This result is better than the result of the algorithm proposed by Changdra et al. [7] with

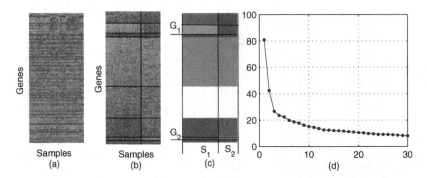

Fig. 5. (a) The unordered gene expression data matrix of Leukemia Microarray dataset. (b) Bi-clustering results of the proposed algorithm for the Leukemia dataset with $l = 2$. (c) Biclusters validation and merging biclusters with $\delta = 1$. S_1 denotes the ALL sample; S_2 denotes the AML sample. (d) The eigenvalues of the first 30 principal components as descending order.

3 misclassifications in this dataset. From Fig. 5(c), the ALL sample have significant difference with AML sample in the 'G_1' and 'G_2' gene group.

4   Conclusion

In this paper, a novel biclustering approach for gene expression data has been proposed. This technique uses singular value decomposition as its framework. Based on the property of SVD, identifying bicluster problem from gene expression matrix is transformed into two global clustering problems. Using this approach, we were able to identify the important co-regulated genes and efficiently group the samples as different classes. The efficacy of the algorithm was tested on three well-known datasets: Human Tissues, Lymphoma and Leukemia. The results demonstrated that this approach has good visualization and interpretation ability.

Acknowledgments

This project is supported in part by NSF of China(60575004, 10231040), NSF of GuangDong(05101817), the Ministry of Education of China(NCET-04-0791).

References

1. Abdullah, A., Hussain, A., et al.: A new biclustering technique based on crossing minimization. Neurocomputing 69, 1882–1896 (2006)
2. Alizadeh, A.A., Eisen, M.B., et al.: Distinct types of diffuse large B-cell lymphoma identified by gene expression profiling. Nature 403, 503–511 (2000)
3. Alter, O., Brown, P.O., Botstein, D.: Singular value decomposition for genome-wide expression data processing and modeling. Proceedings of the National Academy of Sciences 97(18), 10101–10106 (2000)

4. Alter, O., Golub, G.H.: Singular value decomposition of genome-scale mrna lengths distribution reveals asymmetry in rna gel electrophoresis band broadening. Proceedings of the National Academy of Sciences 103(32), 11828–11833 (2006)

5. Ben-Dor, A., Chor, B., Karp, R., Yakhini, Z.: Discovering local structure in gene expression data: the order-preserving sub-matrix problem. In: Proceedings of the 6th Annual International Conference on Computational Biology (RECOMB 2002), New York, USA, pp. 49–57 (2002)

6. Carmona-Saez, P., Pascual-Marqui, R.D., Tirado, F., Carazo, J.M., Pascual-Montano, A.: Biclustering of gene expression data by non-smooth non-negative matrix factorization. BMC Bioinformatics 7(78) (2006)

7. Changdra, B., Shanker, S., Mishra, S.: A new approach: Interrelated two-way clustering of gene expression data. Statistical Methodology 3, 93–102 (2006)

8. Cheng, Y., Church, G.M.: Biclustering of gene expression data. In: Proc. 8th Int. Conf. Intelligent Systems for Molecular Biology (ISMB 2000), San Diego, CA, pp. 93–103 (2000)

9. Dudoit, S., Fridlyand, J., Speed, T.: Comparison of discrimination methods for the classification of tumors using gene expression data. Journal of the American Statistical Association 97, 77–87 (2002)

10. Hartigan, J.A.H.: Direct clustering of a data matrix. Journal of the American Statistical Association 67(337), 123–129 (1972)

11. Ihmels, J., Bergmann, S., Barkai, N.: Defining transcription modules using large-scale gene expression data. Bioinformatics 20, 1993–2003 (2004)

12. Jackson, D.A.: Stopping rules in principal components analysis: a comparison of heuristical and statistical approaches. Ecology 74, 2204–2214 (1993)

13. Kluger, Y., Basri, R., Chang, J.T., Gerstein, M.: Spectral biclustering of microarraydata: co-clustering genes and conditions. Genome Research 13(4), 703–716 (2003)

14. Lazzeroni, L., Owen, A.: Plaid models for gene expression data. Statistica Sinica 12(1), 61–86 (2002)

15. Liu, B., Wan, C., Wang, L.: An efficient semi-unsupervised gene selection method via spectral biclustering. IEEE Transactions on Nanobioscience 5(2), 110–114 (2006)

16. Liu, L., Hawkins, D.M., Ghosh, S., Young, S.S.: Robust singular value decomposition analysis of microarray data. Proceedings of the National Academy of Sciences 100(23), 13167–13172 (2003)

17. Madeira, S.C., Oliveira, A.L.: Biclustering algorithms for biological data analysis: a survey. IEEE/ACM Transactions on Computational Biology and Bioinformatics 1(1), 24–45 (2004)

18. McLachlan, G., Do, K., Ambroise, C.: Analysing microarray gene expression data. Wiley, Chichester (2004)

19. Mirkin, B.: Mathematical Classification and Clustering. Kluwer Academic Press, Norwell (1996)

20. Pomeroy, S.L., Tamayo, P., et al.: Prediction of central nervous system embryonal tumour outcome based on gene expression. Nature 24(415), 436–442 (2002)

21. Tanay, A., Sharan, R., Shamir, R.: Discovering statistically significant biclusters in gene expression data. Bioinformatics 18, 36–44 (2002)

22. Yang, W.H., Dai, D.Q., Yan, H.: Generalized discriminant analysis for tumor classification with gene expression data. In Proceedings of the Fifth International Conference on Machine Learning and Cybernetics (ICMLC 2006), pp. 4322–4327, Dalian, China (2006)

On the Number of Partial Least Squares Components in Dimension Reduction for Tumor Classification

Xue-Qiang Zeng[1,2], Guo-Zheng Li[2], Geng-Feng Wu[2], and Hua-Xing Zou[1]

[1] Computer Center, Nanchang University, Nanchang 330006, China
{xqzeng,hxzou}@ncu.edu.cn
[2] School of Computer Engineering & Science, Shanghai University, Shanghai 200072, China
{gzli,gfwu}@shu.edu.cn

Abstract. Dimension reduction is important during the analysis of gene expression microarray data, because the high dimensionality of data sets hurts the generalization performance of classifiers. Partial Least Squares (PLS) based dimension reduction is a frequently used method, since it is specialized in handling high dimensional data set and leads to satisfying classification performance. This paper investigates the influence on generalization performance caused by the variation of the number of PLS components and the relationship between classification performance and regression quality of PLS on the training set. Experimental results show that the number of PLS components for classifiers can be automatically determined by regression quality of PLS latent variables.

1 Introduction

DNA microarray experiments are used to collect information from tissue and cell samples regarding gene expression differences for tumor diagnosis [1,2,3]. In general, the output of microarray experiments is summarized as an $n \times p$ data matrix, where n is the number of tissue or cell samples, p is the number of genes. Here, p is always much larger than n, which hurts generalization performance of most classification methods. To overcome this problem, we either select a small subset of interesting genes (gene selection) or construct K new components summarizing the original data as well as possible, with $K < p$ (dimension reduction, feature extraction).

Gene selection has been studied extensively in the last few years. The most commonly used procedures of gene selection are based on a score which is calculated for all genes individually and genes with the best scores are selected. Gene selection procedures output a list of relevant genes which may be experimentally analyzed by biologists. This method is often denoted as univariate gene selection, whose advantages are its simplicity and interpretability. However, much information contained in the data set is lost when genes are selected solely according to their individual capacity to separate the samples, since interactions and correlations between genes are omitted, as are of great interest in system biology.

Dimension reduction is an alternative to gene selection to overcome the problem of curse of dimensionality. Unlike gene selection, dimension reduction projects the whole data into a low dimensional space and constructs new dimensions (components) by analyzing the statistical relationship hidden in the data set. Researchers have developed

T. Washio et al. (Eds.): PAKDD 2007 Workshops, LNAI 4819, pp. 206–217, 2007.

different dimension reduction methods in applications of bioinformatics and computational biology [4,5,6], among which Partial Least Squares based Dimension Reduction (PLSDR) is one of the most effective methods [6].

Partial Least Squares (PLS) was firstly developed as an algorithm performing matrix decompositions, and then was introduced as a multivariate regression tool in the context of chemometrics [7,8]. Only in recent years, PLS has been found to be an effective dimension reduction technique for tumor discrimination [9,10].

Nguyen and Rocke proposed to use PLS for dimension reduction as a preliminary step for binary and multi-class classification [9,10]. A numerical simulated study on total predictor variance explained by PLS was also carried out by Nguyen *et al.* [5]. Experiments on microarray data proved that PLSDR is better than Principle Component Analysis (PCA) based dimension reduction. Barker and Rayens explained the relationship between PLS and Canonical Correlation Analysis (CCA) in a formal statistical manner [11], which clarified that PLS is superior to PCA when dimension reduction is needed. Boulesteix compared PLS with some of state-of-the-art classification methods and investigated some interesting properties of PLSDR [12]. Dai *et al.* provided a comparative study of three dimension reduction techniques: PLSDR, Sliced Inverse Regression (SIR) and PCA [6], which evaluated the predictive accuracy and computational efficiency of classification procedures incorporating those methods and suggested PLSDR is most effective among these methods. Zeng *et al.* introduced PLS into the field of text classification and suggested PLS is super to Latent Semantic Indexing (LSI) as a text representation method [13].

The size of the reduced space by PLSDR is critical to generalization performance of classifiers, especially the initial several components of PLSDR contain more information than the others, but it is hard to decide how much tail components are trivial for the aim of discrimination. Nguyen and Rocke fixed the number of components at three in their experiments and suggested the classification accuracy is insensitive to this parameter when it is beyond five [9,10]. Some authors used a more comprehensive way and determined the size of the reduced space by classification performance of cross-validation [6,12]. However, few has studied the influence on classification performance caused by the variation of the number of PLS components. Here, we will investigate the relationship between classification performance and regression quality of PLS components on the training set and propose to employ the PLS regression quality to automatically determine the number of PLS components for classification.

This paper is organized as follows. Some essential notions are given in section 2. In section 3, PLS is shortly introduced and then PLSDR is presented in detail. Data sets, experiment settings and evaluation methods are described in section 4. We show the results and discussions in section 5. Finally, conclusions are given in section 6.

2   Notions

Expression levels of p genes in n microarray samples are collected in an $n \times p$ data matrix $X = (\mathbf{x}_{ij}), 1 \le i \le n, 1 \le j \le p$; of which an entry $\mathbf{x}_{ij}$ is the expression level of the jth gene in the ith microarray sample.

Here we consider binary classification problems, the labels of the n microarray samples are collected in the vector **y**. When the ith sample belongs to class one, the element y_i is 1; otherwise it is -1.

Besides, $\| \bullet \|$ denotes the length of a vector. X^T represents the transpose of X, X^{-1} represents the inverse matrix of X.

Note that X and **y** used in the following are assumed to be centered to zero mean by each column.

3   Partial Least Squares Based Dimension Reduction

PLS is a class of techniques for modeling relations between blocks of observed variables by means of latent variables. It comprises of regression and classification tasks as well as dimension reduction techniques and modeling tools. The underlying assumption of PLS is that the observed data is generated by a system or process which is driven by a small number of latent (not directly observed or measured) variables. Therefore, PLS aims at finding uncorrelated linear transformations (latent components) of the original predictor variables which have high covariance with the response variables. Based on these latent components, PLS predicts response variables **y** and reconstruct original matrix X at the same time.

Let matrix $T = [\mathbf{t}_1, \ldots, \mathbf{t}_K] \in \mathbb{R}^{n \times K}$ represents the n observations of the K components which are usually denoted as latent variables or scores. The relationship between T and X is defined as:

$$T = XV \tag{1}$$

where $V = [\mathbf{v}_1, \ldots, \mathbf{v}_K] \in \mathbb{R}^{p \times K}$ is the matrix of projection weights. PLS determines the projection weights V by maximizing the covariance between the response and latent components.

Based on these latent components, X and **y** are decomposed as:

$$\begin{aligned} X &= TP^T + E \\ \mathbf{y} &= TQ^T + \mathbf{f} \end{aligned} \tag{2}$$

where $P = [\mathbf{p}_1, \ldots, \mathbf{p}_K] \in \mathbb{R}^{p \times K}$ and $Q = [\mathbf{q}_1, \ldots, \mathbf{q}_K] \in \mathbb{R}^{1 \times K}$ are denoted as loadings of X and **y** respectively. Generally, P and Q are computed by Ordinary Least Squares (OLS). E and **f** are residuals of X and **y** respectively.

By the decomposition of X and **y**, response values are decided by latent variables not by X (at least not directly). It is believed that this model would be more reliable than using OLS model on X directly, because the latent variables are coincided with the true underlying structure of original data.

The major point of PLS is the construction of components by projecting X on the weights V. The classical criterion of PLS is to sequentially maximizing the covariance between response **y** and latent components. There are some variants of PLS approaches to solve this problem [8]. Ignoring the miner differences among these algorithms, we demonstrate the most frequently used PLS approach: PLS1 [14].

PLS1 determines the first latent component $\mathbf{t}_1 = X\mathbf{w}_1$ by maximizing the covariance between **y** and $\mathbf{t}_1$ under the constraint of $\| \mathbf{w}_1 \| = 1$. The corresponding objective function is:

$$\mathbf{w}_1 = \arg\max_{\mathbf{w}^T\mathbf{w}=1}(Cov(X\mathbf{w}, \mathbf{y})) \tag{3}$$

The maximization problem of Equation (3) can be easily solved by the Lagrange multiplier method.

$$\mathbf{w}_1 = X^T\mathbf{y}/\parallel X^T\mathbf{y} \parallel \tag{4}$$

To extract other latent components sequentially, we need to model the residual information of X and $\mathbf{y}$ which couldn't be explained by previous latent variables. So, after the extraction of the score vector $\mathbf{t}_1$, PLS1 deflate matrices X and $\mathbf{y}$ by subtracting their rank-one approximations based on $\mathbf{t}_1$. The X and $\mathbf{y}$ matrices are deflated as:

$$
\begin{aligned}
E_1 &= X - \mathbf{t}_1\mathbf{p}_1^T \\
\mathbf{f}_1 &= \mathbf{y} - \mathbf{t}_1\mathbf{q}_1^T
\end{aligned}
\tag{5}
$$

where $\mathbf{p}_1$ and $\mathbf{q}_1$ are loadings determined by OLS fitting:

$$
\begin{aligned}
\mathbf{p}_1^T &= (\mathbf{t}_1^T\mathbf{t}_1)^{-1}\mathbf{t}_1^T X \\
\mathbf{q}_1^T &= (\mathbf{t}_1^T\mathbf{t}_1)^{-1}\mathbf{t}_1^T \mathbf{y}
\end{aligned}
\tag{6}
$$

As an iterative process, PLS1 constructs other latent components in turn by using the residuals as new X and $\mathbf{y}$. Each PLS iteration represents some information which could not explained by previous steps and produces one latent component. The number of components K is the only parameter of PLS which can be fixed by user or decided by a cross-validation scheme. In general, the maximal number of latent components is the rank of matrix X which have non-zero covariance with $\mathbf{y}$.

After projecting and representing each sample in the new space, PLS build models on some latent variables by using the OLS algorithm. When the number of latent components is the same as the rank of matrix X, all the information of X is preserved and PLS exhibits the same as OLS does on X. But, we needn't stick to using OLS in the transformed space. Other statistical learning may also be used in this space, such as Support Vector Machine (SVM), Logistic Discrimination (LD) *etc*. That is, we may just use PLS as a dimension reduction method which denoted as PLSDR instead of a classification/regression model [9], [10]. PLSDR performs dimension reduction by projecting X on the projection weights W or V. Since W are orthogonal project weights and shown in the experiments to be more effective than V for dimension reduction [15], so W instead of V are used in the following.

In PLS, with the increase of the number of latent components, the explained variances of X and $\mathbf{y}$ are extended, and all the information of original data are preserved when K reaches the rank of X. But the full rank is not necessary for classification in most cases. Similar to PCA, tail components are considered as uninformative ones, even noises, which should be removed in dimension reduction. However, noises are difficult to be evaluated, which heavily depend on concrete applications. The previous works set a fixed threshold to reduce noisy information, or determined it by cross-validation. Cross-validation determines sub-optimal threshold for classification, but it is of limited usage because of heavy computation on huge data sets and poor performance or strong bias on small sample data sets. Nguyen and Rocke suggested to fix the number of components at three [9,10], on which the performance convergence of certain classifiers like LD was confirmed. But the threshold of three is an arbitrarily value which is not suitable for all data sets by all classifiers. Therefore, the development of an adaptive method to find the optimal proportion of reduced information is demanded.

We consider linking the threshold with the PLS regression quality, since generalization performance is strongly correlated with the model goodness-of-fit on the training set. PLS is originally designed as a regression method, so, it is reasonable to use the quality of regression R_y^2, whose definition is given in subsection 4.3. At the same time, regression quality has the advantage of little computation and can be embedded into the PLS iteration naturally. Taking the risk of over-fitting into account, the dimension with the top R_y^2 may not be the most suitable one. We propose to use a parameter δ to control the trade-off between the training accuracy and model parsimony. The newly proposed model selection algorithm is described as follows.

1. Start with initial component list $\mathbf{r} = []$, a threshold δ, and initialize $E = X, \mathbf{f} = \mathbf{y}$.
2. Produce the PLS component l on E and $\mathbf{f}$, and compute its corresponding R_y^2.
3. $r = [r, l]$, and If $R_y^2 \geq 1 - \delta$ goto step 5.
4. Update E and $\mathbf{f}$ as residuals, goto step 2.
5. Output component list $\mathbf{r}$.

In the following, we perform experiments to investigate the relationship between the influence of the classification performance and the fitness quality of the training set, then test our newly proposed adaptive model selection algorithm.

4   Experiments

4.1   Date Sets

Four microarray data sets are used in our study which are briefly described as below.

Leukemia
The acute leukemia data set was published by Golub *et al.* [1], which consists of 72 bone marrow samples with 47 ALL and 25 AML. The gene expression intensities are obtained from Affymetrix high-density oligonucleotide microarrays containing probes for 7,129 genes.

Colon
Alon *et al.* used Affymetrix oligonucleotide arrays to monitor expressions of over 6,500 human genes with samples of 40 tumor and 22 normal colon tissues [2]. Expression of the 2,000 genes with the highest minimal intensity across the 62 tissues were used in the analysis.

Breast Cancer
Van't Veer *et al.* used DNA microarray analysis on primary breast tumours and applied supervised classification to identify the significant genes for the disease [16]. The data contains 97 patient samples, 46 of which are from patients who had developed distance metastases within 5 years (labeled as "relapse"), the rest 51 samples are from patients who remained healthy from the disease after their initial diagnosis for interval of at

least 5 years (labeled as "non-relapse"). The number of genes is 24,481 and the missing values of "NaN" are replaced with 100.

Lung Cancer

Gordon *et al.* proposed a data set for the purpose of classifying lung Cancer between malignant pleural mesothe-lioma (MPM) and adenocarcinoma (ADCA) of the lung [17]. There are 181 tissue samples (31 MPM and 150 ADCA). Each sample is described by 12,533 genes.

4.2 Experimental Settings

Each data set is randomly partitioned into a training data set $\mathcal{L}$ containing $n_{\mathcal{L}}$ observations and a test data set $\mathcal{T}$ containing the $n - n_{\mathcal{L}}$ remaining observations are generated. The class distribution of the training and test data set is the same as the original data set. We repeat the process of separation 100 times and do experiments on each partition. This scheme is widely used in the comparative studies of classification methods for microarray data [3]. It is more reliable than leave-one-out cross-validation [18]. We fix the partition ratio $n_{\mathcal{L}}/n$ at 0.5.

For each partition $\{\mathcal{L}, \mathcal{T}\}$, the gene expressions are transformed to have zero mean and standard deviation one across samples on $\mathcal{L}$. In the test set $\mathcal{T}$, data expressions are transformed according to the means and standard deviations of the corresponding training set $\mathcal{L}$. As no gene selection is performed, all genes of the original data set are used in our study.

To make conclusions more sound, we use five commonly used classifiers to predict the observations in $\mathcal{T}$, which are linear version of Support Vector Machine (SVM) with $C = 100$, Logistic Discrimination (LD), Ridge Regression (RR), k Nearest Neighbor (kNN) with $k = 1$ and Artificial Neural Network (ANN) with 10 units in one hidden layer. All the classification models have been applied with the same partitions and data preprocessing.

4.3 Evaluation Methods

Averaged on 100 random partitions, the mean classification success (accuracy) rate (SUC) is used to evaluate the performance of classifiers, whose definition is given by

$$SUC = \frac{1}{100} \sum_{j=1}^{100} \frac{1}{n_{\mathcal{T}j}} \sum_{i=1}^{n_{\mathcal{T}j}} I(\hat{y}_i = y_i) \tag{7}$$

where I is the standard indicator function ($I(A) = 1$ if A is true, $I(A) = 0$ otherwise).

In order to evaluate the information preserved in X and $\mathbf{y}$, the measures R_X^2 and R_y^2 are used, which are similar to the standard regression measure R^2.

$$R_X^2 = 1 - \frac{\sum_{i=1}^{n} \sum_{j=1}^{p} (E_{ij})^2}{\sum_{i=1}^{n} \sum_{j=1}^{p} (X_{ij})^2}$$

$$R_y^2 = 1 - \frac{\sum_{i=1}^{n} (f_i)^2}{\sum_{i=1}^{n} (y_i)^2} \tag{8}$$

where E and $\mathbf{f}$ are the residuals described above. As the number of components used in PLS increases, the square sum of residual variances is decreasing, which represents the information has not been modeled by PLS yet. R_X^2 and R_y^2 stand for the reconstruction quality of X and the regression quality of $\mathbf{y}$ respectively, and measure the proportion of corresponding variances that have been explained.

5   Results and Discussions

We perform two series of experiments here. Firstly, we try to examine how the classification performance varies as the dimension of latent components increases. Secondly, we compared our adaptive model selection algorithm with the fixed-dimension method, and some encouraging results are reported.

5.1   The Influence of Dimensional Size

In order to examine how the classification performance varies as the dimension of latent components increases, we vary the dimension of the reduced space from 1 to 30. At each dimension, we predict the observations in $\mathcal{T}$ using five classical classifiers and record the corresponding R_X^2 and R_y^2. The results on four microarray data sets are showed in Fig. 1~4, which are averaged on 100 random partitions.

It seems that different behaviors are spotted on different classifiers. For the convenience of illustration, we split the SUC lines into two stages as follows.

1. In the first stage of informative components, the most important components of PLS are used by classifiers. Classification performances are improved with the addition of initial informative components in most cases. But the top SUC scores of classifiers are not definitively obtained at this time. The optimal dimension heavily depends on classifiers.

   Furthermore, there are some interesting examples which suggest informative components are not exactly the most discriminative ones. For instance, the SUC scores of all classifiers drop dramatically with the addition of the third components on Breast Cancer data set showed in Figure 3. The SUC scores of SVM, LD and RR increase again when more components are used, and the best scores are obtained when the dimensional sizes reach eight.

2. In the second stage of redundant components, components with little information are constructed by the residuals. By analyzing the exhibited behaviors, we categorize classifiers into two groups, namely sensitive classifiers and insensitive classifiers.

   The performances of sensitive classifiers, such as kNN and ANN, are in the descending order obviously with the increase of the dimensional size. The uninformative components prohibit the effectiveness of these classifiers. On the contrary, insensitive classifiers, such as SVM, LD and RR, show stable performances with the additional tail components, which suggests that tail components are redundant rather than noisy, since they hardly hurt performance of insensitive classifiers. Furthermore, these redundant components may be helpful in certain cases, i.e. the performance of LD on the Leukemia data set is slightly improved by the additional tail components.

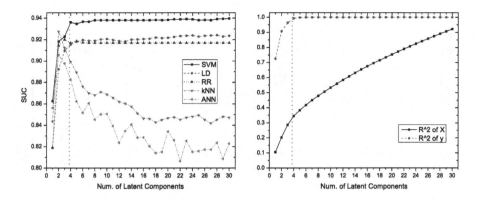

Fig. 1. Statistical results by using five classifiers on the Leukemia data set (training set with 36 samples and 7,129 genes, test set with the same size)

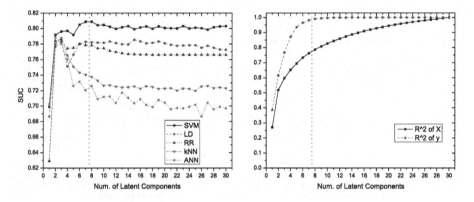

Fig. 2. Statistical results by using five classifiers on the Colon data set (training set with 31 samples and 2,000 genes, test set with the same size)

The exact division point of the above two stages is difficult to be determined, the dot lines plotted in the figures for demonstration have further meaning which will be explained in next subsection.

The results of R_X^2 and R_y^2 coincide with our suppositions, that as the number of latent components increases, the regression and reconstruction qualities on training sets are both promoted. When the dimensional size reach the rank of X, all the information of original data set will be preserved. But the increasing rate is different between R_X^2 and R_y^2. Since only one column is included, the variances of $\mathbf{y}$ are easy to represent with few components, which suggest the further usage of R_y^2 as a model selection criterion of the number PLS components. The reconstruction quality of X is much more comprehensive, since it highly depends on the scale of the training set, the incremental curve of R_X^2 varies with different data sets.

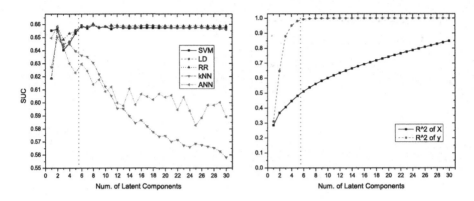

Fig. 3. Statistical results by using five classifiers on the Breast Cancer data set (training set with 49 samples and 24,481 genes, test set with the same size)

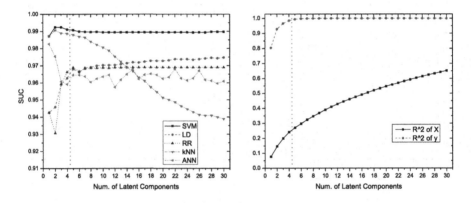

Fig. 4. Statistical results by using five classifiers on the Lung Cancer data set (training set with 91 samples and 12,533 genes, test set with the same size)

5.2   Model Selection by R_y^2

It is clear that there is not a unified optimal dimensional size for all classifiers on each data sets. The best number of latent components for sensitive classifiers hardly exceeds three. On the contrary, insensitive classifiers often obtain the top SUC scores at the stage of redundant components, and the best point depends on the data set heavily. The absolute optimal dimensional size makes sense only when a certain classifier is considered. In the following, we concentrate on the classifier SVM which has outstanding classification performance.

We set δ at 0.015 for the SVM classifier, and select the dimensional size by the R_y^2 criterion on the data sets. We compare our method with the fixed-dimension method which fixes the dimension at three. We also find the optimal dimension for SVM on each data sets by calculating all the SUC values on each dimension. The comparative results are listed in Table 1, where Dim.±std are statistical mean values with its standard

Table 1. Comparative results between the proposed algorithm and the fixed-dimension method for SVM

Data Sets	Proposed		Fixed Dim. at 3		Optimal Dim.	
	Dim.±std	SUC±std	Dim.	SUC±std	Dim.	SUC±std
Leukemia	3.93±0.26	0.936±0.05	3	0.923±0.05	29	0.939±0.05
Colon	7.66±0.69	0.803±0.06	3	0.796±0.06	7	0.809±0.06
Breast Cancer	5.54±0.78	0.658±0.06	3	0.640±0.06	8	0.660±0.06
Lung	4.45±0.50	0.991±0.01	3	0.992±0.01	2	0.992±0.01
Average	5.40±0.56	0.847±0.04	3	0.838±0.04	11.50	0.850±0.04

deviation, and SUC±std have the same meaning, since the experiments are performed 100 times.

From the Table1, it can be seen that our criterion is better than the fixed dimension at three, and much close to the performance of the optimal dimension. Besides, from the point of model complexity, the parsimony space determined by the R_y^2 criterion is favorite than that by the optimal dimension method.

We also plot the obtained dimensional sizes by the R_y^2 criterion as dot lines in Figure 1~4. It is interesting to find that the selected dimensions are nearly the beginning points of the stages of redundant components described above. Although these dimensions are not the global optimal points for SVM, the R_y^2 criterion produce satisfying spaces which have both sub-optimal performance and parsimony model complexity.

The only parameter δ controls the trade-off between fitness of the training set and model parsimony, which does not directly depend on any classifier. The effect of δ at 0.015 for SVM has been examined on four data sets in our study. Due to the paper size, we did not try δ for all the classifiers, we think $\delta = 0.015$ is suitable not only for SVM, but also for LD and RR, but it is not suitable for sensitive classifiers like kNN and ANN, which need more parsimony thresholds. We suggest to choose different fixed δ for different classifiers and apply them on all data sets, which need more investigation, and will be done in an extended work.

6   Conclusions

This work investigates the influence on classification performance of classifiers caused by the variation of dimensional size of Partial Least Squares based Dimension Reduction (PLSDR). Different behaviors of five usually used classifiers are spotted on four microarray data sets. Some classifiers are sensitive to the additional tail components, while others not. For the converged performance of some classifiers, we suggest the tail components are more like redundant ones than noises.

By analyzing the variation of fitness qualities of PLS components on the training set, we propose an adaptive threshold for the determination of dimensional size of PLSDR by using the regression quality R_y^2, which produces satisfying dimensions with high performance on classifiers and the property of parsimony.

Acknowledgment

This work was supported in part by the Nature Science Foundation of China under grant no. 20503015, Nature Science Project of Shanghai Municipal Education Committee under grant no. 05AZ67, open funding by Institute of Systems Biology of Shanghai University and Scientific Research Fund of Jiangxi Provincial Education Departments under grant no. 2007-57.

References

1. Golub, T., Slonim, D., Tamayo, P., Huard, C., Gaasenbeek, M., Mesirov, J., Coller, H., Loh, M., Downing, J., Caligiuri, M., Bloomfield, C., Lander, E.: Molecular classification of cancer: Class discovery and class prediction by gene expression. Bioinformatics & Computational Biology 286(5439), 531–537 (1999)
2. Alon, U., Barkai, N., Notterman, D., Gish, K., Ybarra, S., Mack, D., Levine, A.: Broad patterns of gene expression revealed by clustering analysis of tumor and normal colon tissues probed by oligonucleotide arrays. In: Proceedings of the National Academy of Sciences of the United States of America, pp. 6745–6750 (1999)
3. Dudoit, S., Fridlyand, J., Speed, T.: Comparison of discrimination methods for the classification of tumors using gene expression data. Journal of the American Statistical Association 97(457), 77–87 (2002)
4. Antoniadis, A., Lambert-Lacroix, S., Leblanc, F.: Effective dimension reduction methods for tumor classification using gene expression data. Bioinformatics 19(5), 563–570 (2003)
5. Nguyen, D., David, D.M., Rocke, M.: On partial least squares dimension reduction for microarray-based classification: a simulation study. Computational Statistics & Data Analysis 46(3), 407–425 (2004)
6. Dai, J., Lieu, L., Rocke, D.: Dimension reduction for classification with gene expression data. Statistical Applications in Genetics and Molecular Biology 5(1) (2006) Article 6
7. Wold, S., Ruhe, A., Wold, H., Dunn, W.: Collinearity problem in linear regression. the partial least squares (pls) approach to generalized inverses. SIAM Journal of Scientific and Statistical Computations 5(3), 735–743 (1984)
8. Wold, S., Sjostrom, M., Eriksson, L.: Pls-regression: a basic tool of chemometrics. Chemometrics and Intelligent Laboratory Systems 58(22), 109–130 (2001)
9. Nguyen, D., Rocke, D.: Tumor classification by partial least squares using microarray gene expression data. Bioinformatics 18(1), 39–50 (2002)
10. Nguyen, D., Rocke, D.: Multi-class cancer classification via partial least squares with gene expression profiles. Bioinformatics 18(9), 1216–1226 (2002)
11. Barker, M., Rayens, W.: Partial least squares for discrimination. Journal of Chemometrics 17(3), 166–173 (2003)
12. Boulesteix, A.L.: Pls dimension reduction for classification of microarray data. Statistical Applications in Genetics and Molecular Biology 3(1) (2004)
13. Zeng, X.Q., Wang, M.W., Nie, J.Y.: Text classification based on partial least square analysis. In: 22nd Annual ACM Symposium on Applied Computing, Special Track on Information Access and Retrieval (in press, 2007)
14. Helland, I.: On the structure of partial least squares regression. Communications in statistics. Simulation and computation 17(22), 581–607 (1988)
15. Zeng, X.Q., Li, G.Z.: Orthogonal projection weights in dimension reduction based on partial least squares. Journal of Computational Intelligence in Bioinformatics 1 (in press, 2007)

16. Van't Veer, L., Dai, H., Vijver, M., He, Y., Hart, A., Mao, M., Peterse, H., Kooy, K., Marton, M., Witteveen, A., Schreiber, G., Kerkhoven, R., Roberts, C., Linsley, P., Bernards, R., Friend, S.: Gene expression profiling predicts clinical outcome of breast cancer. Nature 415(6871), 530–536 (2002)
17. Gordon, G., Jensen, R., Hsiao, L.L., Gullans, S., Blumenstock, J., Ramaswamy, S., Richards, W., Sugarbaker, D., Bueno, R.: Translation of microarray data into clinically relevant cancer diagnostic tests using gene expression ratios in lung cancer and mesothelioma. Cancer Research 62(17), 4963–4967 (2002)
18. Ambroise, C., McLachlan, G.: Selection bias in gene extraction on the basis of microarray gene-expression data. In: Fienberg, S. (ed.) Proceedings of the National Academy of Sciences, pp. 6562–6566 (2002)

Mining Biosignal Data: Coronary Artery Disease Diagnosis Using Linear and Nonlinear Features of HRV*

Heon Gyu Lee[1], Ki Yong Noh[2], and Keun Ho Ryu[1,**]

[1] Database/Bioinformatics Laboratory, School of Electrical & Computer Engineering,
Chungbuk National University, Cheongju, 361-763, Korea
{hglee,khryu}@dblab.chungbuk.ac.kr
[2] Korea Research Institutes of Standards and Science, Korea
kyno@kriss.re.kr

Abstract. The main purpose of our study is to propose a novel methodology to develop the multi-parametric feature including linear and nonlinear features of HRV (Heart Rate Variability) diagnosing cardiovascular disease. To develop the multi-parametric feature of HRV, we used the statistical and classification techniques. This study analyzes the linear and the non-linear properties of HRV for three recumbent positions, namely the supine, left lateral and right lateral position. Interaction effect between recumbent positions and groups (normal and patients) was observed based on the HRV indices and the extracted HRV indices used to classify the CAD (Coronary Artery Disease) group from the normal people. We have carried out various experiments on linear and non-linear features of HRV indices to evaluate several classifiers, e.g., Bayesian classifiers, *CMAR*, *C4.5* and *SVM*. In our experiments, *SVM* outperformed the other classifiers.

1 Introduction

Mortality of domestic people from cardiovascular disease ranked second, which followed that of from cancer last year. Therefore, it is very important and urgent to enhance the reliability of medical examination and treatment for cardiovascular disease.

HRV (Heart Rate Variability) analysis has been used extensively to assess autonomic control of the heart under various physiological and pathological conditions, and used as a clinical tool to diagnose cardiac autonomic function [1], [2]. Various measures and explanations have been used to analyze the HRV. For example, simple linear time domain analysis, such as mean, standard deviation, and root mean square of successive RR interval differences have been widely employed in quantification of the overall variability of the Heart Rate (HR) [2]. Frequency domain variable provide

---

* This work was supported by the Korea Research Foundation Grant funded by the Korean Government (The Regional Research Universities Program/Chungbuk BIT Research-Oriented University Consortium).
** Corresponding author.

T. Washio et al. (Eds.): PAKDD 2007 Workshops, LNAI 4819, pp. 218–228, 2007.

markers of the cardiac autonomic regulation, i.e. the sympatho-vagal balance [3], [4]. In addition, there are several nonlinear features. The nonlinear interaction between the various regulatory systems of the heart rate gives rise to clinically useful concepts of variability and regularity. The complexity of the human physiological system, which is reduced in bad health but increased in good health [5], can be analyzed quantitatively by various nonlinear methods. Nonlinear analyses include the Princare plots and complexity estimation. Therefore, we consider it worth-while investigating the linear and nonlinear properties of HRV in patients with CAD, and we evaluate each measured properties.

In this paper, our aim is to propose a quantitative measure for HRV and a suitable prediction model to enhance the reliability of medical examination for cardiovascular disease. To achieve this aim, the proposed method works in two steps as follows. We first analyze several time and frequency domain measures of HRV as linear measures. As nonlinear measures, Princare plots and complexity estimation are used. And then we apply statistical method to select interesting features and use classification techniques to predict the patients with CAD. For the prediction step, we evaluated several supervised learning methods. We tested SVM, associative classifier (CMAR), decision tree induction (C4.5) and Bayesian classifiers and validated their accuracy for diagnosing of cardiovascular disease. Experiments also show that with proper classification methods, the results of diagnosis can be improved.

The rest of the paper is organized as follows. Section 2 describes the process of extracting linear and nonlinear. Section 3 describes statistical analysis to select suitable features. Section 4 describes our multi-parametric HRV data classification methods. Section 5 presents experimental results and Section 6 presents concluding remarks.

2   Linear and Nonlinear Features Extraction

In this section, we describe the process of extracting linear and nonlinear features by HRV analysis from raw biosignal.

2.1   ECG Biosignal Processing

The ECG signals are recorded at Einthoven's Lead- II channel (right arm electrode: negative pole, left leg electrode: positive pole, right leg electrode: ground) with electrocardiograph during 5 minutes for each 3 recumbent postures with random order; supine, right lateral and left lateral positions (see Fig. 1). In this electrocardiograph, the measured analog signal is converted to a digital signal with a sampling frequency of 500 Hz. We extract the R-peaks from the ECG recordings based on Thomkin's algorithm [6]. RR interval data is analyzed during a 5-min baseline period and all RR intervals are edited in order to exclude all ectopic beats or artifacts.

RR intervals time series are re-sampled at a rate of 4 Hz to obtain power spectral density. Fig. 2 shows the RR intervals of ECG biosignal and time series of RR intervals.

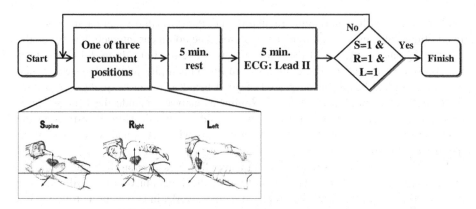

Fig. 1. Measuring ECG biosignal for each position

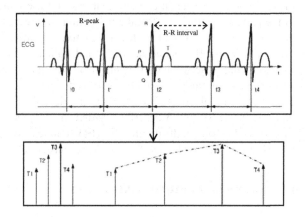

Fig. 2. Time series of RR intervals for ECG biosignal

2.2 Linear Feature

Linear features of HRV can be divided into frequency domain and time domain [7], [8]. In the following these features are defined and discussed briefly.

Frequency domain: After calculating the mean heart rate (beat/min) from the ECG signal, we used *Fast Fourier Transformation (FFT)* to obtain the power spectrum of the RR intervals. We then define the various areas of spectral peaks as follows: The *Total power (TP)*, 0 Hz to 0.4 Hz; *Very Low Frequency (VLF) power*, 0 Hz~0.04 Hz; *Low Frequency (LF) power*, 0.04 Hz to 0.15 Hz; and *High Frequency (HF) power*, 0.15 Hz to 0.4 Hz.

The *TP*, which is a useful index for detecting abnormal autonomic activity, is larger in normal subjects than in patients [9]. Moreover, the *LF* power mainly provides a measure of sympathetic activity with some influence from the parasympathetic nervous system, whereas the *HF* power is responsible solely to the parasympathetic nervous system. We use the normalized *LF (nLF)* as an index of sympathetic modulation, the normalized *HF* (*nHF*) as an index of vagal modulation and the *LF* to *HF* ratio (*LF/HF*) as an index of sympathovagal balance. Above spectral values (*nLF* and *nHF*) are defined as follows and presented in normalized units (nu).

$$nLF = \frac{(TP - VLF)}{LF} \times 100 \tag{1}$$

$$nHF = \frac{(TP - VLF)}{HF} \times 100 \tag{2}$$

Time domain: In the present study, several time-domain measures of HRV were selected. In a continuous ECG record, each QRS complex is detected, and the so-called normal-to-normal intervals (that is all intervals between adjacent QRS complexes resulting from sinus node depolarizations, normal RR intervals), or the instantaneous heart rate is determined. Simple time-domain variables that can be calculated include the mean RR interval (*RRm*), the standard deviation of all RR intervals (*SDRR*), and the standard deviation of differences between adjacent RR intervals (*SDSD*).

2.3 Nonlinear Feature

In this section, we analyze the HRV (RR intervals) by means of nonlinear methods; Proincare plot and approximate entropy (*ApEn*).

Poincare Plot: the Poincare Plot (*PP*) is a scattergram, which is constructed by plotting each RR interval against the previous one. The *PP* may be analyzed quantitatively by fitting an ellipse to the plotted shape [10], [11]. The center of the ellipse is determined by average RR interval. *SD1* means the standard deviation of the distances of points from $y = x$ axis, *SD2* means the standard deviation of the distances of points from $y = -x + \overline{RR}$ axis, where $\overline{RR}$ is the average RR interval [12]. The parameters *SD2/SD1*, and *SD1·SD2* describing the relationship between *SD1* and *SD2* are also computed in our study.

Approximate Entropy (ApEn): *ApEn* quantifies the regularity of time series, so is also called a 'regularity statistic.' It is represented as a simple index for the overall complexity and predictability of each time series. In our study, *ApEn* quantifies the regularity of the RR interval. The more regular and predictable the RR interval series, the lower will be the value of *ApEn* [13].

First of all, we reconstructed the RR interval time series in the *n*-dimensional space using Takens theorem [14]. Takens proposed the following time delay method for the reconstruction of the state space:

$$D_t(n) = \left[RR(t), RR(t + \tau), ..., RR(t + (n-1)\tau) \right], \tag{3}$$

where *n* is the embedding dimension and τ is the time delay. In this paper, the optimal value of τ is 10. The mean of the fraction of patterns with length *m* that resemble the pattern with the same length beginnings at interval *i* is defined by the following equation.

$$\Phi^m(r) = \frac{1}{N-m+1} \sum_{i=1}^{N-m+1} \ln\left[\frac{number\ of\ |D_m(j) - D_m(i)| < r}{N-m-1}\right], \quad j = 1 \sim N-m,\ j \neq i \qquad (4)$$

In the above equation, $D_m(i)$ and $D_m(j)$ are state vectors in the embedding dimension, m. Given N data points, we can define *ApEn* as,

$$ApEn(m, r, N) = \Phi^m(r) - \Phi^{m+1}(r), \qquad (5)$$

where *ApEn* estimates the logarithmic likelihood that the next intervals after each of the patterns will differ. Generally, the embedding dimension, m, and the tolerance, r are fixed at $m=2$ and $r=0.2*SD$ in physiological time series data [15].

All variables used in our study are summarized in Table 1 and Fig. 3 shows the example of feature extraction process from ECG biosignal.

Table 1. Linear and nonlinear features of HRV

	Feature	Description
Linear	nLF	Normalized low frequency power
	nHF	Normalized high frequency power
	LF/HF	The ratio of low- and high-frequency power
	RRm	The mean of RR intervals
	SDRR	Standard deviation of all RR intervals
	SDSD	Standard deviation of differences between adjacent RR intervals
Nonlinear	SD1	Standard deviation of the distance of RR(i) from the line $y = x$ in the Poincare plot
	SD2	Standard deviation of the distance of RR(i) from the line $y = -x + 2RR_m$ in the Poincare plot
	SD2/SD1	The ratio of SD2 and SD1
	SD1SD2	SD1 x SD2
	ApEn	Approximate Entropy

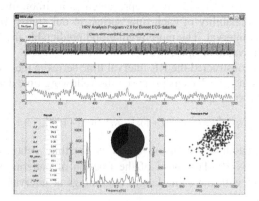

Fig. 3. Development of HRV analysis program

3 Statistical Analysis for Feature Selection

This section describes the feature selection phase. After linear and nonlinear features are extracted, we perform the statistical analysis for selecting the essential features.

For all our analyses, we represent the data in the form of a mean. The significance of differences in the response to different recumbent positions between the normal and patients (CAD group) was determined by ANOVA with repeated measures. Significant effects are followed up by paired t-test to compare normal and patients groups for supine, right and left lateral position separately. Any p-value less than 0.05 are accepted as significant. In patient with CAD group compared with control group, for the right lateral position the *nLF* and *ApEn* are lower moderately (p-value<0.1), *LF/HF, SDRR, SDSD and SD1SD2* features are higher significantly (p-value<0.05), for the left lateral position the *nLF, nHF* and *SDRR* are higher significantly (p-value<0.05), whereas *ApEn* is lower significantly (p-value<0.05). In the normal group, the *SD2/SD1* in the right lateral position is lower than that of the left lateral's significantly (p-value<0.05). In the CAD group, *LF/HF* and *nLF* in the right lateral position are lower than that of the left lateral's significantly (p-value<0.05) and lower than that of the supine moderately (p-value<0.1), whereas *nHF* in the right lateral is higher than that of the left lateral's significantly (p-value<0.05) and higher than that of the supine moderately (p-value<0.1).

From the results of statistical analysis, we select the essential features satisfying values less than 0.05. Eventually, the features selected are as following table.

Table 2. Essential features of HRV

	Features
Linear	*nLF, nHF, LF/HF, SDRR, SDSD*
Nonlinear	*SD2/SD1, SD1SD2, ApEn*

4 Prediction Models for Diagnosing Coronary Artery Disease

In this section we describe the process of preprocessing features and then we introduce several state-of-art classifiers in order to predict coronary artery disease.

4.1 Preprocessing Step

Since the extracted linear and nonlinear features contain continuous variables, those variables also must be made discrete. Therefore, entropy-based discretization has been used because the intervals are selected according to the information they contribute target variable. Due to the decision tree's discretization [16], all continuous contributed variables are cut up into a number of intervals.

Let T partition the set D of examples into the subsets D_1 and D_2. Let there be k classes $C_1,...,C_k$. Let $P(C_i, D_j)$ be the proportion of examples in D_j that have class C_i. The class entropy of a subset $D_j, j=1, 2$ is defined as:

$$Ent(D_j) = -\sum_{i=1}^{k} P(C_i, D_j) \log(P(C_i, D_j)) \qquad (7)$$

Suppose the subsets D_1 and D_2 are induced by partitioning a feature A at point T. Then, the class information entropy of the partition, denoted $E(A, T; D)$, is given by:

$$E(A,T;D) = \frac{|D_1|}{D} Ent(D_1) + \frac{|D_2|}{D} Ent(D_2) \qquad (8)$$

A binary discretization for A is determined by selecting the cut point T_A for which $E(A, T; D)$ is minimal amongst all the candidate cut point. The same process can be applied recursively to D_1 and D_2 until some stopping criteria is reached.

The Minimal Description Length Principle is used to stop partitioning. Recursive partitioning within a set of values D stop if

$$Gain(A,T;D) < \frac{\log_2(N-1)}{N} + \frac{\delta(A,T;D)}{N}, \qquad (9)$$

where N is the number of values in the set D, $Gain(A, T; D) = Ent(D) - E(A, T; D)$, $\delta(A,T;D) = \log_2(3^k - 2) - [k \cdot Ent(D) - k_1 \cdot Ent(D_1) - k_2 \cdot Ent(S_2)]$, and k_i is the number of class labels represented in the set D_i.

4.2 Supervised Methods

SVM (Support Vector Machine): This method is one of the most popular supervised classification methods. *SVM* is basically two-class classifier and can be extended for the multi-class classification (e.g., combining multiple on-versus-the-rest two-class classifiers). In our model each object is mapped to a point in a high dimensional space, each dimension of which corresponds to features. The coordinates of the point are the frequencies of the features in the corresponding dimensions. *SVM* learns, in the training step, the maximum-margin hyper-planes separating each class. In testing step, it classifies a new object by mapping it to a point in the same high-dimensional space divided by the hyper-plane learned in the training step. For experiments, we used the *Radial Basis Function (RBF)* kernel [17].

CMAR (Classification based on Multiple Association Rules): CMAR is a *Classification Association Rule Mining (CARM)* algorithm developed by [18]. CMAR operates using a two stage approach to generating a classifier:

(1) Generating the complete set of CARs according to a user supplied: Support threshold to determine frequent item sets, and confidence threshold to confirm *CRs*.

(2) Prune this set to produce a classifier.

The two stage approach is a fairly common approach used by many CARM algorithms, for example the *CBA (Classification Based on Associations)* algorithm.

The *CMAR* algorithm uses an FP-growth algorithm to produce a set of *CARs* which are stored in data structure referred to as a *CR tree*. CARs are inserted into the *CR tree* if the Chi-Squared value is above a user specified critical threshold (5% significance level; assuming a degree of freedom equivalent to 1, this will equate to a threshold of 3.8415), and the *CR tree* does not contain a more general rule with a higher priority.

NB (Naïve Bayesian classifier): Bayesian classifiers are statistical classifiers and have exhibited high accuracy and speed when applied to a large database. This approach chooses the highest posterior probability class using the prior probability computed from the training data set. *NB* assumes that the effect of an attribute on a given class is independent of the values of the other attributes. This assumption is called class conditional independence. However, attribute values of HRV data may not be entirely independent from each others. In order to address this problem, in this study, we consider a set of extended Bayesian classifier known to work well with correlated data, including *TAN (Tree Augmented Naïve Bayes)* and *STAN (Selective Tree Augmented Naïve Bayes)* [19], [20].

C4.5 (Decision Tree): *C4.5* is a decision tree generating algorithm, based on the *ID3* algorithm [21]. It contains several improvements, especially needed for software implementation. Improvements contain: (1) Choosing an appropriate attribute selection measure. (2) Handling training data with missing attribute values. (3) Handling attributes with differing costs. (4) Handling continuous attributes.

The parameters of the above methods are set as follows. For *SVM*, the soft margin allows errors during training. We set 0.1 for the two-norm soft margin value. All *NB* and *C4.5* parameters are default values. We test both *C4.5* tree method and rule method. For *CMAR*, we set support threshold to 0.05 and confidence threshold to 0.8 and disable the limit on number of rules. Other parameters remain default.

5 Experimental Results

In this section, we evaluate our experiments in building CAD detection model. All the experiments were performed on a 2.4GHz PentiumVI PC with 512MB main memory, running MS Windows server 2000.

Coronary arteriography is performed in patients with angina pectoris, unstable angina, previous myocardial infarction, or other evidence of myocardial ischemia. Patients with stenosis of the luminal narrowing greater then 0.5 were recruited as the CAD group, the others were classified as the normal. By using angiography, 99 patients with abnormal (CAD group) and 94 patients with normal coronary arteries

Table 3. Description of summary results

Classifier	TP	FP	Precision	Recall	Class
SVM	0.909	0.096	0.909	0.909	CAD
	0.904	0.091	0.904	0.904	Normal
CMAR	0.838	0.277	0.761	0.838	CAD
	0.723	0.162	0.81	0.723	Normal
C4.5	0.838	0.277	0.761	0.838	CAD
	0.723	0.162	0.81	0.723	Normal
NB (TAN)	0.818	0.234	0.786	0.818	CAD
	0.766	0.182	0.8	0.766	Normal
NB (STAN)	0.768	0.16	0.835	0.768	CAD
	0.84	0.232	0.775	0.84	Normal

Table 4. The result of attributes ranking

Posture	Attribute Evaluator	Ranked Attributes
Three positions	ChiSquared	*SD2/SD1, nLF, nHF, ApEn, LF/HF, SDSD, SD1SD2, SDRR*
	InfoGain	*SD2/SD1, nLF, nHF, ApEn, LF/HF, SDSD, SDRR, SD1SD2*
Supine	ChiSquared	*nHF, nLF, ApEn, SDSD, SD1SD2, LF/HF, SD2/SD1, SDRR*
	InfoGain	*nHF, nLF, ApEn, SDSD, SD1SD2, LF/HF, SD2/SD1, SDRR*
Right	ChiSquared	*ApEn, SD1SD2, LF/HF, nHF, nLF, SD2/SD1, SDSD, SDRR*
	InfoGain	*ApEn, SD1SD2, LF/HF, nHF, nLF, SD2/SD1, SDSD, SDRR*
Left	ChiSquared	*SDRR, ApEn, nHF, nLF, SD2/SD1 LF/HF, SD1SD2, SDSD*
	InfoGain	*ApEn, SDRR, nHF, nLF, SD2/SD1, LF/HF, SD1SD2, SDSD*

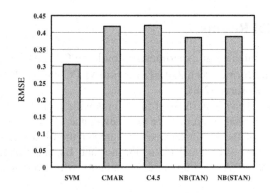

Fig. 4. Comparison of classifier error rate

(normal group) were studied. The accuracy was obtained by using the methodology of stratified *10-fold cross-validation (CV-10)*. We used *TP (True Positive)*, *FP (False Positive)*, *Precision* and *Recall* to evaluate the classifiers performance. The result is shown Table 3 and Fig. 4 shows classifier error rate (*RMSE*: Root Mean Squared Error). As expected, SVM and NB classifiers (TAN and STAN) were more accurate than the other classifiers. Especially, SVM showed the best performance among the tested methods.

In order to compare all features of HRV in terms of their predictability, we performed attributes ranking using Weka tool [22]. The results of attributes ranking for three recumbent positions and for each recumbent position are shown Table 4.

We expected that the use of the features for three recumbent positions (8 features in Table 1) will produce the more effective tool than the use of the features for each recumbent position. To address this expectation, we presented the accuracy of classifiers results for three recumbent positions and for each recumbent position in Fig. 5 and Fig. 6, respectively.

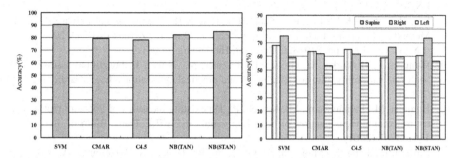

Fig. 5. Accuracy of classifiers for three recumbent position

Fig. 6. Accuracy of classifiers for each recumbent position

6 Conclusion

Most of parameters employ in diagnosing disease have both strong and weak points together. Therefore, it is important to develop multi-parametric indices diagnosing cardiovascular disease in order to enhance the reliability of the diagnosis.

In this paper, we proposed multi-parametric features of HRV from ECG biosignal. The HRV indices proposed in this paper can be used to classify the patients with CAD from normal people. For classification, we employed the proposed multi-parametric features, which allow us to choose a classifier from a large pool of well studied classification methods. We considered several supervised methods including extended Naïve Bayesian classifiers (*TAN, STAN*), decision tree (*C4.5*), associative classifier (*CMAR*) and *SVM*. In experimental results, *SVM* showed the best performance among the tested methods.

References

1. Guzzetti, S., Magatelli, R., Borroni, E., Mezzetti, S.: Heart rate variability in chronic heart failure American Neuroscience. Basic and Clinical 90, 102–105 (2001)
2. Task Force of the European Society of Cardiology and the North American Society of Pacing and Electrophysiology. Guidelines of Heart rate variability: Standards of measurement, physiological interpretation, and clinical use. Eur. Heart J. 17, 354–381 (1996)
3. Lombardi, F., Malliani, A., Pagani, M., Cerutti, S.: Heart rate variability and its sympatho-vagal modulation. Cardiovasc. Res. 32(2), 208–216 (1996)
4. Mezzacappa, E., Kindlon, D., Earls, F., Saul, J.P.: The utility of spectral analytic techniques in the study of the autonomic regulation of beat-to-beat heart-rate-variability. Int. J. Methods Psychiatr. Res. 4, 29–44 (1994)

5. Wagner, C.D., Persson, P.B.: Chaos in blood pressure control. Cardio. Res. 380–387 (1996)
6. Tompkins, W.J.: Bimedical digital signal processing, p. 07458. Prentice Hall PTR, Upper Saddle River, New Jersey (1995)
7. Noh, K.Y., Lee, H.G., Lee, B.J., Shon, H.S., Ryu, K.H.: Associative Classification Approach for Diagnosing Cardiovascular Disease. In: Int. Conf. on Intelligent Computing. LNCIS, vol. 345, pp. 721–727 (2006)
8. Lee, H.G., Noh, K.Y., Lee, B.J., Shon, H.S., Ryu, K.H.: Cardiovascular Disease Diagnosis Method by Emerging Patterns. In: Li, X., Zaïane, O.R., Li, Z. (eds.) ADMA 2006. LNCS (LNAI), vol. 4093, pp. 819–826. Springer, Heidelberg (2006)
9. Pumprla, J., Howorka, K., Groves, D., Chester, M., Nolan, J.: Functional assessment of heart rate variability: physiological basis and practical applications. Int. J. Cardio. 84, 1–14 (2002)
10. Brennan, M., Palaniswami, M., Kamen, P.: Do existing measures of Poincaré plot geometry reflect nonlinear features of heart rate variability? IEEE Trans. Biomed. Eng. 48(11), 1342–1347 (2001)
11. Moraru, L., Tong, S., Malhotra, A., Geocadin, R., Thakor, N., Bezerianos, A.: Investigation of the effects of ischemic preconditioning on the HRV response to transient global ischemia using linear and nonlinear methods. Med. Eng. & Physics 27, 465–473 (2005)
12. Tulppo, M.P., Makikallio, T.H., Takala, T.E.S., Seppanen, T.: Quantitative beat-to-beat analysis of heart rate dynamics during exercise. Am J. Physiol. 271, 244–252 (1996)
13. Tulppo, M.P., Husghson, R.L., Makilkallio, T.H., Airaksinen, K.E.J., Huikuri, H.V.: Effects of exercise and passive head-up tilt on fractal and complexity properties of heart rate dynamics. Am J. Phyisol. Heart Circ. Physiol. 280, H1081–H1087 (2001)
14. Takens, F.: Detecting strange attractors in turbulence. In: Eckmann B.D.A. (ed.). Lectures Notes in Mathematics, vol. 898, pp. 366–381. Springer, Heidelberg (1981)
15. Pincus, S.M., Goldberger, A.L.: Physiological time series analysis: what does regularity quantify? Am J. Physiol. 266, H1643–H1656 (1994)
16. Fayyad, U.M., Irani, K.B.: Multi-Interval discretization of continuous-valued attributes for classification learning. In: Proc. of the Interna'l Joint Conference. on AI, pp. 1022–1027 (1993)
17. Cristianini, N., Shawe-Taylor, J.: An introduction to Support Vector Machines. Cambridge University Press, Cambridge (2000)
18. Li, W., Han, J., Pei, J.: CMAR: Accurate and Efficient Classification Based on Multiple Association Rules. In: Proc. of 2001 Interna'l Conference on Data Mining (2001)
19. Chen, J., Greiner, R.: Comparing Bayesian Network Classifiers. In Proc. of UAI-99, pp. 101–108 (1999)
20. Java Bayesian Network Classifier Toolkit, http://jbnc.sourceforge.net
21. Quinlan, J.: C4.5: Programs for Machine Learning. Morgan Kaufmann, San Mateo (1993)
22. Weka Machine Learning Project, http://www.cs.waikato.ac.nz/ ml/weka/index.html

High Performance Data Mining and Applications Overview

Chao Xie[1] and Jieyue He[2]

[1] Department of Computer Science, Georgia State University, Atlanta, USA
cxie@cs.gsu.edu
[2] Department of Computer Science, Southeast University, Nanjing, China
jieyuehe@seu.edu.cn

Abstract. International workshop on High Performance Data Mining and Applications (HPDMA 2007) was held in conjunction with The 11th Pacific-Asia Conference on Knowledge Discovery and Data Mining (PAKDD 2007), Nanjing, China, May 2007. The workshop aimed at sharing and comparing experiences on high performance data mining methods and applications from both algorithmic and system perspectives. In summary, the workshop gave a discussion forum for researchers working on both data mining and high performance computing where the attendees discussed various aspects on high performance data mining.

1 Introduction

Over the years data is being collected and stored at an unprecedented rate in almost all fields of human endeavor from scientific research to economic activities. To achieve efficient mining of useful information from the data available, parallel hardware platforms, clusters and large-scale distributed computing infrastructures, such as computational grid and peer-to-peer systems, are widely used by data mining communities. This also poses challenges on the design of parallel and distributed algorithms for data mining.

This workshop focused on high performance data mining methods and applications from both algorithmic and system perspectives. The workshop brought together researchers who are interested in both of the areas of data mining and high performance computing, where the attendees discussed various aspects on high performance data mining

The topics of the workshop in call for papers included:

- Parallel or distributed mining
- Cluster-based data mining algorithms and systems
- Grid-based data mining algorithms and systems
- Peer-to-Peer based data mining algorithms and systems
- Data mining algorithms and systems based on parallel hardware platforms, including shared-memory systems (SMPs), distributed-memory systems, etc.
- Resource and location aware data mining algorithms and systems

T. Washio et al. (Eds.): PAKDD 2007 Workshops, LNAI 4819, pp. 229–230, 2007.

- Data mining in mobile and ad hoc environments
- Data mining in sensor networks
- Data mining in distributed security and privacy
- High performance stream data mining and management
- Integration of mining with databases and data warehousing
- Applications of parallel and distributed data mining in business, science, engineering, medicine, and other disciplines

2 Workshop Overview

All submitted papers were carefully peer reviewed by program committee members. We accepted 29 papers (25 regular papers and 4 posters) out of 119 submissions. The acceptance rate is approximately 25%.

We would like to thank all the authors who submitted papers to the workshop and participated in the interesting discussions at the workshop. We would also like to thank the all active program committee members for their efforts in careful reviewing of papers and supporting the success of the workshop.

3 Organization

General Chairs
Mohammed J. Zaki (Rensselaer Polytechnic Institute, US)
Yi Pan (Georgia State University, US)

Program Chairs
Chao Xie (Georgia State University, US)
Jieyue He (Southeast University, CN)

Program Committee Members (Alphabetical)
Phil Chan (Florida Institute of Technology, US)
Benjamin C. M. Fung (Simon Fraser University, CA)
Masaru Kitsuregawa (Tokyo University, JP)
Yuefeng Li (Queensland University of Technology, AU)
Jing Liu (Wuhan University, CN)
Nikos Mamoulis (University of Hong-Kong, HK)
Rosa Meo (Universit di Torino, IT)
Salvatore Orlando (University of Venice, IT)
Yufei Tao (Chinese University of Hong Kong, HK)
Jason Tsong-Li Wang (New Jerseys Science and Technology University, US)
Xintao Wu (UNC Charlotte, US)
Xiaowei Wu (University of Arkansas at Little Rock, US)
Hui Xiong (State University of New Jersey, US)
Ying Zhao (University of Missouri-Rolla, US)

Approximately Mining Recently Representative Patterns on Data Streams*

Jia-Ling Koh and Yuan-Bin Don

Department of Information Science and Computer Engineering
National Taiwan Normal University
Taipei, Taiwan
jlkoh@ntnu.edu.tw

Abstract. Catching the recent trend of data is an important issue when mining frequent itemsets from data streams. To prevent from storing the whole transaction data within the sliding window, the frequency changing point (FCP) method was proposed for monitoring the recent occurrences of itemsets in a data stream under the assumption that exact one transaction arrives at each time point. In this paper, the FCP method is extended for maintaining recent patterns in a data stream where a block of various numbers of transactions (including zero or more transactions) is inputted within each time unit. Moreover, to avoid generating redundant information in the mining results, the recently representative patterns are discovered from the maintained structure approximately. The experimental results show that our approach reduces the run-time memory usage significantly. Moreover, the proposed GFCP algorithm achieves high accuracy of mining results and guarantees no false dismissal occurring.

1 Introduction

The strategies for mining frequent itemsets in static databases have been widely studied over the last decade such as the Apriori[1], DHP[9], and FP-growth[4]. In addition to memory requirement and execution efficiency, the sheer size of mining results is a major challenge in frequent-pattern mining. In many cases, a high minimum support threshold may discover only commonsense patterns but a low one may generate an explosive number of output patterns, which severely restricts its usage. To solve this problem, the closed frequent itemsets [10] provided a lossless compression of the whole collection of patterns; however, its compression power is limited. For providing a general method for high-quality compression, the distance measure between two frequent itemsets was defined in [12] to find a representative pattern for each cluster of patterns. Two greedy algorithms, named RPglobal and RPlocal, respectively, were proposed. The RPglobal algorithm applied the greedy method to find representative patterns among the discovered frequent itemsets. For providing a scalable method, the RPlocal algorithm found the representative patterns locally during the process of pattern-growth. The experimental results showed that the

---

* This work was partially supported by the R.O.C. N.S.C. under Contract No. 95-2221-E-003-011 and 95-2524-S-003-012.

T. Washio et al. (Eds.): PAKDD 2007 Workshops, LNAI 4819, pp. 231–243, 2007.

RPlocal method, whose compression quality is very close to RP-global, is far more efficient than RP-global.

Recently, the data stream, which is an unbounded sequence of data elements generated at a rapid rate, provides a dynamic environment for collecting data sources. Since it is not feasible to store the past data in data streams completely, a method for providing the approximate answers with accuracy guarantees is required. The hash-based approach was proposed in [5], in which each item in a data stream owns a respective list of counters in a hash table, and each counter may be shared by many items. A new novel algorithm, called hCount, was provided to maintain frequent items over a data stream and support both insertion and deletion of items with a less memory space. Lossy-counting is the representative approach for mining frequent itemsets from data streams [8]. Given an error tolerance parameter ε, the Lossy-counting algorithm prunes the patterns with support being less than ε from the pool of monitored patterns such that the required memory usage is reduced. Consequently, the frequency of a pattern is estimated by compensating the maximum number of times that the pattern could have occurred before being inserted into the monitored patterns. It is proved no false dismissal occurs with Lossy-counting algorithm and the frequency error is guaranteed not to exceed a given error tolerance parameter.

In addition to the restriction of memory usage considered in the two works introduced previously, the time sensitivity issue is another important issue when mining frequent itemsets from data streams. It is likely that the embedded knowledge in a data stream will change quickly as time goes by. In order to catch the recent trend of data, the *estDec* algorithm [2] decayed the old occurrences of each itemset as time goes by to diminish the effect of old transactions on the mining result of frequent itemsets in the data steam. However, in certain applications, it is interested only the frequent patterns mined from the recently arriving data within a fixed time period. Under the assumption that exact one transaction arrives at each time unit, the sliding window method [3] defined the current sliding window to consist of the most recently coming w transactions in a data stream according to a given window size w. Consequently, the recently frequent itemsets were defined to be the frequent itemsets mined from the current sliding window. In addition to maintain the occurrence for the new transaction, the oldest transaction has to be removed from the maintained data structure when the window is sliding. However, all the transactions in the current sliding window need to be maintained in order to remove their effects on the current mining result when they are beyond the scope in the window.

To prevent from storing the whole transaction data within the sliding window, we proposed the frequency changing point (FCP) method for monitoring the recent occurrence of itemsets in a data stream [6]. The effect of old transactions on the mining result of recently frequent itemsets is diminished by performing adjusting rules on the monitoring data structure. The experimental results showed that our approach reduces the run-time memory usage significantly by comparing with one of [3]. Moreover, the proposed FCP algorithm achieves high accuracy of mining results and guarantees no false dismissal occurring.

In [7], a time-sensitive sliding window approach was also proposed for mining the recently frequent itemsets within the current sliding window in a data stream. However, a general assumption that a block of various numbers of transactions (zero or more transactions) is inputted into the data stream at each time unit was adopted.

Accordingly, the recently frequent itemsets were discovered from the most recent w blocks of transactions. For each block of transactions, the frequent itemsets in the block were found and all possible frequent itemsets in the sliding window were collected in a PFP (Potential Frequent-itemset Pool) table. For each newly inserted pattern, the maximum number of possible lost counts was estimated. Moreover, a discounting table was constructed to provide approximate counts of the expired data items. However, as the minimum support threshold is reduced, the number of frequent itemsets in a basic block will increase dramatically. Because of the increasing cost of table maintenance, the memory usage of PFP table will increase such that the execution efficiency of the algorithm goes down.

In this paper, the FCP algorithm proposed in [6] is extended for solving the same problem considered in [7]. The provided data structures and adjusting rules in [6] are modified accordingly. Moreover, to avoid generating redundant information in the mining results, the idea of local greedy method is applied to discover recently representative itemsets from the maintained structure. The experimental results show that our approach reduces the run-time memory usage significantly than the one of STW algorithm [7] when the minimum support threshold is low. Moreover, the proposed GFCP algorithm achieves high accuracy of mining results and guarantees no false dismissal occurring.

This paper is organized as follows. The related terms used in this paper are defined in Section 2 first. The provided data structure for monitoring recent patterns in a data stream is shown in Section 3. In Section 4, the proposed algorithm for discovering recently representative patterns from the maintained structure is introduced. The performance evaluation on the proposed algorithms and a related work is reported in Section 5. Finally, in Section 6, we conclude this paper.

2 Preliminaries

Let $I = \{i_1, i_2, \ldots, i_m\}$ denote the set of items in the specific application domain and a transaction is composed of a set of items in I. A **basic block** is a set of transactions arriving within a fixed unit of time. A data stream, $DS = [B_1, B_2, \ldots, B_t)$, is an infinite sequence of basic blocks, where each basic block $B_i=\{T_{i1}, T_{i2}, \ldots, T_{ij}\}$ is associated with an time identifier i and t denotes the time identifier of the latest basic block arriving currently. Let $DS[i,j]$ denote the multi-set of transactions collected from basic blocks in DS from time i to j. Under a predefined window size w, the **current transaction window** at time t, denoted as CTW_t, corresponds to $DS[t-w+1, t]$. The time identifier of the first basic block in CTW_t is denoted as CTW_t^{first}, that is $t-w+1$.

An itemset (or a pattern) is a set consisting of one or more items in I, that is, a non-empty subset of I. If itemset e is a subset of transaction T, we call T **contains** e. The number of transactions in CTW_t which contain e is named the **recent support count** of e in DS, denoted as $RC_t(e)$. The **recent support** of e, denotes as $Rsup_t(e)$, is obtained from $RC_t(e) / |CTW_t|$.

Given a user specified minimum support threshold between 0 and 1, denoted as S_{min}, an itemset e is called a **recently frequent** itemset in DS if $Rsup_t(e) \geq S_{min}$. Otherwise, e is a **recently infrequent** itemset in DS.

For any two patterns p_r and p, let $T(p)$ and $T(p_r)$ denote the sets of transactions which contain p and p_r, respectively. By extending the definition of pattern distances proposed in [12], if $p_r \supseteq p$, the distance between p and p_r in CTW_t is defined to be:

$$RD(p, p_r) = 1 - \frac{|T(p) \cap T(p_r)|}{|T(p) \cup T(p_r)|} = 1 - \frac{|T(p_r)|}{|T(p)|} = 1 - \frac{RC_t(p_r)}{RC_t(p)} = 1 - \frac{Rsup_t(p_r)}{Rsup_t(p)}.$$

Otherwise, $RD(p, p_r) = \infty$. If $RD(p, p_r)$ is less than a given δ value, it is called that p_r δ-covers p and p is *δ-covered by* p_r in CTW_t. A recently frequent pattern p is named a *recently representative pattern* if p is not δ-covered by any recently frequent itemset.

3 Pattern Summarization Method

3.1 Frequency Changing Points

A most compact method to represent the occurrences of a pattern p is to keep the first appearing time, denoted as $p.t_s$, and an accumulated count of the pattern, denoted as $p.f$. Accordingly, $Rsup_t(p)$ is obtained from $p.f/|DS[CTW_t^{first}, t]|$ if $p.t_s$ is within current transaction window. Otherwise, $Rsup_t(p)$ is estimated by $p.f / |DS[p.t_s, t]|$. However, if $p.t_s$ is far earlier than CTW_t^{first}, it is not sure whether the estimated support stands for the recent trend of the pattern p accurately. According to the discussion proposed in [10], it is a critical point to set a new accumulated count of pattern p when the appearing frequency of p becomes infrequent.

Let t_1 denote the first time identifier when a pattern p appeared. The earliest time identifier t', where $t' > t_1$, p appears in $B_{t'}$, and (the support count of p in $DS[t_1, t'-1]$) / $|DS[t_1, t'-1]| < S_{min}$, is named a *frequency changing point(FCP)* of p. If a frequency changing point of p has ever occurred, t_1 is set to be the latest frequency changing point of p. The frequency changing points of a pattern p are going to be used to reset the boundaries of intervals for accumulating the support counts of p.

3.2 Pattern Summarization Tree

For a pattern p, an 8-tuple $(p, t_s, f, f_c, t_e, t_{pre}, C_d, Rqueue)$ summarization record as described as the following is used to represent the summarization of p's occurrences.

(1) p: the corresponding itemset;
(2) t_s: the starting point of support count accumulation for p; initially, it is set to be the identifier of the basic block when p was inserted into the data structure;
(3) f: the accumulated support count of p in $DS[t_s, t]$;
(4) f_c: the support count of p in B_t;
(5) t_e: the time identifier of the latest basic block which contains p;
(6) t_{pre}: the latest time identifier which is a frequency changing point of p;
(7) C_d: the accumulated support count of p in $DS[t_s, t_{pre}-1]$;
(8) $Rqueue$: a queue consists of a sequence of (ct_1, ac_1), (ct_2, ac_2), ..., (ct_n, ac_n) pairs, in which the time identifier ct_i is a frequency changing point of p for $i=1,...,n$. Besides, ac_1 denotes the support count of p in $DS[t_s, ct_1-1]$ and ac_i denotes the support count of p in $DS[ct_{i-1}, ct_i-1]$ for $i=2,...,n$.

A FP-tree-like structure, named pattern summarization tree (PS-tree), is adopted to organize the summarization records of itemsets which appear in current transaction window. However, to prevent from two scans over the data set, the items in a transaction are sorted according to their alphanumeric order instead of their frequency-descending order.

Moreover, an array named TranArray is constructed to maintain the number of transactions in each basic block for the recent blocks, which is used for computing the recent supports of itemsets.

3.3 Maintenance of PS-tree

On the whole, there are four sub-tasks for maintaining the PS-tree structure: (1) insert transactions in the new basic block into PS-tree, (2) remove the expired information of patterns, (3) check and record frequency changing points of patterns, and (4) update the TranArray. The pseudo codes of the whole process are shown in Figure.1.

Program PS-tree_maintenance
Input: DataStream DS, Window_size w
1. TranArray and PS_tree are initialized;
2. While (B_t is inputted) {
3.   New_block_insertion(B_t) ;
4.   For each node N in the PS_tree ;
5.   { Let p denote the associated itemset in N ;
6.   If ($t > w$)
7.   If ($p.t_e < CTW_t^{first}$) Remove_itemset(PS_tree) ;
8.   Else If ($p.t_s < CTW_t^{first}$) Reset_start_time(PS_tree) ;
9.   If ($p.f_s \neq t$ and $p.f_c \neq 0$)
10.   Check_change_point(PS_tree) ;}
11.   TranArray_update(TranArray) ;
12. $t = t+1$; }

Fig. 1. The pseudo codes of PS-tree maintenance

First, each transaction T in the newly inputted basic block is inserted into the PS-tree. The PS-tree is a trie structure, so there is one node for every common prefix of patterns. For each prefix-subset p of T, if the corresponding node exists in the PS-tree, $p.f$, $p.f_c$, and $p.t_e$ in the node are updated accordingly. Otherwise, a new node for storing the summarization record of p with initial setting is constructed.

In the window sliding phase, for each monitored pattern p in the PS-tree, it is necessary to update the summarization information of p within current transaction window. It is indicated that a pattern p does not appear in CTW_t if $p.t_e$ is less than CTW_t^{first}. Therefore, such a pattern is pruned to prevent from storing the unnecessary patterns in the monitoring data structure.

Moreover, it is necessary to adjust the starting point of support count accumulation for the monitored pattern p if $p.t_s$ is less than CTW_t^{first}. If $p.Rqueue$ is empty for such a pattern p, it implies p remains a frequent itemset during the accumulation interval. Therefore, no false dismissal occurs when discovering recently frequent patterns according to the support of p in $DS[p.t_s, t]$ such that it is not necessary to adjust $p.t_s$. On the other hand, it implies there is one or more frequency changing points of p occurring if $p.Rqueue$ is not empty. Therefore, the frequency changing points of p are checked one by one to adjust $p.t_s$ to be a frequency changing point of p as approaching CTW_t^{first} as possible. Let (ct, ac) denoted the first frequency changing point pair got from $p.Rqueue$. It is applicable to adjust $p.t_s$ in the following three cases:

(1) $ct \leq CTW_t^{first}$: it implies the support count accumulated in ac is beyond the scope of CTW_t.

(2) $ct > CTW_t^{first}$ and $ac=1$: it implies the occurring of p before ct is at $p.t_s$ only. Thus, the actual starting point of p within CTW_t is set to be ct.

(3) $ct > CTW_t^{first}$, $ac>1$, and the previous time point when p appears before ct, denoted as t_e', is less than CTW_t^{first}: let X denote the support count of p in $B_{te'}$. The maximum value of t_e', denoted by Max_t_e', must satisfy $(ac-X)/|DS[t_s, Max\_t_e'-1] \geq S_{min}$ and $(ac-X)/|DS[t_s, Max\_t_e'] < S_{min}$ because there is not any changing point occurring from $p.t_s$ to t_e'. Since the support count of p in $B_{te'}$ is not maintained, an upper bound of Max_t_e', denoted by $UB\_Max\_t_e'$, is estimated by $ac/|DS[t_s, UB\_Max\_t_e'-1] \geq S_{min}$ and $ac/|DS[t_s, Max\_t_e'] < S_{min}$. If $UB\_Max\_t_e'$ is less than CTW_t^{first}, it is indicated that all the possible value of t_e' must be less than CTW_t^{first}. Accordingly, ct is the first time p appears within CTW_t.

When satisfying any one of the situations enumerated above, the (ct, ac) pair is removed from $p.Rqueue$, the starting accumulation time $p.t_s$ is adjusted to be ct; and accumulated support counts $p.f$ and $p.C_d$ are modified accordingly.

The situation that $p.t_s$ and $p.f$ are not adjusted occurs when $p.Rqueue$ is not empty for a pattern p and the first changing point ct in $p.Rqueue$ does not satisfy any situation enumerated above. It is implied that $ct > CTW_t^{first}$, $ac>1$, and $t_e' \geq CTW_t^{first}$. In this case, p is frequent in $DS[p.t_s, CTW_t^{first}-1]$ because there is not any frequency changing point appearing between $p.t_s$ and ct. When judging whether p is recently frequent according to its support count in $DS[p.t_s, t]$, even though false alarm may occur, it is certain that no false dismissal will occur.

Moreover, for each pattern p monitored in the SP-tree, if p appears in current basic block $(p.f_c > 0)$ and has ever appeared in any previous basic block $(p.t_s<t)$, the checking of a frequency changing point of p is performed. If t is certified to be a frequency changing point of p, a frequency changing point pair (ct, ac) is inserted into $p.Rqueue$, where $ct=t$ and $ac=((p.f - p.f_c)-p.C_d)$. Besides, $p.t_{pre}$ and $p.C_d$ are updated to be t and $(p.f - p.f_c)$, respectively. After the checking, $p.f_c$ is reset to be 0. Finally, the number of transactions in current basic block is obtained, which is used to update the data in TranArray.

[Example 1]: Table 1 shows an example of data streams. Suppose S_{min} is set to be 0.5 and window size w is 4. The process of constructing the SP-tree is described as the following.

Table 1. An example of data streams

B_1	B_2	B_3	B_4	B_5	B_6	B_7	B_8	B_9	B_{10}	...
ab	b	ab	a	a	c	a		a	cd	...
a	c			b	cd	b				
a	c				c					
					cd					

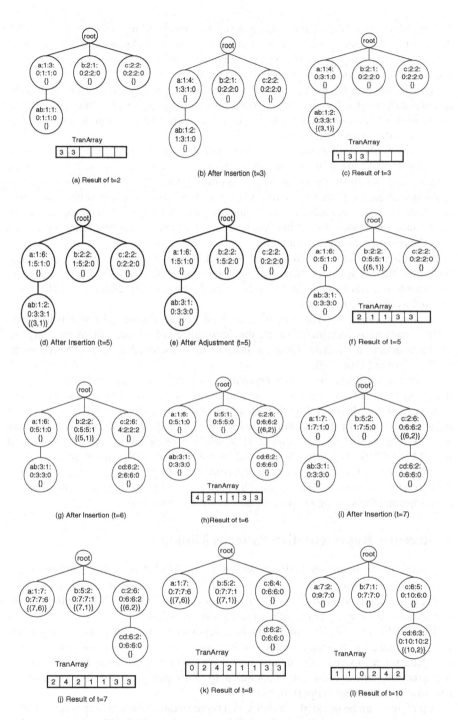

Fig. 2. The maintenance of the monitoring data structure

From time point t_1 to t_2, there was not any frequency changing point occurring for the monitored patterns. The constructed SP-tree and TranArray at t_2 are shown in Figure 2(a). For simplifying the figures, the header table and the links connecting the nodes with same items are now displayed in the figure.

After processing B_3, time point t_3 is a frequency changing point of ab. Therefore, the changing point entry $(3,1)$ is appended into $ab.Rqueue$. Besides, $ab.C_d$ is set to be $ab.f - ab.f_c = 1$, and $ab.t_{pre}$ is set to be 3. Accordingly, the resultant monitoring data structure is shown in Figure 2(c).

Continuing the similar processing, the resultant monitoring data structure at t_5 after inserting the patterns in B_5 into the data structure is shown in Figure 1(d). Then the value of $ab.t_s$ is adjusted because $ab.t_s$ is less than $CTW_5^{first}(=2)$. According to the first frequency changing point pair of ab, which satisfies the first case of adjustment, $ab.t_s$ is reset to be 3, $ab.f$ is updated to be 1, and the pair is removed from $ab.Rqueue$ as the result shown in Figure 2(e). After recording the frequency changing point pair, $(5,1)$, of b, the result is shown in Figure 2(f).

After inserting the transactions in basic block B_6 as the result shown in Figure 2(g), since the first frequency changing point pair of b satisfies the second case of adjustment, $b.t_s$ and $b.f$ are adjusted to be 5 and 1, respectively. Moreover, $c.Rqueue$ is updated as Figure 2(h) shows.

At time point t_7, the value of $ab.t_e$ is 3 (shown in Figure 2(i)) which less than $CTW_7^{first}(=4)$. It is indicated that all the occurrences of ab are out of the range of current transaction window. Thus, the node of ab is removed from the PS-tree as the result shown in Figure 2(j).

In spite of no transaction being inputted at t_8, by satisfying the third case among the three conditions of adjustment, $c.t_s$ is adjusted to be 6 and $c.f$ is reset to be 6. The obtained result is shown in Figure 1(k), which shows the SP-tree correctly catches the recent occurrences of the monitored patterns b, c, d, and cd in the current transaction window. The estimated recent support of a, which is obtained according to its support in $DS[1, 8]$ (7/16), is higher than its real recent support (2/8). However, after B_{10} is processed at time t_{10}, the summarization record of a will be adjusted as the monitoring data structure shown in Figure 1(e). Therefore, the estimation error is compensated after adjusting the starting point of support count accumulation.

4   Recently Representative Patterns Mining

According to the maintained SP-tree structure, the TD-FP Growth algorithm [11] is modified to perform on the structure to find all the patterns p with $Rsup_t(p) \geq S_{min}$ whenever needed. Moreover, the idea of RPlocal algorithm [12] is applied on the mining process to discover recently representative patterns.

The TD-FP Growth algorithm adopts a depth-first approach to perform pattern-growth. During the mining process, our approach constructs a *RP-tree* (Representative Pattern tree) to maintain the discovered representative patterns. The RP-tree is a trie structure, where each node contains a (p, sup) pair to denote a representative pattern p and its recent support, respectively.

A pattern p can be decided whether it is a representative pattern only when no more frequent pattern is grown from p or all the patterns grown from p have been decided. Therefore, a *RP-stack* is used to maintain those patterns which have not been certified to be representative patterns yet. There are five fields in each entry of *RPstack*: (1) *p*: the itemset; (2) *sup*: the recent support of *p*; (3) *covered*: whether *p* is δ-covered by other patterns; (4) *cover_pattern*: the known maximum pattern which δ-covers *p*; (5) *cover_RP*: the known maximal representative pattern which δ-covers *p*.

For a newly discovered recently frequent itemset *p*, each pattern *p'* in *RP-stack* is a prefix of *p*. If *p'* has not been δ-covered by a representative pattern and it is δ-covered by *p*, *p'.covered* is set to be 1. Besides, the pattern in *p'.cover_pattern* is replaced by *p* if *p* is larger than the original one.

In addition, the *RP-tree* is searched to find the representative patterns which δ-cover *p*. If more than one representative pattern δ-covers *p*, let p_r denote the maximum pattern among these satisfying patterns. Accordingly, *p.covered* is set to be 2 and p_r is stored in *p.cover_RP*.

Until the process of generating patterns grown from *p* is complete, the record of *p* is popped from *RP-stack*. According to the value stored in *p.covered*, whether *p* is a representative pattern is certified:

(1) *p.covered*=2: there is a representative pattern δ-covers *p*, *thus*, *p* is not a representative pattern.
(2) *p.covered*=1: *p* is δ-covered by the pattern *p'* stored in *p.cover_patter*. If *p'* is found in the RP-tree, *p* is not a representative pattern because it is δ-covered by a representative pattern. Otherwise, *p* is certified to be a representative pattern and is inserted into the RP-tree.
(3) *p.covered*=0: there is not any pattern found to δ-cover *p*. Pattern *p* is certified to be a representative pattern and is inserted into the RP-tree.

5 Performance Study

The proposed GFCP (Generalized Frequency Changing Point) algorithm and Time-sensitive Sliding Window method (TSW in short) [7] were implemented using Visual C++ 6.0. The experiments have been performed on a 3.4GHz Intel Pentium IV machine with 1G megabytes main memory and running Microsoft XP Professional. Moreover, the datasets were generated from the IBM data generator [1], where each dataset simulates a data stream with each basic block consisting of 1000 transactions.

In the first part of experiments, the accuracy of mining results, execution time, and memory usage were measured to show the effectiveness and efficiency of the proposed GFCP algorithm for mining recently frequent itemsets by comparing with the ones of TSW algorithm. Furthermore, we extended the TSW algorithm by applying a local greedy method [12] to extract representative patterns from the discovered recently frequent itemsets. The performance of GFCP for mining recently representative patterns was observed in the second part of experiments by comparing with the one of the extended TSW algorithm.

[Experiment 1]: To evaluate the effectiveness and efficiency of GFCP and TSW algorithms, the first part of experiments were performed on the dataset T5I4D100K with |N|=100. In this experiment, GFCP algorithm was performed to maintain the monitoring data structures, and TD_FP-Growth algorithm was performed once every 10 time points to find recently frequent itemsets.

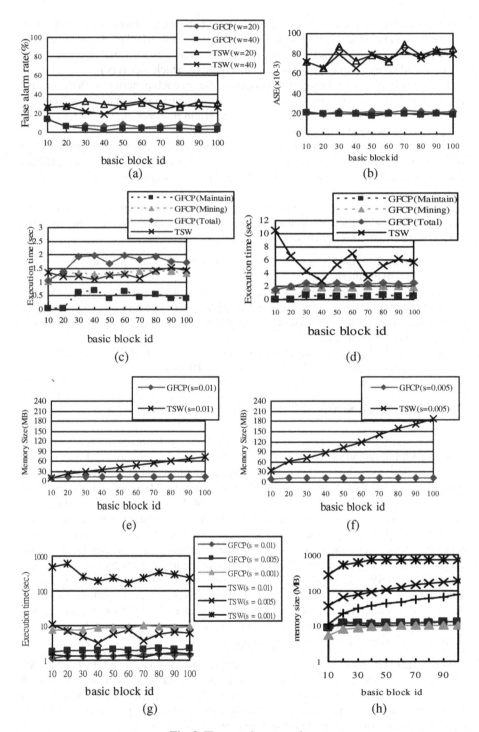

Fig. 3. The experiment results

[Exp.1-1]: By comparing the mining results with the frequent itemsets found by Apriori algorithm on the corresponding CTW_t, the false dismissal rate(FDR), false alarm rate(FAR), and average support error(ASE) of the two algorithms were measured. Both GFCP and TSW guarantee that no false dismissal occurs, thus, the false dismissal rates of two algorithms are zeros. The false alarm rates of the two proposed algorithm at various time points are shown in Figure 3(a). In general, it is reported that the FARs of the executions with window size 40 are less than the ones with window size 20. Besides, the FAR of FCP algorithm is below 10% on average, which is better than the one of TSW. From Figure 3(b) one may note that $ASE(R_{GFCP}|R_{Apriori})$ keeps under 25×10^{-3} at different time points, which is about one third of $ASE(R_{TSW}|R_{Apriori})$.

[Exp.1-2]: In this experiment, GFCP and TSW algorithms are compared on their execution time (including the time of maintenance and mining) and maximum memory requirement. The dataset T5I4D1000K was experimented with window size $=20$. When S_{min} is varied to be 0.01 and 0.005, the results of execution time and maximum memory usages are shown in, Figure 3(c), 3(d), 3(e), and 3(f), respectively.

Although the total execution time of GFCP is more than the one of TSW algorithm slightly when S_{min} is 0.01, the execution efficiency of GFCP is much better than TSW when S_{min} is 0.005. This is due to the fact that the number of frequent itemsets in a basic block is increasing dramatically as the setting of S_{min} is reduced. Accordingly, the execution efficiency of TSW algorithm goes down because of the increasing cost of table maintenance. Regarding GFCP algorithm, little difference of maintenance time is observed between the two different settings of S_{min}. In addition, the effect of changing S_{min} on the mining time of GFCP is not significant. It should be noted that GFCP has better performance efficiency than TSW as S_{min} is smaller. By discarding the mining time, the maintenance time of GFCP is much less than the execution time of STW. These results suggest that, if it was not necessary to get mining results at any time, GFCP algorithm is a very efficient method for monitoring patterns in a data stream to discover recently frequent itemsets on demand. Furthermore, in contrast with TSW, the memory usage of GFCP is not sensitive to the setting of S_{min}, which also keeps stable at different time points. These results clearly demonstrate that GFCP is feasible for the streaming environment with a small memory.

[Experiment 2]: In order to evaluate the GFCP algorithm and extended STW algorithm for mining recently representative patterns, the execution time and maximum memory usage were measured for different settings of S_{min}. The experiment was performed with window size=20 andδ=0.2. By varying the setting of S_{min} to be 0.01, 0.005, and 0.001, Figure 3(g) and 3(h) show the results of execution time and maximum memory usage, respectively.

The result shown in Figure 3(g) is in general agreement with the one of [Exp.1-2]. Especially, when S_{min} is set to be 0.001, the execution time of GFCP is 600 sec. less than the one of extended STW at most, which demonstrates the scalability of our approach. Clearly visible in Figure 3(h) is that the maximum memory usage of extended TSW is increasing as S_{min} is reduced. On the other hand, as a smaller S_{min} is set, the amount of frequency changing points maintained in the FP-tree is decreasing. This is the reason for the maximum memory usage of GFCP under S_{min} =0.001 is only 1/60 of the one required by TSW.

6 Conclusion

The sliding window approach proposes a good solution for providing time-sensitive mining of frequent itemsets in data streams. In this paper, a pattern summarization data structure based on the frequency changing point representation is provided to represent the occurrences of patterns in a data stream under a general input assumption. The effect of old transactions on the mining result of recently frequent itemsets is diminished by performing adjusting rules on the monitoring data structure without needing to keep the whole transactions in the current sliding window physically. Moreover, to avoid generating redundant information in the mining results, the idea of local greedy method is applied to discover the recently representative patterns from the monitoring data structure. The experimental results demonstrate that the proposed GFCP algorithm achieves high accuracy for approximating the supports of recently frequent patterns and guarantees no false dismissal occurring. Not only the memory requirement by GFCP is significantly reduced by comparing with the one of TSW algorithm, but also the maintaining process of the monitoring data structure is very quick under various parameters setting. These results suggest that, if it was not necessary to get mining results at any time, GFCP algorithm is a very efficient method for monitoring patterns in a data streaming environment with a limited memory to discover recently representative patterns on demand.

References

[1] Agrawal, R., Srikant, R.: Fast Algorithms for Mining Association Rules. In: Proc. of Int. Conf. on Very Large Data Bases (1994)
[2] Chang, J.H, Lee, W.S.: Finding Recent Frequent Itemsets Adaptively over Online Data Streams. In: Proc. of the 9th ACM International Conference on Knowledge Discovery and Data Ming (2003)
[3] Chang, J.H., Lee, W.S.: A Sliding Window Method for Finding Recently Frequent Itemsets over Online Data Streams. Journal of Information Science and Engineering 20, 753–762 (2004)
[4] Han, J., Pei, J., Yin, Y., Mao, R.: Mining Frequent Patterns without Candidate Generation: A Frequent-Pattern Tree Approach. Data Mining and Knowledge Discovery 8(1), 53–87 (2004)
[5] Jin, C., Qian, W., Sha, C., Yu, J.X., Zhou, A.: Dynamically Maintaining Frequent Items Over a Data Stream. In: Proc. of the 12th ACM International Conference on Information and Knowledge Management (2003)
[6] Koh, J.L., Shin, S.N.: An Approximate Approach for Mining Recently Frequent Itemsets from Data Streams. In: Tjoa, A.M., Trujillo, J. (eds.) DaWaK 2006. LNCS, vol. 4081, Springer, Heidelberg (2006)
[7] Lin, C.H., Chiu, D.Y., Wu, Y.H., Chen, A.L.P.: Mining Frequent Itemsets from Data Streams with a Time-Sensitive Sliding Window. In: Proc. of SIAM Intl. Conference on Data Mining (2005)
[8] Manku, G.S., Chen Motwani, R.: Approximate Frequent Counts over Data Streams. In: Proc. of the 28th International Conference on Very Large Database, Hong Kong, China (August 2002)

[9] Park, J.S., Chen, M.S., Yu, P.S.: An Effective Hash-based Algorithm for Mining Association Rules. In: Proc. of the ACM SIGMOD International Conference on Management of Data (SIGMOD 1995), May, pp. 175–186 (1995)

[10] Pei, J., Han, J., Mao, R.: CLOSET: An efficient algorithm for mining frequent closed itemsets. In: Proc. of ACM SIGMOD Workshop on Research Issues in Data Mining and Knowledge Discovery (2000)

[11] Wang, K., Tang, L., Han, J., Liu, J.: Top Down FP-Growth for Association Rule Mining. In: Chen, M.-S., Yu, P.S., Liu, B. (eds.) PAKDD 2002. LNCS (LNAI), vol. 2336, pp. 6–8. Springer, Heidelberg (2002)

[12] Xin, D., Han, J., Yan, X., Cheng, H.: Mining Compressed Frequent-Pattern Sets. in Proc. of Int. Conf. on Very Large Data Bases (VLDB 2005) (2005)

Finding Frequent Items in Data Streams Using ESBF

ShuYun Wang, XiuLan Hao, HeXiang Xu, and YunFa Hu

Department of Computing and Information Technology, Fudan University, P.R.C.
{wang_shuyun,hxl2221_cn}@126.com, {051021045,yfhu}@fudan.edu.cn

Abstract. In this paper, we introduce a novel data structure, ESBF (Ex- tensible and Scalable Bloom Filter), and the algorithm FI-ESBF (Finding frequent Items using ESBF) for estimating the frequent items in data streams. FI-ESBF can work with high precision while using much less memory than those of the best reported algorithm does considering the large number of distinct items in the stream. ESBF is the extension of counting Bloom Filter(CBF), By using it, we are allowed to adjust the size of memory used dynamically according to the different data distribution and the number of distinct items in the data streams, therefore the priori knowledge about the data distribution of the streams and the number of distinct elements to be stored is not required.

Keywords: Stream, Frequent Items, Bloom Filter, Algorithm.

1 Introduction

A data stream is an unbounded sequence of data items, which arrive at high speed. Data of many emerging applications takes forms of continuous streams instead of finitely stored data sets, examples include performance measurements in network monitoring and traffic management, call detail records in telecommunications, transactions in retail chains, ATM operations in banks, log records generated by Web Servers and sensor network data, etc.[7].

Finding frequent items plays an important role in database, data mining and computer network etc. For example, iceberg quires [2] generalize the notion of hot items in the relation to aggregate functions over an attribute(or set of attributes) in order to find aggregate values above a specified threshold. Mining association rules firstly requires finding frequent items. Tracking measurement and accounting of IP packets require identifies of flows that exceed a certain fraction of total traffic[9].

In this paper, we present an algorithm using small memory space to generate items above some user-specified threshold in a dynamic environment, where items can be inserted and deleted. It is important to note that our algorithm is based on an efficient data structure, ESBF, which is based on CBF [13]. But ESBF was designed to overcome the CBF's lack of adaptiveness.

T. Washio et al. (Eds.): PAKDD 2007 Workshops, LNAI 4819, pp. 244–255, 2007.

1.1   Related Work

A frequent item is an item whose frequency, in a stream S of a given size N, exceeds a user specified support ϕN, where $0 \leq \phi \leq 1$. There has been a large amount of work reported for finding frequent items in data streams, all involves some combination of sampling, hashing, and counting.

In[3], The algorithm FindApproxTop(S, k, ϵ) was proposed for generating a list of k most frequent items such that every item in the list has frequency larger than$(1 - \epsilon)F_k$, where ϵ is a user-defined error, and F_k is the frequency of the kth most frequent element. In [4], Manku and Motwani developed a sampling based algorithm, *sticky sampling*, which computes an ϵ-deficient synopsis with probability at least $1 - \delta$ using at most $O(\frac{1}{\epsilon} \log(s^{-1}\delta^{-1}))$ expected number of entries, where s is specified support threshold. A counting based algorithm, *lossy counting*, was also introduced in [4], which used $O(\frac{1}{\epsilon} \log(\epsilon N))$ entries for frequent items counting. The algorithm of *Frequent* was proposed in[8] for deterministically finding all categories having a frequency above $1/(m+1)$using m counters, *Frequent* keeps k counters to monitor k items, if a monitored item is observed, its counter is incremented, otherwise, all counters are decremented, in case any counter reaches 0, it is assigned to the next observed item, when the algorithm terminates, the monitored items are the candidate frequent items. In[6], Cormode and Muthukrishman introduced an algorithm called *groupTest*, which is able to output all items with frequency above $1/k+1$ with probability $1-\delta$, the number of counters needed is $O(k(\log k + \log(\frac{1}{\delta})) \log M)$, where M is the number of different items. *hCount* was proposed in[10], which can handle the situation of insertion and deletion, and with probability ρ, $\ln(\frac{-M}{\ln \rho}) \cdot \frac{e}{\epsilon}$ counters are used to estimate the frequency of items in the steam with error parameter no more than ϵ, where M is the number of different items in the stream and e is the the base of the natural logarithm. The algorithms of *sample and hold* and *multistage filters* were proposed in[9], whose errors were proportional to $1/M$, where M is the available memory. Recently, Ahmed Metwally and Divyakant Agrawal introduced the *Space-saving* algorithm in[12], in this algorithm, a list of counters is kept for monitoring the streaming data. A data structure, Stream-Summary, was also introduced in [12] to implement the *Space-saving* algorithm.

It should be noted that the main idea of our work is influenced by *hCount*[10], but ours greatly outperform *hCount* in terms of memory space consumption.

1.2   Our Contribution

In this paper, FI-ESBF algorithm is proposed for finding the frequent items in the data stream, the idea of FI-ESBF is similar to that of *hCount*[10]. However, by using Chernoff Bound, we test that the space bound for FI-ESBF is no more than $\ln(\frac{-M}{\ln \rho}) \cdot \frac{e}{\epsilon} \cdot \frac{1}{\epsilon \cdot M}$, considering the large number of distinct items existing in the stream, this space bound is much tighter than that of *hCount* whose space bound is $\ln(\frac{-M}{\ln \rho}) \cdot \frac{e}{\epsilon}$. In addition, our algorithm relies on a efficient data structure called ESBF, by using the ESBF, it is possible to adjust the size of memory used dynamically according to the different data distribution and the number of distinct items in the data streams, therefore, the priori knowledge about the data distribution of the streams and the number of items to be stored is not required. Experiments show that our algorithm only use a bit more than half of the space needed by *hCount* for processing the same synthetic data, while only using a bit more processing time.

1.3   Paper Outline

The remainder of the paper is organized as follows. In section 2 we describe the basic idea of our data structure ESBF as an extension of the Bloom Filter[1]. The FI-ESBF algorithm is described in section 3 followed by analysis. We present the performance results of FI-ESBF in section 4. Section 5 concludes our study.

2   Extensible and Scalable Bloom Filter(ESBF)

This section reviews the Bloom Filter[1] structure and Counting Bloom Filter (CBF)[13]. The basic implementation of the ESBF relying on CBF is also introduced in this section.

2.1   The Bloom Filter

Bloom Filter was invented by Burton Bloom in 1970 [1], which is capable of representing a set $A = \{A_1, A_2, ..., A_n\}$ of keys from a universe U by using a bite vector V of m = O(n) bits. All the bits in the vector V are initially set to 0. The Bloom Filter uses h hash functions, $h_1, h_2, ..., h_h$, mapping keys from U to the range$\{1...m\}$. For each item $k \in S$, the bits at positions $h_1(k), h_2(k), ..., h_h(k)$ in V are set to 1. Given an item $q \in U$, we check its membership in stream, S, by examining the bits at positions $h_1(q), h_2(q), ..., h_h(q)$, if one (or more) of the bits is equal to 0, then q is certainly not in S, otherwise, we report that q is in S, but there may be false positive error: the bits $h_i(q)$ may be all one even though $q \notin S$, if other keys from S were mapped into these positions. The probability for a false positive error is dependent on the selection of the parameters m, h.

2.2   The Counting Bloom Filter

The Counting Bloom Filters (CBF)[13] is Bloom Filters which replace the bit vector V with a vector of m counters, C. The counters in C roughly represent multiplicities of items, all the counters in C are initially set to 0. When inserting an item s, we increase the counters $Ch_1(s), Ch_2(s), ..., Ch_h(s)$by 1. The CBF stores the frequency of each item, and it also allows for deletions, by decreasing the same counters. We adopt a variant of Counting Bloom filters in this paper, which consists of partitioning the m bits among the h hash functions, thus creating h slices of m' = m/k counters. In this variant, each hash function $h_i()$, with $1 \leq i \leq h$, produces an index over m' for its respective slice. Therefore, each item is always described by exactly h bits, which results in a more robust filter, with no item specially sensitive to false positives.

2.3   The ESBF

Counting Bloom filters (CBF) is a data structure with fast access time, however, their counters are not flexible, consequently, the counters may become saturated resulting in inaccuracy in the stored information. As an alternative, Spectral Bloom Filters (SBF) [4] and DBF[14] have been designed to overcome the lack of adaptiveness, but SBF requires complex indexing structures which make the access to each counter more complex and costly, DBF can access to each counter efficiently, but the counters in the DBF are of the same size, which causes space waste.

In order to access to counter and use the memory efficiently, our data structure, ESBF, is introduced in this paper. ESBF is Bloom Filters whose counters can be extended to larger ones and the number of the filters in it is scalable. ESBF is made up of following three parts: two kinds of counters, basic counters(BC counters)with length of x bits and large counters(LC counters)with length of $2x$ bits, and a bit vector(OF) indicating whether a BC counter has become overflow, the length of the OF is equal to the number of the BC counters. Initially, there are only BC counters and OF in the ESBF, and the values of BC counters and all bits of OF are set to 0. When any of the BC counters become overflow, a LC counter is created, and the value of the overflow counter is turned into a pointer pointing to the newly created LC counter, which is used for counting the corresponding item, which means, the filter is extensible, we called it extensible Bloom Filter(EBF). ESBF is made up of a series of one or more EBF, when the current filters get full due to the limit on its capacity, a new one is added for the newly monitored items arriving in the stream. The size of newly added filter is $M_0 s^i$, where M_0 is the size of the first filter, i is the current number of the filters. In our implementation, s is equal to 2. Querying is made by testing for the presence in each filter.

2.4   ESBF Basic Construction and Maintenance

Let S be a multi-set of keys taken from a universe U. For $x \in U$, let $f_x = min(C[1][h_1(x)], C[2][h_2(x)], ..., C[h][h_h(x)])$ be the frequency of x in S. To add an item $x \in U$ to the ESBF, the counters $C[1][h_1(x)], C[2][h_2(x)], ..., C[h][h_h(x)]$ are increased by 1, if we are to delete a item x in S, the corresponding counters are decreased by 1.

3   Our Algorithm

In this section, we describes our ESBF-based algorithm, FI-ESBF, for memory efficient frequent items mining, theoretical analysis is also presented in this section.

By using the data structure ESBF, our algorithm is straightforward. Fig. 1(a) shows the algorithm, FI-ESBF. The frequency of each item can be estimated from ESBF by the minimum value of its associated counters. Fig. 1(b)shows the

Algorithm : FI-ESBF(k,ktype)

1	if the number of distinct elements to
	be stored is not known
2	FIDA(k,ktype)
3	else
4	if ktype is insert
5	FIIN(k);
6	else
7	FIDE(k);
8	end if of line 4
9	end if of line 1

(a) The FI-ESBF Algorithm

algorithm   Gfreq(M)

for k=1 to M   //M is data range
  fcy=$\min_{1 \leq j \leq h}(c[j][h_j[k]]$
  if fcy >= s*N
    output(k,fcy);
  end if
end for

(b) Gfreq Algorithm

Fig. 1. The FI-ESBF and Gfreq Algorithm

Algorithm: FIIN(k)

1 N=N+1;
2 for j=1 to h
3   mi=$h_j[k]$;
4   If OF[j][mi]==1
5   (*(C[j][mi]))++;
6   else if OF[j][mi]==0 and C[j][mi]==2^x-1
7   creat a larger counter(LC$_{new}$) with size 2x bits;
8   LC$_{new}$=2^x;
9   Let c[j][mi] point to LC$_{new}$;
10   OF[j][mi]=1;
11   else
12   C[j][mi]++;
13   end if
14 end for

(a) Algorithm of FIIN(k)

Algorithm: FIDE(k)

1 T=$2^{x-1}+\lambda(2^x-1-2^{x-1})$
2 N=N-1;
3 for j=1 to h
4   mi=$h_j(k)$;
5   if OF[j][mi]==1
6   if (*(c[j][mi]))>=T
7   (*(c[j][mi]))--;
8   else
9   copyvalue in the *(c[j][mi]) to c[j][mi];
10   c[j][mi]--;
11   deallocate   *(c[j][mi]);
12   OF[j][mi]=0;
13   end if of line 6
14   else
15   c[j][mi]--
16   end if of line 5
17 end for

(b)Algorithm of FIDE(k)

Fig. 2. Subroutines of FIIN and FIDE construction algorithm

algorithm of generating all frequent items(Gfreq) which is similar to that of eFreq in [10]). As we can see from Fig. 1(a), if the number of distinct items to be stored in the filter is known, we simply increase the counters $C[1][h_1(k)], C[2][h_2(k)], ...,$ $C[h][h_h(k)]$ by 1 on observing the item k. if the set only evolves insertion, we can just increase the counter $\min(C[1]h_1(k), C[2]h_2(k), ..., C[h]h_h(k))$, which enable us to have fewer false positives. when a BC counter $C[i][h_i(k)]$, $1 \leq i \leq h$, becomes overflow, we firstly create a larger counter LC_{new}, whose value is set to 2^x(x is the length of a BC counter), and the value of $C[i][h_i(k)]$ is transformed to a pointer pointing to LC_{new}, at the same time, $OF[i][h_i(k)]$ is set to 1, see Fig. 2(a) for details. Upon deleting an item, k, from the data set, $C[1][h_1(k)], C[2][h_2(k)], ..., C[h][h_h(k)]$ are decremented by 1. In the deletion condition, the value of a LC counter may be less than $2^x - 1$, we call such LC counters as small LC counters, whose value is able to be stored in the BC counters. in this case, in order to save memory space, we can copy the value of the small LC counters to the BC counters pointing to them, and deallocate the small LC counters. However, in order to avoid unstable situations of consecutive delete/insert operations, which could result in excessive creation and deletion of LC counters, the process of copy and deallocation is delayed. Therefore, we introduce a threshold, T, between values 2^{x-1} and $2^x - 1$. We define such threshold as $T = 2^{x-1} + \lambda(2^x - 1 - 2^{x-1})$, where λ ranges from 0 to 1(in our algorithm, we set λ to 0.5). Hence, when the value of a LC counter is decreased and its value is less than T, the process of copy and deallocation is carried out. The algorithm of deleting an item, FIDE(k), is described in Fig. 2(b).

In the condition that we don't know the number of distinct items that to be stored, we first build filter with smaller size. when the current filter gets full, a new filter will be created, whose size is twice as large as that of the former one. Fig. 3 outlines the algorithm, FIDA(k,ktype), for handling such situations. A modified Bloom Filter is applied for quick membership queries, which is used to decide whether an item, k, is observed for the first time, and it is also used to tell which EBF that stores the information of item k, we call it MQBF(Membership Queries Bloom Filter). MQBF may contains a series of QBF(Query Bloom Filter). The first QBF(QBF$_1$) is designed to check the item ranging from 1 to R_1(in our system we set R_1 to 20000). when an item to be handled is beyond[1,20000], we create a new QBF to handle items in the range of $[R_1+1,R_2]$ where $R_2=2*R_1$, that is to say the QBF$_{i+1}$ is used to check the item ranging from R_i +1 TO $2*R_i$. suppose QBF$_i$ contains HR$_i$ slices and each slice is composed of MR$_i$ units, from[13][14][15], we know that for any given false positive probability P and the number of stored elements n, the minimum size of filter is $\frac{-n \cdot \ln(P)}{(\ln 2)^2}$, and the optimal number of slices is $\log(\frac{1}{P})$, so HR$_i$=$\log(\frac{1}{P})$, MR$_i$=$\frac{\frac{-n_i \cdot \ln(P)}{(\ln 2)^2}}{\log(\frac{1}{P})}$. In our system, we set P to 0.1% and all the units in the MQBF has 3 bits in length initially. when there are more than 7 EBF in the ESBF, we reconstruct the MQBF by setting more bits to its units. On observing an item k, we first chose the appropriate QBF based on the data range in QBF. suppose that the selected QBF is QBF$_s$, J=min(QBF$_s$[1][h$_1$(k)], QBF$_s$[2][h$_2$(k)], ..., QBF$_s$[HR$_s$][h$_{HR_s}$(k)]). if J is equal to 0, it is sure that k is observed for the first time and the following should be done:if the latest EBF is not full, set QBF$_s$[i][h$_i$(k)]=EBF$_{NUM}$, $0 \le i \le HR_s$; if the latest EBF is full, set QBF$_s$[i][h$_i$(k)]=EBF$_{NUM}$+1 (EBF$_{NUM}$ is the number of EBF in the ESBF); otherwise, we can say that the information of item k is stored in the Jth EBF. however, the Jth EBF may not contains the information of k due to the false positive (J sometimes may bigger than the number of the EBF), under this circumstance, we should check all the EBF until we found the one that contains the item k and update its information there. if there is no EBF that contains k, we are sure that k is observed for the first time. On querying the information about an item k, we also check the MQBF to chose an appropriate EBF in the same way.

Now, we will address the issue of choosing the m and h for ESBF.

Algorithm : FIDA(k, ktype)

```
Chose the appropriate SBF for k
       by hasing k to the MQBF
   if ktype is delete
       FIDE(k);
       if all counters in the current filter is 0
          dellocate the current filter;
       end if
   else
          if k is observed for the first time
          if the current filter is not full
              FIIN(k);
          else
              create a new filter of size m₀2ⁱ,
              // m₀ is the size of first created filter,
              // i is the current filter number
              FIIN(k);
          end if
       else
          FIIN(k);
   end if
```

Fig. 3. Subroutines of FIDA construction algorithm

Proposition 1. *By our algorithm,* $\ln(\frac{-M}{\ln(P)}) \cdot \frac{e}{\epsilon} \cdot \frac{1}{\epsilon \cdot M}$ *counters is enough for estimating each item with error no more than* ϵN *with probability* ρ*, while* $h = \ln(-\frac{M}{\ln \rho})$*, and* $m = \frac{e}{\epsilon} \cdot \frac{1}{\epsilon \cdot M}$

the prove to this Proposition is similar to that of $hCount$[10], but we get much tighter bound by using Chernoff Bound. As seen from what mentioned above, for an arbitrary item k, the respective associated counters are: $c[1][h_1(k)], c[2][h_2(k)]$, ..., $c[h][h_h(k)]$, where each associated counter contains not only net occurrence n_k for k, but also occurrences of some other items that are mapped to the associated counter. let $e_1, e_2, ..., e_h$ denote the errors of counters related with item k, then the value of the associated counters for k are: $n_k+e_1, n_k+e_2, ..., n_k+e_h$. Provided that all the hash functions are well defined, the expected value of each counter would be N/m, where N is stream length, and m is the slice size, so the expected value of each error is no more than N/m. Let random variable Y denote this error, then

$$E[Y] \leq N/m. \tag{1}$$

from Chernoff Bound, we know that:

$$Pr[Y > (1 + \delta)u] < (\frac{e^\delta}{(1 + \delta)^{(1+\delta)}})^u. \tag{2}$$

where $\delta \geq 0$, and $u = E[Y]$, substitute u with N/m, we can get the following equation:

$$Pr[Y > (1 + \delta) \cdot \frac{N}{m}] < (\frac{e^\delta}{(1 + \delta)^{(1+\delta)}})^{\frac{N}{m}}. \tag{3}$$

the above formula shows that for random variable Y, if we try once, the event that the value of Y is greater than $(1 + \delta) \cdot \frac{N}{m}$ happens with probability no more than $(\frac{e^\delta}{(1+\delta)^{(1+\delta)}})^{\frac{N}{m}}$, this probability is denoted as p. if we try h times, the probability of all values are greater than $(1 + \delta) \cdot \frac{N}{m}$ is no more than p^h, in other word, with probability $1 - p^h$, that the value of Y is smaller than $(1 + \delta) \cdot \frac{N}{m}$ happens at least once. Let Y_{min} denote the minimal value of Y among h tries, following formulas can be obtained.

$$Pr[Y_{min} - (1 + \delta) \cdot \frac{N}{m}) > 0] < (\frac{e^\delta}{(1+\delta)^{(1+\delta)}})^{\frac{N}{m} \cdot h}. \tag{4}$$

$$Pr[Y_{min} - (1 + \delta) \cdot \frac{N}{m}) \leq 0] \geq 1 - (\frac{e^\delta}{(1+\delta)^{(1+\delta)}})^{\frac{N}{m} \cdot h}. \tag{5}$$

let ρ denote the probability of event that all M distinct items satisfy the above formula simultaneously. Then,

$$\rho = (1 - (\frac{e^\delta}{(1+\delta)^{(1+\delta)}})^{\frac{N}{m} \cdot h})^M \approx e^{-(\frac{e^\delta}{(1+\delta)^{(1+\delta)}})^{\frac{N}{m} \cdot h} \cdot M} \tag{6}$$

we know that N is the occurrence of all items and Y_{min} is the error part of each estimated value. The user specified error parameter ϵ can be set as follows,

$$\epsilon = \frac{1 + \delta}{m} \tag{7}$$

from(6), we get h as follows

$$h = \frac{\ln(\frac{-M}{\ln(\rho)})}{\ln(\frac{(1+\delta)^{(1+\delta)}}{e^\delta})} \cdot \frac{m}{N} \tag{8}$$

let V denote the size of the hash table, then $V = m \cdot h$, from equation (7) and (8), we get V as follows,

$$V = \frac{\ln(\frac{-M}{\ln(\rho)})}{\ln(\frac{(1+\delta)^{(1+\delta)}}{e^\delta})} \cdot \frac{m}{N} \cdot \frac{1 + \delta}{\epsilon} \tag{9}$$

$$V = \frac{\ln(\frac{-M}{\ln(\rho)})}{\frac{\ln(1+\delta)}{1+\delta} - \frac{\delta}{(1+\delta)^2}} \cdot \frac{1}{\epsilon} \cdot \frac{1}{\epsilon \cdot N} \tag{10}$$

since $N \geq M, \delta >= 0$, in addition, we assume $\delta \ll 1$, so we can get the following formulas,

$$\frac{ln(1 + \delta)}{1 + \delta} - \frac{\delta}{(1+\delta)^2} \approx \frac{ln(1 + \delta)}{1 + \delta} \approx \frac{1}{e}. \tag{11}$$

$$V \leq \ln(\frac{-M}{\ln \rho}) \cdot \frac{e}{\epsilon} \cdot \frac{1}{\epsilon \cdot M}. \tag{12}$$

Therefor, we set m and h as follows,

$$h = \ln(\frac{-M}{\ln \rho}), m = \frac{e}{\epsilon} \cdot \frac{1}{\epsilon \cdot M}. \tag{13}$$

From equation(13) we can see that with M increases, the m decreases. However, it does not mean that m can decrease to 0 when M is very great, because from equation(7) we get $m \geq \frac{1}{\epsilon}$, since $\delta \geq 0$. Proposition 1 guarantees that the maximal error of an estimated value is no more than ϵ with probability ρ. All items whose frequency over threshold will be output, and the items whose frequency less than $(s - \epsilon)$will not be output. the equation (12) shows the space bound in theory, however, in practice, the underlying data set tends to be skewed, and a much smaller space is sufficient with high precision and recall(the experiments in section 4 show our results). From above, we know that only $\ln(\frac{-M}{\ln \rho})$counters need to be updated per transaction when the number of distinct items in the stream is known. and in the condition that the number of distinct items in the stream is not known, the number of counters need to be updated per transaction is: $\ln(\frac{-M}{\ln \rho}) + \log(\frac{1}{P})$.

4   Experiments Study

In this section we study the behavior of the proposed method. The synthetic data set we choose is zipf distribution with range[1...10000], the number of items of the data set varies from 100000 to 10000000. We implemented FI-ESBF and $hCount$[10]in C++, and conducted all experiments on a PC with a 1.7GHz CPU and 256MB main memory.

Fig. 4 (a) shows that FI-ESBF uses much less memory space than hCount does at most cases. when M is 1000, FI-ESBF consume more memory space than hcout does, the reason is that when $M < \frac{1}{\epsilon}$, from equation(13), FI-ESBF needs more counters than Hcount does. as M grows, FI-ESBF needs less counters. when$M > \frac{e}{\epsilon}$, we may get $m < \frac{1}{\epsilon}$ from equation(13), but as mentioned above, it is necessary that $m \geq \frac{1}{\epsilon}$. as a result, when $M > 6000$, $m = \frac{1}{\epsilon}$, and the memory consumed by FI-ESBF only increases slightly due to the overflowed counters. On the other hand, as M grows, the h in the hCount grows(from 10 to 12), and the m remains the same, however, when $4000 \leq M \leq 10000$, the computed h does not change, so the memory consumed by hCount remains the same. It is necessary to allot adequate bits to the counters in the hCount to make sure that the counters will not overflow, on the other hand, we can allocate fewer bits to the counters in the FI-ESBF, since the counters in the FI-ESBF is extensible, which makes hCount uses much more space than FI-ESBF does. Fig. 4(b) shows that FI-ESBF needs a bit more time than hCount does for processing the same data set, since FI-ESBF need to check the counters in the ESBF to see if they are overflowed or about to overflow, which cost more time. Fig. 5 compares the processing time and memory consumption among FI-ESBF and hCount under the circumstance that the data range in the data set is not known. there would be several EBFs and hash tables in the FI-ESBF and hCount respectively

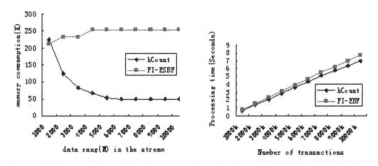

Fig. 4. The memory consumption and processing time between hCount and FI-ESBF when data range is known (a) memory consumption (b) processing time

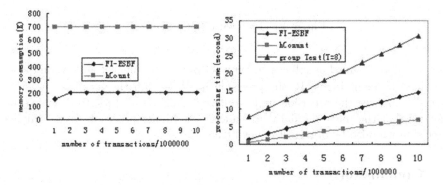

Fig. 5. The memory consumption and processing time between hCount and FI-ESBF when data range is not known (a) memory consumption (b) processing time

under such circumstance. the first constructed EBF in the FI-ESBF is capable of handling first 10000 different items in the data set, and the first constructed hash table in the hCount is capable of handling the item ranging from 1 to 10000. Since hCount processes item based on its range, so in the early stage of processing, new hash table has to be constructed to handle the item that is not in the rang[1,10000], as a result, hCount consume large amount memory space even there are not large number of items processed. As for FI-ESBF, only when the number of distinct item processed in the data set is more than 10000, it needs to construct new EBF. The major space used in FI-ESBF is ESBF and MQBF, and MQBF uses much more space the ESBF does. We also can see from Fig. 5(b) that FI-ESBF needs more time than hCount does to process the same data set, since FI-ESBF needs time to check MQBF to obtain the information about an item, but we can also see that, FI-ESBF needs much less time to process the same data set than groupTest does. Proposition 1 gives the minimal size of a hash table to guarantee the error ϵ, however, in our experiments, we found that a much less memory space is sufficient to achieve high precision and recall is always 1 due to the false positive. For example, when $M = 10000$, and $\epsilon = 0.0001$, we get $\epsilon M = 1$, but when we set $\epsilon M = 10$, the same result is obtained. Precision

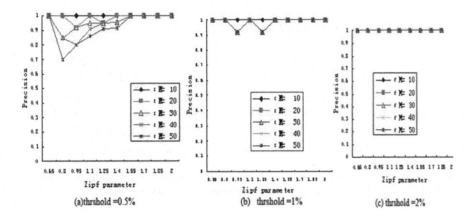

Fig. 6. Precision with Zipf parameter

and recall are defined as follows: given a set of true frequent items, A, and a set obtained frequent items, B, the precision is $\frac{A \cap B}{B}$, and the recall is $\frac{A \cap B}{A}$ [5]. Fig. 6 compares the precision under different zipf parameters, from which, we can see that the more the data set skewed, the higher precision we get, higher precision will also be obtained under higher threshold.

5 Conclusion

In this paper, we propose ESBF-based algorithm FI-ESBF for frequent items mining in streams using much less space than that of the best reported algorithm. FI-ESBF can deal with both insertion and deletion transactions. By using Chernoff Bound, we test that the space bound is no more than $\ln(\frac{-M}{\ln \rho}) \cdot \frac{1}{\epsilon} \cdot \frac{1}{\epsilon \cdot M}$ counters, which is much tighter than what proposed in [10]. Our approach does not require the priori knowledge about the data distribution and length of the distinct items to be stored in the stream, which is more suitable for streaming scenario.

References

1. Bloom, B.: Space/time tradeoffs in hash coding with allowable errors. Commun. ACM 13(7), 422–426 (1970)
2. Fang, M., et al.: Computing iceberg queries efficiently. In VLDB (August 1998)
3. Charikar, M., Chen, K., Farach-Colton, M.: Finding Frequent Items in Data Streams. In: Widmayer, P., Triguero, F., Morales, R., Hennessy, M., Eidenbenz, S., Conejo, R. (eds.) ICALP 2002. LNCS, vol. 2380, pp. 693–703. Springer, Heidelberg (2002)
4. Manku, G., Motwani, R.: Approximate Frequency Counts over Data Streams. In: Proceedings of the 28th International Conference on Very Large Data Bases, pp. 346–357 (2002)

5. Xu Yu, J., Chong, Z., Lu, H., Zhou, A.: False Positive or False Negative:Mining Frequent Itemsets form High Speed Transactional Data Streams. In: Proceedings of the 30th International Conference on Very Large Data Bases, pp. 204–215 (2004)
6. Cormode, G., Muthukrishnan, S.: Whats Hot and Whats Not: Tracking Most Frequent Items Dynamically. In: Proceedings of the 22nd Symposium on Principles of Databse Systems, pp. 296–306 (June 2003)
7. Garofalakis, M., Gehrke, J., Rastogi, R.: Querying and mining data streams: you only get one look. In: the tutorial notes of the 28th Int'l Conference on Very Large Databases, Hong Kong, China (August 2002)
8. Demaine, E.D., Lopez-Ortiz, A., Munro, J.I.: Frequency Estimation of Internet Packet Streams with Limited Space. In: Möhring, R.H., Raman, R. (eds.) ESA 2002. LNCS, vol. 2461, pp. 348–360. Springer, Heidelberg (2002)
9. Estan, C., Varghese, G.: New Directions in Traffic Measurement and Accounting: Focusing on the Elephants, Ignoring the Mice. ACM Trans. Comput. Syst. 21(3), 270–313 (2003)
10. Jin, C., Qian, W., Sha, C., Yu, J.X., Zhou, A.: Dynamically Maintaining Frequent Items over A Data Stream. In: Proceedings of the Twelfth International Conference on Information and Knowledge Management, pp. 287–294. ACM Press, New York (2003)
11. Karp, R., Shenker, S., Papadimitriou, C.: A Simple Algorithm for Finding Frequent Elements in Streams and Bags. ACM Transactions on Database Systems 28(1), 51–55 (2003)
12. Metwally, A., Agrawal, D., El Abbadi, A.: Efficient Computation of Frequent and Top-k Elements in Data Streams. Technical Report 2005-23, University of California, Santa Barbara (September 2005)
13. Fan, L., Cao, P., Almeida, J., Broder, A.Z.: Summary Cache: A Scalable Wide-Area Web Cache Sharing Protocol. IEEE/ACM Transactons on networking 8(3) (June 2000)
14. Aguilar-Saborit, J., Trancoso, P., Muntes-Mulero, V., Larriba-Pey, J.L.: Dynamic Count Filters. SIGMOD Record 35(1) (March 2006)
15. Cohen, S., Matias, Y.: Spectral Bloom Filters. In: SIGMOD 2003, June 912 , San Diego, CA (2003)

A New Decision Tree Classification Method for Mining High-Speed Data Streams Based on Threaded Binary Search Trees*

Tao Wang[1], Zhoujun Li[2], Xiaohua Hu[3], Yuejin Yan[1], and Huowang Chen[1]

[1] Computer School, National University of Defense Technology, Changsha, 410073, China
[2] School of Computer Science & Engineering, Beihang University, Beijing, 100083, China
[3] College of Information Science and Technology, Drexel University,Philadelphia, PA, USA
InsistStar@nudt.edu.cn

Abstract. One of most important algorithms for mining data streams is VFDT. It uses Hoeffding inequality to achieve a probabilistic bound on the accuracy of the tree constructed. Gama et al. have extended VFDT in two directions. Their system VFDTc can deal with continuous data and use more powerful classification techniques at tree leaves. In this paper, we revisit this problem and implemented a system VFDTt on top of VFDT and VFDTc. We make the following three contributions: 1) we present a threaded binary search trees (TBST) approach for efficiently handling continuous attributes. It builds a threaded binary search tree, and its processing time for values inserting is $O(nlogn)$, while VFDT's processing time is $O(n^2)$. When a new example arrives, VFDTc need update $O(logn)$ attribute tree nodes, but VFDTt just need update one necessary node.2) we improve the method of getting the best split-test point of a given continuous attribute. Comparing to the method used in VFDTc, it improves from $O(nlogn)$ to $O(n)$ in processing time. 3) Comparing to VFDTc, VFDTt's candidate split-test number decrease from $O(n)$ to $O(logn)$.Comparing to VFDT, the most relevant property of our system is an average reduction of 25.53% in processing time, while keep the same tree size and accuracy. Overall, the techniques introduced here significantly improve the efficiency of decision tree classification on data streams.

Keywords: Data Streams, VFDT, Continuous Attribute, Threaded Binary Search Tree.

1 Introduction

Decision trees are one of the most used classification techniques for data mining. Tree models have high degree of interpretability. Global and complex decisions can be approximated by a series of simpler and local decisions. Algorithms that construct decision trees from data usually use a divide and conquer strategy. A complex problem is divided into simpler problems and recursively the same strategy is applied to the

---

* This work was supported by the National Science Foundation of China under Grants No. 60573057, 60473057 and 90604007.

T. Washio et al. (Eds.): PAKDD 2007 Workshops, LNAI 4819, pp. 256–267, 2007.
© Springer-Verlag Berlin Heidelberg 2007

sub-problems. The solutions of sub-problems are combined in the form of a tree to yield the solution of the complex problem [3, 20, 22].

More recently, the data mining community has focused on a new model of data processing, in which data arrives in the form of continuous streams [1, 3, 9, 11, 12, 16]. The key issue in mining on data streams is that only one pass is allowed over the entire data. Moreover, there is a real-time constraint, i.e. the processing time is limited by the rate of arrival of instances in the data stream, and the memory and disk available to store any summary information may be bounded. For most data mining problems, a one-pass algorithm cannot be very accurate. The existing algorithms typically achieve either a deterministic bound on the accuracy or a probabilistic bound [21, 23].

Domingos and Hulten [2, 6] have addressed the problem of decision tree construction on data streams. Their algorithm guarantees a probabilistic bound on the accuracy of the decision tree that is constructed. Gama et al. [5] have extended VFDT in two directions: the ability to deal with continuous data and the use of more powerful classification techniques at tree leaves. Wang et al. [28] propose a method to manage continuous data based on binary search trees.

In this paper, we revisit the problem of decision tree construction on streaming data. We propose the VFDTt system, which makes three contributions: 1) we present a threaded binary search trees approach for efficiently handling continuous attributes. It builds a threaded binary search tree, and its processing time for values inserting is $O(nlogn)$, while VFDT`s processing time is $O(n^2)$. When a new example arrives, VFDTc need update $O(logn)$ attribute tree nodes, but VFDTt just need update one necessary node. 2) we improve the method to get the best split-test point of a given attribute. Comparing to the method used in VFDTc, it improves from $O(nlogn)$ to $O(n)$ in processing time. 3) Based on[27], our VFDTt has less candidate split-test point than VFDTc for getting the best one.

The rest of the paper is organized as follows. Section 2 describes the related works that is the basis for this paper. Section 3 presents the technical details of VFDTt. We detail the major options that we implemented and the differences to VFDT and VFDTc. The system has been implemented and evaluated, and experimental evaluation is done in Section 4. Last section concludes the paper, resuming the main contributions of this work.

2   Related Work

In this section we analyze the related works that our VFDTt bases on.

Decision trees support continuous attributes by allowing internal nodes to contain tests of the form $A_i \leq T$ (the value of attribute i is less than threshold T). Traditional induction algorithms learn decision trees with such tests in the following manner. For each continuous attribute, they construct a set of candidate tests by sorting the values of that attribute in the training set and using a threshold midway between each adjacent pair of values that come from training examples with different class labels to get the best split-test point.

There are several reasons why this standard method is not appropriate when learning from data streams. The most serious of these is that it requires that the entire training set be available ahead of time so that split thresholds can be determined.

2.1 VFDT

VFDT(Very Fast Decision Tree) system[2], which is able to learn from abundant data within practical time and memory constraints. In VFDT a decision tree is learned by recursively replacing leaves with decision nodes. Each leaf stores the sufficient statistics about attribute-values. The sufficient statistics are those needed by a heuristic evaluation function that evaluates the merit of split-tests based on attribute-values. When an example is available, it traverses the tree from the root to a leaf, evaluating the appropriate attribute at each node, and following the branch corresponding to the attribute's value in the example. When the example reaches a leaf, the sufficient statistics are updated. Then, each possible condition based on attribute-values is evaluated. If there is enough statistical support in favor of one test over the others, the leaf is changed to a decision node. The new decision node will have as many descendant leaves as the number of possible values for the chosen attribute (therefore this tree is not necessarily binary). The decision nodes only maintain the information about the split-test installed in this node. The initial state of the tree consists of a single leaf: the root of the tree. The heuristic evaluation function is the Information Gain (denoted by $G(\cdot)$. The sufficient statistics for estimating the merit of a discrete attribute are the counts n_{ijk}, representing the number of examples of class k that reach the leaf, where the attribute j takes the value i. The Information Gain measures the amount of information that is necessary to classify an example that reach the node: $G(A_j)=info(examples)-info(A_j)$. The information of the attribute j is given by:

$$\inf o(A_j) = \sum_i P_i (\sum_k -P_{ik} \log(P_{ik})) \quad \text{where} \quad P_{ik} = n_{ijk} / \sum_a n_{ajk} \quad , \quad \text{is the}$$

probability of observing the value of the attribute i given class k and $P_i = \sum_a n_{ija} / \sum_a \sum_b n_{ajb}$ is the probabilities of observing the value of attribute i.

As mentioned in Catlett and others [23], that it may be sufficient to use a small sample of the available examples when choosing the split attribute at any given node. To determine the number of examples needed for each decision, VFDT uses a statistical result known as Hoeffding bounds or additive Chernoff bounds. After n independent observations of a real-valued random variable r with range R, the Hoeffding bound ensures that, with confidence $1\text{-}\delta$, the true mean of r is at least $\bar{r} - \varepsilon$, where $\bar{r}$ is the observed mean of samples and $\varepsilon = \sqrt{\dfrac{R^2 \ln(1/\delta)}{2n}}$. This is true irrespective of the

probability distribution that generated the observations.

Let G(·) be the evaluation function of an attribute. For the information gain, the range R, of G(·) is $log_2 \#classes$. Let x_a be the attribute with the highest G(·), x_b the attribute with second-highest G(·) and $\Delta \overline{G} = \overline{G}(x_a) - \overline{G}(x_b)$, the difference between the two better attributes. Then if $\Delta \overline{G} > \varepsilon$ with n examples observed in the leaf, the Hoeffding bound states with probability 1-δ that x_a is really the attribute with highest value in the evaluation function. In this case the leaf must be transformed into a decision node that splits on x_a.

For continuous attribute, whenever VFDT starts a new leaf, it collects up to M distinct values for each continuous attribute from the first examples that arrive at it. These are maintained in sorted order as they arrive, and a candidate test threshold is maintained midway between adjacent values with different classes, as in the traditional method. Once VFDT has M values for an attribute, it stops adding new candidate thresholds and uses additional data only to evaluate the existing ones. Every leaf uses a different value of M, based on its level in the tree and the amount of RAM available when it is started. For example, M can be very large when choosing the split for the root of the tree, but must be very small once there is a large partially induced tree, and many leaves are competing for limited memory resources. Notice that even when M is very large (and especially when it is small) VFDT may miss the locally optimal split point. This is not a serious problem here for two reasons. First, if data is an independent, identically distributed sample, VFDT should end up with a value near (or an empirical gain close to) the correct one simply by chance. And second, VFDT will be learning very large trees from massive data streams and can correct early mistakes later in the learning process by adding additional splits to the tree.

Thinking of each continuous attribute, we will find that the processing time for the insertion of new examples is $O(n^2)$, where n represents the number of distinct values for the attribute seen so far.

2.2 VFDTc

VFDTc is implemented on top of VFDT, and it extends VFDT in two directions: the ability to deal with continuous attributes and the use of more powerful classification techniques at tree leaves. Here, we just focus on the handling of continuous attributes.

In VFDTc a decision node that contains a split-test based on a continuous attribute has two descendant branches. The split-test is a condition of the form $attrib_j \leq T$. The two descendant branches correspond to the values *TRUE* and *FALSE* for the split-test. The cut point is chosen from all the possible observed values for that attribute. In order to evaluate the goodness of a split, it needs to compute the class distribution of the examples at which the attribute-value is less than or greater than the cut point. The counts n_{ijk} are fundamental for computing all necessary statistics. They are kept with the use of the following data structure: In each leaf of the decision tree it maintains a vector of the classes' distribution of the examples that reach this leaf. For each continuous attribute j, the system maintains a binary attribute tree structure. A node in the binary tree is identified with a value i(that is the value of the attribute j seen in an example), and two vectors (of dimension k) used to count the values that go through that node. Two vectors, *VE* and *VH* contain the counts of values respectively $\leq i$ and $> i$ for the examples labeled with class k. When an example reaches leaf, all the binary trees are updated. In [5], an algorithm of inserting a value in the binary tree is presented. Insertion of a new value in this structure is $O(nlogn)$ where n represents the number of distinct values for the attribute seen so far.

To obtain the Information Gain of a given attribute, VFDTc uses an exhaustive method to evaluate the merit of all possible cut points. Here, any value observed in the examples seen so far can be used as cut point. For each possible cut point, the information of the two partitions is computed using equation 1.

$$\inf o(A_j(i)) = P(A_j \leq i) * iLow(A_j(i)) + P(A_j > i) * i\,High(A_j(i)) \quad (1)$$

Where i is the cut point, $iLow(A_j(i))$ the information of $A_j \leq i$ (equation 2) and $iHigh(A_j(i))$ the information of $A_j > i$ (equation 3).

$$iLow(A_j(i)) = -\sum_K P(K = k \mid A_j \leq i) * \log(P(K=k \mid A_j \leq i)) \quad (2)$$

$$iHigh(A_j(i)) = -\sum_K P(K = k \mid A_j > i) * \log(P(K=k \mid A_j > i)) \quad (3)$$

VFDTc only considers a possible *cut_point* if and only if the number of examples in each of subsets is higher than P_{min} (a user defined constant) percentage of the total number of examples seen in the node. [5] Presents the algorithm to compute #($A_j \leq i$) for a given attribute j and class k. The algorithm's processing time is $O(logn)$, so the best split-test point calculating time is $O(nlogn)$. Here, n represents the number of distinct values for the attribute seen so far at that leaf.

3   Technique Details

We implement a system named VFDTt on top of VFDT and VFDTc. It handles continuous attributes using threaded binary search trees, and uses a more efficient best split-test point calculating method.

For discrete attributes, they are processed using the algorithm mentioned in VFDT [2]. Our VFDTt specially focus on continuous attribute handling.

3.1   Threaded Binary Search Tree Structure for Continuous Attributes

VFDTt maintains a threaded binary search tree for each continuous attribute. The threaded binary search tree data structure will benefit the procedure of inserting new example and calculating best split-test point.

For each continuous attribute i, the system maintains a threaded binary search tree structure. A node in the threaded binary search tree is identified with a value *keyValue* (that is the value of the attribute i seen in the example), and a vector(of dimension k) used to count the values that go through that node. This vector *classTotals[k]* contains the counts of examples which value is *keyValue* and class labeled with k. A node manages *left* and *right* pointers for its left and right child, where its left child corresponds to $\leq keyValue$, while its right child corresponds to $>keyValue$. For the goodness of calculating the best split-test point, a node contains *prev* and *next* pointers for the previous and next node. At most, three nodes` *prev* and *next* pointers will be updated while new example arrives.

VFDTt maintains a *head* pointer for each continuous attribute to traverse all the threaded binary trees.

3.2   Updates the Threaded Search Binary Tree While New Examples Arrives

One of the key problems in decision tree construction on streaming data is that the memory and computational cost of storing and processing the information required to obtain the best split-test point can be very high. For discrete attributes, the number of distinct values is typically small, and therefore, the class histogram does not require much memory. Similarly, searching for the best split predicate is not expensive if number of candidate split conditions is relatively small.

However, for continuous attributes with a large number of distinct values, both memory and computational costs can be very high. Many of the existing approaches are scalable, but they are multi-pass. Decision tree construction requires a preprocessing phase in which attribute value lists for continuous attributes are sorted [20]. Preprocessing of data, in comparison, is not an option with streaming data, and sorting during execution can be very expensive. Domingos and Hulten have described and evaluated their one-pass algorithm focusing only on discrete attributes [2], and in later version they uses sorted array to handle continuous attribute. This implies a very high memory and computational overhead for inserting new examples and determining the best split point for a continuous attribute.

In VFDTt a Hoeffding tree node manages a threaded binary search tree for each continuous attribute before it becomes a decision node.

```
Procedure InsertValueTBSTree(x, k, TBSTree)
  Begin
    while (TBSTree ->right != NULL || TBSTree ->left != NULL )
        if (TBSTree ->keyValue = = x )        then    break;
        Elseif (TBSTree ->keyValue > x )    then TBSTree = TBSTree ->left;
        else TBSTree = TBSTree ->right;
    Creates a new node curr based on x and k;
    If ( TBSTree.keyValue = = x )      then      TBSTree.classTotals[k]++;
    Elesif (TBSTree.keyValue > x)      then      TBSTree.left = curr;
    else              TBSTree.right = curr;
    Threads the tree ;( The details of threading is in figure2)
  End
```

Fig. 1. Algorithm to insert value x of an example labeled with class k into a threaded binary search tree corresponding to the continuous attribute i

In the induction of decision trees from continuous-valued data, a suitable threshold T, which discretizes the continuous attribute i into two intervals: $atrr_i \leq T$ and $atrr_i > T$, is determined based on the classification information gain generated by the corresponding discretization. Given a threshold, the test $atrr_i \leq T$ is assigned to the left branch of the decision node while $atrr_i > T$ is assigned to the right branch. As a new example (x,k) arrives, the threaded binary search tree corresponding to the continuous attribute i is update as Figure 1.

In [5], when a new example arrives, $O(logn)$ binary search tree nodes need be updated, but VFDTt just need update a necessary node here. VFDT will cost $O(n^2)$, and our system VFDTt will just cost $O (nlogn)$ (as presented in Figure 1) in execution time for values inserting, where n represents the number of distinct values for the given attribute seen so far.

3.3 Threads the Binary Tree While New Example Arrives

VFDTt need thread the binary search trees while new example arrives. If the new example's value is equal to an existing node`s value, the threaded binary tree doesn't need be threaded. Otherwise, the threaded binary tree need be threaded as Figure 2.

At most, three relevant nodes need be updated here. This threading procedure mentioned in Figure 2 can be embedded in the procedure presented in Figure 1, and the inserting procedure's processing time is still $O(nlogn)$.

```
                  Procedure TBSTthreads()
                  Begin
                  if (new node curr is left child of ptr)
                          curr->next = ptr;
                          curr->nextValue = ptr->keyValue;
                          curr->prev = ptr->prev;
                          ptr->prev->next = curr;
                          prevPtr->nextValue = value;
                          ptr->prev = curr;
                          if (new node curr is right child of ptr)
                              curr->next = ptr->next;
                              curr->nextValue = ptr->nextValue;
                              curr->prev = ptr;
                              ptr->next->prev = curr;
                              ptr->nextValue = value;
                              ptr->next = curr;
          End
```

Fig. 2. Algorithm to thread the binary search tree while new example arrives

3.4 Best Split-Test Point Selecting

Taking advantage of threaded binary search tree, we use a more efficient method to obtain the information gain of a given attribute.

Assuming we are to select an attribute for a node having a set S of N examples arrived, these examples are managed by a threaded binary tree according to the values of the continuous attribute i ; and an ordered sequence of distinct values a_1, a_2 ... a_n is formed. Every pair of adjacent data points suggests a potential threshold $T = (a_i + a_{i+1})/2$ to create a cut point and generate a corresponding partition of attribute i. Fayyad [27] had proved that only the class boundary points could be the cut points to obtain the maximum information in classification, which implies if a_i and a_{i+1} belong to the same class, a cut point between them can not lead to a partition that has maximum information gain. Therefore, we can generate a smaller set of candidate cut points from the class boundary points. In order to calculate the goodness of a split, we need to compute the class distribution of the examples at which the attribute value is less than or greater than threshold T. The counts *TBSTree.classTotals[k]* are fundamental for computing all necessary statistics.

To take the advantage of threaded binary search tree, we record the *head* pointer of each attribute's threaded binary search tree. As presented in Figure 3, traversing from the *head* pointer to the tail pointer, we can easily compute the information of all the potential thresholds (equation 1, 2, 3).

Here, VFDTc will cost $O(nlogn)$, and our system VFDTt will just cost $O(n)$ in processing time, where n represents the number of distinct values for the given continuous attribute seen so far.

```
Procedure BSTInorderAttributeSplit(TBSTtreePtr ptr,int *belowPrev[])
   Begin
      if ( ptr->next == NULL) then    break;
      for ( k = 0 ; k < count ; k++)
            *belowPrev[k] += ptr->classTotals[k];
      Calculates the information gain using *belowPrev[];
         BSTInorderAttributeSplit( ptr->next,int *belowPrev[]);
   End
```

Fig. 3. Algorithm to compute the information gain of a continuous attribute

4   Evaluation

In this section we empirically evaluate VFDTt. The main goal of this section is to provide evidence that the use of threaded binary search tree decreases the processing time of VFDT, while keeps the same error rate and tree size. The algorithms` processing time is listed in Table 1.

Table 1. Algorithm's processing time

Algorithm Name	Inserting time	Best split-test point calculating time
VFDT	$O(n^2)$	$O(n)$
VFDTc	$O(n \log n)$	$O(n \log n)$
VFDTt	$O(n \log n)$	$O(n)$

We first describe the data streams used for our experiments. We use a tool named *treeData* mentioned in [2] to create synthetic data. It creates a synthetic data set by sampling from a randomly generated decision tree. They were created by randomly generating decision trees and then using these trees to assign classes to randomly generated examples. It produced the random decision trees as follows. Starting from a tree with a single leaf node (the root) it repeatedly replaced leaf nodes with nodes that tested a randomly selected attribute which had not yet been tested on the path from the root of the tree to that selected leaf. After the first three levels of the tree each selected leaf had a probability of f of being pre-pruned instead of replaced by a split (and thus of remaining a leaf in the final tree). Additionally, any branch that reached a depth of 18 was pruned at that depth. Whenever a leaf was pruned it was randomly (with uniform probability) assigned a class label. A tree was completed as soon as all of its leaves were pruned.

VFDTc`s goal is to show that using stronger classification strategies at tree leaves will improve classifier's performance. With respect to the processing time, the use of naïve Bayes classifier will introduce an overhead [5], VFDTc is slower than VFDT. Here we just list the detailed results of VFDT and VFDTt.

We ran our experiments on a Pentium IV/2GH machine with 512MB of RAM, which running Linux RedHat 9.0.

Figure 4 shows the processing (excluding I/O) time of learners as a function of the number of training examples averaged over nine runs. VFDT and VFDTt run with parameters $\delta = 10^{-7}, \tau = 5\%, n_{\min} = 300, noise = 5\%, example\ number = 100000K$, no leaf reactivation, and no rescan. Averagely, comparing to VFDT, VFDTt`s average reduction of processing time is 25.53%.

In this work, we measure the size of tree models as the number of decision nodes plus the number of leaves. We should note that VFDT and VFDTt generate exactly the same tree model. Figure 5 shows the average tree size induced by each of the learner.

Figure 6 shows the error rate curves of VFDT and VFDTt. As for dynamic data stream with 100 million examples, we notice that the two learners similarly have the same error rate. We have done another experiment using 1 million examples generated on disk, the result shows that they have same error rate.

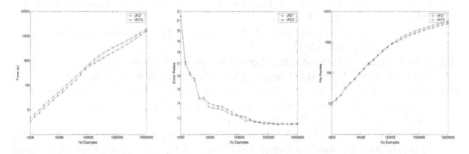

Fig. 4. Processing time as a function of the examples numbers **Fig. 5.** Tree size as a function of the examples numbers **Fig. 6.** Error rate as a function of the examples numbers

5 Conclusions and Future Work

In this paper, we propose a system VFDTt on top of VFDT and VFDTc, one of the most promising algorithms for decision tree induction from data streams. Focusing on continuous attribute, we have developed and evaluated a new technique named TBST to insert new example and calculate best split-test point efficiently. It builds threaded binary search trees, and its processing time for values insertion is *O(nlogn)*. Comparing to the method used in VFDTc, it improves from *O(nlogn)* to *O(n)* in processing time for best split-test point calculating.

In the future, we would like to expand our work in some directions. First, we do not discuss the problem of time changing concept here, and we will apply our method to those strategies that take into account concept drift [4, 6, 10, 14, 15, 19, 24, 25]. Second, we want to apply new fuzzy decision tree methods in data streams classification [8, 13, 17, 18, 26].

References

[1] Babcock, B., Babu, S., Datar, M., Motawani, R., Widom, J.: Models and Issues in Data Stream Systems. In: PODS (2002)

[2] Domingos, P., Hulten, G.: Mining High-Speed Data Streams. In: Proceedings of the Association for Computing Machinery Sixth International Conference on Knowledge Discovery and Data Mining, pp. 71–80 (2000)

[3] Mehta, M., Agrawal, A., Rissanen, J.: SLIQ: A Fast Scalable Classifier for Data Mining. In: Proceedings of The Fifth International Conference on Extending Database Technology, Avignon, France, pp. 18–32 (1996)

[4] Fan, W.: StreamMiner: A Classifier Ensemble-based Engine to Mine Concept Drifting Data Streams. In: VLDB 2004

[5] Gama, J., Rocha, R., Medas, P.: Accurate Decision Trees for Mining High-Speed Data Streams. In: Domingos, P., Faloutsos, C. (eds.) Proceedings of the Ninth International Conference on Knowledge Discovery and Data Mining, ACM Press, New York (2003)

[6] Hulten, G., Spencer, L., Domingos, P.: Mining Time-Changing Data Streams. In: ACM SIGKDD (2001)

[7] Jin, R., Agrawal, G.: Efficient Decision Tree Construction on Streaming Data. In: Proceedings of ACM SIGKDD (2003)

[8] Last, M.: Online Classification of Nonstationary Data Streams. Intelligent Data Analysis 6(2), 129–147 (2002)

[9] Muthukrishnan, S.: Data streams: Algorithms and Applications. In: Proceedings of the fourteenth annual ACM-SIAM symposium on discrete algorithms (2003)

[10] Wang, H., Fan, W., Yu, P., Han, J.: Mining Concept-Drifting Data Streams using Ensemble Classifiers. In: 9th ACM International Conference on Knowledge Discovery and Data Mining, Washington DC, USA. SIGKDD (2003)

[11] Arasu, A., Babcock, B., Babu, S., Datar, M., Ito, K., Nishizawa, I., Rosenstein, J., Widom, J.: STREAM: The Stanford Stream Data Manager Demonstration Description –Short Overview of System Status and Plans. In: Proc. of the ACM Intl Conf. on Management of Data (SIGMOD 2003) (June 2003)

[12] Aggarwal, C., Han, J., Wang, J., Yu, P.S.: On Demand Classification of Data Streams. In: Proc. 2004 Int. Conf. on Knowledge Discovery and Data Mining (KDD 2004), Seattle, WA (2004)

[13] Guetova, M., Holldobter, S.H.-P.: Incremental Fuzzy Decision Trees. In: Jarke, M., Koehler, J., Lakemeyer, G. (eds.) KI 2002. LNCS (LNAI), vol. 2479, Springer, Heidelberg (2002)

[14] Ben-David, S., Gehrke, J., Kifer, D.: Detecting Change in Data Streams. In: Proceedings of VLDB 2004

[15] Aggarwal, C.: A Framework for Diagnosing Changes in Evolving Data Streams. In: Proceedings of the ACM SIGMOD Conference (2003)

[16] Gaber, M.M., Zaslavskey, A., Krishnaswamy, S.: Mining Data Streams: a Review. SIGMOD Record 34(2) (June 2005)

[17] Cezary, Z.J.: Fuzzy Decision Trees: Issues and Methods. IEEE Transactions on Systems, Man, and Cybernetics 28(1), 1–14 (1998)

[18] Utgoff, P.E.: Incremental Induction of Decision Trees. Machine Learning 4(2), 161–186 (1989)

[19] Xie, Q.H.: An Efficient Approach for Mining Concept-Drifting Data Streams, Master Thesis

[20] Quinlan, J.R.: C4.5: Programs for Machine Learning. Morgan Kaufmann, San Mateo, CA (1993)
[21] Hoeffding, W.: Probability Inequalities for Sums of Bounded Random Variables. Journal of the American Statistical Association 58, 13–30 (1963)
[22] Breiman, L., Friedman, J.H., Olshen, R.A., Stone, C.J.: Classification and Regression Trees, Wadsworth, Belmont, CA (1984)
[23] Maron, O., Moore, A.: Hoeffding Races: Accelerating Model Selection Search for Classification and Function Approximation. In: Cowan, J.D., Tesauro, G., Alspector, J. (eds.) Advances in Neural Information Processing System (1994)
[24] Kelly, M.G., Hand, D.J., Adams, N.M.: The Impact of Changing Populations on Classifier Performance. In: Proc. of KDD-99, pp. 367–371 (1999)
[25] Black, M., Hickey, R.J.: Maintaining the Performance of a Learned Classifier under Concept Drift. Intelligent Data Analysis 3, 453–474 (1999)
[26] Maimon, O., Last, M.: Knowledge Discovery and Data Mining,the Info-Fuzzy Network(IFN) Methodology. Kluwer Academic Publishers, Dordrecht (2000)
[27] Fayyad, U.M, Irani, K.B.: On the Handling of Continuous-valued Attributes in Decision Tree Generation. Machine Learning 8, 87–102 (1992)
[28] Wang, T., Li, Z., Yan, Y., Chen, H.: An Efficient Classification System Based on Binary Search Trees for Data Streams Mining. In: ICONS (2007)

Progressive Subspace Skyline Clusters Mining on High Dimensional Data

Rong Hu, Yansheng Lu, Lei Zou, and Chong Zhou

School of Computer Science and Technology
Huazhong University of Science and Technology, Wuhan, 430074, Hubei, China
rhuhust@gmail.com

Abstract. Skyline queries have caused much attention for it helps users make intelligent decisions over complex data. Unfortunately, too many or too few skyline objects are not desirable for users to choose. Practically, users may be interested in the skylines in the subspaces of numerous candidate attributes. In this paper, we address the important problem of recommending skyline objects as well as their neighbors in the arbitrary subspaces of high dimensional space. We define a new concept, subspace skyline cluster, which is a compact and meaningful structure to combine the advantages of skyline computation and data mining. Two algorithms Sorted-based Subspace Skyline Clusters Mining, and Threshold-based Subspace Skyline Clusters Mining are developed to progressively identify the skyline clusters. Our experiments show that our proposed approaches are both efficient and effective.

1 Introduction

Skyline queries are important in many real world applications, such as multi-criteria decision making, data mining, user-preference queries and data visualization. Given a set of multi-dimensional objects, the *skyline* refers to those that are not dominated by any other object. An object p dominates another object q, if p is as good or better in all dimensions and better in at least one dimension. For simplicity, we assume that skylines are computed with respect to *min* conditions on all dimensions; however, the methods discussed here can be applied with any combination of conditions.

The nice property of skyline operation is that it can determine interesting objects from a potentially large set of objects by just specifying query attributes, not exact weightings, which are neither intuitive nor sensible without any prior knowledge about the data. Nevertheless, in a high dimensional space skyline objects no longer offer any interesting insights because of the large number of them. In fact, as claimed in [1], most applications involve only a small subset of attributes. Different users may issue queries regarding various subsets of the dimensions relying on their interests.

On the one hand, the fewer skyline objects in the subspace may not be desirable in some real applications. For instance, if the skyline hotel is occupied, can we find alternatives which are the best choices except the skyline hotel? Can we provide some candidate cell phones that have similar attributes to the skyline so as to help consumers make better choices? On the other hand, the large amount of skylines due to the increasing dimensionality, the large cardinality or the anti-correlated dimensions

T. Washio et al. (Eds.): PAKDD 2007 Workshops, LNAI 4819, pp. 268–279, 2007.

remains an obstacle for obtaining the indeed interesting information. Can we generate high quality patterns reflecting the correlations between the skylines so as to provide users with short cuts to their desired objects?

Fig. 1 gives two typical applications where there are few and many skylines. Take the hotels for example, the left figure shows the situation of too few skylines. In order to offer more suitable candidates, a group of analogous hotels are incorporated as shown in the dashed-lines rectangles. The right figure is an example of organizing numerous nearby skylines into fewer clusters, named *skyline clusters*, to make it convenient to the users for manual processing.

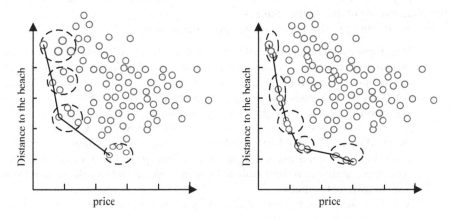

Fig. 1. An example of subspace skyline clusters

The above applications propose a new challenging task for skyline queries: *How to efficiently detect skyline clusters in the ad hoc subspaces.* Most existing work on skyline computation has been focused on efficient queries of skyline objects in large databases. However, these methods are not adapted for finding subspace skyline clusters efficiently because (1) the non-index-based methods require scan the whole data at least once, resulting in expensive overhead; (2) the index-based methods suffer considerable performance degradation in high dimensional space. Furthermore, they are optimized for a fixed set of dimensions, and thus incur poor performance for skyline queries with regard to different attributes.

To deal with the curse of the dimensionality, we adopt the methods motivated by the rank aggregation algorithm over middleware environments studied by Fagin et al [2]. Our model is built on pre-sorting objects separately according to each dimension. This model has at least three advantages. First, it supports the skyline computation in arbitrary subspaces. Second, it guarantees the output is progressive, i.e., the first results should be reported to the user almost instantly and the output size should gradually increase. Third, since this model does not rely on the central index structure, but considers the individual dimension, it makes sense in a number of settings. For example, in distributed environment, each system will sort the objects according to their scores in the specific features. A query retrieves the scores of objects from different related systems and combines them to obtain the final result.

The partial similarity is another focus in this paper. In order to find the top k neighbor of the skyline in the subspace, the partial similarity searches are performed. Most nearest neighbor models consider the similarity over the full dimensions, leaving the partial similarity uncovered. However, the similarity is often affected by a few dimensions with big difference. The probability of encountering this phenomenon becomes greater in high dimensional data. Moreover, users may be interested in a particular aspect of an object instead of the entirety. As a result, the partial similarity search becomes a crucial task. Although we can first compute each skyline, then find the top k neighbor of the skylines, the response time of such approach is not satisfactory. Consequently, it is beneficial to share the computation of the skyline and the determination of the skyline clusters.

Our contributions of this paper can be summaries as follows.

1. We propose the notion of the subspace skyline cluster, which is a compact and meaningful structure to combine the advantages of skyline computation and data mining. Therefore, it not only extends the skyline retrieval in the database, but also finds interesting patterns for data mining tasks.
2. We develop two new progressive skyline clusters identifying algorithms in arbitrary subspaces, namely Sorted-based Subspace Skyline Clusters Mining (SSSCM), and Threshold-based Subspace Skyline Clusters Mining (TSSCM). Instead of one central index designed for the full space. Our methods have the wide applicability into the distributed systems, which collect information from a set of different sources.
3. We show the efficiency and effectiveness of the proposed algorithms through a comprehensive experimental study.

The rest of the paper is organized as follows. In the next section we provide a brief overview of some related research work. Then we formulate the problem and give some analysis in section 3. We describe our algorithms SSSCM and TSSCM in section 4. The experimental evaluation of the proposed algorithms is reported in section 5. Finally, in section 6 we conclude the paper with directions for future work.

2 Related Work

The basic idea of skyline queries originated from some old topics such as maximal vector problem [3], convex hull [4], nearest neighbor search and top k search. The algorithms developed usually suit for a small dataset within the main memory.

Since it was introduced into large database systems by Borzonyi et al. [5], the skyline queries have attracted considerable attention. In the context of databases, the skyline queries can be divided into three categories: the conventional skyline queries, the subspace skyline queries and the skyline variant queries. We will survey the subspace skyline queries since it is closely related to our work.

Yuan et al. [6] proposed to use a complete pre-computation based approach for answering multiple related skyline queries and proposed a new concept named SKYCUBE, which is an union of skyline results over all non-empty subsets of the d dimensions. Two methods that compute a SKYCUBE are: breadth-first and bottom-up manner, depth-first and top-down manner. Another study on computing skylines in subspaces by data cube was introduced by Pei et al. [7]. They investigated the semantics of skylines and introduced new concepts called *skyline group* and *decisive subspaces* to capture the semantics. They also developed an efficient algorithm SKYEY to compute the set of skyline groups and their decisive subspace. We aim at computing the skyline of one particular subspace, as opposed to all subspaces.

Besides the skyline computation in high-dimensional space, our work is also related to optimal aggregation algorithms [2]. To determine the top k objects, the naïve algorithm must access every object in the database, to find its grade under each attribute. Fagin has given two methods, FA and TA. The FA algorithm is optimal with high probability in the worst case. By improving the FA algorithm with setting a threshold, the TA algorithm is essentially optimal, not just for some monotone aggregation functions, but for all of them, not just in a high probability worst case sense, but over every database.

[8] is a quite similar work to ours. A novel concept, called thick skyline is proposed, which recommends not only skyline objects but also their nearby neighbors within ε-distance. Three computation methods, Sampling-and-Pruning, Indexing-and-Estimating and Microcluster-based are developed under three typical scenarios, a single file to represent dataset, general index structure, and special summarization structure of micro-clusters. None of the methods can be applied to detect skyline clusters on any subset of the dimensions because instead of retrieval the skyline in the original space, our algorithm aims at the subspace skyline analysis. Moreover, we consider the problem of discovering the partial similar objects to the skyline object in the subspace, not the ε-neighbor in the original space. The latter is inadequate to capture the most similar objects because it has some drawbacks: First, to compute the ε-neighborhood, we need to specify the value of ε, which is very difficult for there are distinct attributes and sometimes continuous value domains. Second, a fixed ε value for any skyline object may not be the best choice for all skylines. Third, the ε-neighbor of skyline object is not a reasonable structure since it limits the similar objects within a circle region centered at the skyline object. For the inherent characteristic of the skyline object, part of the circle will never cover any object.

Another related study is cooperative database retrieval proposed in [9]. The basic idea is to form an interactive workflow for cooperative retrieval using samples of the skyline to give users an overview of the best possible objects, evaluate their feedback and perform focused, yet meaningful top k retrieval. This approach differs from our topic as they address the problem of utilizing skyline queries to effectively estimate users' compensation functions for top k retrieval.

3   Definitions and Analysis

3.1   Preliminaries

In this section, we give some notations and definitions that will be used throughout the paper.

We assume that the dataset consists of N objects. Associated with each object p are m fields D_1, "", D_m. We may refer to $p.D_i$ as the i field of p. The dataset is thought of as m sorted sets of objects L_1, "", L_m, and each set is stored in an array in non-descending order by the D_i value. We may refer to L_i as list i. Each entry of L_i is of the form (p, D_i).

For objects p, $q \in S$, p is said to *dominate* q, denoted as $p \succ q$, if $p.D_i \leq q.D_i$ for $1 \leq i \leq d$ and there exists at least one dimension D_j such that $p.D_j < q.D_j$. Object p is a skyline object if p is not dominated by any other objects in S.

Definition 1 (Subspace Skyline) The projection of an object p, p_B ($p \in S$) in the subspace $B \subseteq D$ is in the subspace skyline of B, if p_B is not dominated by any q_B in B for any other object $q \in S$. Then p is called a *subspace skyline* object of B. The set of all subspace skyline objects in subspace B is denoted by SKY_B.

To find the cluster members of the skyline object, we compute the Manhattan distance between the skyline object and the data objects on the subspace. If this difference is within a predefined threshold ε, then this object is returned as one nearest neighbor. This leads to a new problem of determining ε. To address this problem, we leave the choice self-adaptive and more flexible. Specifically, for a skyline object p and the potential objects, we sort the Manhattan distances in the subspaces. Next, among the potential objects, we choose the k ones with the smallest difference. Thus ε is this smallest difference. Since ε equals the k smallest match difference on the subspace, k objects well be returned as the answer. In this way, we solve the problem of subspace skyline clusters by utilizing the approach to computing the top k objects.

Definition 3 (Subspace Skyline Cluster) The subspace skyline cluster is composed of two parts: one part is the single skyline object p as the representative sample; as for the case of too many skylines, the other part is the skyline objects which are the k nearest neighbor of p; as for the case of too few skylines, the other part is the non-skyline objects which are the k nearest neighbor of p. The subspace skyline cluster of p is denoted by G_p.

3.2 Analysis

In this section, we will illustrate two important heuristic properties of the subspace skyline clusters.

The first observation is that the nearest neighbor can be applied to skyline computation. This observation has been proven in [1] by contradiction. Obviously, this observation also holds for subspace skyline queries. The reason for choosing this object first is as follows. First, as it is definitely a member of the skyline, there is no dominating object. Second, it is the one with the maximum perimeter of the dominance region among all the skyline objects. Because perimeter and volume are highly correlated, it also gives a large volume on average. Therefore, it is expected to expunge a large portion of the data objects. Third, the object can be computed fast so that the non-skyline objects can be discarded as early as possible, thereby reducing the overhead of maintaining them before they are pruned.

The second observation is based on the effectiveness of the presorting. Our algorithms presort the objects in non-descending order with respect to each dimension. After finding the nearest neighbor, the algorithms sort the potential objects according to their Manhattan distance to the origin. The intuition behind this heuristic is that an object with a smaller sum is likely to have smaller values on more dimensions and is likely to dominate a large amount of objects, thus rending the pruning more effective. The sorting according to the Manhattan distance ensures that an object p dominating another object q must be visited before q. Hence, we can immediately return the skyline object. A similar idea was previously used in [10].

4 Algorithms

In this section, we present two novel algorithms, namely SSSCM and TSSCM to mine the subspace skyline clusters. Each algorithm takes the sorted list in each dimension $L=\{L_i|\ i\in[1,\ m]\}$, the subspace $B=\{D_j|\ j\in[1,\ d]\}$ and a parameter T as input, and outputs the skyline clusters in the subspace G_B.

Inspired by the ideas proposed in [2], we can use two approaches to search the nearest neighbor. The basic idea of the first approach is to access in parallel the sorted list on every dimension. And it will find the top k objects when there is a set of at least k objects such that each of them has been seen in each list. The basic idea of the second idea is to set a threshold by all the smallest values that have seen from all the lists. This approach stops when the current top k objects have the score larger than the threshold. Based on the aforementioned two approaches, we propose two algorithms to mine the subspace skyline clusters.

The algorithms use two basic types of accesses: sorted access, where the system obtains the value of an object in some sorted list by proceeding through the list sequentially from the top; and random access, where the system requests the value of a certain given object in a list, and retrieves it in one step.

4.1 Sorted-Based Subspace Skyline Clusters Mining

In this subsection, we discuss the first algorithm Sorted-based Subspace Skyline Clusters Mining.

Figure 2 illustrates the pseudo-code of SSSCM. It obtains the first skyline p by simultaneous scanning all the lists to find the first match (Line 1). Next, it identifies the subspace skyline cluster G_p of p (Line 2). Then the first skyline cluster can be returned (Line 3). If the object has not been seen before, then it is definitely not the part of the skyline and thus can be eliminated. If the objects have been examined prior to the first match, they may be in the skyline. Consequently, we sort these objects D_r according to the sum of the values in the subspace (Line 4 to 7) and prune the objects in the G_p from this sorted list (Line 8). For the rest of the objects in the D_r, we determine whether it is a skyline. If it is a skyline, we call the algorithm SkylineCluster to find its skyline cluster and output it, then delete the objects in the cluster from D_r(Line 9 to 14).

We now explain the algorithm SkylineCluster. It is used to find the cluster of the given object o. It searches before the position where the given skyline exists in each lists. Since the objects are ordered by the values in that list, the objects nearer to o in the list is more likely to be the top k objects of o. Therefore, we can visit them in the reverse-sorted order, which is in essence similar to the way to find the first skyline. The algorithm proceeds in three steps: a sorted access step (Line 1-2), a random access step (Line 3-5), a final sort step (Line 6). If there are few skylines so that we want to include more non-skyline objects in the cluster, we just need modify Line 2 by not checking whether it is a skyline or not.

Algorithm **SSSCM** (sorted list L, query subspace B)

---

1. do sorted access in parallel to each of the d sorted list L_i. Wait until there is the first match $p(p_1, \text{----}, p_d)$, i.e., the first record to be seen by all the lists

2. $G_p=$**SkylineCluster**(p)

3. output G_p and $G_B=G_B\cup\{\,G_p\,\}$

4. **for** each object r in D_r **do**// D_r is the set of objects that have been seen before

5.   do random access as needed to each of the list L_i to find the ith filed r_i of r

6.   compute the sum of r's dimension values

7. sort D_r in non-descending order

8. **for** each p' in G_p **do**   remove p' from D_r

9. **while** (D_r is not empty) **do**

10.   remove top entry r' from D_r

11.   **if** r is a skyline object **then**

12. $G_r=$**SkylineCluster**(r)

13.   output G_r and $G_B =G_B\cup\{\,G_r\,\}$

14.   **for** each r' in D_r **do**   remove r' from D_r

15. **return** G_B

Algorithm **SkylineCluster** (the representative skyline object o)

---

1. initialize the cursor $C_i=\{P_{o1}-1, \text{----}, P_{od}-1\}$//$P_{oi}$ refers to the pointer to object o in list L_i, and $P_{oi}-1$ refers to the pointer to the object located before o in list L_i

2. do reverse-sorted access in parallel to each of the d sorted list L_i from C_i. Wait until there are k matches $M = (q_1, \text{----}, q_k)$, each of which is a skyline

3. **for** each object q in D_q **do** // D_q contains the objects that have been seen before

4.   do random access to each of the list L_i to find the ith filed q_i of q

5.   compute the Manhattan distance d_q between o and q

6. $G_o =\{\,o\,\}\cup\{$ the top k objects with the smallest d_q in $D_q\,\}$

7. **return** G_o

Fig. 2. SSSCM algorithm

4.2 Threshold-Based Subspace Skyline Clusters Mining

A critical factor that will affect the performance of the SSSCM algorithm is how fast the first match can be found. If a match is found late (which is likely to be the case for large number of dimensions), it will result in a high initial response time. To overcome this limitation, we present our second approach TSSCM which exploits the threshold, the sum of the last reported object to improve the computation of the nearest neighbor (Line 1-6 in Algorithm TSSCM and Line 1-8 in Algorithm T_SkylineCluster). The details of this algorithm are as shown in Figure 3.

Algorithm **TSSCM** (sorted list L, query subspace B)

1. *unfinished*=true

2. **while**(*unfinished*) **do**

3.   do sorted access in parallel to each of the d sorted list L_i. As an object p is seen in some list, find the value p_i in other lists, and compute the sum of p_i, s_p

4.   **if** s_p is the smallest sum we have seen, **then** remember (p, s_p)

5.   Threshold value $\tau=(\underline{x}_1, \text{----}, \underline{x}_d)$, $\underline{x}_i$ is the value of the last object seen

6.   **if** $s_p<\tau$ **then** *unfinished*=false

7. G_p=**T_SkylineCluster**(p)

8.   output G_p and $G_B=G_B\cup\{\ G_p\ \}$

9. **for** each object r in D_r **do** // D_r is the objects appeared before p in each list

10. Line 10-19 refer to Line 5–14 in Algorithm SSSCM

11. **return** G_B

Algorithm **T_SkylineCluster** (the representative skyline object o)

1. initialize the cursor $C=\{P_{o1}-1, \text{----}, P_{od}-1\}$, *unfinished*=true
   //$P_{oi}-1$ refers to the pointer to the object located before o in list L_i

2. **while**(*unfinished*) **do**

3.   do reverse-sorted access in parallel to each of the d sorted list L_i from C_i. As an object q is seen in some list, find the value q_i in other lists

4.   **if** q is a skyline **then**

5.   compute the Manhattan distance d_q between o and q

6.   **if** d_q is the top k smallest distance we have seen, **then** $D=D\cup\{(\ q,\ s_q)\}$

7.   Threshold value $\tau=(\underline{x}_1, \text{----}, \underline{x}_d)$, $\underline{x}_i$ is the value of the last object seen

8.   **if** $(|D|==k)$ and $(\max(s_q)<\tau)$ **then** *unfinished*=false

9. $G_o =\{\ o\ \}\cup\{q|q\in D\}$

10. **return** G_o

Fig. 3. TSSCM algorithm

5 Experimental Evaluation

This section reports the experimental results. As mentioned earlier, there is no existing technique specifically designed to support efficient subspace skyline clusters computation. So we compare the two proposed algorithms. We first discuss the data sets and experimental settings, then show the results.

5.1 Data Sets and Configurations

We use both synthetic and real data sets to evaluate the algorithms we presented. The following are the details of the data sets:

NBA. This dataset can be available from http://www.nba.com. It contains 17k 13-dimensioanl points, where each point corresponds to the statistics of a player in 13 categories, including the number of points scored, rebounds, assists etc.

Color. The dataset can be downloaded at http://kdd.ics.uci.edu. It consists of 9-dimensioanl features of 68k photo images, such as brightness, contrast, saturation, and so on.

We also generate a number of data sets by varying the size of dimensionality and cardinality. We employ two types of distributions popular in the skyline queries: independent and anti-correlated. Each record in an independent data set is a point whose coordinates are obtained uniformly in their respective domains. An anti-correlated record has the property that if its coordinate on one dimension is large, then with a high probability its coordinate on another dimension is small. The details of creating data sets can be found in [5].

We focus on the cost in the computing stage instead of pre-processing stage such as presorting. And we set k at 10. Experiments are conducted on a PC with an AMD Athlon 1700+ CPU and 512M memory running a Microsoft Windows XP operating system.

5.2 Experimental Results

In the section, we conduct three types of experiments to analyze the mining performance with respect to the query dimensionality, the data set cardinality and the effect of parameter k.

Effect of query dimensionality. We first study the impact of query dimensionality on the efficiency of the algorithms by varying the queries dimensions from 2 to 4. Fig. 4(a) and (b) depict the runtime vs query dimensionality for two real data sets. The results on the independent and anti-correlated data sets with 10-dimensionality and 100k-cardinality are shown in Fig. 4(c) and (d). The two algorithms respond within the acceptable time. The runtime increases with the growing number of query dimensionality due to the increase of skyline size. SSSCM outperforms TSSCM on all the four data sets as the effectiveness of the threshold-based method. The stopping rule for the threshold-based method occurs at least as early as that for the sorted-based method, that is, with no more sorted accesses than the sorted-based method.

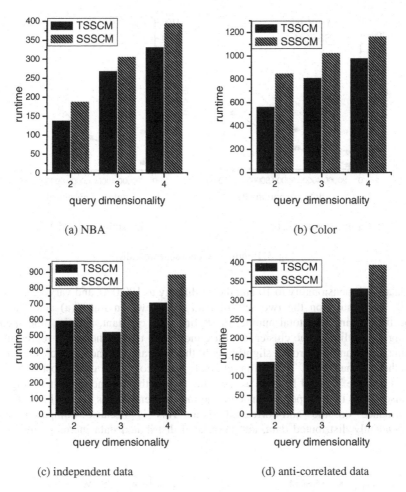

Fig. 4. Runtime vs. query dimensionality

Effect of data set cardinality. Fig. 5 reports runtime versus the data set cardinality on the two synthetic data. On these two data sets, we fix the query dimensionality at 3, set the data set dimensionality to 10 and vary data set cardinality between 10,000 and 100,000. Both algorithms have good scalability. However, TSSCM is consistently superior than SSSCM and thus has better scalability. Moreover, the performance is affected by data set cardinality. Specifically, the larger data set size incurs the more skyline comparisons, and thus the more time. Furthermore, the performance on the independent data is always better than the performance on the anti-correlated data because the inherent characteristic of anti-correlated data incurs more incomparable records and less pruning.

Effect of parameter k. To demonstrate the effect of the parameter k, we conduct experiments on the two synthetic data sets by fixing query dimensionality at 3, setting

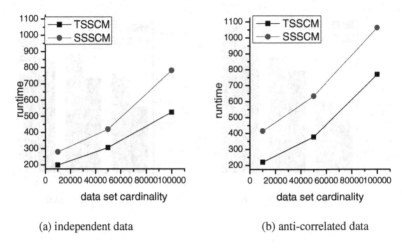

(a) independent data   (b) anti-correlated data

Fig. 5. Runtime vs. data set cardinality

the data set dimensionality to 10 and the cardinality to 100,000, and varying k from 10 to 20. The results on the two sets of data are shown in Fig. 6(a) and Fig. 6(b), respectively. Since the total number of skylines is a constant, and the time spent on finding the skyline in the cluster is a little more than that in the single representative skyline, the runtime grows slightly with the increase of the size of the cluster. Nonetheless, the cost of TSSCM is relatively small compared to the cost of SSSCM. This is because TSSCM has less access time to find the skylines and their neighbors. We also notice that the performance on the independent data is always better than the performance on the anti-correlated data. This is because compared to the independently distributed data, anti-correlated distributed data increases the skyline sizes.

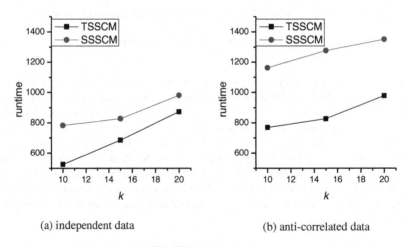

(a) independent data   (b) anti-correlated data

Fig. 6. Runtime vs. k

6 Conclusion and Future Work

In this paper, we address the important problem of recommending skyline objects as well as their neighbors in the arbitrary subspace of high dimensional space. We propose a novel concept, subspace skyline clusters to provide concise representation of the skylines to facilitate users to browse and choose. Two algorithms Sorted-based Subspace Skyline Clusters Mining, and Threshold-based Subspace Skyline Clusters Mining are presented. Experimental evaluations confirm the efficiency and effectiveness of them.

For future work, we will focus on extending our algorithm to support complex environments, such as data streams, peer-to-peer networks and so on.

References

1. Kossmann, D., Ramsak, F., Rost, S.: Shooting Stars in the Sky: An Online Algorithm for Skyline Queries. In: VLDB, Hong Kong, China, pp. 275–286 (2002)
2. Fagin, R., Lotem, A., Naor, M.: Optimal aggregation algorithms for middleware. In: Proc. of ACM Symposium on Principles of Database Systems (PODS 2001), Santa Barbara, CA (2001)
3. Kung, H.T., Luccio, F., Preparata, F.P.: On finding the maxima of a set of vectors. Journal of the ACM 22, 469–476 (1975)
4. Preparata, P.F., Shamos, M.I.: Computational geometry: an introduction. Springer, Heidelberg (1985)
5. Borzsonyi, S., Kossmann, D., Stocker, K.: The Skyline Operator. In: IEEE Conf. on Data Engineering, Heidelberg, Germany, pp. 421–430 (2001)
6. Yuan, Y., Lin, X., Liu, Q., Wang, W., Yu, J.X., Zhang, Q.: Efficient Computation of the Skyline Cube. In: International Conference on Very Large Data Bases (VLDB), Trondheim, Norway, pp. 241–252 (2005)
7. Pei, J., Jin, W., Ester, M., Tao, Y.: Catching the best views of skyline: a semantic approach based on decisive subspaces. In: Proceedings of the 31st international conference on Very large data bases, Trondheim, Norway, pp. 253–264 (2005)
8. Jin, W., Han, J., Ester, M.: Mining Thick Skylines over Large Databases. In: Boulicaut, J.-F., Esposito, F., Giannotti, F., Pedreschi, D. (eds.) PKDD 2004. LNCS (LNAI), vol. 3202, pp. 255–266. Springer, Heidelberg (2004)
9. Balke, W.-T., Zheng, J.X., Güntzer, U.: Approaching the Efficient Frontier: Cooperative Database Retrieval Using High-Dimensional Skylines. In: Zhou, L.-z., Ooi, B.-C., Meng, X. (eds.) DASFAA 2005. LNCS, vol. 3453, Springer, Heidelberg (2005)
10. Chomicki, J., Godfrey, P., Gryz, J., Liang, D.: Skyline with Presorting. In: Proceedings of the 19th ICDE, Bangalore, India, pp. 717–719 (2003)

Efficient Privacy Preserving Distributed Clustering Based on Secret Sharing*

Selim V. Kaya, Thomas B. Pedersen, Erkay Savaş, and Yücel Saygın

Sabanci University
Istanbul, 34956, Turkey
selimvolkan@su.sabanciuniv.edu, {pedersen,erkays,ysaygin}@sabanciuniv.edu

Abstract. In this paper, we propose a privacy preserving distributed clustering protocol for horizontally partitioned data based on a very efficient homomorphic additive secret sharing scheme. The model we use for the protocol is novel in the sense that it utilizes two non-colluding third parties. We provide a brief security analysis of our protocol from information theoretic point of view, which is a stronger security model. We show communication and computation complexity analysis of our protocol along with another protocol previously proposed for the same problem. We also include experimental results for computation and communication overhead of these two protocols. Our protocol not only outperforms the others in execution time and communication overhead on data holders, but also uses a more efficient model for many data mining applications.

1 Introduction

Recent advances in data collection and storage technologies enabled organizations to handle vast amounts of data related to their customers or users. However, this vast amount of plain data needs to be converted into useful knowledge through data mining techniques for high level decision making.

In many cases, databases are distributed among several organizations, which need to collaborate to achieve a more significant and accurate data mining model. Simply sharing the databases is not a feasible approach due to privacy concerns. As a result, Privacy Preserving Distributed Data Mining(PPDDM) techniques are developed for constructing a global data mining model over distributed databases without actually sharing the confidential data.

Efficiency in communication and computation is crucial in PPDDM since databases are often of considerable size. Sample scenarios are sensor networks or RFID applications, where the sensor nodes or RFID readers that contain the data (data holders) have very limited computation and communication capacity. In such scenarios, reducing the communication and computation costs is of utmost importance.

---

* This work was partially funded by the Information Society Technologies Programme of the European Commission, Future and Emerging Technologies under IST-014915 GeoPKDD project.

T. Washio et al. (Eds.): PAKDD 2007 Workshops, LNAI 4819, pp. 280–291, 2007.

In this paper we propose a new protocol for privacy preserving clustering over horizontally partitioned data with only a small constant communication and computation overhead for data holders with no loss of accuracy. We reduce the privacy preserving clustering problem to privacy preserving dissimilarity matrix construction which was proposed by Inan et al. [1]. After the dissimilarity matrix is computed privately, it can be fed into any hierarchical clustering algorithm. Our protocol uses two non-colluding third parties, which receive secret shares of inputs and compute intermediary results, while a data miner performs the actual clustering.

2   Related Work and Background

The pioneering research on privacy preserving data mining was conducted by Agrawal and Srikant[2], and Lindell and Pinkas[3] in 2000. In [2], Agrawal and Srikant use data perturbation for construction of a classification model privately. The basic idea is that original data values can be perturbed in such a way that original distribution of the aggregated data can be recovered but not the individual data values. Perturbation technique is efficient to implement however it has several problems. First of all, even though the distribution of original values can be predicted with a certain confidence level, some accuracy is lost. Secondly, modification of data does not fully preserve privacy of individual values, and may cause privacy breaches as shown in [4,5]. Finally, perturbation has a predictable structure for certain cases and hence may not fully preserve privacy [6]. A different perturbation method is proposed by Saygin et al.[7] in 2001 for association rule hiding, where unknown values are introduced to hide sensitive association rules.

Authors in [3] employ cryptography as its main tool and implements a decision tree learning protocol. However oblivious transfer, which is the main building block of this protocol, causes huge computation and communication overhead due to exponentiation operation for each bit of private data and expansion of each bit of private data as a result of exponentiation respectively. Authors in [8] propose a privacy preserving association rule mining protocol over horizontally partitioned data taking advantage of commutative encryption. Nevertheless the protocol requires encryption and decryption operations to be performed over each private input by all of the participants resulting in a large communication and computation cost.

Several protocols are proposed for privacy preserving clustering. Oliveira and Zaiane [9] introduce geometric data transformation methods(GDTMs) to distort confidential data values. The protocol tries to preserve main features of the confidential data for clustering while perturbing the data to meet privacy requirements. However, perturbation causes accuracy loss in clustering, and privacy of the data is not fully guaranteed. Consequently, Oliveira and Zaiane [10] introduce the notion of Rotation-Based Transformation(RBT). RBT provides confidentiality of attribute values while completely preserving the original clustering results. However RBT method has a computation overhead since attribute values are transformed pairwise, and selection of attribute pairs should be done in such a way that variance between the original and transformed attributes are maximum. In [11], Oliveira and Zaiane propose Object Similarity-Based Representation(OSBR) and Dimensionality Reduction-Based Representation(DRBT) methods for clustering over centralized and vertically partitioned databases. Therefore, OSBR has high computation cost since each data owner sends a dissimilarity matrix to a central party yielding a communication complexity of $O(n^2)$, while DRBT can cause loss of accuracy due to dimensionality reduction in the original data.

Merugu and Ghosh [12], and Klusch, Lodi and Moro [13] propose privacy preserving clustering methods based on sharing models representing the original data instead of sharing the original data itself. Accordingly, clustering can be performed over the model without revealing the original data points. However clustering over low quality representatives of the original data causes loss of accuracy while efforts for high quality representatives means loss of privacy.

Vaidya and Clifton [14] propose a privacy preserving k-means clustering protocol based on secure multi-party(SMC) computation. Nevertheless there is a huge communication and computation cost due to iterative execution of several SMC protocols till a convergence point for the clusters is obtained. Jha, Kruger and McDaniel propose two privacy preserving k-means clustering protocols for horizontally partitioned data in [15]. The protocols use homomorphic encryption and oblivious polynomial evaluation as their building block which are inefficient to be applied over large databases due to cost of modular exponentiation and oblivious transfer respectively.

The most recent study for privacy preserving clustering is proposed by Inan et al. [1] over horizontally partitioned data and the problem is reduced to secure computation of dissimilarity matrix. Each entry of the dissimilarity matrix is computed by a secure difference protocol where confidential data points are disguised by pseudo-random values and the disguise is removed by a trusted third party revealing the final difference. However secure difference protocol leads to privacy breaches because of the way pseudo-random values are used. According to the secure difference protocol, initiator of the protocol creates two disguise factors; one for the follower of the protocol to disguise initiators value and the other for the trusted third party to disguise which participant's input is subtracted from the other. Nevertheless, the latter disguise factor is the same for each entry point within a row of dissimilarity matrix. In other words, trusted third party can guess which site's input is subtracted from the other with a probability of $\frac{1}{2}$ for each row. On the other hand, quadratic communication cost for dissimilarity matrix computation is a huge burden for data holders.

3   Preliminaries

In our scenario we have ℓ *data holders*: $DH_1, \ldots, DH_\ell$ where DH_i has a database with n_i objects (tuples): $o_1^i, \ldots, o_{n_i}^i$. The databases all have the same schema with m numeric attributes (from an algebraic field). Since all databases have the same schema, we can write the union of the databases as $o_1, o_2, \ldots, o_N$, where $N = \sum_{i=1}^{\ell} n_i$, and where object o_i has attribute values $a_1^i, \ldots, a_m^i$. We say that the collective database is *horizontally partitioned* among the ℓ data holders.

The goal of our protocol is to compute the dissimilarity matrix of all objects in all databases, while keeping the actual values secret. At the end, each entry of the dissimilarity matrix will contain the weighted Manhattan distance between two elements from the collective database.

$$D_{ij} = \sum_{k=1}^{m} w_k |a_k^i - a_k^j|, \tag{1}$$

where $i, j = 1, \ldots, N$, and $w_1, \ldots, w_m$ are predefined weights. We introduce the notion of *partial dissimilarity matrices* which contain the numerical distance for a single attribute, so that the dissimilarity matrix can be written

$$D = \sum_{k=1}^{m} w_k D^k, \tag{2}$$

where D^k is the dissimilarity matrix with entries $D^k[i, j] = |a_k^i - a_k^j|$ which results from considering only the kth attribute.

3.1 Homomorphic Secret Sharing

Informally speaking, (m, t)-secret sharing is a method to share a secret among m parties in such a way that $t - 1$ or less colluding parties cannot compute any information about the secret; but t arbitrary parties can recover the secret. A data holder that wishes to share his secret s will create m *secret-shares* $s_1, \ldots, s_m$ and send one share to each party [16,17].

The protocols we present in this paper rely on additive secret sharing. To share a secret integer[1] s between two parties, we choose a random integer r and give the share r to the first party and the share $s - r$ to the second party. Clearly both shares are random when observed alone, so no single party can compute any information about the secret. The secret is revealed by simply two parties adding their shares, hence they can recover the secret together.

Additive secret sharing we used in our protocol is homomorphic with respect to addition since adding shares pairwise gives an additive sharing of the sum of the secrets.

4 Our Protocol

There are two challenges for designing a protocol for computing Manhattan distance: (i)not to reveal private inputs, (ii) to hide which input is the largest. We employ additive homomorphic secret sharing to fulfill the first challenge, with a very small communication and computation overhead for the data holders.

---

[1] Or more precisely: to share an element from an additive group.

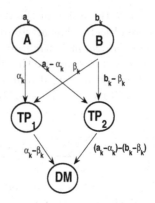

Fig. 1. Overview of the numerical distance protocol

The inputs are shared between two non-colluding third parties, TP_1 and TP_2, who can compute a secret sharing of the difference by using the homomorphic property. To avoid revealing the sign of the difference (which input is larger), TP_1 and TP_2 share a pseudo random number generator. Before the protocol starts, TP_1 and TP_2 will each prepare an $m \times N \times N$ table, whose entries are one-bit binary values from a pseudo random number generator (PRNG), which is initialized with a shared seed.

Let a_k and b_k be the private values for the kth attribute of o_i^A and o_j^B held by DH_A and DH_B, respectively. The (i,j)th entry in the D^k is $|a_k - b_k|$. To compute this Euclidean distance DH_A selects a random number α_k, and sends additive shares α_k and $a_k - \alpha_k$ to TP_1 and TP_2, respectively. Likewise, DH_B creates its own additive shares β_k and $b_k - \beta_k$ and sends them to TP_1 and TP_2, respectively. TP_1 computes $sh_1 = (-1)^{\mathrm{PRNG}(\mathrm{k,i,j})}(\alpha_k - \beta_k)$ and TP_2 computes $sh_2 = (-1)^{\mathrm{PRNG}(\mathrm{k,i,j})}((a_k - \alpha_k) - (b_k - \beta_k))$, and they send the results to the data miner (DM). When DM adds the two received values the result is

$$sh_1 + sh_2 = (-1)^{\mathrm{PRNG}(\mathrm{k,i,j})}(a_k - b_k). \qquad (3)$$

After receiving the numerical values the miner gets the result $|sh_1 + sh_2| = |a - b|$, which is the required (i,j)th entry of D^k. Overview of our Euclidean distance protocol is depicted in Figure 1.

To construct the dissimilarity matrix for the kth attribute, each data holder DH_i computes additive shares of its private values $a_k^1, a_k^2 \dots a_k^{n_i}$, which are stored in arrays $s_1^{i,k}$ and $s_2^{i,k}$. The arrays $s_1^{i,k}$ and $s_2^{i,k}$ are, then, sent to TP_1 and TP_2, respectively. Steps of the protocol for data holders are demonstrated in Algorithm 1.

Receiving $s_1^{1,k}, s_1^{2,k}, \dots, s_1^{\ell,k}$ from all of the data holders, TP_1 merges these arrays into s_1^k. After merge operation, s_1^k contains additive shares of the collective database for the kth attribute. Then, TP_1 initializes an $N \times N$ matrix D_1^k and

---

Algorithm 1. DH_i

Input: private values for attribute k: $a_k^1, a_k^2 \dots a_k^{n_i}$
Output: secret share arrays $s_1^{i,k}$ and $s_2^{i,k}$
1: Initialize secret share arrays $s_1^{i,k}$ and $s_2^{i,k}$ of size n_i
2: **for** $j = 1$ to n_i **do**
3:    $(s_1^{i,k}[j], s_2^{i,k}[j]) = secretshare(a_k^j)$
4: **end for**
5: Sends $s_1^{i,k}$ to TP_1
6: Sends $s_2^{i,k}$ to TP_2

---

fills each entry (i,j) with value $(-1)^{\text{PRNG}[k,i,j]}(s_1^k[a] - s_1^k[b])$. The resulting matrix D_1^k contains additive shares of D^k. TP_1 sends D_1^k to DM. The details of the protocol for TP_1 are depicted in Algorithm 2. TP_2 performs the same steps.

Algorithm 2. TP_1

Input: Secret share arrays $s_1^{1,k}, s_1^{2,k}, \dots, s_1^{\ell,k}$, matrix PRNG shared with TP_2
Output: Secret share matrix D_1^k
1: Initialize secret share array s_1^k of size $N = \sum_{i=1}^{\ell} n_i$
2: Initialize secret share matrix D_1^k of size $N \times N$
3: Merge $s_1^{1,k}, s_1^{2,k}, \dots, s_1^{\ell,k}$ into s_1^k
4: **for** $a = 1$ to N **do**
5:    **for** $b = 1$ to N **do**
6:       $D_1^k[a,b] = (-1)^{\text{PRNG}[k,a,b]}(s_1^k[a] - s_1^k[b])$
7:    **end for**
8: **end for**
9: Sends D_1^k to DM

It is trivial for DM to construct D^k from matrices D_1^k and D_2^k by simply computing $D_1^k[i,j] + D_2^k[i,j]$ for each entry (i,j) of D^k, where $i, j = 1, 2, \dots, N$.

Security of our Protocol: Our security definition reflects that not more than a negligible amount of information is revealed about *any* object in the collective database. One must also note that information leakage is limited to whatever can be deduced from the final result. The following standard definition for security of our protocol relies on negligibly small values of ϵ.

Definition 1. *A protocol for computing partial dissimilarity matrices is ϵ-secure if for all parties, and for all attributes A_j^i*

$$\left| P[A_k^i = x | D^k, M] - P[A_k^i = x | D^k] \right| < \epsilon, \tag{4}$$

where M is a transcript of all messages send to a given party, where P stands for probability.

From this definition, one can conclude that the proposed protocol is ϵ-secure. Since data holders never receive any information, Equation (4) is satisfied for these parties. Since blinding factors α_i are chosen randomly and independently, Equation (4) is also satisfied for TP_1. Since attributes a_i are chosen from algebraic fields and $a_i - \alpha_i$ are also independent of the data, Equation (4) is also satisfied for TP_2. The values received by DM enables to build D, where each entry has a random sign depending on PRNG. If PRNG is secure, no additional information can be computed.

5   Complexity Analysis

In this section, we analyze computation and communication complexities of our protocol for numeric attributes [2]. Each analysis will be performed for DHs, TPs, and DM separately. We also show complexity analysis of the privacy preserving clustering protocol proposed by Inan et al. [1].

Since computation of secret shares of private inputs can be performed in parallel by each DH, computation complexity of our protocol for DHs is $O(n_{max})$, where $n_{max} = max(n_1, n_2, \ldots, n_\ell)$. On the other hand, for the protocol in [1] computation complexity of DHs is $O(N^2)$ (where N is the total number of objects) since data holders compute shared dissimilarity matrices pairwise which requires serial execution.

In our protocol, for TPs, computation of secret share of D_k yields complexity of $O(N^2)$ which is due to computation of the global dissimilarity matrix, if we assume TPs operate in parallel and pseudo random numbers PRN are generated in advance. In [1], there is only one TP and computation complexity of TP is $O(N^2)$. Complexity of our protocol for DH is $O(N^2)$, which is the cost of computing D^k.

In our protocol, each DH sends secret shares of their private inputs to TPs resulting in a total communication complexity of $O(N)$. TPs send secret shares of D^k to DM and the total communication complexity is $O(N^2)$. Since final clustering is done by DM, there is no further communication cost. On the other hand, in [1], each DH sends local and shared dissimilarity matrices to TP where global dissimilarity matrix is computed. Accordingly, communication complexity is $O(N^2)$. Summary of the computation and communication complexity analysis is depicted in Table 1.

Table 1. Computation/Communication Complexities of our Protocol and the protocol in [1]

Attribute Type	DH	TP	DM	Total
Numeric	$O(n_{max})/O(N)$	$O(N^2)/O(N^2)$	$O(N^2)/-$	$O(N^2)/O(N^2)$
Numeric for [1]	$O(N^2)/O(N^2)$	$O(N^2)$	$-/-$	$O(N^2)/O(N^2)$

---

[2] Due to space limitations, we did not include the extension of our protocol to other attributes types.

6   Implementation and Performance Evaluation

In this section, performance evaluation of our protocol is explained and discussed in detail in comparison with the protocol proposed in [1]. Since both protocols do not result in any loss of accuracy, we perform only two tests: communication cost analysis and computation cost analysis. The experiments are conducted on an Intel Dual-Core Centrino PC with 2 MB cache, 2 GB RAM and 1.83 GHz clock speed. We used C# programming language to implement the algorithms.

Experimental Setup: To measure the performance of our protocol and the one in [1], two test cases are identified for varying: i) total number of entities (total database size), and ii) number of data holders.

Each test case is performed over numeric attributes. For each experiment, we measure the communication and computation overhead of our protocol against the protocol proposed in [1]. Comparison of the protocols in the experiments are confined to the formation of the global dissimilarity matrices and clustering is not taken into consideration. For all the experiments, we denote our protocol as "our protocol" and the one in [1] as "protocol" in the figures. In order to measure the effect of the database size, we fixed the number of data holders to four, where each data holder has an equal share of database.

For the test case (i), we used total database sizes of 2K, 4K, 6K, 8K and 10K. In test case (ii), number of data holders, excluding TPs, is 2, 4, 6, 8, and 10. For each test case, we use synthetically generated datasets. Synthetic datasets are more appropriate for our experiments since we try to evaluate scalability and efficiency of our protocol for varying parameters, and synthetic datasets can be generated by controlling the number of entities, number of data holders, and average length of attributes. Data generator is developed in Eclipse Java environment. For the numeric attributes, each entity is chosen from the interval [0, 10000].

In our experiments, we use 128-bit Advanced Encryption Standard (AES) cipher to generate PRNs to disguise data holders' inputs. We use Cryptography namespace of MS .Net platform to perform AES encryption in the implementation of our protocol and [1]. For our protocol, keys and initialization vectors (IVs) are chosen by each data holder independent of the others, while for [1], seeds for pseudo-random number generator shared between data holders are used as keys and an initialization vector (IV) globally known to every data holder is used. Ciphertext as a result of encryption of IV by AES key is used as the pseudo-random number.

Computation Cost Analysis: Comparison of computation costs between our protocol and the protocol in [1] for varying database sizes from $2K$ to $10K$ is depicted in Figure 2. Figure 2 shows that both our protocol and the one in [1] behave quadratically which is due to formation of global dissimilarity matrix. However, our protocol performs better than the one in [1] since data holders operate in parallel in our protocol and the overall computation cost for each data holder is n AES encryptions for computing secret shares of the data, where n is database size of each data holder. However, the protocol in [1] performs n AES encryptions at each data holder to disguise data values and n AES encryptions at TP to remove these disguise factors. As a result, the protocol in [1] performs $k * n$ more AES encryptions than our protocol where there are k data holders. As Figure 2 shows, execution time difference between our protocol and the one in [1] gets larger as database size increases since number of AES encryptions performed at TP also increases. On the other hand, in our protocol no encryption is performed by any party other than data holders.

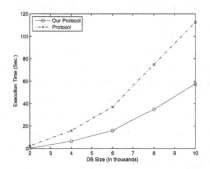 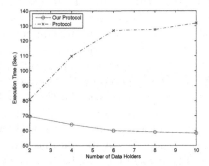

Fig. 2. Computation cost for different database sizes

Fig. 3. Computation cost for different number of data holders

Our protocol scales better than the one in [1] with the number of data holders. For this experiment, we generate a dataset of 10K entities and then horizontally partitioned this dataset by distributing the complete dataset over the data holders so that each party holds the same number of entities. As depicted in Figure 3, execution time for the protocol in [1] increases as number of data holders increases. This is due to the fact that, C_2^k number of pairwise computation between data holders have to be performed to compute shared dissimilarity matrices where k is total number of data holders. However, increase in total execution time for [1] gets smaller as number of data holders increases since amount of data owned by each data holder gets smaller. On the other hand, increase in number of data holders reduces total execution time for our protocol since share of each data holder gets smaller which means fewer numbers of encryptions are performed by data holders in parallel. In our protocol, the computation cost of TPs and DM are not affected by the change in number of DH.

Communication Cost Analysis: Overall communication costs of our protocol and the protocol in [1] for various database sizes are depicted in Figure 4. As seen in the figure, overall communication cost of our protocol is more than the one in [1] due to the secret sharing employed in our protocol where two shares are created for each attribute value. Both protocols behave quadratically since overall communication cost is dominated by communication cost of dissimilarity matrices. On the other hand, communication cost of data holders for our

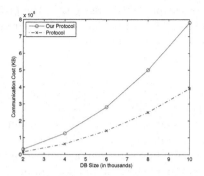

Fig. 4. Overall communication cost for different database sizes

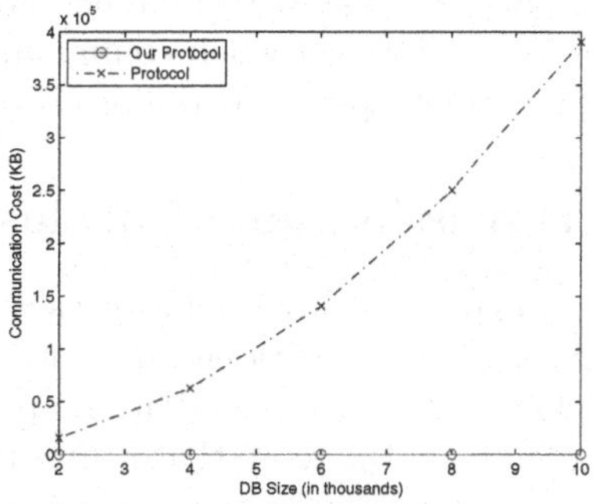

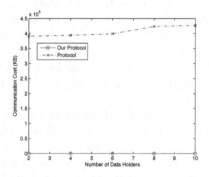

Fig. 5. Communication cost of data holders for different database sizes

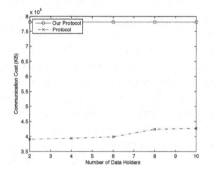

Fig. 6. Overall communication cost for different numbers of data holders

Fig. 7. Communication cost of data holders for different numbers of data holders

protocol and the one in [1] is depicted in Figure 5. Communication cost of our protocol for data holders is linear with respect to the size of each data holder's dataset while the protocol in [1] requires quadratic communication cost for data holders. For that reason, communication cost of our protocol for data holders is negligible compared to the protocol in [1]. Accordingly, our protocol shifts the communication burden to the third parties and requires negligible amount of communication from data holders which are assumed to be resource limited. On the other hand, the protocol in [1] requires all the communication to be performed by data holders, which is not appropriate for our scenario where data holders (e.g. sensor nodes and RFID readers) have limited resources.

Analysis of overall communication costs for different number of data holders are depicted in Figure 6, which shows that communication cost of our protocol remains the same for varying number of data holders. On the other hand, communication cost of the protocol in [1] increases with increasing number of data holders. However, the amount of increase in communication cost gets smaller as number of data holders increase, due to the same reasoning as in Figure 3. As

Figure 6 shows, overall communication cost of our protocol is more than the protocol in [1]. However when communication costs of data holders are compared, our protocol outperforms the protocol in [1] as shown in Figure 7.

7   Discussions and Conclusion

In this paper, we proposed a privacy preserving distributed clustering protocol for horizontally partitioned data using secret sharing scheme, which is homomorphic with respect to addition operation. The model that we adopted is unprecedented in the sense that it uses two non-colluding third parties. The idea of using two third parties that greatly alleviates the computation and communication burden of the data holders is especially useful in applications such as sensor networks and RFID where data holders are resource-limited sensor nodes and RFID readers. When compared to the most efficient former techniques, which exclusively rely on the computation and communication capabilities of the data holders, our protocol can run even on the most simple data holders, such sensor nodes or RFID readers. One can even think that there is no need for the data holders to store the actual data. It is true that the communication overhead between the two third parties is greater than the other protocols. Nevertheless, third parties can easily be equipped with high computation capability and bandwidth.

The use of two third parties and non-collusion property are realistic when they are chosen to have conflicting interests in the data mining results of the actual data. As an example, one third party can be a consumer organization who is interested in the privacy of consumers, while the other is representative of the industry - they both have interests in the right outcome, but will never collude.

We proved information theoretically that the third parties cannot gather any information about the data under the non-colluding assumption. Therefore, our protocol adopts a stronger security model than computational infeasibility, which is used by the majority of other privacy preserving data mining algorithms.

Other two benefits of the proposed protocol are that the data holders do not need to be synchronized or to share keys. Almost all previously proposed methods rely on synchronous and interactive protocols, where data holders must always be on-line during the protocol execution, while the data holders in our protocol can send the shares of their data asynchronously at their convenience. Since there is no communication between the data holders, there is no need for them to share keys; therefore there is no key distribution problem.

And finally, the model based on the use of two third parties and homomorphic secret sharing can be extended to other data types and different dissimilarity metrics.

References

1. Inan, A., Saygın, Y., Savaş, E., Hintoğlu, A.A., Levi, A.: Privacy preserving clustering on horizontally partitioned data. In: Privacy Preserving Clustering on Horizontally Partitioned Data, p. 95. IEEE Computer Society, Los Alamitos (2006)

2. Agrawal, R., Srikant, R.: Privacy-preserving data mining. In: SIGMOD 2000. Proceedings of the 2000 ACM SIGMOD international conference on Management of data, pp. 439–450. ACM Press, New York (2000)
3. Lindell, Y., Pinkas, B.: Privacy preserving data mining. In: Bellare, M. (ed.) CRYPTO 2000. LNCS, vol. 1880, pp. 36–54. Springer, Heidelberg (2000)
4. Evfimievski, A., Gehrke, J., Srikant, R.: Limiting privacy breaches in privacy preserving data mining. In: PODS 2003. Proceedings of the twenty-second ACM SIGMOD-SIGACT-SIGART symposium on Principles of database systems, pp. 211–222. ACM Press, New York (2003)
5. Evfimievski, A., Srikant, R., Agarwal, R., Gehrke, J.: Privacy preserving mining of association rules. Inf. Syst. 29(4), 343–364 (2004)
6. Kargupta, H., Datta, S., Wang, Q., Sivakumar, K.: Random-data perturbation techniques and privacy-preserving data mining. Knowl. Inf. Syst. 7(4), 387–414 (2005)
7. Saygin, Y., Verykios, V.S., Clifton, C.: Using unknowns to prevent discovery of association rules. SIGMOD Rec. 30(4), 45–54 (2001)
8. Kantarcıoğlu, M., Clifton, C.: Privacy-preserving distributed mining of association rules on horizontally partitioned data. IEEE Transactions on Knowledge and Data Engineering 16(9), 1026–1037 (2004)
9. Oliveira, S., Zaiane, O.R.: Privacy preserving clustering by data transformation. In: 18th Brazilian Symposium on Databasesn, pp. 304–318 (2003)
10. Oliveira, S., Zaiane, O.R.: Achieving privacy preservation when sharing data for clustering. In: Jonker, W., Petković, M. (eds.) SDM 2004. LNCS, vol. 3178, pp. 67–82. Springer, Heidelberg (2004)
11. Oliveira, S., Zaiane, O.R.: Privacy-preserving clustering by object similarity-based representation and dimensionality reduction transformation. In: Workshop on Privacy and Security Aspects of Data Mining (PSDM 2004) in conjunction with the Fourth IEEE International Conference on Data Mining (ICDM 2004), pp. 21–30 (2004)
12. Merugu, S., Ghosh, J.: Privacy-preserving distributed clustering using generative models. In: ICDM 2003. Proceedings of the Third IEEE International Conference on Data Mining, Washington, DC, USA, pp. 211–218. IEEE Computer Society, Los Alamitos (2003)
13. Klusch, M., Lodi, S., Moro, G.: Distributed clustering based on sampling local density estimates. In: IJCAI 2003. Proceedings of the 18th International Joint Conference on Artificial Intelligence, pp. 485–490. AAAI Press (2003)
14. Vaidya, J., Clifton, C.: Privacy-preserving k-means clustering over vertically partitioned data. In: KDD 2003. Proceedings of the ninth ACM SIGKDD international conference on Knowledge discovery and data mining, pp. 206–215. ACM Press, New York (2003)
15. Jha, S., Kruger, L.P.M.: Privacy preserving clustering. In: ESORICS'05:10th European Symposium On Research In Computer Security, pp. 397–417 (2005)
16. Asmuth, C., Bloom, J.: A modular approach to key safeguarding. IEEE Transactions on Information Theory 29(2) (1983)
17. Shamir, A.: How to share a secret. Communications of the ACM 22(11), 612–613 (1979)

SePMa: An Algorithm That Mining Sequential Processes from Hybrid Log

Xiaoyu Huang, Huiling Zhong, and Wenxue Cai

School of Electronic Business, South China University of Technology,
Guangzhou, 510006, China
{echxy,hlzhong,ctwxcai}@scut.edu.cn

Abstract. To accommodate ourselves to the changeful and complex business environment, we should be able adjust the business processes within the enterprise whenever changes happen. However, the work to design and redesign the processes is far from trivial, the designers are required to have deep knowledge of the business processes at hand, in traditional approaches it means long term investigation and high cost. To automate the procedure of process discovery, process mining is introduced. Process mining takes the run-time log generated by the process management system as its input, and outputs the process models defined for the system. Unfortunately, current work on process mining often assumes that the input log is generated by the same process, but in many occasions this requisition is hard to be satisfied. In this paper, we propose SePMa, an algorithm that mining sequential processes from hybrid log. SePMa aims at discovering sequential processes from log generated by multiple processes, both of theoretical analysis and experimental results show that SePMa has very high efficiency and effectiveness.

1 Introduction

Capable of promoting business process automation and collaboration, process management systems (PMSes) attract popular attentions in areas such as electrical government, business process engineering and collaborative work. While applying a PMS in enterprise, one main problem the process designer encounters is to make the "well-defined" processes, and, whenever changes happen, the definitions of related processes must be redesigned immediately.

The work to redesign process models is far from trivial [1], it requires the designer to have deep knowledge on the previous processes running in enterprise, in traditional hand-designing approach, it means all the participants, i.e. the process designers, workers and managements have to pay a lot of efforts, and it is also time-consuming and high-cost.

To accelerate the procedure of business process reconstruction, process mining is introduced [2]. Process mining aims at discovering process models automatically, it takes the log generated by the run time process as input, and outputs corresponding process model. Comparing with the hand-design approach, the process mining technology has significant advantages: first, the mining process is completely done by

T. Washio et al. (Eds.): PAKDD 2007 Workshops, LNAI 4819, pp. 292–300, 2007.

computer, so it's more efficient; second, the analyzing process is based on the objective log data, so the output model can be more objective.

Despites its promise, most current work on process mining has the same assumption that the log should be generated by the same process. Unfortunately, in many scenarios this assumption is too rigid. To meet the requisition, the data collectors have to identify the source process for every log record, and sometimes the task is almost impossible to achieve. Being constrained by the presupposition, the application of the process mining technology is limited. On the other hand, in many enterprise applications, most processes are sequential, as a result, it is especially interesting to identify and analyze sequential processes.

In this paper, we propose SePMa, a sequential process mining algorithm. Unlike previous work aiming at discovering the branch structures for a given process, SePMa focuses on mining sequential processes from hybrid log, i.e. the log can be generated by more than one sequential processes. In this paper, both the theoretic analysis and experimental results show that SePMa is of high efficiency and accuracy. The remaining of the paper is organized as follows: in section 2, we review related work on process mining. In section 3, we study the distributional features of hybrid log in depth, and propose SePMa. Section 4 reports on the experiment. Finally, section 5 concludes and ends with a vista of the future work.

2   Related Work

This section provides a brief review on the advance of process mining, a through one can be found in [3].

Process mining was first investigated by Cook [4][5][6], in [5] three kinds of methods are proposed, which are RNet, KTail and Markov. The RNet method is neural networks based, it looks at the past behaviors to predicate current state; the KTail method is finite state machine based, it determines current state by looking at the future k behaviors, where k is a given threshold. While RNet being a purely statistical method and KTail a purely algorithmic, Markov, the third method, is the combination of them. In Markov, current state depends on both the history and the future.

Cook's work mainly focuses on discovering the structure of software process automatically. Recently, it is extended to mining business processes by Aalst [1][3][7]. In [7], the constraint conditions that a minable process should satisfy are deduced, and [7] also proposes the α algorithm, which is a Petri net based approach; in [1], another mining algorithm is presented, which is a purely statistical method.

Although all the algorithms mentioned above have been reported that they perform well in experiments, however, they all have the same assumption that the log should be generated by the same process, which restrict their applications within narrow limit.

3   SePMa

3.1   Definitions and Theorems

Definition 1 (Sequential Process). Given process P composed of activities $a_1, a_2 \ldots a_n$, we say P is a sequential process, if in every execution of P the activities execute in some fixed order and both of the following constraints are satisfied: (1) At any time, there is at most one instance of P running; (2) The intersection of any two different sequential processes is empty, i.e. given two sequential processes P and Q, suppose PS denotes the activity set of P and QS the activity set of Q, then $PS \cap QS = \varnothing$.

For any sequential process with activity set $\{a_i | 1 \leq i \leq n\}$, if the activities execute in the order $a_1, a_2 \ldots a_n$, then we use $P < a_1, a_2 \ldots a_n >$ to denote the process.

Given $P_1 < a_1, a_2 \ldots a_n >$, let $a_i a_{i+1} \ldots a_k$ be one of the executing sequences of the activities in P_1 (for subscript j larger than n, let $a_j = a_{(j \bmod n)+1}$), we can see that there are another (n-1) processes $P_2 < a_2, a_3 \ldots a_n, a_1 >$, $P_3 < a_3 \ldots a_n, a_1, a_2 > \ldots P_n < a_n, a_1 \ldots, a_{n-1} >$ can also generate the same sequence. In fact, given process P_i, if we define the prior for P_i's first activity be the last activity in P_i, and the successor for P_i's last activity be the first one, then we can see all the n processes have the same inner structure, therefore any logs generated by one of them can be also generated by the others. So from the viewpoint of the log, the n processes are unidentifiable. i.e., all the n processes are equivalent.

Definition 2 (Equivalent Processes). Given sequential processes P and Q, if both of their activity sets are the same, and each of their activities has the same prior and successor in them, then we say P and Q are equivalent, and denote $P \sim Q$.

For sequential processes P and Q, if $P \sim Q$, then we take them as the same one.

Definition 3 (Hybrid Sequential Log). Given k sequential processes $P_1, P_2 \ldots P_k$, let s_i^j denote the log record for some activity in P_i and $T_i = s_i^1 s_i^2 \ldots s_i^{is}$ is the log for P_i. We say sequence $HL = s_1 s_2 \ldots s_{i1+i2+\ldots+ik}$ ($s_i \in \bigcup_{i=1}^{K} T_i$) is a hybrid log for $P_1, P_2 \ldots P_k$, iff $\forall i \in [1,N]$, any two elements in T_i has the same relative order in HL i.e. for s_i^r and s_i^t that $r \neq t$, let their positions in HL be r' and t' respectively, then the inequations $r < t$ and $r' < t'$ are equivalent.

In the remaining of this paper, when the context is clear we will use the same symbol A to represent both the activity itself and the log record corresponding to it.

Example 1. Given two sequential processes $P_1 < a_1, a_2, a_3 >$ and $P_2 < b_1, b_2, b_3 >$, then all the sequences $a_1 a_2 a_3$, $b_2 b_3 b_1 b_2 b_3$ and $a_1 b_2 b_3 a_2 b_1 a_3 b_2 b_3$ are legal sequential logs, while sequences $a_2 a_1 a_3 a_1$ and $a_1 b_3 b_2 a_2 b_1 a_3 b_2 b_3$ are illegal.

Definition 4 (Segment and Activity Set). Given hybrid log $HL = s_1 s_2 \ldots s_n$, for any activity a_i, let the positions a_i occurs in HL be $i_1, i_2, \ldots i_k$, where $i_1 < i_2 < \ldots < i_k$. For any $j \in [1,k]$, let $T_{ij} = s_p s_{p+1} \ldots s_{p+q}$, where $p = i_j$ and $q = i_{j+1} - 1 - p$, then we say T_{ij} is a segment for a_i on HL, further more, let TS_{ij} denote the set of all the different elements in T_{ij}, when given i, the intersection of all T_{ij} is said to be the activity set for a_i on HL, and is denoted as $\cap(a_i, HL)$.

Example 2. Given sequence $HL = a_1 b_2 b_3 a_2 b_1 a_3 b_2 b_3 a_1 b_2 b_3 a_2$, then the segment set of a_1 is $\{a_1 b_2 b_3 a_2 b_1 a_3 b_2 b_3 b_1\}$, the activity set for a_1 on HL is $\cap(a_1, HL) = \{a_1, b_2, b_3, a_2, b_1, a_3\}$; the segment set for b_2 is $\{b_2 b_3 a_2 b_1 a_3, b_2 b_3 b_1 a_1\}$, the activity set for b_2 on HL is $\cap(b_2, HL) = \{b_2, b_3, a_2, b_1, a_3\} \cap \{b_2, b_3, b_1, a_1\} = \{b_1, b_2, b_3\}$.

By the definitions above, we have the following theorem:

Theorem 1. Given any activity a and sequential process log HL, if $\cap(a, \text{HL})$ is not empty, then all the other activities in the same process with a also belong to $\cap(a, \text{HL})$.

Proof. By the definition of $\cap(a, \text{HL})$, we have $a \in \cap(a, \text{HL})$, so we just need to prove that the other activities in the same process with a occur at least once in each segment of a. Suppose $a \in P<a_1,a_2...a_n>$, let $a=a_i$ $(1 \le i \le n)$. By definition 1, after one execution of a_i, the new instance of P won't be created until the executions of $a_{i+1}, a_{i+2}...a_n$ are finished, so $a_{i+1}, a_{i+2}...a_n$ must appear earlier than the next a_i in HL. Besides, when the next instance is created, activities $a_1, a_2,...a_{i-1}$ must execute before a_i, so they also occur in front of the next a_i, summarize the conclusions above, we have the theorem.

Corollary 1. Given $P<a_1,a_2...a_n>$ and hybrid log HL, if $\bigcap_{i=1}^{N}(\cap(a_i, HL))$ is not empty, then $\{a_1,a_2...a_n\} \subset \bigcap_{i=1}^{N}(\cap(a_i, HL))$.

Corollary 2. Given hybrid log HL, suppose x and y are different activities in HL, if x and y are in the same process, then the predicate $(x \in \cap(y, \text{HL}) \land y \in \cap(x, \text{HL}))$ must be true.

Corollary 3. Given hybrid log HL, suppose x and y are different activities in HL, if x and y are in the same process, then each of them occurs once in each of the other's segments.

Both of corollary1 and 2 can be directly deduced from theorem1. Here we just prove corollary3: By corollary2, if x and y belong to the same process, then each of them occurs at least for once in each of the other's segments. On the other hand, if there is one segment of x contains more than one ys, notice that in this segment there exists at least one segment of y containing no xs, this is paradoxical to the supposition, so the number of y in each segment of x is exactly one; it is the same to prove that the number of x in each segment of y is one, then we can get the conclusion..

Theorem 2. Suppose different activities a_1, $a_2...a_N$ are in the same hybrid log HL, if $\{a_1, a_2,...a_N\} \subset \bigcap_{i=1}^{N}(\cap(a_i, HL))$, then there exists one sequential process F composed of $a_1, a_2,...a_N$, the executing orders among the activities are the same as their occurrences in HL.

Lemma 1. Given $i \ne j$, then the times a_i occurs in each segment of a_j is 1, and it's also the same to a_j.

The proof of lemma 1 is similar to corollary 3, we omit it here.

Proof of Theorem 2: Given the hybrid log HL, we delete all the activities not in the set $\{a_1, a_2,...a_n\}$ from it, then we get HL', which is purely composed of $a_1, a_2,...a_n$. now we just need to construct a sequential process that can generate the log HL'. Let p_i^k be the position for the kth occurrence of a_i in HL', not lost generality, suppose that if $i<j$ and $a_i \ne a_j$, then $p_i^1 < p_j^1$, now let's compute p_i^k.

By the definition of p_i^k, from the position p_i^1 to p_i^k, the times a_i occurs is k, and the k occurrences of a_i form (k-1) segments of a_i on HL'. By lemma1, for any $j \ne i$, a_j occurs once in each of these segments, so there are (k-1) a_j between the position p_i^1 and p_i^k. Besides this, if j<i, then by the assumption $p_i^1 > p_j^1$ and lemma1, there's just one a_j before the first occurrence of a_i; while if j>i, no a_j in the front of the position p_i^1. Summarize all the conclusions above, we can get : p_i^k =(k-1)*(n-1)+(i-1)+k=(k-1)*n+i, to make it more clear, we rewrite equation in the form $p_i^k \equiv i$ mod n. now we can have HL'=$a_1a_2...a_na_1a_2...a_n$....Let F=<$a_1,a_2...a_n$>, we have the theorem.

Corollary 4. Given activities a_1, $a_2...a_k$, we say these activities can compose a sequential process iff $\{a_1,a_2...a_k\} \subset \bigcap_{i=1}^{N}(\bigcap(a_i, HL))$.

The proof of corollary 4 can be deduced directly from corollary 1 and theorem 2.

3.2 Algorithm

By corollary 4, given a hybrid log HL, if there exists activities $a_1,a_2...a_n$ in HL satisfying the constraint $\{a_1,a_2...a_n\} \subset \bigcap_{i=1}^{N}(\bigcap(a_i, HL))$, then it is possible that $a_1,a_2...a_n$ belong to the same process. By corollary3 and lemma1, the formulation $\{a_1,a_2...a_n\} \subset \bigcap_{i=1}^{N}(\bigcap(a_i, HL))$ is equivalent to the constraint that given $a_i \neq a_j$, each of them occurs just once in each of the other's segments. Based on the analysis above, we propose SePMa.

Corollary 4 shows that each process got by SePMa can be one of the target processes that generated the log, now we will prove that when the length of the log is large enough, then all the processes got by SePMa will be the target processes.

Given process P< $a_1,a_2...a_n$ >, let $a_{n+1}=a_1$ and t_i be the interval between activities a_i and a_{i+1}, we use $C_i=1/t_i$ denote the coupling degree between a_i and a_{i+1}, i.e. the larger C_i, the higher probability that the direct successor of a_i be a_{i+1} in the log.

Given hybrid log HL, let HL be generated by processes $P_1<a_{1,1},a_{1,2},$ $...a_{1,k1}>$,$P_2<a_{2,1},a_{2,2},...a_{2,k2}>$...and $P_n<a_{n,1},a_{n,2}...a_{n,kn}>$. Denote $C_{i,j}$ be the coupling degree between activities $a_{i,j}$ and $a_{i,j+1}$($1 \leq i \leq n$ and $1 \leq j \leq k_i$).Not lost generality, we assume that every activity should be logged before it is called. Suppose the current executing activities of the n processes are $a_{1,j1},a_{2,j2}...a_{n,jn}$, where if process P_i is not running, let j_i be k_i. So the probability that the next record in HL be $a_{i,j+1}$ is $\dfrac{C_{i,ji}}{\sum_{s=1}^{n}C_{s,js}}$.

For any activity $a_{i,j}$, let $a_{i',j'+1}$ be the successor of $a_{i,j}$, if $a_{i',j'+1} \neq a_{i,j+1}$, then $a_{i',j'+1}$ must occur between every couple <$a_{i,j}$, $a_{i,j+1}$>.Suppose there are m $a_{i',j'+1}$s in HL, and let p_a be the probability that each of the m $a_{i',j'+1}$s is between a <$a_{i,j}$, $a_{i,j+1}$> couple, then we

have $p_a < \left(\dfrac{C_{i',j'}}{C_{i',j'} + C_{i,j}}\right)^m = \left(\dfrac{1}{1 + \dfrac{C_{i,j}}{C_{i',j'}}}\right)^m \leq \dfrac{1}{1 + m \cdot \dfrac{C_{i,j}}{C_{i',j'}}}$ .Let $p_a < \varepsilon$ (0<ε<1), we get

Table 1. The algorithm that mining sequential processes from hybrid log

---

Name : SePMa.

Input : Hybrid sequential process log.

Out : The set of processes that generate the input.

Procedure:

ActivitySet←Empty;

E←the first element of the input;

While E<>null

   If E not in ActivitySet then

      Add E into ActivitySet;

      Create an empty queue for E and set a counter for the queue with initial value 0;

   For each element A in ActivitySet

      Let Q_A be the queue of A and C_A the counter for Q_A;

      If E=A then

         Remove all the elements from Q_A whose total occurrences in Q_A not equal to C_A;

         Add E into Q_A;

         C_A←C_A+1;

      Else

         Let C be the number of Es in Q_A;

         If C=C_A-1 then

            Add E into Q_A;

         Else

            Remove all the Es from Q_A;

   E←the next element of the input;

   For each element A in ActivitySet

   Remove all the redundancies in Q_A and output Q_A;

$m \geq \dfrac{1-\varepsilon}{\varepsilon} \cdot \dfrac{C_{i',j'}}{C_{i,j}}$. Notice that m is the count of $P_{i'}$'s occurrences in HL, let

$C_i = \underset{\text{arg } j}{\max}\{C_{i,j}\}(1 \leq i \leq n)$, $C = \sum\limits_{i=1}^{n} C_i$, $p_b = \prod\limits_{j=1}^{ki'} \dfrac{C_{i',j}}{C}$, then the probability that $P_{i'}$ occurs in HL for once is not less than p_b, now let the length of HL be $\dfrac{m}{p_b}$,where m takes the

value $\dfrac{1-\varepsilon}{\varepsilon} \cdot \dfrac{C_{i',j'}}{C_{i,j}}$, i.e. the length of HL is $\dfrac{1}{p_b} \cdot \dfrac{1-\varepsilon}{\varepsilon} \cdot \dfrac{C_{i',j'}}{C_{i,j}}$, then the probability that

each $a_{i',j'+1}$ in HL occurs between every couple $<a_{i,j}, a_{i,j+1}>$ will less than ε.

The deduction above shows that for any activity $a_{i,j}$ and given probability ε, if the length of the log is large enough, then the successor of $a_{i,j}$ got by SePMa will take its correct successor with probability (1-ε). Let ε be small enough and the length of the log large enough, then SePMa will get all their successors for the activities in the log with high probability.

Besides its accuracy, from table 1 it can also be seen that SePMa needn't pre-get the whole information of the log, and the next actions being employed to do just depend on the results that the algorithm has had, so the algorithm is appropriate to dealing with online transactions. Further more, when the length of the input log is n, and suppose n is far more large than the number of different records in the log, i.e. n>>|ActivitySet|, it can be easily deduced that the time complexity of SePMa is $O(n^3)$.

4 Experiments

To evaluate the performance of SePMa, we develop LogGenerator, a synthetic log data generator. The parameters LogGenerator takes are shown in table 2.

Table 2. The parameters of LogGenerator

Parameter	Description
T	Number of the processes
L	Length of the log
S_1	Min steps of each process
S_2	Max steps of each process
C_1	Min coupling degree among the activities in a process ($C_1>0$)
C_2	Max coupling degree among the activities in a process ($C_2>0$)

For example, the input vector $<T,L,S_1,S_2, C_1,C_2>= <5,1000,10,20,0.1,0.85>$ means that the length of the synthetic log is 1000, and it should be generated by 5 different sequential processes. The values of parameters S_1 and S_2 show that all the processes are composed of not less than 10 activities, but none of them has more than 20 activities. For any process, the coupling degree between its neighbor activities would between 0.1and 0.85.

Given hybrid log HL, suppose PS be the process set that generated HL, and PS' be the the set of processes got by SePMa, to evaluate the performance of SePMa, we define the estimated parameter with the **Jaccard coefficient** as

$$\text{Sim(PS, PS')} = \frac{|PS \cap PS'|}{|PS \cup PS'|} \times 100\%$$

Where sim(PS,PS') describes the similarity between the target process set and the result got by SePMa.

In our experiments, the parameters LogGenerator takes and the corresponding results got by SePMa are shown in table 3.

Table 3. Experimental results of SePMa

$<T,S_1,S_2, C_1,C_2>$	L						
	1000	2000	4000	8000	16000	32000	60000
$<10,10,10,0.1,1>$	81.82%	100%	100%	100%	100%	100%	100%
$<10,10,20,0.1,1>$	24.14%	43.75%	100%	100%	100%	100%	100%
$<20,10,20,0.1,1>$	1.95%	10.11%	21.21%	100%	100%	100%	100%
$<20,10,20,0.05,1>$	0.55%	13.64%	100%	100%	100%	100%	100%
$<50,20,30,0.05,1>$	-	0	0	0	2.55%	21.60%	73.02%

Table 3, shows that in the experiments, the number of different activities in the log ranges from 100 to 1500, while the results got by SePMa are kept high accuracy, which shows that SePMa is of high scalability; the experiments also show that the accuracy of SePMa is strong positive relative to the length of the log, when the log's length increases, the output improves significantly.

5 Conclusion and Future Work

Process mining is an important issue in data mining and machine learning, by means of this method, we can rediscover the operating processes within the enterprises automatically, which will accelerate the progress of design and redesign for the enterprise processes significantly.

Most of current works on process mining aims at discovering the branch structures from the log generated by only one process, because of their rigorous requisitions on the log data, the applications of them are limited. Besides this problem, take it into account that most of enterprise processes are sequential, we focus on mining the log generated by multiple sequential processes. In this paper we study the distributional features of hybrid sequential process log in depth, we deduce theorem 1 and 2 and their corollaries, which are profound to describe the log. Further more, based on the conclusions we have, we propose SePMa, the sequential process mining algorithm. Both the theoretic analysis and experimental results show that SePMa is of high efficiency and effectiveness.

There are still more extensions to SePMa, one of the most important issues is to deal with the log generated from multiple multi-branches processes. Taking it into account that every multi-branches process can be decomposed into multiple sequential processes, the results this paper achieves can be helpful.

References

1. Maruster, L., Weijters, A.J.M.M., van der Aalstand, W.M.P., van den Bosch, A.: Process mining: Discovering direct successors in process logs. In: Lange, S., Satoh, K., Smith, C.H. (eds.) DS 2002. LNCS, vol. 2534, pp. 364–373. Springer, Heidelberg (2002)
2. Herbst, J.: A machine learning approach to workflow management. In: López de Mántaras, R., Plaza, E. (eds.) ECML 2000. LNCS (LNAI), vol. 1810, pp. 183–194. Springer, Heidelberg (2000)
3. van der Aalst, W.M.P., van Dongen, B.F., Herbst, J., Maruster, L., Schimm, G., Weijters, A.J.M.M.: Workflow mining: A survey of issues and approaches. Data & Knowledge Engineering 47, 237–267 (2003)
4. Cook, J.E., Wolf, A.A.L.: Automating process discovery through event-data analysis. In: Proceedings of the 17th International Conference on Software Engineering, pp. 73–82. ACM Press, New York
5. Cook, J.E., Wolf, A.L.: Discovering models of software processes from event-based data. ACM Transactions on Software Engineering and Methodology 7(3), 215–249 (1998)
6. Cook, J.E.: Process Discovery and Validation through Event-Data Analysis, University of Colorado (PhD Thesis) (1996)
7. van der Aalst, W.M.P., Weijters, A.J.M.M., Maruster, L.: Workflow Mining: Which processes can be rediscovered? http://tmitwww.tm.tue.nl/staff/wvdaalst/Publications/p169.pdf

Evaluate Structure Similarity in XML Documents with Merge-Edit-Distance

Chong Zhou, Yansheng Lu, Lei Zou, and Rong Hu

College of Computer Science and Technology,
HuaZhong University of Science and Technology,
1037 Luoyu Road, Wuhan, P.R. China
{lys,zoulei,ronghu}@mail.hust.edu.cn,
zhonghuazc@gmail.com

Abstract. XML language is widely used as a standard for data representation and exchange among Web applications. In recent years, many efforts have been spent in querying, integrating and clustering XML documents. Measuring the similarity among XML documents is the foundation of such applications. In this paper, we propose a new similarity measure method among the XML documents, which is based on Merge-Edit-Distance (MED). MED upholds the distribution information of the common tree in XML document trees. We urge the distribution information is useful for determining the similarity of XML documents. A novel algorithm is also proposed to calculate MED as follows. Given two XML document trees A and B, it compresses the two trees into one merge tree C and then transforms the tree C to the common tree of A and B with the defined operations such as "Delete", "Reduce", "Combine". The cost of the operation sequence is defined as MED. The experiments on real datasets give the evidence that the proposed similarity measure is effective.

1 Introduction

XML language is widely used as a standard for data representation and exchange among Web applications. The increasing XML documents have raised some issues on how to manage the heterogeneous XML documents effectively and efficiently. In recent years, many efforts have been spent in querying [1], integrating [2] and clustering [3] XML documents. Measuring the structural similarity among XML documents is a fundamental problem of the mentioned applications.

Some methods such as [4] adopted the (tree) edit distance to identify the similarity among XML documents. To the best of our knowledge, the most commonly cited algorithm[5] to compute general edit distance runs in $O(|T_1||T_2| \min(\text{depth}(T_1), \text{leaves}(T_1)) \min(\text{depth}(T_2), \text{leaves}(T_2)))$ time, where $\text{depth}(T_i)$, $\text{leaves}(T_i)$, $i = 1, 2$ are the depth and the number of leaf nodes of tree T_i respectively. Obviously, the tree-edit distance computation is not scalable for similarity search on large trees in huge datasets. Frequent tree structured patterns is introduced to decide the XML document similarity in [6, 7]. First, their methods represented the documents with full paths (the paths from root to leaves). Secondly, they used a sequential mining algorithm for finding the maximal common paths (that is maximal frequent sequences). Lastly, the

T. Washio et al. (Eds.): PAKDD 2007 Workshops, LNAI 4819, pp. 301–311, 2007.
© Springer-Verlag Berlin Heidelberg 2007

common paths were adopted to determine the similarity. However, since the XML documents are split into some individual paths, some structural information is lost. We urge that the distribution information of the common paths is useful for measuring the similarity between XML document trees. The example in Fig.1 explains our motivation clearly.

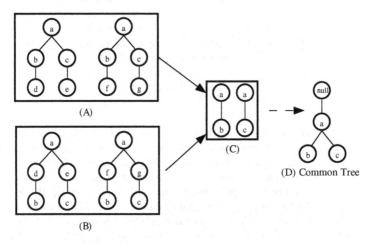

Fig. 1. The distribution information of the common tree

In Fig.1, there are two pairs of XML document trees. The two pairs have the same common paths, labeled (C). According to the method proposed by Lee et al. [5], the two pairs have the same similarity. Although Ho-pong [6] considered the hierarchical information, it still cannot distinguish the two pairs .The common tree (C) consists of two full paths of an induced tree of both trees in (A) ,whereas two paths of (C) make up an embedded tree of both trees in (B). It is obvious the pair in (A) is more similar than the one in (B).

In this paper, we propose a new similarity measure method considering the distribution information of the common tree. Merge-Edit-Distance (MED) is introduced to hold the distribution information. We also propose a high performance algorithm running in $O(|T_1|+|T_2|)$ time to calculate MED.

The rest of the paper is organized as follows. Section 2 presents the framework of measuring the similarity and gives the definitions of MED. An algorithm is proposed to calculate MED and the similarity in Section 3. Section 4 reports experiment results and in the last section we conclude our work.

2 Preliminary

In this section, we first present the framework of our similarity measuring method. Secondly, we give some supporting definition.

2.1 Framework

The whole similarity measure method consists of four steps as follows.

- Step 1 Preprocessing XML documents: Given two XML documents, convert them to tree format. The XML tree only holds the structural information, the element tags of XML documents. We remove repeated and nested nodes in the tree according to the method used in [8]. The tree passed to next step does not have two nodes with the same label.
- Step 2 Compressing trees: Compress the two trees into a Merge tree, labeled as MTree. Section 2.2 describes the Merge Tree structure and the Merge Tree construction Algorithm.
- Step 3 Finding the common tree and MED: Obtain Merge-Edit-Distance(MED) by transforming MTree to the common tree, labeled as CTree, with defined "Delete", "Reduce", "Combine" operations.
- Step 4 Calculating similarity: Calculate the similarity with MED and the common tree according to the similarity formula.

2.2 Basic Definitions

A XML tree T is a tuple T (N, E, r, L, S), where N is a finite set of nodes. $r \in N$ is the root node. E is a binary relation on N^2 and denotes the (acyclic) set of edges. $L : N \rightarrow S$ is a total function and S is the set of all tag names and attribution names in the XML document. The path from the node u to the node v is denoted as $path$ (u, v). $|T|$ stands for the number of nodes in T.

Definition 2.1 [Delete Node]. Let N_1 be a node of T and N_2 be the parent of N_1, Delete$_T(N_1)$ is a node delete operation applied to T, which removes N_1 and let the sons of N_1 be the sons of N_2.

Definition 2.2 [Common Tree]. Given two XML trees T_1 $(N_1, E_1, r_1, L_1, S_1)$, T_2 $(N_2, E_2, r_2, L_2, S_2)$, a tree T_c $(N_c, E_c, r_c, L_c, , S_c)$ is the common tree of T_1 and T_2, where (a) the label of r_c is "null"; (b) all full paths of the subtrees of T_c under the root are common paths of T_1 and T_2;(c) every common path of T_1 and T_2 is a full path of some subtree of T_c under the root.

The common tree is an integration of all common paths. Considering that not all common paths have a common root, we add a root, labeled "null", to construct a tree but not a forest. In fig. 1, the common tree (D) consists of common paths of (C). Note that it has only one subtree under the root since the common paths have a common root 'a'. A common tree is a specific frequent tree pattern containing all common paths.

Definition 2.3 [Merge Tree]. Given two XML trees T_1 $(N_1, E_1, r_1, L_1, S_1)$, T_2 $(N_2, E_2, r_2, L_2, S_2)$, a merge tree is a simple compression of T_1 and T_2, whose data structure is defined as below.

1. It consists of one root r_M labeled as "null", a set of full paths of T_1 and T_2 as the subtrees under the root, and a label header table.
2. Each element in the label table is a five-tuple $(l, node_1, node_2, level_1, level_2)$, where $l \in S_1 \cup S_2$ (l is a tag or attribution name),node $_1$ (node$_2$) points to the corresponding

node of the node carrying the label l in the tree T_1 (T_2), and $level_1$ ($level_2$) registers the level of $node_1$ ($node_2$) in the merge tree. If there is no corresponding node, the link ($node_1$ or $node_2$) is set to be "null" and the level ($level_1$ or $level_2$) is set to be "∞".

Based on this definition, we have the following construction algorithm.

Algorithm 1. Merge tree construction
Input: Two XML document trees T_1, T_2.
Output: the merge tree of T_1 and T_2, M.
Method: The merge tree is constructed in the following steps.
1. Create the root of a merge tree, M, and let its label be "null".
2. Insert the larger tree (i.e., the one with more elements, suppose it is T_1) into M as the subtree under the root. Update the label table as follows: add all labels in T_1 into label table according to the width-first order, let $node_1$ point to the node and $level_1$ be the level of the node, set $node_2$ to be null and set $level_2$ to be "∞".
3. For each full path of T_2, denoted as $path(r_2,k)$, where r_2 and k are the first and last node respectively in the full path, call insert ($path(r_2,k)$, r_M)in the T_2 into M as follows: if r_M has a son m' with the same label of r_2, then let the corresponding $node_2$ and $level_2$ of the label be $node_1$ and $level_1$ respectively; else create a new node m', insert the label of r_2 into the label table, let $node_1$ be "null", $level_1$ be "∞", $node_2$ point to m' and $level_2$ be the level of m'. If r_2 has a son r' in the $path$ (r_2, k), then call insert (path (r',k),m').

Figure 2(A) and Fig.2(B) are the merge trees of the trees in Fig.1(A) and Fig.1(B) respectively. Here, we number the node according to the width-first order and use the unique number of the node instead of the link ("-1" stands for "null"). This presentation shows the merge tree more clearly.

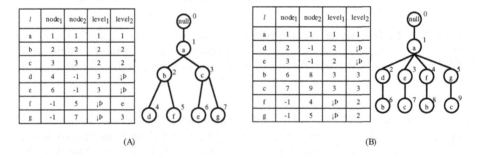

l	$node_1$	$node_2$	$level_1$	$level_2$
a	1	1	1	1
b	2	2	2	2
c	3	3	2	2
d	4	-1	3	¡Þ
e	6	-1	3	¡Þ
f	-1	5	¡Þ	e
g	-1	7	¡Þ	3

(A)

l	$node_1$	$node_2$	$level_1$	$level_2$
a	1	1	1	1
d	2	-1	2	¡Þ
e	3	-1	2	¡Þ
b	6	8	3	3
c	7	9	3	3
f	-1	4	¡Þ	2
g	-1	5	¡Þ	2

(B)

Fig. 2. The merge trees example

3   Transform Amerge Tree to a Common Tree

In this section, we first define several operations to the merge tree in section 3.1. A merge tree can be transformed into a common tree with a series of defined operations. We describe the transform algorithm in section 3.2.

3.1 Operations to a Merge Tree

Definition 3.1 [Delete]. Let $(l, node_1, node_2, level_1, level_2)$ be an element in the label table of the merge tree M. If one of two links, suppose it is $node_2$, is null, $Delete_M(l)$ is a delete operation applied to M.

$Delete_M(l)$ updates M as follows:

(1) $Delete_M(node_1)$.

(2) Remove $(l, node_1, node_2, level_1, level_2)$ from the label table.

(3) Update the level information (e.g. $level_1$ and $level_2$) of remained elements in the label table.

The cost of $Delete_M(l)$ is indicated as $Cost_D(M, l) = 1$.

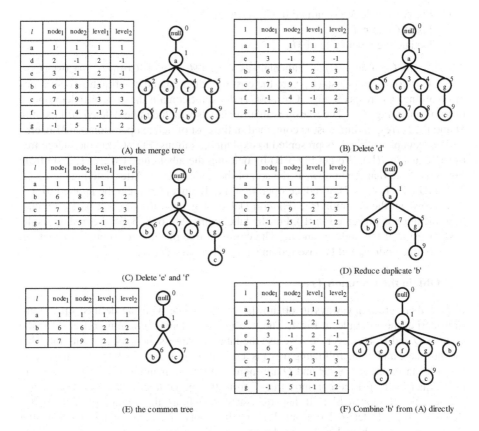

Fig. 3. An example of operations to the merge tree

Definition 3.2 [Reduce]. Let $(l, node_1, node_2, level_1, level_2)$ be an element in the label table of the merge tree M. If $node_1$ and $node_2$ have the same parent node p, $Reduce_M(l)$ is a duplicated node reduction applied to M.

Let T_1 and T_2 be the subtrees rooted at $node_1$ and $node_2$ respectively, M_{12} be the merge tree of T_1 and T_2. $Reduce_M(l)$ updates M as follows:

(1) Remove T_1 and T_2.

(2) Insert M_{12} as the subtree under the node p.

(3) Update the label table of M.

The cost of $\text{Reduce}_M(l)$ is indicated as $\text{Cost}_R(M, l) = 1$.

Definition 3.3 [Combine]. Let $(l, \text{node1}, \text{node2}, \text{level1}, \text{level2})$ be an element in the label table of the merge tree M. If node_1 and node_2 are not null and do not have the same parent node p, $\text{Combine}_M(l)$ is a node combination applied to M. Let T_1 and T_2 be the subtrees rooted at node_1 and node_2 respectively, C_{12} be the common tree of T_1 and T_2, M_{12} be the merge tree of T_1 and T_2. Let node_a be the first ancestor of node_1 and node_2, which does not have any son that is an ancestor of node_1 and node_2. Combine_M (l) updates M as follows:

(1) For each node N_i contained in C_{12}, $\text{Delete}_M(N_i)$.

(2) Insert C_{12} at node_a.

(3) Update the label table of M.

The cost of $\text{Combine}_M(l)$ is indicated as $\text{Cost}_C(M, l) = 1 + \text{Cost}_R(C_{12}) + \text{Cost}_C(C_{12})$, where $\text{Cost}_R(C_{12})$ and $\text{Cost}_C(C_{12})$ are the cost of reduction and combination operations of obtaining C_{12} respectively. $\text{Cost}_C(M, l)$ does not contain the cost of delete operations of obtaining C_{12}. The reason is that the labels which can be deleted still exist in M and the corresponding cost is contained in the cost of delete operations on M itself.

The example in Fig.3. is presented to explain operations. Fig.3(A) is the merge tree as same as Fig.2(B). We get Fig.3(B) by deleting the label entry ('d', 2,-1, 2,-1) from Fig.3(A). Note that the node n_6 is the son of the node n_1 now. After deleting ('e', 3,-1, 2,-1) and ('f',-1, 4,-1, 2), a reduce operation, i.e. Reduce_M('b'), can be applied to (C). At last, we get the common tree (E) with Delete_M('g') and Reduce_M('c'). A combination operation is presented in (F). For the label entry ('d', 6, 8, 3, 3), the node n_1 is the first ancestor of the node n_6 and n_8. The common tree of the trees rooted at n_6 and n_8 has only one node n_6 and be inserted under n_1. $\text{Cost}_C(M,$ 'd')=1.

3.2 Obtain the Common Tree

In fact, there are many ways to obtain the common tree with the defined operations. Considering the example in Fig.4, The tree Q and P have the common tree C as the same as Q and R. However, 'c' (we use the label to express the node carrying the label for short here and in the following part.) is a son of 'b' in R and Q while 'b' is the ancestor of 'c' in the tree P. So R and Q are more similar. As mentioned in Fig.4, there are two sequences to get the common tree C from the merge tree S. One is (Delete_S('g'), Delete_S('e'), Remove_S('b')). It has the same length as the sequence (Delete_T('f'), Remove_T('b') , Delete_T('d')) from T to C. It adopts an operation Delete_S('e') after Delete_S('g') , which makes 'c' be the son of 'b' before 'b' is processed. And then Remove_S('b') and Remove_T('b') have the same effectiveness. However, it ignores the position of 'e' which leads to the key difference between T and S: 'c' is a son of 'b' in T but not in S. The second sequence from S to C is (Delete_S('g'), Remove_S('b'), Delete_S('e'), Remove_S('c')). This sequence processes the node first before processing its children and exposes the information ignored by the first one.

In order to keep the distribution information of nodes as much as possible, a good merge edit sequence should obey the principle that the node closer to the root should be prior to be processed. This can be easily understood since the node closer to the root has more information.

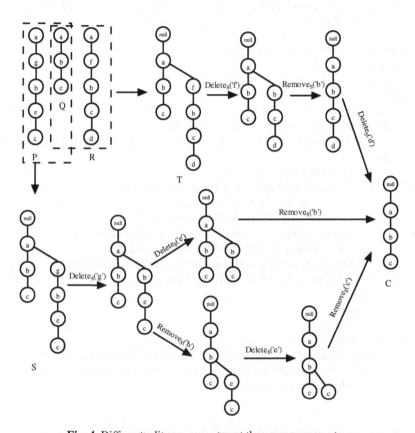

Fig. 4. Different edit sequences to get the same common tree

Definition 3.4 [Valid-Edit-Sequence]. Let L_1 and L_2 be two elements of the label table of the merge tree, suppose $L_1.leve_1 < L_1.leve_2$ and $L_2.leve_1 < L_2.leve_2$. A merge edit sequence is a valid-edit-sequence if and only if L_1 is processed before L_2 under the following conditions:

(1) $L_1.leve_1 < L_2.leve_1$
(2) $L_1.leve_1 = L_2.leve_1$ and $L_1.leve_2 \leq L_1.leve_2$.

Definition 3.5 [Merge-Edit-Distance]. Given a Valid-Edit-Sequence $E(e_1,...,e_n)$ (e_i is a "Delete" or "Reduce" or "Combine" operation for any $i \in [1,n]$.) from a merge tree to its common tree, the Merge-Edit-Distance(MED) of E, denoted MED(E), is defined as the sum of the cost of each operation in the sequence E.

Algorithm2 gives a method to get the good merging edit sequence and its corresponding MED. The function sort(M) ensures that we always first handle the lower label element. In the worst case, Algorithm 2 applies "Delete" operation to delete all the labels from the label table. The time complexity of Algorithm 2 is $O(|M.LT|)=O(|T_1|+|T_2|)$, where $|M.LT|$ is the size of the label table of M which is consist of tree T_1 and T_2.

Algorithm 2. Calculating the Merge-Edit-Distance of the Merging edit sequence
    Input: a merge tree M.
    Output: the common tree C of M, and MED.
    Method:
    1   sort(M);
    2   While the label table of M is not empty do
    3   let $F(l$, node1, node2, level1, level2) be the first element of the label table.
    4   if (node1== node2) then
    5   remove F from the label table and continue.
    6   if (node1==null or node2==null) then
    7   Delete$_M$ (l), MED = MED + Cost$_D(M, l)$;
    8   if (node1!= node2&& node1.parent!=node2.parent) then
    9   Combine$_M$ (l), MED = MED + Cost$_C(M, l)$;
    10   if (node1!= node2&& node1.parent==node2.parent) then
    11   Reduce$_M$ (l), MED = MED + Cost$_R(M, l)$;
    12   sort(M);
    13   end while
    14    $C = M$, return C
    Function sort(M)
        Sort the label table in the ascend order of the minimal level of the node; if some labels have the same minimal level, then sort them in the ascend order of the other level. Let MED be 0.
    End Function

3.3 Measuring the Similarity Between XML Documents

Now we can define the similarity between two XML document trees with MED obtained by Algorithm 2.

Let T_1 and T_2 be two XML document trees, MED(T_1,T_2) be the MED obtained by Algorithm 2. The distance between T_1 and T_2 is defined as follows:

$$D(T_1,T_2)= \text{MED } (T_1,T_2)/(| T_1|+| T_2|), \tag{1}$$

where $| T_1|$ and $| T_2|$ are the number of nodes in T_1 and T_2 respectively. The $D(T_1,T_2)$ value is (a) 0 when the trees are totally same, that is , they have exactly the same structure and the same labels in the matching nodes,(b) 1 when the trees do not have the same label at all, (c) low when the trees have similar structure , and (d) high when the trees have different structure and low percentage of matching nodes.

The similarity between T_1 and T_2 is defined as:

$$Sim(T_1, T_2) = (S_r + 1 - D(T_1, T_2))/2. \tag{2}$$

Where S_r is 1 if the two trees T_1 and T_2 have the same root element; otherwise it is 0.

4 Experiments

We validated the effectiveness of the similarity metric based on MED with real and synthetic data. Our experiments were conducted as follows. We first download two DTDs from NIAGARA [1]: club.dtd and department.dtd. We modified the two DTDs to allow the simple elements are not-mandatory elements. These two DTD both describe person information in club and department respectively. Then we use IBM's XML generator to accept the modified DTDs as input and to create synthetic XML documents. The real data download from NIAGARA and the synthetic data are mixed and used to analyze the similarity of documents. PTR method[7] was chosen for comparisons. The experiment results of XML document similarity between club.dtd and department.dtd are shown in Fig.5 and Table 1, where a_1, a_2, a_3 and a_4 are four documents created with club.dtd and b_1, b_2, b_3 and b_4 are four documents created with department.dtd.

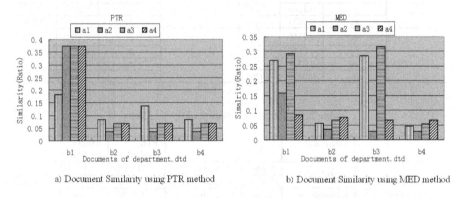

a) Document Similarity using PTR method   b) Document Similarity using MED method

Fig. 5. Results of document similarity between club.dtd and department.dtd

In the following, $Sim_{PTR}(x,y)$(resp. $Sim_{MED}(x,y)$) denotes the similarity between document x from department.dtd and y from club.dtd based on PTR (MED). The following cases are obviously seen:

$Sim_{PTR}(b_1,a_2) = Sim_{PTR}(b_1,a_3) = Sim_{PTR}(b_1,a_4)$, $Sim_{PTR}(b_2,a_3) = Sim_{PTR}(b_2,a_4)$,
$Sim_{PTR}(b_3,a_3) = Sim_{PTR}(b_3,a_4)$, $Sim_{PTR}(b_4,a_3) = Sim_{PTR}(b_4,a_4)$.

However, our MED method can determine the most similar document from them. This is because MED considers structures of both two compared trees. The MED method is sensitive about every element even if the element does not change the common tree. The PTR method chooses the lager document tree to evaluate the similarity,

---

[1] NIAGARA,http://www.cs.wisc.edu/niagara/data.html

so the smaller trees have no effectiveness on the final similarity when their common paths are the same. This brings on a higher stableness of the PTR method. The variance, shown in Table 2, of the PTR method is smaller.

The MED method can identify documents from different DTDs very well. In Table 2, the mean similarity of the MED method is larger than that of the PTR method considering the homogeneous documents but smaller considering the heterogeneous documents. The average similarity of the MED method is larger than that of the PTR method considering documents of department.dtd. The XML document tree of department.dtd has many small level-1 subtrees (direct under the root). Hence, the small subtree (that has few nodes) has a great effectiveness on the PTR method, but little on the MED method. In contrast, the XML document tree of club.dtd has few large level-1 subtrees so that the result of the PTR method is little better than ours. This case indicates that our method is suited for various types of XML documents.

Table 1. Results of XML document similarity between club.dtd and department.dtd

Documents of department.dtd	Documents of club.dtd	Similarity		Documents of department.dtd	Documents of club.dtd	Similarity	
		PTR method	MED method			PTR method	MED method
a_1	b_1	0.1806	0.2693	a_3	b_1	0.3750	0.2917
	b_2	0.0833	0.0556		b_2	0.0926	0.0675
	b_3	0.1389	0.2857		b_3	0.1852	0.3158
	b_4	0.0833	0.0476		b_4	0.0926	0.0526
a_2	b_1	0.3750	0.1591	a_4	b_1	0.3750	0.0853
	b_2	0.0370	0.0357		b_2	0.0667	0.0769
	b_3	0.0370	0.0294		b_3	0.0667	0.0675
	b_4	0.0370	0.0294		b_4	0.0667	0.0675

Table 2. Statistical information of the similarity

	Document-Pair	Statistic	PTR method	MED method
Homogeneous documents	Club-Club	Mean of similarities	0.9516	0.9365
		Variance of similarities	0.002	0.005
	Department-Department	Mean of similarities	0.7192	0.8832
		Variance of similarities	0.072	0.015
Heterogeneous documents	Department-Club	Mean of similarities	0.1433	0.1210
		Variance of similarities	0.004	0.007

5 Conclusions and Future Work

In this paper, we presented a novel similarity metric of XML documents based on Merge-Edit-Distance (MED). We also proposed an effective algorithm to calculate MED between two XML documents. Experiments showed that our method was effective on identifying documents from different DTDs. The similarity value is high between homogeneous documents and low between heterogeneous documents. The experiment results also indicated our method was suited for various types of XML documents.

In the further research, we are focusing on some issues, one of which is to propose an algorithm to calculate MED without eliminating repeated and nested elements first.

References

1. Rege, M., Caraconcea, I., Lu, S., Fotouhi, F.: Querying XML Documents from a Relational Database in the Presence of DTDs. In: Ghosh, R.K., Mohanty, H. (eds.) ICDCIT 2004. LNCS, vol. 3347, pp. 168–177. Springer, Heidelberg (2004)
2. Galhardas, E.H., Florescu, D., Shasha, D., Simon, E., Saita, C.-A.: Declarative Data Cleaning: Language, Model, and Algorithms. In: Apers, P.M.G., Atzeni, P., Ceri, S., Paraboschi, S., Ramamohanarao, K., Snodgrass, R.T. (eds.) Proceedings of 27th International Conference on Very Large Data Bases, pp. 371–380. Morgan Kaufmann, Roma, Italy (2001)
3. Doucet, A., Ahonen-Myka, H.: Naive clustering of alarge XML document collection. In: Norbert Fuhr, N.G., Kazai, G., Lalmas, M. (eds.) First Workshop of the INitiative for the Evaluation of XML Retrieval (INEX), Schloss Dagstuhl, Germany, pp. 81–87 (2002)
4. Nierman, A., Jagadish, H.V.: Evaluating Structural Similarity in XML Documents. In: Mary, F., Fernandez, Y.P. (eds.) WebDB 2002, Madison, Wisconsin, USA, pp. 61–66 (2002)
5. Zhang, K., Dennis, S.: Simple fast algorithms for the editing distance between trees and related problems. SIAM Journal on Computing 18, 1245–1262 (1989)
6. Lee, J.-W., Lee, K., Kim, W.: Preparations for Semantics-Based XML Mining. In: Cercone, N., Lin, T.Y., Wu, X. (eds.) International Conference on Data Mining, pp. 345–352. IEEE Computer Society, California (2001)
7. Leung, H.-p., Chung, F.-l., Chan, S.C.-f.: A New Sequential Mining Approach to XML Document Similarity Computation. In: Whang, K.-Y., Jeon, J., Shim, K., Srivastava, J. (eds.) PAKDD 2003. LNCS (LNAI), vol. 2637, pp. 356–362. Springer, Heidelberg (2003)
8. Dalamagas, T., Cheng, T., Winkel, K.-J., Sellis, T.: A methodology for clustering XML documents by structure. Information Systems 31, 187–228 (2006)

Ensemble Learning Based Distributed Clustering

Genlin Ji and Xiaohan Ling

Department of Computer Science, Nanjing Normal University,
Nanjing 210097, P.R. China
glji@njnu.edu.cn, nolen0@163.com

Abstract. Data mining techniques such as clustering are usually applied to centralized data sets. At present, more and more data is generated and stored in local sites. The transmission of the entire local data set to server is often unacceptable because of performance considerations, privacy and security aspects, and bandwidth constraints. In this paper, we propose a distributed clustering model based on ensemble learning, which could analyze and mine distributed data sources to find global clustering patterns. A typical scenario of the distributed clustering is a 'two-stage' course, i.e. firstly doing clustering in local sites and then in global site. The local clustering results transmitted to server site form an ensemble and combining schemes of ensemble learning use the ensemble to generate global clustering results. In the model, generating global patterns from ensemble is mathematically converted to be a combinatorial optimization problem. As an implementation for the model, a novel distributed clustering algorithm called DK-means is presented. Experimental results show that DK-means achieves similar results to K-means which clusters centralized data set at a time and is scalable to data distribution varying in local sites, and show validity of the model.

Keywords: Distributed clustering, ensemble learning, data mining.

1 Introduction

Most of the machine learning and data mining algorithms work with a rather basic assumption that all the training data have been pooled in a centralized data repository. Recently, there exist a growing number of cases that the data have to be physically distributed, due to either their huge volumes or privacy concern. Relevant examples include distributed medical data analysis, intrusion detection, data fusion in sensor networks, customer record analysis, etc. Analyzing and mining these distributed sources require distributed machine learning and data mining techniques because the wanted patterns should be mined on a global view of these distributed sources.

Clustering is a significant issue of data mining. So far, many algorithms coping with different aspects of clustering have been proposed such as in [1], [2], [3]. All of them process centralized data, but we only focus on distributed clustering topic in this paper. As described in [4], a distributed data mining solution consists of the following steps. First, a data mining algorithm is locally applied to each of the k sites separately and independently. The results are k local sets of patterns called *local models*. Second,

T. Washio et al. (Eds.): PAKDD 2007 Workshops, LNAI 4819, pp. 312–321, 2007.

the local models are transferred to a central server. The central server combines the local models to generate a *global model*. Third, the global model may optionally be sent back to local sites. In [5] a distributed version of DBSCAN [3] is presented. The local clusters are represented by special objects that have the best representative power. This representative power is based on two quality measures that take the density-based clustering concepts into account. For each representative, a covering radius and a covering number are aggregated for the global merge step. In [6] a distributed model-based clustering algorithm that uses EM for detecting local models in terms of mixtures of Gaussian distributions is proposed. To get meaningful global model, an effective algorithm is used to merge local Gaussian distributions.

On the other hand, research of ensemble learning that trains multiple classifiers and combines their predictions has evolved remarkably in machine learning communities. Since this technique can improve accuracy of classification significantly, it is also applied in clustering called clustering ensembles. Clustering ensembles produce multiple partitions, each of which is generated in the same data set by different methods, and combine them to find a new partition with better quality than any known one. For more details, we refer readers to [7].

In this paper, we propose a distributed clustering model for analyzing and mining distributed data sources to obtain the global clustering patterns. The key idea of the model is to regard distributed clustering as clustering ensembles. Alike bagging, one famed ensembling scheme, resampling technique [8] is used to generate multiple partitions to form an ensemble. While data volume is dramatically huge, sampling subset of the entire data repository in each time can prominently reduce computational complexity and does not lose quality of eventual clustering ensembles results. Actually, local sites collect subset of global data in distributed environment, which resembles sampling subset in clustering ensembles. Therefore, based on clustering ensembles, we derive a function that represents relationship of local clustering and global clustering. The function is, exactly, a general model that one can employ it to resolve concrete distributed clustering cases. As an implementation for the model, we also present a novel distributed version of K-means named DK-means. Empirical results indicate that DK-means achieves similar results to K-means and is more scalable to data distribution varying in local sites than algorithm proposed in [9] which also uses K-means for distributed clustering.

The rest of the paper is organized as follows. In Section 2, we firstly review research work of ensemble learning and then derive the general distributed clustering model. In Section 3, a novel distributed clustering algorithm based on the model is presented. And empirical study on the algorithm is given in Section 4. Finally, Section 5 concludes this paper.

2 Modeling Distributed Clustering Via Ensemble Learning

2.1 Ensemble Learning

In general, the whole course of ensemble learning can be obviously divided into two steps, i.e. constructing a number of component classifiers sequentially or simultaneously and then combining the predictions of those components.

For constructing component classifiers, a great number of methods can be employed. The most prevailing approaches are Bagging and Boosting. Bagging is based on bootstrap sampling, which generates several training sets from the original data set and then trains a component classifier from each of those training sets. Boosting sequentially produces some component classifiers of which training sets are determined by the performance of former ones. In order to combine the predictions, plurality voting or majority voting for classification and simple averaging or weighted averaging for regression are mostly employed.

In unsupervised learning, though hundreds of clustering algorithms exist, it is difficult to find a single clustering algorithm that can handle all types of cluster shapes and sizes or even decide which algorithm would be the best one for a particular data set [10]. Since ensemble learning technique can improve accuracy of classification significantly, it is also applied in clustering. In [11] Fred and Jain introduce the concept of evidence accumulation clustering that maps the individual data partitions in a clustering ensemble into a new similarity measure between patterns, summarizing interpattern structure perceived from these clusterings. The final data partition is obtained by applying the single-link method to this new similarity matrix. The results of this method show that, the combination of "weak" clustering algorithms such as the K-means, which impose a simple structure on the data, can lead to the identification of true underlying clusters with arbitrary shapes, sizes and densities.

2.2 Derivation of Distributed Clustering Model

Suppose there are multiple data sources distributed over several local sites and our aim is to obtain the global clustering patterns combined by local clustering results. This distributed clustering problem like ensemble learning can be divided into two subproblems — (i) generating local clustering results and choosing appropriate parameters sent to central site, and (ii) combining the local clustering information effectively to obtain an exact and meaningful global clustering result.

For convenience of following discussion, we define *local cluster model* and *global cluster model*, which represent abstract description to one cluster of all in a local site and abstract description to one cluster of all of the final global clustering, respectively. Moreover we assume there exist n local sites denoted by $\{S_1, S_2, ..., S_n\}$ and one central site S in the distributed environment.

Definition 1 (local cluster model). In a local site S_i ($1 \leq i \leq n$), there are K_i clusters generated by a certain clustering algorithm such as K-means, and then call each cluster a local cluster model denoted by λ_i^j ($1 \leq j \leq K_i$).

Note that above model λ_i^j is an abstract description to a cluster. It has variant forms as using different algorithms for clustering. For instance, when applying EM clustering algorithm, each cluster is described by Gaussian distribution. Thus, λ_i^j is a tuple (μ, Σ) in this case, where μ is the mean value of all points in the cluster and Σ is the $d \times d$ covariance matrix of all points in the cluster.

Definition 2 (global cluster model). The underlying true data distribution comprising data in all local sites is a partition, which involves L clusters. Then one cluster of all is called a global cluster model denoted by $\lambda^j (1 \le j \le L)$.

Obviously, the problem of distributed clustering is to obtain the unknown global cluster models. Inspired by practice of ensemble learning, data in each local site can be considered as subset sampled from original data set, hence a local clustering result described by local cluster models is a biased version of global clustering result.

Therefore, global clustering result described by global cluster models can be obtained by employing combining schemes such as simple averaging or weighted averaging. Stating it formally by simple averaging (also can use weighted averaging) with respect to the j-th global cluster model as follows:

$$\lambda^j = \frac{1}{n} \sum_{i=1}^{n} \lambda_i^j , \qquad (1)$$

where n is the number of local sites. Before using Eq. (1), we need to find out approximately which local cluster model in which local site corresponds to which local cluster model in other sites. Eq. (1) assumes that cluster labels have already been corresponded. This consensus function problem is studied in [7].

So distributed clustering is accomplishable by computing each global cluster model via Eq. (1). Despite Eq. (1) seems brief and simple, a number of distributed clustering approaches are based on it in essence. Taking [9] for example, local cluster labels are firstly corresponded by a matrix-linking algorithm, then global value is computed as the weighted mean of all linked local centroids in matrix.

However, there exist two drawbacks with Eq. (1). First, cluster labels must be corresponded one-to-one among local sites so that each partition in local sites need contain the same number of clusters as well as the global clustering. Second, Eq. (1) assumes correspondence of clusters must occur between different sites. In some cases, clusters in the same site may belong to a same global cluster, so they ought to be corresponded. Thereby we derive a second formula that enhances Eq. (1) to overcome the limits and generally represents relationship between local cluster models and global cluster models.

Suppose central site S takes all local cluster models and then the total number of local cluster models in S can be computed as $\sum_{i=1}^{n} K_i$ while i-th local site includes K_i clusters. In fact, each local cluster model is a biased estimator of a certain global cluster model of all. In other words, an underlying global cluster model λ^j is approached by some local cluster models and these local cluster models form an ensemble. So the distance from the underlying global cluster model λ^j to those local cluster models that estimate it, i.e. $D(\lambda^j, \Theta_j)$ where Θ_j is the set of satisfied local cluster models, is minimal when compared with distance to other local cluster models. It is worth mentioning that $D(\cdot,\cdot)$ is a suitable distance measure for the models in terms of actual form of the model. The rest global cluster models may be deduced by analogy, thus, we can derive the following objective function for all global cluster models, i.e.,

$$\{ \lambda^1,...,\lambda^L \} = \arg \min_{\lambda^j} \sum_{j=1}^{L} D(\lambda^j, \Theta_j), \tag{2}$$

where Θ_j is a nonempty subset of $\Theta = \{ \lambda_i^j \mid 1 \leq i \leq n, \ 1 \leq j \leq K_i \}$ and $\Theta_1 \cap ... \cap \Theta_L = \phi$, $\Theta_1 \cup ... \cup \Theta_L = \Theta$. Although the underlying global cluster models are unknown, they can be obtained by optimizing Eq. (2). So distributed clustering is mathematically converted to be a combinatorial optimization problem.

3 DK-Means

In this section, we propose a new distributed clustering algorithm based on K-means called DK-means (Distributed K-means). The algorithm is an implementation for the model described in previous section. Local clustering uses K-means, intuitively, local cluster model then can be represented by the mean value of points in the cluster. The remain problem is how to combine all local mean values to obtain the global mean values of global clustering.

Since employing K-means, Euclidean distance is naturally the distance measure for the model. Let $\{O_1, O_2, ..., O_L\}$ be global mean values, and $m_i \ (1 \leq i \leq \sum_{i=1}^{n} K_i)$ be each local mean value. Applying Eq. (2), i.e.,

$$\begin{aligned}
\{ \lambda^1,...,\lambda^L \} &= \arg \min_{\lambda^j} \sum_{j=1}^{L} D(\lambda^j, \Theta_j) \\
&= \arg \min_{\lambda^j} \sum_{j=1}^{L} D(O_j, \{m_i \mid m_i \in \Theta_j\}) \\
&= \arg \min_{\lambda^j} \sum_{j=1}^{L} \sum_{m_i \in \Theta_j} D(O_j, m_i) \\
&= \arg \min_{\lambda^j} \sum_{j=1}^{L} \sum_{m_i \in \Theta_j} \| O_j - m_i \|^2.
\end{aligned} \tag{3}$$

As we do not know which and how many local mean values form an ensemble for obtaining a global mean value, it is difficult to directly solve the optimization of Eq. (3). Viewing (3) again, one can observe that while regarding O_j as mean of a cluster and m_i as point in the same cluster, it is surprisingly similar to the criterion function of K-means. Therefore, we employ K-means for approximately optimizing Eq. (3) and synchronously obtaining global mean values.

DK-means firstly does clustering in each local site using K-means, then sends all mean values to central site, finally global mean values of underlying global clustering are obtained by using K-means again. The pseudo code of DK-means is described in Fig. 1.

Input: D_i – data in the i-th site
   K_i – number of clusters in the i-th site
   K – number of clusters of global clustering
Output: C – the set of global mean values

In local site S_i :

   1. C_i=k-means(K_i,D_i);   //C_i is the set of means in the i-th site
   2. Send C_i to global site S;

In global site S :

   1. for i=1 to n do
       Receive C_i;

   2. $D=\bigcup_{i=1}^{n}C_i$;

   3. C=k-means(K,D);
   4. Optionally send C back to local sites;

Fig. 1. Description of DK-means algorithm

4 Experimental Results

We have tested DK-means on both synthetic data set and five real world data sets. Synthetic data set is 2-dementional with three clusters and five real world data sets from the UCI Machine Learning Repository [12] are *Adult, Heart, Iris, Image Segmentation* and *Wine*. The detailed information of real world data sets is shown in Table 1. We prepare 4 computers as local sites, and another one as central site for producing global clustering results.

Table 1. Real world data sets used in experiments

Data Set	NO.attributes	NO.classes	NO.instances
Adult	14	2	48842
Heart	14	5	303
Iris	4	3	150
Image Segmentation	19	7	2310
Wine	13	3	178

Test results on synthetic data set are illustrated by Figs. 2 and 3. In Fig. 2, it is obvious that local site (a), (b) and (d) collect part data that only contains one cluster, and site (c) has data including two clusters. Then we apply DK-means, setting number of clusters in each local site with 2, 3, 3, 2 respectively and number of clusters of global clustering with 3, and get result of global clustering shown in Fig. 3. Intuitively, position of global means is quite accurate and meaningful. Though local

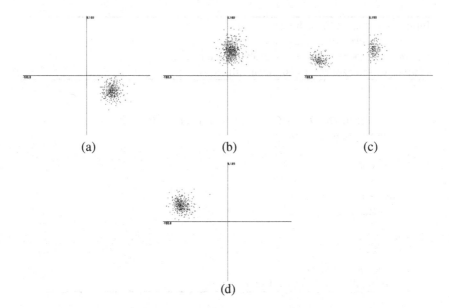

Fig. 2. Data distributed in 4 local sites

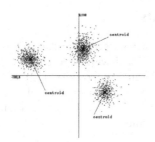

Fig. 3. Result of global clustering

view of global data in local sites is incomplete and inaccurate, DK-means combining information of local clustering generates exact result.

To evaluate performance of DK-means more well, we compare it with centralized clustering using K-means and distributed clustering algorithm proposed in [9] named D-Combining on real world data sets. Data in each local site is randomly sampled from the original data set and probability theory indicates that in this case data distribution in local sites is similar to each other in the most of time. For centralized clustering, data in all local sites is pooled in central site after distributed clustering has done. And for DK-means, the number of clusters of both local clustering and global clustering directly uses the number of classes of each data set. In [13] Modha and Spangler present a kind of measure for evaluating clustering quality taking class labels of original data set into account. The so-called Micro-precision is defined as follows:

$$\text{Micro-precision} = \frac{1}{m} \sum_{i=1}^{k} \alpha_i , \qquad (4)$$

where m is the total number of data samples, k is the number of clusters and α_i is the number of data samples that are correctly assigned to i-th cluster. Note that the greater Micro-precision is, the better is quality of clustering result. Test on each data set runs 10 times and average results are shown in Table 2.

As shown in Table 2, clustering quality is best on *Iris* with average lower bound not less than 0.89 while on data set *Image Segmentation* all three employed algorithms gain Micro-precision not more than 0.7. Small quantity of data samples and distinct parted clusters might lead to the high quality of clustering on data set *Iris*. All in all, our algorithm, DK-means, is most consistent with K-means about Micro-precision and is better than D-Combining on four data sets except on *Heart*. In order to investigate the influence of data distribution different among local sites on distributed clustering quality, we assign data in different class to each local site instead of randomly sampling. Thus, data in each local site not only reveals part view of global data set but also falls in different subspace of global data set. Results are the average of 10 tests depicted in Fig. 4. As Fig. 4 shows, in this case clustering quality of D-Combining falls down a lot while DK-means falls down a little compared with results in Table 2. The reason of why performance of D-Combining fluctuates dramatically between these two experiments is that the algorithm is mostly suitable to the case that all local sites have data with almost identical distribution.

To sum up, experimental results show that DK-means can achieve similar results to the centralized clustering and is much less susceptible to data distribution varying in local sites. Therefore, it is feasible and applicable.

Table 2. Micro-precision on 5 data sets with uniform distribution in local sites

Data Set / Algorithm	Adult	Heart	Iris	Image-S	Wine
K-means	0.804	0.784	0.913	0.678	0.704
D-Combining	0.757	0.778	0.89	0.663	0.681
DK-means	0.782	0.771	0.907	0.681	0.695

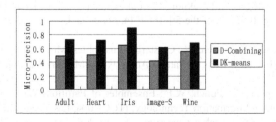

Fig. 4. Micro-precision on 5 data sets with nonuniform distribution in local sites

5 Conclusion

In this paper, we propose a distributed clustering model based on ensemble learning. In the model, distributed clustering is mathematically converted to be a combinatorial optimization problem. As an implementation for the model, we also present a novel distributed version of K-means called DK-means. While obtaining global results, DK-means uses clustering for indirectly optimizing the objective function. Empirical results on both synthetic and real world data sets demonstrate that DK-means can achieve similar results to the centralized clustering and is scalable to data distribution varying in local sites. Furthermore, efficiency of DK-means shows our model is valid and applicable in a certain extent.

For future work we will apply more clustering algorithms based on the model to distributed environment and explore the influence of factors, such as distribution of data appearance, distribution of data size, clustering algorithm used in local sites and the number of clusters, on accuracy of distributed clustering.

Acknowledgments. This research is supported by the Natural Science Foundation of Jiangsu Province, P.R.China, under grant number BK2005135.

References

1. MacQueen, J.: Some methods for classification and analysis of multivariate observations. In: LeCam, L.M., Neyman, J. (eds.) Proc. of the 5th Berkeley Symp. on Mathematical Statistics and Probability, Berkeley, vol. 1, pp. 281–297. University of California Press (1967)
2. McLachlan, G., Basford, K.: Mixture Models: Inference and Application to Clustering, Marcel Dekker, New York (1988)
3. Ester, M., Kriegel, H.P., Sander, J., et al.: A density based algorithm of discovering clusters in large spatial databases with noise. In: Simoudis, E., Jiawei, H., Fayyad, U.M. (eds.) Proc. of the 2nd Int. Conf. on Knowledge Discovery and Data Mining, Portland, Oregon, pp. 226–231. AAAI Press, Stanford, California, USA (1996)
4. Park, B.H., Kargupta, H.: Distributed Data Mining: Algorithms, Systems, and Applications. In: Ye, N. (ed.) The Handbook of Data Mining, Lawrence Erlbaum Associates Publishers, Mahwah, NJ (2003)
5. Januzaj, E., Kriegel, H.P., Pfeifle, M.: Scalable Density-Based Distributed Clustering. In: Boulicaut, J.-F., Esposito, F., Giannotti, F., Pedreschi, D. (eds.) PKDD 2004. LNCS (LNAI), vol. 3202, Springer, Heidelberg (2004)
6. Kriegel, H.-P., Kröger, P., Pryakhin, A., et al.: Effective and Efficient Distributed Model-based Clustering. In: Proc. of the 5th IEEE International Conference on Data Mining, pp. 258–265 (2005)
7. Topchy, A., Jain, A.K., Punch, W.: Clustering Ensembles: Models of Consensus and Weak Partitions. IEEE Transactions on Pattern Analysis and Machine Intelligence 27(12), 1866–1881 (2005)
8. Minaei, B., Topchy, A., Punch, W.F.: Ensembles of Partitions via Data Resampling. In: Proc. Intl. Conf. on Information Technology, ITCC 2004, Las Vegas (2004)
9. Hore, P., Hall, L.O.: Scalable Clustering: A Distributed Approach. IEEE International Conference on Fuzzy Systems 1, 143–148 (2004)

10. Dubes, R., Jain, A.K.: Clustering Techniques: The User's Dilemma. Pattern Recognition 8, 247–260 (1976)
11. Fred, A., Jain, A.K.: Evidence Accumulation Clustering Based on the k-Means Algorithm. In: Caelli, T., et al. (eds.) Proc. Structural, Syntactic, and Statistical Pattern Recognition, pp. 442–451 (2002)
12. Newman, D.J., Hettich, S., Blake, C.L., Merz, C.J.: UCI Repository of machine learning databases, Irvine, CA. University of California, Department of Information and Computer Science (1998), http://www.ics.uci.edu/ mlearn/MLRepository.html
13. Modha, D.S., Spangler, W.: Feature weighting in k-means clustering. Machine Learning 52(3), 217–237 (2003)

Deploying Mobile Agents in Distributed Data Mining

Xining Li and JingBo Ni

State Key Laboratory of Novel Software Technology (Nanjing University)
Department of Computing and Information Science, University of Guelph, Canada
xli@cis.uoguelph.ca, jni@uoguelph.ca

Abstract. Mining information from distributed data sources over the Internet is a growing research area. The introduction of mobile agent paradigm opens a new door for distributed data mining and knowledge discovery applications. In this paper, we present the design of a mobile agent system, which couples a logic language based application programming interface for service discovery and database access. Our proposal aims at implementing system tools to enable intelligent mobile agents to search for distributed data services, to roams the Internet for accessing distributed data sites, to discover patterns and extract useful information from facts recorded in databases, to generate global data model through the communication and aggregation of local results, and to overcome the barriers posed by network congestion, poor security and unreliability.

Keywords: Mobile agent, distributed data mining, service discovery, database service.

1 Introduction

Mobile agent systems bring forward the creative idea of moving user defined computations – agents, towards network resources, and provide a whole new architecture for designing distributed systems. Distributed data mining (DDM) is one of the important application areas of deploying intelligent mobile agent paradigm [1], [2]. Most existing DDM projects focus on approaches to apply various machine leaning algorithms to compute descriptive models of the physically distributed data sources. Although these approaches provide numerous algorithms, ranging from statistical model to symbolic/logic models, they typically consider homogeneous data sites and require the support of distributed databases. As the number and size of databases and data warehouses grow at phenomenal rates, one of the main challenges in DDM is the design and implementation of system infrastructure that scales up to large, dynamic and remote data sources. The objective of our research is to equip mobile agents with system tools such that those agents can search for data sites, move from hosts to hosts, gather information and access databases, carry out complex data mining algorithms, and generate global data model or pattern through the aggregation of the local results.

To deploy mobile agents in DDM, a mobile agent system must provide languages and various programming interfaces for fast and easy development of applications. Different languages, such as C and Java, have been chosen as agent-programming

T. Washio et al. (Eds.): PAKDD 2007 Workshops, LNAI 4819, pp. 322–331, 2007.

languages for variety of reasons. Among them, logic-programming languages prove to be an alternative tool of building mobile agents. Benefiting from their powerful deductive abilities, complex calculations can often be represented by a set of compact logic predicates, which make agents more suitable to migrate around the network. In addition, mobile agents must interact with their hosts in order to use data services or to negotiate services with service providers. Discovering services for mobile agents came form two considerations. First, the agents possess local knowledge of the network and have a limited functionality, since only agents of limited size and complexity can efficiently migrate in a network and have little overhead. Hence specific services are required which aim at deploying mobile agents efficiently in system and network management. Secondly, mobile agents are subject to strong security restrictions, which are enforced by the system security mechanism. Thus, mobile agents should find services that help to complete security-critical tasks, other than execute code, which might jeopardize remote servers. Following this trend, it becomes increasingly important to give agents the ability of finding and making use of data services that are available in a network [3]. A variety of Service Discovery Protocols (SDPs) are currently under development by some companies and research groups. The most well known schemes are Sun's Java based Jini^{TM} [4], Salutation [5], Microsoft's UPP [6], IETF's draft Service Location Protocol [7] and OASIS UDDI [8]. Some of these SDPs are extended and applied by several mobile agent systems to solve the service discovery problem.

In a DDM environment, data may be stored among physically distributed sites and may be artificially partitioned among different sites for better scalability. Therefore endowing mobile agents with the ability of accessing remote databases is the basis for DDM applications. This encourages us to explore the strategies of coupling a mobile agent programming language with database access facilities. In recent years numerous approaches have been made under the topic of designing a coupled system that integrates a relational database and a logic programming language. They all enable programmers to access large amounts of shared data for knowledge processing, manage data efficiently as well as process data intelligently. Generally speaking, coupling a logic programming language with database access facility can be roughly divided into two categories. A loosely coupled system embeds a non-logic language, such as the Structure Query Language (SQL), within the logic-programming context and a closely coupled system provides database query interface as a subset or an extension of the language for featuring dynamic query formulation and view access. For example, SWI-Prolog [9] and KB-Prolog [10] belong to the first and CGW [11] and Quintus [12] fall into the second category.

In this paper we present the design and implementation of a mobile agent system, which couples a logic language based application programming interface for DDM. Two important system modules, namely, service discovery and database access, have been encapsulated and installed inside of the existing IMAGO (Intelligent Mobile Agent Gliding On-line) system. The IMAGO system is an infrastructure for mobile agent applications. It includes code for the IMAGO server - a Multi-threading Logic Virtual Machine, the IMAGO-Prolog - a Prolog-like programming language extended with a rich API for implementing mobile agent and DDM applications, and the IMAGO IDE, a Java-GUI based program from which users can do editing, compiling, and invoking an agent application. The remainder of the paper is organized as follows. Section 2 gives an overview of the design of service discovery module and integration with the IMAGO system. In Section 3, we discuss definitions of the database predicates and how agents to use them in DDM applications. In Section 4, we present the database connection management and briefly explain the way of integrating the database interface into the IMAGO system. Finally, we give the concluding remarks as well as future work.

2   Data Service Discovery

The general idea of distributed data services is that a DDM application may be separated from the data sites needed to fulfill a task. These data sites are mostly modeled by local and centralized databases, which are independent of, or sometimes unknown to the application. A commonly used DDM approach is to apply traditional data mining algorithms to an aggregation of data retrieved from physically distributed data sources. However, this approach may be impractical to a large scale of data sets distributed over the Internet. Deploying mobile agent paradigm in DDM offers a possible solution because the application may decompose data mining problems to scale up to a large distributed data sources and dispatch agents to carry out distributed data processing. This in turn leads us to the data service discovery problem, that is, how to find data sites available to a DDM application.

Clearly, the number of services that will become available in the Internet is expected to grow enormously. Examples are information access via the Internet, multi-media on demand, Web services and services that use computational infrastructure, such as P2P computing and Grid computing. In general, the service usage model is role-based. An entity providing services that can be utilized by other requesting entities acts as a provider. Conversely, an entity requesting the provision of a service is called a requester. To be able to offer services, a provider in turn can act as a requester making use of other services. In a distributed system, requesters and providers usually live on physically separate hosts. Providers should from time to time advertise services by broadcasting to requesters or registering their services on third party servers.

In the IMAGO system, we have implemented a new data service discovery model DSSEM (Discovery Service via Search Engine Model) for mobile agents [13]. DSSEM is based on a search engine, a global Web search tool with centralized index and fuzzy retrieval. This model especially aims at solving the database service location problem and is integrated with the IMAGO system. Data service providers manually register their services in a service discovery server. A mobile agent locates a specific service by migrating to the service discovery server and subsequently submitting requests with the required data description. The design goal of DSSEM is to provide a flexible and efficient service discovery protocol in a mobile agent based DDM environment.

Before a service can be discovered, it should make itself public. This process is called service advertisement. The work can be done when services are initialized, or every time they change their states via broadcasting to anyone who is listening. A service advertisement should consist of the service identifier, plus a simple string describing what the service is, or a set of strings for specifications and attributes.

The most significant feature of DSSEM is that we enrich the service description by using web page's URL to replace the traditional string-set service description in mobile agent systems. That is, service providers use web pages to advertise their services. Because of the specific characteristics, such as containing rich media information (text, sound, image, *etc.*), working with the standard HTTP protocol and being able to reference each other, web pages may play a key role acting as the template of the service description. On the other hand, since the search engine is a mature technology and offers an automated indexing tool that can provide a highly efficient ranking mechanism for the collected information, it is useful for acting as the directory server in our model. Of course, DSSEM also benefits from previous service discovery research in selected areas but is endowed with a new concept by combining some special features of mobile agents as well as integrating service discovery tool with agent servers.

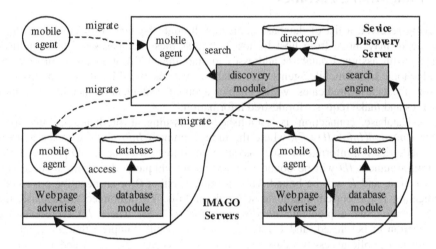

Fig. 1. An Example of Service discovery and Data Mining Process

In principle, data service providers register the URLs of their websites that advertise all the information concerning services. As a middleware on the service discovery server, the search engine will periodically retrieve web pages indicated in URLs and all their referencing documents, parse tags and words in documents and set up the relationship between the keywords and the host address of these service providers. On the other hand, a mobile agent can move to a service discovery server, utilize the system interface to access the search engine's database and obtain an itinerary that includes a list of ranked host addresses of the data service providers. Based on the given itinerary, the mobile agent may travel from host to host to carry out a DDM application. Figure 1 gives an example of service discovery and data mining process.

The application programming interface of data service discovery for mobile agents is a built-in predicate, namely, *web_search(Query, Number, Result)*, where *Query* is a compound term specifying characteristics of the search, *Number* is an integer indicating what is the maximum number of results expected, and *Result* is a variable to hold return values. For example, suppose a food company wants to analyze the customer transaction records for quickly developing successful business strategies, its DDM agent may move to a known IMAGO service discovery server and then issue a query predicate requesting up to 10 possible food industry database locations:

web_search(locate("food", "customer transaction", "imago data server"), 10, R)

The agent is blocked and control is transferred to the service discovery module of the hosting IMAGO server. The discovery module will communicate with the searcher, wait for search results, and resume the execution of the blocked agent. Search results will be delivered to the requesting agent in the form of list, where entries of the list are ranked in priorities from high to low.

3 Data Mining Facilities

Having received a list of database addresses through the service discovery module, the agent may move from host to host to access these databases or clone multiple agents with assigned database addresses to start the DDM application. In order to bridge logic based mobile agents with database systems, the IMAGO system provides a set of database predicates, which enables agents to establish connection with data sources and make requests for desired information.

A database connection is established by issuing a predicate of the form *db_connection(Info, ID, E)*, where the input argument *Info* binds with the address and authorization information, such as database address, user name and password, the output argument *ID* will be bound with an integer uniquely identifying a successful connection, and *E* will be bound with an integer, either 0 or an negative error code indicating if the current predicate succeeds or fails. To disconnect a database, the agent may issue a predicate *db_disconnection(ID, E)*, where *ID* represents the connection to be closed, and *E* returns execution result. Suppose that an agent has obtained a public-accessible database server address *xxx.yyy.zz.ww*, and it wants to access data anonymously, the agent may simply execute the following code:

> *Info = connect("xxx.yyy.zz.ww", "anonymous", _),*
> *db_connection(Info, ID, E1),*
> *// Access and apply data mining algorithm through connection ID*
> *db_disconnection(ID, E2).*

To facilitate DDM applications, the IMAGO system provides two different ways of database access operations, *i.e.*, the set retrieval and the tuple retrieval. The former returns the entire matching data set to the requesting agent, whereas the latter allows the requesting agent to consume the matching data on the tuple by tuple basis. A set retrieval operation is defined by predicate *db_search_set(ID, DB, Q, Dataset, E)*, where *ID* indicates a pre-established database server connection, *DB* is bound with a string representing the target database name, *Q* gives the searching query, *Dataset* is a variable holding the retrieval result, and *E* is a variable to be bound with a positive integer, which gives the number of records in the data set if the operation succeeds, or a negative integer if the operation fails. For example, suppose that an agent has established a database server connection *ID*, it can search for the whole set of customer records from a public database by issuing the following predicates:

db_search_set(ID, "public_database", "select * from customer", Dataset, E),
data_mining_algorithm(Dataset).

Obviously, the set retrieval is not practical if the return result involves a huge amount of matching records. The operation of tuple retrieval is introduced to handle this case. There are two predicates to implement tuple retrieval operations, *i.e.*, *db_search_tuple* and *db_tuple_next*. The first predicate shares the same syntactic form as the set retrieval predicate. Its behavior is to initialize a system manipulated searching buffer and return the first marching record. The second predicate consumes the next available data record from the system buffer orderly. Remaining records are maintained by the IMAGO database module and linked with the connection identified by *ID*. Therefore through this linkage the matching data records can be sequentially consumed by a sequence of *db_tuple_next* predicates, which are probably invoked recursively. An example of using tuple retrieval is given by the following example:

db_search_tuple(ID, "public_database", "select * from customer", DataRec, E),
data_mining_algorithm(ID, DataRec), // invoke data mining algorithm

where the procedure *data_mining_algorithm* may be defined as follows:

data_mining_algorithm(ID, DataRec):-
    // Processing the current DataRec,
    db_tuple_next(ID, DataRecNext, E),
    E<0 →
    terminate;
    data_mining_algorithm(ID, DataRecNext).

Through the recursive invocation of *db_tuple_next* predicate, the entire result set is accessed on the record-by-record basis. The value of *E* in the *db_tuple_next* predicate can be used as a recursion termination condition because it will be bound with a negative integer when the bottom of the record set is reached.

In a DDM application, agents are not working alone and they need to communicate with each other to cooperate and generate a global data aggregation for further analysis. Most existing mobile agent systems adopt some kind of communication models/ protocols from traditional distributed systems. However, the IMAGO system adopts a different strategy to cope with this issue. The idea is to deploy intelligent mobile messengers for inter-agent communication [14]. Messengers are thin agents dedicated to deliver messages. Like normal agents, a messenger can move, clone, and make decisions. Unlike normal agents, a messenger is anonymous and its special task is to track down the receiving agent and reliably deliver messages in a dynamic, changing environment. The IMAGO system provides a set of built-in messengers as a part of its programming interface. A data-mining agent at any remote sites and at any time may dispatch messengers to deliver data to designated receivers. For example, suppose that a mobile agent has completed its data mining work at a remote database server, it can either call *move('home')* to carry results and migrate back to its stationary server, or invoke *dispatch($oneway_messenger, 'home', Results)* to create a messenger which is responsible to deliver *Results* to the *home* server.

Communication among agents takes place by means of an Agent Communication Language (ACL). The essence of an ACL is to make data mining agents understanding the purpose and meaning of their exchanged data. In general, a message consists of two aspects, namely, performative and content. The performative specifies the purpose of a message and the content gives a concrete description for achieving the purpose. In order to facilitate open standards of ACL's, the IMAGO agent-based communication model is in compliance with the FIPA ACL message structure specification [15]. Of course, the performative and content of a message often determine the reaction of the receiver. In addition to the various types of system built-in messengers for sending agents, the IMAGO system provides a set of predicates for receiving agents. The predicate which is similar to an unblocking receive is *accept(Sender, Msg)*. An invocation to this procedure succeeds if a matching messenger is found, or fails if either the caller's messenger queue is empty or there is no matching messenger in the queue. Likewise, the predicate which implements the concept of blocking receive is *wait_accept(Sender, Msg)*. A call to this procedure succeeds immediately if a matching messenger is found. However, it will cause the caller to be blocked if either the caller's messenger queue is empty, or no matching messenger can be found. In this case, it will be automatically re-executed when a new messenger attaches to the caller's messenger queue. Pragmatically, the semantics of matching messengers is implemented by logic unification.

4 Database Connection Management

In the design of a logic based DDM system, an important consideration is efficiency. Several proposals on efficiently manipulating databases have been made at both technical level and logic level. On the technical level, a database connection management scheme is proposed to reduce the cost of network establishments by reusing existing database connections [16]. In addition, the data set retrieved from databases can be handled efficiently by a cache management system, which stores pre-fetched data and enables most queries to be fulfilled in these local caches at least partially [17]. On the logic level, a method of delayed evaluation is introduced by [18], which prevents redundant information being retrieved in a single logic program context. Obviously, establishing, maintaining, or terminating a database connection is always regarded as an expensive job, which not only occupies network resources but also takes significant amount of user/system overhead. The idea of connection management is to let all database operations issued by agents share a number of pre-generated database connections, and therefore many unnecessary establishment and termination operations can be avoided.

To simply our discussion, let *C* denote a set of database connections, where each entry of *C* is a quadruple *(Id, Status, Name, Link)*. As before, *ID* is an integer identifying a connection, *Status* holds the current connection states, such as *FREE, CONNECTED*, or *OCCUPIED, Name* gives the symbolic name of the database and *Link* indicates the address of the database associated with the connection. As an example, let us assume the following set of database connections:

C = [(1, FREE, _, NULL), (2, CONNECTED, "public_database", "131.104.49.49"), (3, OCCUPIED, "localhost", "127.0.0.1")].

It is clear that the first entry of *C* represents a *FREE* connection, the second entry holds a *CONNECTED* state indicating that the connection has been established and linked to a target database but is not being used currently, whereas the last entry shows an *OCCUPIED* connection, which is connected to the database installed on the *localhost* and an agent is accessing data through this connection. Initially all entries in *C* are *FREE*. Whenever an agent issues a connection request, the IMAGO database module will first try to find a *CONNECTED* entry with the condition that whose target database is the same as that requested by the *db_connection* predicate. The selected connection is ready for data retrieval because it has already connected to the desired database. However, if such an entry cannot be found, a *FREE* entry will be picked up and connection establishment operation is invoked. A worse case is that set *C* has reached its maximum capacity and none of the above two types of entries exist. Therefore, the system must find a *CONNECTED* entry, disconnect it, and set it *FREE* to establish a new connection to the required database. Of course, the state of a selected entry needs to be changed from either *CONNECTED* or *FREE* to *OCCUPIED* before being used and a new pair of username and password provided by the *db_connection* predicate needs, if necessary, to be re-authorized by the database system. Finally, if all entries of *C* are of *OCCUPIED* state, the subsequent connection requests will be blocked until some data mining agents release their connections explicitly.

Having finished the desired data mining, an agent may call *db_disconnection* predicate to return the database connection by simply changing its state from *OCCUPIED* to *CONNECTED* without actually terminating the connection with the target database. Other agents would probably reuse this non-terminated connection sooner or later and therefore save network resources and reduce system/user overhead.

5 Conclusion

In this paper, we discussed the scheme of deploying mobile agents in DDM applications. The advantage of adopting mobile agents for DDM is to scale up to large, dynamic and remote data sources, such as various databases distributed over the Internet. We presented the design of data service discovery module and database management module. The programming interface of these modules is a set system built-in predicates capable to couple a logic programming language with functionalities of locating data services and accessing remote databases. Equipped with those system tools, mobile agents may search for suitable data sites, roam the Internet to collect useful information, and communicate with each other to generate a global view of data through the aggregation of distributed computations. In order to verify the feasibility and efficiency of the mobile agent based DDM proposal, experimental service discovery module and database management module have been implemented and integrated with the IMAGO system. The service discovery module is based on the search engine technology and concentrates on locating database services. It uses web pages as a medium to advertise services, and runs an independent search engine to gather and index service provider's information, such as service types, database names, URLs, access modes, as well as possible verification information. The database management module not only provides flexible interface for accessing data, but also manipulates database connections efficiently. At the current stage, the database model in the IMAGO system is MySQL, the most popular open source DBMS system in the world [19].

Research on the agent based DDM involves further extensions of the IMAGO system. First, the current implementation of service discovery module deals with only a limited number of logical relationships. To be able to offer more precise discovery service, this module could be enhanced to parse some complex search criteria, such as conditional expressions and sub-string matching. Secondly, since databases may contain multi-dimensional data, retrieving such kind of information from flat web pages is a pending problem. We are looking to use XML meta-data to solve the database dimensional problem. In addition, we are making investigations on adding more programming languages to the IMAGO system, as well as introducing more flexible and efficient communication tools, such as mobile socket, to facilitate DDM applications.

Acknowledgments. We would like to express our appreciation to the Natural Science and Engineering Council of Canada and State Key Laboratory of Novel Software Technology (Nanjing University) for supporting this research.

References

[1] Klusch, M., Lodi, S., Moro, G.: The Role of Agents in Distributed Data Mining: Issues and Benefits. In: IEEE/WIC International Conference on Intelligent Agent Technology (IAT 2003) (2003)

[2] Park, B., Kargupta, H.: Distributed Data Mining: Algorithms, Systems, and Applications, Data Mining Handbook. In: Ye, N. (ed.) (2002)

[3] Bettstetter, C., Renner, C.: A Comparison of Service Discovery Protocols and Implementation of the Service Location Protocol. In: Proc. of EUNICE 2000, sixth EUNICE Open European Summer School, pp. 13–15 (2000)

[4] Hashman, S., Knudsen, S.: The Application of Jini Technology to Enhance the Delivery of Mobile Services, White Paper (2001), http://wwws.sun.com/

[5] Salutation Consortium, Salutation Architecture Overview, White Paper (1998), www.salutation.org/whitepaper

[6] Universal Plug and Play Forum, Universal Plug and Play Device Architecture, Version 0.91, White Paper (2000)

[7] Guttman, E., Perkins, C.: Veizades, Service Location Protocol, Version 2, White Paper, IETF, RFC 2608 (1999)

[8] OASIS UDDI, UDDI White Paper (2005), http://www.uddi.org

[9] Wielemaker, J.: SWI-Prolog ODBC Interface, technical report University of Amsterdam, The Netherlands (2000)

[10] Bocca, J., Dahmen, M., Macartney, G.: KB-Prolog User Guide, Technical Report, ECRC Munich (1989)

[11] Ceri, S., Gottlob, G., Wiederhold, G.: Efficient Database Access from Prolog. IEEE Transactions on Software Engineering 15(2), 153–164 (1989)

[12] Quintus Inc., Quintus Prolog Database Interface Manual, technical report (1998), http://www.sics.se/

[13] Song, L., Li, X., Ni, J.: A Database Service Discovery Model for Mobile Agents. International Journal of Intelligent Information Technologies 2(2), 16–29 (2006)

[14] Li, X., Autran, G.: Inter-agent Communication in IMAGO Prolog, Lecture Notes in Artificial Intelligence. In: Bordini, R.H., Dastani, M., Dix, J., Seghrouchni, A.E.F. (eds.) Programming Multi-Agent Systems. LNCS (LNAI), vol. 3346, pp. 163–180. Springer, Heidelberg (2005)

[15] FIPA ACL, Agent Communication Language Specifications, FIPA (2005), http://www.fipa.org

[16] Mckay, D., Finin, T., O'Hare, A.: The Intelligent Database Interface: Integrating AI and Database Systems. In: Proceedings of the 8th National Conference on Artificial Intelligence, pp. 677–684 (1990)

[17] Sheth, A., O'Hare, A.: The Architecture of BrAID: A System for Bridging AI/DB Systems. In: Proceedings of the Seventh International Conference on Data Engineering, pp. 570–581 (1991)

[18] Cuppens, F., Demolombe, R.: A Prolog-relational DBMS Interface Using Delayed Evaluation. In: Proc. 3rd Int. Conf. On Data and Knowledge Bases, pp. 135–148 (1988)

[19] MySQL AB MySQL Reference Manual, user's manual (2005), http://dev.mysql.com/doc/mysql/en/

ODDC: Outlier Detection Using Distance Distribution Clustering

Kun Niu[1], Chong Huang[2], Shubo Zhang[3], and Junliang Chen[1]

[1] Dept. of Computer Sci. and Eng., Beijing University of Posts and Telecommunications,
Beijing 100876, China
niukun2006@gmail.com
[2] Graduate University, Chinese Academy of Sciences, Beijing 100039, China
chuang@jdl.ac.cn
[3] Dept. of Strategy Research, China Telecom Beijing Institute, Beijing 100035, China
zhangshb@ctbri.com.cn

Abstract. Outlier detection is an important issue in many industrial and financial applications. Most outlier detection methods suffer from two problems: First, they need parameter tuning in accord to domain knowledge. Second, they are incapable to scale up to high dimensional space. In this paper, we propose a distance-based outlier definition and a detection algorithm ODDC (Distribution Clustering Outlier Detection). We redefine the problem by clustering in the distribution difference space rather than the original feature space. As a result, the new algorithm is stable regardless of different input and scalable to the dimensionality. Experiments on both synthetic and real datasets show that ODDC outperforms the counterpart both in effectiveness and efficiency.

1 Introduction

Outlier detection is an alternative to supervised learning methods, particularly for applications where label information is either hard to obtain or unreliable. Typical examples of such application include network intrusion detection, fraud detection and fault detection in manufacturing, and so on [1]. In data mining, detection of anomalous patterns is of more interest than detecting inlier clusters. For example, a breast cancer detection system might consider inlier observations to represent healthy patient and outlier observation as a patient with breast cancer [3].

The existing approaches of outlier detection can be broadly classified into the following categories [12]: (1) Distribution-based approach. This approach assumes a distribution or probability model for the given data set and then identifies outliers with respect to the model using a discordancy test. (2) Depth-based approach. This approach computes different layers of k-d convex hulls and flags objects in the outer layer as outliers [13]. (3) Distance-based approach. Distance-based methods define outlier as an observation that is d_{min} distance away from p of observations in the dataset. (4) Density-based approach. In density-based methods, outlier is detected from abnormal local density of observations.

One major drawback of the existing detection approaches is that though they usually work well with low dimensionality, they can not achieve good performance in

T. Washio et al. (Eds.): PAKDD 2007 Workshops, LNAI 4819, pp. 332–343, 2007.

very large and highly dimensional datasets. Moreover, they require users to input parameters. Finding suitable parameters is laborious and error-prone [2].

In this paper, we present a new distance-based outlier detection method with distance distribution clustering. This method transits the feature space to the new space by discretizing the distance distribution of each object. With observations clustered in the new feature space, we take objects in small clusters as outliers. Our contributions are: (1) we propose a new outlier definition which improves the speed of outlier detection process. (2) We also present a stable, scalable and effective detection algorithm – ODDC (an *O*utlier *D*etection algorithm based on distance *D*istribution *C*lustering). ODDC takes the distance distribution difference as the criteria of detecting outliers and uses clustering results to describe the difference, so ODDC is scalable to high dimensionality. Also, ODDC is insensitive to input parameters, which outperforms most other approaches. Experiments on both synthetic and real datasets show that ODDC outperforms two traditional outlier detection methods significantly in terms of the ability to find outliers and the scalability to dimensionality of datasets.

The remainder of this paper is organized as follows: In section 2, we discuss the new definition of outlier and presents the proposed method in detail. Section 3 reports the extensive experimental evaluation on both synthetic and real datasets. Finally, section 4 concludes the paper.

2 Efficient Outlier Detection

In the following sections, we assume that D is a database of n points in a d-dimensional feature space, i.e. $D \subseteq \mathbb{R}^d$.

2.1 General Idea

Many algorithms have been proposed in recent years for outlier detection, most of which are either distance based or density based, but the curse of high dimensionality is still daunting. Thus, in this paper, we focus on highly dimensional datasets.

The key idea of outlier detection algorithm using distance distribution clustering is: for each object in a cluster, the distance distribution of its neighborhoods is different from outliers. Accordingly, we redefine the problem by clustering in the distribution difference space rather than the original feature space.

Our method ODDC consists of the following four steps:

1. **Preprocessing:** First, all distances between each two point in the original feature space are computed. Then the *distance spheres* (defined in section 2.2) of each point are generated.
2. **Distance Distribution Discretization:** In the second step, for each point p in original feature space, the density[1] difference of neighboring distance spheres are calculated to get the distance distribution of each point in order to complete the process of distance distribution discretization.

---

[1] The density of a distance sphere is defined to be the fraction of total data points contained in the sphere.

3. **Feature Space Transformation:** The third step is to transform original feature space to the distribution difference space. In the new space, each point in the original space is represented as a new vector, and each dimension of the new vector is a distance distribution from step 2.

4. **Transformation Clustering:** The last step is clustering on the new feature space and the points in small clusters are defined outliers.

In what follows, we give the formal definition of the new definition of outlier and its related concepts.

2.2 Preprocessing

From a knowledge discovery standpoint, outliers are often more interesting than the common ones since they contain useful information underlying the abnormal behavior. Despite the important differences between various outlier detection algorithms, it is widely acknowledged that the detection performance critically depends on the quality of outlier definition. In [7], Knorr and Ng's defined distance based outlier as:

An object O in a dataset T is an outlier if at least fraction pct of the objects in T lies at a distance greater than or equal to threshold distance, d_{min}, from O. We denote it as $DB(pct, d_{min})$.

From this definition we can think of (According to this definition, we view) *distance-based outliers* as objects that do not have "enough" neighbors, where *neighbors* are defined based on distance from the given object. This definition avoids the excessive computation of fitting the observed distribution into some standard distribution. However, it requires the user to set both pct and d_{min}. Finding suitable settings for these parameters can involve many efforts and errors[1].

The purpose of this step is to get the distances between each object to the other objects in data sets. First, ODDC takes distance sphere to partition the normalized original feature space.

Definition 1 (Distance Sphere). Let DB be the given datasets and $m \in N^+$ be the dimensionality of DB. Given an object p in DB, $\forall \varepsilon \in N^+$, the distance sphere of p with regard to ε, is the set of concentric hyper spheres with p as the center and $\sqrt{m}/\varepsilon$ as the step length, which is denoted by $DS_1(p), DS_2(p), ..., DS_\varepsilon(p)$.

Note that in m-dimensional normative feature space, the maximal distance between each two points is $\sqrt{m}$. As a result, for each point p in it, the hyper sphere with p as the center and $\sqrt{m}$ as the radius is able to contain all points in the feature space consequentially. ODDC takes p as the center and expand the hyper sphere with $\sqrt{m}/\varepsilon$ as the step length to form the distance spheres of p. Figure 1 shows a point p in feature space and its distance spheres.

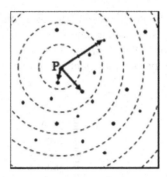

Fig. 1. The distance sphere of object p

2.3 Distance Distribution Discretization

We propose to use the distance spherical shell to complete the step of distance distribution discretization.

Definition 2 (Distance Spherical Shell). Given an object p in DB, $\forall \varepsilon \in \mathbb{N}^+$, the distance sphere shell of p with regard to ε, is the set of spherical shells between each two neighboring distance spheres of p, which is denoted by $DSS_1(p)$, $DSS_2(p)$,..., $DSS_{(\varepsilon-1)}(p)$.

The objects that lie within the distance spherical shells of p are the foundation to form transformation vector.

2.4 Feature Space Transformation

The purpose of this step is to map all objects into the new feature space, where the transformation vector capturing the distance distribution of each point.

Definition 3 (Transformation Vector). Given an object p in DB, $\forall \varepsilon \in \mathbb{N}^+$, let $DSS_1(p)$, $DSS_2(p)$, ..., $DSS_{(\varepsilon-1)}(p)$ be the distance spherical shells of p and $|DSS_1(p)|$, $|DSS_2(p)|$, ..., $|DSS_{(\varepsilon-1)}(p)|$ be the number of objects within $DSS_1(p)$, $DSS_2(p)$,..., $DSS_{(\varepsilon-1)}(p)$ respectively. The transformation vector of p with regard to ε, denoted by TV(p), is the vector of $(|DSS_1(p)|, |DSS_2(p)|, ..., |DSS_{(\varepsilon-1)}(p)|)$. That is,

$$TV(p) = (|DSS_1(p)|, |DSS_2(p)|, ..., |DSS_{(\varepsilon-1)}(p)|).$$

As Fig. 2 shows, the numbers of objects within the distance spherical shells of each object form a vector which embodies the isolation of the object. In Fig. 2(a), there are several points in the first distance sphere, which means the point is not an outlier because it has enough neighbors. When it comes to the outlier in Fig. 2(b), there are not any points in both the first and the second distance sphere. However, it is worth to note that when ε has a too small value, the property makes no sense. As a result, instead of direct comparison, ODDC takes clustering to recognize the difference, which could depress the improper judgment according to the unsuitable values of parameters effectively.

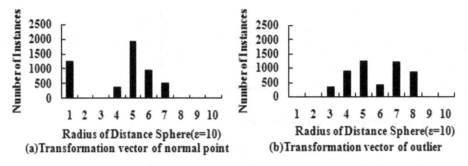

Fig. 2. The transformation vector of normal point and outlier

Transformation vector is the discrimination expression of the distance distribution of each object, which has the similar mathematical meaning as histogram to describe the distribution feature of distances between objects.

2.5 Transformation Clustering

We collect the transformation vectors of all objects in original datasets to get the new datasets called transformation datasets which include the distance distribution information of all objects. Accordingly, the clustering operation on transformation datasets is called transformation clustering.

So far, ODDC takes uses distance sphere to partition the normalized feature space and get the distance distribution of each object by distance sphere shell. Furthermore, it takes transformation vector to denote the discrimination result to get transformation datasets. After clustering on transformed datasets, we get outliers, including clusters in small scale.

Finally, we give the new outlier definition as follows:

Definition 4 (Outlier). Let DB' be the transformation dataset of DB, $\delta \in N^+$, $C_1, C_2, ..., C_k$ be the clustering results of transformation clustering on DB'. If $|C_i| \leq \delta$ $(1 \leq i \leq k)$ is satisfied, we define outliers as the set of each object o in cluster C_i.

In contrast to clustering on original datasets, transformation clustering has several advantages as follows: the input vector of direct clustering is the values in each dimension of data points. When there are clusters with different densities and granularities in feature space, the clustering results are easy to be influenced by the difference on different dimensions. However, the input vector of transformation clustering is the relative location information which reflects the isolation of each point. As a result,

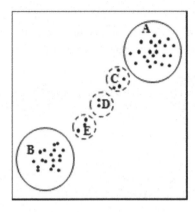

Fig. 3. Transformation clustering

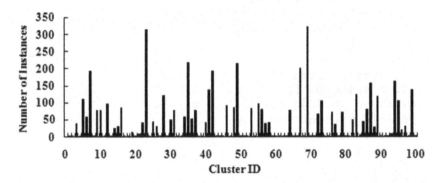

Fig. 4. The results of transformation Clustering

transformation clustering is more effect than direct clustering. To demonstrate the difference between transformation clustering and direct clustering, we give an example in Fig. 3. In Fig. 3, if we take direct clustering method, considering the sensitivity to input parameters of clustering algorithm, cluster C, D and E are all possible to be merged by cluster A or cluster B, even they are all in the same cluster. While after transformation, the object attributes in cluster C, D and E lie on the relative location of other objects from cluster C, D and E, which ensures cluster C, D and E detach.

According to the property of outliers, there are fewer neighbors around them, which results in fewer objects with similar transformation vector. On the other hand, for normal points, because they are located in clusters, there are many objects with similar transformation vector. That is, the distance spheres of objects in the same clusters are almost overlapped, which results in better coherence of transformation vector of each objects in the cluster during transformation clustering. Fig. 4 is the clustering results of transformation clustering. There are 100 clusters in Fig. 4. Some clusters are small scale, which just include few objects. ODDC defines objects in such clusters as outliers.

The ODDC algorithm, based on the discussion in the previous section, is illustrated as follows.

```
Algorithm ODDC
Input: a data set DB of m-dimension; ε,specifying the
number of distance sphere; δ,the threshold of small-
scale clusters
Output: outliers;
/*Distance Distribution Discretization Step */
for each object p∈ DB do
  for i=1 to ε do
    count_i(p):=|DS_i(p)|
  end for
end for
for j=1 to (ε-1) do
  |DSS_j(p)|:=Count_(j+1)(p)-Count_j(p)
end for
/*Feature Space Transformation Step */
for each object p∈ DB do
  /* to get the transformation vector of object p */
  TV(p):=(|DSS_1(p)|,|DSS_2(p)|,...,|DSS_(ε-1)(p)|)
  DB':=DB'∪(TV(p))
end for
/* Clustering Step */
C:=Clustering(DB');
for each c in C do
  if |c|≤δ then
    for each object o∈ C do
      flag(o):=1;
    end for
  end if
end for
Return the objects whose flags are 1.
```

3 Experiments

We compare the accuracy of ODDC with the well-known outlier detection methods called the LOF method [8], which is based on a modified nearest-neighbor approach. In the first set of experiments, we evaluate the accuracy of the proposed method against LOF on a synthetic data set. In the second set of experiments, we compare the two methods on three real data sets from UCI [11]. In the third group of experiments, we focus on the scalability of two algorithms. The last set of experiments demonstrates the impaction on ODDC of three input parameters. The tests are run on 1.3GHZ AMD processor, with 512MB of main memory, under Windows 2000 advanced-server operating system. All algorithms are implemented in Java. We use the same parameter setting and pruning process of LOF as in paper [8], and we choose the best experimental result.

3.1 Evaluation

Outlier detection algorithms are typically evaluated using the detection rate, the false alarm rate, and the ROC curves [9]. In order to define these metrics, we take the confusion matrix in [10], described in Table 2.

Table 1. Confusion matrix defines four possible scenarios when classifying class C

	Predicted Outliers – Class C	Predicted Normal class NC
Actual Outliers- Class C	True Positives(TP)	False Negatives(FN)
Actual Normal Class - NC	False Positives(FP)	True Negatives(TN)

From Table 1, detection rate and false alarm rate may be defined as follows:

$$Detection\ rate = TP/\ (TP+FN)$$
$$False\ alarm\ rate = FP/\ (FP+TN)$$

Detection rate gives information about the number of correctly identified outliers, while the false alarm rate reports the number of outliers misclassified as normal data records (class NC). The ideal instance has 0% false alarm rate, while having 100% detection rate.

3.2 Synthesis Datasets

In this section, we evaluate the ability to find outliers of the ODDC algorithm and compare it with LOF. In this experiment, we take $\varepsilon =10$, k=100 and $\delta =2$. We use a data generator to produce a 2-dimensional datasets with 5000 objects, with 53 outliers included. In our datasets, there are four clusters with mixture Gaussian distributions. The original datasets are shown in Fig. 5.

At the same time, we run LOF on the same datasets. Taking the method in [8], we take k=4 and MinPts=6. The results are shown in Table 2.

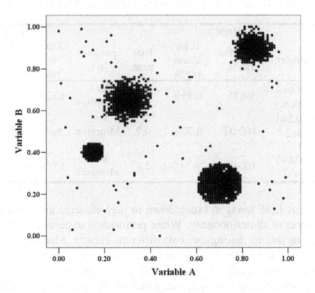

Fig. 5. The original datasets

Table 2. Detecting outliers in synthesis datasets (ODDC vs. LOF)

Algorithm	LOF	ODDC
Run Time	25	21
Detection Rate	92.45%	100.00%
False Alarm Rate	0.08%	0.10%

Table 2 demonstrates that ODDC has the detection rate of 100%, which means that it finds all outliers, while LOF just has the detection rate of 92.45%. Moreover, the run time of ODDC is 4 seconds lesser than that of LOF.

3.3 Real Life Datasets

All real life datasets used in our experiments are from UCI [11]. Similarly, we run both ODDC and LOF on the three datasets in Table 3. The results are shown in Table 4.

Table 3. Summary of datasets used in experiments

Datasets	Size of datasets	Number of features	Number of outliers(rare class records)	Percentage of outliers
Breast Cancer	699	9	241	34.48%
Lymphography	148	18	6	4.05%
Ionosphere	350	34	126	36.00%

Table 4. The detection results on real life datasets(ODDC vs. LOF)

Datasets	parameter	ODDC Detection rate%	False alarm rate%	Run time	parameter	LOF Detection rate%	False alarm rate%	Run time
Breast cancer	ε =10,k=2, δ inefficacy	98.76	6.55%	0.4	k=4, Minpts=6	62.24%	0%	0.8
Lym-pho-grphy	ε =10,k=10, δ=2	100.00	0.70%	0.2	k=4, Minpts=6	failure	failure	0.3
Ionosph-ere	ε =10,k=10, δ=32	100.00	10.71%	0.4	k=4, Minpts=6	15.08%	6.25%	0.4

Table 4 tells us LOF tends to break down in high-dimensional feature spaces, because of the curse of dimensionality. When performing experiments on Lymphography, it is even unable to distinguish two different classes. Meanwhile, dealing with high-dimensional sparse datasets is the advantage of ODDC. It transforms the feature space of absolute coordination to the relative coordinate by clustering. The dimensionality of transformed feature space depends on the parameter of ε instead of the dimensionality of the original feature space, which results in better performance on three real life datasets than LOF.

3.4 Efficiency

In this section, we present an experiment to evaluate the efficiency of our approach on synthetic datasets. First, we consider a set of synthetic datasets with the same distribution to study the influence of factors such as the size of datasets, the dimensionality and the percentage of outliers. Then, we use the same datasets to study the influence of the three parameters on the method.

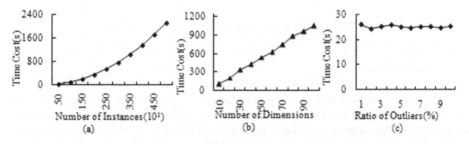

Fig. 6. The scalability of ODDC

Fig. 6(a) shows the scalability as the size of the datasets is increased from 5,000 to 50,000. The data space has 2 dimensions and the percentage of outliers is 1%. As expected, the curve exhibits quadratic behavior.

Fig. 6(b) shows the scalability as the dimensionality of the data space is increased from 2 to 20. The datasets has 5000 instances and the percentage of outliers is also 1%. The run time scales linearly with the dimensionality of the datasets.

Fig. 6(c) shows the scalability as the percentage of outliers on the 2-dimensional datasets with 5000 instances. Due to the computational complexity of deciding a normal point is the same as deciding an outlier, the run time has no relation to the percentage of outliers.

3.5 Parameters

The parameters involved in the algorithm are 1) ε, the specifying the number of distance sphere, 2) k, the number of clusters, 3) δ, the threshold of small-scale clusters. We evaluated the impact of parameterization by changing one of the three parameters at a time, while the other two are fixed at an optimized value. The results are illustrated in Fig. 7.

The parameter ε is not critical as far as it is not too small. If it is too low, some outliers are still found with high precision but some of the outliers are not found at all. Roughly the same can be said about the parameter k. If k is reduced to a value less than 100, the detection rate drops significantly, whereas the false alarm rate rises significantly if k is increased. The third parameter δ turned out to be stable during a large range of tested settings, i.e. the precision of ODDC is high at any tested δ value. However, when the δ value is less than 5, the false alarm rate decreased significantly, whereas the detection rate is nearly stable unless δ is less than 2.

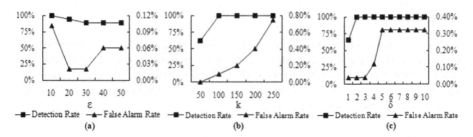

Fig. 7. Impact of parameterization on the accuracy of ODDC

4 Conclusion

In this paper, we proposed a new outlier detection algorithm ODDC. It overcomes problems of existing methods including weak scalability with respect to data dimensionality and the sensitivity to input parameters. Experiments on both synthetic datasets and real life datasets have shown that ODDC is significantly more effective than LOF. In addition, ODDC has good performance on scalability and running time with respect to data dimensionality. In the near future, we would like to apply this approach to applications in telecommunication.

References

1. Abe, N., Zadrozny, B., Langford, J.: Outlier Detection by Active Learning. In: Proceedings of the ACM SIGKDD International Conference on Knowledge Discovery and Data Mining, Philadelphia, USA (2006)
2. Han, J., Kamber, M.: Data Mining: Concepts and Techniques, 2nd edn. Morgan Kaufmann Publishers, San Francisco (2006)
3. Hautamäki, V., Kärkkäinen, I., Fränti, P.: Outlier Detection using k-Nearest Neighbour Graph. In: Proceedings of the 17th International Conference on Pattern Recognition, Cambridge, United Kingdom (2004)
4. Angiulli, F., Basta, S., Pizzuti, C.: Distance-Based Detection and Prediction of Outliers. IEEE Transactions on Knowledge and Data Engineering 18(2) (2006)
5. Angiulli, F., Pizzuti, C.: Outlier Mining in Large High-Dimensional Data Sets. IEEE Transaction on Knowledge and Data Engineering 2(17), 203–215 (2005)
6. Ren, D.M., Rahal, I., Perrizo, W., et al.: A Vertical Distance-based Outlier Detection Method with Local Pruning. In: Proceedings of the 13th Conference on Information and Knowledge Management, Washington, D.C., USA (2004)
7. Knorr, E.M., Ng, R.T.: A Unified Notion of Outliers: Properties and Computation. In: Proceedings of the 3rd International Conference on Knowledge Discovery and Data Mining Proceedings, pp. 219–222 (1997)
8. Breunig, M.M., Kriegel, H., Ng, R.T., et al.: LOF: Identifying Density-Based Local Outliers. In: Proceedings of ACM SIGMOD International Conference on Management of Data, Dalles, TX, pp. 93–104 (2000)
9. Provost, F., Fawcett, T.: Robust Classification for Imprecise Environments. Machine Learning 42, 203–231 (2001)

10. Lazarevic, A., Kumar, V.: Feature Bagging for Outlier Detection. In: Proceedings of the ACM SIGMOD International Conference on Knowledge Discovery and Data Mining, Chicago, USA (2005)

11. Merz, C., Murphy, P., Aha, D.: UCI Repository of Machine Learning Databases. At, http://www.ics.uci.edu/ mlearn/MLRepository.html

12. Zhu, C., Kitagawa, H., Papadimitriou, S., et al.: OBE: Outlier by Example. In: Proceedings of the 8th Pacific-Asia Conference on Knowledge Discovery and Data Mining, Sydney, Australia, pp. 222–234 (2004)

13. Johnson, T., Kwok, I., Ng, R.T.: Fast computation of 2-dimensional depth contours. In: Proceedings of the 4th International Conference on Knowledge Discovery and Data Mining, New York, USA, pp. 224–228 (1998)

Spatial Clustering with Obstacles Constraints Using Ant Colony and Particle Swarm Optimization

Xueping Zhang[1,2,3], Jiayao Wang[2], Zhongshan Fan[4], and Bin Li[2]

[1] School of Information Science and Engineering, Henan University of Technology,
Zhengzhou 450052, China
[2] School of Surveying and Mapping, PLA Information Engineering University,
Zhengzhou 450052, China
[3] Geomatics and Applications Laboratory, Liaoning Technical University, Fuxin 123000, China
[4] Henan Academy of Traffic Science and Technology, Zhengzhou 450052, China
zhang_xpcn@yahoo.com.cn

Abstract. Spatial clustering is an important research topic in Spatial Data Mining (SDM). This paper proposes an Ant Colony Optimization (ACO) and Particle Swarm Optimization (PSO) method for solving Spatial Clustering with Obstacles Constraints (SCOC). In the process of doing so, we first use improved ACO to obtain the shortest obstructed distance, which is an effective method for arbitrary shape obstacles, and then we develop a novel PKSCOC based on PSO and K-Medoids to cluster spatial data with obstacles. The PKSCOC algorithm can not only give attention to higher local constringency speed and stronger global optimum search, but also get down to the obstacles constraints and practicalities of spatial clustering. The experimental results demonstrate the effectiveness and efficiency of the proposed method, which performs better than Improved K-Medoids SCOC (IKSCOC) in terms of quantization error and has higher constringency speed than Genetic K-Medoids SCOC (GKSCOC).

Keywords: Spatial Clustering, Obstacles Constraints, Ant Colony Optimization, Particle Swarm Optimization, K-Medoids Algorithm.

1 Introduction

Spatial clustering has been an active research area in the data mining community. Spatial clustering is not only an important effective method but also a prelude of other task for Spatial Data Mining (SDM). As reported in surveys on data clustering, clustering methods can be classified into Partitioning approaches, Hierarchical methods, Density-based algorithms, Probabilistic techniques, Graph theoretic, Grid-based algorithms, Model-based approaches, Genetic Algorithms, Fuzzy methods, Rough Set methods etc. Some algorithms have also integrated two or three kinds of clustering methods. As pointed out earlier, these techniques have focused on the performance in terms of effectiveness and efficiency for large databases. However, few of them have taken into account constraints that may be present in the data or constraints on the clustering. These constraints have significant influence on the results of the clustering process of large spatial data.

T. Washio et al. (Eds.): PAKDD 2007 Workshops, LNAI 4819, pp. 344–356, 2007.

Spatial clustering with constraints has two kinds of forms [1]. One kind is Spatial Clustering with Obstacles Constraints (SCOC), such as bridge, river, and highway etc. whose impact on the result should be considered in the clustering process of large spatial data. In a GIS application studying the movement of pedestrians to identify optimal bank machine placements, for example, the presence of a highway hinders the movement of pedestrians and should be considered as an obstacle. As an example, Fig.1 shows clustering data objects in relation to their neighbors as well as the physical obstacle constraints. Ignoring the constraints leads to incorrect interpretation of the correlation among data points. The other kind is Spatial Clustering with Handling Operational Constraints [2], it consider some operation limiting conditions in the clustering process. SCOC is mainly discussed in this paper.

To the best of our knowledge, only three clustering algorithms for SCOC have been proposed very recently, that is COD-CLARANS [3], AUTOCLUST+ [4], and DBCluC [5]-[8], and many questions exist in them. COD-CLARANS computes obstructed distance using visibility graph costly and is unfit for large spatial data. In addition, it only gives attention to local constringency. AUTOCLUST+ builds a Delaunay structure for solving SCOC costly and is also unfit for large spatial data. DBCluC cannot run in large high dimensional data sets etc. We developed GKSCOC (Genetic K-Medoids SCOC) based on Genetic algorithms (GAs) and IKSCOC (Improved K-Medoids SCOC) in [9]. The effectiveness and efficiency of GKSCOC is better than IKSCOC. But the drawback of GKSCOC is a comparatively slower speed in clustering.

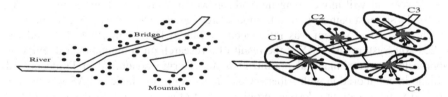

(a) Data objects and obstacles constraints (b) Clusters ignoring obstacle constraints

Fig. 1. Clustering data objects with obstacles constraints

Ant Colony Optimization (ACO) is an evolutionary computation technique. ACO is initiated by Colorni and Dorigo in the beginning of 1990s, it attracted much attention from many scholars and it has been applied widely in all kinds of fields because of its excellent ability of optimization. ACO algorithm is fit for solving large-scale combination optimization problems, such as Traveling Salesman Problem (TSP), Quadratic Assignment Problem (QAP), Job Scheduling Problem (JSP), and so on, especially [10].

Particle Swarm Optimization (PSO) is a population-based optimization method and it is developed by Dr. Eberhart and Dr.Kennedy in 1995. Compared to GAs, the advantages of PSO are that it is easier to implement and there are fewer parameters to be adjusted, and it can be efficiently used on large data sets. Recently, PSO has been applied to data clustering [11-14].

In this paper, we explore the applicability of ACO and PSO for SCOC. In the process of doing so, we first use improved ACO to obtain the shortest obstructed distance, which is an effective method for arbitrary shape obstacles, and then we develop a novel PKSCOC based on PSO and K-Medoids to cluster spatial data with obstacles. The experimental results demonstrate the effectiveness and efficiency of the proposed method, which performs better than Improved K-Medoids SCOC (IKSCOC) in terms of quantization error and has higher constringency speed than Genetic K-Medoids SCOC (GKSCOC).

The remainder of the paper is organized as follows. Section 2 introduces ACO, and using improved ACO to get the optimal obstructed path is discussed in Section 3. Section 4 introduces PSO, and PKSCOC is presented in Section 5. The performances of PKSCOC implementation on real datasets are showed in Section 6, and Section 7 concludes the paper.

2   Ant Colony Optimization

ACO has been inspired by the observation on real ant colony's foraging behavior, and on that ants can often find the shortest path between food source and their nest. Biologist finds some obvious distinctive features of real ants in the process of looking for food. For instance, ant will release some substance called pheromone in the process of moving; the released pheromone will lessen gradually along with the passing of time; ant can detect the existence of the interclass pheromone trail within a given area; ant will move along the path on which pheromone trail is plentiful. Ant can find a shortest path from nest to food source based on the above basic features.

The first ACO system was introduced by Marco Dorigo in 1992 [15], and was called Ant System (AS). AS is the result of a research on computational intelligence approaches to combinatorial optimization that Dorigo conducted at Politecnico di Milano in collaboration with Alberto Colorni and Vittorio Maniezzo. AS was initially applied to the traveling salesman problem, and to the quadratic assignment problem. Since 1995 Dorigo, Gambardella and Stützle have been working on various extended versions of the AS paradigm. Dorigo and Gambardella have proposed Ant Colony System (ACS) [16], while Stützle and Hoos have proposed MAX-MIN Ant System (MMAS) [17]. They have both have been applied to the symmetric and asymmetric traveling salesman problem, with excellent results. Dorigo, Gambardella and Stützle have also proposed new hybrid versions of ant colony optimization with local search.

ACO algorithm has two important features. One is the positive feedback process shown in the group behavior of ant colony. With the help of the positive feedback, the whole system can evolve gradually. In this process, the feedback information is global. Adjusted via the feedback mechanism, the better solutions get the help of self-enhancement effect, which makes the solutions evolve in the direction of global optimization and obtains the globally relatively optimal solutions effectively at the very end. This also is the most important feature of ACO algorithm. The other feature is the excellent distributed parallel computing ability. This ability enables the searching for solutions can be implemented in global many points simultaneously, which avoids the possibility of getting the local optimal solutions effectively. When ACO algorithm is used to solve the nonlinear problems, this ability also ensures that it has unexampled excellent effect and high robustness. In many references, it has been proved that ACO algorithm has very preferable ability in finding the better solutions. The reason is just that this algorithm not only uses positive feedback principle to quicken the evolutionary process to a certain extent, but also can be regarded as a kind of parallel algorithm that promotes the continuous communication and transmission of information between various individuals to ensure colony collaboration to find the better solutions.

3   Using Improved ACO to Get the Shortest Obstructed Distance

Definition 1 (Obstructed distance). Given point p and point q, the obstructed distance $d^o(p,q)$ is defined as the length of the shortest Euclidean path between two points p and q without cutting through any obstacles.

Let W be the sum of ants; n be the number of partition modules in two dimensions; $d_{ij}^o(i, j = 1,2,\cdots,n)$ be the obstructed distance between point i and point j; $b_i(t)$ be the number of ants on point i at moment t ; $\tau_{i,j}(t)$ be the quantity of the residual pheromone path $<i, j>$ at moment t . Then we have equation $w = \sum_{i=1}^{n} b_i(t)$. The transition probability which ant $k(k = 1,2,\cdots,W)$ wants to transit from point i to j at moment t can be defined as:

$$p_{i,j}^k(t) = \begin{cases} \left(\tau_{i,j}^\alpha(t) \bullet \eta_{i,j}^\beta(t)\right) \Big/ \left(\sum_{r \in S_i^k} \left(\tau_{i,r}^\alpha(t) \bullet \eta_{i,r}^\beta(t)\right)\right) & if \quad j \in S_i^k \\ 0 & otherwise \end{cases} \tag{1}$$

where $\eta_{i,j}(t)$ is a local heuristic function and it can be defined as $1/d_{ij}^o$ in our discussed problem; α and β determine the relative weightiness of the quantity of pheromone trail $\tau_{i,j}(t)$ and the heuristic function $\eta_{i,j}(t)$, respectively; S_i^k denotes the feasible neighboring region of ant k on point i . Here, α and β will be modified based on the time variation instead of numerical constants so as to obtain the better solutions [10]:

$$\alpha = \begin{cases} (4q)\Big/m & 0 \le q < m \\ 4 & m \le q \le u \end{cases} \tag{2}$$

$$\beta = \begin{cases} (3m-1.5q)\Big/m & 0 \le q < m \\ 1.5 & m \le q \le u \end{cases} \tag{3}$$

where m is the critical moment and u is the pre-established terminative moment.

After h moments, ant colony finishes the moving in a circle. Then, the quantity of pheromone trail on every path will be modified based on the following equations:

$$\tau_{i,j}(t+h) = \rho \bullet \tau_{i,j}(t) + \sum_{k=1}^{W} \Delta \tau_{i,j}^{k} \qquad (4)$$

where ρ denotes the degree of redundancy of pheromone trial after it volatilized on certain path; $\Delta \tau_{i,j}^{k}$ is the quantity of pheromone trail released by ant k on path $<i, j>$ in this circle, and its definition is as the following:

$$\Delta \tau_{i,j}^{k} = \begin{cases} Q/L_k & \begin{array}{l} \textit{if} \text{ ant } k \text{ passes the} \\ \text{path } <i, j> \text{ in this circle} \end{array} \\ 0 & \textit{otherwise} \end{cases} \qquad (5)$$

where Q is a numerical constant that denotes the intensity of pheromone trail and L_k denotes the length of all paths visited by ant k in this circle. At initial time, $\tau_{i,j}(0) = C$ (C is a numerical constant) and $\Delta \tau_{i,j}^{k} = 0 (i, j = 1, 2, \cdots, n; k = 1, 2, \ldots, W)$.

Let $\varphi_{i,j}^{k}$ be the heuristic distance information probability of every point on the moving path, and it can be defined as:

$$\varphi_{i,j}^{k} = \begin{cases} \dfrac{\left(\left(Maxd_{A(i),e}^{o} - d_{j,e}^{o} \right) \cdot \omega + \mu \right)^{\lambda}}{\sum\limits_{j \in Available(i)} \left(\left(Maxd_{A(i),e}^{o} - Dist\,an\,ce_{j,e} \right) \cdot \omega + \mu \right)^{\lambda}} & \text{if } j \in Available(i) \\ 0 & \textit{otherwise} \end{cases} \qquad (6)$$

where $Available(i)$ is the set of points that are around the point i, which are not in obstacle area and within one unit distance. Here, we define that the ant can only move in four directions: front, back, left, and right. Then we have inequation $0 < Available(i) \le 4$. $d_{j,e}^{o}$ is the obstructed distance from point j to point e . $Maxd_{A(i),e}^{o}$ is the maximum value of all $d_{j,e}^{o}$ .In order to avoid that the heuristic probability (7) become a random function and the heuristic effect cannot be realized, we adopt three calibration parameters ω, μ, and λ [10], and let ω, μ, and λ be 10, 2, and 2.

The improved ACO algorithm used to get the shortest obstructed distance is adopted as follows.

1. Initialize feasible paths from point i to point j according to equation (6) and calculate the shortest path;
2. For $t = 1$ to t_{max} do {
3. For every path $<i, j>$ do {
4.   Calculate $length(<i, j>)$;

5.   Calculate $\Delta \tau_{i,j}^{k}$ according to equation (5);

6.   Calculate $\tau_{i,j}(t+h)$ according to equation (4);

7.   Straighten path $< i, j >$;

8.   Calculate $shortest(< i, j >)$ and adjust $\tau_{i,j}(t+h)$ accordingly}

9.   Generate feasible paths from point i to point j according to equation (6) }

10. Output the shortest obstructed distance $d^{o}(i, j)$.

where $t_{\max}$ is the maximum number of iterations. STEP 7 is to improve the speed of the improved ACO algorithm, which can be used in the large number of picture element points in bitmap environment, by decreasing the length of moving path artificially. The simulation result is in Fig.2 and Fig.3, and the red solid line represents the optimal obstructed path obtained by ACO.

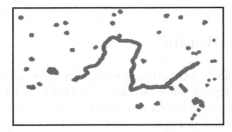 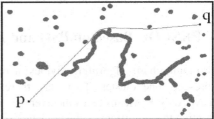

Fig. 2. Layer of obstacle constraints   **Fig. 3.** Optimal obstructed path

4 Particle Swarm Optimization

The Particle Swarm Optimization (PSO) is a population-based optimization method first proposed by Kennedy and Eberhart [18, 19]. In order to find an optimal or near-optimal solution to the problem, PSO updates the current generation of particles (each particle is a candidate solution to the problem) using the information about the best solution obtained by each particle and the entire population. The mathematic description of PSO is as the following. Suppose the dimension of the searching space is D, the number of the particles is n. Vector $X_i = (x_{i1}, x_{i2},..., x_{iD})$ represents the position of the ith particle and $pBest_i = (p_{i1}, p_{i2},..., p_{iD})$ is its best position searched by now, and the whole particle swarm's best position is represented as $gBest = (g_1, g_2,..., g_D)$. Vector $V_i = (v_{i1}, v_{i2},..., v_{iD})$ is the position change rate of the ith particle. Each particle updates its position according to the following formulas:

$$v_{id}(t+1) = w * v_{id}(t) + c_1 * rand() * [p_{id}(t) - x_{id}(t)] + c_2 * rand() * [g_d(t) - x_{id}(t)] \quad (7)$$

$$x_{id}(t+1) = x_{id}(t) + v_{id}(t+1) , \quad 1 \le i \le n, 1 \le d \le D \quad (8)$$

where w is the inertia weight, c_1 and c_2 are positive constant parameters, and $Rand()$ is a random function with the range [0, 1]. Equation (7) is used to calculate the particle's new velocity, then the particle flies toward a new position according to equation (8).The various range of the dth position is $[XMINX_d, XMAXX_d]$ and the various range $[-VMAXX_d, VMAXX_d]$. If the value calculated by equations (7) and (8) exceeds the range, set it as the boundary value. The performance of each particle is measured according to a predefined fitness function, which is usually proportional to the cost function associated with the problem. This process is repeated until user-defined stopping criteria are satisfied.

PSO is effective in nonlinear optimization problems and it is easy to implement. In addition, only few input parameters need to be adjusted in PSO. Because the update process in PSO is based on simple equations, PSO can be efficiently used on large data sets. A disadvantage of the global PSO is that it tends to be trapped in a local optimum under some initialization conditions [20].

5   PKSCOC Based on PSO and K-Medoids

Partitioning-base algorithm divides n objects into $k(k < n)$ parts, and each part represents one cluster. There are there typical types: K-Means, K-Medoids and CLARANS. K-Means takes the average value of a cluster as the cluster centre. While adopting this algorithm, a cluster center possibly just falls on the obstacle (Fig. 4), and it cannot be implemented in reality. On the other hand, K-Medoids takes the most central object of a cluster as the cluster centre, and the cluster center cannot fall on the obstacle. In view of this, K-Medoids algorithm is adopted for SCOC.

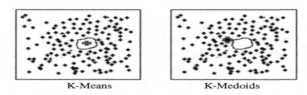

K-Means   K-Medoids

Fig. 4. K-Means vs. K-Medoids

In order to overcome the disadvantage of partitioning approach which only gives attention to local constringency, and keep the advantage of PSO which has stronger global optimum search at the same time, we propose a novel Spatial Clustering with Obstacles Constraints based on PSO and K-Medoids (PKSCOC).

This section first introduces IKSCOC in section 5.1, and then presents the PKSCOC algorithm in section 5.2.

5.1 IKSCOC Based on K-Medoids

K-Medoids algorithm selects the most central object of a cluster as the cluster centre. The clustering quality is estimated by an object function. Square-error function is adopted here, and its definition can be defined as:

$$E = \sum_{j=1}^{N_c} \sum_{p \in C_j} (d(p,m_j))^2 \tag{9}$$

where N_c is the number of cluster C_j, m_j is the cluster centre of cluster C_j, $d(p,q)$ is the direct Euclidean distance between the two points p and q.

To handle obstacles constraints, accordingly, criterion function for estimating the quality of spatial clustering with obstacles constraints can be revised as:

$$E^o = \sum_{j=1}^{N_c} \sum_{p \in C_j} (d^o(p,m_j))^2 \tag{10}$$

where $d^o(p,q)$ is the obstructed distance between point p and point q.

The method of Improved KSCOC (IKSCOC) is adopted as follows [9].

1. Select N_c objects to be cluster centers at random;
2. Distribute remain objects to the nearest cluster center;
3. Calculate E according to equation (10);
4. Do {let current $E = E^o$;
5. Select a not centering point to replace the cluster center m_j randomly;
6. Distribute objects to the nearest center;
7. Calculate E according to equation (9);
8. If $E >$ current E , go to 5;
9. Calculate E^o ;
10. If $E^o <$ current E , form new cluster centers;
11.} While (E^o changed).

While IKSCOC still inherits two shortcomings because it is based on standard partitioning algorithm. One shortcoming is that selecting initial value randomly may cause different results of the spatial clustering and even have no solution. The other is that it only gives attention to local constringency and is sensitive to an outlier.

5.2 PKSCOC Based on PSO and K-Medoids

In the context of clustering, a single particle represents the N_c cluster centroid. That is, each particle x_i is constructed as follows:

$$x_i = (m_{i1},...,m_{ij},...,m_{iN_c}) \tag{11}$$

where m_{ij} refers to the jth cluster centroid of the ith particle in cluster C_{ij}. Here, the objective function is defined as follows:

$$f(x_i) = \frac{1}{J_i} \qquad (12)$$

$$J_i = \sum_{j=1}^{N_c} \sum_{p \in C_{ij}} d^o(p, m_j) \qquad (13)$$

The lower J_i , the higher fitness value.

Spatial Clustering with Obstacles Constraints based on PSO and K-Medoids (PKSCOC), which consults the K-means PSO hybrid [12], is adopted as follows.

1. Execute the IKSCOC algorithm to initialize one particle to contain N_C selected cluster centroids;
2. Initialize the other particles of the swarm to contain N_C selected cluster centroids at random;
3. For $t = 1$ to t_{max} do {
4.   For each particle x_i do {
5.   For each object p do {
6.   Calculate $d^o(p, m_{ij})$;
7.   Assign object p to cluster C_{ij} such that $d^o(p, m_{ij}) = \min \forall c = 1, ..., N_c \{d^o(p, m_{ic})\}$;
8.   Calculate the fitness according to equation (12) ;}}
9.   Update $gBest$ and $pBest_i$;
10.   Update the cluster centroids according to equations (7) and (8);
11.   If $\|v\| \le \varepsilon$, terminate;
12.   Optimize new individuals using the IKSCOC algorithm ;}

where t_{max} is the maximum number of iterations. STEP 1 is to overcome the disadvantage of the global PSO which tends to be trapped in a local optimum under some initialization conditions. STEP 11 is to improve the local constringency speed of the global PSO.

The population-based search of the PKSCOC algorithm reduces the effect that initial conditions has, as opposed to the IKSCOC algorithm; the search starts from multiple positions in parallel. Section 6 shows that the PKSCOC algorithm performs better than the IKSCOC algorithm in terms of quantization error and has higher constringency speed than the GKSCOC algorithm.

6 Results and Discussion

This section presents experimental results on synthetic and real datasets. We have made experiments separately by K-Medoids, IKSCOC, GKSCOC and PKSCOC. The number of particles $n = 50, w = 0.72, c_1 = c_2 = 2, V_{max} = 0.4, t_{max} = 100, \varepsilon = 0.001$.

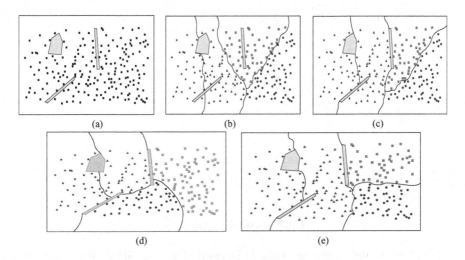

(a)   (b)   (c)

(d)   (e)

Fig. 5. Clustering dataset Dataset1

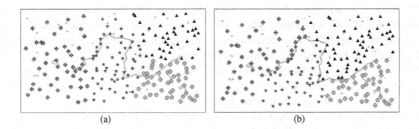

(a)   (b)

Fig. 6. Clustering dataset Dataset2

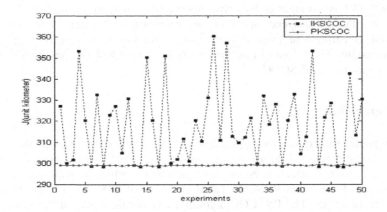

Fig. 7. PKSCOC vs. IKSCOC

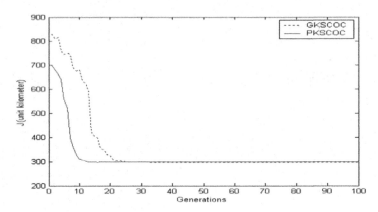

Fig. 8. PKSCOC vs. GKSCOC

Fig. 5 shows the results on synthetic Dataset1. Fig. 5 (a) shows the original data with simple obstacles. Fig. 5 (b) shows the results of 4 clusters found by K-Medoids without considering obstacles constraints. Fig. 5 (c) shows 4 clusters found by IKSCOC. Fig. 5 (d) shows 4 clusters found by GKSCOC. Fig. 5 (e) shows 4 clusters found by PKSCOC. Obviously, the results of the clustering illustrated in Fig. 5(c), Fig. 5 (d) and Fig. 5 (e) have better practicalities than that in Fig. 5 (b), and the ones in Fig. 5 (e) and Fig. 5 (d) are both superior to the one in Fig. 5 (c).

Fig. 6 shows the results on real Dataset2. Fig. 6 (a) shows 4 clusters found by K-Medoids. Fig. 6(b) shows 4 clusters found by PKSCOC. The one in Fig. 6(b) is superior to the one in Fig. 6 (a), obviously.

Fig. 7 is the value of J showed in every experiment on Dataset1. It is showed that IKSCOC is sensitive to initial value and it constringes in different extremely local optimum points by starting at different initial value while PKSCOC constringes nearly in the same optimum points each time. Therefore, we can draw the conclusion that PKSCOC has stronger global constringent ability comparing with IKSCOC.

Fig. 8 is the constringency speed in one experiment on Dataset1. It is showed that PKSCOC constringes in about 12 generations while GKSCOC constringes in nearly 25 generations. So, we can draw the conclusion that PKSCOC has higher constringency speed than GKSCOC.

7   Conclusions

In this paper, we explore the applicability of ACO and PSO based K-Medoids for SCOC. In the process of doing so, we first use improved ACO to obtain the shortest obstructed distance, which is an effective method for arbitrary shape obstacles, and then we develop a novel PKSCOC based on PSO and K-Medoids to cluster spatial data with obstacles. The PKSCOC algorithm can not only give attention to higher local constringency speed and stronger global optimum search, but also get down to the obstacles constraints and practicalities of spatial clustering. The experimental results demonstrate the effectiveness and efficiency of the proposed method, which performs better than IKSCOC in terms of quantization error and has higher constringency speed than GKSCOC.

Acknowledgments. This work is partially supported by the Natural Sciences Fund Council of China (Number: 40471115), the Natural Sciences Fund of Henan (Number:0511011000, Number: 0624220081), and the Open Research Fund Program of the Geomatics and Applications Laboratory, Liaoning Technical University (Number: 2004010).

References

1. Tung, A.K.H., Han, J., Lakshmanan, L.V.S., Ng, R.T.: Constraint-Based Clustering in Large Databases. In: Van den Bussche, J., Vianu, V. (eds.) ICDT 2001. LNCS, vol. 1973, pp. 405–419. Springer, Heidelberg (2000)
2. Tung, A.K.H., Ng, R.T., Lakshmanan, L.V.S., Han, J.: Geo-spatial Clustering with User-Specified Constraints. In: MDM/KDD 2002, pp. 1–7 (2000)
3. Tung, A.K.H., Hou, J., Han, J.: Spatial Clustering in the Presence of Obstacles. In: Proceedings of International Conference on Data Engineering (ICDE'01), Heidelberg Germany, pp. 359–367 (2001)
4. Estivill-Castro, V., Lee, I.J.: AUTOCLUST+: Automatic Clustering of Point-Data Sets in the Presence of Obstacles. In: Proceedings of the International Workshop on Temporal, Spatial and Spatial-Temporal Data Mining, Lyon France, pp. 133–146 (2000)
5. Zaïane, O.R., Lee, C.H.: Clustering Spatial Data When Facing Physical Constraints. In: Proceedings of the IEEE International Conference on Data Mining (ICDM 2002), Maebashi City Japan, pp. 737–740. IEEE Computer Society Press, Los Alamitos (2002)
6. Wang, X., Hamilton, H.J.: A DBRS Density-Based Spatial Clustering Method with Random Sampling. In: Proceedings of the 7th PAKDD, Seoul Korea, pp. 563–575 (2003)
7. Wang, X., Rostoker, C., Hamilton, H.J.: DBRS+: Density-Based Spatial Clustering in the Presence of Obstacles and Facilitators. Ftp.cs.uregina.ca/Research/Techreports/2004-09.pdf (2004)
8. Wang, X., Hamilton, H.J.: Gen and SynGeoDataGen Data Generators for Obstacle Facilitator Constrained Clustering. Ftp.cs.uregina.ca/Research/Techreports/2004-08.pdf (2004)
9. Zhang, X., Wang, J., Wu, F., Fan, Z., Li, X.: A Novel Spatial Clustering with Obstacles Constraints Based on Genetic Algorithms and K-Medoids. In: Proceedings of the Sixth International Conference on Intelligent Systems Design and Applications (ISDA 2006), Jinan Shandong China, pp. 605–610 (2006)
10. Habib, M.K., Asama, H.: Optimal Path Planning for Mobile Robots Based on Intensified Ant Colony Optimization Algorithm. In: Proceedings of the 2003 IEEE International Conference on Robotics, Intelligent Systems and Signal Processing, Changsha China, pp. 131–136 (2003)
11. Xiao, X., Dow, E.R., Eberhart, R., Miled, Z.B., Oppelt, R.J.: Gene Clustering Using Self-Organizing Maps and Particle Swarm Optimization. In: Proceedings of the International Conference on Parallel and Distributed Processing Symposium (IPDPS) (2003)
12. Van der Merwe, D.W., Engelbrecht, A.P.: Data Clustering Using Particle Swarm Optimization. In: Proceedings of IEEE Congress on Evolutionary Computation 2003 (CEC 2003), pp. 215–220. IEEE Computer Society Press, Los Alamitos (2003)
13. Omran, M.G.H.: Particle Swarm Optimization Methods for Pattern Recognition and Image Processing. Ph.D. thesis, University of Pretoria (2005)

14. Cui, X., Potok, T.E., Palathingal, P.: Document clustering using particle swarm optimization. In: Proceedings of IEEE on Swarm Intelligence Symposium (SIS 2005), pp. 185–191 (2005)
15. Dorigo, M.: Optimization, Learning and Natural Algorithms. Ph.D.Thesis, Politecnico di Milano, Italy (1992)
16. Dorigo, M., Gambardella, L.M.: Ant Colony System: A Cooperative Learning Approach to the Traveling Salesman Problem. IEEE Transactions on Evolutionary Computation 1(1), 53–66 (1997)
17. Stützle, T., Hoos, H.H.: MAX-MIN Ant System. Future Generation Computer Systems, 16(8), 889–914 (2000)
18. Eberhart, R.C., Kennedy, J.: A new optimizer using particle swarm theory. In: Russ, C. (ed.) Proceedings of the Sixth International Symposium on Micro Machine and Human Science, Nagoya Japan, pp. 39–43 (1995)
19. Kennedy, J., Eberhart, R.C.: Particle Swarm Optimization. In: Proceedings of IEEE International Conference on Neural Networks, Perth Australia, vol. IV, pp. 1942–1948. IEEE Computer Society Press, Los Alamitos (1995)
20. van den Bergh, F.: An Analysis of Particle Swarm Optimizers. Ph.D. thesis, University of Pretoria (2001)

A High Performance Hierarchical Cubing Algorithm and Efficient OLAP in High-Dimensional Data Warehouse*

Kongfa Hu[1,2], Zhenzhi Gong[2], Qingli Da[2], and Ling Chen[1]

[1] Department of Computer Science and Engineering, Yangzhou University, 225009 China
[2] School of Economics & Management, Southeast University, 210096 China
kfhu05@126.com

Abstract. Data cube has been playing an essential role in fast OLAP (online analytical processing) in many data warehouses. The pre-computation of data cubes is critical for improving the OLAP response time of in large high-dimensional data warehouses. However, as the sizes of data warehouses grow, the time it takes to perform this pre-computation becomes a significant performance bottleneck. In a high dimensional data warehouse, it might not be practical to build all these cuboids and their indices. In this paper, we propose a hierarchical cubing algorithm to partition the high dimensional data cube into low dimensional cube segments. It permits a significant reduction of CPU and I/O overhead for many queries by restricting the number of cube segments to be processed for both the fact table and bitmap indices. Experimental results show that the proposed method is significantly more efficient than other existing cubing methods.

Keywords: Data cube, hierarchical cubing algorithm, high-dimensional data warehouse.

1 Introduction

Data warehouses integrate massive amounts of data from multiple sources and are primarily used for decision support purposes. Data warehouses integrate massive amounts of data from multiple sources and are primarily used for decision support purposes. They have to process complex online analytical processing (OLAP). OLAP refers to the technologies that allow users to efficiently retrieve data from the data warehouse for decision support purposes[1]. A lot of research has been done in order to improve the OLAP query performance and to provide fast response times for queries on large data warehouses. A key issue to speed up the OLAP query processing is efficient indexing and materialization of data cubes [2,3,4]. Many efficient cube computation algorithms have been proposed recently, such as BUC [5], H-cubing [6], Quotient cubing [7], and Star-cubing [8]. However, in the large data warehouse applications, such as bioinformatics, the data usually has high dimensionality with more than 100 dimensions.

---

* The research in the paper is supported by the National Natural Science Foundation of China under Grant No. 70472033 and 60673060; the National Facilities and Information Infrastructure for Science and Technology of China under Grant No. 2004DKA20310; the National Tenth-Five High Technology Key Project of China under Grant No. 2003BA614A; the Natural Science Foundation of Jiangsu Province under Grant No. BK2005047 and BK2005046; the 'Qing Lan' Project Foundation of Jiangsu Province of China.

T. Washio et al. (Eds.): PAKDD 2007 Workshops, LNAI 4819, pp. 357–367, 2007.
© Springer-Verlag Berlin Heidelberg 2007

Since data cube grows exponentially with the number of dimensions, it is generally too costly in both computation time and storage space to materialize a full high-dimensional data cube. For example, a data cube of 100 dimensions, each with 100 distinct values, may contain as many as 101^{100} cells. If we consider the dimension hierarchies, the aggregate cell will increase even more tremendously. Although condensed cube[9], dwarf cube[10], or star cubes [8] can delay the explosion, it does not solve the fundamental problem[11]. In this paper, we develop a feasible cubing algorithm that supports dimension hierarchies for high-dimensional data cubes and answers OLAP queries efficiently. The algorithm decomposes a multi-dimensional hierarchical data cube into smaller segments. For each segment, we encode their dimension hierarchies using a prefix bitmap indexing. Such an approach leads to significant reduction of processing and I/O overhead for many queries by restricting the number of cube segments to be processed for both the fact table and bitmap indices. The bitmap index is designed to support efficient OLAP by allowing fast look-up of relevant tuples.

2   Cube Segmentation

We proposed a decomposition scheme that partitions a high dimensional data cube into low dimensional shell cube segments with support for dimensional hierarchies.

In the data warehouse, a cube C is formally defined as the following (n+m)-tuple: $C=(D_1,...,D_n, M_1, ...,M_m)$ where D_i, for 1<= i <=n, is a dimension and M_j, for 1<= j <=m, is a measure. Each dimension D_i containing the hierarchical attributes $\{ L_1^i, L_2^i, \cdots, L_h^i \}$ (L_1^i being the most aggregated level and L_h^i the most detailed one), where L_j^i is the level j dimension hierarchy of the dimension D_i.

To illustrate the method ,a tiny warehouse, Table 1, is used as a running example.

Table 1. A sample warehouse

TID	DimProduct			DimRegion			DimTime			Measure	
	Category	Class	Product	Country	Province	City	Year	Month	Day	Count	SaleNum
1	Office	OA	Computer	China	Jiangsu	Nanjing	1998	1	1	1	20
2	Office	OA	Computer	China	Jiangsu	Nanjing	1998	1	2	1	60
3	Office	OA	Computer	China	Jiangsu	Yangzhou	1998	1	2	1	40
4	Office	OA	Computer	China	Jiangsu	Yangzhou	1998	1	3	1	20
...	...	...	...	...	...	...	...	...	...	...	...
367	Office	OA	Computer	China	Jiangsu	Nanjing	1999	1	2	1	60
...	...	...	...	...	...	...	...	...	...	...	...

For a cube of d dimensions, it will create 2^d cuboids. If we consider the dimension hierarchies of each dimension, the cube will create $\prod_{i=1}^{d} (h_i + 1)$ cuboids (where h_i is the number of hierarchy levels of dimension D_i).But in a high-dimensional warehouse, there is a substantial I/O overhead for accessing a fully materialized data cube. A partial solution which has been implemented in some commercial data warehouses is to compute a thin shell cube. For example, one might compute all the cuboids with 3 dimensions or less in a 30-dimension data cube. There are two disadvantages of this approach. First, it still needs to compute $C_{30}^3 + C_{30}^2 + C_{30}^1 = 4525$ cuboids if there is no hierarchies and it needs to compute $2^h*(C_{30}^3 + C_{30}^2 + C_{30}^1) = 2^3*4525 = 36200$ cuboids when each dimension has h=3 levels dimension hierarchies. Second, it does not support OLAP in a large portion of the high-dimensional cube space.

In this paper, we propose an orthogonal way to partition the cube space. We partition all the dimensions of a cube into subsets called the cube segments. For example, for a warehouse of 30 dimensions, $D_1, D_2,...,D_{30}$, we first partition the 30 dimensions into 10 Cube segments of size 3: $(D_1,D_2,D_3), (D_4,D_5,D_6),...,(D_{28},D_{29},D_{30})$. For each cube segment, we compute its full data cube. For example, in Cube segment (D_1,D_2,D_3), we compute the eight cuboids:$\{(D_1,D_2,D_3), (D_1,D_2,All), (D_1,All,D_3), (All, D_2,D_3), (D_1,All,All), (All,D_2,All), (All,All,D_3), (All,All,All)\}$. If we consider that each dimension of the 3-D cube (D_1,D_2,D_3) has three hierarchy levels as $D_1(L_1^1,L_2^1,L_3^1)$, $D_2(L_1^2,L_2^2,L_3^2)$, $D_3(L_1^3,L_2^3,L_3^3)$, we will compute 64 cuboids:$\{(L_3^1,L_3^2,L_3^3), (L_3^1,L_3^2,L_2^3),...,(All,All,All)\}$.

Lemma 1. Given a warehouse of T tuples and d dimensions, the entire shell Cube segment will create $\sum_{i=1}^{f} C_f^i * \lceil d/f \rceil = (2^f * \lceil d/f \rceil)$ cuboids and needs $O(T* \sum_{i=1}^{f} C_f^i * \lceil d/f \rceil) = O(T* (2^f * \lceil d/f \rceil))$ storage space, while the partial cube will create $\sum_{i=1}^{f} C_d^i$ cuboids and needs $O(T* \sum_{i=1}^{f} C_d^i)$ storage space, and the full cube will create 2^d cuboids and needs $O(|T|*2^d)$ storage space.

Rational. In the shell Cube segment method, the cube partition into $\lceil d/f \rceil$ cube segments. For each cube segment will create C_f^1 cuboids of 1-dimension, C_f^2 cuboids of 2-dimension,..., C_f^f cuboids of f-dimension and one cuboid(All,All, ...,All). Thus each cube segments will create $(C_f^1 + C_f^2 + \cdots + C_f^f + 1)$ $= \sum_{i=1}^{f} C_f^i$ cuboids. So the entire shell Cube segment will create $\sum_{i=1}^{f} C_f^i * \lceil d/f \rceil$ $= (2^f * \lceil d/f \rceil)$ cuboids and needs $O(T* \sum_{i=1}^{f} C_f^i * \lceil d/f \rceil) = O(T*(2^f * \lceil d/f \rceil))$ storage space.

In partial cube, we select f dimensions from the d dimensions to create the partial cube. It will create C_d^1 cuboids of 1-dimension, C_d^2 cuboids of 2-dimension,..., C_d^f cuboids of f-dimension. Thus the f partial cube will create $C_d^1 + C_d^2 + \cdots + C_d^f = \sum_{i=1}^{f} C_d^i$ cuboids and needs $O(T* \sum_{i=1}^{f} C_d^i)$ storage space.

In full cube, for each dimension D, the dimension of its aggregate cuboids is D or All. For every dimension $\{ D_1,...,D_d \}$, the dimension of its aggregate cuboids is chosen form the 2-values $\{D_i, \text{All}\}$. So for the entire full cube, it will create $\prod_{i=1}^{d} 2 = 2^d$ cuboids and needs $O(|T|*2^d)$ storage space.

Lemma 2. If we consider each dimension has h hierarchies, our prefix-index cubing method will create $\prod_{i=1}^{f}(h_i+1)*\lceil d/f \rceil = ((h+1)^f * \lceil d/f \rceil)$ cuboids, while the minimal cubing method will create $\sum_{i=1}^{f} C_f^i (\sum_{j=1}^{h_i} C_{h_i}^j)*\lceil d/f \rceil = (2^{(f+h)} * \lceil d/f \rceil)$ cuboids.

Rational. In prefix-index cubing method, each dimension D_i has h_i hierarchies, the dimension hierarchies of its aggregate cuboids is chosen form the (h_i+1)-values $\{L_1^i, L_2^i, \cdots, L_h^i, \text{All}\}$. So it will create $\prod_{i=1}^{f}(h_i+1)*\lceil d/f \rceil = ((h+1)^f * \lceil d/f \rceil)$ cuboids.

In the minimal cubing method, the cube partition into $\lceil d/f \rceil$ cube segments. For each cube segments will create $\sum_{i=1}^{f} C_f^i$ cuboids for the f dimensions cube segments. For each dimensions of these cube segments have h_i hierarchies and create $\sum_{j=1}^{h_i} C_{h_i}^j$ dimensional hierarchy cuboids. So the entire minimal cubing will create $\sum_{i=1}^{f} C_f^i (\sum_{j=1}^{h_i} C_{h_i}^j)*\lceil d/f \rceil = (2^{(f+h)} * \lceil d/f \rceil)$ cuboids.

The benefit of this model can be seen by a simple calculation. For a cube of 30 dimensions without hierarchy, if we partition it into 10 segments, each with 3 dimensions, each segment will have 8 cuboids and there are only 8×10 = 80 cuboids to be computed. If each dimension has three hierarchy levels, then each segment will have 64 cuboids as shown above, and there are in total 64×10 = 640 cuboids to be computed. Comparing this to the 36200 cuboids needed by the shell cube technique, the savings in cubing time and space are significant.

3 Prefix Bitmap Indexing

To index the data, we employ a bitmap join index called dimension hierarchical encoding on the higher-cardinality dimensions *dimProduct*, *dimRegion* and *dimTime*.

For each dimension D_i, each hierarchy level L_j^i is encoded as $B^{L_j^i}:dom(L_j^i)$ $\rightarrow \{<b_{k-1}... b_i...b_0>|b_i \in \{0,1\}, i=0,...,k-1\}$. The dimension hierarchical encoding B^{D_i} of each member on the dimension D_i is defined as follows.

$$B^{D_i} = (...((B^{L_1^i} <<Bit\ L_2^i\ |B^{L_2^i})<<Bit\ L_3^i\ |B^{L_3^i})...)<<Bit\ L_h^i\ |B^{L_h^i}$$

Where k is the bit number of the hierarchy level L_j^i on the dimension D_i. k can be computed by $k = Bit\ L_j^i = \lceil \log_2 m \rceil$, where $m = \max(|L_j^i|)$ is the max number of the distinct member of the hierarchy L_j^i. $B^{L_j^i}$ is the encoding of the level of every hierarchy on the dimension D_i and $Bit_{L_j^i}$ is the bit number of the $B^{L_j^i}$. Bit_{D_i} ,the bit number of the member of the dimension D_i ,can be computed with $Bit_{D_i} = \sum_{j=1}^{h} BitL_j^i$.

We can create the *dimTime, dimRegion* and *DimProduct* dimension hierarchy encoding shown in Table 2, Table 3 and Table 4.

Table 2. The *dimTime* dimension hierarchy encoding

TimeID	Year	Month	Day	B^{TimeID}
	yyy	mmmm	ddddd	yyymmmmddddd
1	98	1	1	001000100001
2	98	1	2	001000100010
3	98	1	3	001000100011
...	...	...	...	...
367	99	1	2	010000100010
...	...	...	...	...

Table 3. The *dimRegion* Dimension Hierarchy Encoding

RegionID	Country	Province	City	$B^{RegionID}$
	uuuuuuuu	vvvvv	cccc	uuuuuuuuvvvvvcccc
1	China	Jiangsu	Nanjing	0000001000010001
2	China	Jiangsu	Yangzhou	0000001000010010
...	...	...	...	...

Table 4. The *dimProduct* Dimension Hierarchy Encoding

ProductID	Category	Class	Product	$B^{ProductID}$
	gggg	aaaaa	ppppppp	ggggaaaaappppppp
1	Office	OA	Computer	0001000010000001
2	Office	OA	Printer	0001000010000010
...	...	...	...	...

In Table 2, the dimension hierarchical encoding of the member "1999.1.2" on the dimension *DimTime* is 010000100010. It consists of $B^{Year=1999}B^{Month=1}B^{Day=2}$ and is computed by $B^{1999.1.2}=(B^{Year=1999}<<Bit_{Month}|B^{Month=1})<<Bit_{Day}|B^{Day=2}$ =(010<<4|0001) <<5|00010 =010000100010.

Lemma 3. Our prefix-index cubing method needs $O(T*(h+1)^f *\lceil d/f \rceil)*\lceil \log_2 m \rceil/8)$ storage space, while the minimal cubing method needs $O(T*(2^{(f+h)} *\lceil d/f \rceil)* \lceil \log_{10} m \rceil)$ storage space.

Rational. In prefix-index cubing method, the member of each dimension needs $\lceil \log_2 m \rceil$ bits dimension hierarchical encoding. So the storage space of prefix-index cubing method needs $O(T*((h+1)^f * \lceil d / f \rceil)* \lceil \log_2 m \rceil /8)$ bytes. In the minimal cubing method, the member of each dimension needs $\lceil \log_{10} m \rceil$ integer indices. So the storage space of the minimal cubing needs $O(T*(2^{(f+h)} * \lceil d / f \rceil) \lceil \log_{10} m \rceil *)$ bytes.

By using dimension hierarchical encoding, we can register a list of tuples IDs (tids) associated with the dimension members for each dimension. For example, the TID list associated with the *DimProduct*, *DimRegion* and *DimTime* dimension are shown in Table 5, Table 6 and Table 7 in turn. To compute a data cube for this database with the measure avg() (obtained by sum()/count()), we need to have a tid-list for each cell: $\{tid_1,..., tid_n\}$. Because each tid is uniquely associated with a particular set of measure values, all future computations just need to fetch the measure values associated with the tuples in the list. In other words, by keeping an array of the ID-measures in memory for online processing, one can handle any complex measure computation. Table 8 shows what exactly should be kept, which is substantially smaller than the database itself.

Table 5. *dimProduct* dimension TID

$B^{ProductID}$	TID List
0001000010000001	1-2-3-4-367
...	...

Table 6. *dimRegion* dimension TID

$B^{RegionID}$	TID List
0000001000010001	1-2-367
0000001000010010	3-4
...	...

Table 7. *dimTime* dimension TID

B^{TimeID}	TID List
001000100001	1
001000100010	2-3
001000100011	4
...	...
010000100001	367
...	...

Table 8. TID-measure array of Table 2

tid	Cnt	Num
1	1	20
2	1	60
3	1	40
4	1	20
...	...	...
367	1	60
...	...	...

4   Hierarchical Cubing Algorithm and Efficient OLAP

4.1   Parallel Construction of Shell Cube Segments

The data cube can be distributed across a set of parallel computers by parallel constructing the Cube segments. Therefore, for the end-user and other potential applications, we consider this data cube as one large virtual cube, which is distributed across a set of parallel computers, which manage the creation, updates and querying of the associated cube portions. To develop appropriate scheduling mechanisms for these management tasks, we consider that the virtual cube is split into several smaller parts, called Cube segments. But a Cube segment could furthermore also be split into smaller segments and so on, till we achieve the level of chunks. They can then be assigned to parallel computers, having sequential or parallel computing power, which are responsible for their management. The algorithm for shell cube segment parallel computation can be summarized as follows.

```
Algorithm 1(Parallel computation of shell cube
segments)
{ partition the set of dimensions :(D₁; ... ;Dₙ) into a
    set of k Cube segments {P₁;..., Pₖ};

  scan base cuboid BC once and do the following with
    parallel processing

    {insert each <tid, measure> into ID-measure array;

     for each attribute value aᵢ of each dimension Dᵢ;

       build an dimension hierarchy encoding index
         entry: <B: TID list>;}

  parallel processing all segment partition Pᵢ as follows

     build a local  Cube segments CSᵢ by intersecting
       their corresponding tid-lists and computing
       their measures;}
```

We can parallel construct the high dimensional cube with the Cube segments parallel construction. The system architecture of these shell Cube segment parallel construction is shown in Figure 1.

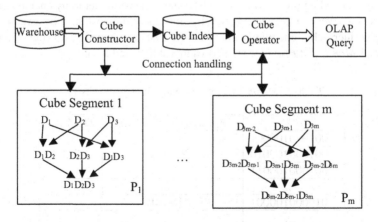

Fig. 1. The System Architecture of Parallel Construction of Shell Cube Segments

The Cube *Constructor* reads one tuple after the other, passes over the items to the index warehouse, retrieves its global index and then passes the (raw) measure and its associated global index to the data cube structure. The *Querying Cube operator* is some kind of highly sophisticated, recursively nested loops for aggregation of measures. Because the number of computational operations of nested aggregation depends on the size of the dimensions and thereby on the order in which dimensions are aggregated, the engine uses a kind of query plan optimization to select dimensions in a "good" way.

4.2 Efficient OLAP Query Handling

Based on the bitmap indexing, we can efficiently retrieve the matching hierarchy levels of each dimension, evaluate the set of query ranges for each dimension, and improve the efficiency of OLAP queries. A key property of our encoding is that it is a prefix indexing scheme that allows one to quickly retrieve a path prefix for each dimension.

The path prefix of the member d_k^i of the hierarchy level L_j^i is defined as

$$DMPrefixpath(DTree, \ d_k^i \ )= \ \bigcup_{j=i}^{1} \ DMPrefixpath(DTree, Parent(\ d_k^j \ ))$$

$=\{ \ Ancestor(d_k^i)\}$, where $Ancestor(d_k^i)$ is the all *ancestors* of the member d_k^i according to its dimension hierarchy tree. The encoding prefix of the member d_k^i is defined as $Bprefix(B^{d_k^i}, L_{m-1}^i)= B^{d_k^i} >> \sum_{l=m}^{j} \ (Bit \ L_l^i)$, where $m=\{1, ...j\}$.

In the above example, the path prefix of the member "1998.1.2" on the dimension *DimTime* is *DMPrefixpath(DTimeTree*, 1998.1.2)={1998.1, 1998}.Its encoding prefix on the hierarchy *Month* is $BPrefix(B^{1998.1.2}, Month) = B^{1998.1} = B^{1998.1.2} >> Bit_{Day} = 001000100010>>5=0010001$. Its encoding prefix on the hierarchy *Year* is $BPrefix(B^{1998.1.2}, Year)=B^{1998}=B^{1998.1.2} >> (Bit_{Day}+Bit_{Month})=001000100010>>(5+4)=001$.

By using encoding prefix, we can register the dimension hierarchy encoding and its TID list for every dimension hierarchy for each dimension. For example, the dimension hierarchy encoding and its TID list associated with the dimension hierarchies *Month* and *Province* are shown in Table 9, and so on.

Table 9. *Month* hierarchy encoding *Prefix* AND its TID

B^{TimeID}	$Bprefix(B^{TimeID}, Month)$	TID List
001000100001		
001000100010	0010001	1-2-3-4
001000100011		
...	...	...
010000100001	0100001	367
...	...	...

For each fragment, we compute the complete data cube by intersecting the TID-lists in the dimension and its hierarchies in a bottom-up depths-first order in the cuboid lattice (as seen in [8]). For example, to compute the cell {0001000010000001, 0000001000010001, 0010001}, we intersect the TID lists of $B^{ProductID}$ =0001000010000001, $B^{RegionID}$ =0000001000010001, and $Bprefix(B^{TimeID},Month)$= 0010001 to get a new list of {1,2}. The algorithm of efficient OLAP query can be summarized as follows:

```
Algorithm 2 (OLAP Query)
    { ascertain all Cube Segment CS_i according as the
      each dimension attribute of the query
      Q<a_1,...,a_n,M>;

    for each CS_i

      {compare the CS_i with query Q<a_1,...,a_n,M> using the
       Lattice and find the all dimension D_i of
       CS_i•{a_1,...,a_n} with parallel processing;

       compute the TID List of the all BC_i cells of
       CS_i•{a_1,...,a_n} in D_i and its aggregate Cuboids;

       intersect the TID List of th BC_i and compute the
       query result set RQ{TID List};

    compute the aggregate with each TID of the TID-
    measure array from the set RQ{TID List};}}
```

This method uses the small dimension hierarchical encoding and their prefix path, so it can rapidly retrieve the matching dimension member hierarchical encoding and evaluate the set of query ranges for each dimension. It rapidly aggregated the clustered fact data that is clusteringly stored by the dimension hierarchical encoding, so that it can drastically reduce the multi-table join effort and so much as could remove completely one or more join operations. As a result, the algorithm can greatly reduce the disk I/Os and highly improve the efficiency of OLAP queries.

5 Performance Study

In this section, we perform a thorough analysis of these costs. In our experimentation we generated a large number of synthetic data sets which in terms of the following parameters: d— number of dimensions, h_i — number of hierarchy levels of dimension D_i, m— maximum number of distinct members of the hierarchy L^i_j, T— number of tuples , f— size of the shell cube segment.

Figure 2 shows the cuboids created by the Cube segment and the other cube (such as partial cube) and their storage size shows in the Figure 3. Figure 4 shows the storage size of the prefix-index cubing and the minimal cubing methods on the cube had T=10^6 tuples and h=3 level hierarchy and with shell fragment size f=3 and their average I/O page access for online query shows in the Figure 5.

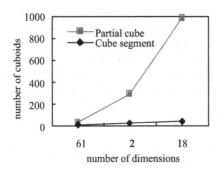

Fig. 2. Cuboids of Cube Segment with f=3

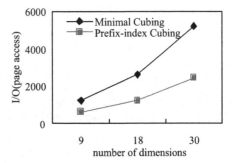

Fig. 3. Storage size of Cube Segment with f=3

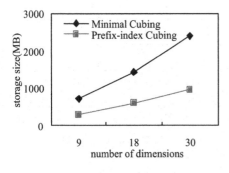

Fig. 4. Sorage Size of Shell Segment h=3   **Fig. 5.** Average I/Os with f=3

Figure 2 and Figure 3 show the method of Cube segments has more efficient than other cube such as partial cube and full cube. The Figure 4 and Figure 5 show the performance of the prefix-index cubing method is more efficient than the other existed leading cubing algorithms such as minimal cubing.

6   Conclusions

Data cube has been playing an essential role in fast OLAP in many multi-dimensional data warehouses. The pre-computation of data cubes is critical to improving the OLAP response time in large data warehouses. But in high-dimensional warehouse, it might not be practical to build all these cuboids and their indices. We have proposed a novel hierarchical cubing algorithm to partition a high dimensional cube into a set of disjoint low dimensional cubes (i.e., shell cube segments). This cubing algorithm supports not only high-dimensional data cubes but also hierarchical data cubes with multiple levels in a dimension. The decomposition of the data cube space leads to significant reduction of processing and I/O overhead for many queries by restricting the number of cube segments to be processed for both the fact table and bitmap indices. Using a prefix bitmap indexing and pre-aggregated results, OLAP queries are computed online by dynamically constructing the cuboids from these cube segments. We have experimentally compared the proposed method with the other existing cubing algorithms such as partial cubing and minimal cubing. The analytical and experimental results show that the proposed hierarchical cubing algorithm is significantly more efficient in time and space than the other leading cubing methods in large high-dimensional data warehouses.

References

1. Chauduri, S., Dayal, U.: An overview of data warehousing and OLAP technology. SIGMOD Record 26(1), 65–74 (1997)
2. Wu, K., Otoo, E.J., Shoshani, A.: A performance comparison of bitmap indexes. In: CIKM(2001), pp. 559–561 (2001)
3. Mistry, H., Roy, P., Sudarshan, S.: Materialized view selection and maintenance using multi-query optimization. In: SIGMOD(2001), pp. 307–318 (2001)
4. Gray, J., Chaudhuri, S., Bosworth, A., Layman, A., Reichart, D., Venkatrao, M., Pellow, F., Pirahesh, H.: Datacube: A relational aggregation operator generalizing group-by, cross-tab and subtotals. Data Mining and Knowledge Discovery 1, 29–54 (2001)
5. Beyer, K., Ramakrishnan, R.: Bottom-up computation of sparse and iceberg cubes. In: ACM SIGMOD, pp. 359–370. ACM Press, New York (1999)
6. Han, J., Pei, J., Dong, G., Wang, K.: Efficient computation of iceberg cubes with complex measures. In: ACM SIGMOD, pp. 1–12. ACM Press, New York (2001)
7. Lakshmanan, L.V.S., Pei, J., Han, J.: Quotient cubes: how to summarize the semantics of a data cube. In: Bressan, S., Chaudhri, A.B., Lee, M.L., Yu, J.X., Lacroix, Z. (eds.) CAiSE 2002 and VLDB 2002. LNCS, vol. 2590, pp. 778–789. Springer, Heidelberg (2003)
8. Xin, D., Han, J., Li, X., Wah, B.W.: Star-cubing:computing iceberg cubes by top-down and bottom-up integration. In: Aberer, K., Koubarakis, M., Kalogeraki, V. (eds.) Databases, Information Systems, and Peer-to-Peer Computing. LNCS, vol. 2944, pp. 476–487. Springer, Heidelberg (2003)
9. Sismanis, Y., Deligiannakis, A., Kotidis, Y., Roussopoulos, N.: Hierarchical dwarfs for the rollup cube. In: VLDB 2004, pp. 540–551 (2004)
10. Lakshmanan, L.V.S., Pei, J., Zhao, Y.: QC-trees: An efficient summary structure for semantic OLAP. In: ACM SIGMOD, pp. 64–75. ACM Press, New York (2003)
11. Li, X., Han, J., Gonzalez, H.: High-dimensional OLAP: A minimal cubing approach. In: VLDB 2004, pp. 528–539 (2004)

Grid-Based Clustering Algorithm Based on Intersecting Partition and Density Estimation

Bao-Zhi Qiu[1], Xiang-Li Li[1], and Jun-Yi Shen[2]

[1] School of Information & Engineering, Zhengzhou University,
Zhengzhou, 450052, China
[2] School of Electronic Information & Engineering, Xi'an Jiaotong University,
Xi'an, 710049, China
{bzqiu,iexlli}@zzu.edu.cn

Abstract. In order to solve the problem that traditional grid-based clustering techniques lack of the capability of dealing with data of high dimensionality, we propose an intersecting grid partition method and a density estimation method. The partition method can greatly reduce the number of grid cells generated in high dimensional data space and make the neighbor-searching easily. On basis of the two methods, we propose grid-based clustering algorithm (GCOD), which merges two intersecting grids according to density estimation. The algorithm requires only one parameter and the time complexity is linear to the size of the input data set or data dimension. The experimental results show that GCOD can discover arbitrary shapes of clusters and scale well.

Keywords: Grid-based clustering, intersecting partition, density estimation, algorithm.

1 Introduction

Clustering has been an important field of data mining, which groups a large amount of data samples into several clusters, maximizing the similarity of data samples in the same cluster and minimizing the similarity between data samples in different clusters [1] . Currently, clustering algorithms can be categorized into partition-based, hierarchical, density-based, grid-based and model-based. Grid-based clustering algorithms are drawing great attention for its advantages of discovering clusters of different shapes and sizes with high efficiency.

Traditional grid-based clustering approaches can be divided into two categories according to partition method: fix-up grid partition method and adaptive grid partition method. Certain clustering algorithms for example CLIQUE[2], STING [3], WaveCluster [4], DENCLUE [5], GDILC [6], shifting grid clustering[7], SCI [8]and Dclust[9] adopt fix-up grid partition technology; while some other clustering algorithms adopt adaptive grid partition method, for example MAFIA [10], OptiGrid [11], MMNG [12], DESCRY [13], CBCM [14], GCHL [15] and so on.

A typical grid-based clustering algorithm using fix-up grid partition technology is CLIQUE, whose basic idea is to partition every dimension of data space

T. Washio et al. (Eds.): PAKDD 2007 Workshops, LNAI 4819, pp. 368–377, 2007.

into intervals with equal length, and then non-intersecting rectangular cells with equal size are formed. Because points fallen into the same grid are probably to belong to a cluster, they can be processed as one object. All the clustering processes are operated on these grid cells, so they have nothing to do with the size of input data, but the number of grid cells. However, fix-up grid partition clustering approach suffers problems as follows: (1) The position of fix-up grid may cause the loss of small clusters. If certain cluster is partitioned into several adjacent grids and none of these grid cells are dense, clusters wouldn't be formed. On the other hand, this partition technology may split the inherent connection of data. (2) The precision of clustering isn't high [8]. When the boundary points of clusters fall into sparse grid, these points will be discarded in the process of clustering. (3) Fix-up grid partition technology is very feeble to clustering in high dimension. For example, a dataset with 100 dimensions would produce enormous 10^{100} grids if we partition every dimension into 10 intervals. So the efficiency of clustering algorithms adopting fix-up grid partition technology will degenerate rapidly with increasing dimensions.

Another typical grid-based clustering algorithm using adaptive grid partition technology is MAFIA, which partitions data space into non-intersecting grid cells with different sizes according to distributional feature of data. The total number of grid cells is greatly reduced in comparison of those algorithms adopting fix-up grid partition technology, but selecting spitting points still needs much computation.

Clustering algorithms adopting fix-up and adaptive grid partition method use non-intersecting grids to form clusters, which need neighbor seeking. Because searching for neighbors usually cost much more time, the latter clustering algorithms mostly build index for grids to improve the efficiency of searching neighbors. Those clustering algorithms usually use depth-first search to connect grids and use growing pattern of greedy algorithm to form clusters.

Generally speaking, fix-up grid partition technology would greatly affect the results of clustering. When the number of dimension is high, there would be enormous grid cells, which make neighbor-seeking a time-consuming process. Although adaptive grid partition technology could decrease the total number of grids, selecting optimal cutting points still need much computation. To solve these problems above, this paper proposes a new clustering algorithm, named GCOD, based on intersecting partition and density estimation, which can decrease the number of grids generated and improves the efficiency of neighbor seeking and precision of clustering.

2   Related Concepts

To sum up, clustering methods based on traditional grid partition technology have low precision and efficiency. So these clustering methods are not suitable to cluster in high dimensional datasets. We propose a new clustering algorithm to solve these problems.

Definition 1. *Given a d-dimensional data point* $x = (x_1, x_2, \cdots, x_d)$, *a grid cell is a hyper-rectangle with* x *as its central point and* Eps_i *as its length. It could be described as:*

$$\prod_{i=1}^{d} [x_i - Eps_i/2, x_i + Eps_i/2] \tag{1}$$

and Eps_i *is defined as*

$$Eps_i = (max_{x \in D}\{x_i\} - min_{x \in D}\{x_i\})/k \tag{2}$$

in which k *is the number of intervals of partition on every dimension, and* x_i *is attribute value of points* x *on its ith dimension.*

From definition 1, we can see that each grid cell should have at least one point. If a point $y = (y_1, y_2, \cdots, y_d)$ falls into the grid cell whose central point is x, the following condition must be satisfied and vice versa.

$$x_i - Eps_i/2 \le y_i \le x_i + Eps_i/2, i = 1, 2, \cdots, d \tag{3}$$

Traditional technology dealing with grid cell has a right-open interval in every dimension; the grid cell we defined here has a full-close interval in every dimension.

Definition 2. *Suppose* C_1, C_2 *are denoted as two grid cells respectively. We define* C_1 *intersects with* C_2, *if* $C_1 \cap C_2 \neq \emptyset$. *Otherwise we define* C_1 *and* C_2 *are separated. The two intersecting grids are adjacent to each other.*

If two intersecting or tangent grids don't share any common point physically, we also take them as separated, because grid-based clustering algorithms suppose points in the same grid are similar and they should belong to the same cluster. These grids which share no points and intersect or tangent to each other physically also can be partitioned by border-lines of grid cells. Because they have different property, we don't take them as neighbors. Therefore, it is reasonable in this paper to partition grids into intersection or separation.

Definition 3. *Given two grids* C_1 *and* C_2, x_i *is attribute value of point* x *on its ith dimension. The density of a grid cell and intersecting block* $(C_1 \cap C_2)$ *are defined respectively as follows:*

$$density(C_1) = \frac{|C_1|}{1/d * \sum_{i=1}^{d} EPs_i} \tag{4}$$

$$density(C_1 \cap C_2) = \frac{|C_1 \cap C_2|}{1/d * \sum_{i=1}^{d} (max_{x \in C_1 \cap C_2}(X_i) - min_{x \in C_1 \cap C_2}(X_i))} \tag{5}$$

Definition 3 uses the length of edge of hyper rectangular to estimate its density. Three circumstances need to be deal with:

(1) If following condition:

$$density(C_1 \cap C_2) \geq max\{density(C_1), density(C_2)\} \tag{6}$$

is satisfied, then the intersecting block must be the center of a cluster, so points in the two grids should belong to the same cluster.

(2) If following condition:

$$min\{density(C_1), density(C_2)\} \leq density(C_1 \cap C_2) < max\{density(C_1), density(C_2)\} \tag{7}$$

is satisfied, then the density is decreasing from one grid cell to another and getting closer to that of the border of the cluster, so points in the two grid cells should belong to the same cluster.

(3) If following condition is satisfied:

$$0 \leq density(C_1 \cap C_2) < min\{density(C_1), density(C_2)\} \tag{8}$$

then the intersecting block is the border of two clusters. Since the $density(C_1 \cap C_2)$ is closer to the grid cell with lower density, shared points should be removed from the grid cell with higher density. For the case that $|C_1 \cap C_2| = 1$, assign the point in the intersecting block to the grid cell with higher density and remove its related information from the grid cell with lower density. If the condition $|C_1 \cap C_2| = 0$ is satisfied, it means that the two grids are separated or share no points physically. Therefore, these points in the two grids shouldn't be assigned to the same cluster. In the former two cases, the process of connecting intersecting grid cells is called density estimation connection.

Definition 4. *We take a grid cell as an isolated grid cell, if it doesn't intersect with any other grid cell.*

Definition 5. *A cluster is the maximal region with density estimation connection.*

3   GCOD Algorithm

The idea of clustering algorithm based on intersecting grids and density estimation is: firstly, compute the length of edge of grid cell on each dimension, and read a point from the data set. If it belongs to certain existing grid cells, record its related information. Otherwise, if it doesn't belong to any existing grid cell, create a new grid cell with itself as center. Then we get clusters by connecting intersecting grids according to density estimation, until every point has a cluster ID. The algorithm is described in details as follows:

Algorithm GCOD:
Inputs: k, // k is the number of interval on each dimension.
1: Computing length of edge: compute the length of each edge of grid;
2: Creating grids: create grid cells according to input data, and record the information of shared points of intersecting grid cells;
   2.1 read dataset;
   2.2 generating grids;
   2.3 record information of shared points.
3: Generating clusters: Forming clusters according to density estimation connection, and deal with the clusters that isolated grid cells belong to.
4: Output clusters.

Computing the Length of Edge Stage: We compute the length of each edge of hyper rectangle Eps_i ($1 \leq i \leq d$, d is the number of dimension)according to the value of k and definition 1, and prepare for generating hyper rectangle (grid cell) next.

Creating Grids Stage: Read a data point from dataset, and judge whether it belongs to certain existing grid cell or not. If it does belong to certain existing grid cells, assign the point to this grids cell. If the point belongs to several grid cells simultaneity, that means these grid cells intersects each other, then record the information of intersecting grids and shared points. Otherwise, if the point doesn't belong to any existing grid cell, create a new grid cell with itself as the central point according to definition 1 and set its length of ith edge of hyper rectangular as Eps_i. The process doesn't stop until all the data points are processed.

The stored information of a grid includes grid-ID of the grid cell, central point, points fallen in the grid, grid-ID of shared grids, the shared points and cluster-ID of clusters. Thus the total number of grids ($Gnum$) is much less than that of points in the dataset, and it is a cover of dataset D. The former input points may belong to grids that will be created latterly during grid creating process, so we need to add shared points in order to prevent shared information from being lost. Fig 1 is 2-dimension intersecting grid sketch map.

Theorem 1. *Given a d-dimensional space D, n is the total number of points, m is the number of grids generated according to definition 1, so $1 \leq m \leq n$.*

Proof. we can see from definition 1 that each grid should contain at least one point. The best situation is that all the data points belong to one grid, while the worst situation is that each data point belongs to the grid cell with itself as the central point and it's impossible that two grids have different central points while they share the same point, which means there aren't two different grids which contain same points. So at most the number of grids is n, and theorem 1 is proved.

Theorem 2. *Supposed d is the number of dimension of data point, n is the total number of data point, m is the number of grids and each grid has q data points in average. If $d \geq lg(n/q)/lgk$, then $m \leq k^d$.*

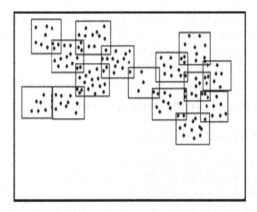

Fig. 1. 2-dimension intersecting grids sketch map

Proof. because $d \geq lg(n/q)/lgk$, and $k \geq 1$, we would get $(lgq + d*lgk) \geq lgn$, viz. $lgk^d \geq lg(n/q)$, so $k^d \geq n/q = m$.

Theorem 2 states that when the number of dimension of data points is high enough, the number of grids generated by intersecting grid partition method is less than that of traditional fix-up grid partition method.

Generating Clusters Stage: clustering begins with a random grid cell and merges the grid cells intersecting according to density estimation connection with broad-first search algorithm. Compute the density of two intersecting grid cells and the intersecting block respectively. If the condition $density(C_1 \cap C_2) \geq min\{density(C_1), density(C_2)\}$ is satisfied, the two grid cells are merged. Factually, the process of merging grids is to modify cluster-ID of grid cells. If the condition $0 \leq density(C_1 \cap C_2) < min\{density(C_1), density(C_2)\}$ is satisfied, points in the intersecting block are merged with the grid of lower density, and delete points as well as their related information from grid with higher density. Special treatment is needed when the condition $|C_1 \cap C_2| = 1$ is satisfied, assign the share point to the higher density grid and delete its relation information from the lower one. The merging process continues until grids of density estimation connection don't exist, and then a cluster is generated. Randomly select a grid from the remainder grids and repeat the process above until all the grids are clustered. Finally, output the results of clustering.

In the process of clustering, there will be some isolated grids, which may either belong to certain cluster or contain outliers merely. The processing method in the paper is to enlarge length of each edge of grids by $\alpha\%$ (In our experiments we set $\alpha = 10$). If the enlarged grid still doesn't intersect with any other clusters, all the points belonging to the grid are taken as outliers. If the enlarged grid contains points of certain existing cluster, merge the grid with the cluster, which prevents points in the cluster from being lost.

Because clusters generated by clustering method above contain small clusters and single-point clusters, we can set a threshold according to clustering environment. Points in the cluster are taken as outliers if the number of points contained in the cluster is less than the given threshold.

4   Experimental Results and Analysis

4.1   Analysis of Time Complexity

The time complexity of GCOD is $O(ndm)$. Here, n is the number of total data points, d is the number of dimensions, and m is the number of grids. The time complexity is linear to the size of data dimension, while the time complexity of clustering algorithms adopting fix-up and adaptive grids partition (for example CLIQUE and MAFIA) grow exponentially with size of dimensions.

4.2   Experimental Results

All experiments are run on PC with Pentium2.4G CUP, 256M memory, windows 2000 professional operation system. The algorithm is programmed and compiled in Borland C++ builder6.0.

Synthesized datasets usually are used to evaluate effectiveness and clustering precision of clustering algorithms. The primitive dataset of Fig 2 comes from CHAMELEON which contains 8000 data points and consists of 6 clusters, some outliers and streaks running across clusters. Fig2 (a) is the clustering result of GCOD. Although the dataset contains clusters of different shapes and sizes, GCOD can identify six clusters correctly, which demonstrates effectiveness of the clustering algorithm (in Fig 2a, the clusters whose total number of points is less than 300 are denoted by one color). Fig 2b is the clustering result of CLIQUE. We can see from Fig 2b that clustering result of CLIQUE loses many boundary points, while GCOD doesn't. Obviously clustering precision of GCOD is higher than that of CLIQUE.

Using a real dataset KDD $cup99$[6], we select 1000 records and set $k = 3$. Table 1 describes the number of grids created by CLIQUE and GCOD. From

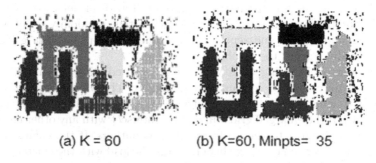

(a) K = 60   (b) K=60, Minpts= 35

Fig. 2. The comparison of clustering results of GCOD and CLIQUE

Table 1. The number of grids at different dimensions using CLIQUE and GCOD

	2	10	20	30	40
CLIQUE	3^2	3^{10}	3^{20}	3^{30}	3^{40}
GCOD	20	83	125	231	332

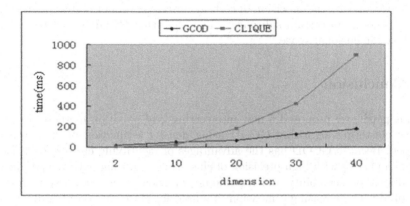

Fig. 3. Scalability with the number of dimensions

Table 2. The precision of clustering and Recall on Iris and Wine

	Iris data set			Wine data set		
	i=1	i=2	i=3	i=1	i=2	i=3
$|C_i^0|$	50	50	50	59	71	48
$|C_i^5|$	50	19	81	51	43	42
$|C_i^5 \cap C_i^0|$	50	18	49	51	41	39
Precision (%)	100	94.7	60.5	100	95.4	92.9
Recall(%)	100	36	98	86.8	57.8	81.3

table 1, we can see that the number of grids generated by GCOD doesn't grow exponentially with increasing dimensions. So GCOD is suitable for clustering in high dimension. Fig 3 shows the relation between running time and the number of dimension, from which we can see that execution efficiency of CLIQUE is higher than that of GCOD when the size of dimension is lower, while running time of GCOD is much less than that of CLIQUE with increasing dimension.

4.3 Correctness Test

We will use two datasets Iris and Wine that come from UCI for classification to evaluate clustering precision of GCOD. The dataset Iris contains 150 instances of 4 dimensions and 3 classes, and each class contains 50 instances. The dataset Wine contains 178 instances of 13 dimensions and 3 classes, and the number of instances contained in each class is $59, 71, 48$ respectively. We denote factual instances contained in ith cluster by $|C_i^0|$ and denote instances detected by clustering algorithm by $|C_i^s|$, then the precision of clustering (Precision) and Recall can be denoted by$|C_i^0 \cap C_i^s|/|C_i^s|$ and $|C_i^0 \cap C_i^s|/|C_i^0|$ respectively. Table 2 shows clustering results of GCOD on Iris and Wine data sets. Compared to the results in reference [16], the precision of clustering algorithm GCOD is higher than that of CURE and clustering method based on shrinking.

The reason why the precision of certain cluster is still not high enough is that this clusters lie in certain subspace, while algorithm GCOD generates clusters in whole dimensional space.

5   Conclusion

In this paper, we proposed a new intersecting grid partition method, on which a new clustering algorithm GCOD using density estimation of grids was developed. Because GCOD has the advantages of identifying clusters of different shapes and sizes with high precision of clustering, requiring only one parameter, well dimension scalability and the number of grids generated doesn't grow exponential with increasing dimension, it's used for clustering on high dimension. Our future work is how to introduce the technology of clustering to subspace clustering to improve precision of clustering.

Acknowledgments. This research is partially supported by the National Science Foundation of China under grants No.60673087.

References

1. Xu, R., Wunsch, D.: Survey of clustering algorithms. IEEE transaction on neural networks 16(3), 645–678 (2005)
2. Agrawal, R., Gehrke, J., Gunopulos, D., et al.: Automatic Subspace Clustering of High Dimensional Data for Data Mining Applications. In: Proc. of the ACM SIGMOD Int'l Conference on Management of Data. Seattle, Washington, pp. 94–105 (June 1998)
3. W., Yang, J., Muntz, R.: STING: A statistical information grid approach to spatial data mining. In: Proc. 23rd Int. conf. on very large data bases, pp. 186–195. Morgan Kaufmann, San Francisco (1997)
4. Sheikholeslami, G., Chatterjee, S., Zhang, A.: WaveCluster: A multiresolution clustering approach for very large spatial databases. In: Proc. 1998 Int. conf. very large data bases, New York, pp. 428–439 (1998)
5. Hinneburg, A., Keim, D.A.: An Efficient Approach to Clustering in Large Multimedia databases with Noise. In: KDD 1998, pp. 58–65.
6. Yanchang, Z., Jjunde, S.: GDILC: A Grid-based Density-Isoline Clustering Algorithm. In: Proc. of 2001 Int'l Conferences on Info-tech and Info-net, Beijing, China, October 2001, pp. 140–145
7. Ma, E.W.M., Chow, T.W.S.: A new shifting grid clustering algorithm. Pattern Recognition 37, 503–514 (2004)

8. Hsu, C.-M., Chen, M.-S.: Subspace clustering of high dimensional spatial data with noises. In: Dai, H., Srikant, R., Zhang, C. (eds.) PAKDD 2004. LNCS (LNAI), vol. 3056, pp. 31–40. Springer, Heidelberg (2004)
9. Zhang, J., Hsu, W., Lilee, M.: Clustering in dynamic spatial database. Journal of intelligent information systems 24(1), 5–27 (2005)
10. Goil, S., Nagesh, H., Choudhary, A.: Mafia: Efficient and scalable subspace clustering for very large data sets. Technical Report CPDC-TR-9906-010, Northwestern University, 2145 Sheridan Road, Evanston IL 60208 (June 1999)
11. Hinneburg, A., Keim, D.A.: Optimal Grid-Clustering: Towards breaking the curse of Dimensionality in high-dimensional Clustering. In: Proceedings of the 25th VLDB Conference, Edinburgh, pp. 506–517 (1999)
12. Rickard, J.T., Yager, R.R., Miller, W.: Mountain clustering on non-uniform grids using P-tree. Fuzzy optimization and decision making 4, 87–102 (2005)
13. Angiulli, F., Pizzuti, C., Ruffolo, M.: DESCRY: A Density Based Clustering Algorithm for Very Large Data Sets. In: Yang, Z.R., Yin, H., Everson, R.M. (eds.) IDEAL 2004. LNCS, vol. 3177, pp. 203–210. Springer, Heidelberg (2004)
14. Chang, J.-W., Kim, Y.-K.: An Efficient Clustering Method for High-Dimensional Data Mining. In: Bazzan, A.L.C., Labidi, S. (eds.) SBIA 2004. LNCS (LNAI), vol. 3171, pp. 276–285. Springer, Heidelberg (2004)
15. Pilevar, A.H., Sukumar, M.: GCHL: a grid-clustering algorithm for high-dimensional very large spatial data bases. Pattern recognition letters 26, 999–1010 (2005)
16. Shi, Y., Song, Y., Zhang, A.: A shrinking-based clustering approach for multi-dimensional data. IEEE transaction on knowledge and data engineering 17(10), 1389–1403 (2005)
17. Qiu, B., Zhang, X., Shen, J.: Grid-based clustering algorithm for multi-density. In: Proceedings of 2005 international conference on machine learning and cybernetics, pp. 1509–1512.
18. Qiu, B., shen, J.: GBCBE:Grid-based clustering algorithm with boundary point extraction. Intelligent information management systems and technologies 1(2), 271–276 (2005)

Depth First Generation of Frequent Patterns Without Candidate Generation

Qunxiong Zhu and Xiaoyong Lin

School of Infromation Science and Technology, Beijing University of Chemical Technology,
15 Beisanhuan East Road, ChaoYang District, 100029, Beijing, China
{zhuqx,linxy}@mail.buct.edu.cn

Abstract. Mining frequent patterns has been studied popularly in data mining research. Most of the current studies adopt an FP_growth-like approach which does not bring the candidate generation. However, the cost of recursively constructing each frequent item's conditional frequent pattern tree is high. In this paper, we propose a depth first algorithm for mining frequent patterns. Efficiency of mining is achieved with the following techniques: large database is compressed into a frequent pattern tree with a children table but not a header table, which avoids costly repeated database scans, on the other hand the mining algorithm adopts a depth first method which takes advantage of this tree structure and dynamically adjusts links instead of generating a lot of redundant sub trees, which can dramatically reduces the time and space needed for the mining process. The performance study shows that our algorithm is efficient and scalable for mining frequent patterns, and is an order of magnitude faster than Trie, FP_growth, H-mine and some recently reported new frequent patterns mining methods.

Keywords: Frequent pattern mining, Data mining, Depth first, Top-down.

1 Introduction

Data mining has recently attracted tremendous amount of attention in database research. The attention is motivated by the large amount of data which many organizations now have about their own business. For instance, supermarkets store electronic records of millions of receipts, banks and credit card companies maintain extensive transaction records and factories collect a lot of flow parameters. The goal in data mining is to analyze these large data and discover patterns, rules and trends that are useful for decision support.

Mining association rules, an important subfield of data mining, was firstly introduced in [1]. So far there have been considerable researches on designing fast algorithms for this task [1,2,3,4,5,6,7,8,9,10,11,12,13,14]. The problem of mining all association rules can be decomposed into two sub problems. Firstly all frequent patterns that have a support above a minimum support defined by the user are mined. Secondly, rules are generated by using the frequent patterns mined. Comparing with the first problem, the solution to the second one is rather straightforward [1], so we only consider the first problem in detail.

T. Washio et al. (Eds.): PAKDD 2007 Workshops, LNAI 4819, pp. 378–388, 2007.

Frequent patterns mining was firstly presented in [1]. Reference [3] is considered as one of the most important contributions to this subject. Its main algorithm, Apriori, not only influenced the association rules mining community, but affected other data mining fields as well. Most of the previous studies, such as [5, 6, 11, 12], adopt an Apriori-like approach, which is based on an anti-monotone Apriori heuristic: if any length k pattern is not frequent in the database, its all length (k+1) super patterns can never be frequent. The essential idea is to iteratively generate the set of candidate patterns of length (k+1) from the set of frequent patterns of length k (for k ≥ 1), and then check their corresponding occurrence frequencies in the database [3].

The Apriori heuristic achieves food performance gain by reducing the size of candidate set. However, the generation of candidate set is still costly, especially there are a mass of patterns or long patterns. Jiawei Han [2] proposed a frequent pattern tree (FP_tree) structure and developed an efficient algorithm, FP_growth, which didn't bring the candidate set and was scalable for mining both short and long patterns. Whereas, the cost of recursively constructing each frequent item's conditional frequent pattern tree is still high.

Jian Pei [4] presented a simple hyperlinked data structure, H-struct, and a new mining algorithm, H-mine, which takes advantage of the data structure and dynamically adjusts links in the mining process. Because it does not generate a lot of conditional frequent pattern trees, high time and space performance in various kinds of data is achieved. However, the structure still suffers from the low compactness; especially the datasets are very large and dense.

After many careful examinations, we believe that if one can get a highly condensed structure and avoid recursively constructing the conditional frequent pattern trees, the mining performance can be substantially improved.

This problem is settled in the following three aspects.

First, we proposed a novel frequent pattern tree with a children table (not a header table in [2]), which is an extended prefix-tree structure storing complete information of all frequent patterns. Only frequent length-1 items will appear in the tree, and the tree nodes are arranged in such a way that more frequently occurring items will have better chances of sharing nodes than less frequently occurring ones.

Second, we present a depth first method (namely top-down strategy) of frequent patterns mining, DF_Miner, which starts from the node which represents most frequent length-1 item (as an initial prefix pattern) and examines only the subtree from this node. After achieving all frequent patterns with this item, we consider the second frequent length-1 item and its subtree, and so on. In the subsequent process, by the children table, we maintain the contribution of the subtree that was aleady considered for the following mining process. Since the frequent itemset in any transaction is always encoded in the corresponding path of the frequent pattern tree, DF_Miner ensures the completeness of the result. In this paper, we hold the compactness by using the FP_tree [2] and take top-down strategy instead of bottom-up strategy in FP_grown. The major operations of mining are count accumulation and the local adjustment for tree, which are usually much less costly than recursively constructing all conditional frequent pattern trees.All these techniques contribute to substantial reduction of search costs.

A performance study has been conducted to compare the performance of DF_Miner with Trie (a fast Apriori in [6]), FP_growth and H-mine. The study shows that DF_Miner achieved good performance.

The rest of the paper is organized as follows. Section 2 formally introduces the problem. Section 3 proposes our new algorithm.. Section 3.3 analyzes the efficiency of the algorithm. Section 4 gives our performance study. Section 5 summarizes our study and points out the future work.

2 Problem Statement

In this section, we will state basic definition of frequent patterns mining.

2.1 Mining Frequent Patterns

Let $I = \{x_1, x_2, x_3 \ldots x_n\}$ be a set of items. An itemset X that is also called pattern is a subset of I, denoted by $X \subseteq I$. A transaction $T_X = (TID, X)$ is a pair, where X is a pattern and TID is its unique identifier. A transaction T_X is said to contain T_Y if and only if $Y \subseteq X$. A transaction database, named TDB, is a set of transactions. The number of transactions in DB that contain X is called the support of X. A pattern X is a frequent pattern, if and only if its support is larger than or equal to s, where s is a threshold called minimum support.

Given a transaction database, TDB, and a minimum support threshold, s, the problem of finding the complete set of frequent patterns is called the frequent patterns mining problem.

3 Depth First Generation of Frequent Patterns

In this section, we first introduce the frequent pattern tree with a children table and present an algorithm for constructing such a tree. Then we develop a depth first algorithm, DF_Miner, for mining all frequent patterns from the tree constructed. Finally we discuss the efficiency of DF_Miner.

3.1 Frequent Pattern Tree with a Children Table

An FP_tree with a children table can be designed based on the following observations.

To avoid repeatedly scanning the database, one stores the set of frequent items of each transaction in a tree structure. It is necessary to perform one scan of database to identify the set of frequent items with its frequency count, because only the frequent items will appear in the FP_tree. Then one need perform another scan of database to construct the FP_tree.

All transactions which share an identical frequent items set can be merged into one branch with the numbers of occurrences registered as count. It is easy to check whether two sets are identical if the frequent items in all of the transactions are ordered according to a given order.

If the frequent items are ordered in their support descending order, there are better chances that more prefix strings can be shared. If two transactions share a common prefix, according to the descending order, the shared parts can be merged using one prefix structure as long as the count is registered properly [2].

Each node in an FP_tree consists of four fields: item_name, count, next link, and child links, where item_name registers which item this node represents, count registers the number of transactions represented by the portion of the path reaching this node, next link links to the next node in the FP_tree carrying the same item_name, and child links point its children.

With all observations that mentioned above, one can design a highly condensed frequent pattern tree structure for efficient frequent pattern mining. Let's examine an example.

Example 1. Let the database, TDB, be the first two columns of Table 1, and the minimum support threshold, s, be 3.

One may construct an FP_tree with a children table as follows:

First, one may perform a scan of TDB to get all frequent items, {(a:5), (i:4), (k:3), (b:3), (f:3)} (The number after":" indicates the item's support in database TDB.), in which items ordered in the support descending order, each path of the frequent pattern tree will follow this order. For convenience of later discussions, the frequent items in each transaction are listed in this order in the third column of Table 1.

Table 1. The transaction database TDB and the frequent items in a given s=3

TID	Items	Frequent Items(ordered and s=3)
100	i, k, a, h, c, f, d, g	a, i, k, f
200	a, b, f, i, j	a, i, b, f
300	b, a, e, l, m	a, b
400	a, h, j	a
500	b, i, k, l, n, o	i, k, b
600	a, k, i, f	a, i, k, f

Second, one may scan the TDB again to build the FP_tree. At the beginning, one creates the root of the tree. After scanning the first transaction, one construct the first branch of the tree: {(a: 1), (i: 1), (k: 1), (f: 1)}. Notice that the frequent items in the transaction are ordered according to the order in the list of frequent items. For the second transaction, since its frequent items (ordered) list {a, i, b, f} shares a common prefix {a, i } with the existing path {a, i, k, f}, the count of each node along the prefix is incremental by 1 and a new node (b:1) is created and linked as a child of (i:2), and another new node (f:1) is created and linked as a child of (b:1). After all transactions are mapped into the tree, the FP_tree of TDB is constructed. For using a top-down strategy, a children table of the root is built in which each child of the root is listed to the relevant child link field. In the whole process of construction, one need not link all nodes with the same item, which happen in [2], in order to save more construction time. In this example, after scanning all the transactions in TDB, one finishes the FP_tree with a children table, which is shown in Fig. 1(except for C_a and C_{ai}).

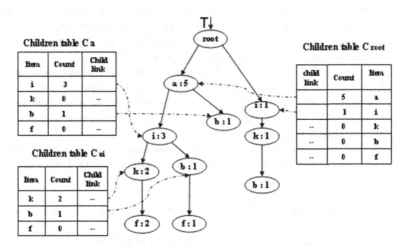

Fig. 1. The frequent pattern tree with a children table of TDB (s=3)

Based on the observations and example mentioned above, The FP_tree with a children table construction algorithm is in the following step (The construction algorithm of an FP_tree is proposed in [2]. Readers are referred to the cited paper for detailed information.).

Algorithm 1.   FP_tree_Construction(TDB, s, FP_tree)

    F= {all frequent items with their supports in the support descending order};
    create the root of FP_tree, T   //T is null now.
    construct an FP_tree;   //see reference [2].
    create the children table of the root, C_{root} //for the top-down strategy.
    call Initialize (C_{root}, T);
end

The procedure Initialize is shown as follow.

procedure Initialize (C_α, α) {
    for each child c_i of α do begin
        append c_i into the c_i's child link field in C_α;
        modify the relevant local-count fields;
    end
}

3.2 Depth First Generation of Frequent Patterns Algorithm

In this section, we present a novel mining algorithm, called depth first generation of frequent patterns (DF_Miner for short), which uses an FP_tree with children tables. We use a prefix to identify a subtree. For instance, in Fig 1, a-prefix represents the leftmost subtree of root, and ab-prefix identifies the second subtree rooted at a-prefix. For simplicity, we use i_j and i_k for a single item, and use α and β for a pattern which contains multi items or empty.

Given an FP_tree with a children table for a set of frequent items $I = \{i_1, i_2, i_3 \ldots i_n\}$, where a total order (in the support descending order) is defined on I. The top-down strategy of DF_Miner adopts an Apriori heuristic (however it doesn't bring the candidate set generation.): A pattern represented by $\alpha i_j i_k$ may be frequent only if the pattern represented by αi_j is frequent. Namely when we find that the pattern αi_j is not frequent, we need not test the whole $\alpha i_j i_k$-prefixes ($j < k \leq n$). Another strategy (divide-and-rule) of DF_Miner is based on the following observation.

Mining a pattern-frefix represented by α-prefix is to mine all the union of prefixes whose form is $\beta\alpha$-prefix. For example, Fig 1 shows a FP_tree with a children table, which contains five items $\{a,i,k,b,f\}$. The union of i-prefix is {i-prefix and ai-prefix} and the union of b-prefix is {ab-prefix, aib-prefix, and ikb-prefix}.

Based on the above two strategies. We present the DF_Miner algorithm (which will be explained at all in next sub section) for mining frequent patterns using an FP_tree with a children table.

Algorithm 2.   DF_Miner(α, C_α, F) { //depth first generation of frequent patterns
　　　　　　　　//where α represents a pattern, C_α is a children table of the α
　　　　　　　　//and F is the set of all frequent patterns, which is an output.
　　for all c_i in the children table C do begin
　　　　if c_i.support $\geq$ s then
　　　　　　{αc_i.support $= c_i$.support and F=F $\cup$ αc_i
　　　　　　generate the children table for αc_i, $C_{\alpha ci}$
　　　　　　call Initialize ($C_{\alpha ci}$, αc_i);
　　　　　　call DF_Miner(αc_i, $C_{\alpha ci}$, F)}　　　　　　//depth first strategy
　　　　　call LinkforUnion(c_i, $C_{\alpha ci}$,C_α)　　　　　//divide-and-rule strategy
　　end
}
　　procedure LinkforUnion(c_i, $C_{\alpha ci}$,C_α)
　　　　//link all items in $C_{\alpha ci}$ into C_α
　　　　for each item in $C_{\alpha ci}$ do begin
　　　　　　append the child link behind the child link of the same item in C_α
　　　　end
end

3.3 Analysis of DF_Miner

We observe some interesting properties of an FP_tree structure which will accelerate frequent patterns mining.

Property 1 (Next link property). For all prefixes which form is αi_j-prefix, their appropriate appearances can be exactly obtained by i_j's next link.

This property assures that we will not miss any appropriate prefixes which form is αi_j-prefix.

Property 2 (Child link property). All the possible frequent patterns that contain i_j can be obtained completely by traversing all αi_j's next link and following all their child links.

This property ensures that we can accurately find all frequent patterns which contain i_j.

The two properties which are mentioned above are based directly on the designing and mining process, which will commendably accelerate the traversal of all appropriate prefixes related to i_j.

Example 2. Let us analyze the mining process based on the FP_tree shown in Fig. 1. Based on Property 1, we collect all the patterns that a node i_j participates by starting from i_j's next link and the child-links. We examine the mining process by starting from the top of children table (a top-down strategy) instead of the bottom of the header table in FP_growth (a bottom-up strategy).

For node a, because its count field is 5, which is larger than the minimum support (s=3), it gets a frequent pattern (a: 5) and two children in a-prefix. At first we create the children table of pattern a, C_a, as shown in Fig. 1. The first child-link points the ai-prefix and the second one points the ab-prefix. In this matter, we take the top-down (depth first) strategy for generating all frequent patterns containing a. For pattern ai, because i's count in this node is equal to s, we derives a frequent pattern (ai: 3) and two children in ai-prefix. We also create the children table of pattern ac, C_{ai} as shown in Fig. 1. The first child link points the aik-prefix and the second one points the aib-prefix. For pattern aik, because the relevant count field of the node k in aik-path is less than s, we stop generating any frequent pattern containing aik (Because any length k+1 pattern can never be frequent if its length k sub-pattern is not frequent). Before considering the aib-prefix, we call LinkforUnion procedure to adjust the children tables C_a, C_{ai} and their relevant next links (a divide-and-rule strategy). The adjusted children tables C_a and C_{ai} is shown in Fig. 2. Similarly, the pattern ak and ab are not frequent, we adjust the relevant children tables and their next links only. When we finished mining a-prefix, we get {(a: 5), (ai: 3), (aif: 3), (af: 3)}.

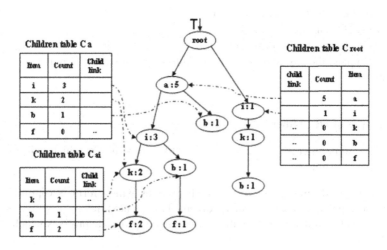

Fig. 2. The frequent pattern tree with a children table after getting pattern (ai: 3)

Similarly using the depth first strategy, node i drives {(i: 4), (ik: 3), (if: 3)} and node k and b get {(k: 3), (b: 3)}. Finally, there is only remaining one frequent item, f, which doesn't bring the generation of children table any more, because there are no frequent items in the ordered frequent item list . At that time, we get the last frequent pattern, {(f: 3)}.When we finish the mining process, the final FP_tree with a children table is shown in Fig. 3.

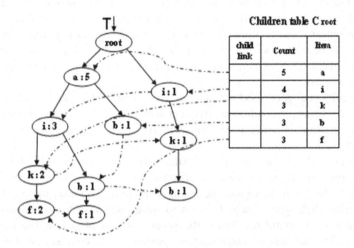

Fig. 3. The frequent pattern tree with a children table after finishing the mining process

4   Performance Study

In this section, we present a performance study of our algorithm, DF_Miner, and give the comparison of DF_Miner with Trie [6] (a fast Apriori implementation), FP_growth [2] (a classical frequent patterns mining algorithm) and H-mine [4] (a fast and clever algorithm). It shows that DF_Miner outperforms Trie, FP_growth and H-mine and is efficient and highly scalable for mining very large databases.

All the experiments are performed on a 1.8GHz Pentium PC machine with 256 Megabytes main memory and 60G hard disk, running on Microsoft Windows/NT. All codes were written in Visual C++6.0 and run in the same environment, while the version of Trie that we used is a well-known version, available at http://fimi.cs.helsinki.fi/. All reports of the runtime of DF_Miner include both the time of constructing tree and mining frequent patterns. They also include both CUP and I/O time.

We have tested various data sets, with consistent result. Limited by space, only the results on two typical datasets, Gazelle (a real and sparse data set) and T25I15D10k (a synthetic and dense data set), are reported here.

Gazelle is a web store visit (click stream) data set from Gazelle.com. It is a sparse data set. It contains 59,602 transactions, while where per transaction are up to 267 items.

Figure 4 shows that the run times of DF_Miner, Trie, FP_growth and H-mine on this data set. Obviously, DF_Miner outperforms the other three algorithms, and the gaps (in terms of run time) become larger as the support threshold goes lower.

Trie works well in such sparse data sets since most of the candidates that it generates turn out to be frequent patterns. However, it has to construct a hash tree for the candidates and match them in the tree and update their counts each time when scanning a transaction that contains the candidates. That is the major cost for Trie.

FP_growth gets a similar performance as Trie and sometime is even slightly worse. This is two reasons, firstly when the data set is sparse; it can not compress data as effectively as what it does on dense data sets, on the other hand, recursively constructing every frequent item's conditional frequent pattern tree over sparse data set has its overhead.

H-mine achieved good performance in such sparse data sets since it avoids recursively constructing conditional frequent pattern tree by taking advantage of a simple data structure, H-struct. It only needs to dynamically adjust links in the mining process.

Figure 5 gives the space usage of DF_Miner, Trie, FP_growth and H-mine during the mining process. To make the comparison clear, the space usage (axis Y) is in logarithmic scale. From the figure, the memory usage of Trie does not scale well as the support threshold goes down. Trie has to store level-wise frequent patterns and generate next level candidates. When the support threshold is low, the number of frequent patterns and that of candidates are nontrivial. In contrast, the DF_Miner, FP_growth and H-mine are very scalable in terms of space usage with respect to support threshold. Even when the support threshold reduces to very low, the memory usage is still stable, since they do not need to store any frequent patterns or candidates, which are output immediately and never read back.

T25I15D10k is generated by a synthetic data generator described in [3]. It is a relatively dense data set, which contains 10,000 transactions and each transactions and each transaction has up to 25 items. There are 1,000 items in the data set and the average longest potentially frequent pattern is 15 items.

Figure 6 shows the run time of the four algorithms on this data set. When the support threshold is high, most patterns are short; Trie and FP_growth have similar performance, at the same time H-mine and DF_Miner have the similar performance too. When the support threshold becomes low, DF_Miner is the fastest one.

Figure 7 gives the space usage of the four algorithms in mining process. The space usage is drawn in logarithmic scale too. As support threshold goes down, the number of the frequent patterns goes up dramatically. Trie requires an exponential amount of space. FP_growth, H-mine, and DF_Miner use stable amount of space. In dense data set, a frequent pattern tree is smaller than the set of all frequent items projections (H-mine), so DF_Miner uses less space than H-mine. However, long patterns mean more recursive frequent pattern trees, on the other hand, since the number of frequent items is large in this data set, the FP_tree may have many branches in various levels and becomes bushy, which make FP_growth require more space than H-mine in this case.

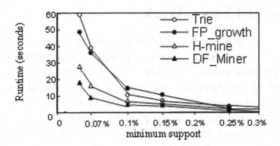

Fig. 4. Runtime on data set Gazelle

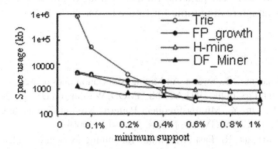

Fig. 5. Space usage on data set Gazelle

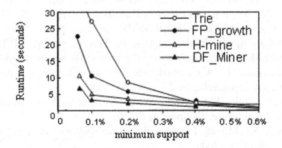

Fig. 6. Runtime on data set T25I15D10k

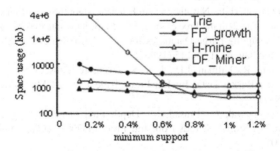

Fig. 7. Space usage on data set T25I15D10k

5 Conclusions and Future Work

In this paper, we present a novel algorithm for mining frequent patterns, which take depth first strategy to generate all frequent patterns. This method avoids repeatedly constructing conditional pattern tree for every frequent item. The algorithm was implemented, and its performance is studied and compared with three of best algorithms for mining frequent patterns so far. The study shows that our algorithm has superior performance on mining frequent patterns from large database.

Recently, there have been some interesting studies at mining frequent patterns with a changed minimum support threshold or in such databases which allow new data is added to or old data are deleted from. The extension of the depth first technique, for maintenance of the frequent patterns mined as both database and support threshold being updated, is an interesting topic for future research.

References

1. Agrawal, R., Imielinski, T., Swami, A.: Mining Association rules between Sets of Items in Large Databases. In: Intl Proc of the 1993 ACM SIGMOD, Washington D.C, pp. 207–216 (1993)
2. Han, J., Pei, J., Yin, Y.: Mining Frequent Patterns without Candidate Generation. In: Intl Proc of the 2000 ACM SIGMOD, Dallas, pp. 1–12 (2000)
3. Agrawal, R., Strikant, R.: Fast Algorithms for mining association rules. In: 20th Intl Proc of the 1994 VLDB, Santiago, pp. 487–499 (1994)
4. Pei, J., Han, J., Lu, H., et al.: H-Mine: Hyper-Structure Mining of Frequent Patterns in Large Databases. In: Intl Proc of the 2001 ICDM, San Jose, pp. 441–448 (2001)
5. Park, J., Chen, M., Yu, P.S.: An Effective Hash-Based Algorithm for Mining Association rules. In: Intl Proc of the 1995 ACM SIGMOD, San Jose, pp. 175–186 (1995)
6. Bodon, F.: A Fast Apriori Implementation. In: Proc of the IEEE ICDM Workshop on Frequent Itemset Mining Implementations Repository. Melbourne, Florida (2003)
7. Rozenberg, B., Gudes, E.: Association Rules Mining in Vertically Partitioned Databases. Data & Knowledge Engineering 59(2), 378–396 (2006)
8. Palashikar, G.K., Kale, M.S., Apte, M.M.: Association Rules Mining Using Heavy Itemsets. Data & Knowledge Engineering (2006)
9. Chen, T., Hsu, S.: Mining Frequent Tree-Like Patterns in large datasets. Data & Knowledge Engineering (2006)
10. Xin, D., Han, J., Yan, X., Cheng, H.: On Compressing Frequent Patterns. Data & Knowledge Engineering (2006)
11. Savasere, A., Omiecinski, E., Navathe, S.: An Efficient Algorithm for Mining Association Rules in Large Databases. In: 21th Intl Proc of the 1995 VLDB. San Francisco, pp. 432-444 (1995)
12. Toivonen, H.: Sampling Large Databases for Association Rules. In: 22nd Intl Proc of the 1996 VLDB, Bombay, pp. 134–145 (1996)
13. Brin, S., Motwani, R., Ullman, J.D., Tsur, S.: Dynamic Itemset Counting and Implication Rules for Market Basket Data. In: Intl Proc of the 1997 ACM-SIGMOD, pp. 255–264. New York (1997)
14. Agarwal, R., Aggarwal, C., Prasad, V.: Depth First Generation of Long Patterns. In: Ramakrishnan, R., Stolfo, S. (eds.) 6th Intl Proc. of 2000 ACM SIGKDD, Boston, pp. 108–118 (2000)
15. Goethals, B., Zaki, M.J.: Advances in frequent itemset mining implementations. Report on FIMI 2003. SIGKDD Explorations 6(1), 109–117 (2003)

Efficient Time Series Data Classification and Compression in Distributed Monitoring*

Sheng Di[1], Hai Jin[1], Shengli Li[1], Jing Tie[2], and Ling Chen[1]

[1] Cluster and Grid Computing Lab
Services Computing Technology and System Lab
Huazhong University of Science and Technology, Wuhan, 430074, China
[2] Department of Computer Science, University of Chicago
1100 E 58th Street Chicago, USA
hjin@hust.edu.cn

Abstract. As a key issue in distributed monitoring, time series data are a series of values collected in terms of sequential time stamps. Requesting them is one of the most frequent requests in a distributed monitoring system. However, the large scale of these data users request may not only cause heavy loads to the clients, but also cost long transmission time. In order to solve the problem, we design an efficient two-step method: first classify various sets of time series according to their sizes, and then compress the time series with relatively large size by appropriate compression algorithms. This two-step approach is able to reduce the users' response time after requesting the monitoring data, and the compression effects of the algorithms designed are satisfactory.

1 Introduction

Efficiently storing monitoring data is a crucial function to a distributed monitoring system, however, many current systems [1, 2, 3] have not considered it well enough, especially on how to classify several sets of time series and compress them appropriately. In practice, it is very important to design a way of reducing historical monitoring data before transmitting them in large-scale distributed systems. As a matter of fact, if there is not a processing module, the size of the transmission data may be so large that clients' memories can not afford such high costs (Some exceptions may be thrown out, such as "OutOfMemoryException" [4]), or the corresponding transmission time will be so long that users can not endure.

There are several candidate ways to solve the problem mentioned above. The first way is that we can make a buffer on the client device that retrieves historical monitoring data. The size of this buffer can be set as an appropriate value to avoid being out of memory. As soon as the buffer is full, the monitoring data retrieved will be displayed on GUI (*Graphic User Interface*) or stored into external storage. This way can certainly avoid being out of memory, but it unwillingly enhances the loads of client devices and cannot reduce transmission time. Moreover, it is against to the

---

* This paper is supported by National Science Foundation of China under grant 90412010, ChinaGrid project from Ministry of Education, and CNGI projects under grant CNGI-04-15-7A.

T. Washio et al. (Eds.): PAKDD 2007 Workshops, LNAI 4819, pp. 389–400, 2007.

principle of designing a thin client instead of a fat one in the grid field [5]. The other way is to compress historical monitoring data before transmitting them. It can reduce not only the loads of clients but also the transmission time.

From clients' views, the historical monitoring data are displayed as several sets of time series to be compared. In our monitoring system, as a use case, each set of time series stands for the performance on some metric of one node inside a cluster. Thus, the compression issue could be transformed to how to compress time series. Moreover, it is apparent that as long as we choose appropriate compression algorithms, the fluctuation of time series compressed can still be displayed clearly on clients' computers.

This paper mainly studies a strategy in processing time series data. Not only do we develop a kind of classification algorithm to organize all the time series to be compared (such as, the monitoring data about all the nodes inside one cluster with some metrics), but we also study some algorithms used to compress a specific set of time series (i.e. to reduce dimensionality of time series).

In our practical distributed monitoring platform, CGSV [15], these approaches mentioned above are processed in a server called Archive [6] which mainly provides two services. One service is used to receive requests from *Target System Service* [15] (Target System Service is a web service deployed on the front node of a cluster and used to notify monitoring data to Archive) and execute corresponding operations (including "create table", "insert data", etc.), and the other one is used to publish interfaces for clients.

The rest of this paper is organized as follows: in section 2, we present the related work. In section 3, we introduce our classification algorithm (i.e. the preparation task of compressing). We give some detailed compression algorithms and relative analyses in section 4. We analyze the results of our practical testing in section 5 and conclude and present the future work in section 6.

2 Related Work

2.1 Storage Modules in Grid Monitoring System

The field on distributed monitoring system has a long history in literatures, but the issue about how to effectively process monitoring data is still underway.

- VisPerf [1]: VisPerf is a monitoring tool for grid computing. There is no storage module in it.
- MonALISA [2]: MonALISA has a special storage module used to compress monitoring data. As data are becoming older over time, the values stored will be compressed by evaluating the mean values on large time intervals. In our paper, we provide several better compression algorithms than this mean value algorithm.
- GridView [3]: GridView is a dynamic and visual grid monitoring system. It has a Grid Information Manager used to compress monitoring data periodically. Compared with it, our approach does not rewrite the original data so as to keep them lossless. The Archive module compresses the corresponding time series just as clients request them.

2.2 Compression Algorithms

In recent years, there has been an explosion of interests in mining time series databases. Several high level dimensionality reductions of time series have been proposed, including *Singular Value Decomposition* (SVD) [7], *Discrete Fourier Transform* (DFT) [8], *Discrete Wavelets Transform* (DWT) [9], *Symbolic Mappings* [10, 11, 12], *Piecewise Linear Representation* (PLR) [13], and *Piecewise Aggregate Approximation* (PAA) [14]. As a matter of fact, SVD, DFT and DWT are mainly designed to match patterns, so they are relatively intricate to be implemented to compress data. In this paper, we adopt the clearest approach, *Piecewise Linear Representation* (PLR), to compress the monitoring data. The goal of this approach is to index some special points in a time series and link them in a series. The most straightforward index algorithms used in PLR are *Random Walk Algorithm* (RWA) and *Mean Value Algorithm* (MVA). However, neither are they effective when encountering many white noises [16] (i.e. where each value x_t is completely independent to its neighbors x_{t-1}, x_{t+1}). In addition, although *Height Sorting Reduction* (HSR) algorithm and *Angle Sorting Reduction* (ASR) algorithm can adapt to white noises well, neither of them is suitable to the other cases without white noises. Facing this issue, we design two new algorithms (PFR and A-ISR) which are able to effectively index special points of time series in both cases (with white noises and without white noises). Finally, we test these two algorithms and prove that PLR can get a very high efficiency in compression.

3 Classification Algorithm of Historical Monitoring Data

In our monitoring platform (CGSV) [6], all the monitoring data are stored into the database deployed on the Archive. In order to reduce the complexity and keep database legible, all the time series data are saved together. We provide a use case: the monitoring performance data about all the nodes inside one cluster. In this case, Archive has to classify these time series out from the original data before compressing them.

3.1 Classification Algorithm

We assume the original data have been organized as n sets of time series shown in Fig.1. S is defined as the total remaining free space (i.e. the maximum number of monitoring data). With monitoring data being indexed over time, this free space will keep shrinking. The initial space is assumed to be S_0 (such as $S=S_0=10000$), and it means confining the maximum number of data transmitted at one time to *10000*. The main purposes of the classification algorithm are to justify which time series should be compressed and calculate the appropriate compression ratios for them.

It is obvious that $\bar{s} = S/n$ is the mean length of all the target time series. We define a rule as follows: If the length of some original time series is less than or equal to $\bar{s}$, it will not be compressed. Hence, these data will be indexed immediately. We call this kind of time series *Raw-Index Time Series*, while we call the time series which should be compressed *Compress-Index Time Series*. Apparently, if the number of *Raw-Index Time Series* is m, then the number of *Compress-Index Time Series* will be $(n-m)$ and the remaining data space is: $S-\sum_m$(the size of the *Raw-Index Time Series*). After executing Formula (3) and Formula (4), one search round will be done, and then, the work flow will be recursively returned back to Formula (1). In each round, $\overline{S}$ may be greater than that of last round, so the length of a time series whose length was greater than that of $\overline{S}$ in previous rounds may become shorter than that of $\overline{S}$ in the current or future rounds.

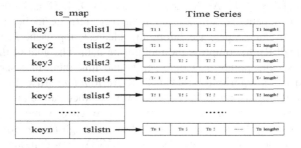

Fig. 1. Time Series of Ts_map

$$\overline{S} = \frac{S}{n} \tag{1}$$

$$l \begin{cases} \leq \overline{S} & \text{Raw-Index Time Series} \\ > \overline{S} & \text{Compress-Index Time Series} \end{cases} \tag{2}$$

$$S \leftarrow S - \sum_m (\text{the size of Raw-Index Time Series}) \tag{3}$$

$$n \leftarrow (n - m) \tag{4}$$

3.2 Pseudo-code of Classification Algorithm

All the variables used in the classification algorithm are listed in Table 1. The pseudo-code of classification algorithm is shown in Fig.2.

4 Compression Algorithms

4.1 Characteristics of Monitoring Data

Monitoring data own a number of characteristics. We study them carefully in our distributed monitoring platform: CGSV (*ChinaGrid SuperVision*) [6, 15]. In this system, all the sensors deployed on monitored nodes are used to retrieve monitoring data every β seconds (β can be modified by users). In other words, as soon as the value of a metric is changed over a revisable threshold (its default value is 0.1 in CGSV), the node's sensor will receive a corresponding notification and forward it upwards. The characteristics of monitoring data are as follows:

- Small amplitude fluctuations may inevitably exist in high frequencies.
- It is possible to generate abrupt fluctuations at some time stamps in a time series.
- The randomicity of monitoring time series data is very high.
- There are huge discrepancies among the monitoring data belonging to different metrics.

Table 1. Variables in the Pseudo-code of Classification Algorithm

Variable	Explanation
ts_map	A map that consists of time series (such as Fig.1)
ts	One element in *ts_map* (i.e. a set of time series)
cutNumber	The number of the sets of time series not to be compressed in this round
mark	It is used to mark if there is at least one set of time series which need not be compressed in a round

```
1.     recursion(ts_map, S)
2.     {
3.         m ←the length of ts_map //m is the number of the elements in ts_map.
4.         S̄ ←S/m      // the mean length of time series
5.         cutNumber←0// the number of the sets of time series not to be compressed in this round
/*As follows is to traverse ts_map and check whether the lengths of all time series are greater than S̄ .*/
6.         mark ←false
7.         do for tranverse ts_map
8.             do ts ← one set of time series
9.             do if the length of ts ≤ S̄
10.                then do if mark = false
11.                    then do mark ←true
12.                do cutNumber ← cutNumber + the length of ts
13.                    Remove this ts from ts_map;
14.         if mark = false
15.             then do return S;
16.         else
17.             do S₁ ← S- cutNumber,
18.                 return recursion(ts_map, S₁);
19.     }
```

Fig. 2. The Pseudo-code of Classification Algorithm

The area magnified in Fig.3 shows the small amplitude fluctuations. In practice, when the metric is set to be CPU_Idle, most of monitoring data may often fluctuate from 99.8 to 100.0. Fig.3 also shows the abrupt fluctuations and the randomicity of monitoring data. Fig 4 shows the monitoring data of free memory. Plus the different units of diverse metrics or other changeful cases, the fluctuations of different time series may be extremely disparate.

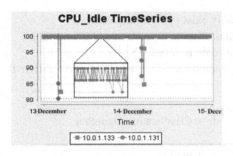 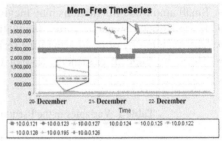

Fig. 3. Small Amplitude Fluctuation and Abrupt Fluctuations at Time Stamp

Fig. 4. Monitoring Data of Some Free Memory

There have been many existing outstanding solutions to compress large-scale time series data [7, 8, 9, 10, 11, 12, 14]. However, all of them are relatively complicated to implement. For instance, as for DFT [8], programmers have to implement the intricate DFT algorithm, choose appropriate coefficients, and implement the R*-Tree algorithm [17].

In order to adapt to the features of our monitoring data mentioned above and also facilitate the implementation, we adopt PLR [13] and design a couple of new algorithms to index special points.

4.2 Time Series Index Algorithms in PLR

In order to facilitate the implementation of compression, we adopt the PLR method and put forward two new index algorithms in it. Let a set of values as $\{x_0, x_1, ..., x_n\}$ and the corresponding time stamps as $\{t_0, t_1, ..., t_n\}$. We assume the time series with n original points will be extracted into a sequence with $\bar{s}$ ($\bar{s}$ is the $\bar{s}$ after executing classification algorithm) special points and the points (t_0, x_0) and (t_n, x_n) are always being indexed.

Peak Focus Reduction (PFR)
The focus of this algorithm is all the peaks of a time series, shown in Fig.5. First, sort the points in the given time series according to d_i calculated in Formula (5) by descend.

$$d_i = x_i - \bar{v}, \bar{v} = \frac{\sum_{i=1}^{n} x_i}{n} \tag{5}$$

And then, pick up these points in the time series by descend one by one. As for each point selected, estimate its neighbor points in time series toward two sides (left and right). Once the warp between its value and that of its neighbor points surpasses a threshold, this neighbor point will be indexed and this estimation section will be ended. Otherwise, estimate its neighbor's neighbor point analogically. The whole algorithm will not be done until the number of points indexed is saturated (i.e. up to $\bar{s}$ points).

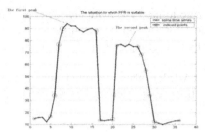 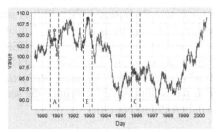

Fig. 5. The Special Points Indexed with PFR Algorithm   **Fig. 6.** The Special Points Indexed with A-ISR Algorithm

Adaptable Interzone Sampling Reduction (A-ISR)

Define K as an adaptable coefficient determining the value of the threshold TH and we take $K=10$ in our practical test. In Formula (6), the variables max and min denote the maximal value and the minimal value, respectively, during the whole time series.

$$TH = \frac{1}{K} \mid max - min \mid \tag{6}$$

As long as Condition (7) is met, the corresponding fluctuation will be viewed as distinct. In contrast, if it is not met, the fluctuation will be ignored (i.e. we treat this tiny fluctuation as no fluctuation).

$$\Delta x > TH \quad , \quad \Delta x = \mid x_{i+1} - x_i \mid \tag{7}$$

We use a variable p to mark the current trend of the segment of time series.

$$p = \begin{cases} 1 & \text{The current value is larger than the last one} \\ 0 & \text{The two values' discrepance should be ignored} \\ -1 & \text{The current value is smaller than the last one} \end{cases} \tag{8}$$

We use n_{turn} to record the number of the turning points. By the end of the algorithm, n_{turn} will be checked. $n_{turn}=0$ means there is no turning point during the specific period of this time series. In this case, we index a random point. $n_{turn}=1$ means there exists one and only one extreme point during the specific period, and this special extreme point will be indexed. $n_{turn} \geq 2$ means there are more than one turning points during this specific period, and in this case, we will calculate the mean time stamp of all the special extreme points in this period as the time stamp of the special point and the mean value. Fig.6 shows the points indexed with this algorithm when processing a time series. In section A, $n_{turn}=3$, so the calculated blue point will be indexed. In section B, $n_{turn}=1$, so the exclusive special point will be indexed. In section C, $n_{turn}=0$, therefore we index a random point.

4.3 Analysis of Algorithms

As to the two index algorithms mentioned above, they have their own features and their time complexities are acceptable.

- *Peak Focus Reduction* (PFR): PFR is suitable to the cases with and without white noises. This algorithm is a little similar to HSR (it will be mentioned in section 5.2), but it is more reasonable than HSR. Its time complexity is $O(n \log n)$ because the sort algorithm we use is Heapsort [18].
- *Adaptable Interzone Sampling Reduction* (A-ISR): This algorithm does not sort time series but just scans through the set of time series. Hence, the time complexity is not $O(n \log n)$ but $O(n)$.

5　Performance Analysis

5.1　Performance of Classification Algorithm

The whole time consumed is immensely reduced because of the Classification Algorithm. In our practical testing, when the total number of raw data is *12000*, the time consumed by Classification Algorithm is still less than 1ms. However, the transmission time is much longer than it. So the time cost by this algorithm can be ignored. Fig.7 shows the difference of the response time consumed between with and without our approaches. From this figure, we could find that when users request *12000* data, the total time consumed is mainly in transmitting data. However, when we compress them into *600* data with our algorithms before transmission, the whole response time consumed will be cut down greatly.

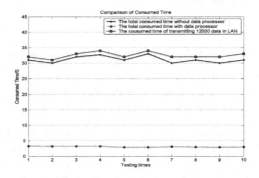

Fig. 7. Comparison of Consumed Time between with and without Our Approaches

5.2　Performance and Effect of Index Algorithm in PLR

We test all the index algorithms mentioned above with two groups of original monitoring data: one group with white noises and the other without them. In addition, we compare our index algorithms with traditional ones: *Random Walk Algorithm, Mean Value Algorithm, Height Sorting Reduction* (HSR), and *Angle Sorting Reduction* (ASR). *Random Walk Algorithm* and *Mean Value Algorithm* choose random point and mean value point to replace the values in one segment, respectively. HSR sorts all monitoring data in terms of their values' discrepancies to mean value and indexes the $\overline{s}$ biggest ones. ASR is similar to HSR, and their exclusive difference is ASR's criterion of sorting data is angles of time series lines.

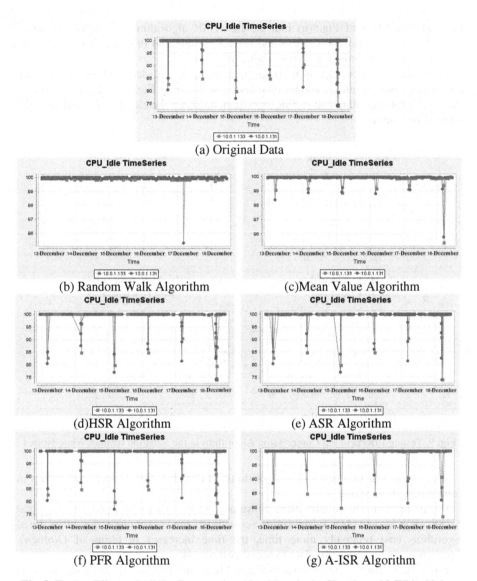

Fig. 8. Testing Effects of All the Compression Algorithms in the Situation with White Noises

Fig 8 (a) presents the original time series and its size is *6076*. The result size of time series processed in each figure is *300*. So each compression ratio is up to (6076-300)/6076*100%=95%.

The other sub-figures in Fig.8 show the different compression effects of the various algorithms when it comes to many white noises in the original time series. Fig.8(b) and Fig.8(c) show that the fluctuations generated by *Random Walk Algorithm* or *Mean Value Algorithm* are highly different from the original time series. Fig.8(d),

Fig.8(e), Fig.8(f), and Fig.8(g) indicate that ASR algorithm and A-ISR algorithm generate better compression effects and HSR algorithm and PFR algorithm come into being the closest fluctuations to the original time series.

Fig 9 shows the compression effects with the group of data without many white noises (the corresponding original time series is shown in Fig.4). They reveal that in this case, PFR algorithm and A-ISR algorithm still keep the closest fluctuations to the original time series.

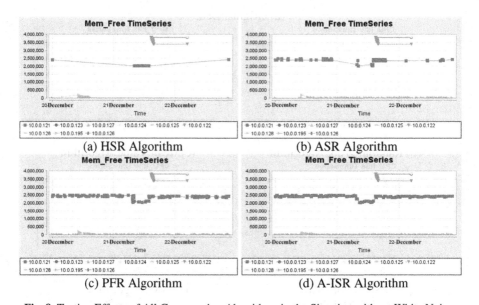

(a) HSR Algorithm   (b) ASR Algorithm

(c) PFR Algorithm   (d) A-ISR Algorithm

Fig. 9. Testing Effects of All Compression Algorithms in the Situation without White Noises

As a result, our two new index algorithms, PFR and A-ISR, indeed own values in compressing time series.

Fig.10 presents the relative time consumed. In this figure, it is easy to find that A-ISR, RWA, and MVA algorithms cost least time. Although the other three algorithms cost relatively more time, the time increases in terms of $O(n\log n)$. Apparently, these costs still adapt to users' commands.

From the testing results shown in Fig.8, Fig.9, and Fig.10, we can summarize as follows:

In the white noise situations, PFR algorithm and HSR algorithm have the best compression effects, and *Random Walk Algorithm*'s compression effect is the worst. In the situation without white noises, PFR algorithm and A-ISR algorithm generate better compression effects than the others. In addition, *Random Walk Algorithm*, *Mean Value Algorithm* and A-ISR algorithm consume the shortest time because their time complexities are all $O(n)$. In fact, the time consumed by each algorithm mentioned above could be ignored compared with the time cost via network. So, we recommend PFR algorithm and A-ISR algorithm in most situations.

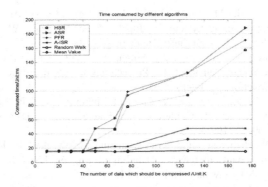

Fig. 10. Time Consumed by Different Algorithms

6 Conclusion and Future Work

In order to solve the problem of transmitting massive data in the distributed monitoring field, we studied how to classify and compress time series to be compared. The classification algorithm is a recursive algorithm that can make full use of memory space of clients' devices. This algorithm is also a key step of preparing for the following time series compression. As to the time series compression method, we adopted *Piecewise Linear Representation* (PLR) method to process data. Based on this method, we designed two kinds of fresh index algorithms and analyzed their performance. The testing results present that both of them are suitable to the cases with and without white noises and consume acceptable time.

In the future, we still have some things to study, such as:

- We are supposed to make progress on index algorithms, especially on compression effect of A-ISR algorithm. In addition, we should try to exploit other better ones.
- Different index algorithms are adapted to different kinds of time series. So, we will enhance the adaptability of archive in CGSV by adopting various algorithms to process diverse kinds of time series.

References

1. Lee, D., Dongarra, J., Ramakrishna, R.: Visperf: Monitoring Tool for Grid Computing. In: Proceeding of ICCS, pp. 1–12 (2003)
2. Newman, H.B., Legrand, I.C., Galvez, P., Voicu, R., Cirstoiu, C.: MonALISA: A Distributed Monitoring Service Architecture. In: Proceedings of CHEP, La Jolla, CA, pp. 1–8 (2003)
3. Guangbo, N., Jie, M., Bo, L.: GridView: A dynamic and visual grid monitoring system. In: Proceedings of HPC Asia, pp. 89–92 (2004)
4. OutOfMemoryException and other pathological cases, http://haacked.com/archiv e/2004/02/11/189.aspx
5. Foster, I., Kesselman, C., Tuecke, S.: The Anatomy of the Grid: Enabling Scalable Virtual Organizations. International J. Supercomputer Applications 15(3) (2001)

6. Zheng, W., Liu, L., Hu, M., Wu, Y., Li, L., He, F., Tie, J.: CGSV: An Adaptable Stream-Integrated Grid Monitoring System. In: Jin, H., Reed, D., Jiang, W. (eds.) NPC 2005. LNCS, vol. 3779, pp. 22–31. Springer, Heidelberg (2005)

7. Wu, D., Angrawal, D., Abbadi, A.E., Singh, A., Smith, T.R.: Efficient Retrieval for Browsing Large Image Databases. In: Proceedings of 5th International Conference on Knowledge Information, pp. 11–18 (1996)

8. Agrawal, R., Faloutsos, C., Swami, A.: Efficient Similarity Search in Sequence Databases. In: Proceedings of the 4th Conference on Foundations of Data Organization and Algorithms, pp. 69–84 (1993)

9. Chan, K., Fu, W.: Efficient Time Series Matching by Wavelets. In: Proceedings of the 15th IEEE International Conference on Data Engineering, pp. 126–133 (1999)

10. Agrawal, R., Lin, K.I., Sawhney, H.S., Shim, K.: Fast Similarity Search in the Presence of Noise, Scaling, and Translation in Times-series Databases. In: Proceedings of 21th International Conference on Very Large Data Bases, pp. 490–500 (1995)

11. Das, G., Lin, K., Mannila, H., Renganathan, G., Smyth, P.: Rule Discovery from Time Series. In: Proceedings of the 3rd International Conference on Knowledge Discovery and Data Mining, pp. 16–22 (1998)

12. Perng, C., Wang, H., Zhang, S., Parker, S.: Landmarks: a New Model for Similarity based Pattern Querying in Time Series Databases. In: Proceedings of 16th International Conference on Data Engineering, pp. 33–42 (2000)

13. Keogh, E.J., Pazzani, M.J.: Scaling up Dynamic Time Warping to Massive Datasets. In: Żytkow, J.M., Rauch, J. (eds.) Principles of Data Mining and Knowledge Discovery. LNCS (LNAI), vol. 1704, pp. 1–11. Springer, Heidelberg (1999)

14. Keogh, E., Chakrabarti, K., Pazzani, M., Mehrotra, S.: Dimensionality Reduction for Fast Similarity Search in Large Time Series Databases. In: Proceedings of the ACM SIGMOD International Conference on Management of Data, pp. 151–162 (2001)

15. CGSV Manual, http://www.chinagrid.edu.cn/CGSV/doc/CGSV-Manual/en/html/ book.html

16. Schroeder, M.: Fractals, Chaos, Power Laws: Minutes From an Infinite Paradise. W.H. Freeman and Company, New York (1991)

17. Bechmann, N., Kriegel, H.P., Schneider, R., Seeger, B.: The R*-tree: An Efficient and Robust Access Method for Points and Rectangles. In: Proceedings of ACM SIGMOD International Conference on Management of Data, pp. 322–331 (1990)

18. Cormen, T.H., Leiserson, C.E., Rivest, R.L., Stein, C.: Introduction To Algorithms, 2nd edn., pp. 127–128. MIT Press, Cambridge

Best-Match Method Used in Co-training Algorithm

Hui Wang, Liping Ji, and Wanli Zuo

College of Computer Science and Technology, Jilin University,
Key Laboratory of Symbolic Computation and Knowledge
Engineering of the Ministry of Education,
Changchun 130012, China
whui05@yahoo.com.cn

Abstract. Since 1998 there has been significant interest in supervised learning algorithms that combine labeled and unlabeled data for text learning tasks. The co-training algorithm applied to datasets which have a natural separation of their features into two disjoint sets. In this paper, we demonstrate that when learning from labeled and unlabeled data using co-training algorithm, selecting those document examples first which have two parts of best matching features can obtain a good performance.

Keywords: Best-match, co-training, expectation-maximization, Naive Bayes, text classification, labeled data, unlabeled data.

1 Introduction

Combining information from labeled and unlabeled data for text classification is very important now. Some approaches have been proposed since 1998, e.g. Expectation Maximization (EM) estimates maximum posteriori parameters of a generative model [1]; generative model built from unlabeled data performs a discriminative classification [2]; transductive inference for support vector machine optimizes performance on a specific test set [3]. All these results have shown that making use of unlabeled data can significantly decrease classification error, especially when labeled training data are short of supplied.

This paper organized as follows: we recall related works in section 2. Section 3 relates with Best-Match method proposed in this paper. Experimental results are elaborated in section 4. Section 5 is concerned with conclusion and future works.

2 Related Works

Blum and Mitchell [4] proposed the co-training algorithm for classifying web pages that builds two classifiers: one over the words that appear on a page, and the other over the words appearing in hyperlinks pointing to that page. Nigam and Ghani [5] perform further extensive experiments to demonstrate that

T. Washio et al. (Eds.): PAKDD 2007 Workshops, LNAI 4819, pp. 401–409, 2007.

when learning from labeled and unlabeled data, algorithms explicitly leveraging a natural independent split of the features outperform algorithms that do not leverage it. When a natural split does not exist, co-training algorithms that manufacture a feature split may out-perform algorithms not using the split.

Naive Bayes is a simple but effective text classification algorithm for learning only from labeled data [6,7]. The parameterization given by naive Bayes defines an underlying generative model assumed by the classifier. If supervised learning setting is extended to include unlabeled data, the naive Bayes is no longer adequate to find maximum posteriori parameter estimates.

Traditional text classification techniques require labeled training examples of all classes to build a classifier [8,9]. As for learning from unlabeled data, there are some researchers who do a lot of researches on this respect in recent years [10]. A theoretical study of Probably Approximately Correct (PAC) learning from positive and unlabeled examples was done in [11]. Sample complexity results from learning by maximizing a number of unlabeled examples labeled as negative while constraining the classifier to label all the positive examples correctly were presented in [12]. Text classification techniques can be used in a very broad area [13]. Expectation-Maximization technique can be used to find maximum parameter estimates locally. In addition, EM is an iterative statistical technique for maximum likelihood estimation in problems with incomplete data [14].

3   Best-Match Method Used in Co-training Algorithm

The co-training algorithm explicitly uses a feature split when learning from labeled and unlabeled data. Its approach is to incrementally build classifiers over each of the feature sets. Each classifier is initialized using just a few labeled documents supplied. At each iteration of co-training algorithm, each classifier chooses some unlabeled documents per class to put them into the labeled data set of examples. The documents selected are those which have the highest classification confidence given by the underlying classifier. Then, each classifier rebuilds from the augmented labeled data set, and the progress repeats.

Blum and Mitchell made a few assumptions about the data used by the co-training algorithm:

- The first assumption is that the instance distribution is compatible with the target function; that is, for most examples, the target functions over each feature set can predict the same label.
- The second assumption is that the features in one set of instances are conditionally independent of the features in the other set, given the classes of the instances.

This seems to be a somewhat plausible, as Blum and Mitchell said, since the page itself is constructed by a different user than the one who made the hyperlink [4]. Enlightened by Gale Shapely [15] algorithm, we propose the Best-Match method which selecting those instances first which have two parts of best matching features to add them into the labeled data set.

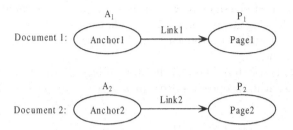

Fig. 1. Two document examples, the features of each one are consisted by two parts, the anchor part and the page part

More specifically, suppose the features of document 1 are consisted by A_1 and P_1, and the features of document 2 are consisted by A_2 and P_2 (see Fig. 1). Here A_1 and P_1 stand for an anchor text and a web page respectively. Of course, this anchor text's hyperlink points to the web page. In the same way, A_2 and P_2 stand for the similar meanings of document 2. For simplicity, we also use A_1 to stand for A_1 anchor text feature score marked by a anchor-based classifier. The same procedure is also applied to A_2, P_1 and P_2. In fact, A_1 is not a feature score, but an order given by an anchor-based classifier. Similarly, A_2, P_1 and P_2 stand for orders either. If formula (1) and (2) are met, then we prefer adding document 1 into the labeled data set to document 2.

$$A_1 + P_1 < A_2 + P_2. \tag{1}$$

$$\left| \frac{A_1}{A_1 + P_1} - \frac{1}{2} \right| < \left| \frac{A_2}{A_2 + P_2} - \frac{1}{2} \right|. \tag{2}$$

Obviously, we can derive from formula (2) and attain formula (3):

$$\frac{|A_1 - P_1|}{A_1 + P_1} < \frac{|A_2 - P_2|}{A_2 + P_2}. \tag{3}$$

According to formula (1) and (3), documents which meet the following two conditions will be added into the labeled data set first and foremost:

- First, it has a small total order number (anchor text order plus page order).
- Second, its two part orders are more likely equal.

Two probability values, one $Pr(v^T|f(v))$ from a text-based classifier and the other $Pr(v^A|f(v))$ from an anchor-based classifier may have widely different ranges of values. Generally, because the dimensionality of the textual sub-problem is higher than the anchor sub-problem, $Pr(v^T|f(v))$ tends to be lower in magnitude than $Pr(v^A|f(v))$. Here, $f(v)$ stands for the class assignment to node v. Just because of this, we figure out formula (1) and (3). We do not want these scores to be compared with or compounded directly. We simply use their orders given by a classifier. For example, suppose having five anchor texts, we use an anchor-based classifier to give orders for them. The final result maybe look like the following sequence: 2 1 5 3 4. That is to say, the third anchor text has the fifth order and it has the least similarity with the anchor-based classifier among the five ones.

Applying Best-Match method to the co-training algorithm, we finally get the Best-Match algorithm proposed in this paper (see Table 1).

Table 1. The outline of Best-Match algorithm. Experiments reported here were trained using a Naive Bayes algorithm, and algorithm parameters were set to $p = 1$, $n = 3$, $k = 30$ and $u = 75$. All these parameters are the same as the co-training algorithm proposed by Blum and Mitchell.

Inputs:

- an initial set L of labeled training examples
- an initial set U of unlabeled examples

Create a pool U' of examples by choosing u examples randomly from U

Loop for k iterations:

- Use L to train a classifier f_1 that considers only the Anchor portions of x
- Use L to train a classifier f_2 that considers only the Page portions of x
- Let f_1 label $2p$ positive and $2n$ negative examples from U'
  ◇ Select p positive examples from $2p$ and let them subject to formula (1) and (3)
  ◇ Select n negative examples from $2n$ and let them subject to formula (1) and (3)
- Let f_2 label $2p$ positive and $2n$ negative examples from U'
  ◇ Select p positive examples from $2p$ and let them subject to formula (1) and (3)
  ◇ Select n negative examples from $2n$ and let them subject to formula (1) and (3)

Add these labeled examples to L

Randomly choose $2p + 2n$ examples from U to replenish U'

Outputs:

- Two classifiers, f_1 and f_2, which can assign class labels for new documents

---

For each positive candidate document example selected by an anchor-based classifier, we put it into one of four categories according to the following conditions:

(1) A(positive conclusion made by an anchor-based classifier) + P(no valid page content).
(2) A(positive conclusion made by an anchor-based classifier) + P(positive conclusion made by a page-based classifier).

(3) A(positive conclusion made by an anchor-based classifier) + P(negative conclusion made by a page-based classifier).

(4) Other.

We give case 2 the highest priority, next case 1, case 3 and give case 4 the lowest priority. The similar procedure also applies to a page-based classifier. The reason why we select candidate document examples this way is that: the anchor-based classifier and the page-based one should cooperate with each other and reinforce each other to find a balance point. The candidate document example selecting policy applies to co-training algorithm outlined in Table 1. More specifically, when selecting p positive examples from $2p$ candidate positive ones or n negative examples from $2n$ candidate negative ones using an anchor-based or a page-based classifier, we apply these document example selecting strategies to the co-training algorithm.

The main difference between our Best-Match and the co-training algorithm is that: when selecting an example from an unlabeled data set to add it into the labeled one, decisions are made based on the anchor-based classifier as well as the page-based one. That is to say, examples are given the highest priority only when they are considered as positive or negative by both classifiers. In this paper, we call those examples as consistent candidate document examples. On the contrary, in co-training algorithm, documents selected are those with the highest classification confidence given by either the anchor-based classifier or the page-based one.

4   Experimental Results

Blum and Mitchell [4] present experimental results that compare the co-training algorithm with labeled and unlabeled data to the naive Bayes classifier with labeled data alone. Their experimental domain is the WebKB-Course dataset, a collection of 1051 web pages collected from computer science departments at four universities. The task is to identify those that are home pages of academic courses. Each example consists of the words that occur on the web page, as well as the words occurring in the anchor texts of hyperlinks pointing to that page.

Experiment One: We run naive Bayes, SMO [16], co-training and Best-Match on this dataset for ten paired trials of randomly selected train/test splits either. Three course documents and 9 non-course documents form the training set. 25% of the documents are held aside as a test set(230 documents or so), and the remaining documents consist of the unlabeled data set.

Table 2 shows classification error rates for the co-training and our Best-Match, in comparison with baselines provided by naive Bayes and SMO. Numbers shown here are the minimum error rates among the 30 iterations averaged over the ten random train/test splits. Both the co-training algorithm and our Best-Match lower classification errors considerably compared with a naive Bayes and SMO classifiers using the labeled data alone. Our method, Best-Match algorithm, has the lowest error rate among the four methods. From here, we think that unlabeled data can have some positive effects if they can be used properly in text

Table 2. Minimum classification error rates for the naive Bayes, SMO, the co-training and our Best-Match on the WebKB-Course dataset among the 30 iterations averaged over the ten random train/test splits

Algorithm	#labeled	#unlabeled	Error
Naive Bayes	12	-0-	11.8%
SMO	12	-0-	9.1%
Co-Training	12	776	8.0%
Best-Match	12	776	6.6%

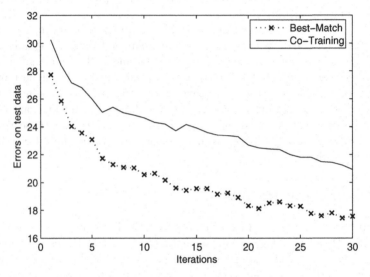

Fig. 2. Error versus number of iterations for our Best-Match and the co-training on the WebKB-Course dataset averaged over the ten random train/test splits

classification. Ordinarily, SMO is one of superior traditional text classification techniques, but here, it is not because of its not making full use of unlabeled data. Figure 2 gives a plot of error versus number of iterations averaged over the ten random train/test splits.

Experiment Two: In this experiment, we want to validate the effectiveness of formula (1) and (3) proposed in this paper and investigate in details about the learning behavior of the co-training algorithm and our Best-Match method. The experimental result is shown in Table 3. Note that, the numbers in Table 3 denote the count of consistent candidate document examples selected by the two classifiers. From Table 3, we can draw some conclusions:

– First, page-based classifier has more reliable performance when compared with anchor-based classifier. On average, page-based classifier can obtain 0.74

positive and 2.59 negative consistent candidate document examples, whereas, anchor-based classifier can obtain 0.68 positive and 2.5 negative ones only when using Co-training algorithm. If our Best-Match method is applied to Co-training algorithm, page-based classifier can obtain 0.76 positive and 2.7 negative consistent candidate document examples; anchor-based classifier can obtain 0.71 positive and 2.54 negative ones. Usually, A page has longer text than an anchor and this maybe the reason for the experimental results,
– Second, formula (1) and (3) do extract some different candidate document examples. On one hand, when using anchor-based classifier, our Best-Match method can obtain 0.03(0.71-0.68) more positive and 0.04(2.54-2.5) more negative documents. On the other hand, when using page-based classifier, our Best-Match method can obtain 0.02(0.76-0.74) more positive and 0.11(2.7-2.59) more negative documents.

Table 3. The count of consistent candidate document examples extracted by the co-training algorithm and our Best-Match method based on page and anchor text separately averaged over the ten random train/test splits

Methods	Co-training				Best-Match			
Classifiers	anchor-based		page-based		anchor-based		page-based	
Documents	Pos.	Neg.	Pos.	Neg.	Pos.	Neg.	Pos.	Neg.
1	0.8	2.7	0.8	2.7	0.8	2.8	0.8	2.8
2	0.5	2.2	0.6	2.4	0.6	2.3	0.6	2.4
3	0.6	2.4	0.6	2.5	0.6	2.4	0.6	2.5
4	0.7	2.7	0.8	2.8	0.7	2.7	0.8	2.8
5	0.5	2.3	0.6	2.5	0.5	2.4	0.6	2.5
6	0.6	2.5	0.6	2.5	0.7	2.5	0.7	2.5
7	0.7	2.2	0.8	2.3	0.7	2.3	0.9	2.5
8	0.7	2.3	0.8	2.4	0.7	2.3	0.8	2.4
9	0.8	2.8	0.9	2.9	0.9	2.8	0.9	2.9
10	0.9	2.9	0.9	2.9	0.9	2.9	0.9	2.9
Average	0.68	2.5	0.74	2.59	0.71	2.54	0.76	2.7

5   Conclusion and Future Works

In this paper, we propose Best-Match method used in co-training algorithm. In order to expand the labeled data set, we simply select more reliable documents which have two parts of the best matching features using our Best-Match method from the unlabeled data set to put them into the labeled one. Through experiments, we demonstrate that our Best-Match method has desirable performance when compared with the co-training algorithm, as well as SMO and naive Bayes which use labeled data alone.

In the near future, we want to test our Best-Match method on other data sets and more than one target functions, besides WebKB-Course dataset, to test its performance and robustness extensively.

Acknowledgement

This work is sponsored by the Natural Science Foundation of China under grant number 60373099.

References

1. Nigam, K., McCallum, A., Thrun, S., Mitchell, T.: Text classification from labeled and unlabeled documents using EM. Machine Learning 39(2/3), 103–134 (2000)
2. Jaakkola, T., Haussler, D.: Exploiting generative models in discriminative classifiers. In: Advances in NIPS 11 (1999)
3. Joachims, T.: Transductive inference for text classification using support vector machines. In: Proceedings of ICML 1999 (1999)
4. Blum, A., Mitchell, T.M.: Combining labeled and unlabeled data with co-training. In: COLT 1998, Madison, WI, USA,
5. Nigam, K., Ghani, R.: Analyzing the effectiveness and applicability of co-training. In: 9th International Conference on Information and Knowledge Management (CIKM), 2000. Computational Learning Theory, pp. 92–100 (1998), www.cs.cmu.edu/knigam
6. Lewis, D.D.: Naive (Bayes) at forty: The independence assumption in information retrieval. In: Nédellec, C., Rouveirol, C. (eds.) Machine Learning: ECML-98. LNCS, vol. 1398, Springer, Heidelberg (1998)
7. McCallum, A., Nigam, K.: A comparison of event models for naive Bayes text classification. In: AAAI-98 Workshop on Learning for Text Categorization, Tech. rep. WS-98-05, AAAI Press (1998), http://www.cs.cmu.edu/mccallum
8. Sebastiani, F.: Machine learning in automated text categorization. ACM Computing Surveys 34(1), 1–47 (2002)
9. Yang, Y., Liu, X.: A re-examination of Text Categorization Methods. In: SIGIR-99
10. Liu, B., Dai, Y., Li, X., Lee, W.S., Yu, P.S.: Building text classifiers using positive and unlabeled examples. In: Proc. of the 3rd IEEE Int'l Conf. on Data Mining. Melbourne (ICDM-03), pp. 179–188. IEEE Computer Society, Los Alamitos (2003)
11. Denis, F.: PAC learning from positive statistical queries. In: Proc. 9th International Conference on Algorithmic Learning Theory-ALT 987, pp. 112–126 (1998)
12. Liu, B., Lee, W.S., Yu, P., Li, X.: Partially supervised classification of text documents. In: ICML-02
13. Shih, L.K., Karger, D.R.: Using URLs and table layout for Web classification tasks. In: Feldman, S., Uretsky, M., Najork, M., Wills, C.E. (eds.) Proc. of the 13th Int'l Conf. on the World Wide Web (WWW-2004), pp. 193–202. ACM Press, New York (2004)

14. Dempster, A.P., Laird, N.M., Rubin, D.B.: Maximum likelihood from incomplete data via the EM algorithm. Journal of the Royal Statistical Society, Series B 39(1), 1–38 (1977)
15. Brualdi, R.A.: Introductory Combinatorics, 3rd edn., pp. 200–300. Prentice Hall Inc, Englewood Cliffs (1999)
16. Keerthi, S.S., Shevade, S.K., Bhattacharyya, C., Murthy, K.R.K.: Improvements to Platt's SMO Algorithm for SVM Classifier Design. Neural Computation 13(3), 637–649

A General Method of Mining Chinese Web Documents Based on GA&SA and Position-Factors

Xi Bai[1,2], Jigui Sun[1,2], Haiyan Che[1,2], and Jin Wang[3]

[1] College of Computer Science and Technology, Jilin University,
Changchun 130012, China
[2] Key Laboratory of Symbolic Computation and Knowledge Engineering of Ministry
of Education, Jilin University, Changchun 130012, China
[3] Institute of Network and Information Security, Shandong University,
Jinan 250100, China
xibai@email.jlu.edu.cn, {jgsun,chehy}@jlu.edu.cn, wangjin06@mail.sdu.edu.cn

Abstract. Clustering and classification are two important techniques of mining Web information. In this paper, a new adaptive method of mining Chinese documents from the internet is proposed. First, we give an algorithm of clustering documents which combines Genetic Algorithm(GA) and Simulated Annealing(SA) based on Boolean Model. This Algorithm avoids the disadvantage of clustering documents by using pure GA which can not be utilized accurately since GA converges too early and bogs the local optimum. Then, considering that the effect of classification with traditional Vector Space Model(VSM) is not satisfying enough since it is not related to the grades of importance of words, we add the position-factors of key words into VSM and set up a new classifier model to classify Chinese Web documents. Experimental results indicate that this adaptive method can make the process of clustering and classification more accurate and reasonable comparing to the methods which does not have the positions of words considered.

1 Introduction

With the development of the Web technology, the internet has become one of the most important tools for people to gather information. Information on the Web is updated almost in every second and increases geometrically. In order to get useful and valuable information from those mountainous data, the research on searching engines, document mining, information filter and retrieval has reached an unprecedented tide. The present widely used method makes use of the document classification models and can correctly identify the category of the documents inputted by users.

Boolean Model[1], Vector Space Model(VSM)[2], and Naive Bayes Model[3] are among the most widely used ones. Boolean Model is the simplest in structure but strict in its expression. VSM is a widely used document classification model which takes the appearance frequencies of words in documents as the element of

T. Washio et al. (Eds.): PAKDD 2007 Workshops, LNAI 4819, pp. 410–420, 2007.
© Springer-Verlag Berlin Heidelberg 2007

the feature vector. However,if the document is short in length or its key words are replaced by pronouns, the frequencies of the key words which can reflect the category of the document will become low. In recent years many clustering algorithms have been proposed, but few of them are dedicated to the clustering of documents in Chinese[4,5]. A new method is using Genetic Algorithm(GA), since it is capable of searching globally and can find the feature vectors of documents[6]. However, it might converge too early and bog the local optimum and the clustering result is unsatisfying. On the contrary, Simulated Annealing(SA) which is capable of searching locally has also been used to documents mining and it has limitation too[7].

In this paper, we propose a new adaptive method of mining Chinese Web documents. In the following section we are going to design a new method which is a combination of GA and SA based on Boolean Model first of all. By this way, we can identify the correct feature vector efficiently which can best represent the feature of the documents belonging to the same category. Secondly we are going to set a new classifier model which is an expert of classifying documents in Chinese. It adds the position-factor of the key words into the feature vector based on VSM and also optimizes the category's features vector dynamically. To some extent, it maintains the actual weighted value of the key words and helps us choose the threshold value of similarities. Then, we present experimental results which show that the effect of this new adaptive method is more effective and accurate comparing to the clustering methods which use pure GA only or the classification methods which do not have the positions of words considered. VSM, Boolean Model, GA and SA which were extensively studied in [1,2,8,9], and will not be described here for brevity.

2   Document-Clustering Algorithm Based on GA&SA

GA was proposed by John Holland from the University of Michigan at the beginning of the 60s and SA was first tested by Kirkpatrick and Gelatt on the travelling salesman problem. They are both adaptive method of serching for the optimum solution. In this section, we combine GA and SA and propose a new clustering method.

2.1   Design of Fitness Function

After fixing the individuals' number of the group, which is denoted by n, we take the n random benches of 0 and 1 as the initial individuals, denoted by $chrom_i$, which indicates the code of the ith chromosome. Since Boolean model derives from VSM, we still design the fitness function in the way of getting similarity according to the VSM's calculation of the similarity. Obviously, the best individual should represent the feature in a training set of documents which belong to the same category. Therefore, the average value of similarities between the best individual and the feature vectors of these documents should be the highest. Suppose the number of the documents in the training set is denoted by N and their feature vectors are denoted by $Tchrom_1, Tchrom_2, ..., Tchrom_N$ respectively. And then we define the fitness function f_i of the ith individual as following:

$$f_i = \frac{\sum\limits_{k=1}^{N} sim(chrom_i, Tchrom_k)}{N} \qquad (1)$$

Here

$$sim(C, P) = \cos(C, P) = \frac{\sum\limits_{i=1}^{n}(F_{W_i} \times f_{W_i})}{\sqrt{\sum\limits_{i=1}^{n}(F_{W_i})^2} \times \sqrt{\sum\limits_{i=1}^{n}(f_{W_i})^2}} \qquad (2)$$

F_{W_i} denotes the value(0 or 1) of the word W_i in feature vector C. f_{W_i} denotes the value of the word W_i in feature vector P. n denotes the dimension quantities of the feature vector. $W_1, W_2, W_3, ..., W_n$ denotes the words which appear in the feature vector.

2.2   Genetic Operation

−Selection

Here we use the Monte Carlo Selection[8,9]. The selecting probability P_i of the ith individual in each selection forms a direct ratio with its relative fitness, so we have:

$$P_i = \frac{f_i}{\sum\limits_{t=1}^{N} f_t} \qquad (3)$$

The higher the fitness, the more chances of the individual being chosen, which best reflects the principal *survival of the fittest*. The pseudo code of selection is described as follows:

Algorithm 1. Selection

```
get n individuals' fitness in the InitialGroup ;
for(int i = 0; i < n; i++){
  get the accumulative total S_i of the fitness ;
  create random integer R_i between the interval [0, S_i] ;}
for(int j = 0; j < n; j++)
  for(int k = 0; k < n; k++){
    find out the first S_k which is bigger than R_j ;
    copy the corresponding individual chrom_k into the SelectedGroup ;}
return SelectedGroup ;
```

−Crossover

Crossover is a vital part of the GA. There are two steps in crossover: first, match the selected chromosome in the current group and then deciding whether to crossover them or not according to crossover probability P_c; second, choose the cross point randomly and then crossover. A chromosome with the length of l has $l - 1$ crossover-points. Here we use the *single-point crossover*[8,9]. The pseudo code of crossover described as follows:

Algorithm 2. Crossover

```
for(int i = 0; i < n; i++) {
    get n random integers Rᵢ between the interval [0, n − 1] ;}
for(int k = 0; k < n; k++) {
    newchromₖ = chromₖ ;}
int j = 0 ;
While (j < [n/2]){
    get a random integer r from the interval [l, l − 1] as a cross point ;
    get a random number P from the interval [0, 1] ;
    if(P > Pc) continue ;
    else {change the latter part of the cross point r between newchrom₂ⱼ and
    newchrom₂ⱼ₊₁ ;}
    put newchrom₂ⱼ and newchrom₂ⱼ₊₁ into XoveredGroup ;}
return XoveredGroup ;
```

–Mutation

Here we choose the *basic mutation* to change the values of chromosome in 3 places randomly according to the mutation probability P_m[8,9]. So we get the pseudo code of mutation as follows:

Algorithm 3. Mutation

```
get 20 different random integers R₀, R₁, ..., R₁₉ between the interval[0, n − 1];
get 3 different random integers P₀, P₁, P₂ from the interval [0, n − 1] ;
get a random number r from the interval [0, 1) ;
if(Pm > r){MutatedGroup = XoveredGroup ;}
else {
    for(int i = 0; i < 3; j++)
        for(int j = 0; j < 3; j++) {
            change the value in Rⱼ of chromⱼ ;
            put the mutated chrom into the MutatedGroup ; } }
return MutatedGroup ;
```

2.3   Process of Simulated Annealing

As regard to the disadvantage of GA, that is, it bogs in the local optimum before a complete and mature convergence, we add the periodical annealing process into it. Suppose the best solution as a whole is denoted by *BestIndividual*, when the genetic process reaches a preseted upper limit of the times of the generation, then the algorithm will regard the *BestIndividual* in the present group as the initial value and begin Simulated Annealing. Then we compare the *TempIndividual* in a balanceable state after annealing with the *BestIndividual* and execute the *seeding*

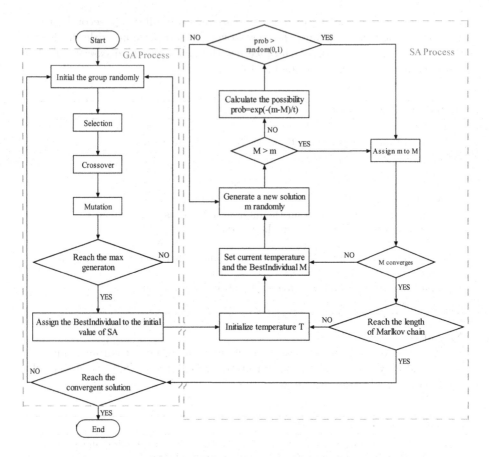

Fig. 1. The flowchart of GA&SA

operation: if the *TempIndividual* is superior, then we displace S individuals in the present group with S copies of the *BestIndividual*; Otherwise, we use S copies of the *TempIndividual* and assign the value of the *TempIndividual* to the *BestIndividual*. Here the chosen of the displaced individuals is random. After that, the updated groups will be genetically operated and then annealed in the next round until the *BestIndividual* converges. We can see the while process of GA&SA is in Figure 1.

3   A New Weighted Vector Space Model for Classifying Documents

The traditional VSM simply takes the words' appearing frequencies as the feature vector which is improved by TFIDF nowadays[10]. But using this model to classify documents is not satisfying enough. Two documents belonging to the same category may share low similarity. This make the selection of threshold value difficult. This is mainly because that the frequencies of the key words which can reflect the category of the text are too low. Maybe the text is short in length so that the total number of the words is too small; or maybe the key words appear in the title or the beginning of a document and then don't appear any more,since they are replaced by pronouns.

To avoid the condition that the appearance time of the keywords is too small, we can add weighted variable to keywords. The words appearing in a text have different important degrees. For instance, the titles always summarize the content of an article and do their best to attract readers' attention. Titles can direct the readers to do their reading. When a reader get an article, what does she or he look at first? It must be the title. People won't be attracted by the articles whose title is obscure. Besides the title, the first sentence of an article is also very important. It is the window of the article and can help reader to get the meaning of the following content. we can make use of structured information in the Web including the important marks in HTML files such as: title, subtitle, over-striking, underline, etc to extract keyword-sets. For example, we can extract the words from documents with the structure such as $\langle title \rangle$ and $\langle /title \rangle$, and add the weighted values to these words which appear in the feature vector of a certain document. We add the title-factor T_{W_i} and the words in the first-sentence-factor B_{W_i}:

$$T_{W_i} = \begin{cases} \alpha, & \text{word } W_i \text{ appears in the title,} \\ 1, & \text{otherwise.} \end{cases} \tag{4}$$

$$B_{W_i} = \begin{cases} \beta, & \text{word } W_i \text{ appears in the beginning sentance,} \\ 1, & \text{otherwise.} \end{cases} \tag{5}$$

α and β are parameters greater than 1. Their magnitude can be adjusted by the effect of classifying. If we add the position-factor, then the feature vector of the document $P = (tf_{W_1}, tf_{W_2}, tf_{W_3}, ..., tf_{W_n})$, which is to be classified, will be changed into $P' = (tf'_{W_1}, tf'_{W_2}, tf'_{W_3}, ..., tf'_{W_n})$, where $tf'_{W_i} = T_{W_i} \times B_{W_i} \times tf_{W_i}$. Here tf_{W_i} and tf'_{W_i} both denote the appearing frequencies of word W_i. Then the similarity between this document and the compared category whose feature vector is denoted by C is as following:

$$sim(C, P') = \cos(C, P') = \frac{\sum\limits_{i=1}^{n}(TF_{W_i} \times tf'_{W_i})}{\sqrt{\sum\limits_{i=1}^{n}(TF_{W_i})^2} \times \sqrt{\sum\limits_{i=1}^{n}(tf'_{W_i})^2}} \tag{6}$$

TF_{W_i} denotes the appearing frequency of word W_i in C. This improvement helps the key words regain their true weighted value in documents, and the difference of the similarity between the same category of documents and different categories of documents will be obvious. Therefore, it is easy for us to choose the threshold value and classify the documents.

4   A Clustering Method Based on the Dynamic Update of Words' Frequencies

Comparing to the feature vector formed by 0 and 1, the feature vector based on the words' frequencies has more information. However, if we don't update the feature vector, the words' frequencies will not change. Sometimes the number of the collected texts in a particular category is limited, and with the adding of new texts, the feature vector of the category may not accurately represent the category of texts any more.

After deciding the category of a particular text, we must update the feature vector of the category timely. This can endow the classifier with the function of learning and in its applying process the feature vector can be fine adjusted so that it can better represent the category's feature again and lay the foundation for classifying other texts in the future. As regard to the close relationship between the feature vectors before and after update and the feature vectors of new added texts, we can update those of the category by means of getting the average frequencies of the words in the classifier.

If we only get the average value of the old one's feature vector and the new added one's, the updated feature vector would be unstable due to the new added texts' short length or some other reasons. To avoid such condition, we can add an accommodating coefficient to the old feature vector so as to strengthen its influence. Thus, we take the number of the texts which has been collected in a certain category as the accommodating coefficient. As the number of classified texts increases, the updating will be more effective and the feature vectors will be more reasonable. Suppose we have already know that a text With P' as its feature vector in section 3.1 exceeds the preset threshold value(When two documents' similarity is equal to or more than the threshold value, we deem that they are belong to the same category). Supposed that the number of the collected texts is k, then the feature vector of the compared category after update is C'. We have a formula to calculate C' as following:

$$C' = avg(P', C, k) = (\frac{tf'_{W_1} + TF_{W_1} \times k}{k+1}, \frac{tf'_{W_2} + TF_{W_2} \times k}{k+1}, ..., \frac{tf'_{W_n} + TF_{W_n} \times k}{k+1})$$

(7)

5   Experiments and Results

We test the performance of the method of using GA&SA and compare it with the method of using traditional GA only. As expected, GA&SA got a better feature vector and cost less time. Then we test our new classifier which uses the weighted vector space model. The results indicate that this classifier can make the classifying result more accurate than some other ones. If the position-factor is 5, and the threshold value is 0.65, then the max difference of the similarities between the same and different category will below 0.58. The above experiments were carried out on a PC with a Celeron 1.2 GHz cpu and 256M of RAM.

Study 1:

We take 70 documents in Chinese from the Web as the tested samples of which 45 are about securities and 25 of other subjects. We plan to choose 10 documents from the 45 and get the feature vector. After the word classifying, we compress the word aggregates reasonably and keep the 35 meaningful verbs and nouns which are translated into English as following: *rise, national, bond, debt, market, bargain, index, bankroll, bank, fall, dollar, increase, interest, dividend, trafficking, risk, stock, fund, exchange, economy, quotation, finance, business, income, policy, ratio, quote, invest, ledger, bourse, currency, price, investor, delivery, merger.*

In the test, we set the parameters of Simulated Annealing & Genetic Algorithm as: the max generation is 80, crossover probability P_c is 80%, mutation probability P_m is 1%, the number of mutated bits is 10, the number of mutated individual is 20, maximum anneal times is 20, initial temperature T is 100 degree centigrade, temperature decay coefficient is 0.95, allowance is 10^{-3} , and *seeding* number is 3. To fully realize SA's advantage of jumping out of the local optimum, we preset that the minimum temperature after annealing should be at least 0.05, and absolute value of the best-individual's fitness after annealing minus that of the best individual before annealing should be smaller than the allowance. We set the threshold value of the similarity 0.65, if the calculated similarity reaches or exceeds the threshold value, then the documents will be defined as the same category. After operation, we get the best feature vector as: (1,1,1,1,1,1,1,1,1,1,1,0,1,1,1,1,1,0,0,0,1,1,0,1,1,1,0,1,0,0,0,0,0,0,1), and the best similarity is 0.768. We get the best solution after 8 times annealing which costs 84.882 seconds.

Table 1. GA&SA versus GA

	GA	GA&SA
Best Similarity	0.669	0.768
Search Time	67.382s	84.882s
Recall	78%	92%
Precision	81%	100%
percentage of exclusion	76%	100%

From Table 1, we can see that in the process of getting the similarity between the best feature vector and the other 35 documents of the same category, 32 of them exceed 0.65(the threshold value of the similarity)which account for 92% in total, see in Table 2. The similarity between the best feature vector and the 25 documents of different categories are all smaller than 0.65, described in Table 3. The accuracy of discarding documents of different categories reaches 100%. It is obvious that the classifying effect is satisfying. We also use the GA to generate the best feature. We find that the solution bogs the local optimum early, and it can not get out of this local optimum any more. The final best feature vector is (1,0,1,1,1,1,1,0,0,0,1,0,1,1,0,1,1,1,0,0,1,1,0,1,1,1,1,1,0,0,0,1,0,0,1),

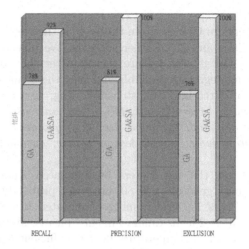

Fig. 2. Classification effect of using GA and using GA&SA

Table 2. The similarities between *BestIndiviual* and 35 documents of the same category

0.783	0.677	0.685	0.745	0.702	0.674	0.730
0.609	0.827	0.730	0.783	0.713	0.776	0.714
0.722	0.796	0.757	0.656	0.724	0.639	0.792
0.809	0.740	0.625	0.713	0.625	0.714	0.776
0.685	0.612	0.722	0.702	0.644	0.802	0.730

Table 3. The similarities between *BestIndiviual* and 25 documents of different categories

0.386	0.250	0.354	0.408	0.309
0.408	0.118	0.309	0.546	0.433
0.408	0.354	0.399	0.379	0.0
0.0	0.144	0.0	0.0	0.0
0.0	0.0	0.0	0.156	0.204

and the best similarity is 0.669. Table 1 and Figure 2 indicate that using GA&SA to mining documents is better than using traditional GA.

Study 2:
We collect 80 sets of Chinese documents. Each set contains four documents, three Chinese documents belonging to the same sort and one belonging to another sort, totally 320 documents. Then we calculate the similarities of documents in each set. For instance, in a set, there are three documents about space navigation and one document of another sort. They will serve as the sample. Firstly, we do the word segment and get the feature vector of each document by means of traditional VSM(vector without weighted value) and the number of the words chosen as feature vector is 25. Then, we add the position-factors to the words and do tests in three cases which are $\alpha = 1$, $\alpha = 3$ and $\alpha = 5$ respectively.

We preset the threshold value of the similarity as 0.8 and calculate the similarity between each two documents in above three case. In the second and the third test, document 1 and document 3 are of the same category. Their feature vectors changed into a common feature vector after the process in equation (7).

Table 4. The similarities between every two documents in a test set

Fitness function	Result of the calculation of the similarities		
	un-weighted	weighted($\alpha = 3$)	weighted($\alpha = 5$)
$sim(A_1, A_2)$	0.622	0.668	0.695
$sim(A_1, A_3)$	0.661	0.822	0.869
$sim(A_2, A_3)$	0.726	0.738	0.795
$sim(A_1, A_4)$	0.400	0.318	0.298
$sim(A_2, A_4)$	0.565	0.535	0.468
$sim(A_3, A_4)$	0.537	0.318	0.298

During the processing of our program, we can get Table 4 where A_i denotes the feature vector of the ith document. The data in this table and the similarities in other document sets show that the similarities which are calculated with traditional VSM between the same and different category of documents are close to each other. Therefore, it is hard to choose a threshold value. After we add the weighted value, the similarities between the same documents are magnified and those between different categories decrease. Meanwhile, the dynamically updated feature vector can reflect the category feature as the increasing of the collected documents. However, if α is too big, the contents of the documents can less influence the similarities so that the titles can define the categories of the documents, which is not what we expected. Tests show that when $\alpha = 5$, we set the threshold value as $\mu = 0.65$, then we can get a better result.

6   Conclusion

With the rapid development of the web searching technology, clustering method now seriously related to the classification of documents. However, the efficiency and quality of searching are not satisfying enough due to huge searching spaces and availabilities of information. Therefore, In this paper we developed a new clustering and classification method. We give a new adaptive method including the clustering method based on GA&SA and the classification method which weights the feature vector according to the positions of the key words in a document. The experimental results indicate that it is a hopeful method of mining Chinese Web documents.

This is a temporary study of the weighting method of documents' feature vector. In our future work, we will add suitable weighted values to other words which appear in the structure information of HTML pages(sub-title, over-striking and underline etc), and apply the position-factors on the mining model using TFIDF.

Acknowledgements

This work is supported by the National Science Foundation of China (60496321, 60473003), Ministry of Education Program for New Century Excellent Talents in University(NECT) and Doctor Point Founds of Educational Department (20050183065).

References

1. Melucci, M.: Context modeling and discovery using vector space bases. In: Proceedings of the 14th ACM international conference on Information and knowledge management, October 31-November 05 (2005)
2. Goncalves, A., Jianhan, Z., Dawei, S., Uren, V., Pacheco, R.: LRD: Latent relation discovery for vector space expansion and information retrieval. In: Yu, J.X., Kitsuregawa, M., Leong, H.V. (eds.) WAIM 2006. LNCS, vol. 4016, pp. 122–133. Springer, Heidelberg (2006)
3. Schneider, K.M.: On word frequency information and negative evidence in Naive Bayes text classification. In: Vicedo, J.L., Martínez-Barco, P., Muñoz, R., Saiz Noeda, M. (eds.) EsTAL 2004. LNCS (LNAI), vol. 3230, pp. 474–485. Springer, Heidelberg (2004)
4. Tang, C., Lau, R.W.H., Li, Q., Li, T., Yu, Z.: Distance courseware discrimination based on representative sentence assaying. In: Proceedings of Seven-th International Conference of Advanced Database Applications, pp. 92–99. IEEE Publishing, Hong Kong (2001)
5. Li, T., Tang, C.J., Zuo, J.: Web document filtering technique based on natural language understanding. International Journal Computer Processing of Oriental Language 14(3), 279–291 (2001)
6. Riyaz, S., Selwyn, P.: Efficient genetic slgorithm based data mining using feature selection with hausdorff distance. Information Technology and Management 6(4), 315–331 (2005)
7. Metropolis, N., Rosenbluth, A.W., Rosenbluth, M.N., Teller, A.H., Teller, E.: Equation of state calculations by fast computing machines. Journal of Chemical Physics, 1087–1092 (1953)
8. Casillas, A., de Lena, M.T.G., Martínez, R.: Document Clustering into an Unknown Number of Clusters Using a Genetic Algorithm. In: Matoušek, V., Mautner, P. (eds.) TSD 2003. LNCS (LNAI), vol. 2807, pp. 43–49. Springer, Heidelberg (2003)
9. Xu, X.S., Ma, J., Wang, H.: An improved simulated annealing algorithm for the maximum independent set problem. In: Huang, D.-S., Li, K., Irwin, G.W. (eds.) ICIC 2006. LNCS, vol. 4113, pp. 822–831. Springer, Heidelberg (2006)
10. Kang, Y.H.: Representative term based feature selection method for svm based document classification. In: Khosla, R., Howlett, R.J., Jain, L.C. (eds.) KES 2005. LNCS (LNAI), vol. 3681, pp. 56–61. Springer, Heidelberg (2005)

Data Management Services in ChinaGrid for Data Mining Applications*

Song Wu, Wei Wang, Muzhou Xiong, and Hai Jin

Services Computing Technology and System Lab
Cluster and Grid Computing Lab
School of Computer Science and Technology
Huazhong University of Science and Technology, Wuhan, 430074, China
wusong@hust.edu.cn

Abstract. Grid systems, as large-scale distributed computing environments, are widely used by data mining communities. This paper proposes a set of system-level Grid services to form an infrastructure supporting data-intensive applications and data mining. ChinaGrid, aiming at integrate heterogeneous massive resources distributed on China Education and Research Network (CERNET), is a national-wide Grid project supported by the Chinese government. ChinaGrid Supporting Platform (CGSP) is a Grid middleware developed for the ChinaGrid. It provides a series of system-level services of the ChinaGrid, helps to build application portals and integrate Grid resources, and supports the secondary development of Grid services. The Data Management Services (DMS) is a group of Grid services in CGSP to manage storage and data resources, support transparent data access, and guarantee high-performance data transfer on the Grid. It consists of metadata management service, storage resource management service, replication management service, storage agent and transfer client. It offers the fundamental support for data mining applications on ChinaGrid. In this paper, we introduce the design principle and implementation of DMS.

1 Introduction

The term "Grid" was coined in the mid-1990s to denote a proposed distributed computing infrastructure for advanced science and engineering [1]. Researchers have since made much progress in constructing such an infrastructure and extending and applying it to a broad range of computing problem. As a result, "Grid" has entered the computer science vocabulary to denote middleware infrastructure, tools, and applications concerned with integrating geographically distributed computational resources.

Researchers first developed Grid concepts and technologies to enable resource sharing within scientific collaborations – first in early gigabit test beds, and then on increasingly larger scales. At the root of these collaborations was the need for participants to share not only data sets but also software, computational resources. More

---

* Supported by the National Science Foundation of China under Grant Nos. 60673174, 90412010 and the National High-Tech Research and Development Plan of China under Grant No. 2006AA01A115.

T. Washio et al. (Eds.): PAKDD 2007 Workshops, LNAI 4819, pp. 421–432, 2007.

generally, scientists needed technologies to support coordinated resource sharing and problem solving in dynamic, multi-institutional collaborations.

ChinaGrid, a Grid project funded by the Chinese government, forms a national-wide Grid infrastructure consisting of the top 20 Chinese universities [2][3]. Through developing corresponding grid middleware and various Grid applications, ChinaGrid can integrate heterogeneous massive resources on China Education and Research Network (CERNET) effectively. It avoids resource islands on CERNET and provides useful Grid applications.

ChinaGrid Supporting Platform (CGSP) is a Grid middleware developed for ChinaGrid [4]. It provides a series of system-level services of ChinaGrid, helps to build application portals and integrate Grid resources, and supports the secondary development of Grid services. Data Management Services (DMS) plays an important role in CGSP. It provides the basic data support for the common users of the CGSP and the requirements of the Grid applications running on the CGSP. By means of DMS, the users can have a virtualized data space, use it for storing files and run data mining applications on it.

In this paper, we present the prototype of DMS in CGSP and discuss the design principle and the details of implementation of DMS. The rest of the paper is organized as follows: Section 2 describes related work; in Section 3, we present the design of DMS; we give performance results in Section 4; finally, we conclude and discuss future work in Section 5.

2 Related Work

There are several similar systems to realize the data management in a large-scale distributed computing environment or Grid. We list and compare the typical systems as follows.

- Globus [5]: Globus Toolkit is an open source software toolkit used for building Grid systems and applications. It is being developed by the Globus Alliance and many others all over the world. It uses GridFTP, OGSA-DAI, and Metadata Catalog Service to provide uniform Grid interfaces to various types of data in the Grid space.
- SRB [6]: Storage Resource Broker (SRB) is a Data Grid Management System (DGMS) or simply a logical distributed file system based on a client-server architecture which presents the user with a single and global logical namespace or file hierarchy.
- OceanStore [7]: OceanStore is a global persistent data store designed to scale to billions of users. It provides a consistent, highly-available, and durable storage utility atop an infrastructure comprised of un-trusted servers.

Comparing with Globus and OceanStore, our DMS can integrate heterogeneous storage resources on the Grid and provide a virtual storage system to users. Comparing with SRB, DMS realize the data management functions in the form of Grid service according to Web-Service Resource Framework (WSRF)[8]. The WSRF is a set of six Web services specifications that is used to modeling and managing state in a Web services context. We don't have a convenient method that manages the user's state in a common Web service until the come of WSRF. Every client that has logged in the DMS is treated as a resource in the DMS. We can do many things such as querying its state, managing its lifetime, etc. The WSRF preserves the advantages of Web service and can do many complicate operations that a Grid System needs. Every operation in DMS is realized as a WSRF service, which makes DMS more suitable for the Grid environment.

3 Data Management Services

3.1 Architecture

CGSP is organized into many domains. The concept of domain in CGSP is an autonomy organization. Every domain can interact with each other with the help of the Information Center of CGSP. In one domain, DMS traditionally consists of a single master and multiple storage nodes and is accessed by multiple clients, as shown in Figure 1. In the current preliminary implementation, each storage node is typically a commodity Linux machine running a user-lever server process. We can also replace it with a high capability storage device with a IP. This type of storage device is called "smart bricks" [9].

Although there are many distributed architectures or even P2P-style architectures, the central architecture has its advantages. It can simplify the design of DMS and make flexible strategy through the efficient usage of global knowledge. Meanwhile, DMS has to minimize the master's involvement into read and write operations so that it can avoid a system bottleneck. In fact, clients never read and write file data through the master. A client first query service running on the master to get the information of storage nodes which have the destination files and then contacts and interacts with the corresponding node directly. The traffic of the communication between clients with the service is very small.

Let us explain the interactions for a simple read with reference to Figure 1. First, the client sends the User Space service running on the master a request containing the file name. The service replies with the corresponding file's handle, stripes, and replica locations. The client can download files from one replica or many replicas. The file can be striped in many stripes and in our preliminary implementation, each file only have one stripe (It can be extended in the later versions to fasten the speed of the transfer). The client uses the GridFTP[10] protocol to communicate with the GridFTP Server running on the storage node to transfer the file.

There are three core services running on the single master: User Space Service (USS), Storage Resource Service (SRS) and Replication Management Service (RMS). But these services will not need to running on a single master and can be distributed on different machines to support more flexibility. Each service is running as a WSRF service in the container of CGSP.

The User Space Service (USS) maintains all file system metadata. This includes the namespace, the mapping from logical files to physical files, users' information, etc. The Storage Resource Service communicates with each storage node in HearBeat messages to give it instructions and collect its state. The Replica Management Service controls a third-party file transfer between two different storage nodes to complete the functions of replication.

Files are distributed on the different storage nodes and are identified by a 128-bit length UUID look like this: 550e8400-e29b-41d4-a716-446655440000. The UUID is the abbreviation of the Universally Unique Identifier. It is widely used in distributed systems to uniquely identify information without significant central coordination. With the help of UUID, we can ensure that there are no same name files in a storage node.

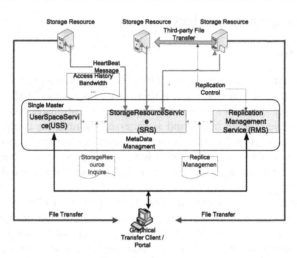

Fig. 1. DMS Architecture

3.2 User Space Service

The User Space Service running on the master stores three types of metadata: the users' file system information, the mapping from the logical files to physical files, and the locations of each file's replicas. All types of metadata are kept persistently in the database. The database stores the file's metadata in a flatten structure but we should present the users in a tree-style structure on the portal. We use a XML file to act as the broker between these two styles. In the database, every file's record contains two important fields: nodeID and parentID. We use these two fields to organize the user's information into a XML tree-style to present on the portal. This XML file also plays the role of cache to speed up the operation in the user's file namespaces. For example, if we want to delete a directory in the CGSP just using the database. As we just store every file's id and its parentID in the database, it will need many times to access database to find out which files are belong to this directory. But it is very simple in the XML file just using a simple XPath expression. We first deal with XML file for the intermediate complicate operations and write the result to the database permanently.

The USS is an important service in DMS. It virtualizes the metadata of all geographically distributed storage resources to a unique view. With the help of the User Space Service, it is easy to achieve the transparency about the naming of the file, location of the file, and the transfer protocol etc. Every user in CGSP has its own data space, and can upload, search, download files in this space.

3.3 Storage Resource Service

The state of the storage resources is an important basis to make out the strategies about data placement, replica creation and so on. The Storage Resource Service collects and maintains the state of each storage node. There is a property file in the storage node that specifies the address of the Storage Resource Service. There is also a GridFTP server running on the storage node.

There is a user-level process called Storage Agent running on each storage node. This process collects information such as the disk storage, frequency of the CPU, bandwidth of the network card, and reports them to the Storage Resource Service by a HeartBeat message periodically. If the Storage Resource Service doesn't get the reply message of a storage node within a threshold time period, it will set the state of this node to failure. The interaction between Storage Agent and the Storage Resource Service is a light weight protocol. The HeartBeat messages will not bring the heavy traffic to the network. The storage resource's state stored in the Storage Resource Service will be used for determining the physical locations when the users upload files to the CGSP.

After a file is deleted, DMS does not immediately reclaim the available physical storage until regular garbage collection at the physical level. This makes the system much simpler and more reliable.

When a file is deleted by the client, the logical files are just moved to another table in the database. After the come of next HeartBeat message of the storage node, the Storage Resource Service sends the storage agent the information about the deleted file's information located on that storage node. The physical file is renamed to a file with the deletion timestamp. The Storage Agent also removes any such files if they have existed for more than 24 hours (the interval is configurable by a property file located on the storage node).

The delete task will not be performed by the master but the storage agent, because the basic transfer protocol of the DMS is based on GridFTP. If the master wants to delete a file located on the storage node directly and immediately, it will connect to the storage node using GridFTP protocol to complete this task. The cost of this task is very large since FTP is not a light weight protocol, especially the client wants to delete a complicate directory.

This strategy offers several advantages over eager deletion. First, it is simple and reliable in a large-scale distributed system.. The GridFTP connections that are used to delete files may be lost due to unreliable network, and the Storage Resource Service has to remember to reconnect to those storage nodes. It is not robustness and may bring large overhead to the whole network. Let the files be deleted by the storage agent can overcome these problems.

The main disadvantage of this strategy is that the delay sometimes hinders user effort to fine tune usage when storage is tight. Applications that repeatedly create and delete temporary files may not be able to reuse the storage right away. We can address these issues by shorten the interval time to delete physical files.

3.4 Replication Management Service

Data replication is one of the best known strategies to achieve high levels of availability and fault tolerance, as well as minimal access times for large, distributed user communities. Replication Management Service (RMS) manages the replicas located in the CGSP.

The replica's creation in DMS is done by the third-party file transfer. We can transfer two files directly between two different GridFTP servers located on two storage nodes.

CGSP is highly distributed at more levels than one. So, DMS typically has hundreds of storage nodes spread geographically. These storage nodes in turn may be accessed from hundreds and thousands of clients. Using the replication, we can maintain the high availability and the efficiency of the system. By default, we have three replicas per file. Users can also designate the replica degree for each file if they have enough privileges.

In our current implementation, the files are distributed randomly over the underlying storage nodes.

File replicas may become dirty if the storage node where they reside has failed and missed operations to them while it is down. For each file, the master maintains a file version number to distinguish between up-to-date and dirty replicas.

Whenever the master grants a new operation on a file, it increases the file's version number and informs the up-to-date replicas. If some replica is currently unavailable, its version number will not be advanced. The RMS will detect that this storage node has a dirty replica when the storage node restarts and reports its set of files and their associated version numbers

The RMS stores the replica's information in the database permanently. The RMS has the global view of all replicas in the system, and as the SRS has the global view of all storage resources in the system. We can develop many optimize strategy with the help of these two services. For example, we can choose the most suitable storage node to store the replica.

3.5 Transfer Client

Users need an easy and friendly method to access the data in DMS. Except for the standard web portal, we offer a dedicated transfer client to the users. The client consists of two parts: a graphical client for the common users and a jar-style module for the other parts of CGSP such as GRS[11] and GridPPI[12].

Fig. 2. The Graphical Client for CGSP

As the limitations of HTTP, users can't upload or download a directory through the web portal. We developed a graphical client, as shown in Figure 2, to support the directory's transfer, and also support other features including the breakpoint resume transportation, graphical statistic, detail logging, auto start, and task group.

The graphical client is based on SWT[13], a great graphical library used in Eclipse. It is an open source widget toolkit for Java designed to provide efficient, portable access to the user-interface facilities of the operating systems on which it is implemented. We choose it for developing the client as it runs more efficiently than Swing.

The graphical client can support the breakpoint resume transportation. It uses a file to log the current position of the transfer and resume it at the next transmission. It also support the group of tasks, users can drag the task into a specifically group to manage it.

When the client wants to access the data in DMS, it first contacts the User Space Service running on the master and gets the mapping information from the logical file names to the physical file names. The information that the User Space Service returns is packaged as a XML-based file named "GridTorrent", as shown in Figure 3. The "GridTorrent" file borrows the concept of the "torrent file"[14] used in P2P network.

```
<?xml version="1.0" encoding="UTF-8"?>

<TORRENT>
    <TORRENTFLAG>true</TORRENTFLAG>
    <TORRENTNAME>Azureus_2.5.0.0_source.zip</TORRENTNAME>
    <INFOCENTER_URL>http://211.69.198.208:9090/wsrf/services/UserSpaceService</INFOCENTER_URL>
    <CREATION_DATE>1168158348305</CREATION_DATE>
    <COMMENT>CGSP comment</COMMENT>
    <CREATED_BY>admin</CREATED_BY>
    <LOCALFILENAME>2</LOCALFILENAME>
    <LOCALDIR>D:\CGSP\tmp\</LOCALDIR>
    <REMOTEDIR>/</REMOTEDIR>
    <REMOTEFILENAME>test.wmv</REMOTEFILENAME>
    <SLICEINFO FILENUM="1" LENGTH="5856328" TYPE="file">
      <FILE LENGTH="5856328" NAME="Azureus_2.5.0.0_source.zip" SLICENUMBER="1" SN="0">
        <SLICE BEGINPOS="0" CURPOS="1843200" ENDPOS="5856327" LENGTH="5856328" SN="0">
          <SLICEURL DIR="/usr/store/" IP="211.69.198.208" NAME="31F03D16-60E6-8134-B28D-2F845DBB17ED" NETWORK="0" PORT="2811"/>
        </SLICE>
      </FILE>
    </SLICEINFO>
</TORRENT>
```

Fig. 3. Grid Torrent File

The "GridTorrent" file plays the role of a broker between the client and the service. In a traditional P2P network, the clients can use the torrent file to interact with each other with the help of the tracker server. In our implementation, users can only interact with the service using the "GridTorrent" file. We also want to introduce a P2P network protocol based on the GridFTP to let every client can interact with each other and improve the transfer experience remarkably.

In the common sense, the "GridTorrent" file contains the address of the User Space Service, the user's login name, the creation date. If clients upload file or directory to DMS, the "GridTorrent" file also contains the destination folder in the virtual space, and users fill in the address of the local file or directory. If clients download a file or directory from DMS, they fill in the destination folder's name.

Another problem in the file transfer of DMS is the channel reuse [15] of the GridFTP. It is very slow when we transfer a lot of small files as we need to establish connection many times between the client and the server, and disconnect it when we finish the transfer of a file. This kind of overhead can not be omitted as the GridFTP protocol is based on the FTP protocol. The solution is to introduce the concept of "channel reuse". We can package a set of transfer tasks into a single interaction and just establish and disconnect once during the transfer. The cost will reduce much by this method. We have implemented it in the GDSS (Global Distributed Storage System)[16], a virtualized storage system. The transfer protocol of the GDSS is modified on the base of FTP; we added some new instructions to FTP and implemented the "channel reuse". As we adopt the GridFTP protocol as the transfer protocol in DMS, the development work for "channel reuse" is very hard and we will collaborate with the Globus development team to target this issue.

4 Data Transfer Performance

We test the performance of DMS in a single domain. The test environment is described below in table 1. And The files we used in the test are list in the table 2.

Table 1. Test Environment of domain

	Client	Single Master	Storage Node
Operating System	Red Hat Linux 9	Red Hat Linux 9	Red Hat Linux 9
CPU	Pentium 4 1.60GHZ	Pentium 4 2.40 GHZ	Pentium 4 2.40GHZ
Memory	1G	2G	2G
Disk	40G	40G	80G
Network Card	10Mbit/s	100Mbit/s	100Mbit/s
Additional Software	DMS Transfer Client	CGSP Container, DMS Service	GridFTP Server, Storage Agent
Numbers	5	1	5

Table 2. Test Case Information

Sn	Type	Name	Size	File Numbers
1	File	1.rar	1.2M	1
2	File	2.rar	5M	1
3	File	3.rar	11.3M	1
4	File	4.rar	21.8M	1
5	File	5.rar	54.6M	1
6	File	6.rar	103M	1
7	Directory	test1	0.495k	10
8	Directory	test2	2.41k	50
9	Directory	test3	4.83k	100
10	Directory	test4	24.1k	500
11	Directory	test5	48.3k	1000

First, we test the DMS with a single client. The test case consists of two parts: First is the single file. Second is the directory with multiple small files. For each file, we test the three values: the time for the service consume, the time to connect to the remote storage node and the time to transfer file. The test results can be shown in Figure 4, 5, 6, 7.

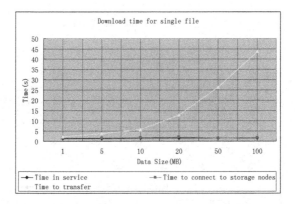

Fig. 4. The Download Time for Single File

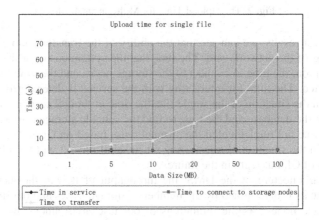

Fig. 5. The Upload Time for Single file

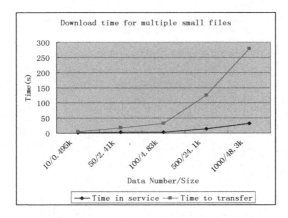

Fig. 6. The Download Time for Multiple small files

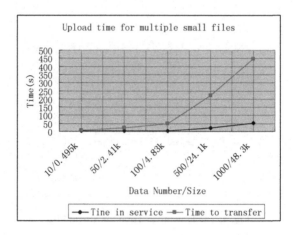

Fig. 7. The Upload Time for Multiple small files

From the test results, we can see that the ability of the transfer almost has reached the limitations of the network card. The time that the system costs consists of the time for transportation, the time to connect to the storage nodes and the time for interacting with the service. The time for interacting with the service is determined by the number of files. In the situation of the single file, the time is about 2 seconds. It consists of the look up in the database and the time to find the suitable storage resource.

As DMS have not implemented the channel reuse based on GridFTP, it is not performed well in the situation of transferring much small files. It will be solved with the collaboration of Globus team in the future.

From the architecture of the DMS, the throughput of it with the increase of the number of clients is determined by the number of storage nodes and the spread of data in these storage nodes. As the limitations of our test environment, we just test the response time of the service in the DMS with the increase of number of client's

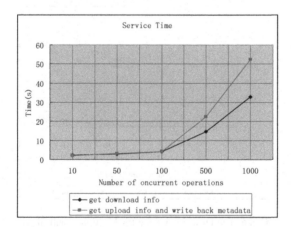

Fig. 8. The Service Response Time for many concurrent operations

concurrent operations but the transfer speed. These main concurrent operations conclude the "get download file info","write upload file info and write back the metadata". The test result can be shown in Figure 8.

The replication of CGSP is most used to maintain the reliability of data. It also helps to speed up the transfer of data. As mentioned above, the speed is determined by the spread of replica in the underlying storage nodes. We don't test it in our test environment.

In our current CGSP implementation, DMS meet the needs of storage and the needs of the Grid jobs running on CGSP. It achieves the goal of its design.

5 Conclusions and Future Work

DMS (*Data Management Services*) is the core module of CGSP, the Grid middleware of ChinaGrid. It is based on WSRF and supply system-level data service supporting common users and Grid jobs including data mining applications running on CGSP.

We design three WSRF services to complete the functions of the data management in CGSP. The User Space Service is in charge of the management of metadata. The Storage Resource Service manages all the geographically storage nodes. The Replica Management Service manages the replicas to offer high availability.

The future work will first focus on the more efficient data structure for storing metadata. The current implementation for storing metadata is to store these metadata in the database and use a XML file to act as a broker between the tree-style structure and the flatten structure of database. There are two drawbacks for this method: first, the cost to interact with database is very large although we use a database connection pool technology; second, the cost to parse the XML file is also very large. Though, in our current environment, it is not a bottle-neck of the system, we want to design a file-based or even memory-based structure to manage the metadata to speed up the management of metadata.

In the future, we will also have a further study on the channel reuse of the GridFTP and the integrity of the P2P and the GridFTP protocol.

References

1. Foster, I., Kesselman, C., Tuecke, S.: The anatomy of the Grid: Enabling scalable virtual organization. International Journal of Supercomputer Applications 15(3), 200–222 (2001)
2. Jin, H.: ChinaGrid: Making Grid Computing a Reality. In: Chen, Z., Chen, H., Miao, Q., Fu, Y., Fox, E., Lim, E.-p. (eds.) ICADL 2004. LNCS, vol. 3334, pp. 13–24. Springer, Heidelberg (2004)
3. ChinaGrid, http://www.chinagrid.edu.cn
4. ChinaGrid Supporting Platform, http://www.chinagrid.edu.cn/cgsp
5. Foster, I., Kesselman, C.: Globus: A Metacomputing Infrastructure Toolkit. International Journal of Supercomputer Applications 11(2), 115–129 (1998)
6. Rajasekar, A., Wan, M., Moore, R.: MySRB and SRB - components of a Data Grid. In: Proceedings of the 11th IEEE International Symposium on High Performance Distributed Computing, Edinburgh, pp. 301–310. Institute of Electrical and Electronics Engineers Inc, San Francisco (2002)

7. Kubiatowicz, J., Bindel, D., Chen, Y., et al.: OceanStore: An Architecture for Global-Scale Persistent Storage. ACM SIGPLAN Notices 35(11), 190–201 (2000)

8. The Web Services Resource Framework, http://www.globus.org/wsrf

9. Ganger, G.R., Strunk, J.D., Klosterman, A.J.: Self-* Storage: brick with automated administration. Technical Report CMU-CS-03-178. Carnegie Mellon University (August 2003)

10. Allcock, W., Bresnahan, J., Kettimuthu, R., Link, M., Dumitrescu, C., Raicu, I., Foster, I.: The Globus Striped GridFTP Framework and Server. In: Gschwind, T., Aßmann, U., Nierstrasz, O. (eds.) SC 2005. LNCS, vol. 3628, Springer, Heidelberg (2005)

11. Liu, L.K., Wu, Y.W., Yang, G.W., et al.: General Running Service: An Execution Framework for Executing Legacy Program on Grid. In: Fifth International Conference on Grid and Cooperative Computing Workshops, pp. 522–529.

12. He, F., Wu, Y.W., Yang, G.W., et al.: Grid Programming Environment over ChinaGrid Support Platform. In: Fifth International Conference on Grid and Cooperative Computing Workshops, pp. 530–535.

13. SWT, http://www.eclipse.org/swt

14. BitTorrent Protocol Specification, http://www.bittorrent.org/protocol.html

15. Guan, X.S., Jin, H., Xie, C., Wang, Q.C.: An Adaptive transfer Algorithm in GDSS. In: Second International Conference on Knowledge Economy and Development of Science and Technology(KEST2004), BeiJing, China, September 17-19 2004, pp. 226–233.

16. Jin, H., Ran, L., Wang, Z., et al.: Architecture Design of Global Distributed Storage System for Data Grid. High Technology Letters 9(4), 1–4 (2003)

Two-Phase Algorithms for a Novel Utility-Frequent Mining Model

Jieh-Shan Yeh[1], Yu-Chiang Li[2], and Chin-Chen Chang[3,4]

[1] Department of Computer Science and Information Management,
Providence University, Taichung 433, Taiwan
jsyeh@pu.edu.tw

[2] Department of Computer Science and Information Engineering,
Southern Taiwan University, Yung-Kang City, Tainan 71005, Taiwan
lyc@cs.ccu.edu.tw

[3] Department of Information Engineering and Computer Science,
Feng Chia University, Taichung 40724, Taiwan

[4] Department of Computer Science and Information Engineering,
National Chung Cheng University, Chiayi 62102, Taiwan
ccc@cs.ccu.edu.tw

Abstract. When companies seek for the combination of products which can constantly generate high profit, the association rule mining (ARM) or the utility mining will not achieve such task. ARM mines frequent itemsets without knowing the producing profit. On the other hand, the utility mining seeks high profit items but no guarantee the frequency. In this paper, we propose a novel *utility-frequent mining* model to identify all itemsets that can generate a user specified utility in transactions, in which the percentage of such transactions in database is not less than a minimum support threshold. A utility-frequent itemset indicates that such combination of products can constantly generate high profit. For finding all utility-frequent itemsets, there is no efficient strategy due to the nonexistence of "downward/upward closure property". In order to tackle such challenge, we propose a bottom-up two-phase algorithm, **BU-UFM**, for efficiently mining utility-frequent itemsets. We also introduce a novel concept, *quasi-utility-frequency*, which is upward closed with respect to the lattice of all itemsets. In fact, each utility-frequent itemset is also quasi-utility-frequent. A top-down two-phase algorithm, **TD-UFM**, for mining utility-frequent itemsets is also presented in the paper.

1 Introduction

Data Mining has made a profound impact on business practices and knowledge management in recent years. Association Rule Mining (or market basket analysis), finding interesting association or correlation relationships among data items, is one of the most important data mining strategies. Since the concept of association rules was introduced by Agrawal et al. [2] in 1993, many algorithms

T. Washio et al. (Eds.): PAKDD 2007 Workshops, LNAI 4819, pp. 433–444, 2007.
© Springer-Verlag Berlin Heidelberg 2007

and techniques for mining association rules have been proposed in the literature. Traditional association rule mining (ARM) model treats all the items in the database equally by only considering if an item is present in a transaction or not. ARM focuses on deriving correlations among a set of items and their association rules.

Although finding correlations of itemsets is very important, frequent itemsets identified by ARM may only contribute a small portion of the overall utility. In many situations, people may be more interested in finding out how a set of items support a specific objective that they want to achieve. Recently, a utility mining model was defined [17]. The goal of utility mining is to identify high utility itemsets that drive a large portion of the total utility. Utility mining is useful in a wide range of practical applications. However, it does not indicate how often such itemsets appear in the database. There may be some full priced items or high margin items which are high utility, but only appear in a small number of transactions. Due to the infrequency, such items may not be beneficial to the companies consistently.

When companies seek for the combination of products which can constantly generate high profit, clearly, the association rule mining or the utility mining will not achieve such task. ARM mines frequent itemsets without knowing the producing profit. On the other hand, the utility mining seeks high profit items but no guarantee the frequency. In this paper, we propose a novel *utility-frequent mining* model to identify all itemsets that can generate a user specified utility in each of certain transactions, in which the percentage of such transactions in database is not less than a minimum support threshold. An utility-frequent itemset indicates that the combination of items (or products) can constantly generate high utility (or profit). For finding all utility-frequent itemsets, there is no efficient strategy due to the nonexistence of "downward/upward closure property" (anti-monotone property). In order to tackle such challenge, we propose a bottom-up two-phase algorithm, called Bottom-Up Utility-Frequent Mining algorithm (**BU-UFM**), for efficiently mining utility-frequent itemsets. We also introduce a novel concept, *quasi-utility-frequency*, which is upward closed with respect to the lattice of all itemsets. In fact, each utility-frequent itemset is also quasi-utility-frequent. Therefore, a top-down approach can be applied to identify utility-frequent itemsets as well. **TD-UFM**, a top-down two-phase algorithm, for mining utility-frequent itemsets is also presented in the paper.

The rest of this article is organized as follows. Section 2 overviews the related work. In Section 3, we propose the utility-frequent mining model. Section 4 presents the algorithms. Section 5 provides experimental results. Finally, we conclude in Section 6 with a summary of our work.

2   Related Works

In this section, we review the Support-Confidence Framework and the utility mining model introduced in [17].

2.1 Support-Confidence Framework

Association rule mining attempts to discover the interesting relationships among items in a given a transaction database. The presence of some items in a transaction implies the high possibility of other items also appear in the same transaction. The formal definition is as follows.

Let $I = \{i_1, i_2, \ldots, i_m\}$ be a set of items. Let $DB = \{T_1, T_2, \ldots, T_n\}$, the task-relevant data, be a set of database transactions where each transaction T_j is a set of items, that is, $T_j \subseteq I$. A set of items is also referred as an *itemset*. An itemset that contains k-items is called a k-itemset. Each transaction is associated with an identifier, called TID. Let X be an itemset, a transaction T is said to contain X if and only if $X \subseteq T$. An *association rule* is an implication of the form $X \Rightarrow Y$, where $X \subseteq I$, $Y \subseteq I$, and $X \cap Y = \emptyset$. The rule $X \Rightarrow Y$ holds in the transaction set DB with **support** s, where s is the percentage of transaction in DB that contain $X \cup Y$ (i.e. both X and Y). This is taken to be the probability, $P(X \cup Y)$. The rule $X \Rightarrow Y$ has **confidence** c in the transaction set DB, if c is the percentage of transactions in DB containing X that also contain Y. This is taken to be the conditional probabilities, $P(Y|X)$. That is, $support(X \Rightarrow Y) = P(X \cup Y)$ and $confidence(X \Rightarrow Y) = P(Y|X)$.

A rule that satisfy a minimum support threshold (min_sup) and a minimum confidence threshold (min_conf) is called *strong*. An itemset is called a *frequent* itemset, if it satisfies minimum support.

Apriori, a multiple passes algorithm [3], is the most famous method to discover frequent itemsets. The Apriori Principle indicates that each subset of a frequent itemset must be frequent; otherwise the itemset is infrequent. This property is also called *downward closure property* or *anti-monotone property*. In each pass, Apriori scans a database once and employs the downward closure property to filter out many useless candidates.

Numerous efficient methods have been proposed to discover frequent itemsets, such as level-wise algorithms [3], [5], [6], [7] and pattern-growth methods [1], [9], [8], [14].

2.2 Utility Mining Model

A utility mining model has been proposed to measure how "useful" an itemset is [17]. It overcomes the shortcomings of traditional association rule mining, which ignores the sale quantity and price (or profitability) among items in a transaction. The traditional association rule ming becomes a special case of utility mining [16].

The following is the definition of a set of terms, given in [17], that leads to the formal definition of utility mining problem.

- The *item count of item* $i_p \in I$ in transaction T_q, $c(i_p, T_q)$, is the number of item i_p purchased in transaction T_q. For example, $c(A, T_1) = 0$ $c(B, T_1) = 0$, and $c(C, T_1) = 18$, in Table 1 (a).
- Each item i_p has an associated set of transactions $\mathcal{T}_{i_p} = \{T_q \in DB | i_p \in T_q\}$.

Table 1. An example of transaction database [15]

(a) The transaction table

TID	A	B	C	D	E
T_1	0	0	18	0	1
T_2	0	6	0	1	1
T_3	2	0	1	0	1
T_4	1	0	0	1	1
T_5	0	0	4	0	2
T_6	1	1	0	0	0
T_7	0	10	0	1	1
T_8	3	0	25	3	1
T_9	1	1	0	0	0
T_{10}	0	6	2	0	2

(b) The external utility table

Item	$eu(i_p)$
A	3
B	10
C	1
D	6
E	5

(c) The transaction utility table

TID	$TU(T_q)$
T_1	23
T_2	71
T_3	12
T_4	14
T_5	14
T_6	13
T_7	111
T_8	57
T_9	13
T_{10}	72

- A k-itemset $X = \{x_1, x_2, \ldots, x_k\}$ is a subset of I, where $1 \leq k \leq m$, $x_i \in I$ for all $i = 1, 2, \ldots, k$.
- Each k-itemset X has an associated set of transactions $T_X = \{T_q \in DB| \ X \subseteq T_q\}$.
- The *external utility* of item $i_p \in I$, $eu(i_p)$, is the value associated with item i_p in the external utility table. This value reflects the importance of an item, which is independent of transactions. For example, in Table 1 (b), the external utility of item A, $eu(A)$, is 3.
- The *utility of item* $i_p \in I$ *in transaction* T_q, $u(i_p, T_q)$, is the quantitative measure of utility for item i_p in transaction T_q, defined as $eu(i_p) \times c(i_p, T_q)$. For example, $u(A, T_8) = 3 \times 3 = 9$.
- The *utility of itemset* X *in transaction* T_q, $u(X, T_q)$, is $\sum_{i_p \in X} u(i_p, T_q)$, where $X \subseteq T_q$. For example, let $X = \{A, C\}$, $u(X, T_8) = 3 \times 3 + 25 \times 1 = 34$.
- The *utility of itemset* X, $u(X)$, is defined as $\sum_{X \subseteq T_q \in DB} u(X, T_q)$. For example, let $X = \{A, C\}$, $u(X) = u(X, T_3) + u(X, T_8) = 7 + 34 = 41$.
- The *transaction utility* of T_q, $TU(T_q)$, is equal to $u(T_q, T_q) = \sum_{i_p \in T_q} u(i_p, T_q)$. For example, $TU(T_1) = 18 \times 1 + 1 \times 5 = 23$.

Utility mining is to find all the itemsets whose utility values are beyond a user specified threshold. An itemset X is a *high utility itemset* if $u(X) \geq \epsilon$, where $X \subseteq I$ and ϵ is the minimum utility threshold, otherwise, it is a *low utility itemset*. For example, in Table 1, $u(\{A, D, E\}) = u(\{A, D, E\}, T_4) + u(\{A, D, E\}, T_8) = 14 + 32 = 46$. If $\epsilon = 120$, $\{A, D, E\}$ is a low utility itemset.

There is no efficient strategy to find all the high utility itemsets due to the nonexistence of "downward closure property" in the utility mining model. The challenge of utility mining is in restricting the size of the candidate set and simplifying the computation for calculating the utility.

An exhaustive search method can be applied to identify all high utility itemsets. However, such method is too time-consuming and space-consuming to work well in a large dataset environment. Several heuristic methods have been proposed to accelerate discovering high utility itemsets (or share frequent itemsets), such as MEU [17], SIP, CAC, and IAB [4] methods. Nevertheless, they may loss some high utility itemsets.

Recently, Li, et al. developed some efficient approaches, including the FSM, SuFSM, ShFSM, and DCG methods for share mining [11], [12], [13]. In the meanwhile, Liu, et al. [15] also presented a Two-Phase (TP) algorithm for fast discovering all high utility itemsets. In fact, under appropriate adjustment on item count and external utility of items, share mining is equivalent to utility mining.

3 Utility-Frequent Mining Model

As defined in the previous section, the utility of itemset X in transaction T_q, $u(X, T_q)$, is defined as $\sum_{i_p \in X} u(i_p, T_q)$, where $X \subseteq T_q$.

For a given utility μ, each itemset X is associated with a set of transactions $\mathcal{T}_{(X,\mu)} = \{T_q \in DB|\ X \subseteq T_q \text{ and } u(X, T_q) \geq \mu\}$.

$\mathcal{T}_{(X,\mu)}$ can be seen as the set of transactions that contain X and generate at least utility μ on X. The ratio of the size of $\mathcal{T}_{(X,\mu)}$ and the total number of transactions is denoted as

$$support(X, \mu) = \frac{|\mathcal{T}_{(X,\mu)}|}{|DB|} \qquad (1)$$

Definition 1. *For a given utility μ and a given minimum support threshold s, an itemset X is utility-frequent (U-frequent), if $support(X, \mu)$ is not less than s. Otherwise, X is utility-infrequent.*

For example, in Table 1, let $X = \{A, C\}$, $\mu = 20$ and $s = 20\%$, then $\mathcal{T}_{(X,\mu)} = \{T_8\}$, where $X \subseteq T_8$ and $u(X, T_8) = 34 > 20$. $support(X, \mu) = 10\% < s$, therefore, X is utility-infrequent.

Utility-frequent mining is to obtain all itemset X, in which the percentage of transactions, containing X and generating utility value on X beyond a user specified threshold μ, is greater than or equal to a user specified threshold s.

However, U-frequency is neither upward nor downward closed with respect to the lattice of all itemsets. For example, in Table 1, let $\mu = 20$ and $s = 20\%$, $\mathcal{T}_{(\{C\},\mu)} = \{T_8\}$, $\mathcal{T}_{(\{C,E\},\mu)} = \{T_1, T_8\}$ and $\mathcal{T}_{(\{C,D,E\},\mu)} = \{T_8\}$. Since $|DB| = 10$, $support(\{C\}, \mu) = support(\{C, D, E\}, \mu) = 10\%$ and $support(\{C, E\}, \mu) = 20\%$. $\{C, E\}$ is utility-infrequent, however, neither $\{C\}$ nor $\{C, D, E\}$. Therefore, U-frequency is neither upward nor downward closed.

Similar to the utility mining model, the challenge of utility-frequent mining is also in restricting the size of the candidate set and simplifying the computation for calculating the supports.

First, we observe the following important property on utility-frequent itemsets.

Theorem 1. *If itemset X is utility-frequent, X is also frequent.*

Proof: For a given utility μ and a threshold s, clearly, $T_{(X,\mu)} = \{T_q \in DB |\ X \subseteq T_q$ and $u(X, T_q) \geq \mu\} \subseteq \{T_q \in DB|\ X \subseteq T_q\} = T_X$. $support(X) \geq support(X, \mu)$. If X is U-frequent, $support(X, \mu) \geq s$, so, $support(X) \geq s$. Therefore, X is frequent. ☐

On the other hand, we give an extended definition of utility-frequent, called *quasi-utility-frequent* in the follows.

For a given utility μ, each itemset X is associated with a set of transactions $T'_{(X,\mu)} = \{T_q \in DB|\ \sum_{i_p \in X} u(i_p, T_q) \geq \mu\}$. Note that, X contained in T_q is not necessary. In Table 1, consider $X = \{A, C, E\}$ and $\mu = 20$, $\sum_{i_p \in X} u(i_p, T_1) = u(A, T_1) + u(C, T_1) + u(E, T_1) = 0 \times 3 + 18 \times 1 + 1 \times 5 = 23 > \mu$. Accordingly, $T_1 \in T'_{(\{A,C,E\},20)}$. $T'_{(X,\mu)}$ can be seen as the set of transactions that generate at least utility μ on X. The ratio of the size of $T'_{(X,\mu)}$ and the total number of transactions is denoted as

$$Qsupport(X, \mu) = \frac{|T'_{(X,\mu)}|}{|DB|} \qquad (2)$$

Definition 2. *For a given utility μ, an itemset X is quasi-utility-frequent (QU-frequent), if $Qsupport(X, \mu) \geq s$, where s is the user specified threshold.*

Theorem 2. *QU-frequency is upward closed with respect to the lattice of all itemsets.*

Proof: For a given utility μ and a threshold s, let X and Y be subsets of I, $X \subseteq Y$ and X be QU-frequent. For any T_q in $T'_{(X,\mu)}$, $\sum_{i_p \in Y} u(i_p, T_q) = \sum_{i_p \in Y-X} u(i_p, T_q) + \sum_{i_p \in X} u(i_p, T_q) \geq \mu$, then T_q is also in $T'_{(Y,\mu)}$. Therefore, $|T'_{(Y,\mu)}| \geq |T'_{(X,\mu)}|$, $Qsupport(Y, \mu) \geq Qsupport(X, \mu) \geq s$. By definition, Y is QU-frequent. ☐

Moreover, we have the following theorem.

Theorem 3. *If itemset X is utility-frequent, X is also quasi-utility-frequent.*

Proof: Given an utility μ and a threshold s, for any $T_q \in T_{(X,\mu)}$, by definition, $T_q \supseteq X$, so that $\sum_{i_p \in X} u(i_p, T_q) = u(X, T_q) \geq \mu$. Accordingly, $T_q \in T'_{(X,\mu)}$. In addition, $Qsupport(X, \mu) \geq support(X, \mu)$. If X is U-frequent, $support(X, \mu) \geq s$, then $Qsupport(X, \mu) \geq s$. Therefore, X is QU-frequent. ☐

For example, in Table 1, let $\mu = 20$ and $s = 20\%$, $T'_{(\{C,D,E\},\mu)} = \{T_1, T_8\}$, $T'_{(\{C,E\},\mu)} = \{T_1, T_8\}$ and $T'_{(\{C\},\mu)} = \{T_8\}$. $Qsupport(\{C, D, E\}, \mu) = Qsupport(\{C, E\}, \mu) = 20\%$ and $Qsupport(\{C\}, \mu) = 10\%$. $\{C, D, E\}$ and $\{C, E\}$ is QU-frequent, however, $\{C\}$ is not.

Figure 1 and Figure 2 illustrate the inclusion relationships between frequent itemsets and utility-frequent itemsets, and between quasi-utility-frequent itemsets and utility-frequent itemsets, respectively. The set of utility-frequent itemsets, $\{B, BD, BE, CE, BDE\}$, is a subset of both the set of frequent itemsets

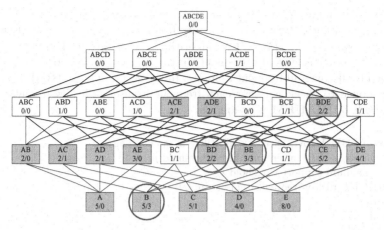

Fig. 1. Itemsets lattice related to the example in Table 1 with $\mu = 20$ and $s = 20\%$. Itemsets in gray-shaded boxes are frequent. Itemsets in circles are utility-frequent. Numbers in each box are "support count/utility support count $(|\mathcal{T}_X|/|\mathcal{T}_{(X,\mu)}|)$".

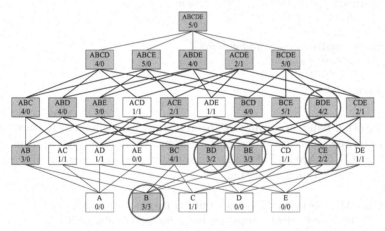

Fig. 2. Itemsets lattice related to the example in Table 1 with $\mu = 20$ and $s = 20\%$. Itemsets in gray-shaded boxes are quasi-utility-frequent. Itemsets in circles are utility-frequent. Numbers in each box are "quasi-utility support count/utility support count $(|\mathcal{T}'_{(X,\mu)}|/|\mathcal{T}_{(X,\mu)}|)$".

and the set of quasi-utility-frequent itemsets. Based on the above observation, we are able to develop a bottom-up and a top-down two-phase utility mining algorithms in the next section.

4   Algorithms

Intuitively, an exhaustive search algorithm can extract all utility-frequent itemsets. For a transaction database with n distinct items, the algorithm must generate 2^n possible itemsets that is impractical.

To avoid generating too many candidates, this study proposes a bottom-up two-phase algorithm to discover utility-frequent itemsets called Bottom-Up Utility-Frequent Mining **BU-UFM** algorithm. According to Theorem 1, each utility-frequent itemset is also frequent. In the first phase, Apriori algorithm is employed to discover all frequent itemsets. In the second phase, **BU-UFM** scan database once to check whether each frequent itemset is utility-frequent. The pseudo-code of **BU-UFM** is as follows:

Algorithm: BU-UFM. Discover all utility-frequent itemsets.
Input: Database DB; minimum utility threshold μ;
        minimum support threshold s.
Output: UFI, utility-frequent itemsets in DB.
Method:
//Phase I
(1) $UFI = \emptyset$;
(2) $CandidateSet = $ **Apriori**(DB, s);
//Phase II
(3)   **foreach** candidate $c \in CandidateSet$ {
(4)   **foreach** transaction $T \in DB$ { //scan database
(5)   **if**($c \subseteq T$ and $u(c,T) < \mu$){
(6) $c.count--$;} } }
(7)   **foreach** candidate $c \in CandidateSet$ {
(8)   **if** $(c.count \geq s)$ {
(9) $UFI := UFI + c$;} }
(10) **return** UFI;

On the other hand, according to Theorem 2, we can utilize the upward closure property on the quasi-utility-frequent itemsets. We first propose an Apriori-like algorithm, **QUF-Apriori**, for the quasi-utility-frequent itemset mining. The following is the detailed algorithm.

Algorithm: QUF-Apriori. Find quasi-utility-frequent itemsets using an
        iterative level-wise approach based on candidate generation.

Input: Database DB; minimum utility threshold μ;
        minimum support threshold s.
Output: $QUFI$, quasi-utility-frequent itemsets in DB.
Method:
(1) $L_{m-1} = $ **find_quasi-utility-frequent_$(m-1)$-itemsets**(DB, μ, s);
        // m is the size of I.
(2)   **for**$(k = 2; L_{m-k+1} \neq \emptyset; k++)$ {
(3) $C_{m-k} = $ **apriori_gen**(L_{m-k+1});
(4)   **foreach** candidate $c \in C_{m-k}$ { //scan DB for count
(5)   **foreach** transaction $T \in DB$ {
(6)   **if**($u(c,T) \geq \mu$) {
(7) $c.count++$; } } }
(8) $L_{m-k} = \{c \in C_{m-k} \mid c.count \geq s\}$; }
(9)   **return** $QUFI = \bigcup_{m-k} L_{m-k}$;

procedure find_quasi-utility-frequent_$(m-1)$-itemsets (
$\quad\quad\quad\quad DB$: database; μ : utility; s : minimum support threshold)
(1)   **foreach** item $i \in I$ {
(2) $\overline{\{i\}} = I - \{i\};$ //$(m-1)$-itemset
(3)   **foreach** transaction $T \in DB$ {
(4)   **if**($u(\overline{\{i\}}, T) \geq \mu$) {
(5) $\overline{\{i\}}.count + +;$ } } }
(6) $L_{m-1} = \{\overline{\{i\}} \mid \overline{\{i\}}.count \geq s\};$
(7)   **return** L_{m-1};

According to Theorem 3, the complete set of utility-frequent itemsets is a subset of quasi-utility-frequent itemsets. Now, we propose Top-Down Utility-Frequent Mining algorithm (**TD-UFM**). In Phase I, we utilize **QUF-Apriori** algorithm to discover the complete set of quasi-utility-frequent itemsets. We prune those overestimated itemsets in Phase II. The detailed algorithm is as follows.

Algorithm: TD-UFM. Discover all utility-frequent itemsets.
Input: Database DB; minimum utility threshold μ;
          minimum support threshold s.
Output: UFI, utility-frequent itemsets in DB.
Method:
//Phase I
(1) $CandidateSet = $ **QUF-Apriori**(DB, μ, s);
//Phase II
(2)   **foreach** candidate $c \in CandidateSet$ { //scan DB for count
(3)   **foreach** transaction $T \in DB$ {
          //calcuate the utility value of c in T for $c \subseteq T$
(4)   **if**($c \subseteq T$ and $u(c, T) \geq \mu$){
(5) $c.count + +;$ } } }
(6) $UFI = \{c \in CandidateSet \mid c.count \geq s\};$
(7)   **return** UFI;

5   Experimental Results

All the experiments were performed on an AMD K8 3500+ (2200 MHz) PC with 1 GB main memory, running the Windows XP Professional operating system. The **BU-UFM** algorithm was implemented in Visual C++ 6.0 and applied to several synthetic datasets. The two datasets T10.I6.D1000k.N1000 and T20.I6.D1000k.N1000 was generated from the IBM synthetic data generator [10].

The item count of each item in the two datasets were randomly generated between one and four. Observed from real world databases, most items are in the low profit range. Therefore, the external utility of each item was heuristically chosen between 0.01 and 10 and randomly generated with a log-normal distribution.

To choose appropriate minimum utility thresholds in the experiments, instead of randomly selecting μ, we pick different ratios of the average of transaction

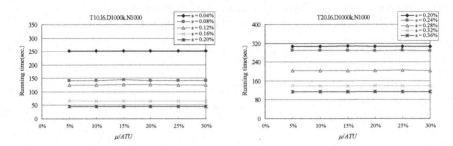

Fig. 3. Running times of **BU-UFM** on T10.I6.D1000k.N1000 and T20.I6.D1000k.N1000 for five minimum support thresholds

Table 2. Itemsets number comparison between utility-frequent itemsets and frequent itemsets

Dataset	T10.I6.D1000k.N1000						
Support threshold (s)	0.16%						
u/ATU Frequent itemset size	5% UF_k	10% UF_k	15% UF_k	20% UF_k	25% UF_k	30% UF_k	F_k
$k = 1$	616	402	255	157	109	84	848
$k = 2$	1575	830	416	212	90	45	2106
$k = 3$	38	18	5	1	0	0	49
$k = 4$	5	5	4	2	0	0	5
$k \geq 5$	0	0	0	0	0	0	0
Dataset	T20.I6.D1000k.N1000						
Support threshold (s)	0.28%						
u/ATU Frequent itemset size	5% UF_k	10% UF_k	15% UF_k	20% UF_k	25% UF_k	30% UF_k	F_k
$k = 1$	428	185	93	54	29	16	859
$k = 2$	4587	1573	503	176	45	14	9037
$k = 3$	109	23	1	0	0	0	205
$k = 4$	253	62	4	0	0	0	339
$k = 5$	422	161	15	0	0	0	462
$k = 6$	462	217	14	0	0	0	462
$k \geq 7$	0	0	0	0	0	0	0

utility (ATU) in database. For example, in Table 1, $\mu = 20$ is equal to 50% of the average of transaction utility (ATU), where the total utility of the database is 400 and the average of transaction utility is 40.

Figure 3 appears the running time performances of **BU-UFM** on both datasets T10.I6.D1000k.N1000 and T20.I6.D1000k.N1000. The x-axis value indicates the percentage of utility threshold to the average of transaction utility in database. A high minimum support threshold resulted in a short running time. For a certain minimum support threshold, the process of Phase I was identical over different utility thresholds, since Phase I is a traditional frequent itemset mining. The parameter of utility threshold almost had no influence the performance, since Phase II checked which candidates were utility-frequent. The running time of Phase II was relatively small. Therefore, the distinction of running time performances over various utility thresholds was insignificant.

Table 2 lists the number of utility-frequent itemsets and frequent itemsets with different item sizes on both T10.I6.D1000k.N1000 and T20.I6.D1000k.N1000 for several several distinct percentages of utility thresholds to *ATU*.

6   Conclusion

Utility measures the total utility derived from itemsets in a database. It does not indicate how often such itemsets appear in the database. We proposed a novel *utility-frequent mining* model to identify all itemsets that can generate a user specified utility in transactions, in which the percentage of such transactions in database is greater than or equal to a user specified threshold. We proposed a bottom-up two-phase algorithm, **BU-UFM**, for efficiently mining utility-frequent itemsets. We also introduced *quasi-utility-frequent* which is upward closed with respect to the lattice of all itemsets. Since each utility-frequent itemset is also quasi-utility-frequent, therefore, a top-down approach can be applied to identify utility-frequent itemsets as well. In the future, we will investigate more on the relationship among frequent, utility-frequent, and quasi-utility-frequent itemsets. We believe that a combination of top-down and bottom-up approach can more accelerate the mining process on utility-frequent itemsets.

References

1. Agarwal, R.C., Aggarwal, C.C., Prasad, V.V.V.: A tree projection algorithm for generation of frequent item sets. Journal of Parallel and Distributed Computing 61(3), 350–371 (2001)
2. Agrawal, R., Imieliński, T., Swami, A.: Mining association rules between sets of items in large databases. In: Proceedings of the 1993 ACM SIGMOD International Conference on Management of Data, Washington, D.C., USA, pp. 207–216 (1993)
3. Agrawal, R., Srikant, R.: Fast algorithms for mining association rules. In: Proceedings of 20th International Conference on Very Large Data Bases, pp. 487–499 (1994)
4. Barber, B., Hamilton, H.J.: Extracting share frequent itemsets with infrequent subsets. Data Mining and Knowledge Discovery 7(2), 153–185 (2003)
5. Berzal, F., Cubero, J.C., Marín, N., Serrano, J.M.: Tbar: An efficient method for association rule mining in relational databases. Data and Knowledge Engineering 37(1), 47–64 (2001)
6. Brin, S., Motwani, R., Ullman, J.D., Tsur, S.: Dynamic itemset counting and implication rules for market basket data. In: Peckham, J. (ed.) Proceedings of ACM SIGMOD International Conference on Management of Data, pp. 255–264 (1997)
7. Chang, C.C., Lin, C.Y.: Perfect hashing schemes for mining association rules. The Computer Journal 48(2), 168–179 (2005)
8. Grahne, G., Zhu, J.: Fast algorithms for frequent itemset mining using fp-trees. IEEE Transactions on Knowledge and Data Engineering 17(10), 1347–1362 (2005)

9. Han, J., Pei, J., Yin, Y., Mao, R.: Mining frequent pattern without candidate generation: A frequent pattern tree approach. Data Mining and Knowledge Discovery 8(1), 53–87 (2004)
10. IBM. Synthetic data generation code for associations and sequential patterns (accessed Jan. 2007), http://www.almaden.ibm.com/cs/projects/iis/hdb/Projects/data_mining/mining.shtml
11. Li, Y.C., Yeh, J.S., Chang, C.C.: Direct candidates generation: A novel algorithm for discovering complete share-frequent itemsets. In: Wang, L., Jin, Y. (eds.) FSKD 2005. LNCS (LNAI), vol. 3614, pp. 551–560. Springer, Heidelberg (2005)
12. Li, Y.C., Yeh, J.S., Chang, C.C.: Efficient algorithms for mining share-frequent itemsets. In: Fuzzy Logic, Soft Computing and Computational Intelligence - 11th World Congress of International Fuzzy Systems Association (IFSA 2005), pp. 534–539 (2005)
13. Li, Y.C., Yeh, J.S., Chang, C.C.: A fast algorithm for mining share-frequent itemsets. In: Zhang, Y., Tanaka, K., Yu, J.X., Wang, S., Li, M. (eds.) APWeb 2005. LNCS, vol. 3399, pp. 417–428. Springer, Heidelberg (2005)
14. Liu, J., Pan, Y., Wang, K., Han, J.: Mining frequent item sets by opportunistic projection. In: Proceedings of the 8th ACM-SIGKDD International Conference on Knowledge Discovery and Data Mining, pp. 229–238 (2002)
15. Liu, Y., Liao, W.-K., Choudhary, A.: A two-phase algorithm for fast discovery of high utility itemsets. In: Ho, T.-B., Cheung, D., Liu, H. (eds.) PAKDD 2005. LNCS (LNAI), vol. 3518, pp. 689–695. Springer, Heidelberg (2005)
16. Yao, H., Hamilton, H.J., Geng, L.: A unified framework for utility based measures for mining itemsets. In: Proceedings of ACM SIGKDD 2nd Workshop on Utility-Based Data Mining, pp. 28–37 (2006)
17. Yao, H., Hamilton, H.J., Butz, C.J.: A foundational approach to mining itemset utilities from databases. In: Proceedings of the 4th SIAM International Conference on Data Mining, pp. 428–486 (2004)

Top-Down and Bottom-Up Strategies for Incremental Maintenance of Frequent Patterns

Qunxiong Zhu and Xiaoyong Lin

School of Infromation Science and Technology, Beijing University of Chemical Technology,
15 Beisanhuan East Road, ChaoYang District, 100029, Beijing, China
{zhuqx,linxy}@mail.buct.edu.cn

Abstract. Mining frequent patterns has been studied popularly in data mining research. For getting the real useful frequent patterns, one must continually adjust a minimum support threshold. Costly and repeated database scans were done due to not maintaining the frequent patterns discovered. In this paper, we first propose a top-down algorithm for mining frequent patterns, and then present a hybrid algorithm which takes top-down and bottom-up strategies for incremental maintenance of frequent patterns. Efficiency is achieved with the following techniques: large database is compressed into a highly condensed and dynamic frequent pattern tree structure, which avoids repeated database scans, the top-down mining approach adopts a depth first method to avoid the recursive construction and materialization of conditional frequent pattern trees, which dramatically reduces the mining cost. The performance study shows that our algorithm is efficient and scalable for mining frequent patterns, and is an order of magnitude faster than FP_growth and Re-mining.

Keywords: Frequent patterns mining, Incremental maintenance, Data mining, Top-down, Bottom-up.

1 Introduction

Mining association rules, an important subfield of data mining, was firstly introduced in [1]. So far there have been considerable researches on designing fast algorithms for this task [2,3,4,5,6,7,9,10,14,15]. The problem of mining all association rules can be decomposed into two sub problems. Firstly all patterns that have a support above a minimum support defined by the user are mined. We call these frequent patterns. Secondly, rules are generated by using the frequent patterns mined. Comparing with the first problem, the solution to the second one is rather straightforward [1].

According to many studies in mining association rules, mining frequent patterns from databases has the following characteristics:

- The size of the database is significantly large, it could be gigabytes, terabytes, or even large.
- As data is inserted into, deleted from or modified in a database, the previous frequent patterns may lose their interestingness, or new frequent patterns appear.

T. Washio et al. (Eds.): PAKDD 2007 Workshops, LNAI 4819, pp. 445–456, 2007.
© Springer-Verlag Berlin Heidelberg 2007

- When users want to get the real useful frequent patterns, they have to continually adjust a minimum support threshold. This process is interactive. The set of all frequent patterns may change too.

These observations show that the algorithms of frequent patterns mining can be divided into three aspects.

1.1 Frequent Patterns Mining

The problem was firstly presented in [1]. Reference [3] is considered as one of the most important contributions to this subject. Its main algorithm, Apriori, not only influenced the association rules mining community, but affected other data mining fields as well. Most of the previous studies, such as [5, 6, 7, 8, 9, 11, 12], adopt an Apriori_like approach, because the Apriori heuristic achieves good performance by reducing the size of candidate set. However, candidate set is still costly, especially when there are a mass of patterns or long patterns. Reference [2] proposed a frequent pattern tree (FP_tree) structure and developed an efficient algorithm, FP_growth, which didn't bring the candidate set and was scalable for mining both short and long patterns. Some other algorithms based new structures were also proposed in [13].

1.2 Incremental Frequent Patterns Mining on Updated Database

As data is inserted into, deleted from or modified in the original database, the previous frequent patterns may lose their support or new frequent patterns may appear in the updated database. The process of generating frequent patterns using the updated part of the database and the previous frequent patterns mined is called incremental frequent patterns mining. The FUP algorithm proposed in [5] and the MAAP algorithm presented in [6] are efficient incremental mining methods.

1.3 Incremental Frequent Patterns Mining with the Change of Minimum Support Threshold

For getting the real useful frequent patterns, one need continually tune a minimum support threshold. The process of generating frequent patterns using the new minimum support and the frequent patterns mined is called incremental frequent patterns mining too. The IUA algorithm proposed in [7] is an efficient incremental mining algorithm. However, it still suffers from the candidate set due to basing on Apriori_like heuristic. The Re-mining algorithm proposed in [4] is an incremental mining algorithm through materializing old conditional frequent pattern trees. Whereas, there are still a lot of I/O operations needed, especially in some sparse database. Comparing with the incremental mining issue on updated database, little work has been done on this problem.

In this paper, a dynamic frequent patterns tree (DFP_tree) structure is proposed to store the information about all frequent patterns, and an efficient algorithm, DFP_Miner, is presented for computing all frequent patterns when the minimum support threshold is being changed constantly. All previous mined frequent patterns and the constructed DFP_tree can be reused in this interactive process. Two optimization techniques, children table, for depth first generation of frequent patterns, and trailer table, for reducing the size of database during the incremental mining process will be discussed. All these techniques contribute to substantial reduction of search costs.

A performance study has been conducted to study the capability of DFP_Miner, which is compared against FP_growth (a classical and fast algorithm) and Re-mining (a fast incremental algorithm). The study shows that DFP_Miner is at least an order of magnitude faster than these algorithms.

The rest of the paper is organized as follows: In Section 2, we formally introduce the problem. In Section 3, we propose our new algorithm. Our performance study is presented in Section 4. Finally, the conclusions and the future work are stated in Section 5.

2   Problem Statement

In this section, we first provide basic definitions of frequent patterns mining and the incremental frequent patterns mining with the change of minimum support threshold. Then we discuss the incremental problems in detail by using an example.

2.1   Frequent Patterns Mining

Let I={x_1, x_2, x_3 ... x_n} be a set of items. An itemset X that is also called pattern is a subset of I, denoted by $X \subseteq I$. A transaction T_X = (TID, X) is a pair, where X is a pattern and TID is its unique identifier. A transaction T_X is said to contain T_Y if and only if $Y \subseteq X$. A transaction database, named TDB, is a set of transactions. The number of transactions in TDB that contain X is called the support of X. A pattern X is a frequent pattern, if and only if its support is larger than or equal to s, where s is a threshold called minimum support. The frequent patterns mining problem is to find the complete set of frequent patterns in a given transaction database with respect to a given support threshold, s.

2.2   Incremental Frequent Patterns Mining with the Change of Minimum Support Threshold

Let s is the original support threshold. The number of frequent patterns with s may be less or larger than the number which users want. Then the minimum support threshold is continually adjusted until appropriate frequent patterns are achieved. Let s' is the new threshold, the process of mining frequent patterns using the previous mined frequent patterns and s' is called incremental frequent patterns mining. The goal in this paper is to solve the efficient incremental problem of frequent patterns mining after the minimum support is changed.

One approach to this problem is to rerun the frequent patterns mining algorithm on the whole database. This approach, though simple, has some obvious disadvantages. All computation previously done at finding out the old frequent patterns is wasted and all frequent patterns have to be computed again from scratch.

Example 1. Let the first two columns of Table 1 be our transaction database, denoted by TDB. Let the original minimum support threshold be s, and let the set of frequent patterns be F, we can get F through any algorithm of mining frequent patterns. For example, set s = 3, the frequent items are shown in the forth column of Table 1, and the F = {(a: 4), (c: 4), (d: 4), (e: 3), (ac: 3), (ad: 3), (cd: 3), (de: 3).}(The number after ":" indicates the support of the pattern before ":"). Considering the first matter (s'>s), in this matter, we don't rescan the database, e.g. if s' is changed to 4 (s' = 4), the new frequent items are shown in the third column of Table 1, and the new set of frequent pattern, F' = {(a: 4), (c: 4), (d: 4).}, which can be got straightway only using F and s' through the following formula:

$$F' = \{ f \in F \mid f.\text{sup } port \geq s' \} \tag{1}$$

Considering the second one (s'<s), e.g. if s' is changed to 2 (s'=2), the frequent items are shown in the fifth column of Table 1, the final complete frequent patterns, F'={(a: 4), (c: 4), (d: 4), (e: 3), (g: 2), (ac: 3), (ad: 3), (ae:2), (cd:3), (ce:2), (de: 3), (dg: 2), (eg:2), (acd:2), (ade:2), (cde:2), (deg: 2)}, and the set of new frequent patterns, F^+= { (g: 2) , (ae:2), (ce:2) , (dg: 2), (eg:2), (acd:2), (ade:2), (cde:2), (deg: 2)}.

The goal of this paper is to discuss the problem of incremental frequent patterns mining with the change of minimum support threshold, and introduce an efficient method to solve the problem. We will firstly construct a novel DFP_Tree, which can be neatly updated when the minimum support is changed. Then we present an algorithm for getting all frequent patterns through only mining new frequent patterns (F^+) without candidate generation from the DFP_Tree reconstructed.

Table 1. The transaction database TDB and different frequent items with different support

TID	Items	Frequent items		
		s=4	s=3	s=2
T1	c, d, e, f, g, i	c, d	c, d, e	c, d, e, g
T2	a, c, d, e, m	a, c, d	a, c, d, e	a, c, d, e
T3	a, b, d, e, g, k	a, d	a, d, e	a, d, e, g
T4	a, c, h	a, c	a, c	a, c
T5	a, c, d	a, c, d	a, c, d	a, c, d

3 Top-Down and Bottom-Up Strategies for Incremental Maintenance of Frequent Patterns

In this section, we firstly present an algorithm for constructing a DFP_tree and a top-down algorithm for mining frequent patterns. Then we propose an algorithm of reconstruction of DFP_tree when the minimum support is changed. Finally we provide a bottom-up algorithm for incremental maintenance of frequent patterns.

3.1 Dynamic Frequent Patterns Tree with a Children Table and a Trailer Table

To design a Dynamic frequent patterns tree, let's first observe an example.

Example 2. Let the database, DB, be the first two columns of Table 2, and the minimum support threshold, s, be 3.

Table 2. The transaction database DB, the frequent items and new databases

TID	Items	(Ordered)Frequent Items (s=3)	DB'	DB''
100	i, k, a, h, c, f, d, g	a, i, k, f	h, c, d, g	c, d, g
200	a, b, f, i, j	a, i, b, f	j	
300	b, a, e, l, m	a, b	l, e, m	e, m
400	a, h, j	A	h, j	
500	b, i, k, l, n, o	i, k, b	l, n, o	n, o
600	a, k, i, f	a, i, k, f		

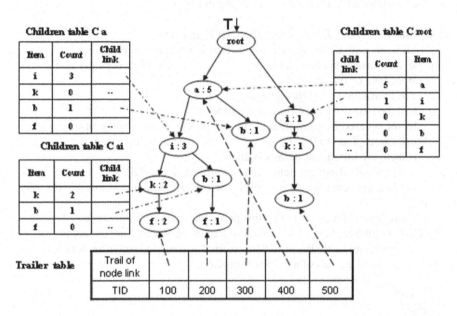

Fig. 1. The DFP_tree with a children table and a trailer table of database DB (s=3)

We first scan the database to collect the set of all frequent items with their supports and order them in the support descending order, P = {(a:5), (i: 4), (k:3), (b:3), (f:3), (h:2), (j:2), (l:2),(c:1), (d:1), (e:1), (g:1), (m:1), (n:1), (o:1)}. Then we create the root of the tree, T. Then we perform another scan of the DB to build the DFP_tree: After scanning the first transaction of DB, {100, (i, k, a, h, c, f, d, g)}, we construct the first branch of the tree with the frequent items in the transaction, {(a: 1), (i: 1), (k: 1), (f: 1)}, then we insert a new transaction (h, c, d, g), which is deleted all frequent items in the first transaction of the original DB, into the new database, DB' , At the same time, we add the first transaction's TID, 100, to the trailer table and let the new transaction's trail of node link pointing to the last node (f:1) in this path. If all items in a transaction are frequent, after we add these items to the DFP_tree, we directly delete this transaction through not adding any item to the new database. E.g. when we consider the transaction {600, (a, k, i, f)}, because all items are frequent, we don't add any item in this transaction to DB', i.e. we need not add {600, trial} to the trailer table (the size of database is reduced gradually by deleting frequent items or whole transaction). After we finish the second scan, the new database DB' that is smaller than DB is shown in the forth column of Table 2. For using a top-down strategy in the mining process, a children table of the root is built in which each child of the root is be listed, and the relevant link field points to its first occurrence in the tree by the child links of the root. We don't link all nodes with the same item, which happen in [2]. In this example, when we finished the construction, the DFP_tree with a children table and a trailer table is finished and shown in Fig. 1.

Algorithm 1.   DFP_Insert_Tree(DB, s, DFP_tree, DB')
   P={All items with their supports are ordered in support descending order}
   create the transaction trailer table   //the table is empty now
   create the root of IFP_Tree, T
   for each transaction t_i in DB do begin
       sort the items in t_i according to the order of P
       let the ordered frequent item list in t_i be [x|X] //where x is the first item and
       call Insert_Tree ([x|X], T)   // X is the remaining item list.
       if there are infrequent items in t_i then
            {add all infrequent items as a new transaction to DB'
            link the new transaction's trail   to the last node just inserted to the tree}
   end
   procedure Insert_Tree([x|X], T) //this procedure cites from function
     if T has children then   //insert_tree of algorithm 1 in [2], readers
       for each child c do begin   //are referred to the cited paper for detail
         if c.item_name=x then   //information.
             {c.count++, break}
     end
     else
         create a new node c and set the count field be 1
         link the closet node with the same item name to c via the next link field
         let T become c's parent
     if X is not empty then
         call Insert_Tree(X,c)
   end

3.2 Top-Down Strategy for Mining Frequent Patterns

In this section, we present a novel mining algorithm, TopDown_Miner, which takes a top-down strategy. We use a prefix to identify a subtree. For instance, in Fig. 1, a-prefix represents the leftmost subtree containing a, and ab-prefix identifies the second subtree rooted at a-prefix. For simplicity, we use i_j or i_k for a single item, α or β for a pattern which contains multi items or empty.

Given a set of frequent items $I = \{i_1, i_2, i_3 \ldots i_n\}$, where a total order (in the support descending order) is defined on I. The TopDown_Miner adopts an Apriori heuristic (however it doesn't bring the candidate generation.): A pattern $\alpha i_j i_k$ never be frequent if the pattern represented by αi_j is not frequent. Namely when we find that the pattern αi_j is not frequent, we need not test the whole $\alpha i_j i_k$-prefixes ($j < k \leq n$). The divide-and-rule strategy of TopDown_Miner is based on the following observation: Mining a pattern-frefix represented by α-prefix is to mine all the union of prefixes whose form is $\beta\alpha$-prefix. For example, Fig 1 shows an DFP_tree with a children table, which contains five items $\{a, i, k, b, f\}$. The union of i-prefix is $\{$i-prefix, ai-prefix$\}$ and the union of b-prefix is $\{$aib-prefix, ab-prefix and ikb-prefix$\}$.

Based on the above two strategies. We present the TopDown_Miner algorithm for mining frequent patterns by using a DFP_tree with a children table.

Algorithm 2.   TopDown_Miner(α, C_α, F)   //where α represents a prefix C_α
                    // is a children table of the α-prefix and F is the frequent patterns.
    for all c_i in the children table C do begin
        if c_i.support $\geq$ s then
            $\{$F=F $\cup$ αc_i
            generate the children table for αc_i-prefix, $C_{\alpha c i}$,
            call TopDown_Miner (αc_i, $C_{\alpha c i}$, F)$\}$ //depth first strategy
            call LinkforUnion(c_i, $C_{\alpha c i}$, C_α)   //divide-and-rule strategy

    end
    procedure LinkforUnion(c_i, $C_{\alpha c i}$, C_α)
    //link all items in $C_{\alpha c i}$ into C_α
    for each item in $C_{\alpha c i}$ do begin
            append the child link behind the child link of the same item in C_α
    end for
end

3.3 Reconstruction of DFP_tree

In this section, one will study how to reuse the previous DFP_tree and the new database, DB', reforming the new DFP_tree when the minimum support is changed.

Let s is the old minimum support threshold, and F is the set of all frequent patterns mined. When one changes the minimum support to s', there are two probabilities: if s'>s, some frequent patterns are not frequent any more, otherwise, some infrequent patterns will be frequent. Set the new frequent patterns set as F'. In the first matter, F' will be got directly from F as in formula (1). In the following part of this section, one will study the second matter.

Because s' is smaller than s, some patterns that were not frequent will be frequent. But the information about new frequent patterns is not in the original DFP_Tree. The following algorithm uses the original DFP_Tree and new database DB' built in algorithm 1 to finish the reconstruction of DFP_Tree. After the reconstruction, the new DFP_Tree will contain all information about new frequent patterns.

Algorithm 3.   DFP_tree Reconstruction(DFP_tree, DB', s')
set all old frequent items' count fields to 0 //which can help us avoid repeatedly
                                                            //computing the frequent items mined
add all new frequent items into children table in their support descending order
for each transaction tᵢ in DB' do begin
    let the ordered frequent item list in tᵢ be [x|X]   //where x is the first
                                                            // item and X is the remaining item list.
    trail = the trail of node link of tᵢ
    call Insert_Tree ([x|X], trail)
    if there are infrequent items in tᵢ then
        add them as a new transaction to DB'' or delete them in DB'
            link the new transaction's trail to the last node just inserted to the tree
end

After scanning all transactions in DB', the new DFP_tree is reconstructed. E.g. as
the minimum support threshold is changed to 2, after running this algorithm the new
DFP_tree is reconstructed, which is shown in Fig. 2, and a far smaller database DB''
is shown in the fifth column of Table 2.

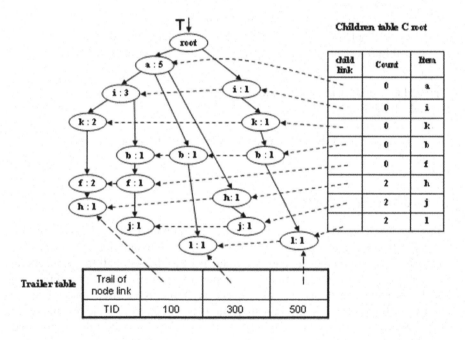

Fig. 2. The DFP_tree of database DB after the minimum support is changed to 2

3.4 Bottom-Up Strategy for Incremental Maintenance of Frequent Patterns

Based on the DFP_Tree constructed by Algorithm 1 or reconstructed by Algorithm 3,
the algorithm for mining frequent patterns with this tree can be performed as follow.

Algorithm 4.   DFP_Miner(DFP_tree, s, s' DB, F, DB')
  if (s'=0) then   //first time mining
    call DFP_Insert_Tree(DB,s,DFP_tree, DB')
    call TopDown_Miner(null, C_{root}, F)
  else
    if (s<s')
        F' = {f∈ F I f.support ≥ s'}
    else{
        call DFP_tree_Reconstruction(DFP_tree, DB',s')
        for each new item a_i from the bottom of C_{root} do begin //bottom-up strategy
           construct a_i's conditional FP_tree $Tree_{ai}$ //with a children table
          if $Tree_{ai} \neq \varnothing$ then   // instead of a header table
             call TopDown_Miner($Tree_{ai}$, C_{Treeai}, F^+)
             $F = F \cup F^+$ and a_i.count=0
    end
        s=s'}
  end

When the user firstly run the algorithm, the s'=0. If s'=0, one can call TopDown_Miner directly. Otherwise s is compared with s'. If s<s', the new frequent pattern F' can be got easily using (1). If s>s', some patterns that were not frequent may become frequent. By the DFP_tree_Reconstrucion algorithm, the new DFP_tree that contains new frequent patterns is achieved. Only all frequent patterns containing new frequent items need be computed. The process of mining is not from scratch. All work done can be recycled substantially.

4 Performance Study

In this section, we present a performance study of our algorithm, DFP_Miner, and give the comparison of DFP_Miner with FP_growth (a classic frequent pattern mining algorithm) and Re-mining (an effective incremental algorithm). It shows that DFP_Miner outperforms FP_growth and Re-mining and is efficient and highly scalable for mining very large databases.

All the experiments are performed on a 1.8GHz Pentium PC machine with 256 Megabytes main memory and 60G hard disk, running on Microsoft Windows/NT. All codes were written in Visual C++6.0 and run in the same environment. All reports of the runtime of DFP_Miner include both the time of reconstructing tree and mining frequent patterns. They also include both CUP and I/O time.

We have tested various data sets, with consistent result. Limited by space, only the results on two typical datasets, Gazelle (a real and sparse data set) and T25I15D10k (a synthetic and dense data set), are reported here. We only consider the case s'<s.

Gazelle is a web store visit (click stream) data set from Gazelle.com. It is a sparse data set. It contains 59,602 transactions, while where per transaction are up to 267 items. We start the mining with threshold 0.25%, and then tune the threshold to 0.2%, 0.15%, 0.1%, 0.05% (There are exponentially numerous frequent patterns as the support threshold goes down in such sparse data set).

Figure 3 shows that the run time of the three algorithms on this data set. Obviously, DFP_Miner outperforms the other two algorithms, and the gaps (in terms of run time) become larger as the support threshold goes lower. FP_growth works not very well in such sparse data sets since the data set is sparse. Because it can not compress data as effectively as what it does on dense data sets, on the other hand, recursively constructing every frequent item's conditional frequent pattern tree over sparse data set has its overhead.

Re-mining achieved better performance than FP_growth in such sparse data sets except for the first time mining. Because at the beginning, it needs more time to integrate and materialize the conditional frequent pattern trees for the purpose of the later extension. In the following mining, it avoids recursively constructing them by loading them directly from disk.

DFP_Miner achieved best performance in such sparse data sets since it saves all frequent patterns discovered and avoids recursively constructing conditional frequent pattern tree by taking advantage of the Top-Down strategy (which only needs to dynamically adjust links in the mining process). At the same time, DFP_Miner never need to store any further more information like in Re-mining, which can dramatically reduces the time and space needed.

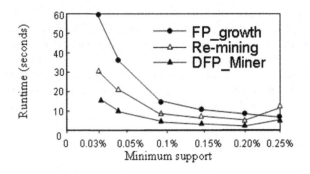

Fig. 3. Runtime on data set Gazelle

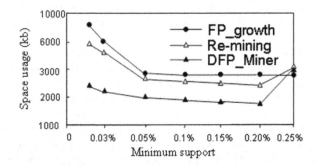

Fig. 4. Space usage on data set Gazelle

Figure 4 gives the space usage of DFP_Miner, FP_growth and Re-mining during the mining process. To make the comparison clear, the space usage (axis Y) is in logarithmic scale. From the figure, DFP_Miner is more scalable than FP_growth and

Re-mining since it need not any sub conditional frequent pattern trees in the mining process.

T25I15D10k is generated by a synthetic data generator described in [3]. It is a relatively dense data set, which contains 10,000 transactions and each transactions and each transaction has up to 25 items. There are 1,000 items in the data set and the average longest potentially frequent pattern is 15 items. We start the mining with threshold 0.6%, and then tune the threshold to 0.5%, 0.4%, 0.3%, 0.2%, 0.1% (There are pretty long frequent patterns as well as a large of short ones in them).

Figure 5 shows the run time of the three algorithms on this data set. At the first time mining, the support threshold is high, most patterns are short, and they have similar performance. When the support threshold becomes low, DFP_Miner is the fastest one.

Figure 6 gives the space usage of the three algorithms in mining process. The space usage is drawn in logarithmic scale too. As support threshold goes down, the number of the frequent patterns goes up dramatically. FP_growth need generate and Re-mining need load more sub conditional frequent patterns trees. In contrary, DFP_Miner use stable amount of space. In dense data set, a DFP_tree is smaller than the sum of all frequent pattern trees used in the other two mining processes, so DFP_Miner uses less space than the other algorithms.

According to the experimental results, the DFP_Miner is at least about an order of magnitude faster than FP_growth and Re-mining, and it need less space too.

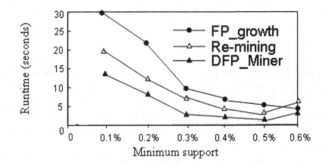

Fig. 5. Runtime on data set T25I15D10k

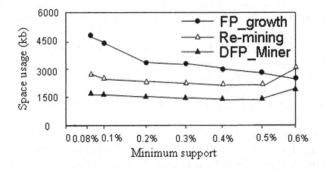

Fig. 6. Space usage on data set T25I15D10k

5 Conclusions and Future Work

One studied an efficient, incremental updating frequent pattern mining technique for maintenance of the frequent patterns mined. This method uses the DFP_tree to avoid repeatedly scanning, reconstructing, and computing. The algorithm is implemented, and its performance is studied and compared with two of best algorithms: FP_growth and Re-mining. The study shows that the DFP_Miner algorithm has superior performance on support threshold updating.

Recently, there have been some interesting studies at mining frequent patterns in databases which allow new data are added to or old data are deleted from. The extension of the DFP technique, for maintenance of the frequent patterns mined as both database and support being updated, is an interesting topic for future work.

References

1. Agrawal, R., Imielinski, T., Swami, A.: Mining Association rules between Sets of Items in Large Databases. In: Intl Proc of the 1993 ACM SIGMOD, Washington D.C, pp. 207–216 (1993)
2. Han, J., Pei, J., Yin, Y.: Mining Frequent Patterns without Candidate Generation. In: Intl. Proc. of the 2000 ACM SIGMOD, Dallas, pp. 1–12 (2000)
3. Agrawal, R., Strikant, R.: Fast Algorithms for mining association rules. In: 20th Intl. Proc. of the 1994 VLDB, Santiago, pp. 487–499 (1994)
4. Ma, X., Tang, S., Yang, D., et al.: Towards Efficient Re-mining of Frequent Pattern upon Threshold Changes. In: Meng, X., Su, J., Wang, Y. (eds.) WAIM 2002. LNCS, vol. 2419, pp. 80–88. Springer, Heidelberg (2002)
5. Cheung, D.W., Han, J., Ng, V.T., Wong, C.Y.: Maintenance of Discovered Association Rules in Large Databases: An Incremental Updating Technique. In: Intl Proc of the 1996 Data Engineering. New Orleans (1996)
6. Zhou, Z., Ezeife, C.: A Low-Scan Incremental Association Rule Maintenance Method Based on the Apriori Property. In: Proc of the 2001 on AI. Canadian (2001)
7. Feng, Y., Feng, J.: Incremental Updating Algorithms for mining Association Rules. Journal of Software 9(4), 301–305 (1998)
8. Park, J., Chen, M., Yu, P.: An Effective Hash-Based Algorithm for Mining Association rules. In: Intl. Proc. of the 1995 ACM SIGMOD, San Jose (1995)
9. Bodon, F.: A Fast Apriori Implementation. In: Proc of the IEEE ICDM Workshop on Frequent Itemset Mining Implementations Repository. Melbourne, Florida (2003)
10. Rozenberg, B., Gudes, E.: Association Rules Mining in Vertically Partitioned Databases. Data & Knowledge Engineering 59(2), 378–396 (2006)
11. Palashikar, G., Kale, M., Apte, M.: Association Rules Mining Using Heavy Itemsets. Data & Knowledge Engineering, Articles in Press
12. Chen, T., Hsu, S.: Mining Frequent Tree-Like Patterns in large datasets. Data & Knowledge Engineering (2006)
13. Xin, D., Han, J., Yan, X., et al.: On Compressing Frequent Patterns. Data & Knowledge Engineering (2006)
14. Savasere, A., Omiecinski, E., Navathe, S.: An Efficient Algorithm for Mining Association Rules in Large Databases. In: 21th Intl Proc of the 1995 VLDB, San Francisco (1995)
15. Toivonen, H.: Sampling Large Databases for Association Rules. In: 22nd Intl Proc of the 1996 VLDB. Bombay (1996)

GC-Tree: A Fast Online Algorithm for Mining Frequent Closed Itemsets

Junbo Chen and ShanPing Li

Department of Computer Science, ZheJiang University
Dorm 10, 5035, Zhejiang University YuQuan Campus,
Hangzhou City, Zhejiang Province, China
chjb@zju.edu.cn,
shan@cs.zju.edu.cn

Abstract. Frequent closed itemsets is a complete and condensed representaion for all the frequent itemsets, and it's important to generate non-redundant association rules. It has been studied extensively in data mining research, but most of them are done based on traditional transaction database environment and thus have performance issue under data stream environment. In this paper, a novel approach is proposed to mining closed frequent itemsets over data streams. It is an online algorithm which update frequent closed itemsets incrementally, and can output the current closed frequent itemsets in real time based on users specified thresholds. The experimental evaluation shows that our proposed method is both time and space efficient, compared with the state of art online frequent closed itemsets algorithm FCI-Stream [3].

1 Introduction

Frequent closed itemsets is a complete and condensed representation for all the frequent itemsets. Therefor, the study of the Frequent closed itemsets has arisen a lot of interest in the data mining community. Extensive researches have been carried out in this area, in the following they are split to four categories:

Both A-Close [12] and TITANIC [8] exploit a level-wise process to discover closed itemsets through a breadth-first search strategy. In each iteration, they try to search for candidates of MGs (Minimum Generators) with the help of search space pruning technique, and then verify them. Finally the MGs are used to generate all the closed itesmsets . Usually these algorithms are required to scan the whole dataset many times.

For CLOSET [13] and CLOSET+ [6]. With the help of high compact data structure FP-Tree, they try to project the global extraction context to some smaller sub-contexts, and then apply FCI mining process recursively on these sub-context. Better performance can be achieved than the adoption of A-close and TITANIC algorithms.

CHARM [7] and DCI-Closed [4] exploit hybrid techniques which try to use the properties of both previous mentioned techniques. Due to a data structure called IT-Tree, CHARM simultaneously explores both the closed itemset space and transaction space, with the depth-first search strategy, and generates one candidate each time, then with tidset intersection and subsumption checking, it will see whether the candidate is closed. DCI-Closed could be considered as an improvement of CHARM.

T. Washio et al. (Eds.): PAKDD 2007 Workshops, LNAI 4819, pp. 457–468, 2007.

FCI_Stream [3] is an online algorithm which performs the closure checking over a data stream sliding window. It uses a in memory data structure called DIU-tree to store all the Closed Itemsets discovered so far. With a specific search space pruning technique, it tries to perform the time consuming closure checking operation only when it's really needed.

Among all the algorithms, DCI-Tree has the best performance. Inspired by this algorithm, in this paper, we try to adapt it as an online algorithm under data stream sliding window environment, called GC-Tree.This algorithm is named after the in memory data structure it used, which is also called GC-Tree.

The rest of this paper is organized as follows. Section 2 formally defines the concept of closed itemsets, describes the notations to be used throughout the paper and introduces related works. Section 3 presents our proposed GC-Tree algorithm. The performance evaluation is depicted in Section 4. Finally, comes conclusion of this paper, Section 5.

2   Problem Definition

Let $\mathcal{I} = \{i1, i_2, ..., i_n\}$ be a set of items, $\mathcal{D} = \{t_1, t_2, ..., t_n, ...\}$ be a set of infinite stream of transactions, and $\mathcal{S} = \{t_{x_0}, t_{x_1}, ..., t_{x_M}\}$ be a sliding window of $\mathcal{D}$ which contains the recent M transactions. A subset $I \subseteq \mathcal{I}$ is called an itemset. Each transaction $t \in \mathcal{D}$ is a set of items in $\mathcal{I}$. There's an unique $transactionid$ for every t. Given a set of transactions $T \subseteq \mathcal{D}$, it can be represented by a tid_list, and the support of an itemset I in T is the percentage of transactions that contain I.

The concept of a closed itemset is based on the two following functions:

$f(T) = \{i \in \mathcal{I} \mid \forall t \in T, i \in t\}$
$g(I) = \{t \in \mathcal{D} \mid \forall i \in I, i \in t\}$

Definition 1. *An itemset X is said to be closed if and only if $C(I) = f(g(I)) = f \circ g(I) = I$ where the composite function $C = f \circ g$ is called a Galois operator or a closure operator.*

As a closure operator, C have the following properties:

Property 1. $C(X) \supseteq X$

Property 2. $Y \subseteq X \Rightarrow C(Y) \subseteq C(X)$

Property 3. $C(C(X)) = C(X)$

The number of transactions in the dataset including an itemset I is defined as the *support* of I, denoted as $supp(I)$. Mining all the frequent closed itemsets from the current sliding window of data stream $\mathcal{D}$ requires to discover all the closed itemsets which have higher support than a given threshold *min_supp* in the sliding window.

The closure operator, C defines a set of equivalence classes. All elements belong to the same equivalence class share the same *closed itemset*. Our algorithms tries to find exactly one generator in every equivalence class, then calculates the closure of them. There're 2 popular lemmas which are wildly exploited:

Lemma 1. *Given two itemsets X and Y, if $X \subset Y$ and $supp(X) = supp(Y)$, then $C(X) = C(Y)$.*

Lemma 2. *Given an itemset X and an item i, if $g(X) \subseteq g(i) \Rightarrow i \in c(X)$.*

3   The GC-Tree Algorithm

3.1   Algorithm Overview

Our algorithm is inspired by 'DCI_Closed' algorithm [4], It exploits a total lexicographic order relation $\prec$ among all the itemsets of the search space[1]. The usage of a *closure climbing* technique to obtain *closure generators* on the basis of the total lexicographic order, can efficiently detect the duplicate generators in the same equivalence class.

In this paper, we propose an algorithm which can work under the data stream sliding window environment. This algorithm uses an in memory data structure called GC-Tree (Generator and frequent Closed itemsets Tree) to store all the frequent closed itemsets in the current sliding window. Each element in The GC-Tree has the following format: $< gen, eitem, clo >$[2]. Where gen is the *closure generator*, $eitem$ is the extension item with which gen used to extend another closed itemset[3], clo is the corresponding *closed itemset*[4]. We also define two functions on the gen, *preset(gen)* and *posset(gen)* which are defined as:

Definition 2. *Let $gen = Y \cup i$ be a generator of a closed itemset where Y is a closed itemset and $i \notin Y$. preset(gen) is defined as $\{j \prec i | j \notin gen\}$. posset(gen) is defined as $\{j \in \mathcal{I} | j \notin preset(gen) \text{ and } j \notin C(gen)\}$.*

Definition 2 along with Definition3, Theorem 1 and Lemma3 provided by Claudio Lucchese *et al.* [4] are the base of our GC-Tree algorithm:

Definition 3. *A generator $X = Y \cup i$, where Y is a closed itemset and $i \notin Y$, is said to be order preserving one   iff $i \prec (C(X) - X)$.*

Theorem 1. *For each closed itemset $Y \neq C(\emptyset)$, there exists one and only one sequence of $n(n \geq 1)$ items $i_0 \prec i_1 \prec \ldots \prec i_{n-1}$ such that $\{gen_0, gen_1, \ldots, gen_{n-1}\} = \{Y_0 \cup i_0, Y_1 \cup i_1, \ldots, Y_{n-1} \cup i_{n-1}\}$ where the various gen_i are order preserving generators, with $Y_0 = C(\emptyset), Y_{j+1} = C(Y_j \cup i_j) \forall j \in [0, n-1]$ and $Y = Y_n$.*

Lemma 3. *Let $gen = Y \cup i$ be a generator of a closed itemset where Y is a closed itemset and $i \notin Y$. gen is not order preserving   iff $\exists j \in preset(gen)$, such that $g(gen) \subseteq g(j)$.*

GC-Tree is lexicographically ordered, for each closed itemset, there exists a path in GC-Tree from the root. All the elements in this path compose the order preserving generator sequence $\{gen_0, gen_1, \ldots, gen_{n-1}\}$ mentioned in Theorem 1. It has the following properties:

---

[1] This lexicographic order is induced by an order relation between single item literals, according to which each k-itemset I can be considered as a sorted set of k distinct items $\{i_0, i_1, \ldots, i_k\}$.

[2] For the concision of representation, we omit the links between the parents and children, and the support of the closed itemsets.

[3] *eitem* is the item i in Definition 2.

[4] The GC-Tree Node may be written in the compact way in the rest of this paper: $< gen, clo >$.

Property 4. *Let N_p < gen, clo > be parent node of N_c < gen, clo > in GC-Tree, Then we have: (1) $N_c.gen = N_p.clo \cup j, j \in N_p.pos$ and gen is order preserving in every Node. (2) $N_c.gen \supset N_p.gen$. (3) $N_c.clo \supset N_p.clo$. (4) posset($N_c.gen$) $\subset$ posset($N_p.gen$).*

Property 5. *Every Frequent Closed Itemset in the current sliding window can be found in the GC-Tree, the path from the root to the specific FCI composes the sequence $\{gen_0, gen_1, \ldots, gen_{n-1}\}$ in Theorem 1.*

Property 6. *Let N_r < gen, clo > be the root in the GC-Tree, then we have $N_r.gen = \emptyset$, $N_r.clo = C(\emptyset)$, preset($N_r.gen$)=$\emptyset$ and posset($N_r.gen$)=$\mathcal{I} - C(\emptyset)$.*

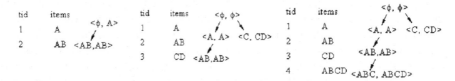

Fig. 1. GC-Tree

When a transaction arrives or leaves the current sliding window, the algorithm update GC-Tree incrementally. Figure 1 illustrates the GC-Tree when the data stream {A, AB, CD, ABCD} arrives.

3.2 Add a Transaction to the Sliding Window

In this section, we discuss how to maintain the GC-Tree when a new transaction arrives. Let t be the arrived transaction, T_1 be the set of transaction in the sliding window before t comes, T_2 be the set of transaction in the sliding window after t comes, then we have: $T_2 = T_1 \cup t$.

Lemma 4. *Let C_{T_1} be all the closed closed itemsets in T_1, and C_{T_2} be all the closed closed itemsets in T_2, then we have: $C_{T_1} \subseteq C_{T_2}$.*

Proof: $\forall c \in C_{T_1}$. if $c \notin t$, then, $g(c)_{T_2} = g(c)_{T_1} \Rightarrow f(g(c)_{T_2}) = f(g(c)_{T_1}) = c \Rightarrow c$ is closed. If $c \in t$, then, $g(c)_{T_2} = g(c)_{T_1} \cup t \Rightarrow f(g(c)_{T_2}) = f(g(c)_{T_1}) \cap t = c \cap t = c \Rightarrow$ c is closed.

According to Lemma 4, when a new transaction is added to the sliding window, the only action we should concern is to add new nodes(closed itemsets) into FCI-Tree.

3.2.1 The Conditions to Add New Closed Itemsets

Lemma 5. *Let $Y \nsubseteq t \Rightarrow C(Y)_{T2} = C(Y)_{T1}$.*

Proof: $g(Y)_{T_2} = g(Y)_{T_1} \Rightarrow C(Y)_{T_2} = f(g(Y)_{T_2}) = f(g(Y)_{T_1}) = C(Y)_{T_1}$.

According to Lemma 5, If $Y \nsubseteq t$, Y will remain closed/un-closed in T2 as in T1. So, we simply omit this situation.

Lemma 6. *Let* $Y \subseteq t, t \in T_1 \Rightarrow C(Y)_{T2} = C(Y)_{T1}$.

According to Lemma 6, If $Y \nsubseteq t$, and $t \in T_1$ Y will remain closed/un-closed in T2 as in T1. So, we simply omit this situation, too.

3.2.2 How to Add New Closed Itemsets to GC-Tree

First, we give the pseudo-code algorithm in Table 1, then we explain how it works.

Table 1. GC-Tree−Addition

```
1:    procedure Add(t, curr)
2:       if(t ∈ T₁)
3:          for all Y ∈ t, Y is closed itemset: s(Y)++;
4:          return;
5:       else
6:          if(curr.gen ⊈ t) return;
7:          newclo ⟵ curr.clo ∩t
8:          if(newclo = curr.clo )
9:             s(curr.clo)++;
10:            for all (i ∈ posset*(curr))
11:               checkChild(curr.clo ∪ i, curr, t)
12:         else
13:            m ⟵ min≺{curr.clo − newclo}
14:            newnode ⟵ <new-clo ∪ m, curr.clo>
15:            s(newnode.clo) ⟵ s(curent.clo);
16:            for all (child of curr node)
17:               if(child.eitem ∈ preset(newnode))
18:                  moveChild(child, curr, new-node);
19:               else
20:                  child.gen ⟵ newclo ∪ child.eitem;
21:            addChild(curr, new-node);
22:            curr.clo ⟵ newclo;
23:            for all (i ∈ posset*(curr))
24:               checkChild(curr.clo ∪ i, curr, t)
```

In Table 1, $preset(node)$ is defined as the preset of $node.gen$, $posset(node)$ is the posset of $node.gen$, $posset^*(gen)$ is defined as $posset(gen) \cap t$, moveChild(child,old-parent,new-parent) detaches child from old-parent and add it as a child of new-parent, checkChild($curr.clo \cup i$, curr, t) searches the children of curr node: if any of them have $gen = curr.clo \cup i$, the algorithm invokes Add() recursively; if there's no such child found, the algorithm invokes closureCheck(curr, t) function[5], and closureCheck() is defined in Table 2.

According to Lemma 6, it's unnecessary to perform any operation if $t \in T_1$ except to increase the supports, see Table 1, line 2-4.

Lemma 7. *Let* C_{T_1} *be all the closed closed itemsets in* T_1, *and* C_{T_2} *be all the closed closed itemsets in* T_2, $\forall c_i \in (C_{T_2} - C_{T_1})$, *we have:* $c_i \subseteq t$.

---

[5] Similar to the function closureCheck() in DCI-Closed, devised by Claudio Lucchese *et al.* [4].

Table 2. GC-Tree–Closure Checking

```
1:    procedure closureCheck(parent, item)
2:        gen ⟵ parent.clo ∪ item
3:        if(is_dup(gen) = false)
4:            newclo ⟵ gen;
5:            for all j ∈ posset*(parent.gen), j ≻ itemm, g(gen) ∈ g(j)
6:                newclo ⟵ newclo ∪j
7:            new-node ⟵ < gen, newclo>;
8:            addChild(parrent, new-node);
9:            for all k, k ∈ posset*(gen)
10:               closureCheck(this, k)
```

Proof: If $\exists c_j \not\subseteq t$, then we have, $g(c_j)_{T_2} = g(c_j)_{T_1} \Rightarrow C(c_j)_{T_2} = C(c_j)_{T_1}$, which means c_j will remain closed/un-closed in T2 as in T1, so that $c_j \notin (C_{T_2} - C_{T_1})$.

According to Lemma 7, we have, $gen \not\subseteq t \Rightarrow C(gen) \not\subseteq t \Rightarrow C(gen)_{T_2} = C(gen)_{T_1}$, and notice that if $gen \not\subseteq t$, then for any offspring of the current node, we have $gen \not\subseteq t$, that means there should be no new closed itemset in the current node and its sub-tree, see Table 1, line 6.

Lemma 8. *For any node n in the GC-Tree, if $n.gen \subseteq t$, we have:* $C(n.gen)_{T_2} = C(n.gen)_{T_1} \cap t$.

Proof: $C(n.gen)_{T_2} = f(g(n.gen)_{T_2}) = f(g(n.gen)_{T_1} \cup t) = C(n.gen)_{T_1} \cap t$.

According to Lemma 8, we have: if $curr.gen \subseteq t$ and $curr.clo = curr.clo \cap t$, then the closed itemset in the current node remains closed, so we leave the current node alone and checking it's children. See Table 1, line 8-11.

Lemma 9. *For any node n in the GC-Tree, if $n.gen \subseteq t$ and $C(n.gen)_{T_2} \neq C(n.gen)_{T_1} \cap t$, then $\exists i \in \mathcal{I}$ which makes $gen^* = C(n.gen)_{T_2} \cup i$ order preserving and $C(n.gen^*)_{T_2} = C(n.gen)_{T_1}$.*

Proof: First, we prove gen^* is order preserving: According to Lemma 8, it's easy to see $C(n.gen)_{T_1} \supseteq C(n.gen)_{T_2}$. Let $i = min_{\prec}\{C(n.gen)_{T_1} - C(n.gen)_{T_2}\}$, obviously, $gen^* = C(n.gen)_{T_2} \cup i$ is order preserving.

Second, we prove $C(n.gen^*)_{T_2} = C(n.gen)_{T_1}$:

$$
\begin{align}
C(n.gen^*)_{T_2} &= f(g(n.gen^*)_{T_2}) \tag{1}\\
&= f(g(C(n.gen)_{T_2} \cup i)_{T_2}) \tag{2}\\
&= f(g(C(n.gen)_{T_2} \cup i)_{T_1}) \tag{3}\\
&= C(C(n.gen)_{T_1}) \tag{4}\\
&= C(n.gen)_{T_1} \tag{5}
\end{align}
$$

Equation (3) holds because $i \notin t$ (see Lemma 8), equation (4) holds because in T_1, $C(n.gen)_{T_2} \cup i$ and $C(n.gen)_{T_1}$ belongs to the same equivalence class.

According to Lemma 9, if the *clo* of the current node is changed after t arrives, the new $clo = C(gen)_{T_2}$ is simply $C(gen)_{T_1} \cap t$, and a new node $< gen^*, C(gen)_{T_1} >$ is required as a child of the current node. To make sure the children of current node are order preserving in the updated GC-Tree: if j is not in the *preset* of the new node, we should move them from the current node to the new node as children; else, we set the $child.gen$ as $C(gen)_{T_1} \cup j^6$. See Table 1, line 12-24.

When the first transaction t arrives, GC-Tree is initialized as a single node $< \emptyset, t >$. After that, every upcoming transaction t would cause $Add(t, GC - Tree.root)$ to be invoked.

3.3 Delete a Transaction from the Sliding Window

In this section, we discuss how to maintain the GC-Tree when an old transaction leaves the sliding window. Let t be the leaving transaction, T_1 be the set of transactions in the sliding window before t leaves, and T_2 be the set of transactions in the sliding window after t leaves, then we have: $T_1 = T_2 \cup t$.

Lemma 10. *Let C_{T_1} be all the closed closed itemsets in T_1, and C_{T_2} be all the closed closed itemsets in T_2, then we have: $C_{T_1} \supseteq C_{T_2}$.*

Proof: $\forall c \in C_{T_2}$. if $c \notin t$, then, $g(c)_{T_2} = g(c)_{T_1} \Rightarrow f(g(c)_{T_2}) = f(g(c)_{T_1}) = c \Rightarrow c$ is closed in T_1. If $c \in t$, then, $g(c)_{T_1} = g(c)_{T_2} \cup t \Rightarrow f(g(c)_{T_1}) = f(g(c)_{T_2}) \cap t = c \cap t = c \Rightarrow c$ is closed in T_1.

According to Lemma 10, when an old transaction leaves the sliding window, the only action needed is to delete some existing nodes(closed itemsets) from GC-Tree whose itemsets are not closed any longer.

3.3.1 The Conditions to Delete Existing Closed Itemsets

Lemma 11. *Let $Y \not\subseteq t \Rightarrow C(Y)_{T2} = C(Y)_{T1}$.*

Proof: $g(Y)_{T_2} = g(Y)_{T_1} \Rightarrow C(Y)_{T2} = f(g(Y)_{T_2}) = f(g(Y)_{T_1}) = C(Y)_{T_1}$.

According to Lemma 11, If $Y \not\subseteq t$, Y will remain closed/un-closed in T2 as in T1. So, we could simply omit this situation.

Lemma 12. *Let $Y \subseteq t, t \in T_2 \Rightarrow C(Y)_{T2} = C(Y)_{T1}$.*

3.3.2 How to Delete Expired Closed Itemsets from GC-Tree
First, we give the pseudo-code algorithm in Figure 4, then we explain how it works.

In Table 3, moveChild(child,old-parent,new-parent) detaches child from old-parent and add it as a child of new-parent, moveChild(children, old-parent,new-parent) detaches all the children from old-parent and add it as a child of new-parent, removeChild(p-node, c-node) is detach c-node as child of p-node.

According to Lemma 12, we do not need to perform any operation if $t \in T_2$ except decreases supports, see Table 3, line 2-4.

---

[6] This is acceptable because it's easy to proof if $j \notin t$, then the closure of the new generator equals to the closure of the old generator, if $j \in t$, then the recursive checking would find the closure of the new generator, pushing down the closure of the old generator, meanwhile keeping the GC-Tree order preserving.

Table 3. GC-Tree−Deletion

```
1:    procedure Delete(t, curr)
2:      if(t ∈ T₂)
3:        for all Y ∈ t, Y is closed itemset: s(Y)− −;
4:        return;
5:      else
6:        if(curr.clo ⊈ t) return;
7:        for all (child of curr node)
8:            Delete(t, child)
9:        if(isClosed(curr))
10:           s(curr)− −;
11:           return;
12:       else
13:         if(curr node has more than one child)
14:             newClo ⟵ ∩j{∀j|j = child.clo};
15:             curr.clo ⟵ newClo;
16:             for all(chl of curr)
17:                 chl.gen ⟵ newClo ∪ chl.eitem;
18:             find mchl where mchl.gen=newClo;
19:             moveChild(mchl.children, mchl, curr);
20:             curr.deleteChild(mchild);
21:         else if(curr has only one child)
22:             child.gen ⟵ curr.gen;
23:             moveChild(child, curr, curr.parent);
24:             removeChild(curr.parent, curr)
25:         else
26:             removeChild(curr.parent, curr)
```

Table 4. GC-Tree−Is Closed

```
1:    procedure IsClosed(curr)
2:      M ⟵ I
3:      c ⟵ curr.clo
4:      for all (U ⊃ c , U ∈ GC-Tree and U ≺ c)
5:          M ⟵ M ∩ U
6:      if(M = curr.clo)
7:          return true;
8:      else
9:          return false;
```

Lemma 13. *Let C_{T_1} be all the closed closed itemsets in T_1, and C_{T_2} be all the closed closed itemsets in T_2, $\forall c_i \in (C_{T_1} - C_{T_2})$, we have: $c_i \subseteq t$.*

The proof is similar with Lemma 7. According to this Lemma, if $curr.clo \nsubseteq t$, all closed itemsets of the current node and all it's offsprings would remain closed. See Table 3, line 6.

Lemma 14. *If X and Y are 2 closed itemsets, and $Z = X \cap Y \neq \emptyset$, then Z is closed itemset too.*

Proof: To prove Z is closed itemset, we only need to prove C(Z)=Z, in turn, we need to prove $C(Z) \subseteq Z$ and $C(Z) \supseteq Z$.

$$g(Z) \supseteq g(X) \cup g(Y)$$
$$\text{For } \subseteq : \Rightarrow f(g(Z)) \subseteq f(g(X)) \cap f(g(Y))$$
$$\Rightarrow C(Z) \subseteq X \cap Y$$
For $\supseteq$: According to Property 1, $C(Z) \supseteq Z$.

Lemma 15. *Transaction t is leaving the sliding window, given n is a node in the GC-Tree, it has more than one child remains after all its offsprings are checked for closure, and $n.clo$ is not closed any longer. Then there exists a child of n, denoted as $mchl$, that $mchl.clo = newClo$ (newClo is the intersection of all the chl.clo, where chl is the child of n).*

Proof: On one hand, $newClo$ is a closed itemset according to Lemma 14. Then there should be a GC-Node in GC-Tree which represents this closed itemset, denoted as $ncNode$. On the other hand, since $n.clo$ is not closed, then we have, $newClo \supset n.clo$, and after that $newClo$ could be represented as $n.clo \cup L$, where $L = \{i \in \mathcal{I} | \forall chl, i \in chl.clo, i \notin n.clo\}$, that means $L \subset posset(n.gen)$. It's easy to see that such a closed itemset $newClo = n.clo \cup L$ could only appear under n as its child.

According to Lemma 15, if the current node is no longer closed, and it has more than one child which are still closed, there exists one child, having a closed itemset the same with the intersection of all the children of the current node. Then this specific child should be promoted to replace the current node. See Table 3, line 13-20.

Lemma 16. *Transaction t is leaving the sliding window, given n, which is a node in the GC-Tree, it has only one child (denoted as c) remains after all its offsprings are checked for closure, and $n.clo$ is not closed any longer. Then, we have: n and c should be merged as one node $< n.gen, c.clo >$.*

Proof: If n is closed in T_1 and not closed in T_2, from Lemma 13, we have $n.clo \subseteq t$, Let $c.gen = n.clo \cup i$, we have:

$$\left. \begin{array}{l} g(n.clo)_{T_1} \supseteq t \\ g(n.clo)_{T_1} \supseteq g(c.gen)_{T_1} \end{array} \right\} \Rightarrow g(n.clo)_{T_1} \supseteq g(c.gen)_{T_1} \cup t \qquad (6)$$

Assume $\exists s \in g(n.clo)_{T_1}, s \notin g(c.gen) \cup t$, we have, $s \supseteq n.clo$, so that $\forall j \in c.clo - n.clo, j \notin s$, because if $j \in s$, we have $s \supseteq n.clo \cup j$, since $n.clo \cup j$ is in the equivalence class of $c.clo$, we have $s \supseteq c.clo$, it's contradictory with the assumption $s \notin g(c.gen)$.
There're 2 possibilities:

1. $s = n.clo$, it's impossible because if it's true, then n.clo would remain closed after t leaves.
2. $\exists k \notin c.clo, k \in s$, so that $\exists C(s) \supseteq s$, which satisfies $C(s) \cap c.clo = n.clo$, according to Lemma 14, $n.clo$ should be closed too in T_2, contradiction.

So, we proved $\nexists s, s \in g(n.clo)_{T_1}, s \notin g(c.gen) \cup t$, in another word:

$$g(n.clo)_{T_1} \subseteq g(c.gen)_{T_1} \cup t \qquad (7)$$

Form (1) and (2), we have:

$$g(n.clo)_{T_1} = g(c.gen)_{T_1} \cup t \qquad (8)$$

So, that:
$$\begin{aligned} g(n.gen)_{T_2} &= g(n.gen)_{T_1} - t \\ &= g(n.clo)_{T_1} - t \\ &= g(c.gen)_{T_1} \\ &= g(c.gen)_{T_2} \\ &= c.clo \end{aligned}$$

According to Lemma 16, we can merge the current node with its only child once the current node is proved not closed in T_2. See Table 3, line 21-24.

3.4   Some Optimization

We adopted some optimizations to reduce the runtime of our algorithm.

First, we cached the hashcode of all the transactions in the sliding window with a hash table. In this way the condition check whether t belongs to T_1 could be done in $\mathcal{O}(1)$.

Second, the function $preset()$ and $posset()$ of the GC-Node would be recalculated whenever a new transaction arrives. To store the result in every GC-Node will improve the performance. However, the GC-Nodes might be large enough to occupy a lot of memory. We address this issue in the following item.

Third, compact representation of itemsets. For dense dataset, we represent the itemsets with a bits vector, if the ith bit is 1 in an itemset, then, the itemset contains the ith item, vice versa. This could reduce both the time and the space complexity. For sparse dataset, it's a different story, the bitwise representation is not suitable. So we devised an inversed representation of itemset which contains 4 elements: $< its, inversed, f, t >$, its is an array of short, which stores the ids of items. $inversed$ is a flag indication whether the itemset is represented inversely: if $inversed$ is $false$, the itemset is all the items in its; if $true$, the itemset is all the items from f to t, except for the items in its.

4   Performance Evaluation

We compare our algorithm with CFI-Stream[3], which is the state of-the-art algorithm. A series of synthetic datasets are used. Each dataset is generated by the same method as described in [14], an example synthetic dataset is T10.I6.D100K, where the three numbers denote the average transaction size (T), the average maximal potential frequent itemset size (I) and the total number of transactions (D), respectively.

There are 7 datasets used in our experiments: T4.I3.D100K, T5.I4.D100K, T6.I4.D100K, T7.I4.D100K, T8.I6.D100K, T9.I6.D100K, T10.I6.D100K. The main

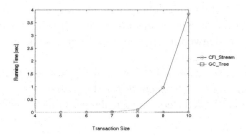

Fig. 2. Runtime Performance

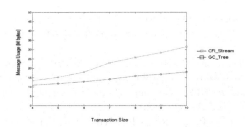

Fig. 3. Memory Usage

difference between them are the average transaction size(T). Figure 2 shows the average processing time for CFI-Stream and GC-Tree over the 100 sliding windows under different average transaction size.

With the increment of the average transaction size, the running time for CFI-Stream increases exponentially since CFI-Stream will check every subset of the current transaction. For GC-Tree, the running time increases slightly since it only checks all the GC-Nodes with $gen \in t$, where t is the current transaction. And the traverse of the tree is likely to stop in the first several levels, because the length of the generators of the GC-Node grows quickly according to the depth of the node.

Figure 3 shows the memory usage of CFI-Stream and GC-Tree according to the transaction size. CFI-Stream requires almost one third less than GC-Tree, because GC-Tree stores all of the generators, the closed itemsets, the presets and the possets in the GC-Nodes. However, by the inversed representation of itemsets, we reduced the memory usage of GC-Tree to be acceptable comparing to CFI-Stream.

From the above discussion, we could see that with a slightly more memory usage, GC-Tree accelerates the online sliding window closed frequent itemsets mining algorithm CFI-Stream dramatically.

5   Conclusions and Future Work

In this paper we proposed a novel algorithm, GC-Tree, to discover and maintain closed frequent itemsets in the current data stream sliding window. The algorithm checks and maintains closed itemsets online in an incremental way. All closed frequent itemsets in data streams can be output in real time based on users specified thresholds. Our performance studies show that with slightly more memory usage, this algorithm is much faster than the state-of-the-art algorithm CFI-Stream when mining data streams online. The algorithm also provide two representations of itemsets so that it could be adaptive for both dense and sparse datasets.

In the future, the algorithm could be improved by only storing closed itemsets with a user specified lower bound of frequent thresholds. In this way, the number of nodes stored in GC-Tree could be reduced dramatically, and this is beneficial for both time and space efficiency.

References

1. Ben Yahia, S., Hamrouni, T., Mephu Nguifo, E.: Frequent closed itemset based algorithms: a thorough structural and analytical survey. ACM SIGKDD Explorations Newsletter 8(1), 93–104 (2006)
2. Lin, C.-H., Chiu, D.-Y., Wu, Y.-H., Chen, A.L.P.: Mining Frequent Itemsets from Data Streams with a Time-Sensitive Sliding Window. In: Proc. of SDM Conf. (2005)
3. Jiang, N., Gruenwald, L.: CFI-Stream: mining closed frequent itemsets in data streams. In: Proc. of KDD Conf., pp. 592–597 (2006)
4. Lucchese, C., Orlando, S., Perego, R.: DCI Closed: A Fast and Memory Efficient Algorithm to Mine Frequent Closed Itemsets. In: Proc. of FIMI Conf., (2004)
5. Lucchese, C., Orlando, S., Perego, R.: Fast and Memory Efficient Mining of Frequent Closed Itemsets. IEEE Journal Transactions of Knowledge and Data Engineering (TKDE) 18(1), 21–36 (2006)
6. Wang, J., Han, J., Pei, J.: CLOSET+: Searching for the Best Strategies for Mining Frequent Closed Itemsets. In: Proc. of KDD Conf., (2003)
7. Zaki, M.J., Hsiao, C.-J.: CHARM: An Efficient algorithm for closed itemsets mining. In: Proc. of SIAM ICDM Conf. (2002)
8. Stumme, G., Taouil, R., Bastide, Y., Pasquier, N., Lakhal, L.: Computing iceberg concept lattices with TITANIC. Journal of Knowledge and Data Engineering(KDE) 2(42), 189–222 (2002)
9. Zaki, M.J., Gouda, K.: Fast vertical mining using diffsets. Technical Report 01-1, Computer Science Dept., Rensselaer Polytechnic Institute (March 2001)
10. Lucchese, C., Orlando, S., Palmerini, P., Perego, R., Silvestri, F.: KDCI: a multistrategy algorithm for mining frequent sets. In: Proc. of ICDM Conf., (2003)
11. Orlando, S., Palmerini, P., Perego, R., Silvestri, F.: Adaptive and resource-aware mining of frequent sets. In: Proc. of ICDM Conf., (2002)
12. Pasquier, N., Bastide, Y., Taouil, R., Lakhal, L.: Discovering frequent closed itemsets for association rules. In: Beeri, C., Bruneman, P. (eds.) ICDT 1999. LNCS, vol. 1540, Springer, Heidelberg (1998)
13. Pei, J., Han, J., Mao, R.: CLOSET: An efficient algorithm for mining frequent closed itemsets. In: Proc. of DMKD Conf. (May 2000)
14. Agrawal, R., Srikant, R.: Fast algorithms for mining association rules. In: Int'l. Conf. on Very Large Databases (1994)

Integration of Distributed Biological Data Using Modified K-Means Algorithm*

Jongil Jeong, Byunggul Ryu, Dongil Shin, and Dongkyoo Shin**

Department of Computer Science and Engineering, Sejong University
98 Kunja-Dong, Kwangjin-Ku, Seoul 143-747, Korea
{jijeong,baramryu}@gce.sejong.ac.kr, {dshin,shindk}@sejong.ac.kr

Abstract. The goals of bioinformatics are the solving of biological questions and the active driving of the work of biologists by offering search and analysis methods for research data. The internet brings us distributed environments in which we can access the databases of various research groups. However, a very large quantity of data always causes trouble, creating crucial problems, such as problems with the search for and analysis of data in these distributed environments. Data clustering can be a solution when searching for data. However, this task is very tedious because its execution time is directly proportional to the volume of data. In this paper we propose a distributed clustering scenario and a modified K-means algorithm for the efficient clustering of biological data, and demonstrate the enhancement in performance that it brings.

Keyword: Distributed clustering, K-means algorithm, Bioinformatics.

1 Introduction

Thanks to the Internet, it is easy to transmit and merge data in distributed environments. Particularly in the bioinformatics field, the Internet has opened up many possibilities for solving biological questions by allowing meaningful bioinformatics data to be shared among research groups [1]. The goals of bioinformatics are the solving of biological questions and the active driving of the work of biologists. Nevertheless, there are difficulties that we must overcome to achieve these ideal goals. The very large quantity of data in bioinformatics creates crucial problems, such as delays when searching for and transmitting data in the distributed environment. In this paper we propose a distributed clustering scenario and modified K-means algorithm that decrease execution time for the clustering of huge quantities of biological data. We then demonstrate the enhancement in performance that these generate.

This paper consists of five sections. In section 2, we introduce compression techniques for XML data and the K-means algorithm for data clustering. In section 3, we consider a scenario and modified K-means algorithm for exchanging and integrating

---

* This study was supported by a grant of the Korea Health 21 R&D Project, Ministry of Health & Welfare, Republic of Korea. (0412-MI01-0416-0002).
** Corresponding author.

T. Washio et al. (Eds.): PAKDD 2007 Workshops, LNAI 4819, pp. 469–475, 2007.

bioinformatics data. In section 4, we evaluate the scenario. Finally, we conclude this paper in section 5.

2 Background

A variety of research has been performed in the quest to overcome the classical problems encountered in achieving the goal of the bioinformatics grid. Representative research has proposed parallel versions of the clustering algorithms and application to distributed databases [2], [3]. Although such researches are actively being performed in the e-business field, this approach is also valuable in the bioinformatics field. As distributed systems become stronger when more resources cooperate, the goals of bioinformatics can be fully achieved as the technology to overcome the problems is put into place. In the following subsections, we introduce the K-means algorithm, which is used as the main algorithm for clustering bioinformatics data, and MAGE-ML (MicroArray Gene Expression Markup Language) [4], which is a language designed to describe and communicate information about microarray based experiments.

2.1 K-Means

Clustering is used to search for or analyze data existing separately in a distributed environment. K-means is a very popular clustering algorithm. The execution steps of the K-means algorithm are as follows [5]:

− Step 1:
Initial central set $[Y_1...Y_K]$ is made by randomly selecting a vector of K from a data set $[X_1...X_N]$.
− Step 2:
If X_N is nearest to Y_i then X_N is so labeled to belong to X_i. Finally, a data set is divided in to $[X_1...X_N]$ that is the K number of clusters.

$$X_i = \{X_n \mid d(X_n, Y_i) \leq d(X_n, Y_j), j = 1,...K\}$$

− Step 3:
This step updates each of the cluster centers (that is, **centroid**) in new clusters generated in Step 2.

$$Y_i = c\,(X_i),\, i = 1,...K$$

− Step 4:
By calculating the sum of the distances between the data and the nearest centers of clusters, calculates the total distortion.

$$D = \sum_{m=1}^{N} d(x_n, y_{i(n)})$$

where $i(n) = k$, if $x_n \in X_k$

– Step 5:
Until distortion does not change properly or reaches the number of repetitions that is set, Steps 2~4 are iterated.

2.2 MAGE-ML

MAGE (MicroArray Gene Expression) consists of three components: MAGE-OM (Object Model), MAGE-ML (Markup Language), and MAGE-STK (MAGE-Software Tool Kit). MAGE-OM is a standard object model for microarray data and consists of 132 classes and 17 packages of MIAME. MAGE-ML, which is produced based on MAGE-OM, is a standard data format based on XML for exchanging microarray data with others. MAGE-STK is an implementation package set for MAGE-OM in various programming languages [6].

A MAGE-ML document has only one markup <MAGE-ML> as a root element and the root element has the following 13 child elements [1], [4], [6]. Each of these elements also has a number of sub-elements. The root element can have all the elements or some of the elements selectively.

- Experiment: information related to a microarray experiment.
- ArrayDesign: information related to array design
- DesignElement: information related to the elements constituting the design
- BioMaterial: information related to materials for a microarray experiment
- BioAssayData: information related to factors from a microarray
- BioSequence: information related to a sequence
- QuantitionType: information related to measurement after a microarray experiment
- Array: information related to each array
- Protocol: information related to the hardware and software needed to perform a microarray experiment
- AuditAndSecurity: information related to a laboratory or researcher
- Description: information related to an added file, including comments about an experiment by a researcher
- HigherLevelAnalysis: information related to an analysis of experimental data

The design purpose of MAGE-OM/ML is to share microarray data with other research groups. Therefore, if each research group is using a MAGE compliant database system, it can collect target microarray data from other groups through various mining techniques. In this case, it is easy to merge the target data from other groups into a repository in a specific research group.

3   Clustering and Merging Large Quantities of Bioinformatics Data in Distributed Environments

In this section, we present an efficient scenario for integrating bioinformatics data in distributed environments.

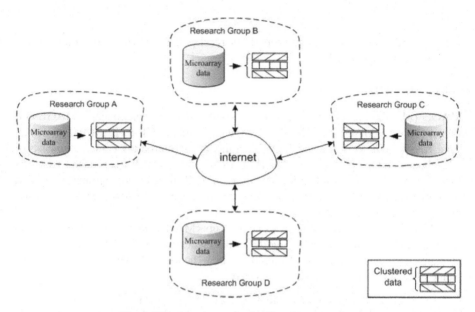

Fig. 1. Distributed repositories for microarray data

Figure 1 shows an experimental environment in which several research groups are linked via the Internet. The basic scenario is to get bioinformatics data from other research groups and merge it into one research group. In this scenario, data clustering and XML compression are the main techniques in which we are interested. According to the order of applying the main techniques, several scenarios can be designed. The probable scenarios are as follows:

- Scenario 1: Each research group (A, B, and C) transmits bioinformatics data to research group D. Research group D clusters data from each research group using the K-means algorithm. That is, clustering is performed only in research group D.
- Scenario 2: Each research group (A, B, and C) clusters the bioinformatics data that is requested by research group D and transmits its centroid (that is, its cluster center) information and data to research group D. Research group D then averages each group's centroid and clusters integrated data by using the averaged centroid. We modified step 1 of the K-means clustering algorithm, which is introduced in section 2.1, and used the averaged centroid as the initial central set. We performed the remaining steps of the algorithm as stated in section 2.1.

By evaluating the scenarios, we will show that Scenario 2 improves performance when clustering, transmitting, and merging very large quantities of bioinformatics data in distributed environments.

4 Evaluation

We did experiments to evaluate the performance of the two scenarios illustrated in the previous section. We experimented with measured bioassay data whose attribute values were extracted from a physical bioassay. A physical bioassay, measured bioassay, and derived bioassay are generated by a BioAssay, which is a stage of a MicroArray experiment. Measured bioassay data has 40 types of attributes and we performed K-means clustering to make two clusters with the selected 7 attributes: CH1I_MEAN, CH2I_MEAN, CH1B_MEDIAN, CH2B_MEDIAN, CH1D_ MEAN, CH2D_MEAN, and CH1I_MEDIAN [6]. (We named these Attr1, ⋯, Attr7 in Table 1).

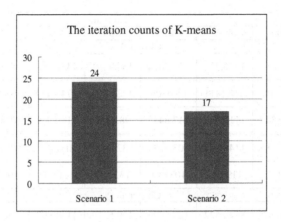

Fig. 2. A comparison of iteration counts in research group D

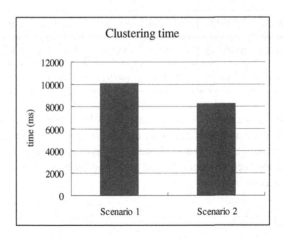

Fig. 3. A comparison of clustering time in research group D

Figure 2 shows a comparison of the iteration counts for the K-means algorithm for the two scenarios. In Scenario 1, where research group D requests data from the other groups (A,B and C) and each research group sends the data to research group D, the number of iterations is 24. In Scenario 2, where research group D requests data and centroid information from the other groups (A,B, and C) and each research group sends data and its centroid to research group D, the number of iterations is 17, by using the averaged centroid. Therefore, we see that using the modified k-means clustering algorithm in Scenario 2 improves performance. We also measured the execution time in milliseconds for K-means clustering in research group D for the two scenarios. As we can see in Figure 3, the execution time is reduced by 20% with Scenario 2.

Table 1. Centroids of attributes in Scenario 2

Data	Centroid						
	Att1	Att2	Att3	Att4	Att5	Att6	Att7
Cluster 0							
Research Group A	1597.866	1686.334	125.5074	154.9082	1472.358	1531.426	1522.416
Research Group B	642.7346	641.1411	68.6306	142.7207	574.1039	498.4204	605.5098
Research Group C	1258.544	1113.168	117.6433	98.3927	1140.911	1014.776	1189.647
Averaged centroid	1166.385	1146.881	103.9271	132.0072	1062.458	1014.874	1105.857
Final centroid	1221.97	1194.27	103.641	133.294	1118.32	1060.97	1162.92
Cluster 1							
Research Group A	11178.88	8743.522	129.8076	160.5319	11049.07	8582.99	11132.37
Research Group B	3677.535	2489.578	75.0644	171.0307	3602.47	2318.547	3660.108
Research Group C	10799.72	8389.727	129.0404	106.001	10670.68	8283.726	10599.97
Averaged centroid	8552.042	6540.942	111.3041	145.8545	8440.738	6395.088	8464.15
Final centroid	10181	7352.93	121.441	140.915	10059.6	7212.02	10080.5

Table 1 shows the centroids that are calculated by clustering from each research group(A,B, and C), the averaged centroid, and the final centroid in research group D. We can see that in both cluster 0 and cluster 1 the averaged centroid is very similar to the final centroid for each attribute. So we get a good initial central set by using each cluster's averaged centroids.

5 Conclusion

This paper proposed an efficient distributed clustering scenario and modified K-means algorithm for clustering biological data in distributed environments. The proposed method reduces the iteration time for clustering and achieves an improvement in clustering time. We expect that this achievement will contribute to the development of bioinformatics and distributed clustering algorithms.

Acknowledgement

We express gratitude to Mr. Junyoung Jeong for his assistance in the experimentation and for his helpful ideas.

References

1. Crignon, I., Grzybek, S., Staedtler, F., Masiello, A., Dressman, M., Taheri, F., Stock, R., Lenges, E., Pitarelli, R., Genesio, F., Reinhardt, M.: An Architecture for Standardization and Management of Gene Expression Data in a Global Organization. In: ECCB (European Conference on Computational Biology), Paris, France, September 27-30 (2003)
2. Zhang, B., Formaml, G.: Distributed Data Clustering Can Be Efficient and Exact, Software Technology Laboratory HPLaboratories, Palo Alto HPL-2000-158 (December 4, 2000)
3. Albert, O., Zomaya, Y.: Tarek E1-Ghazawi.:Parallel and Distributed Computing for Data Mining. IEEE Concurrency 7(4), 11–13 (1999)
4. MAGE-ML, http://www.mged.org/Workgroups/MAGE/mage-ml.html
5. Zhang, Y.-F., Mao, J.-L., Xiong, Z.-Y.: An efficient clustering algorithm. 2003 International Conference on Machine Learning and Cybernetics 1, 261–265 (2003)
6. Martin, W., Horton, R.M.: MageBuilder: A Schema Translation Tool for Generating MAGE-ML from Tabular Microarray Data. In: 2003 IEEE Proceedings of Computational Systems Bioinformatics (CSB 2003), pp. 431–432 (2003)

A Parallel Algorithm for Enumerating All the Maximal *k*-Plexes*

Bin Wu and Xin Pei

Beijing Key Laboratory of Intelligent Telecommunications Software and Multimedia, Beijing
University of Posts and Telecommunications, Beijing 100876, China
wubin@bupt.edu.cn, peixin11@gmail.com

Abstract. Finding and enumerating subgraphs of different structures in a graph
or a network is one of the fundamental problems in combinatorics. One of the
earliest subgraph models is clique. However, the clique approach has been
criticized for its overly restrictive nature. *k*-plex is one of the models which are
introduced by weakening the requirement of clique. The problem to enumerate
all the maximal *k*-plexes is NP complete. We consider this problem and propose
an algorithm *Pemp* (Parallel Enumeration of all Maximal *k*-Plexes) for
enumerating all the maximal *k*-plexes. We also propose a strategy to accelerate
the pruning. A diameter pruning strategy is proposed. This strategy reduces the
number of small maximal *k*-plexes and improves the performance greatly. We
also state the parallel edition of our algorithm to analysis large networks and a
load balancing strategy is given. In addition, we evaluate the performance of
Pemp on random graphs.

1 Introduction

Graph-based data mining (GDM) becomes a hot-spot research of KDD (Knowledge
Discovery in Databases). Finding and enumerating subgraphs of different structures in
a graph or a network is one of the fundamental problems in combinatorics, which has
many important applications in GDM.

One kind of structures is termed community or cohesive subgroups. Given a graph,
a community is a subgraph, whose nodes are tightly connected, i.e. cohesive.
Members of a cohesive subgroup tend to share information, have homogeneity of
thought, identity, behavior, even food habits. Modeling a cohesive subgroup
mathematically has long been a subject of interest in social network analysis. The
earliest graph models used for studying cohesive subgroups was the clique model.
However, the clique approach has been criticized for its overly restrictive nature [5].
In real world, there is a lot of noise. Thus a strict definition of clique is not very
helpful for real applications. Researchers extend the definition of clique by weakening
the requirements of different aspects. The models *k-clique* and *k-clan* are defined by
weakening the requirement of reachability (the diameter of clique is 1), and the

* This work is supported by the National Science Foundation of China under grant number
60402011and the co-sponsored project of Beijing Committee of Education SYS100130422.

models *k-core*, *quasi-clique* and *k-plex* are defined by weakening the requirement of closeness(all pairs of vertices are connected).

k-plex is introduced in [1] and k-plexes with low k values ($k = 2, 3$) provide good relaxations of clique that closely resemble the cohesive subgroups that can be found in real life social networks. There are few works about finding all the maximal k-plexes in a graph as we have known.

The main contribution of this paper is to present a parallel algorithm for enumerating all the maximal k-plexes in graphs and describe the strategy to reduce the number of small maximal k-plexes. The rest of this paper is organized as follows. Basic definitions, notations and related works are presented in Section 2. Section 3 presents our algorithm, the optimization methods and the parallel edition of *Pemp*. Finally the paper is concluded with a summary and directions for future work in Section 5.

2 Definitions, Notations and Related Works

Let $G = (V, E)$ be a simple undirected graph with vertices $V = \{v_1, v_2, ... v_n\}$ and edges $E = \{e_1, e_2, ... e_n\}$. $|G|$ denotes the number of vertices in G; $d_G(u,v)$ denotes the length of the shortest path between vertices u, v; $\deg_G(v)$ denotes the degree of vertex v in graph G and $diam(G) = \max_{u,v \in V} d_G(u,v)$ denotes the diameter of G. Denote by $G[S] = (S, E \cap S \times S)$, the subgraph induced by S.

A subset of vertices $S \subseteq V$ is a clique if all pairs of its vertices are adjacent. A clique is maximal if it is not a subset of another clique. The problem to enumerate all the maximal cliques is termed "Maximal Clique Problem". The base algorithm about Maximal Clique Problem called BK [4] was first published in 1973.

The formal definition of k-plex is stated as follows:

Definition 1. A subset of vertices $S \subseteq V$ is a k-plex if $\deg_{G[s]}(v) \ge |S| - k$ for every vertex v in S.

A k-plex is a maximal k-plex if it is not a subgraph of another k-plex and it is maximum if it is the largest maximal k-plex. "Maximal k-plex Problem" is the generalization of "Maximal Clique Problem" because a 1-plex is a clique. So it is more difficult for its weak restriction. [2] proves that the decision edition of Maximum k-plex Problem (finding the maximum k-plex) is NP complete. Because the Maximal k-plex Problem is the general edition of Maximal Clique Problem, we use the basic framework of BK algorithm to find all the maximal k-plexes. However, *Pemp* is more complex than BK because of the weak requirement of k-plex.

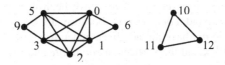

Fig. 1.

We should point out the fact that Definition 1 dose not emphasize the requirement of the connectivity. The triangles $\{10, 11, 12\}$ and $\{0, 1, 6\}$ in Fig.1 can also form a k-plex ($k \geq 4$). This kind of disconnected k-plex is trivial. The number of trivial k-plex is tremendous while it is useless to analyze these k-plexes. So the object of *Pemp* is just to find all the non-trivial k-plexes (hereinafter called k-plexes unless noted).

3 Algorithm Pemp

3.1 Basic Idea and Method

We use depth first search method to find all the maximal k-plexes. [1] states a property of k-plex as follows:

Theorem 1. Any vertex-induced subgraph of a k-plex is a k-plex.

According to Theorem 1, we can expand a k-plex until we get a maximal k-plex. If a vertex v is adjacent to at least one vertex in a k-plex S, we say that v is connected to S or v and S is connected.

Definition 2. Assume that a vertex v in graph G is connected to a k-plex S of G. v can expands S if the vertex-induced subgraph of G, $S \cup \{v\}$, is a k-plex.

Theorem 2. A vertex v can expand a k-plex S if v satisfies the two conditions as follows:

1. v is adjacent to at least $|S| - k + 1$ vertices in S
2. Any vertex u in S is adjacent to at least $|S| - k + 1$ vertices in $S \cup \{u\}$

Assume that v is associated with a counter, counting the number of vertices in S that v is not connected to. v can satisfy condition 1 if the counter of v is not greater than $k - 1$. A definition of critical vertex is introduced for condition 2.

Definition 3. A vertex u in a k-plex S is a critical vertex if u is not connected to $k - 1$ vertices in $\{S - u\}$.

To satisfy condition 2 the vertex to expand S should be selected in the neighborhood of critical vertex if there is a critical vertex in S. Otherwise there will be k vertices which the critical vertex u disconnected to in $S \cup \{u\}$. The candidate vertex used to expand S should be chosen in the intersection of neighborhoods of all the critical vertices if there are n critical vertices.

Like BK algorithm, *Pemp* maintains three vertices sets: *compsub*, *candidate* and *not*. The vertices in *compsub* are found k-plex. The vertices in *candidate* are used to expand the *compsub* set. The vertices in *not* have been used in previous search and can not be used to expand *compsub* set because all the extensions of *compsub* containing any vertex in *not* have already been generated. The vertices set of *compsub* should always be a k-plex and the vertices in *candidate* and *not* should always be able to expand *compsub*. There are another two vertices sets are generated from previous three sets. *connected_candidate* (*connected_not*) is the subset of *candidate* (*not*) and it contains all the vertices connected to *compsub* in *candidate* (*not*). Every vertex in every set is associated to a counter. The counter of vertex u in *compsub* counts the number of points that u is disconnected to in $S - \{u\}$. The counter of vertex u in *candidate* and *not* counts the number of disconnected *compsub* vertices of u.

Input: A Graph
Output: All the Maximal k-plex
Method:
1: Read the Graph and Copy all the vertices to
candidate set. Let *compsub* and not be empty;
2: **while** there are vertices in *candidate*:
3:         Call *FindAllMaxKplex* (*compsub, candidate, not*);
4:         Move the used vertex to *not* upon the return;
5: **endwhile;**
Function: *FindAllMaxKplex* (*inputCompsub, nputCandidate,*
inputNot)
6: Copy *inputCompsub* to *compsub*. Copy *inputCandidate* to
candidate. Copy *inputNot* to *not*;
7: Select a vertex v in *connected_candidate* of
inputCandidate in lexicographic order;
8: Move v to *compsub* and Update the counters of the
vertices in *compsub*;
9: **if** there are n critical vertices in *compsub* ($n > 0$):
10:         Compute the intersection C of the
neighborhoods of the n critical vertices and Remove all
the vertices in *candidate* and *not* which are not in C;
11: Update the counters of the vertices in *candidate*
and *not* and Remove the vertices of *candidate* and *not* if
the vertices can not expand *compsub* (the counter is
greater than k-1);
12: Generate *connected_candidate* of *candidate* and
connected_not of *not*;
13: **if** *connected_candidate* and *connected_not* are empty:
14:         *compsub* is a maximal k-plex, Return v;
15: **if** *connected_candidate* is empty and *connected_not*
is not empty:
16:         Return that there are no vertices which can
expand *compsub*;
17: **if** *connected_candidate* and *connected_not* is not
empty:
18:         **while** there are vertices in *candidate*:
19:         Call *FindAllMaxKplex*(*compsub, candidate, not*);
20:         Move the used vertex to *not* upon the return;
21:         **endwhile;**
22: Return v;

Fig. 2. Algorithm *Pemp*: the basic case

The found k-plex (*compsub*) becomes larger by moving the vertices which can expand the found k-plex from *candidate* to *compsub* until *compsub* is a maximal k-plex. It is easy to find out that the combination of any m vertices in candidate is a k-plex when $m \leq k$. There will be great cost and many trivial k-plexes will be found. Taking the Fig.1 as an example, there is no vertex which can expand compsub if compsub is {6, 9} by Theorem 2. So vertex 6 and 9 form a trivial maximal k-plex without analysis significance. That is why we use the *connected_candidate* set. In order to prevent trivial k-plex we only use the vertices connected to found k-plex to expand *compsub*. Copy the *candidate* vertices which are connected to *compsub* into *connected_candidate*. Then we select vertices from *connected_candidate* to expand *compsub*. This method does not only prevent the algorithm from finding trivial k-plexes but also improve its performance because it will not traverse all the combinations of any m ($m \leq k$) vertices in *candidate*. The basic algorithm is summarized in Fig. 2.

A necessary condition for having created a maximal k-plex is that the *connected_candidate* is empty. This condition, however, is not sufficient, because if now *not* is not empty, we know from the definition of *not* that the present configuration of *compsub* has already been contained in another configuration and is therefore not maximal. We may now state that the *compsub* is maximal k-plex if and only if *connected_candidate* and *connected_not* is empty.

3.2 Acceleration of Pruning

The pruning condition is that *connected_candidate* is empty while *connected_not* is not empty. We can bring forward the pruning if we can predict that the *connected_not* will not be empty finally at an early stage. If a vertex v in *connected_not* is connected to all the vertices in *compsub* and *candidate*, *connected_not* will not be empty. So we can prune as soon as we find such a vertex. For this object another counter is added to the vertex in *not*. The counter defined in section 3.1 is termed counter1 and the new counter is called counter2, which counts the number of vertices in both *candidate* and *compsub* that the *not* vertex disconnected to. The pruning condition is satisfied when the counter2 equals 0.

Pruning can be accelerated if we can make the counter2 of vertices in *not* decrease faster. We introduce a vertices set termed *prunable_not*. The entire *not* vertices which connected to all the vertices in *compsub* (Because the counter2 of vertex will not be 0 if the vertex is not adjacent to all the vertices in *compsub*.) are copied to *prunable_not*. When expanding the *compsub* we will not select the next *candidate* vertex in lexicographic order but the vertex which is disconnected to the vertex with minimum counter2 in *prunable_not* set. Note that no counter2 is ever decreased by more than one when a vertex is moved to the *compsub*. Let v be the vertex which has minimum counter2 in *prunable_not*. If we keep selecting *candidate* vertices disconnected to v, the counter2 of v will be decreased by one at every repetition. Obviously no other *not* vertices's counter2 can go down more rapidly than v.

It should be pointed out that the method to accelerate the pruning can not be used during the whole process of the algorithm. Before there are critical vertices in *compsub*, *connected_candidate* may not contain the vertex which is disconnected to the vertex in *prunable_not* which has minimum counter2. So the vertex can only be selected in lexicographic order during this period. Additionally, this method can be used when *compsub* is empty. In conclusion, we can replace line 7 in Fig. 2 by calling the function in Fig. 3 to select a better vertex.

Function: *SelectCandiate (compsub, candidate, not)*
1: **if** *compsub* is not empty and there are no critical vertices in *compsub*:
2:        Return next vertex in *connected_candidate* of *candidate* in lexicographic order
3: **endif;**
4: **else:**
5: Copy all the vertices in *connected_not* of *not*, which are adjacent to all the vertices in *compsub*, to *prunable_not* set;
6:        **if** *prunable_not* is not empty:
7:        **if** the minimum counter2 of the vertices in *prunable_not* is 0:
8:            Return that there is no vertex can expand *compsub;*
9:        **endif;**
10:        **else:**
11:            Return the *connected_candidate* vertex which is not connected to the vertex in *punable_not* which has minimum counter2;
12:        **endelse;**
13:        **endif;**
14   **else:**
15:            Return next vertex in *connected_candidate* of *candidate* in lexicographic order;
16:        **endelse;**
17: **endelse;**

Fig. 3. The function to find a vertex to expand *compsub*

Although the method can bring forward the pruning, the overhead increases because of the extra counter. The additional cost may even countervail the effect of this method, so further optimization is necessary. We can consider the process of the algorithm as traversing the backtracking trees whose nodes correspond to the vertex which is selected to expand the found *k*-plex. Let d_1 be the degree of the root of the first tree to be traversed. We know that the algorithm needs to traverse $|G| - d_1 + 1$ trees if we want to find all the maximal *k*-plexes. Obviously the tree rooted by the vertex which has maximum degree should be traversed first.

3.3 Diameter Pruning

Clearly it is not all the vertices in the graph can form a non-trivial *k*-plex with a certain vertex *v* but only the vertices within finite hops of *v*. The size of *k*-plex has something with its diameter. We find the general relationship between the size of *k*-plex and its diameter as follows.

Theorem 3. Let *G* be a *k*-plex. If |G|>2k-d, then $diam(G) \leq d$.

We can get the follow corollary from Theorem 3.

Corollary 1. The maximal k-plex, whose size is greater than m, and which contains vertex v can be found within $2k - m$ hops of v.

We only need to find all the non-trivial maximal k-plexes whose sizes are larger than k if we want to find all the maximal k-plexes because the maximal k-plexes whose sizes are smaller than k will be found inevitably by *Pemp*. By Corollary 1, we can find all the maximal k-plexes which contain vertex v within k hops of v and the vertices in *candidate* and *not* which are not in k hops of v can be removed. The next *candidate* vertex should be within k hops of both u and v when *compsub* is $\{u, v\}$. In our experiment, this method is only used when the size of *compsub* is 1. And we can also only find the maximal k-plexes with sufficient size by Corollary 1.

3.4 Parallel Model of *Pemp*

The algorithm traverses all the trees whose roots are the vertices of graph G to find all the maximal k-plexes. Every vertex will be moved to the *not* set after the backtracking of the tree rooted by it. The backtracking of every tree only depends on the *candidate* and *not*. If the order in which the vertices are selected is $\{v_1, v_2...v_n\}$ when *compsub* is empty (We can know the order before the algorithm begins.), the traverse of the tree rooted by the vertex v_m begins with *not* set $\{v_1, v_2, ..., v_{m-1}\}$ and *candidate* set $\{v_m, v_{m+1}, ..., v_n\}$. It is unnecessary to begin after the first m-1 trees have been traversed. Thus, we can traverse the trees collaterally. Practically, the number of processing elements of the parallel platform is often much smaller than the number of vertices in graph G, so mapping techniques are required.

Suppose that the order in which vertices are selected is $\{v_1, v_2...v_n\}$ when *compsub* is empty. A simple mapping scheme is to assign the n tasks to P processing elements sequentially. However, this scheme suffers from load unbalancing. Obviously the size of the tree rooted by v_1 is larger than v_n because the vertices $v_1, v_2, ..., v_{n-1}$ have been moved to *not* when the tree rooted by v_n are traversed. Let $I_p \in \{1, 2, ..., P\}$ denotes the index of processing elements. We can solve this problem by defining the tasks set on a single processing element I_p as follows:

$$\left\{ v_i \middle| i = I_p + j \times P \vee i = n - \left(I_p + j \times P\right), I_p + P \times j \le \frac{n+1}{2} j = 0, 1, ..., \left\lfloor \frac{n}{P} \right\rfloor \right\} \qquad (1)$$

4 Experiment Result

To evaluate the performance of *Pemp*, we implement it in C++ and run it on random graphs. Our experiments are performed on a Cluster (84 3.2GHz processors with 2Gbytes memory on each node, Linux AS3). The number of vertices and edges of the random graphs and the runtime (in seconds) on 1 CPU (Sequential) and on 20 CPUs (Parallel) when k=1, 2, 3, 4 are shown in Table 1. Many other results of experiments, witch show the strategy of acceleration of Pruning and diameter pruning are very effective, can not be covered because of space constraints.

Table 1. Runtime on 1 CPU (Sequential) and on 20 CPUs (Parallel) when k=1, 2, 3, 4

|V|	|E|	Sequential (1 CPU)				Parallel (20 CPUs)			
		k=1	k=2	k=3	k=4	k=1	k=2	k=3	k=4
1000	14432	3.60	26.7	595	12158	0.43	2.34	47.2	977
2000	28709	10.5	70.6	1416	28601	3.15	8.18	105	2000
4000	58063	60.2	307	4896	89099	22.7	45.1	418	6568
8000	116276	436	1860	23804	-	161	269	1944	29810
16000	231622	3340	13550	-	-	1214	1976	13101	186938

5 Conclusion and Future Work

By allowing some strangers or noise in a cohesive subgroup, k-plex provides a more realistic alternative to model cohesive subgroups in a real network. So we propose *Pemp* algorithm and we hope most of the applications which use maximal clique can obtain better performance if clique is superseded by k-plex. We also propose the optimization for the pruning and the size of maximal k-plex and the parallel model of *Pemp* to analysis large networks. The future work mainly focuses on improving the performance. We are currently studying the strategy of pruning and load balancing to improve *Pemp*.

References

[1] Seidman, S.B., Foster, B.L.: A graph theoretic generalization of the clique concept. Journal of Mathematical Sociology 6, 139–154 (1978)
[2] Balasundaram, B., Butenko, S., Hicks, I.V., Sachdeva, S.: Clique Relaxations in Social Network Analysis: The Maximum k-plex Problem (2006), http://ie.tamu.edu/people/faculty/Hicks
[3] Hanneman, R.: Introduction to Social Network, Methods (1998), http://faculty.ucr.edu/~hanneman/networks/nettext.pdf
[4] Bron, C., Kerbosch, J.: Algorithm 457, Finding all cliques of an undirected graph. Proceedings of the ACM 16(9), 575–577 (1973)
[5] Alba, R.: A graph-theoretic definition of a sociometric clique. Journal of Mathematical Sociology 3, 113–126 (1973)

A Multi-dependency Language Modeling Approach to Information Retrieval

Keke Cai, Chun Chen, Jiajun Bu, Guang Qiu, and Peng Huang

College of Computer Science, Zhejiang University, Hangzhou, Zhejiang 310027, China
{caikeke,chenc,bjj,qiuguang,huangpeng}@zju.edu.cn

Abstract. This paper presents a multi-dependency language modeling approach to information retrieval. The approach extends the basic KL-divergence retrieval approach by introducing the hybrid dependency structure, which includes syntactic dependency, syntactic proximity dependency and co-occurrence dependency, to describe dependencies between terms. Term and dependency language models are constructed for both document and query. The relevant between a document and a query is then evaluated by using the KL-divergence between their corresponding models. The new dependency retrieval model has been compared with other traditional retrieval models. Experiment results indicate that it produces significant improvements in retrieval effectiveness.

Keywords: Term dependency, language model, information retrieval.

1 Introduction

Language Modeling (LM) is a new retrieval approach that has been used in many recent studies in information retrieval. Experiments have proved its promising retrieval effectiveness compared with traditional retrieval models. However, most existing LM approaches usually assume independence between terms, which means that terms are statistically independent from each other. Although this assumption makes the retrieval models easier to implement, it is not the truth in actual textual data.

The term "dependency" can be understood from two different perspectives: dependency between words within a query or within a document; dependency between query words and document words [1]. In this paper, the study will concentrate on the first kind of dependency. Some researches have been carried out to study term dependencies in LM retrieval approaches. To realize the effect of term dependency in the practice of LM retrieval model, two primary problems should be taken into account: how to define the dependencies between terms and how to apply term dependencies to LM retrieval model.

- Different approaches adopt different criteria to define the dependencies between terms, which, however, are always assumed from just one point of view. A question that one may raise is, besides exclusively using one form of dependency, can we integrate multi forms of dependency together?
- Most dependency LM approaches to IR base their implementations on query-likelihood model. However, the lacking of query model always makes them inconvenient to deal with valued dependency information. It is considered a necessary to introduce query model into retrieval.

T. Washio et al. (Eds.): PAKDD 2007 Workshops, LNAI 4819, pp. 484–491, 2007.

The two problems above outline the main tasks of this paper. To solve them, a novel dependency retrieval model is proposed. This retrieval model suggests an extension to the existing LM approach to IR by introducing dependency models for both query and document. Relevance between document and query is then evaluated by reference to the Kullback-Leibler divergence (KL-divergence) between their dependency models [2]. This paper also introduces a novel hybrid dependency structure for describing dependencies between terms. This hybrid dependency structure allows integration of various sources of dependency, including syntactic dependency, syntactic proximity dependency and co-occurrence dependency, within a single framework. Moreover, instead of estimating term dependencies statistically on the scale of whole documents, our model concludes the hybrid dependency structure on the level of sentence, which therefore limits the dependencies to the most important relationships that are useful for retrieval.

The rest of this paper is organized as follows. Section 2 introduces the related works. Section 3 presents our proposed dependence model and the methods used for parameter estimation. In section 4, a set of experiments is presented. At the end, section 5 concludes the whole paper.

2   Related Works

Since the proposal of the first language modeling approach to IR [3], various IR approaches based on language modeling have been proposed. The retrieval model based on KL-divergence is one of the popular ones. In this retrieval model, language models are associated with both query and document. Relevance between query and document is then evaluated by the divergence between models of document and query:

$$D(\theta_Q \parallel \theta_D) = \sum_{w \in V} P(w \mid \theta_Q) \log \frac{P(w \mid \theta_Q)}{P(w \mid \theta_D)} . \tag{1}$$

where V is a vocabulary, θ_Q and θ_D are language models of query Q and document D. In this paper, KL-divergence retrieval model is adopted as the underlying retrieval framework. The relevance of D with respect to Q is then measured by $-D(\theta_Q \parallel \theta_D)$.

The concept of term dependency has been discussed in many studies. In some traditional approaches, term dependency is always in the form of phrases [4] [5] or term co-occurrences [6]. For early works of LM based approaches, term dependency is formed in the style of bigram or biterm [7] [8]. Two terms that are adjacent are assumed to be related with each other. Recently, statistical and linguistic information is also utilized in the exploration of term dependency. In [9], Nallapati and Allan represent term dependencies in a sentence using a maximum spanning tree. In [10], an existing dependency parsing is utilized to detect the linkage of a term sequence. Experiments have illustrated the usefulness of these identified dependencies in the succeeding retrieval. Although these approaches have different definitions for term dependency, they commonly consider term dependency in the context of sentence. As illustrated in [11], "the most significant dependencies between terms occur within a sentence". In our work, term dependency is also viewed in such a way. What is different is that in our method more than one forms of dependency are considered. Moreover, just dependencies that are most important to retrieval are noticed.

3 Multi-dependency Language Modeling Approach to IR

3.1 Model of the Dependency Based Retrieval

KL-divergence retrieval model allows using any language model to describe query and document. Therefore, our method evaluates relevance of document through two components. The first component is to evaluate the similarity of term distributions between query and document. Specifically, given the term distributions of query Q and document D: $\theta_{Q,T}$ and $\theta_{D,T}$, relevance between Q and D can be measured by:

$$score_w(D,Q) = \sum_w P(w \mid \theta_{Q,T}) \log P(w \mid \theta_{D,T}) - \sum_w P(w \mid \theta_{Q,T}) \log P(w \mid \theta_{Q,T}) \ . \tag{2}$$

The second part of formula 2 is identical for all documents, it can be removed.

The second component measures the similarity between query and document based on term dependency. Let $\theta_{Q,P}$ and $\theta_{D,P}$ be dependency models of query Q and document D, L be a set of dependencies, relevance between Q and D is evaluated by:

$$score_{DP}(D,Q) = \sum_{l \in L} P(l \mid \theta_{Q,P}) \log P(l \mid \theta_{D,P}) - \sum_{l \in L} P(l \mid \theta_{Q,P}) \log P(l \mid \theta_{Q,P}) \ . \tag{3}$$

Similarly, the second part is document-independent constant and can be dropped.

The above two components observe the relevance between query and document from different perspectives. To incorporate them together, the simplest strategy, i.e., linear interpolation method, is adopted.

$$score(D,Q) = (1 - \lambda) * score_w(D,Q) + \lambda * score_{DP}(D,Q) \ . \tag{4}$$

Term dependency in this paper is explored in the context of sentence. According to different degrees of closeness between terms, three forms of dependency are formed: syntactic dependency, proximity dependency, and co-occurrence dependency. Syntactic dependency describes the direct syntactic relation between terms, while proximity dependency describes the relation between terms within a certain syntactic distance. Co-occurrence dependency relaxes the requirement of syntactic relationship between terms and defines the dependency based on co-occurrence among terms. Considering term dependency in the ways above, formula 3 is then reformulated as:

$$score_{DP}(D,Q) = \sum_{R_i} \alpha_i * \sum_{l \in R_i} P(l \mid \theta_{Q,P}) \log P(l \mid \theta_{D,P}) \ . \tag{5}$$

where R_i represents the dependencies of the i-th form and the coefficient α_i is to control the influence of each form of dependency, $i \in [1,3]$.

Syntactic dependency and proximity dependency are constructed on the basis of syntactic relationships between terms. Here, a special kind of syntactic relationship, namely dependency syntactic [12], is concerned. Term dependency is consequently described in the form of dependency parse tree [13].

3.2 Parameter Estimations

As shown in formula 3, four parameters should be estimated in our proposed retrieval model, i.e., $P(w \mid \theta_{Q,T})$, $P(w \mid \theta_{D,T})$, $P(l \mid \theta_{Q,P})$ and $P(l \mid \theta_{D,P})$.

(1) $P(l \mid \theta_{Q,P})$; Dependency proposed in this paper is defined in the context of sentence. However, most web queries are not grammatical sentences. To identify the dependency structure of a query, one may suggest applying statistical analysis over sentences in the document collection. However, performing such an analysis is computationally expensive. This paper adopts a solution based on pseudo relevance feedback. The idea underneath is to utilize feedback to obtain the sentences relevant to the given query. Relations between query terms are then statistically evaluated. Here, relevance of sentence is simply evaluated by the number of common terms contained in query and sentence.

Given the sentences relevant to the given query, syntactic dependencies between terms in each sentence are obtained. Dependency probabilities between query terms therefore can be generalized by statistical method. In this paper, the backoff scheme [10] is adopted. The probability of $P(l \mid \theta_{Q,P})$ is then computed as:

$$P(l \mid \theta_{Q,P}) = \lambda_1 * E_1(w_i, w_j) + (1 - \lambda_1) * (\lambda_2 E_2(w_i, w_j) + (1 - \lambda_2) E_3(w_i, w_j)) \ . \tag{6}$$

where
$$\lambda_1 = \frac{\delta_1}{\delta_1 + 1} \quad \lambda_2 = \frac{\delta_2 + \delta_3}{\delta_2 + \delta_3 + 1}$$

$$E_1(w_i, w_j) = \frac{\eta_1}{\delta_1} \quad E_2(w_i, w_j) = \frac{\eta_2 + \eta_3}{\delta_2 + \delta_3} \quad E_3(w_i, w_j) = \frac{\eta_4}{\delta_4} \ . \tag{7}$$

$$\eta_1 = c(w_i, w_j, l) \quad \delta_1 = c(w_i, w_j) \quad \eta_2 = c(w_i, *, l) \quad \delta_2 = c(w_i, *)$$
$$\eta_3 = c(*, w_j, l) \quad \delta_3 = c(*, w_j) \quad \eta_4 = c(*, *, l) \quad \delta_4 = c(*, *)$$

where $c(w_i, w_j, l)$ is the number of w_i and w_j co-occurring in the same sentences with dependency relation l and $c(w_i, w_j)$ is the number of w_i and w_j co-appearing in the same sentences.

(2) $P(l \mid \theta_{D,P})$; This paper uses the linear interpolation smoothing approach to estimate the value of $P(l \mid \theta_{D,P})$. Based on the dependency models of document D and the entire document collection C, $P(l \mid \theta_{D,P})$ is finally defined as:

$$P(l \mid \theta_{D,P}) = (1 - \gamma) * P(l \mid D) + \gamma * P(l \mid C) . \tag{8}$$

where $P(l \mid D)$ and $P(l \mid C)$ is computed by using the same approach of $P(l \mid \theta_{Q,P})$. The only difference is that the computation of $c(w_i, w_j, l)$, $c(w_i, w_j)$ etc are implemented in the context of document D and collection C.

(3) $P(w \mid \theta_{Q,T})$; The generation probability of term from the query Q is evaluated by the maximum likelihood estimate. Formally, $P(w \mid \theta_{Q,T})$ is evaluated by:

$$P(w \mid \theta_{Q,T}) = \frac{c(w, Q)}{\sum_w c(w, Q)} . \tag{9}$$

where $c(w, Q)$ denotes the number of times term w occurring in the query Q.

(4) $P(w \mid \theta_{D,T})$; Experiments in [14] show that the smoothing methods of Jelinek-Mercer and Dirichlet clearly have a better average precision than absolute discounting. In this paper, we select Dirichlet smoothing method to compute the probability $P(w \mid \theta_{D,T})$. The model is given by:

$$p(w \mid \theta_{D,T}) = \frac{c(w,D) + \mu P(w \mid C)}{u + \sum_{w \in D} c(w,D)} .$$ (10)

where μ is the parameter of the Dirichlet distribution and always is set to around 2000, $P(w \mid C)$ denotes the probability of w occurring in the collection C and is computed by the maximum likelihood estimate.

4 Experiments

This section will present the experimental results of our proposed retrieval model. To verify the applicability of term dependency to the actual web retrieval, our experiments are specially performed on a web document collection. To construct such a document collection, 46 queries are selected from the query log of AlltheWeb[1] 2002. For each of these queries, the top 100 pages are selected by using Meta-search engine.

In this paper, the proposed dependency retrieval model (EKLM) is compared with the probabilistic retrieval model BIR (BIRM) and some state-of-the-art language modeling approaches including unigram model (UGM), bigram model (BGM) and KL-divergence model (BKLM). These retrieval models are performed by using the Lemur toolkit[2]. The result of unigram language model provides the baseline from which to compare other retrieval models. In our experiments, to solve the problem of data sparseness, some smoothing methods are implemented. The unigram model is smoothed by Dirichlet method. The Bigram model involves a linear interpolation of the bigram probability with the unigram probability and the KL-divergence model is implemented with Dirichlet smoothing.

In these retrieval models, several parameters need to be determined, such as λ and α_i in formulas 4, 5 and the other smoothing parameters. For simplicity, our experiments determine these parameters empirically. This paper will narrates the parameter tuning process of the proposed models in detail and just reports the best effectiveness of other retrieval models.

Table 1. Average precision of each retrieval model

	AvgP	%change	AvgR	%change
BIRM	0.2343	+5.0%	0.5352	-1.27%
UGM (μ=2000)	0.2232	--	0.5421	--
BGM (χ=0.5)	0.2475	+10.9%	0.5879	+8.4%
BKLM (μ=2000)	0.2298	+3.0%	0.5215	-3.8%
EKLM (λ= 0.8) ($\alpha1$=0.1,$\alpha2$=0.1$\alpha3$=0.8)	0.2581	+15.6%	0.6087	12.3%

---

[1] http://www.alltheweb.com
[2] http://www.lemurproject.org

Table 1 shows the average precision obtained by each experimental model when implemented on three document collections. The percentages in the table are the relative changes with respect to the baseline approach. By comparing EKLM with other retrieval models, we can see the proposed dependency model achieve clear improvements in retrieval effectiveness. The precision and recall gain over the unigram model are respectively 15.6% and 12.3%. By incorporating term dependency into KL-divergence retrieval model, the average precision changes from 0.2298 to 0.2581.

As introduced above, the proposed retrieval can be decomposed into two components. The influence of these two components is controlled by the parameter λ, as in equation 5. To make clear the effect of each component on retrieval, we change the value of λ in the series of experiments. In these experiments, parameters $\alpha1$, $\alpha2$ and $\alpha3$ are respectively set to 0.1, 0.1 and 0.8. Table 2 shows the average precision with different value of λ. As shown in Table 2, when λ is set to 0.8, the performance is optimal.

Table 2. P@k of the proposed retrieval model with the change of λ

λ	0.9	0.8	0.7	0.6	0.5	0.4
P@10	0.4031	0.4161	0.3933	0.3511	0.3317	0.3038
P@20	0.2272	0.2886	0.2586	0.2372	0.2221	0.2016
P@30	0.1487	0.1789	0.1582	0.1461	0.1252	0.1206

Three kinds of dependency are considered in this retrieval model. To explore the effect of each form of dependency on retrieval, a series of experiments are implemented with the change of parameters α_1, α_2 and α_3. The results are shown in Fig. 1. In Fig.1, P@10, P@20 and P@30 represent precisions at 10, 20 and 30 ranked documents respectively. It is observed that the model achieves a significant performance when the parameters are set to 0.1, 0.1 and 0.8. As aforementioned, parameter α_i is used to control the influence of each form of dependency. This shows that the relative loose dependency relationship is more suitable to web queries. However, this does not negate the effect of other forms of dependency. As shown in Fig. 1, P@10 of each model will change along with the changes of parameters, but P@20 and P@30 change slightly. It suggests that every form of dependency can provide valued dependency information for retrieval. Our experiments show that each of these three forms of dependency can be used independently. However, when they work together can produce more improvement. It can be learned from Table 3.

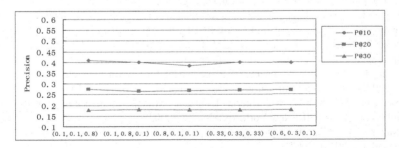

Fig. 1. Average precision of each expanded retrieval model with the change of values of $\alpha1$, $\alpha2$ and $\alpha3$

Table 3 illustrates the precision comparisons between the model *HM* that synthetic-cally consider three forms of dependency and models *SR, PR, CR* that respectively consider syntactic relevancy, proximity relevancy, and co-occurrence relevancy between terms. As shown in Table 3, a significant increase in performance is visible when all three forms of dependency are considered together.

Table 3. Average precision of each retrieval model with or without consideration of all forms of dependency

	HM	*SR*	*PR*	*CR*
AvgP	0.2842	0.2725	0.2713	0.2682

5 Conclusions

In this paper, a novel multi-dependency language modeling approach for information retrieval is proposed. It extends the state-of-the-art KL-divergence approach to information retrieval by introducing dependency models for both query and document. This paper also introduces a hybrid dependency structure, which views term dependency respectively from three perspectives, including syntactic dependency, syntactic proximity dependency, and co-occurrence dependency. Experiments have proved the effectiveness of the proposed dependency model. It outperforms substantially over both the probabilistic retrieval model BIR and the language models with and without consideration term dependency.

References

1. Cao, G., Nie, J.-Y., Bai, J.: Integrating word relationships into language models. In: SIGIR, pp. 298–305 (2005)
2. Lafferty, J., Zhai, C.: Document Language Models, Query Models, and Risk Minimization for Information Retrieval. In: Proc. of SIGIR (2001)
3. Ponte, J.M., Croft, W.B.: A Language Modeling Approach to Information Retrieval. In: Proc. of the 21st Intl. Conf. on Research and Development in Information Retrieval (1998)
4. Croft, W.B., Turtle, H.R., Lewis, D.D.: The Use of Phrases and Structured Queries in Information Retrieval. In: Annual International ACM/SIGIR Conference on Research and Development in Information Retrieval, Chicago, pp. 32–45 (October 1991)
5. Fagan, L.: Automatic phrase indexing for document retrieval: an examination of syntactic and non-syntactic methods. In: Proc of SIGIR 1987, pp. 91–101 (1987)
6. Van Rijsbergen, C J.: A Theoretical Basis for the Use of Co-occurrence Data in Information Retrieval. Journal of Documentation 33, 106–119 (1977)
7. Song, F., Croft, W.: A general language model for information retrieval. In: Proc. of Eighth Intl. Conf. on Information and Knowledge Management (1999)
8. Srikanth, M., Srihari, R.: Biterm Language Models for Document Retrieval. In: Proceedings of SIGIR, New York, pp. 425–426 (2002)
9. Nallapati, R., Allan, J.: Capturing term dependencies using a sentence tree based language model. In: CIKM (2002)

10. Gao, J., Nie, J.-Y., Wu, G., Cao, G.: Dependence language model for information retrieval. In: SIGIR, pp. 170–177 (2004)
11. Nallapati, R., Allan, J.: An Adaptive Local Dependency language Model: Relaxing the Naïve Bayes Assumption. In: Workshop on Mathematical and Formal Models in IR, ACM Special Interest Group in Information Retrieval (2003)
12. Hays, D G.: Dependency theory: A formalism and some observations. Language 40, 511–525 (1964)
13. Lin, D.: Principar—an efficient, broadcoverage, principle-based parser. In: Proceedings of COLING–94, Kyoto, Japan, pp. 482–488 (1994)
14. Zhai, C., Lafferty, J.: A Study of Smoothing Methods for Language Models Applied to Ad Hoc Information Retrieval. In: Proc. of SIGIR (2001)

Factoid Mining Based Content Trust Model for Information Retrieval

Wei Wang, Guosun Zeng, Mingjun Sun, Huanan Gu, and Quan Zhang

Department of Computer Science and Technology, Tongji University,
Shanghai 201804, China
Tongji Branch, National Engineering & Technology Center of High Performance Computer,
Shanghai 201804, China
willtongji@gmail.com

Abstract. Trust is an integral component in many kinds of human interactions and the need for trust spans all aspects of computer science. While most prior work focuses on entity-centered issues such as authentication and reputation, it does not model the information itself, which can be also regarded as quality of information. This paper discusses content trust as a factoid ranking problem. Factoid here refers to something which can reflect the truth of the content, such as the definition of one thing. We extracts factoid from documents' content and then rank them according to their likehood as a trustworthy ones. Learning methods for performing factoid ranking are proposed in this paper. Trust features for judging the trustworthiness of a factoid is given, and features for constructing the Ranking SVM models are defined. Experimental results indicate the usefulness of this approach.

Keywords: Content trust, factoid, information quality, ranking, SVM, text mining.

1 Introduction

Trust is an integral component in many kinds of human interactions allowing people to act under uncertainty and with the risk of negative consequences. The need for trust spans all aspects of computer science. Human users, software agents, and increasingly, the machines that provide services all need to be trusted in various applications or situations. The communication channels between computers and users, and the content exchanged between computers and users also require trust, in both directions, for real world use. Trust can be used to protect data, to find accurate information, to get the best quality service, and even to bootstrap other trust evaluations.

Recent work in trust is motivated by applications in security, electronic commerce, peer-to-peer (P2P) networks, and the Semantic Web, which all may use trust in different ways. Traditional trust mechanisms were envisioned to address authentication, identification, reputation, authentication and proof checking [1-3], but did not mention trust in the information content itself. Such trust representations and metrics do not take into account content trust [4]. But they only considered exploring the content

T. Washio et al. (Eds.): PAKDD 2007 Workshops, LNAI 4819, pp. 492–499, 2007.

trust through the transfer of trust using a resource's associations, and not taking the information content into account actually.

How can we distinguish accurate information from inaccurate or untrustworthy information is a big challenge. We live in a time when millions of people are adding information to the Web. All kinds of content are published on Web pages by ordinary citizens. Blogs, wikis, message boards, shared spreadsheets, and new publishing forums seem to appear almost monthly. As it becomes easier for people to add information to the Web, it is more difficult, and also more important, to distinguish reliable information and sources from those that are not.

In this paper we consider the acquisition of content trust as a problem of what we call 'factoid learning' on the web. Factoid here refers to something which can reflect the truth of the content, such as the definition of one thing which describe the problem of "what" in the context of information retrieval, and the other kinds of factoid may include the information related to "when", "where", "who", "why", and "how" about something. More specifically, given a topic term, we automatically extract all likely factoid candidates about the terms (paragraphs or sentences) from documents and rank the factoid candidates according to the degrees of trusted factoids. We identify key factors in modeling content trust in open sources, and we then describe a novel model that integrates a set of trust features to model content trust by using ranking support vector machine. Experimental results indicate the usefulness of this approach.

The rest of this paper is organized as follows. We review some related work in Section 2. Section 3 introduces the proposed content trust model. We first describe the basic concept of this model, and then describe some key factors of this model. We evaluate our approach and analyze the experiments results in Section 4. Section 5 concludes the paper.

2 Related Work

Recent related work described so far focuses on trusting entities rather than trusting content. Trust is an important issue in distributed systems and security.

To trust that an entity is who it says it is, authentication mechanisms have been developed to check identity, typically using public and private keys. To trust that an entity can access specific resources or perform certain operations, a variety of access control mechanisms generally based on policies and rules have been developed [5].

Popularity is another important factor correlated with trust, but not necessarily. One measure of popularity in the Web is the number of links to a Web site, and is the basis for the widely used PageRank algorithm [6].

Authority also has great impact on content trust. Authoritative sources on the Web can be detected automatically based on identifying bipartite graphs of 'hub' sites that point to lots of authorities and 'authority' sites that are pointed to by lots of hubs [7].

Reputation of an entity can result from direct experience or recommendations from others. Reputation may be tracked through a centralized authority or through decentralized voting [1]. Varieties of trust metrics have been studied, as well as algorithms for transmission of trust across individual webs of trust, including ours previous research [8].

On the other hand, current search engines, such as Google or MSN search, do not capture useful information about whether or not the content provided by a given web resource is reliable. We wish to capture some information about why any content provided by a resource is trusted. This information can be used to decide what resources should be more highly ranked in terms of trust. This is very useful in some real world information retrieval tasks. For example, we can utilize a baseline of topic and popularity to rank search results, and we can use additional trust factors to rerank the results so that more trustworthy web pages appear higher in the results list.

In summary, the challenge is how to enable trust on the web based on the content itself, or how to evaluate the quality of the information.

3 Content Trust Model Based on Factoid Learning

3.1 Basic Concept

We first distinguish between entity trust and content trust. *Entity trust* is a trust judgment regarding an entity based on its identity and its behavior, and is a blanket statement about the entity. *Content trust* is a trust judgment on a particular piece of information or some specific content provided by an entity in a given context [4]. This kind of content can give users some useful information and knowledge which is very valuable in the context of information retrieval.

Content trust is often subjective, and there are many factors that determine whether content could or should be trusted, and in what context. Some sources are preferred to others depending on the specific context of use of the information. Some sources are considered very accurate, but they are not necessarily up to date. Content trust also depends on the context of the information sought. Sources may also specify the provenance of the information they provide, and by doing so may end up being more trusted if the provenance is trusted in turn.

In reference [4], the authors list many salient factors that affect how users determine trust in content provided by Web information sources, such as topic, context and criticality, and appearance of the web pages.

We discuses content trust as a factoid ranking problem in this paper, and learning methods for performing factoid ranking are proposed. The following two sections describe the detail of this method.

3.2 Trustworthiness of Factoid

When we decide whether to trust the content of a document, the definitions in the content can help us to make decision. The more good definitions in the content, the more trustworthy the content may be regarded. Definition can also be regarded as one kind of factoid which related to "what". The other types of factoid may be related with when, where and how about something.

Normally, it is hard to judge whether a factoid is trustworthy or not in an objective way. However, we can still provide relatively objective guidelines for the judgment. Here, we first consider definitions because this kind of definitions describe the meanings of terms and thus belong to the type of frequently accessed factoid. In light of this, based on previous work [9], we develop the following rules to definitions as well as factoid evaluation.

Without losing generalization, we create three categories for factoids which represent their trustworthiness as factoid: 'good factoid/definition', 'normal factoid/definition' and 'bad factoid/definition'.

A *good factoid/definition* must contain the general notion of the term (i.e., we can describe the term with the expression "是一种" (is a kind of)) and several important properties of the term. From a trusted factoid, one can understand the basic meaning of the term.

A *normal factoid/definition* is one that between trusted and distrusted factoid.

A *bad factoid/definition* neither describes the general notion nor the properties of the term. It can be an opinion, impression, or feeling of people about it. One cannot get the meaning of the term by reading an unreliable factoid.

The following example (showed in figure 1), which is crawled in web pages by term "Data mining" illustrates the concept above. Most people would like to say that the content in box A is more reliability which is more valuable to the users.

A	**B**
Data mining (sometimes called data or knowledge discovery) is the process of analyzing data from different perspectives and summarizing it into useful information - information that can be used to increase revenue, cuts costs, or both.	**Data mining** and knowledge discovery is one of the fast growing computer science fields, but it seems difficult to study for me.

Fig. 1. Example of trust definitions

3.3 Factoid Ranking

The rest questions are how to extract definition from information content and how to evaluate their level of trustworthiness. We can regard the second question as a ranking problem. In fact, our method is largely based on the idea from Microsoft, where Xu et al. [9] developed a supervised learning approach to search of definitions.

In factoid ranking problem, we extract from the entire collection of documents <Term, Candidate, Score> triples. They are respectively term, a factoid of the term, and its score representing its likelihood of being a good factoid.

First, we collect factoid/definition candidates (paragraphs) using heuristic rules. That means that we filter out all unlikely candidates. Second, we calculate the score of each candidate as factoid using a Ranking SVM. As a result, we obtain triples of <Term, Candidate, Score>. The Ranking SVM is trained in advance with labeled instances. The whole process can be illustrated in Figure 2.

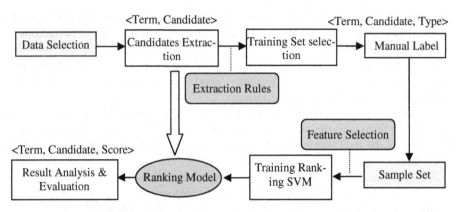

Fig. 2. Process of factoid/definition mining from content

We first parse all the sentences in the paragraph and identify <Term> using the following rules first.

- – <Term> is the first Base NP of the first sentence.
- – Two Base NPs separated by 'of' or 'for' are considered as <Term>.

In this way, we can identify not only single word <Term>s, but also more complex multi-word <term>s.

Next, we extract factoid candidates with the following patterns,

- – <Term>是+   (<Term> is a|an|the +)
- – <Term>是指+   (<Term>, +, a|an|the +)
- – <Term>是一个/一种+   (<Term> is one of +)
- – 所谓<Term>是+   (The notion of <Term> is +)
- – …

Here, '+' denotes a word string containing one or more words and '|'denotes 'or'.

It is possible to uses other rules for candidate extraction, but they are not essential for the problem of factoid ranking. So, we can skip this step or reinforce it by using more 'clever' rules.

Thirdly, we rank factoid candidates to determine the likehood of a candidate as a good factoid. We take a statistical machine learning approach to address the ranking problem. We label candidates in advance, and use them for training.

Given a training data set $D = \{x_i, y_i\}_1^n$, we construct a model that can minimize error in prediction of y given x. Here $x_i \in X$ and $y_i \in \{$trusted, normal, distrusted$\}$ represent a factoid candidate and a label, respectively. When applied to a new instance x, the model predicts the corresponding y and outputs a value of the candidate, which can be regarded as the trust value of it.

We employ Ranking SVM discussed above for ordinal regression. Ranking SVM assigns a value to each factoid candidate. The higher the value, the better the candidate is as a trusted factoid.

Classifying instances into the categories: 'trusted, 'normal' and 'distrusted' is a typical ordinal regression (or ranking) problem, because there is an order between the three categories, and the cost of misclassifying a good instance into 'bad' should be larger than that of misclassifying the instance into 'normal'.

Like reference [9], we also employ Ranking SVM as the model of ranking. Given an instance x (factoid candidate), Ranking SVM assigns a value to it based on the following formula:

$$f(x) = w^T x \tag{1}$$

where w represents a vector of weights. The higher the value of $f(x)$ is, the better the instance x is as a good factoid.

Ranking SVM is a method which formalizes learning to rank as learning for classification on pairs of instances and tackles the classification issue by using SVM.

Assume that there exists an input space $X \in R^n$, where n denotes number of features. There exists an output space of ranks, or categories, represented by labels $Y = \{c_1, c_2, \cdots, c_q\}$ where q denotes number of ranks. Further assume that there exists a total order between the ranks $c_q \succ c_{q-1} \succ \cdots \succ c_1$, where $\succ$ denotes a preference relationship. A set of ranking functions $f \in F$ exists and each of them can determine the preference relations between instances:

$$\vec{x}_i \succ \vec{x}_j \Leftrightarrow f(\vec{x}_i) \succ f(\vec{x}_j) \tag{2}$$

Assume that we are given a set of ranked instances $T = \{(\vec{x}_i, y_i)\}_{i=1}^m$ from the space $X \times Y$. The task here is to select the best function f' from F that minimizes a given loss function with respect to the given ranked instances.

Herbrich et al. [10] propose formalizing the above learning problem as that of learning for classification on pairs of instances, and constructing the Ranking SVM model is equivalent to solving a novel Quadratic Optimization problem. More detail can be found in reference [10].

Ranking SVM utilizes the set of features. Table 1 shows the list of some features.

Table 1. Features used in ranking models

1. *<Term>* occurs at the beginning of the paragraph
2. *<Term>* begins with none.
3. All the words in *<term>* followed by '所谓' (so-called).
4. Paragraph contains predefined negative words, e.g. '他' (he), '她' (she), etc
5. *<Term>* contains pronouns.
6. *<Term>* contains '具有, 和, 或' ('of', 'for', 'and', 'or').
7. *<Term>* re-occurs in the paragraph.
8. *<Term>* is followed by '是指/是一个' (is a/an/the).
9. Number of sentences in the paragraph.
10. Number of words in the paragraph.
11. Number of the adjectives in the paragraph.
12. Bag of words: words frequently occurring within a window after *<term>*

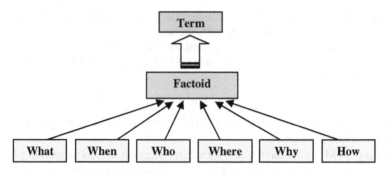

Fig. 3. Process of factoid/definition mining from content

Other factoid factors, besides the definitions, can also be defined in a similar way. We generalize the following six factors when mining the factoid: what, when, who, where, why, and how about a given term. This concept can be described in Figure 3.

The reliability of the factoid about a given term can be determined by the above six factors, and definition mining can be regarded as a kind factoid about the problem of "what". Other factors can be treated as the same way with different extraction rules and features. Some of these factors are also under our research, and how to extend the framework for general purpose information quality evaluation is one of our major works in the future.

4 Experiments and Results Analysis

In the problem of mining of factoids, given a query term, we retrieve all the triples matched against the query term and present the corresponding factoids in descending order of the scores. Our data set was collected from both Google and MSN search engines, and followed the whole process showed in Figure 2.

Table 2. Ranked list of factoids for 'Data Mining'

1. **Data mining** (sometimes called data or knowledge discovery) is the process of analyzing data from different perspectives and summarizing it into useful information - information that can be used to increase revenue, cuts costs, or both.	0.846
2. **Data mining** is a rapidly growing field that is concerned with developing techniques to assist managers to make intelligent use of these repositories.	0.781
3. **Data mining** allows you to find the needles hidden in your haystacks of data.	0.628
4. **Data mining** uncovers patterns in data using predictive techniques	0.520
5. **Data mining** software is one of a number of analytical tools for analyzing data.	0.496
6. **Data mining** is primarily used today by companies with a strong consumer focus	0.448
7. **Data Mining** and Knowledge Discovery is one of the fast growing computer science fields.	0.355
8. **Data Mining** is a prospective science that focuses on the discovery of previously unknown relationships among existing data.	0.321
9. **Data mining** has a promising future.	0.171
10. **Data mining** is difficult to study for me.	0.121

All the data necessary for factoids search is stored in a database table in advance. The data is in the form of <Term, Candidate, Score> triples.

During search, we retrieve the sorted factoid candidates with regard to the search term by table lookup. We retrieved both Chinese and English web pages, and here we only give the results of English ones. Given the query term 'Data mining', we retrieve the ranked list of the factoid candidates as those in Table 2. The results indicate the usefulness of this approach.

5   Conclusions

In this paper, we propose a content trust model based on factoid learning to solve the problem of evaluation trustworthiness through web content. We identify key factors of factoid learning in modeling content trust in open sources, and use Ranking SVM to construct a novel content trust model. We hope that this method will help to move the content trust closer to fulfilling its promise.

Acknowledgements

This research was partially supported by the National Natural Science Foundation of China under grant of 60673157, the Ministry of Education key project under grant of 105071 and SEC E-Institute: Shanghai High Institutions Grid under grant of 200301.

References

[1] Resnick, P., Zeckhauser, R., Friedman, R., et al.: Reputation systems. Communications of the ACM 43(12), 45–48 (2000)
[2] Golbeck J., Hendler J.: Inferring reputation on the semantic web. In Proceedings of the 13th International World Wide Web Conference (2004)
[3] Wang, W., Zeng, G.S., Yuan, L.L.: A Semantic Reputation Mechanism in P2P Semantic Web. In: Mizoguchi, R., Shi, Z., Giunchiglia, F. (eds.) ASWC 2006. LNCS, vol. 4185, pp. 682–688. Springer, Heidelberg (2006)
[4] Gil, Y., Artz, D.: Towards content trust of web resources. In: Proceedings of the 15th International World Wide Web Conference (2006)
[5] Miller, S.P., Neuman, B.C., et al.: Kerberos authentication and authorization system. Tech. rep., MIT (1987)
[6] Brin, S., Page, L.: The anatomy of a large-scale hypertextual web search engine. In: Proceedings of the 7th International World Wide Web Conference (1998)
[7] Kleinberg, J.M.: Authoritative sources in a hyperlinked environment. Journal of the ACM 46(5), 604–632 (1999)
[8] Wang, W., Zeng, G.S., Yuan, L.L.: A Reputation Multi-Agent System in Semantic Web. In: Shi, Z.-Z., Sadananda, R. (eds.) PRIMA 2006. LNCS (LNAI), vol. 4088, pp. 211–219. Springer, Heidelberg (2006)
[9] Xu, J., Cao, Y.B., Li, H., et al.: Ranking Definitions with Supervised Learning Methods. In: Proceedings of the 14th International World Wide Web Conference (2005)
[10] Herbrich, R., Graepel, T., Obermayer, K.: Large Margin Rank Boundaries for Ordinal Regression. Advances in Large Margin Classifiers, 115–132 (2000)

Service, Security and Its Data Management
for Ubiquitous Computing – Overview

Jong Hyuk Park[1] and Deqing Zou[2]

[1] School of Computer Engineering, Kyungnam University,
Masan-si, Kyungnam-do, Korea
parkjonghyuk@gmail.com
http://parkjonghyuk.wo.ro
[2] School of Computer Science and Technology,
Huazhong University of Science and Technology, Wuhan, 430074, China
Deqingzou@hust.edu.cn
http://grid.hust.edu.cn/deqingzou/

Abstract. International workshop on Service, Security and its Data management for Ubiquitous Computing (SSDU-07) was held in conjunction with The
11th Pacific-Asia Conference on Knowledge Discovery and Data Mining
(PAKDD 2007), Nanjing, China, May 22-25, 2007. The main purpose of this
workshop (SSDU-07) was to foster research in the areas of security and intelligence integrating into Ubi-com and data management technology. In addition,
the workshop provided an opportunity for academic and industry professionals
to discuss the latest issues and progress in the area of Ubi-com, security, data
management.

1 Introduction

Recently, the new computing paradigm - Ubiquitous Computing (Ubi-com) is just
around the corner for us. Ubi-com is a user-centric environment where various kinds
of computers is embodied into person, object and the environment which are connected each others and the seamless services are provided to users anytime, anywhere
with any device. However, the many security risks and problems can be occurred
because data resources can be connected and accessed by anyone in Ubi-com. Therefore, it will be needed to more secure and intelligent mechanism in Ubi-com. The
main purpose of this workshop (SSDU-07) is to foster research in the areas of security
and intelligence integrating into Ubi-com and data management technology. In summary, the workshop brought together researchers who are interested in the area of
Ubi-com, security, data management, where the attendees discussed intensively various aspects on this area.

The topics of the workshop included, but were not limited to:

- Context-Awareness and its Data mining for Ubi-com service;
- Human-Computer Interface and Interaction for Ubi-com;
- Smart Homes and its business model for Ubi-com service;
- Intelligent Multimedia Service and its Data management for Ubi-com;
- USN / RF-ID for Ubi-com service;

T. Washio et al. (Eds.): PAKDD 2007 Workshops, LNAI 4819, pp. 500–501, 2007.

- Network security issues, protocols, data security in Ubi-com;
- Database protection for Ubi-com;
- Privacy Protection and Forensic in Ubi-com;
- Multimedia Security in Ubi-com;
- Authentication and Access control for data protection in Ubi-com;
- Service, Security and its Data management for U-commerce;
- New novel mechanism and Applications for Ubi-com;

2 Workshop Overview

There was a large number of paper submissions (124), representing more than 10 countries and regions, not only from Asia and the Pacific, but also from Europe, and North and South America. It was extremely difficult to select the presentations for the conference because there were so many excellent and interesting submissions. In order to keep the high quality of the conference, we finally decided to accept 16 papers for presentations. We believe that all of these papers and topics not only provided novel ideas, new results, work in progress and state-of-the-art techniques in this field, but also stimulated the future research activities in the area of Ubi-com.

The exciting program for this conference was the result of the hard and excellent work of many others, such as program and technical committee members under a very tight schedule. We were also grateful to workshop Chair, Takashi Washio, and program chair, Zhi-Hua Zhou, of PAKDD 2007. They provided much help for SSDU-07.

3 Program Committee

Steering Chair
  Laurence T. Yang, St. Francis Xavier University, Canada

General Chairs
  Jong Hyuk Park, Kyungnam University, Korea
  Han-Chieh Chao, National Ilan University, Taiwan
  Sajal K. Das, University of Texas at Arlighton, USA

Program Chairs
  Deqing Zou, Huazhong University of Sci. & Tech., China
  Kuan-Ching Li, Providence University, Taiwan
  Mario Freire, Department of Informatics, Portugal

Program Vice-Chair
  Yuh-Shyan Chen, National Taipei University, Taiwan
  Jieh-Shan George Yeh, Providence University, Taiwan

Publicity Chairs
  Hakan Duman, University of Essex, UK
  Tsutomu Terada, Osaka University, Japan
  Nicolas Sklavos, Technological Educational Institute of Messolonghi, Greece

Study on Trust Inference and Emergence of Economical Small-World Phenomena in P2P Environment*

Yufeng Wang[1], Yoshiaki Hori[2], and Kouichi Sakurai[2]

[1] College of Telecommunications and Information Engineering,
Nanjing University of Posts and Telecommunications, Nanjing 210003, China
[2] Department of Computer Science and Communication Engineering, Kyushu University,
Fukuoka 812-0053, Japan

Abstract. With the increasing popularity of self-organized communication systems, distributed trust and reputation systems in particular have received increasing attention. By formalizing trust relationships, trust between parties within the community can be derived by analyzing the trust paths linking the parties together. This paper treats trust network as an emergent property. The emergence interpretation concerns both the maintenance and usage of trust network inference. Specifically, in P2P system, trust network is maintained by individual agents at micro level, and it is used (i.e., queried) as one entity at macro level. In this paper, we first discuss micro level activities, that is, we classifies trust into functional trust and referral trust to facilitate trust inference, and extend the referral trust to include factors of similarity and truthfulness, so that our approach can accommodate the personalized feature of reputation rating, and reduce trust inference error effectively; then we discuss macro level properties of trust network. Specifically, we investigate the emergence of network structural properties of trust and reputation system in terms of efficiency and cost. That is, efficiency measures how well information propagates over trust system, and cost measures how expensive it is to build this system. Preliminary simulation results show the performance improvement of P2P community and the emergence of economical small-world trust network, namely relatively high efficiency and low cost.

1 Introduction

Self-organized computer and communication systems have received increasing attention over the last few years, in terms of both deployment and research. They are typically organized according to the peer-to-peer (P2P) organization principle. That is, participants in the system are equal in that they have equivalent capabilities and responsibilities – they are peers. Such P2P systems can also be found in a variety of other networks, such as social or biological networks. One of the major problems in self-organized communication systems is that of cooperation. Typically, users are

* Research supported by the NSFC Grants 60472067, JiangSu education bureau (5KJB510091) and State Key Laboratory of Networking and Switching Technology, Beijing University of Posts and Telecommunications (BUPT).

T. Washio et al. (Eds.): PAKDD 2007 Workshops, LNAI 4819, pp. 502–514, 2007.

concerned primarily about their own benefits and thus cooperation and fairness cannot be guaranteed. This selfish behavior is called free-riding and is a well-known phenomenon in economics. Malicious attacks and random failures are other important problems. A promising approach to all of these is that of using trust and reputation systems. Trust and reputation has been considered in many disciplines other than computer and communication science, including economics, sociology, psychology, management science as well as marketing. Generally, in society, we accord trust to only small numbers of people (see [1] for detailed description about reputation and trust). These people, again, trust another limited set of people, and so forth. The network structure emanating from our very person, composed of trust statements linking individuals, constitutes the basis for trusting people we do not know personally. The structure has been dubbed "Web of Trust" [2], which means that each user explicitly specifies a (possibly small) set of users she trusts. By formalizing trust relationships, trust between parties within the community can be derived by analyzing the trust paths linking the parties together.

There exists a great deal of work on trust and reputation systems in self-organized communication system. Ref. [3] provides an overview of research on distributed reputation systems in the various computer science communities as well as the social sciences. Ref. [4-5] maps trust inference approaches into scalar metrics and group metrics, particularly, scalar metrics analyze trust assertions independently, while group trust metrics evaluate groups of assertions "in tandem". PageRank and Eigentrust (and other variations) [6-7] belong to group-based trust inference algorithms that allow to calculate a global emergent reputation from locally maintained trust values. But those work ignore that, intuitively, trust is not always transitive in real life [8]. In order to address the problem of trust transitivity, Jøsang classifies the trust into functional trust, that is trust on peer's ability to provide service, and referral trust, that is trust on peer's ability to recommend, and provides a subjective logic-based method to analyze the trust inference[9][10]. Different propagation schemes for both trust score and distrust score are studied in Ref. [11] based on a network from a real social community website. Ref. [12] also elaborates on trust network representation, articulates different types of trust, and provides the concept of emergent property in trust and reputation system. But this paper did not describe how to characterize the structure of trust network.

Trust and reputation system is a special kind of social network. One important characteristic of society is the emergence as patterns developed from relative simple interactions in a network of individuals. Ref. [13-14] make some research on social trust networks. But there still exist a great deal of exciting research in this direction. Generally, the following aspects in trust and reputation system should be considered thoroughly: How to find statistical properties in trust and reputation networks, to characterize the structure and behavior of trust network, and to suggest appropriate ways to measure these properties.

In brief, this paper treats trust network as an emergent property, which is maintained by individual agents at micro level, and it is used (i.e., queried) as one entity at macro level. Then, from micro-level, we classify the trust into functional trust and referral trust, and extend the referral trust to include two factors: similarity and truthfulness, which can effectively reduce the trust inference error. Particularly, similarity is used to characterize the personalized reputation feedback, and truthfulness is adopted to measure the credibility of reputation feedback; from the macro-level, we adapt the economic concepts like efficiency and cost to characterize the emergence of social properties of trust network.

The paper is organized as follows: section 2 briefly discusses the research motivations of this paper. From micro-level interaction, section 3 provides several basic trust categories and calculation methods including functional trust, referral trust composed of similarity and truthfulness properties, and then designs the trust evaluation process based on the above concepts. In section 4, we adapt the concepts of efficiency and cost borrowed from economic field to characterize the emergent properties of social trust network. The simulation settings and preliminary results are provided in section 5. Finally, we briefly conclude the paper in section 6.

2   Research Motivations

The scientific research in the area of computational mechanisms for trust and reputation in virtual societies is a recent discipline oriented to increase the reliability and performance of electronic communities. In any sizeable social networks, members are unlikely to have interacted with (rated) every other members. Humans rely on gossip or word-of-mouth to propagate their opinions about each other. In evaluating a stranger's trustworthiness, we weight those of our friends' opinions about this stranger by how much we trust our friends and come to a conclusion on whether we are going to trust this stranger. Hence, propagation of opinions (of which ratings is one) in human society is a very natural phenomenon. Our paper explicitly adopts the separation of trust into two categories: trust as a participator in the service, functional trust, and trust as a recommender of other participants, referral trust. This avoids the inherent difficulty associated with having to make an assumption about a principal's ability to recommend, based on their ability to participate. Without this separation, reputation networks are open to the colluding nodes attack. For example, peer j and k form collusion group. Peer j builds up the trust of another node i (by participating), then makes false recommendations about k to 'transfer' i's trust into a set of malicious nodes, such as k, which i now trusts transitively on peer k via j. Ref. [15-16] proposes to alleviate the above attacks through the separation between functional trust and referral trust. Another advantage of this separation is precisely what allows trust to become transitive, and referral trust chain always assumes the existence of a functional trust scope at the end of the transitive path. Note that we argue that ratings should be context- and individual-dependent quantities. Clearly, sharing approval ratings in multiple contexts need to take into account the similarities and differences among the various contexts. In this paper, we assume all peers in same context. Even in same context, due to the fact that trust and reputation is subjective, heterogeneous peers might have different preference and criteria to judge an interaction. For example, some peers might be picky, and some might be generous. So, they might have different evaluations on an identical specific peer's behaviors. If two peers i and j are similarity in their evaluation criteria, peer i can trust peer j's recommendations, if it knows that peer j is truthful. Thus, Referral trust should include the following two aspects:

- Similarity – Whether peer i is similar to peer j in preferences and ways of judging issues.
- Truthfulness – Whether peer j is truthful in telling its information;

Ref. [17] uses the same classification, namely, similarity and truthfulness, but this paper does not provide the formal description, and furthermore, avoids the truthfulness problem through assuming that all peers are truthful in telling their evaluations. In brief, Peer j's referral trust depends on both being truthful and similar in its preferences to the peer requesting the recommendation. In the next section, we will introduce how to evaluate the similarity and truthfulness in referral trust.

Many people all over the world participate online in established social networks (complex networks). Most social networks show "community structure", i.e., groups of vertices that have a high density of edges with in them, with a lower density of edges between groups, like small world: highly clustered like regular lattices and small characteristic path lengths like random graphs. The small-world effect has obvious implications for the dynamics of process taking place on networks. But, to our knowledge, few work attempts to investigate the structural properties of trust and reputation system, which make us try to do some preliminary research in this field. Inspired by the ideas in [18], we adopt two leading concepts, efficiency and cost to model and analyze the characteristics of trust network, which show the features of small-world. More specifically, efficiency measures how well information propagates over trust system, and cost measures how expensive it is to build this system. The combination of these factors formalizes the idea of trust system that are "cheap" to build, and nevertheless efficient in propagating information, both at global and local scale.

The reputation of a peer is generally associated with the identity of the peer. Identity can be a digital certificate or pseudonym that can uniquely identify a peer. In fact, many security threats arise from the anonymous identities in distributed networks, like whitewashing and Sybil attack. In our paper, we simply assume that nodes should be correctly identified to one another, for example, using a decentralized PKI architecture like explained in [19] or any other mechanism to ensure that the information comes from the location we are expecting it from.

3 Trust Categories and Inference

3.1 Functional Trust ρ_{ij}^{F}

We use the Bayesian probability to represent the functional (direct) trust between peers. Bayesian systems take binary ratings as input (i.e. positive or negative), and are based on computing reputation scores by statistical updating of beta probability density functions (PDF). A posteriori (i.e. the updated) reputation score is computed by combining the a priori (i.e. previous) reputation score with the new rating. The reason to use beta probability density function for evaluating the functional trust lies in that the beta function is the conjugate prior for the binomial likelihood distribution. This implies that if the prior distribution is the beta distribution and the new observations follow a binomial distribution, than the posterior distribution will also be a beta distribution. We use this typical property of beta distribution to derive a closed form expression for reputation updates when a node makes some direct observations. Specifically, the reputation score can be represented in the form of the beta PDF parameter tuple (α, β) (where α and β represent the amount of positive and negative ratings respectively). The beta PDF denoted by $beta(p|\alpha, \beta)$ can be expressed using the gamma function:

$$beta(p \mid \alpha, \beta) = \frac{\Gamma(\alpha + \beta)}{\Gamma(\alpha)\Gamma(\beta)} p^{\alpha-1}(1-p)^{\beta-1} \quad where \, 0 \leq p \leq 1, \alpha, \beta > 0 \quad (1)$$

The probability expectation value of the beta distribution is given by: $E(p) = \alpha/(\alpha+\beta)$.

So, after observing α_j positive and β_j negative outcomes (from the perspective of node i) the reputation of node j maintained at node i is given by: $\rho_{ij}^F = beta\ (\alpha_j +1, \beta_j +1)$, which corresponds to the probability that peer j with cooperate with peer i in next event. Then, assume peer i again interacts with peer j for $r+s$ more events. r cooperative and s non-cooperative, then the updated $\rho_{ij} = beta(\alpha_j+r+1, \beta_j+s+1)$. This clearly shows the flexibility associated with using beta reputation system, that is, the reputation update is equivalent to just updating the value of the two parameters α_j and β_j: $\alpha_j^{new} = \alpha_j + r, \beta_j^{new} = \beta_j + s$. Note that the main reason to use Beta distribution to represent the functional trust lies in that the Beta distribution has the following features: flexibility, simplicity as well as its strong foundations on the theory of statistics.

3.2 Similarity ρ_{ij}^S and Truthfulness ρ_{ij}^T in Referral Trust

Heterogamous peers might have different preference and criteria to judge an interaction. So, when inferring transitive trust, Peer j's referral trust should partially depends on being similar in its preferences to the peer requesting the recommendation (peer i). The similarity between peer i and peer j can be inferred from their ratings on common transaction set. Let $S(i)$ denote the set of peers that have interacted with peer i, Let $S(j)$ denote the set of peers that have interacted with peer j. The common set of peers that have interacted with both peer i and j, is denoted by $CS(i,j)=S(i)\cap S(j)$. To measure the similarity of peer i's referral trust on peer j, peer i computes the feedback similarity between i and j over the common set $CS(i,j)$. We model the feedback by peer i and j as two vector respectively. Particularly, we use the standard deviation (dissimilarity) of the two feedback vector to characterize the similarity. So

$$\rho_{ij}^S = 1 - \sqrt{\frac{\sum_{k \in CS(i,j)} \left(\rho_{ik}^F - \rho_{jk}^F\right)^2}{|CS(i,j)|}} \quad (2)$$

This notion of local or personalized similarity measure provides great deal of flexibility and stronger predictive value as the feedback from similar raters are given more weight. Now we consider how peer i computes the truthfulness of peer j.

Essentially, what peer i want to do is to compare the recommendations peer j makes with its view of the same. The measurement is evaluated as follows:

$$\rho_{ij}^T = 1 - \sqrt{\frac{\sum_{k \in TS(i,j)} \left(\rho_{ik}^F - \rho_{jk}'\right)^2}{|TS(i,j)|}} \qquad (3)$$

where $TS(i,j)$ denotes the set of pees that peer j recommend to peer i. Note that ρ_{jk}' denotes that trust value of peer k declared by peer j. we explicitly model the possibility that peer j may lie about the trust value of peer k.

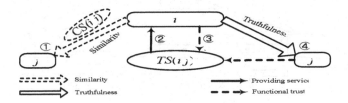

Fig. 1. The procedure of trust evaluation

Just illustrated in Fig.1, firstly, peer i evaluates its similarity with peer j (step 1); if the similarity exceeds certain threshold, then peer i selects the peers in $TS(i,j)$ to ask for service (step 2); after finishing the transaction, peer i updates the functional trust on peers in $TS(i,j)$ (step 3), and, meanwhile, updates the truthfulness of referral trust on peer j (step 4). In this paper, we let $\rho_{ij}^R = \rho_{ij}^S \cdot \rho_{ij}^T$, where ρ_{ij}^R denotes the peer i's general referral trust on peer j. Note that, our paper explicitly use similarity to characterize the personalized reputation feedback, and use truthfulness to measure the credibility of reputation feedback so that we can identify malicious peers effectively.

4   Emergence of Properties of Social Trust Network

Generally, trust and reputation system is a virtual network built on top of P2P system. We represent this system by a directed referral trust graph T. The vertices in T correspond exactly to the peers in the trust and reputation system. Each directed edge has a weight ranging from 0 to 1, the larger value implies stronger trust relationship. This represents a trust value assigned by the peer that requests information (source of the link) to the peer that provides it (destination of the link). When a peer cannot assess the trustworthiness of another, then the corresponding edge is missing from the graph. Note that, before any of trust inference algorithms can be used, the trust graph has to be realized in a scalable manner. We use the storage approach described in NICE systems for its scalability [20]. Assume that a current version of the trust graph T is available to a certain peer, and suppose the peer wish to compute a trust value for another peer. If those two peers have had prior transactions (functional trust, the last hop in referral trust graph), then the former can just look up the value in trust graph. However if they have never had a prior transaction, the former could infer trust value for the later by following directed paths on the trust graph. We employ the multiplication operation to infer trust for distant sources of information (note that, in the inference path, the referral trust appears from the source to the penultimate hop, and the last hop is functional trust). This operation is motivated as follows: for each edge, weight expresses the probability of a successful transaction, then the total probability of successful transaction over a path of two consequent edges equals their product (illustrated in Fig.2).

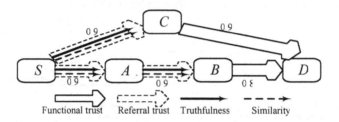

Fig. 2. Illustration of trust inference

The following matrices are used to describe the trust graph: referral trust between neighbors { ρ_{ij}^R }. If peer i and j without direct referral link, we set $\rho_{ij}^R = 0$. ρ_{ij}^R represents the peer i's trust on j as referrer, then $-\log \rho_{ij}^R$ can reasonably denote the cost that i provide the trust and reputation feedback for peer j. This implies that if peer i provides lower rating on peer j, than peer i bears more cost. This assumption is based on the following observations:

- Peers tend to provide good rating for the partner, for good rating will not annoy the partner;
- The truthful low rating is based on the long-term interaction, so the effort to gather this rating is huge;
- If the peer i provides the truthful low rating about peer j, than peers i may bear the danger that peer j threaten it, in this setting, the psychological cost of peer i is huge.

Based on above consideration, we use the $-\log \rho_{ij}^R$ as peer i's cost for providing referral trust rating for peer j. Similarly, in the last hop of trust inference path, there is functional trust matrix { ρ_{ij}^F }, than $-\log \rho_{ij}^F$ is the peer i's cost for maintaining functional trust on peer j.

Corollary 1: Based on the above definition of cost in weighted trust graph, and the multiplication operation used to infer the maximal trust between pair of peers, the shortest cost path between any peer i and j, denoted by c_{ij}, corresponds to the maximum trust path between peer i and j.

Suppose that each peer uses the trust graph to discover trusted peer to conduct transactions. We assume that $\varepsilon_{ij} = \exp(-c_{ij}) \ \forall i, j$ represent the efficiency between any peer i and j. When there is no path in the graph between i and j, we get $c_{ij} = \infty$, and consistently $\varepsilon_{ij} = 0$.

Corollary 2: The efficiency ε_{ij} equals the inferred maximal trust value between pair of peer i and j.

So, it is reasonable that ε_{ij}, denotes the efficiency in trust and reputation network, for this variable represents the inferred maximum trust valued between peer i and j.

In a recent seminal paper [21], Watts and Strogatz have identified two main attractive properties of a small-world network are: a highly clustered like regular lattices; small characteristic path length like random graph. Ref. [18] proposes a generalization of the theory of small world based on two leading concepts, efficiency and cost. In detail, efficiency measures how well information propagates over the network, and cost measures how expensive it is to build network. The combination of these factors formalizes the idea of networks that are "cheap" to build, and nevertheless efficient in propagating information, both at global and local scale. We use the idea and adapt the definition of cost and efficiency to make it more appropriate for trust network.

The global efficiency in trust network E_{global} is defined as follows:

$$E_{global} = E(T) = \frac{1}{N(N-1)} \sum_{i \neq j \in G} \varepsilon_{ij} \tag{4}$$

where N is the number of peers in trust and reputation system.

The efficiency $E(T)$ belongs to the range $[0,1]$. The maximum value $E_{max}=1$ is typically reached when there is edge between each couple of peers, and trust value of each edge is 1. E_{global} is the real variable to be considered when we want to characterize the efficiency of a trust and reputation system finding trusted parts in parallel.

On the other side, the same measure, the efficiency, can be evaluated for any subgraph of T, and therefore it can be used also to characterize the local properties of the graph. Specifically, the local properties of T can be characterized through evaluating the efficiency of T_i for each peer i, the subgraph of the neighbors of i (the set of peer i's neighbors, except peer i). Then, the local efficiency is defined as follows:

$$E_{local} = \sum_{i \in T} E(T_i) \Big/ N \text{ where } E(T_i) = \frac{1}{k_i(k_i-1)} \sum_{i \neq j \in T_i} \varepsilon_{ij} \tag{5}$$

where k_i is the number of peer i's neighbors. E_{local} is an average of the local efficiency and plays a role similar to the clustering coefficient [21]. Since $I \notin T_i$, the local efficiency shows how much the system is fault tolerant, thus how efficient is the trust inference between the first neighbors of i, when i is removed.

Another important variable to consider is the cost of a trust and reputation system, which is defined as follows:

$$Cost(\boldsymbol{T}) = \frac{1}{N(N-1)} \sum_{i \neq j \in G} \gamma \left(- \log \rho_{ij} \right) \tag{6}$$

Here, γ is the so-called cost evaluator function, which calculates the cost needed to build up a trust connection. Here we let $\gamma(x) = 1 - \exp(-x)$. Note that the reason to use the above cost evaluator function is to map the cost of trust network into range [0,1]. Of course, other appropriate cost evaluation functions can be used.

It is widely assumed that most social networks show "community structure", i.e., groups of vertices that have a high density of edges within them, with a lower density of edges between groups. It is a matter of common experience that people do divide into groups along lines of interest, which is also very reasonable phenomenon in social trust network, for each peer have incentive to keep touch with other peers with similar high reputation (that is "short-range" link), but, for the global efficiency, trust network needs some long-range link to infer the trustworthiness of every possible peer. So, in reality, trust system has to somehow be a compromise between efficiency of trust inference, and the need for economy. The cost of a trust system nicely couples with the efficiency to provide a meaningful description of the "good" behavior of a trust system. For the dynamics and scalability, we use gossiping to build trust system, which combines the active spreading of peer reputation profiles (gossiping) with trust system restructuring [22]. Specifically, in order to discover potentially "good" or "bad" peers efficiently, each peer keeps a preference list (which includes the peers with high reputation value), and blacklist (which includes the peers with low reputation value). In our gossip-based trust system, each peer periodically selects some other random peers from the preference list to communicate with the local trust view. Its receiver may use his own knowledge about the target of the opinion to form a new opinion about the sender's ability to recommend other service providers or other recommenders, and form the trust relation.

5 Simulation Results

We consider a peer-to-peer system where certain service is exchanged among peers. After a transaction, the involved peer sends feedback about the service provider. Initial experiments are performed to show the benefits of social trust network and its structural properties.

The simulation settings include three kinds of peers: malicious peers, strategic peers and honest peers, and their behavior patterns are given in Table 1. To model the subjective feature of reputation ratings, the honest peers are divided into three categories: picky, moderate and generous. Two transaction settings are simulated, namely, random setting and referral trust setting. The experiments start as peers perform random transactions with each other. After 1000 direct transactions, this simulation conducts referral trust-based transactions. Specifically, peers initiate transactions by issuing a transaction request. For each request, a certain percentage of peers respond. The initiating peer then uses the trust inference method to select peer with highest trust value to perform the transaction, and update the referral trust and functional trust on the partner. Each simulation result is the mean of ten simulations.

Table 1. Summary the main parameters used and default values

	Parameter	Value range
Peer Model	Number of peers in the network Percentage of malicious peers Percentage of strategic peers Percentage of honest peers	50 ~ 200 20% 20% ~ 50% 50% ~ 100%
Behavior patterns	Malicious peers provide incorrect service and incorrect feedback; Strategic peers do not provide any feedback without incentive mechanism.	
Categories of honest pees	Honest peers truthfully provide service and feedback Three different rating criteria: • Picky standard: the lowest rating on partner; • Moderate standard: the middle rating on partner; • Generous standard: the high rating on partner.	1/3 percentage 1/3 percentage 1/3 percentage

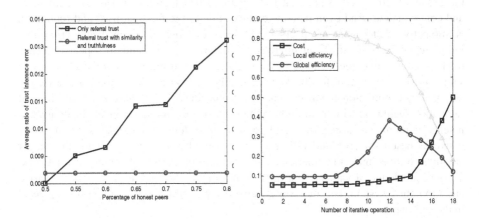

Fig. 3. Effect of truthfulness on trust inference error

Fig. 4. Relations among cost, global and local efficiency

Fig. 3 illustrates the effect of separation of similarity and truthfulness in referral trust on improving the trust inference accuracy. Note that, in our paper, the ratio of trust inference error is defined as the probability that the honest peer is mistakenly regarded as malicious peer for different evaluation criteria. As shown in Fig. 3, if trust and reputation system only uses the single scalar of referral trust, as the percentage of honest peers increases, this approach may mistakenly identify some honest peers with different rating criteria as malicious peers. But our approach that adopts the truthfulness feature to identify whether a peer is malicious can alleviate this mistake.

Finally, we attempt to show the structural properties of social trust network through the relationship among cost, global efficiency, and local efficiency (illustrated in Fig. 4). Using the formed trust graph, this simulation first makes the trust value less than 0.8 zero, and, each following iteration adds the trust links between peers in descending order of trust value, then calculates the cost, global efficiency and local efficiency. Initially, peers connect to other peers with high reputation, just like regular lattice, but in this way, the global efficiency is relative low, the main reason is that there exists no path to large part of peers. So, adding a few links to relative low reputation peers (that is, relatively high cost link) may increase the global efficiency. But adding too many high cost links may also decrease the global efficiency. This also illustrates, maybe in trust and reputation system, the completely connected trust graph is not optimal.

6 Conclusion and Future Works

This paper treats trust network as an emergent property, which concerns both the maintenance and usage of trust network inference. In P2P system, trust network is maintained by individual agents at micro level, and it is used (i.e., queried) as one entity at macro level. Macro level properties associating with a trust network could be the graph structure. We argue that two distinguished features of trust are subjective and contextualize. And even in same context, due to the fact that trust and reputation is subjective, heterogeneous peers might have different preference and criteria to judge an interaction. So, from micro-level viewpoint, we maintain a peer's referral trust as multiple component scores: similarity and truthfulness, to accommodate subjectivity of reputation feedback. Specifically, we use similarity to characterize the personalized reputation feedback, and use truthfulness to measure the credibility of reputation feedback, which can reduce the ratio of trust inference error; From macro-level viewpoint, we argue trust and reputation system is a special kind of social network, based on cost and efficiency borrowed from economic field, this paper attempts to investigate the emergence of structural properties of social trust network. To our knowledge, it was the first time to investigate the structural properties of trust and reputation system. But our works was very preliminary, which can be improved in the following aspects:

- It is imperative to investigate how to combine the scalar trust metrics and group trust metrics to better characterize the trust network, and integrate the propagation of trust and distrust to solve or alleviate various attacks in P2P environment.
- A great deal of excellent outcomes are achieved on the characterization and modeling of network structure. But, studies of the effects of structure on behaviors of trust and reputation systems are still in their infancy.

References

[1] Wang, Y., Hori, Y., Sakurai, K.: On securing open networks through trust and reputation-architecture, challenges and solutions. In: Proceeding of the 1st Joint Workshop on Information Security (JWIS), Seoul Korea (2006)

[2] Golbeck, J., Parsia, B., Hendler, J.: Trust networks on the semantic web. In: Proceedings of Cooperative Intelligent Agents, Helsinki, Finland (2003)

[3] Mundinger, J., Le Boudec, J.-Y.: Reputation in self-organized communication systems and beyond. In: Proceedings of Workshop on Interdisciplinary Systems Approach in Performance Evaluation and Design of Computer & Communications Systems (Inter-Perf 2006) (2006)

[4] Levien, R.: Attack resistant trust metrics, PhD thesis, UC Berkeley (2003)

[5] Ziegler, C.-N., Lausen, G.: Spreading activation models for trust propagation. In: Proceedings of the IEEE International Conference on e-Technology, e-Commerce, and e-Service, Taipei, Taiwan (2004)

[6] Page, L., Brin, S., Motwani, R., Winograd, T.: The pagerank citation ranking: Bringing order to the web. Technical report, Stanford Digital Library Technologies Project (1998)

[7] Kamvar, S.D., Schlosser, M.T., Garcia-Molina, H.: The eigentrust algorithm for reputation management in P2P networks. In: Proceedings of the 12th International World Wide Web Conference, Budapest, Hungary (2003)

[8] Christianson, B., Harbison, W.S.: Why isn't trust transitive? In: Proceedings of the Security Protocols International Workshop. University of Cambridge (1996)

[9] Jøsang, A., Hayward, R., Pope, S.: Trust network analysis with subjective logic. In: Australasian Computer Science Conference (2006)

[10] Jøsang, A., Ismail, R., Boyd, C.: A survey of trust and reputation systems for online service provision, Decision Support Systems (2006)

[11] Guha, R., Kumar, R., Raghavan, P., Tomkins, A.: Propagation of trust and distrust. In: Proceedings of the 13th International World Wide Web Conference, New York City (2004)

[12] Ding, L., Kolari, P., Ganjugunte, S., Finin, T., Joshi, A.: Modeling and evaluating trustnetwork inference. In: Proceedings of the Seventh International Workshop on Trust in Agent Societies (2004)

[13] Zhou, R., Hwang, K.: PowerTrust: A robust and scalable reputation system for trusted P2P Computing. IEEE Transactions on Parallel and Distributed Systems (2006)

[14] Marti, S., Ganesan, P., Garcia-Molina, H.: SPROUT: P2P routing with social networks. In: Proceedings of the International Workshop on Peer-to-Peer Computing & DataBases (P2P&DB 2004) (2004)

[15] Moreton, T., Twigg, A.: Enforcing collaboration in Peer-to-Peer routing services. In: Proceedings of the First International Conference on Trust Management, Heraklion, Crete (2003)

[16] Swamynathan, G., Zhao, B.Y., Almeroth, K.: Decoupling service and feedback trust in a Peer-to-Peer reputation system. In: Proceedings of the 1st International Workshop on Applications and Economics of Peer-to-Peer Systems (AEPP) (2005)

[17] Wang, Y., Vassileva, J.: Trust and reputation model in Peer-to-Peer networks. In: Proceedings of the 3rd International Conference on Peer-to-Peer Computing (2003)

[18] Latora, V., Marchiori, M.: Economic small-world behavior in weighted networks. The European Physical Journal B 32, 249–263 (2003)

[19] Datta, A., Hauswirth, M., Aberer, K.: Beyond 'Web of Trust': Enabling P2P e-commerce. [Online], available: www.citeseer.ist.psu.edu/article/datta03beyond.html

514   Y. Wang, Y. Hori, and K. Sakurai

[20] Lee, S., Sherwood, R., Bhattacharjee, B.: Cooperative peer groups in NICE. In: Proceedings of INFOCOM 2003
[21] Watts, D.J., Strogatz, S.: Collective dynamics of 'small-world' networks. Nature 93, 440–444 (1998)
[22] Jelasity, M., Montresor, A., Babaoglu, O.: Gossip-based aggregation in large dynamic networks. ACM Transactions on Computer Systems 23, 219–252 (2005)

A Secure Time Synchronization Protocol for Sensor Network*

Hui Li, Yanfei Zheng, Mi Wen, and Kefei Chen

Department of Computer Science and Engineering
Shanghai Jiao Tong Univeristy
200030, Shanghai, China P.R.C
{li_hui,yfzheng,superwm,kfchen}@sjtu.edu.cn

Abstract. Clock synchronization is a critical issue in most wireless sensor network applications. Although the synchronization is well studied in last few years. Most of these synchronization protocols assume benign environments, but cannot survive malicious attacks in hostile environments, especially when there are compromised nodes. In this paper, we propose a secure synchronization protocol for sensor network. Our protocol combine the sender-receiver model and receiver-receiver model to verify the synchronization process between each synchronizing pair. The approach guarantees that normal nodes can synchronize their clocks to global clock even if each normal node has up to t colluding malicious neighbor nodes during synchronization phase.

Keywords: Synchronization, security, sensor network.

1 Introduction

Recent advances in micro-electro-mechanical systems and low power and highly integrated digital electronics have led to the development of micro-sensors, which are capable of wireless communication and data processing [1]. Wireless sensor networks (WSNs) are distributed networks of such sensors, dedicated to closely observing real-world phenomena. Such sensors may be embedded in the environment or enabled with mobility; they can be deployed in inaccessible, dangerous, or hostile environments.

Clock synchronization is a critical issue in most wireless sensor network applications, such as measuring the time-of-flight of sound; distributing an acoustic beam forming array; forming a low-power TDMA radio schedule; integrating a time-series of proximity detections into a velocity estimate; suppressing redundant messages by recognizing duplicate detections of the same event by different sensors; integrating multi sensor data; or coordinating on future action. However, due to the resource constraints of sensors, traditional clock synchronization protocols (e.g. network time protocol (NTP) [2]) can not be directly applied in sensor networks.

* This work is supported by Specialized Research Fund for the Doctoral Program of Higher Education of China under grant 20050248043.

T. Washio et al. (Eds.): PAKDD 2007 Workshops, LNAI 4819, pp. 515–526, 2007.

In recently years, much literatures [7, 8; 9, 10] has fully study the synchronization issues in sensor network. But all these clock synchronization protocols have not built with security in mind. However, the sensor network is so vulnerable to attacks that both the internal and external adversarial can easily corrupt the synchronization mechanism in sensor networks. Although authentication [3, 4] can be used to defend against external attacks, an attacker may still attack clock synchronization through compromised nodes. In this paper, we proposed a secure clock synchronization protocol for sensor networks. The idea behind our scheme is using node's neighbors as verifiers to check if the synchronization is processed correctly so that it can detect the attacks launched by compromised node.

The rest of this paper is organized as follow: Section 2, 3 discusses related works, the system setting and attack model for our protocol. Section 4 presents our secure synchronization protocol. Section 5 and 6 provides a security and performance analysis about our protocol. Section 7 give a simulation result about our protocol. Section 8 concludes our work.

2   Related Works

Several clock synchronization protocols (e.g., [7,8]) have been proposed for sensor networks to achieve pairwise and/or global clock synchronization. All these clock synchronization techniques are all based on single-hop pairwise clock synchronization, which discovers the clock difference between two neighbor nodes that can communicate with each other directly. Two approaches have been proposed for single-hop pairwise clock synchronization: receiver-receiver synchronization (e.g., reference broadcast synchronization (RBS) [7]), and sender-receiver synchronization (e.g., timing-sync protocol for sensor networks (TPSNs) [8]). In receiver-receiver model, it synchronizes receivers by recording the arrival time of a reference message in a broadcast medium, and then exchanging this arrival time to determine the discrepancy between the respective clocks of the receivers. This technique is shown in *Fig. 1*. The approach avoids various sources of synchronization error because a broadcast message is received at exactly the same time by all receivers, at least in terms of the granularity of their clocks.

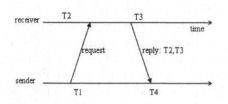
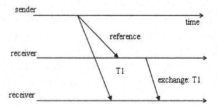

Fig. 1. Sender to receiver synchronization   **Fig. 2.** Receiver to receiver synchronization

In sender-receiver synchronization model, synchronization is performed by a handshake protocol between a pair of nodes. As showing in *Fig 2*, $T1, T4$ represent the time measured by the local clock of node A. Similarly $T2, T3$ represent the time measured by B's local clock. At time $T1$, A sends a synchronization pulse packet to B. Node B receives this packet at $T2$, where $T2$ is equal to $T1 + \delta + d$. Here, δ and d represent the offset between the two nodes and end-to-end delay respectively. At time $T3$, B sends back an acknowledgement packet. This packet contains the values of $T2$ and $T3$. Node A receives the packet at $T4$. Similarly, $T4$ is related to $T3$ as $T4 = T3 - \delta + d$. Node A can now calculate the clock offset and the end-to end delay as below:

$$\delta = \frac{(T2 - T1) - (T4 - T3)}{2} \tag{1}$$

$$d = \frac{(T2 - T1) + (T4 - T3)}{2} \tag{2}$$

Some recent works [3, 4, 5] attempt to deal with malicious attacks against clock synchronization in sensor networks. In [4], LS linear regression is used by each node to calculate the skew of its clock which can reduce the synchronization impact introduced by corrupted nodes. [4] also employ authentication techniques to deal with the external attacks. [3] provide resiliency to internal attacks by checking triangle consistency. Though this approach can prevent incorrect clock synchronization due to malicious attacks, it will also lead to denial of clock synchronization in such situations. Furthermore, [3] only provide internal synchronization, which is not suitable for those time sensitive applications. The basic idea in [5] is to provide redundant ways for each node to synchronize its clock with the common source, so that it can tolerate partially missing or false synchronization information provided by the malicious nodes. However the each node requires $2t + 1$ parents' synchronization messages which introduce high communication overhead to power limited sensors.

3   System Setting and Attack Model

3.1   System Setting

In a typical sensor networks, sensors are densely deployed; Each sensor has multiple neighbor sensors within its communication range. We assume there exist some pairwise key scheme for sensors to establish pairwise key with its neighbors [11, 12]. We also note that there are always many sensors located in the intersection of a pair of sensor's communication range as shown in *fig 3*. As we will show in later sections, we exploit this observation to develop a protocol for detecting compromised node and ensuring secure synchronizaiton.We also mentioned that in sensor networks there are one or more base stations, which are sinks and aggregation points for the information gathered by the nodes. Since base stations are often connected to a larger and less resource-constrained network, we assume a base station is trustworthy as long as it is available. Another

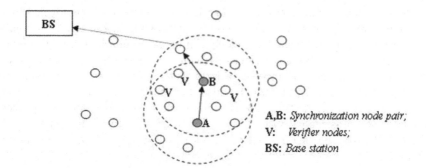

Fig. 3. System model of sensor network

assumption is coming from the fact that intruders need take some time to capture a sensor in sensor network. So we assume there is no comprised sensor during network initializing phase. It is clear that base station can well synchronized to external clock. We would like to synchronize the clocks of all the sensor nodes in the network to the base station.

3.2   Attack Model

In sensor networks, intruders can launch variety of attacks to corrupt the synchronization in sensor networks. For external attackers, they may affect the single-hop pairwise clock synchronization between two normal nodes by launching, for example, pulse-delay attacks [3], message replay, faulty message injection and so on. Fortunately, it is easy to prevent those attacks from external intruders by authentication techniques. For the internal attacks, they are much more complicated to be prevent. The internal adversaries can fully control a few of compromised sensor and therefore also has access to the secret keys for authorized communication with other nodes. Take TPSN as example, the compromised node can effect its children by sending incorrect time stamps for the reception time and the transmit time of the time-synchronization request. The adversary can propagate this de-synchronization down the spanning tree toward the leaves on its own branch.

4   Secure Clock Synchronization

In this paper we develop a secure clock synchronization protocol for sensor network. Our goal is to detect the compromised sensor and ensure the synchronization security of the sensor network. Our protocol is adapt from sender-receiver model and combine with the RBS protocol. The basic idea is based on the observation that there are many neighbor sensors located in the communication range of each synchronizing sensor pair. Due to the broadcast nature of wireless sensors, those neighboring sensors can overhear all the synchronization messages between each synchronization pairs, hence can verify if the synchronization process is correctly performed.

4.1 Protocol Detail

Our protocol consists of three phases: *level discovery phase, synchronization phase and verification phase*. The level discovery phase is to organize sensor nodes into a hierarchy rooted at the base station so that two nodes connected in the hierarchy are neighbors. Each node except for the base station has a parent node in the hierarchy. Each node is also associated with a *level*, which is the number of hops in the longest path from the base station to itself. The nodes beside *level*1 nodes also have a verifiers set to ensure the correctness of the synchronization. We refer to this hierarchy as the level hierarchy. In synchronization phase, each node is synchronized with its parent using sender-receiver model. The verification phase is performed during synchronization phase. We denote the nodes have same or less level and locates in each other communication range as sibling nodes. The sibling nodes of a parent node who are also located in the communication range of its child nodes are formed as verifier set for this synchronization pair. Due to the verifiers can overhear all the synchronization process between synchronization pair, they can judge whether the sender-receiver synchronization is correctly performed, hence detect the compromised node.

Level Discovery Phase. The level discovery phase can be run when sensor network initializing. To establish the level hierarchy, each node maintains three variables: *level, parent* and *verifiers*. The variable records the level of the node, record the parent of the node in the level hierarchy and verifiers set of the node respectively. We assume all the sensor nodes has discovered its neighbors before level discovery and established the unique pairwise key among them [12].Consider a node x. Initially, $x.level = 0$, $x.parents = Null$ and $x.verifiers = NULL$. Base station initials the level discovery phase by unicast a *level discovery message* to each of its neighbors. A level discovery message contains the senders identity and its level number, authenticated with the pairwise key shared between the sender and the receiver. After receiving an authenticated level discovery message from base station, each node i sets as $i.level = 1$ and $i.parents = BastStation$. It then unicasts a level discovery message to each of its neighbor nodes except for base station.

For node i that are multiple hops away from base station, it may receive more than one level discovery messages from their neighbors. Node i may wait for a period of γ time units after getting the first level discovery message and then chooses least level number from those messages and set $i.level = leastlevelnumber + 1$. Then node unicast the level discovery message to the neighbors who do not send level discovery message to itself.

At the end of level discovery phase, each node declares its level to all its neighbors by encrypted unicast. Then node adds those neighbors whose level is lower than itself to verifiers set and set one of them as its parent.

Synchronization Phase. In this phase, pair wise synchronization is performed along the edges of the hierarchical structure established in the earlier phase. We use the classical approach of sender-receiver synchronization [8] for doing this handshake between a pair of nodes. The synchronization is performed level by level. Each node synchronize with its parent, specially level 1 nodes synchronize with base station. Due to the sensor clock drift, base station periodically initials the synchronization phase by broadcast a *synchronization start message*. On receiving *synchronization start message*, node start a timer $u * l$ where u is the period for one pairwise synchronization and l is the node's level. Once node timer is expired, the node launch the pairwise synchronization with its parent as shown in table 1.

Table 1. Synchronization Phase

Synchronization Phase:
1. Base station broadcast *synchronization start message* in network;
2. On receiving *synchronization start message*, node set timer $u * l$
3. **while** (timer is not expired) do nothing;
4. Node A send a *sync message* to its parent B:
$\qquad$ A (T1)$\rightarrow$ B (T2): $A, B, N_A, sync$
5. B received *sync message* and reply a *ack message*:
$\qquad$ B (T3)$\rightarrow$ * (T4): $\{\ B, A, N_A, T2, T3, ack,$
$\qquad\quad MAC_{K_{AB}}(B, A, N_A, T2, T3, ack)\}$
6. A calculates clock offset
$\qquad \delta = (T2 - T1) - (T4 - T3)/2;$
7. After waiting for a period, node A correct its own clock by the offset calculated in above step if no *warning message* received.

In this protocol, message integrity and authenticity are ensured through the use of Message Authentication Codes (MAC) and a pairwise key K_{AB} shared between A and B. This prevents external attackers from successfully modifying any values in the synchronization pulse or in the acknowledgement packet. Furthermore, the attacker cannot assume an identity of node B as it does not hold the secret key K_{AB}. An attacker can hear the packet over the wireless channel and can use the MAC in the future to generate authenticated packets. Using a random nonce, N_A, during the handshake safeguards the protocol against such replay attacks.

Verification Phase. The verification phase is taking place at the same time as synchronization phase. Due to the broadcast nature of wireless sensor network, each verifiers of node A can overhear the synchronization messages in step 4 and 5. Meanwhile, the verifiers are also formed as a receiver set during step 4. We use this nice property to detect malicious node in above phase. Since all verifiers have less level than node A, they all have been synchronized with base station. In the step 4 of synchronization phase, each verifier also records the message receiving time $C_v(T2)$, where $C_v(T2)$ denotes the verifier's local

Table 2. Verification Phase

Verification Phase:

A. On receiving *sync message*, verifier v records the receiving time $C_v(T2)$;

B. On receiving *ack message*, verifier v records local time $C_v(T4)$ and check

C. *if* $(|C_v(T2) - T2| < \varepsilon$ && $d_{min}^* < C_v(T4) - T3 < d_{max}^*)$
    return;

D. *else*
    verifier v unicast a *warning message* to A:
    $v \rightarrow A : Enc_{K_{VA}}(v, A, C_v(T4), N_A, ack\ mesage)$;

E. On receiving *warning message* from a verifier, say v^*, node A firstly check
if the *ack mesage* received by v^* is valid. If so, node A requests at least
$2t-1$ verifiers send their clock time of receiving *ack message* and calculates
the clock offsets for these verifiers. Then node A sorts these $2t + 1$ offsets
(including the offset calculated from its parent)and selects a middle one
δ_{mid} as its offset to global time, and correct its clock accordingly.

F. Node A compares δ_{mid} and δ (calculated in synchronization phase):
    if $(|\delta_{mid} - \delta| > \varepsilon)$
    set $B.credit = B.credit - 1$;
    else
    set $v^*.credit = v^*.credit - 1$;
    if $(B.credit < 0)$
    delete the pair wise key K_{AB}, and set $A.parent = v^*$;
    if $(v^*.credit < 0)$
    delete the pairwise key K_{AV^*};
    delete v^* from verifier set;

---

clock time. According to receiver-receiver model, verifiers and Node B received
sync message at the same time, that is $|C_v(T2) - T2| < \varepsilon$, where ε is the error
bound of clock synchronization[1]. On the other hand, T3 can be also bounded as
$d_{min}^* < C_v(T4) - T3 < d_{max}^*$, where $C_v(T4)$is the local clock time when receiving
ack message, d_{min}^* and d_{max}^* is the minimum and maximum propagation delay
between neighbor nodes. d_{min}^* and d_{max}^* can be estimated before network work
deployed with high accuracy [3], hence T3 can be bounded in tens of μs. We
give a formal description about *verification phase* as shown in table 2.

Once detected abnormal $T2$ or $T3$ value in *ack message*, verifier can notify
node A to re-synchronize with verifiers. Since verifiers have formed a receiver set
with node A in *step 5*, node A can calculates offset by exchanging the receiving
time of *ack message*; Note that all verifiers have already been synchronized with
base station, node A can also synchronize with global time by this receiver-
receiver synchronization. To prevent the attacks launched by malicious verifiers,
node A calculates at least $2t + 1$ clock offset with verifier and selects middle one
as its reference offset. t is a threshold value for the number of verifiers and will
be discussed in the following sub-section. Variable *credit* is used to evaluate the
neighbor's credits. Nodes maintain a *credit* variable for each of its neighbor and

---

[1] ε is normally only tens of μs [8].

initially sets as T. Once the neighbor's credit value falls below zero, node deems it as compromised node and delete the pairwise key between them to isolate the compromised node from itself.

4.2 Number of Verifiers

The robustness of our protocol is determined by the threshold value t. And the value of t depends on the average number of verifiers for each node. We present a method to estimate the number of verifiers as follow. As shown in fig 3, we assume that nodes are uniformly distributed in target area with density ρ. We also assume the average distance between two nodes is d, and the communication range of the node is r. Then we can calculate the size of the intersection area of the neighbor nodes' communication circle as:

$$S = 2 \times \frac{2 \times \arccos(d/2r)}{2\pi} \times \pi \times r^2 - d \times \sqrt{r^2 - d^2/4} \qquad (3)$$

the average distance d can be also estimated as $d = \frac{2}{\sqrt{\pi\rho}}$. It is reasonable to suppose that $1/3$ of the nodes in the intersection region can be act as verifiers. Combine with the equation (3), the threshold t can be estimated by following equation:

$$3 \times (2t + 1) = S \times \rho$$

$$t = \frac{(2 \times \arccos(\frac{1}{\sqrt{\pi\rho} \times r}) \times r^2 - \frac{2 \times \sqrt{\pi\rho r^2 - 1}}{\pi\rho}) - 3}{6} \qquad (4)$$

Fig. 4 shows the threshold t's value given with different communication range r and the network density ρ. It is clear that t is almost linear with network density ρ, and can be as large as several tens in most typical network depolyments. In practice, t should select lower bound of the valid range to ensure each node can achieve enough verifiers to complete the synchronization process.

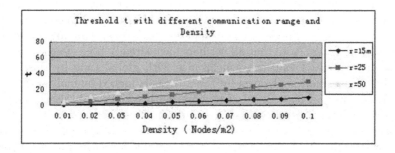

Fig. 4. Number of Verifiers given different communication range and network density

5   Security Analysis

Our synchronization protocol is secure against internal attacks as well as external attacks.

Lemma 1. *Assume a normal node has at least $2t + 1$ verifier nodes and parent node, among which there are at most t colluding malicious nodes. Node can obtain a correct clock offset to global clock from verifiers and parent.*

Proof. According to our model, there are two cases for a normal node to obtain its clock offset:

Case 1: Parent node is normal node, and no malicious verifier node sends wrong *warning message*. In this case, the synchronization process is same as *sender-receiver* synchronization protocol, hence the node can synchronize with its parent node;

Case 2: Parent node or verifier node are malicious nodes, node must collect $2t + 1$ offsets from those verifiers and parent . Because there are at most t malicious nodes, the median offset δ_{mid} must be between two clock offsets obtained through two normal nodes. Therefore, the median offset δ_{mid} is trustable [13], and can help node to synchronize itself to global clock correctly.

Attackers may attempt to launch resource consumption attacks to deplete the limited battery power of sensor nodes in a period of short time. For example, a malicious verifier node can interfere each correct synchronization process by sending *warning message*. Therefore, node has to request messages from verifiers to check the synchronization process and waste more power. However, this attack's impact can be limited by isolating such malicious node after attack performed several times in our protocol.

As for external intruders, the intruder may forge the identity in the network to inject faked messages. By using unique pairwise keys to authenticate messages, our schemes can prevent such attacks. The intruder may also launch replay attacks during the synchronization process. Specifically, a intruder may record a synchronization message during before synchronization phase, and replay it to normal nodes in later. As a result, the normal nodes may accept the replayed message, and derive a false source clock offset. This attack can be prevented by a random nonce, N_A, during the synchronization handshake.

6   Performance Analysis and Simulation Result

We discuss the performance of the proposed protocol using communication overhead, synchronization precision, and storage requirement.To better explain our protocol's performance, we simulate our protocol on NS2. In our settings, there are 100 nodes uniformly distribute on a $50m \times 50m$ simulation area. We assume the communication range of each node is 25m and the bandwidth of each physical link is 250 kb/s, as for MICAz motes. We also simulate a local clock C_i for

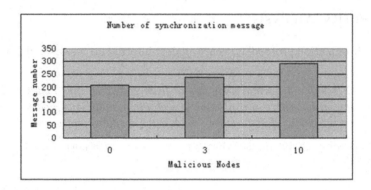

Fig. 5. Number of synchronization message

each node as $C_i = (1 + \rho_i)C_g$, where C_g is the global time, and ρ_i is the node i's clock drift rate. Each ρ_i is randomly generated using a uniform distribution between $-10\mu s/s$ and $10\mu s/s$. We also set the one pairwise synchronization time u as $2s$.

Communication Overhead. The communication overhead can be divided into two parts. In level discovery phase, after a node determines its level, it unicasts level discovery messages to the neighbors that have not sent level discovery messages to it. Assuming there is no communication failure and all the nodes are included in the level hierarchy. To be clear, we consider the network is as a graph $G = V, E$, in which each vertex in V stands for a node in the network, and each edge in E represents that the two vertices of the edge are neighbor nodes. Therefore, the communication overhead in level discovery phase is $|E|$. In synchronization phase, there are two cases. If no attacks during the synchronization process, the communication overhead is same as normal sender-receiver synchronization, two messages for each node, that is $2|V|$ for whole network synchronization. If the synchronization process is under attack, the node has to request additional $2t$ verify messages. However, the attacks may only occur in a small part of synchronization pairs, hence these additional communication overhead is insignificant.Fig 5. shows the the number of messages sent during synchronization under different situation in our simulation.

Synchronization Precision. A high precision pairwise clock synchronization scheme is critical for our schemes, since the synchronization error may accumulate for nodes that are multiple hops away from the base station. Our protocol is based on sender-receiver model or receiver-receiver model if under attack. As in TPSN [8], we can also use MAC layer timestamp to minimize the clock error for sender-receiver synchronization. And hence the synchronization precision can be restrict in tens of μs [8]. Even under attacks, median offset is used to synchronize the node and global time which can effectively eliminate the impact of attacks and achieve almost the same synchronization precision.

Fig 6. show the average synchronization error and maximum synchronization error in different level. As shown in fig 6. the error does not blows up with hop distance.This is because synchronization error between any pair of nodes will probabilistically take different clock drift from the normal distribution as said above. The randomness of clock drift prevents the error from blowing up. In fact from our simulation result it seems that error almost becomes a constant beyond 3-4 hops.

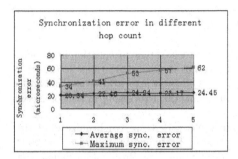
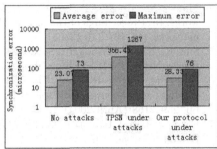

Fig. 6. Synchronization error in different hop count

Fig. 7. Synchronization error

We also simulate the situation that the network is under attacks. We place 10 malicious nodes in different level, and each malicious nodes randomly generate a synchronization error between $200us$ and $1000us$. Fig 7. shows the synchronization precision comparison among different situation. From the fig 7, it is clear that even under attack our protocol can achieve almost the same synchronization accuracy as no attack.

Storage Overhead. Memory usage is a critical issue for resource constrained sensor nodes. In our protocol, each node needs to record node's parent, level, verifiers set and credits of its neighbors. During synchronization, each node may store at most $2t + 1$ clock offsets. However, these storage overhead is not exceed $1K$ bytes in general, which is relative small for the sensor's memory.

7 Conclusion

In this paper, we present a secure global clock synchronization techniques for sensor networks. The idea behind our protocol is to monitor synchronization process by the verifiers who located in both communication range of synchronization pair. Our approach guarantees that normal nodes can synchronize their clocks to global clock even if each normal node has up to t colluding malicious nodes during synchronization phase. Our protocol is also very efficient. The communication overhead is same as traditional sender-receiver model like NTSP protocol in normal environment. Even under attack, there are only $2t + 1$ more messages need to be transmitted.

References

1. Akyildiz, I.F., et al.: Wireless sensor networks: a survey. Computer Networks 38(4), C393–C422 (2002)
2. Mills, D.: Internet time synchronization: The network time protocol. IEEE Trans. Commun., 39(10), C1482–C1493 (1991)
3. Ganeriwal, S., Capkun, S., Han, C., Srivastava, M.B.: Secure time synchronization service for sensor networks. In: Proc. ACM Workshop on Wireless Security, September 2005, pp. C97–C106.
4. Manzo, M., Roosta, T., Sastry, S.: Time Synchronization Attacks in Sensor Networks. In: Proceedings of the 3rd ACM workshop on Security of ad hoc and sensor networks, pp. 107–116 (2005)
5. Sun, K., Ning, P., Wang, C.: Secure and Resilient Clock Synchronization in Wireless Sensor Networks. IEEE Journal on selected areas in communications 24 (2006)
6. Newsome, J., Shi, R., Song, D., Perrig, A.: The sybil attack in sensor networks: Analysis and defenses. In: Proc. IEEE Int. Conf. Inf. Processing in Sensor Netw., pp. 259–2168 (April 2004)
7. Elson, J., Girod, L., Estrin, D.: Fine-Grained Network Time Synchronization Using Reference Broadcasts. ACM SIGOPS Operating Systems Rev. 36, 147–163 (2002)
8. Ganeriwal, S., Kumar, R., Srivastava, M.B.: Timing-Sync Protocol for Sensor Networks. In: Proc. First Intl. Conf. Embedded Networked Sensor Systems (SenSys) (2003)
9. Hu, A., Servetto, S.D.: Asymptotically Optimal Time Synchronization in Dense Sensor Networks. In: Proc. Second ACM Intl. Workshop Wireless Sensor Networks and Applications (WSNA) (September 2003)
10. Li, Q., Rus, D.: Global Clock Synchronization in Sensor Networks. In: Proc. IEEE INFOCOM 2004 Conf. (March 2004)
11. Liu, D., Ning, P.: Establishing pairwise keys in distributed sensor networks. In: Proc. 10th ACM Conf. Compute. Commun. Security, pp. 52–61 (October 2003)
12. Chan, H., Perrig, A., Song, D.: Random key predistribution schemes for sensor networks. In: Proc. IEEE Symp. Res. Security and Privacy, pp. C197–C213 (2003)
13. Coleri, S., Varaiya, P.: Fault tolerance and energy efficiency of data aggregation schemes for sensor networks. In: IEEE VTC (September 2004)

On Grid-Based Key Pre-distribution: Toward a Better Connectivity in Wireless Sensor Network*

Abedelaziz Mohaisen**, YoungJae Maeng, and DaeHun Nyang***

Information Security Research Laboratory - INHA University
253 YongHyun-dong, Nam-Gu, Incheon 402-751, Korea
{asm,brendig}@seclab.inha.ac.kr, nyang@inha.ac.kr

Abstract. In this paper, we revisit Grid-Based Key Pre-Distribution Scheme in Wireless Sensor Network to investigate improving the connectivity of the network and maintain both the security level and communication overhead. Both of the original work and our modification are based on using symmetric bivariate polynomials for generating cryptographic keys. In addition, their work relies on the usage of multi-dimensional grid to assign the polynomials on the sensor nodes allocated on the intersections of the grid and provide a needed connectivity. In this work we consider the simplification of the key establishment logic, the enhancement of connectivity in what we call the plat polynomial assignment. We present detailed discussion on the connectivity, resources usage, and security features that shows better results on the side of the connectivity, intermediate node discovery and security measurement. Finally, we provide a comparison between our results and other existing solutions including the revisited scheme.

Keywords: Sensor Networks, Key Distribution, Bivariate Symmetric Polynomials, Network Connectivity.

1 Introduction

Wireless Sensor Network is a resulting successful coalescence of different technologies that includes microelectronics and semiconductors, networking, signal processing, and others [1]. This network consists of large number of sensor nodes which are inexpensive devices with limited resources that work in cooperative method to perform some sensing tasks [2]. Sensors communicate in peer-to-peer fashion in an open air environments that provide opportunities for man-in-the-middle (MITM) [3], Sybil [3, 4], or node replication attacks [5]. The growth of WSN applications brings the necessity to provide security rules to guard the

---

* This research was supported by the MIC (Ministry of Information and Communication), Korea, under the ITRC (Information Technology Research Center) support program supervised by the IITA (Institute of Information Technology Advancement)(IITA-2006-C1090-0603-0028).
** Mohaisen's work was done when he was graduate student at INHA University.
*** Corresponding author.

communication traffic. The symmetric key cryptography which is based on using the same key in both of the two communication entities to encrypt and decrypt messages is very efficient on the typical sensor node's platform [1, 6]. Due to the weak infrastructure of the WSN, the bottleneck challenge in the security is the distribution of keys on sensor nodes [7, 6, 8]. Traditional key distribution mechanisms such like the Trust Third Party (TTP) or the Key Distribution Center (KDC) are impossible solutions as well [9]. Therefore, it is required to be wisely distributed on the nodes in pre-deployment phases. In the next subsection, we provide a survey on the key pre-distribution schemes in WSN followed by our contribution and the structure of this paper.

1.1   Researches on Key Pre-distribution: Survey

General Schemes: Two of the early works in [10, 11] are widely known for its novelty. In the first work by Blom *et al.* [10] a symmetric matrix of size $N \times N$ is required to store the different N^2 keys for securing communication within the entire network of N nodes. Node $s_i \in N$ has row and column in the matrix. If two nodes s_i, s_j would like to communicate, they use the entries $\mathbf{E_{ij}}$ in s_i side and $\mathbf{E_{ji}}$ in s_j side which are equal (*i.e.* $\mathbf{E_{ij}} = \mathbf{E_{ji}}$ since the matrix is symmetric). To reduce the memory requirements, a slight modification is introduced by Du *et al.* [12]. The following are defined, a public matrix $\mathbf{G}$ of size $(\lambda+1) \times N$ and a private symmetric matrix $\mathbf{D}$ of size $(\lambda+1) \times (\lambda+1)$ where $\mathbf{D}$ entries are randomly generated. Also, $\mathbf{A} = (\mathbf{D} \cdot \mathbf{G})^{\mathbf{T}}$ of size $N \times (\lambda+1)$ is defined. For a node s_i, row $\mathbf{R_i}$ in $\mathbf{A}$ and column $\mathbf{C_i}$ in $\mathbf{G}$ is selected. When two nodes s_i, s_j would eventually communication securely, they exchange their $\mathbf{C_i}, \mathbf{Cj}$ and compute $k_{ij} = \mathbf{R_i} \cdot \mathbf{C_j}$ in the side of s_i and $k_{ji} = \mathbf{R_j} \cdot \mathbf{C_i}$ in the side of s_j. The second work by Blundo *et al.* [11] uses Symmetric Bivariate Polynomial (SBP) to distribute keys for N nodes. The SBP is in the form of $f(x, y) = \sum_{i,j=0}^{t} a_{ij} x^i y^j, (a_{ij} = a_{ji})$ of degree $t \leq N$. For a node s_i with identifier i, the share $g^i(y) = f(i, y)$ is calculated and loaded to its memory generate secure keys. Similarly, for two nodes s_i, s_j, $k_{ij} = g^i(j), k_{ji} = g^j(i)$ are are evaluated locally and used respectively.

Random Key Pre-distribution Schemes: The early scheme of key pre-distribution specifically for WSN is introduced by ESCHENAUER-GLIGOR (EG) [13]. Each node is let to randomly pick a key ring S_k of size k from big keys pool of size P. The picking process provides a probabilistic connectivity $p_{actual} = 1 - \frac{((P-k)!)^2}{(P-2k)!P!}$. If two nodes s_i, s_j share a key $k : k \in S_{k_i} \cap S_{k_j}$ they use it a secret key. Otherwise, a path discovery phase via intermediate nodes is performed. In [13], the usage of memory is reduced, however, a weak resiliency is resulted. To improve the resiliency, Chan *et al.* proposed the Q-COMPOSITE scheme [14]. Using the same procedure of EG, a key between two nodes s_i, s_j is available **iff** $S_{k_i} \cap S_{k_j}$ is a set of q keys. If $\{k_1, \ldots, k_q\} \in \{S_{k_i} \cap S_{k_j}\}$, $\mathbf{hash(k_1 || k_2, \ldots, || k_q)}$ is used as k_{ij}, k_{ji}. Otherwise, intermediate node(s) are used. More analytical analysis on the probabilistic schemes is shown in by Kwang and Kim in [15].

Symmetric Matrices Based Key Pre-distribution: In addition to improving Blom's scheme in [10], Du *et al.* proposed two schemes [16, 12]. In the early one they introduced a deployment knowledge based scheme that improves Blom's [10] by avoiding the unnecessary memory, communication, and computation with reasonable connectivity [16]. In [12], a multi-space matrix scheme based on [10, 13] is introduced. A τ number of private matrices $\mathbf{D}$ is selected randomly out of ω pre-constructed matrices providing connectivity of $p_{actual} = 1 - \frac{((\omega-\tau)!)^2}{(\omega-2\tau)!)\omega!}$. Different $\mathbf{A}$'s are created using the different $\mathbf{D}$s. τ rows of the different $\mathbf{A}$s are selected and assigned for each node. For (s_i, s_j), If they have a common space $\tau_{i,j} : \tau_{i,j} \in \tau_i, \tau_j$, the rest of Blom's is performed, else, an intermediate space is used to construct a key path. Even though much memory and communication are required and smaller connectivity is generated, this work provides a higher resiliency than in [13, 14]. For more accuracy, different deployment structures with practical error measurements and the probability distribution functions pdf based on [16] is used by Ito *et al.* in [17].

Symmetric Bavariate Polynomial Based Schemes: At the same time, Liu *et al.* proposed several schemes [9, 18] for key distribution and mainly based on [11]. In [19], Blundo's scheme is used by assigning more than polynomial for each node similar as in EG scheme [13]. Another scheme is introduced where for a network of size N a two dimensional deployment environment constructing a grid of $N^{1/2} \times N^{1/2}$ is constructed. The different nodes are deployed on different intersection points of the grid and different polynomials are assigned for the different rows and columns of the grid. For any two nodes s_i, s_j, if $R_i = R_j$ or $C_i = C_j$, (*i.e.* they have the same SBP share), a direct key establishment as in [11] is performed. Else (*i.e.* $R_i \neq R_j$ and $C_i \neq C_j$), an intermediate node is used in indirect key establishment phase. Even if a big fraction of nodes p_c as 50% of N is compromised, the network still has the ability to be connected via alternative intermediate nodes. an n-dimensional scheme is introduced in [18].

1.2   Our Contribution and Paper Organization

We improve the three dimensional grid based key pre-distribution scheme in [9, 18] using an extended order of grid with a plat polynomial assignment. Our contribution includes the nodes/polynomials assignment in a three dimensional grid, performance analysis to support the improvement of the connectivity and security performance. We provide a detailed study on the usage of the different resources and study the impact of the communication traffic modeling on required resources. On the structure of the paper, section 2 introduces our detailed contribution, section 3 introduces an analysis of the connectivity, resources usage, performance and security. Section 4 introduces a comparison between our scheme and other possibly comparable schemes and section 5 introduces a conclusion and set of remarks.

2   Grid-Based KPD with Plat Polynomial Assignment

Consider a network of size N, we use a three dimensional grid as in Fig. 1(a). The three dimensional grid has x, y, z axis with each of $N^{1/3}$. We denote the columns

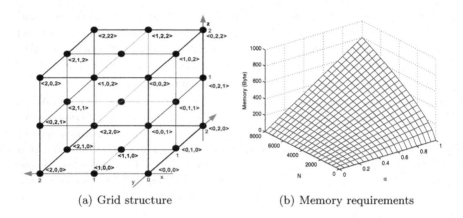

(a) Grid structure   (b) Memory requirements

Fig. 1. (a) The structure of the grid with plat-polynomial assignment (b) Memory required to store the polynomial for different network and security parameters

of the different dimensions as C_x, C_y, C_z. The different nodes are deployed on the intersections of the grid. Each node s_i has a three axis coordinate as $\langle C_x, C_y, C_z \rangle$. For each axis of the grid, different symmetric polynomials are assigned in a way that all of the plat with the same axis value owns the same corresponding polynomial. The total number of the polynomials in the network is $3(N^{1/3})$. Each node in the network has three polynomials shares of the corresponding axis. In the following we provide details on the keying material generation and secure key establishment.

2.1 Keying Material Assignment and Identifiers Structure

On a key management server, the following procedure is performed for one time:

- To construct the grid introduced earlier as of Fig. 1(a), the server picks an integer $m = \lceil N^{1/3} \rceil$ where N is a larger integer than the real network size to provide a flexible extendibility for the network size that provides the ability for other nodes to join and leave the network freely.
- Each sensor node s_i with identifier (i) of size $\log_2 N$ bits is mapped to the proper position on the grid of Fig. 1(a) for which the node will have the ID structure of $i = \langle c_x || c_y || c_z \rangle$.
- The server construct $3 \times m$ different symmetric polynomials. Each polynomial satisfies the condition that $f(x, y) = f(y, x)$ where $f(x, y)$'s coefficients are randomly picked in a finite field F_q with $q > N$ to avoid the collision and accumulate a reasonable security stand.
- The different polynomials are grouped in triples that constructs all of the possible outcomes for three groups of m size. Each group also has the notation $< f_{c_x}, f_{c_y}, f_{c_z} >$.
- Unlike [19], each node with the identifier $\langle c_x || c_y || c_z \rangle$ picks three polynomials with the indices (c_x, c_y, c_z) that are equal to the node's identifier coordinates (*i.e.* all nodes with same plat has same polynomial).

- For each sensor node s_i with identifier i and polynomials $< f_{c_x}, f_{c_y}, f_{c_z} >$, the server evaluate the shares $g_{c_x} = f_{c_x}(i, y), g_{c_y} = f_{c_y}(i, y), g_{c_z} = f_{c_z}(i, y)$ and load these shares into s_i's memory.

Note that this will guarantee that all of the nodes with the same x-axis (*i.e.* that belongs to the same plat of the same dimension) have the same polynomial.

2.2   Key Establishment

Assuming that limited number of nodes is compromised, the assigned polynomial for each node provides actual connectivity that enables two different nodes to communicate directly using their polynomials shares. Other cases require using intermediate nodes to establish keys. In the following we show the different cases.

Direct Key Establishment: For two nodes s_i, s_j that has the IDs $i = \langle c_x || c_y || c_z \rangle$, $j = \langle c_x || c_y || c_z \rangle$, if $i.c_x = j.c_x$ or $i.c_y = j.c_y$ or $i.c_z = j.c_z$; that means that both s_i, s_j belong to at least one common dimension and have a shared polynomial share, use the common shared polynomial's share $g^*(y)$ to establish a common key. If $i.c_x = j.c_x$ and $i.c_y = j.c_y$, or $i.c_x = j.c_x$ and $i.c_z = j.c_z$, or $i.c_y = j.c_y$ and $i.c_z = j.c_z$ use the common polynomial with least compromised shares to establish the key (Note that the last case where all of the coordinates are equal is concerned with the node itself). Finally, If neither of the node's coordinates is equal, an intermediate node or more are used to establish a key using the Indirect Key Establishment.

Indirect Key Establishment: If the two nodes don't belong to the same plat of dimension, they should establish a key path that consists of one or more intermediate nodes. For nodes s_i, s_j as communication parties with i, j as identifiers, they pick s_ϕ with ϕ as the identifier such that any of the following is satisfied:

- $\phi.c_x = i.c_x$ and $\phi.c_y = j.c_y$ or $\phi.c_z = j.c_z$.
- $\phi.c_y = i.c_y$ and $\phi.c_x = j.c_x$ or $\phi.c_z = j.c_z$.
- $\phi.c_z = i.c_z$ and $\phi.c_x = j.c_x$ or $\phi.c_y = j.c_y$.

Based on the matching of the parts, the corresponding shares in ϕ are used (at least two) to make the node s_ϕ as an intermediate node. For example, the first matching leading to the following key generations $k_{\phi i} = g_{cx}^{\phi}(i)$, $k_{i\phi} = g_{cx}^{i}(\phi)$, $k_{\phi j} = g_{cx}^{\phi}(j)$, and $k_{j\phi} = g_{cx}^{j}(\phi)$ and so on for other matchings.

3   Analysis

3.1   Distribution of Communication

In the wireless network, the communication activities are distributed and modeled using a communication traffic function with a probability distribution function f_R defined on the area R. To use this advantage, we introduce $f_R(n)$ that is defined on the area and the plats to which nodes are related. In other words, n would mean the number of hop under the non-adversary condition.

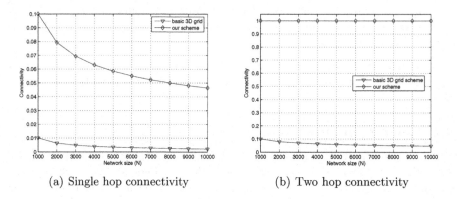

(a) Single hop connectivity   (b) Two hop connectivity

Fig. 2. The connectivity comparison for our scheme and the grid based scheme [18] for both the single and two hop operation

3.2 Connectivity

The connectivity in general is defined as the fraction of the nodes that can communicate with each other using its own keying material in one hop manner. In our scheme, the directly guaranteed actual connectivity is due to the three plats' ploynomials that represents the axis that the node resides on. Let $m = N^{1/3}$, the connectivity of our scheme is $C_{actual} = 3(\frac{m^2-1}{m^3-1})$ which is approximately equal to $\frac{3}{m+1}$. For the basic grid scheme, the single hope connectivity is $C_{actual} = 3(\frac{m-1}{m^3-1}) = (\frac{3}{m^2+m+1})$ [18], which is smaller than the corresponding in our scheme. Fig. 2 shows a comparison for the connectivity between our modified scheme and the original work where Fig. 2(a) is for the single hop connectivity and Fig. 2(b) is for the two hop connectivity. Note that the connectivity of our scheme is always better than the corresponding of [18] since $\frac{3}{m+1} > \frac{3}{m^2+m+1}$ for any $N > 0$. Practically, the minimum network size under which the grid construction is valid is $N = 8$ which leads to $m = 2$ that guarantees the validity of the connectivity advantage in our scheme.

3.3 Memory Overhead

The memory is required mainly dependent on the desirable security level. Let (α) be a security parameter such that $0 \leq \alpha \leq 1$ which determines the level of the security for the nodes which hold the shares of a given polynomial [15], the required memory for storing the coefficients $a_0, a_1, \ldots, a_t$ of the polynomial terms $x^0, x^1, \ldots, x^t$ is $(t+1)\log_2(q)$ bits which can be written as $(\alpha \times m + 1) \times \log_2(q)$. Let N_c be the number of compromised nodes, the required memory M is as in Eq. 1. The required memory for different parameters is shown in Fig. 1(b)

$$M = 3\left((N_c + 1)\left\lceil \log_2(N^{1/3})\right\rceil + (t+1)\log_2(q)\right) \tag{1}$$

3.4   Communication Overhead

The security-related communication overhead is required to exchange two nodes identifiers. For the key establishment under the condition that small fraction of nodes is compromised, there are two different cases: The direct key establishment that requires single ID exchange and the KPE through one intermediate node that requires two identifiers exchange. Based on the identifier structure presented earlier, $3\log_2(N^{1/3})$ bits are required to represent it. At average, the required communication overhead in bits is the average to exchange the IDs in the two cases as follows:

$$C_{cm_{avg1}} = \frac{1+2}{2} \times 3 \left\lceil \log_2(\sqrt[3]{N}) \right\rceil = 4.5 \left\lceil \log_2(\sqrt[3]{N}) \right\rceil \qquad (2)$$

For practical models that consider the usage of the communication traffic function, the communication overhead is defined as:

$$C_{cm_{avg2}} = 3 \left\lceil \log_2(\sqrt[3]{N}) \right\rceil \sum_{i=1}^{n} i f_R(i) \qquad (3)$$

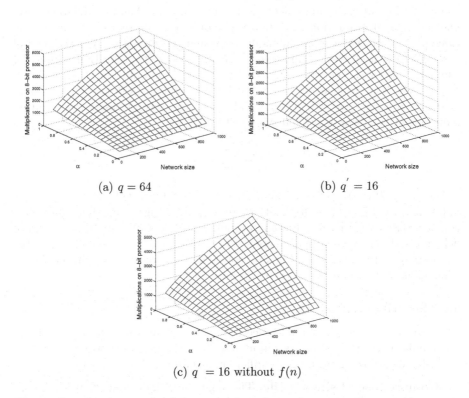

(a) $q = 64$

(b) $q' = 16$

(c) $q' = 16$ without $f(n)$

Fig. 3. Computation overhead for network and security parameters: varying q, α, n with and without communication traffic function $f(n)$ where $f(n) = \frac{c}{2^{n-1}}$

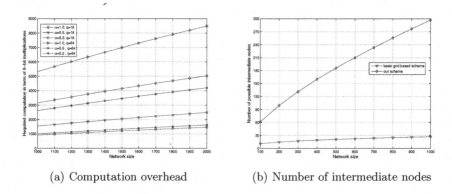

(a) Computation overhead   (b) Number of intermediate nodes

Fig. 4. (a) The computation overhead in term of required multiplications on 8-bit word processor using different security parameters (b) Possible intermediate nodes for two hops secure communication in our scheme and in the basic grid-based scheme

3.5 Computation Overhead

The same like in the communication, the computation has two different case: one polynomial evaluation of degree t is required if the two nodes are in the same plat. Else, two or more polynomials evaluation is required. For the general case, $f_R(n)$ is used to determine the required average computation within the running time of the network life. The first case's computation requirement is $C_m = 2t - 3$ multiplications on a big integer field to evaluate a polynomial of degree $t = \alpha \times m^2$ where $m = N^{1/3}$ and the second case's computation requirement is shown in the following in Eq. 4.

$$C_c = C_m \sum_{i=1}^{n} i f_R(i) \qquad (4)$$

Based on [19], two integers' multiplication on a finite field of 16 or 64 bits requires 16 or 27 8-bit multiplications respectively. Fig. 3 and Fig. 4(a) show comparisons between the required computations on both the 16 an 64 finite fields for different security parameters. Fig. 3(a) considers $q = 64$ and Fig. 3(b) considers $q' = 16$ with $F_R(n)$. Fig. 3(c) considers the computation without $F_R(n)$ which is notably greater than the early one.

3.6 Security Analysis

The security of any polynomial-based scheme is driven from the fact that the polynomial is secure as long as the number of compromised nodes is less than $t + 1$. In the following we measure different situations of our scheme:

- **Compromise of single node:** The single node holds one share for a concerned polynomial, even if it holds two other shares for two different polynomials. Thus, the compromise of single node will not reveal more than the share of the sensor node and other internal information.

- **Compromise of single plat:** We define the single plat by the nodes which holds the same polynomial's shares which are required to recover a polynomial when $\alpha = 1$. The probability for this event to happen when the number of the compromised nodes N_c is given is p_c as shown in Eq. 5.

$$p_c = 1 - \sum_{i=0}^{t} \left(\frac{(N^{\frac{2}{3}})!}{i!(N^{\frac{2}{3}} - i)!} \right) \left(F_c \right)^i \left(1 - F_c \right)^{N^{\frac{2}{3}} - i} \tag{5}$$

  where F_c is the fraction of compromised nodes, i is the number of compromised shares of a given polynomial, and N is the network size. An example that illustrates the difference between our scheme and the scheme in [18], for $N = 1000$ and $F_c = 0.5$, $p_c \approx 0.2$ in our scheme while $p_c \approx 0.4$ for the same network size for the construction of [18].

- **Attack against the network:** The attack against the whole network to break the security of the pair-wise communication between the sensor nodes is possible by compromising every single polynomial using the above approach. Even though there is a big overlap between the nodes which hold shares for different polynomials, still the network can resist up to a big fraction of compromised nodes and the revealed shares will be useless till a determined threshold fraction determined by α.

- **Possible intermediate nodes for key path establishment:** Unlike the basic grid scheme in [18] which provides $3\sqrt[3]{N}$ possible intermediate nodes for any key path establishment through two hops, our scheme provides $3\sqrt[3]{N}^2$ for the same process under the assumption of limited nodes are compromised. Fig. 4(b) shows a comparison between our scheme and [18].

4   Comparison with Other Schemes

The security of our scheme is typically same like the original work of Liu-Ning's [9]. The comparison between our scheme and other schemes is defined in the

Table 1. Comparison between our scheme and a set of other schemes in term of the resources usage and the resulting connectivity. The connectivity for the probabilistic key pre-distribution schemes is probabilistic while it is certain for other schemes including ours. Also, the polynomial degree t differs as shown earlier based on α.

Scheme	Communication	Computation	Memory	Connectivity
GBS [18]	c	SBP Evaluation	ID+2 SBP	$\frac{2}{N^{1/2}-1}$
3D-GBS [18]	c	SBP Evaluation	ID+3 SBP	$\frac{3}{N^{2/3}+N^{1/3}+1}$
our scheme	c	SBP Evaluation	ID+3 SBP	$\frac{3}{N^{1/3}}$
EG [13]	$c \log_2(S_k)$	$\frac{(2C+p-p_k)}{2} \log_2(C)$	S_k keys	$1 - \frac{((P-k)!)^2}{(P-2k)!P!}$
CPS [14]	c	c	S_k keys	$\frac{m}{N}$
DDHV [12]	$c \log_2(n \times \tau)$	2 vectors mult.	$\tau + 1$ vectors	$1 - \frac{((\omega-\tau)!)^2}{(\omega-2\tau)!\omega!}$
HGBS[20]	c	SBP Evaluation	ID+n SBP	1

resources usage shown in section 3. Table 1 shows a list of resources comparison with other possible schemes for key distribution in wireless sensor networks. This includes grid and three dimensional grid based schemes [9], hierarchical grid scheme. [20], Du et al. [12], and our scheme. The comparison shows that relatively, the usage of resources is relatively comparable with the corresponding in [9, 18]. As well, considering the advantageous connectivity shown earlier.

5   Conclusions

In this paper, we revisited the grid-based key pre-distribution in sensor networks which uses symmetric bi-variate polynomials to generate symmetric keys for set of sensor nodes deployed in a grid construction that provides certain connectivity. To improve such connectivity, we introduce the plat deployment method that guarantees a higher connectivity and maintains the security. We analyzed the performance of our modification on the side of the required computation, communication, and memory. Our modification provides a scheme for promising applications that require much connectivity for larger networks.

References

[1] Culler, D., Estrin, D., Srivastava, M.B.: Overview of sensor networks, pp. 41–49. IEEE Computer Society Press, Los Alamitos (2004)
[2] Akyildiz, I., Su, W., Sankarasubramaniam, Y., Cayirci, E.: A survey on sensor networks (2002)
[3] Newsome, J., Shi, E., Song, D.X., Perrig, A.: The sybil attack in sensor networks: analysis & defenses. In: IPSN, pp. 259–268 (2004)
[4] Zhang, Q., Wang, P., Reeves, D.S., Ning, P.: Defending against sybil attacks in sensor networks. In: ICDCS Workshops, pp. 185–191 (2005)
[5] Parno, B., Perrig, A., Gligor, V.D.: Distributed detection of node replication attacks in sensor networks. In: IEEE Symposium on Security and Privacy, pp. 49–63 (2005)
[6] Pietro, R.D., Law, Y.W., Etalle, S., Hartel, P.H., Havinga, P.: State of the art in security of wireless sensor networks. IEEE Computer 35, 1–10 (2002)
[7] Perrig, A., Stankovic, J., Wagner, D.: Security in wireless sensor networks. Commun. ACM 47, 53–57 (2004)
[8] Tillet, J., Ziobro, J., Sharma, N.K.: Secure wireless sensor networks: Problems and solutions. Journal on Systemic, Cybernetics and Informatics 1, 1–11 (2004)
[9] Liu, D., Ning, P.: Establishing pairwise keys in distributed sensor networks. In: ACM CCS, pp. 52–61 (2003)
[10] Blom, R.: An optimal class of symmetric key generation systems. In: Beth, T., Cot, N., Ingemarsson, I. (eds.) EUROCRYPT 1984. LNCS, vol. 209, pp. 335–338. Springer, Heidelberg (1985)
[11] Blundo, C., Santis, A.D., Herzberg, A., Kutten, S., Vaccaro, U., Yung, M.: Perfectly-secure key distribution for dynamic conferences. In: Brickell, E.F. (ed.) CRYPTO 1992. LNCS, vol. 740, pp. 471–486. Springer, Heidelberg (1993)
[12] Du, W., Deng, J., Han, Y.S., Varshney, P.K., Katz, J., Khalili, A.: A pairwise key predistribution scheme for wireless sensor networks. ACM Trans. Inf. Syst. Secur. 8, 228–258 (2005)

[13] Eschenauer, L., Gligor, V.D.: A key-management scheme for distributed sensor networks. In: ACM CCS, pp. 41–47 (2002)

[14] Chan, H., Perrig, A., Song, D.X.: Random key predistribution schemes for sensor networks. In: IEEE Symposium on Security and Privacy, p. 197 (2003)

[15] Hwang, J., Kim, Y.: Revisiting random key pre-distribution schemes for wireless sensor networks. In: SASN, pp. 43–52 (2004)

[16] Du, W., Deng, J., Han, Y.S., Chen, S., Varshney, P.K.: A key management scheme for wireless sensor networks using deployment knowledge. In: INFOCOM (2004)

[17] Ito, T., Ohta, H., Matsuda, N., Yoneda, T.: A key pre-distribution scheme for secure sensor networks using probability density function of node deployment. In: SASN, pp. 69–75 (2005)

[18] Liu, D., Ning, P., Li, R.: Establishing pairwise keys in distributed sensor networks. ACM Trans. Inf. Syst. Secur. 8, 41–77 (2005)

[19] Liu, D., Ning, P.: Establishing pairwise keys in distributed sensor networks. In: ACM CCS, pp. 52–61 (2003)

[20] Mohaisen, A., Nyang, D.: Hierarchical grid-based pairwise key pre-distribution scheme for wireless sensor nets. In: EWSN, pp. 83–98 (2006)

A Distributed and Cooperative Black Hole Node Detection and Elimination Mechanism for Ad Hoc Networks

Chang Wu Yu[1], Tung-Kuang Wu[2], Rei Heng Cheng[3], and Shun Chao Chang[1]

[1] Department of Computer Science and Information Engineering,
Chung Hua University, HsinChu, Taiwan, R.O.C.
cwyu@chu.edu.tw
[2] Department of Information Management,
National Changhua University of Education, ChangHua, Taiwan, R.O.C.
tkwu@mail.tkwu.net
[3] Department of Computer Science,
Hsuan Chuang University, HsinChu, Taiwan, R.O.C.
rhc@hcu.edu.tw

Abstract. A mobile node in ad hoc networks may move arbitrarily and acts as a router and a host simultaneously. Such a characteristic makes nodes in MANET vulnerable to potential attacks. The black hole problem, in which some malicious nodes pretend to be intermediate nodes of a route to some given destinations and drop any packet that subsequently goes through it, is one of the major types of attack. In this paper, we propose a distributed and cooperative mechanism to tackle the black hole problem. The mechanism is distributed so that it can fit with the ad hoc nature of network, and nodes in the protocol work cooperatively together so that they can analyze, detect, and eliminate possible multiple black hole nodes in a more reliable fashion. Simulation results show that our method achieves a high black hole detection rate and good packet delivery ratio, while the overhead is comparatively lower as the network traffic increases.

Keywords: Ad Hoc Networks, Black Hole, Wireless Networks, Security.

1 Introduction

Ad hoc network is a wireless network without fixed infrastructure. Each mobile node in an ad hoc network may move arbitrarily and acts as both a router and a host. When a node needs to send data to a destination, it broadcasts a *Route Request* (RREQ) packet to find a route. Every node in the network that receives this RREQ packet then rebroadcasts the RREQ packet until it reaches the destination. In response, the destination node sends back a *Route Reply* (RREP) packet to the source node right after it receives the RREQ packet. To accelerate the route finding process, many routing protocols allow nodes that have fresh enough routes to the destination reply with RREP packets. Unfortunately, some malicious nodes may take advantage of such routing characteristic and reply with a RREP to any RREQ received, claiming that it

T. Washio et al. (Eds.): PAKDD 2007 Workshops, LNAI 4819, pp. 538–549, 2007.

has a route to a given destination. However, once the data packets begin flowing through this route, they may just be dropped without being relayed. In this case, the node acts like a "black hole", which consumes any incoming data packets. The *black hole attack* is considered a kind of Denial of Service (DoS) attack [1, 2, 3].

In this work, we propose a distributed and cooperative mechanism that would detect potential multiple black hole nodes. Through collection of some local information, nodes in the network evaluate if there exists any suspicious node among their one-hop neighbors. Once a node is found suspicious, a cooperative procedure will be initiated to further check the potential "black hole" nodes. Simulation results show that our method achieves a high black hole detection rate and good packet delivery ratio, while the overhead is comparatively lower as the network traffic increases.

The rest of the paper is organized as follows. Section 2 reviews a few previous studies that were related to our work. The details of our proposed distributed and cooperative mechanism are given in Section 3. In Section 4, the simulation results, in terms of data packet delivery rate, control overhead, and "black hole" node detection rate are presented and discussed. Finally, some concluding remarks and direction of future study are listed in Section 5.

2 Related Work

A number of protocols were proposed to solve the black hole problem [4, 5, 6], which require a source node initiates a checking procedure to determine the reliability of any intermediate node claiming that it has a fresh enough route to the destination.

Deng, Li, and Agrawal [4] assume that the black hole nodes do not work as a group and propose a solution to identify a single black hole. With their protocol, the *source node* (SN) does not send data packets right away upon receiving the RREP packet from an *intermediate node* (IN). Instead, SN extracts the *next hop node* (NHN) information from the RREP packet and then sends a *Further-Request* (FRq) to the NHN (of IN) to verify if it has a route to the node (IN) that sends back the RREP message, or if it has a route to the destination node. The requested NHN can send back a *FurtherReply* (FRp) message, which includes the check result. If the NHN has no route to IN, but has a route to the destination node, it discards the reply packets from IN and uses the new route via the NHN to the destination. In the mean time, SN also sends out an alarm message to the whole network to isolate the malicious node. However, if the NHN has no route to IN and the destination node, SN would initiate another routing discovery process, together with an alarm message to isolate the malicious node.

Unfortunately, the above protocol would not be able to identify the scenario that the NHN works cooperatively with IN and sends back a false FRp. Accordingly, Ramaswamy, Fu, Sreekantaradhya, Dixon and Nygard [5] developed a solution to such a problem in which multiple black hole nodes cooperating as a group. Their protocol requires all nodes to maintain an additional *Data Routing Information* (DRI) table. Within the DRI table of a node, e.g., node X, information regarding to whether node X has ever routed packet from or through some other node is recorded. When an IN replies a RREP to a given SN, the NHN and DRI entry of NHN should also be sent together. The SN will then use the information together with its own DRI table to check whether the IN is a reliable node.

Lee, Han, and Shin also proposed a solution to eliminate a single black hole node that works with the DSR (Dynamic Source Routing) routing protocol [6]. In their protocol, an intermediate node (IN) needs to send a *Route Confirmation Request* (CREQ) to its NHN toward the destination in addition to sending a RREP back to the source. Upon receiving a CREQ, the NHN looks up its cache for a route to the destination and sends a *Route Confirmation Reply* (CREP) to the source with its route information in case it does have one. The source is then able to learn whether the path in RREP is valid by verifying the information with that provided in CREP. However, if the RREP is initiated by the destination, additional CREQ and CREP are not necessary since the destination should give correct route if it wishes to receive data packets.

3   The Distributed and Cooperative Mechanism

The proposed distributed and cooperated "black hole" node detection mechanism composes of four sub-steps as shown in Figure 1.

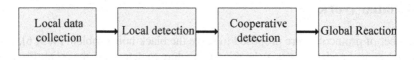

Fig. 1. Working procedures

In local data collection, each node collects information through overhearing packets to evaluate if there is any suspicious node in its neighborhood. If finding one, the detecting node would initiate the local detection procedure to analyze whether the suspicious one is a malicious black hole node. Subsequently, the cooperative detection procedure is initiated by the initial detection node, which proceeds by first broadcasting and notifying all the one-hop neighbors of the possible suspicious node to cooperatively participate in the decision process confirming that the node in question is indeed a malicious one. As soon as a confirmed black hole node is identified, the global reaction is activated immediately to establish a proper notification system to send warnings to the whole network. The details of each sub-step are described in the following four sub-sections.

3.1   Local Data Collection

With the local data collection step, each node in the network is required to evaluate if there is any suspicious node in its neighborhood by collecting information through overhearing packets and using the collected information to construct an estimation table. The estimation table, which is maintained by each node, contains information regarding to nodes that are within its power range (*Node* field), whether there is any data packet sent from the neighboring nodes to itself (*From* field), whether there is any data packet routed from itself to the neighboring node and received corresponding

ACK reply (*Through* field), ratio of received and transmitted packets of the neighboring node (*RTS/CTS* field), and a field indicating whether that neighboring node has been verified as being "suspicious" (*Suspicious* field). Table 1 is an example of such estimation table that is maintained by node 7 in scenario depicted in Figure 2.

Table 1. Estimation table maintained by node 7 of Figure 2

Node	From	Through	RTS/CTS	Suspicious
4	False	False	10	True
5	False	True	5	True
6	True	False	2	False
8	True	True	1	False

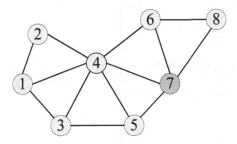

Fig. 2. An example for local data collection

Table 2. The inspection items of the estimation table

From	Through	Action
False	False	The node need to be inspected further
True	False	The node may possibly need to be inspected further (should take into consideration the RTS/CTS ratio)
False	True	The node may possibly need to be inspected further (should take into consideration the RTS/CTS ratio)
True	True	The node may not need to be inspected further

The estimation table can be used to identify suspicious black hole nodes. If a node has not successfully routed any data packets from or through some neighboring node, that particular node may be a potential black hole node and needs to be inspected further. On the other hand, if a node has already routed data packets from and through some neighboring node, that node should be a normal one. In case the information derived from the *From* and *Through* fields is not decisive, it is then up to the RTS/CTS field to determine whether it is urgent to examine further a particular node. A node with high receiving/transmission ratio should be inspected with higher priority. The combinations of criteria and their corresponding actions are summarized in Table 2.

3.2 Local Detection

When a suspicious black hole node is identified according to the estimation table in the local data collection phase, the detecting node, referred to as the initial detection node, would initiate the local detection procedure to analyze whether the suspicious one is indeed a malicious black hole node. The local detection procedure, as presented

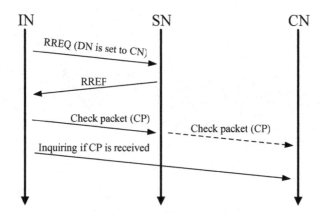

Fig. 3. Local detection flow chart

Local detection algorithm.
{*IN*: initial detection node, *CN*: cooperative detection node, *SN*: suspicious node, *DN*: destination node, *S1HN*: *SN*'s 1 hop neighboring node(s), *ET*: estimation table, *wt*: wait time}
Begin
IN selects a *CN* from *ET* and broadcasts a RREQ (with *DN* being set to *CN*).
IN waits for a *wt* time to collect RREPs from *SN*.
IF (the received RREP is from *SN*)
{ Establish a route to *DN* via *SN*.
   Send a check packet to *DN*.
   Wait for a *wt* time and then ask *CN* if it receives check packet.
   IF (YES)
     Mark suspicious field of *SN* entry in the *ET* to be "FALSE".
   ELSE
     Broadcast to all *S1HN*s and call for a "Cooperative Detection".   }
ELSE
   Ignore RREP.
End

Fig. 4. Local detection algorithm

in Figures 3 and 4, starts with the initial detection node selecting a one-hop neighbor as its cooperative node, and then sends out a Route Request (RREQ) to find a route to the cooperative node. Among the many received Route Reply (RREP) messages, which originated from the cooperative node and arrive from various intermediate nodes, the initial detection node would usually use the information contained in the first received RREP to construct the main route. However, in our case, the initial detection node waits for the RREP that is coming from the suspicious node.

That is, our protocol intentionally builds a route to the cooperative node that goes through the suspicious node. The initial detection node then sends a check packet containing an inspection value onto the route. After a pre-determined time, that initial detection node asks the cooperative node directly if it has received the check packet. If the response is positive, the suspicious node is confirmed to be a normal node. Otherwise, the cooperative detection procedure will be activated to further test on the suspicious node.

3.3 Cooperative Detection

Once the local detection procedure has detected a possible black hole node, the cooperative detection procedure (its flow chart and algorithm presented in Figure 5 and Figure 6) is activated. The cooperative detection procedure is initiated by the initial detection node, which proceeds by first broadcasting and notifying all the one-hop neighbors of the possible suspicious node to cooperatively participate in the decision process confirming that the node in question is indeed a malicious one.

Since the cooperative detection procedure uses broadcast to notify all the neighbors of the suspicious node, it inevitably increases the network traffic. In order to reduce the impact to the bandwidth-limited network, we also propose a constrained broadcasting algorithm that can limit the range of notification message to within a

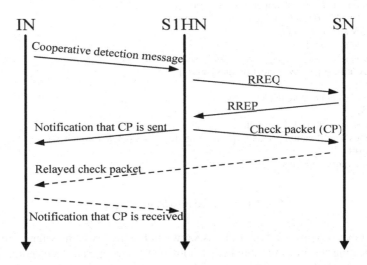

Fig. 5. Cooperative detection flow chart

Cooperative detection algorithm.
{*IN*: initial detection node, *SN*: suspicious node, *DN*: destination node, *S1HN*: *SN*'s 1 hop neighboring node(s), *ET*: estimation table, *VT*: voting table, *wt*: wait time}
Begin
  IN broadcasts cooperative detection message to all *S1HN*s.
  For each *S1HN*
  { Broadcast RREQ (with *DN* being set to *IN*)
   Upon receiving RREPs
   IF (received RREP is from *SN*)
  Send a check packet to *IN* via this route.
  ELSE
  Select a route and send a notification packet to *IN*.   }
  IN waits for two *wt* time to collect the check and notification packets from *S1HN*s.
  { IF a notification message is received
    Add source node *ID* of the message to *VT*'s Voter field.
    Mark FALSE to the corresponding "checked" field.
    ELSE IF a check message is received
         IF source *ID* of the message not in *VT*, add this node *ID* to *VT*'s Voter field.
       Mark TRUE to the "checked" field of the corresponding *ID* entry in the *VT*.   }
  Wait for another *wt* time and then check *VT*.
  IF ("checked" fields of all entries in *VT* are FALSE)
    Add *SN*'s *ID* to black hole list.
  Else
    Mark FALSE to the suspicious field of *SN*'s entry in *ET*.
End

Fig. 6. Cooperative detection algorithm

Constrained broadcasting algorithm.
{*IN*: initial detection node, *CN*: cooperative detection node, *SN*: suspicious node, *ET*: estimation table, *hr* : threshold}
Begin
  For each node receives cooperative detection message from *IN*
  { Extract *SN*'s *ID* and hop field from the message.
   IF (*SN*'s *ID* is not in *ET*)
  { IF ($hop <= thr$)
  { *hop++*
  Forward cooperative detection message.   }
  ELSE
   Ignore cooperative detection message.   }
  ELSE
  { Send a check packet through *SN* to *IN*.
   Sends a notification packet to *IN* via route that does not go through *SN*.
   hop++.
   Forward cooperative detection message.   }
End

Fig. 7. Constrained broadcasting algorithm

fixed number of hops. The algorithm is shown in Figure 7, which assigns a threshold, *thr*, indicating the maximal number of hops that the cooperative detection message would go. With the above mechanism, the control overhead can be greatly reduced.

However, if a lower *thr* value is set, some of neighbors of the suspicious node may not be found.

3.4 Global Reaction

As soon as a confirmed black hole node is identified, the fourth procedure–the global reaction–is activated to establish a proper notification system to send warnings to the whole network. An easy way to do this is simply broadcast the message to all the mobile nodes in the network. However, the simple mechanism may be exploited by other kind of malicious node to falsely charge some node as being a black hole, which constitutes a new form of attack. To preclude the above possibility, we propose two global reaction modes of operation that will improve the black hole notification mechanism.

With the first reaction mode, all mobile nodes on the network will be notified. The procedure begins with the initial detection node notifying all the neighboring nodes of the suspicious node in the same way as the cooperative detection process. The notified nodes then send warning messages accordingly. When all the nodes on the network receive enough warning messages, the malicious node would be added into the black hole list. All later data transmission will not go through nodes in the black hole list. Such a mode may consume more network resource, but achieve a higher reliability and would not be easily exploited by malicious attack.

On the other hand, many on demand routing protocols do not require mobile nodes to be aware of all the situation of the network. For example, it is sufficient that nodes in AODV concern only their one-hop neighbors [7]. As a result, with the second reaction mode, each mobile node only needs to manage its own black hole list and adjust its own transmission route accordingly. Such a mode may take more time and potentially be exploited by malicious attack, but would save lots of communication overhead by not sending warning messages to the network.

Table 3. The voting table

Voters	Checked
1	TRUE
2	FALSE
3	FALSE
4	TRUE
5	FALSE

4 Simulation Results

To demonstrate the performance of our proposed protocol, we run simulations with various setup environments using ns-2 and compare our distributed cooperative mechanism (DCM) with the original AODV protocol (AODV) and the AODV extension proposed in [5] (FRq-FRp). The simulation environment is specified by a 1000m×500m rectangular region with 50 or 100 mobile nodes moving around. Each of the mobile nodes is placed randomly inside the region. Once the simulation begins, each node moves toward a randomly selected direction with a random speed ranges from 0 to 10 meters per second. Upon reaching some randomly determined location, the node pauses for a fixed time, *pause time*, and proceeds again in a similar manner (in our simulation, the pause time is set to 200 seconds). The radio coverage region of each mobile node is assumed to be a circular area of diameter 150 meters. The transmission time for a hop takes 0.002 seconds, and the beacon period is one second. We assume the source node sends a 512-byte data packet every 0.33 second (with constant bit rate), and use UDP as the transport protocol. Each node has a queue, provided by network interface, for packets awaiting transmission that holds up to 50 packets and managed in a drop-tail fashion. A totally 2500 data packets are transmitted in each simulation run. Different parameters, such as the number of nodes, number of black hole nodes and number of data flows, are considered.

Our simulation is composed of two parts. The first part of simulation is designed to measure how our proposed protocol performs in terms of the following three metrics: (1) data delivery rate, which is the percentage of data packets delivered to destination with respect to the total number of packets sent; (2) control overhead, which is the ratio of the total number of control packets (including all AODV routing related traffic and black hole node detection messages) sent with respect to the total number of packet transmissions. Each hopping of a packet is counted as one transmission; (3) detection rate, which is the ratio of the number of detected black hole nodes and the total number of actual black hole nodes.

In the second part of the simulation, we measure the effectiveness of the constrained broadcasting algorithm, which is described in Section 3.3, in terms of (1) notification traffic ratio, which is defined as the ratio of the number of cooperative detection messages sent using the constrained broadcasting algorithm with respect to its broadcasting counterpart; (2) notification messages delivery rate, which is the ratio of total number of cooperative detection messages that are received by the cooperative nodes using the constrained broadcast algorithm with respect to the broadcast-based mechanism.

Figure 8 shows data packet delivery rate as a function of the number of black hole nodes. All the three methods perform well in case of no black hole nodes. With the presence of black hole nodes, the data packet delivery rate of AODV drops sharply, and it becomes worse as the number of black hole nodes increase. On the other hand, both of our method and FRq_FRp seem to be effective in eliminating the black hole effect. However, our proposed method performs better in the case of lower node density (number of node=50), while having slight edge over FRq-FRp in the case of higher node density (number of node=100).

Figure 9 shows the effect of the black hole detection procedure on control overhead with respect to the data packets transmitted. Standard AODV is used as a baseline for comparison. We can see that our method, in most cases, has comparatively lower control overhead than the FRq-FRp method in both lower and higher node density conditions. In addition, our method is more like a proactive detection mechanism, which means it generates constant number of control packets even though there is no data transmission in the network. However, the busier the network, the less the control overhead is (in percentage). On the other hand, the FRq-FRp method tends to

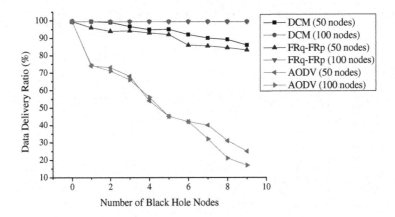

Fig. 8. Data packet delivery rate

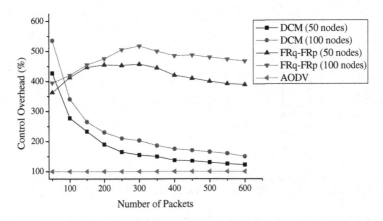

Fig. 9. Control overhead

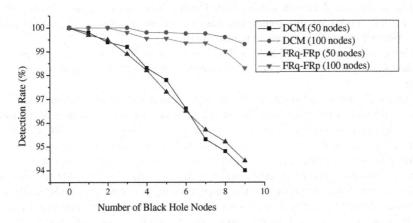

Fig. 10. Detection rate

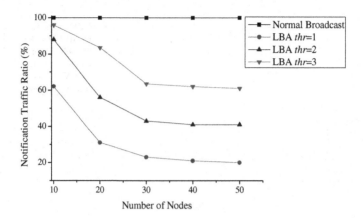

Fig. 11. Comparison of notification traffic ratio using different threshold values

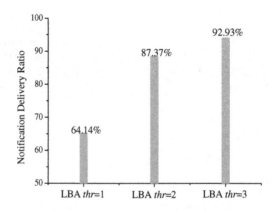

Fig. 12. Comparison of notification delivery ratio using different threshold values

be like a reactive detection mechanism. Upon a new discovery of a route, all the intermediate nodes on this route start the inspection procedure, which then adds a large number of control packets to the network.

Figure 10 shows the black hole nodes detection rate. We can see that our proposed method has higher black hole detection rate than the FRq-FRp method in most of the cases.

Figure 11 shows the notification messages ratio with respect to the number of nodes within the simulation area. The standard broadcast method is used as a baseline for comparison to our proposed constrained broadcasting method with various *thr* values. We can see that our proposed constrained broadcasting method does lower the traffic resulted from the notification procedure. But according to Figure 12, which presents the notification messages delivery rate, setting the *thr* value too low may results in that some neighbors of the suspicious node not receiving the notification message. However, a *thr* value of 3 seems to be a good tradeoff between saving of notification messages and getting most of the nodes aware of their suspicious neighbor.

5 Conclusion and Future Work

We propose a new solution against the black hole attack problem for ad hoc networks. Our method works in a distributed manner, and all mobile nodes cooperate together to analyze and detect possible multiple black hole nodes in a more reliable fashion. Simulation results show that our method has higher black hole detection rate and achieves better packet delivery rate than the notable FRq-FRp method [5] in most cases. And better yet, our method achieves that with much less overhead, especially when the network is busier. In addition, we also propose a constrained broadcasting algorithm to effectively reduce the overhead further, while maintain a competitive performance at the same time.

For further research, we will work on improving the voting mechanism, which may provide a solution to the gray hole and drop random attacks.

References

1. Zhang, Y., Lee, W.: Intrusion Detection in Wireless Ad Hoc Networks. In: Proceedings of the 6th Annual International Conference on Mobile computing and networking, pp. 275–283 (2000)
2. Gupta, V., Krishnamurthy, S., Faloutsos, M.: Denial of Service Attacks at the MAC Layer in Wireless Ad Hoc Networks. In: MILCOM, pp. 1118–1123 (2002)
3. Kumar, A., Koput, J.K., Chancham, S., Kim, Y.M.: Denial of Service Attacks in Ad Hoc Networks. A capstone paper submitted as partial fulfillment of the requirements for the degree of Masters in Interdisciplinary Telecommunications at the University of Colorado, Boulder, May 2, 2003. Project directed by Professor Timothy X Brown (2003)
4. Deng, H., Li, W., Agrawal, D.P.: Routing Security in Wireless Ad Hoc Networks. IEEE Communications Magazine 40, 70–75 (2002)
5. Ramaswamy, S., Fu, H., Sreekantaradhya, M., Dixon, J., Nygard, K.: Prevention of Cooperative Black Hole Attack in Wireless Ad Hoc Networks. In: International Conference on Wireless Networks, pp. 570–575 (2003)
6. Lee, S., Han, B., Shin, M.: Robust Routing in Wireless Ad Hoc Networks. In: International Conference on Parallel Processing Workshops, pp. 18–21 (2002)
7. Perkins, C.E., Royer, E.M., Das, S.: Ad-hoc On-demand Distance Vector (AODV) Routing. RFC 3561 (2003)

A Novel Adaptive and Safe Framework for Ubicomp*

Xuanhua Shi[1,3] and Jean-Louis Pazat[2]

[1] IRISA/INRIA, Campus de Beaulieu, 35042 Rennes Cedex, France
[2] IRISA/INSA de Rennes, Campus de Beaulieu, 35042 Rennes Cedex, France
[3] Huazhong University of Science and Technology, Wuhan 430074, China
{xuanhua.shi,jean-louis.pazat}@irisa.fr

Abstract. Although ubiquitous computing is more and more attractive, ubicomp doesn't become really pervasive. There are many reasons for this, lack of distributed infrastructure is one of them. Grid Computing is a new promising and powerful distributed infrastructure, so the merge of these two technologies is going to make ubicomp really ubiquitous. This paper presents an adaptive framework, Dynasa, which handles the safety problems for high performance computing applications in ubicomp environment based on the adaptive replication management.

1 Introduction

Marc Weiser presented his vision of ubiquitous computing(ubicomp) about 15 years ago, and he expected ubicomp to be a dominant form of computer interaction within 20 years [22]. However, despite numerous lab-based prototypes, ubiquitous computing technology failed to be commonplace [18]. Many reasons have been cited for this slow progress [7], one of them is lacking of common distributed systems infrastructure. Grid computing is a new promising and powerful research area which has the potential to completely change the way of computing as well as the data acquisition and dissemination process [9]. The grid promises a world where access to computational and storage resources across institutional boundaries is standardized, uniform, inexpensive,and ubiquitous. The next logical step in grid computing is a shift from a closed user group to a ubicomp environment for the benefits of geographically distributed computing requirements [2]. The merge of the ubicomp and the grid makes the ubicomp really ubiquitous [11].

The features of the ubiquitous infrastructure is as follows: (1) distributed, the computational resources, the storage resources and the users are dispersed geographically; (2) dynamic, the network status and the computational resources status change dynamically; (3) large scale, there is a large scale of computational

---

* This work is partly funded by the ARA SSIA SafeScale project of ANR, and this work is partly supported by the National Science Foundation of China under Grant No. 60603058.

T. Washio et al. (Eds.): PAKDD 2007 Workshops, LNAI 4819, pp. 550–561, 2007.

and storage resources involved in this infrastructure, and there are a large number of users involved in it, too. Running distributed applications on this kind of infrastructure has many threats, for example when a user in domain A runs a application in domain B, the network between these domains may have terrible congestion, or the machines in domain B maybe overloaded. For the dynamic and large-scale features, there are lots of threats for the application, such as, some malicious users the infrastructure, make parts of the infrastructure denial of services. Upon such kinds of situations, how to make a reliable and safe ubiquitous environment for the applications is what this paper targets.

Most applications over the grid are high performance computing(HPC) applications, such as numerical simulations, image processing [12]. As mentioned in [20], to integrate mobile devices into the computational grid is challenging. The best way for the combination between the mobile devices and the grid is probably to have the mobile devices work as clients for the applications. For the HPC applications, the mobile client doesn't has much power to control the applications, so the backend server should take some actions to deal with the threats the applications meet. Because of the large number of mobile devices that could be connected to the Grid, even with strong authentication mechanisms the Grid may not resist to some malicious attacks, such as breaking the access control to intrude the system, injection flaws to the application, or traditional Denial of Service attacks at the backend servers.

This paper describes a framework, Dynasa, to make a safe ubiquitous infrastructure at the application level, using the adaptive technology [1]. In this framework, the application at the backend server can be adaptive to create replicas, to increase replica numbers, reduce replicas and switch replication styles according to the environment changes.

The rest of this paper is organized as follows: in section 2, the related work is presented, and the framework is illustrated in section 3. In section 4, we will give a use case study scenario about this framework. The performance of this framework is evaluated by simulations in section 5, and we conclude and present some future work in section 6.

2   Related Work

The ubiquitous computing environment is most suitable for adaptive technology for the dynamic features; lots of research work has been done in this field. In [19], a protocol is presented which allows mobile devices to keep their own address in spite of mobility. F. Mouel, et al. [16] described an adaptive framework for dynamic data distribution over mobile environments based on a kind of data cache method. Most of the adaptive issues in mobile environments are related to data, such as the data distribution and live streaming adaptation, but there is not so much research work that has been conducted on the adaptive issues for HPC applications in ubiquitous environments.

The adaptive idea in grid computing is not a new concept, in [4] an adaptive scheduling method is presented, while this adaptive method doesn't give full control of applications to the developers, only the grid itself can make adaptive action according to the applications and the environments. J. Buisson, et al. [6] provides Dynaco for adaptive parallel components, which can be used to handle the adaptive requirements of the parallel and distributed application, giving the application developers full-control of the applications. In [1], an adaptive version of Nas parallel benchmark application is demonstrated with adaptation of the MPI processing nodes at the application runtime.

Security issues in ubicomp are hot topics. Frank Stajano [21] gives a whole description about the the security issues in ubicomp Systems, such as authentication, authorization, and availability. Alastair R. Beresford introduce a mix zone to protect the mobile users' privacy, which is a construction inspired by anonymous communication techniques [3]. Srdjan Capkun, et al. [5] introduce a fully self-organized public-key management for mobile Ad Hoc networks, which allows users to generate their public-private key pairs, to issue certificates, and to perform authentication regardless of the network partitions and without any centralized services. Most of research on ubiquitous computing security is about security at the system level. In this paper, we do not target authentication nor try to detect malicious behavior at the system level. We target the safety of applications not the safety of the Grid infrastructure itself and we rely on "safety sensors" at the application level to detect malicious behavior.

3   Adaptive Framework for Safe Ubiquitous HPC Applications

In this section, we present an adaptive framework for safe HPC applications. For illustration, we first give a brief view about the ubiquitous environment for the HPC applications. then the adaptive frame will be presented with the illustration of the framework components and the related methods involving in it.

3.1   Ubiquitous Environments of HPC Applications

The ubiquitous environments for HPC applications is shown in Fig. 1. Mainly, the ubiquitous environments are composed with two parts, one is the grid infrastructure, which includes the grid resources and the resource management middle-ware, the other are the mobile devices, such as cellphones, car stereos, televisions, VCRs, watches, GPS receivers, digital cameras. The grid infrastructure provides the HPC computational resources and the storage resources. The mobile devices provide the pervasive access to the grid resources.

3.2   Dynasa: Adaptive Framework

The Dynasa framework for HPC applications is composed of two parts, as shown in Fig. 2. One is the safety sensor which monitors the status of the environment and notifies the applications to be adaptive to different event. The other

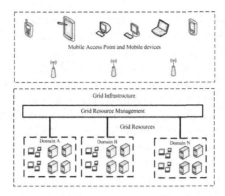

Fig. 1. Ubiquitous HPC environments

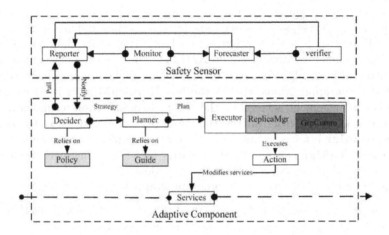

Fig. 2. Dynasa: Adaptive framework for safe HPC applications

part is the adaptive component, which is adaptive to create replication jobs, decrease replication number, and change replication styles according to the system and application situations. This part is handled by means of the Dynaco system [6]. An illustration of the components in the framework about their functions and how to be adaptive to different situations is presented in following subsections.

Components of the Adaptive Framework. The adaptive component is the component part of the framework, which is deployed in one component, and which includes Decider, Planner, Executor, Action and Service, a brief illustration is as follows:

– The **Decider** decides what kinds of strategies should be made for the planner according to the information given by the safety sensor. The information given by the safety sensor can be gathered in two ways: the safety sensor can notify the decider or the decider can pull information from the safety sensor. The decider makes the strategy based on the **Policy**, which describes component-specific information required to make the decisions. In this paper, the policy is the set of rules for safety applications with several kinds of techniques, such as resubmission of jobs, create replications. The decider has been implemented in Dynaco, Dynasa just add new policies which target the safety problems. For example if a safety sensor detects denial of service on some nodes in one domain, the decider could decide to enforce the security level of the application by "moving the application" to a "safer place" if stated by the application policy.

– The **Planner** is a program that modify the components. The planner makes an adaptation plan that makes the component adopt a given strategy by the decider, and the plan is passed to the **Executor**. The **Planner** has been implemented in Dynaco, Dynasa just add new plans to target the safety problems. For example the plan to "move the application" maybe stop the application; checkpoint process; redeploy application in domain B; restart.

– The **Executor** is a kind of virtual machine to interpret the plan provided by the planner by the mean of **Actions**. For example checkpoint is an action. Additionally, the executor handle the replica management sematic for the safe applications, that is the box **ReplicaMgr** in it. Because the replica management is based on the group communication technology, there is a box named **GrpComm** in the executor. Compared with Dynaco, the **ReplicaMgr** is a new component which is not implemented in Dynaco, and the **ReplicaMgr** can handle the security and fault tolerance problems in ubicomp, while the Afpac in Dynaco can't deal with.

– The **Service** is the application program wrapped as a component.

The safety sensor is deployed outside the component, which includes Reporter, Monitor, Verifier and Forecaster. A brief introduction is as follows:

– The **Reporter** is a service which provides the subscription and notification interface for the adaptive components to get the information about the status of the applications and the environments. The **Reporter** generates **Events** to trigger the adaptive components to be adaptive. For example, event can be system overload, malicious attacks existing, network congestion, etc.

– The **Monitor** is a set of tools for monitoring the environment, such as the network status, CPU load of the computational nodes, malicious attack. The **Monitor** sends what it monitors to the **Reporter** to generate events, sends to **Forecaster** for forecasting what may happen in the future.

– The **Verifier** is a tool which verifies whether there is any modifications of the applications results by malicious users [15]. The **Verifier** sends what it gets to the **Reporter** for adaptive events, and sends to the **Forecaster** for pre-adaptation.

– The **Forecaster** is a set of services which forecasts the status of the applications and the environment. For example the network in domain A may have terrible congestion within 30 minutes, the domain A is attacked by malicious users and the domain B will be attacked in the following 3 hours for the connection between domain A and domain B has weak security protection mechanisms.

4   Use Case Scenario

In this section, we take NAS Parallel Benchmark [17] FFT codes as an example to illustrate how the adaptive framework is suited for different unsafe situations. As illustrated in section 3, the mobile users submit job request to the grid infrastructure, and wait for the answer from the grid, here is an overview of the processing of the job.

- **Mobile user point of view:** The user logs in with his certificate, and submit job request to the grid resources which he has been authorized. When the user moves to another place, the connection session expired, and the user re-logins in to check the results of what he requests before.
- **Backend server point of view:** The grid infrastructure receives the job request, reserves the available and suitable resources for this job. The grid resource management software monitors the processing of the job.
- **Adaptive framework point of view:** The safety sensor monitors the application and the environment, when it finds some event happens, notifies the adaptive component to handle the event. The adaptive component makes replication adaptation based on the policies and the guides.

4.1   Situations for Adaptation

The NAS Parallel Benchmark FFT codes are MPI programs [17], and for MPI programs over the grid, there are five kinds of threats, which are depicted in Fig. 3:

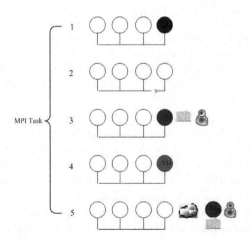

Fig. 3. Situations for adaptation

- **Nodes crashed:** For HPC application, it is very common that the computational nodes crash during job execution, because these application often run for days or more (as shown in situation 1). For an MPI program without fault-tolerance mechanisms, if one nodes crashes, the whole job crashes.
- **Network partition:** Also, it is very common that the grid infrastructure has the problem of network partition, (as shown in situation 2), because the grid is build upon a bandwidth-shared network. Just like what we mention above, the network partition problem will leads to the whole MPI program crashing.
- **Malicious attack:** As shown in Fig. 3, the third kind of threat for the safety of the MPI application is that there are some malicious attack in the job execution domain. These kinds of malicious attack may lead to a denial of service of the backend server or change the results of one of the process of the program.
- **System overload:** The fourth threat for MPI applications on the grid is that the end server or the network may become overloaded. Because the overload of the system will lead to the system crash in the near future.
- **System unsafe in the near future:** One of the potential threat for the application is that the system will be unsafe in the near future, but currently it is safe. For example, there are malicious attack in neighbor domain, while the execution domain always trust the users in the neighbor domain, as depicted for situation 5 in fig. 3. If there is no actions made for this kind of situation, there will be some mistakes for the application, such as changing the application result, making the backend server crashing.

For the unsafe situations presented above, if the system or the application doesn't make any actions, the mobile users can't get what they want, the ubicomp environment will be non-ubiquitous. In next subsection, a set of policies and plans are given to target these threats.

4.2   Adaptive Policies and Plans

The main idea of the Dynasa framework is to make application replications adaptive, the replication of an application is the same program running on another site and the state of the program is synchronized by the application checkpoints. The checkpoint of a application is a serial context files which save the running state of the application program.

For situation 1 and situation 2, a fail-stop error occurs. What the application can do is to restart the application from the checkpoint, but if the nodes or the network can't be repaired in time, the restart action will be useless. So for Dynasa, the application will be restarted on other resources, or just switch the application to another replica which is created by the adaptive component before. The policy for situation 1 and 2 is shown in table 1.

For situation 3, different kinds malicious attacks will lead to different consequences, for example, the traditional DOS attack will lead to the Denial of Service of the nodes, the malicious input attack will lead to the total wrong result of the computation. Upon situation 3, the polices are shown in Table 2.

Table 1. Policies for fail-stop threat

Situation	Stategy
{upon nodes crash}	{switch replica} and {new replica}
{upon network partition}	{switch replica} and {new replica}

Table 2. Policies for malicious attacks

Situation	Stategy
{upon DoS attack}	{new replica}
{upon malicious input attack}	{switch replica} and {new replica}
{upon session hijack attack}	{switch replica} and {new replica }
{upon user credential attack}	{switch replica} and {new replica }

Table 3. Policies for system unhealthy

Situation	Stategy
{upon one nodes unhealthy}	{switch replica} and {new replica} or {terminate process}

Table 4. Policies for anticipated threat

Situation	Stategy
{upon threat in the near future}	{new replica}

For situation 4, one nodes is unhealthy, it is better to create a replica or to kill one MPI process on this nodes, so the adaptive policy could be either of them, as shown in Table 3. The *terminate process* strategy has been implemented in Dynaco system.

For situation 5, there are no threats for the application currently, so the policy for safety is just to create one more new replica, as shown in Table 4.

For the policies presented above, there are four plans to execute, that is, *switch replica, new replica, terminate process,* and *reduce replica number.* We take *new replica* as an example to illustrate what the plan looks like. For the consideration

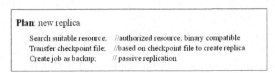

Fig. 4. Plan for *new replica*

of the usage of resources, generally the replica creation uses the warm passive style, that is the new replica serves as a backup, the plan for *new replica* is show in Fig. 4.

5   Simulations

In this section, simulation evaluations for the adaptive framework are presented. The simulation parameters are based on the work of Dynaco [1]. From the application point of view, the performance of the HPC application may have lots of metrics, such as execution time, execution cost, application response latency, but the execution time is one of the most important parameters for it, so we take execution time as the performance metric for the performance evaluation of Dynasa.

Two simulations are presented in this paper, the first one is to evaluate execution time between the adaptive replication and the no-adaptive performance for DoS attacks. The DoS attack will lead to the denial of service of the backend server, which can be regarded as failure to the application. For the adaptive replication, as explain in [13], the execution time for the application is shown in equation 1. We assume that if there is any problem happened, the user will resubmit the job, so the execution time for no-adaptive is shown in equation 2. T_F is the failure free execution time of the application, T_I is the interval of the checkpoint, λ is the fail rate of the backend server for the massive attacks, T_D is the down time of the end server, T_C is the checkpoint overhead, T_R is the recovery time of the application from passive replica.

$$T_{Adaptive} = \left(\frac{T_F}{T_I}\right) \frac{e^{\frac{\lambda}{N_R}(T_D - T_C + T_R)}(e^{\frac{\lambda}{N_R}(T_I + T_C)} - 1)}{\lambda/N_R} \tag{1}$$

$$T_{non-adaptive} = \frac{e^{\lambda T_D}(e^{\lambda T_F} - 1)}{\lambda} \tag{2}$$

We take the parameter values as follows: T_D is 60s, T_C is 0.5s, T_R is 0.5s, T_I is 6s, T_F is 60s, and the adaptive replica number are 3, 4, and 5. The simulation results are depicted in Fig. 5. Results show that the adaptive framework has much better performance than the non-adaptive one, and Fig. 5 also shown that the replication number has less impact to the performance for the DoS attack.

For simulation 2, we evaluate the performance between the adaptive framework and the non-adaptive for malicious input attack. There is a result verifier which verifies the result periodically. Just as illustrated in simulation 1, when the verifier finds the malicious attack, the non-adaptive method will resubmit the job. While for the adaptive framework, it is just switch to another replica and verify the result. As explained in [13], the execution time for the adaptive framework is shown in equation 3. Here the λ means the malicious attack rate. We take the same parameters in simulation 1, and the simulation results are depicted in Fig. 6. Fig. 6 shows that the adaptive framework has better performance, especially when the malicious attack is very common, just as shown in

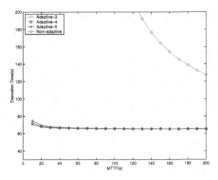

Fig. 5. Performance evaluation for DoS attacks

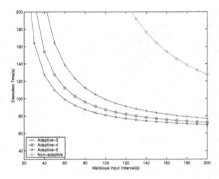

Fig. 6. Performance evaluation for malicious input attacks

Fig. 6 that the malicious attack interval is less than 30s; We can also conclude that the replica number is another important parameter for malicious input attack, especially when the malicious attack interval is less than 60s, For the malicious attack interval is 60s, the execution time between the 3 replica and 5 replica is about 60s, which is almost the failure free run time of the application.

$$T_{Adaptive} = \frac{e^{\frac{\lambda}{N_R}T_D}(e^{\frac{\lambda}{N_R}T_F} - 1)}{\lambda/N_R} \tag{3}$$

6 Conclusion and Future Work

In this paper, we have presented Dynasa, a new adaptive framework, This framework can handle the safety problems for HPC applications in ubicomp environments composed of a grid infrastructure and the mobile devices. Dynasa can make adaptive decisions according to the threats, and the simulation evaluation shows that Dynasa performance are quite good for such threats. In the near future, we will deploy the Dynasa component over the Grid5000 testbed [10], and make experimental evaluations for some HPC applications, such as the NPB, MCell, and GadGet 2. Also, we will make more policies and plans to handle different kinds of malicious attacks in ubicomp environments. A more constrained programming approach based on tasks graphs is being studied by another group in the SafeScale project [14]. We will compare Dynasa with their approach in a near future.

Acknowledgments. Experiments presented in this paper were carried out using the Grid'5000 experimental testbed, an initiative from the French Ministry of Research through the ACI GRID incentive action, INRIA, CNRS and RE-NATER and other contributing partners (see https://www.grid5000.fr).

References

1. Buisson, J., André, F., Pazat, J.: In: Proceeds of ParCo 2005 (September 2005)
2. Bellavista, P., Corradi, A., Stefanelli, C.: The Ubiquitous Provisioning of Internet Services to Portable Devices. IEEE Pervasive Computing 1(3), 81–87 (2002)
3. Beresford, A.R., Stajano, F.: Location Privacy in Pervasive computing. IEEE Pervasive Computing 2(1), 46–55 (2003)
4. Berman, F., Wolski, R., Casanova, H., et al.: Adaptive Computing on the Grid using AppLes. IEEE Trans. on Parallel and Distributed Systems (TPDS) 14(4), 369–382 (2003)
5. Capkun, S., Buttyan, L., Hubaux, J.: Self-Organized Public-Key Management for Mobile Ad Hoc Networks. IEEE Trans. on Mobile Computing 2(1), 52–64 (2003)
6. Dynaco, http://gforge.inria.fr/projects/dynaco
7. Davies, N., Gellersen, H., Prototypes, B.: Challenges in Deploying Ubiquitous Systems. IEEE Pervasive Computing 1(1), 26–35 (2002)
8. Dumitras, T., Srivastava, D., Narasimhan, P.: Architecting and Implementing Versatile Dependability, Architecting Dependable Systems, vol. III (2005)
9. Foster, I.: The Grid: A New Infrasture for 21st Century Science. Physics Today 55(22), 42–47 (2002)
10. Grid5000, https://www.grid5000.fr
11. Hingne, V., Joshi, A., Finin, T., et al.: Towards a Pervasive Grid. In: Proceedings of IEEE International Parallel and Distributed Processing Symposium (April 2003)
12. Jin, H., Shi, X., Qi, L.: Uses Case Study of Grid Computing with CGSP. In: Shimojo, S., Ichii, S., Ling, T.-W., Song, K.-H. (eds.) HSI 2005. LNCS, vol. 3597, pp. 94–103. Springer, Heidelberg (2005)
13. Jin, H., Shi, X., Qiang, W., Zou, D.: DRIC: Dependable Grid Computing Framework. IEICE Transactions on Information and Systems E89-D(2), 612–623 (2006)
14. Krings, A.W., Roch, J., Jafar, S.: Certification of large distributed computations with task dependencies in hostile environments. In: IEEE Electro/Information Technology Conference(EIT 2005), Lincoln, Nebraska (May 2005)
15. Krings, A., Roch, J., Jafar, S., Varrette, S.: A Probabilistic Approach for Task and Result Certification of Large-Scale Distributed Applications in Hostile Environments. In: Sloot, P.M.A., Hoekstra, A.G., Priol, T., Reinefeld, A., Bubak, M. (eds.) EGC 2005. LNCS, vol. 3470, pp. 323–333. Springer, Heidelberg (2005)
16. Mouël, F.L., André, F., Segerra, M.: AeDEn: An Adaptive Framework for Dynamic Distribution over Mobile Environments. Annales de télécommunications 57(11-12) (December 2002)

17. NAS Parallel Benchmark, http://www.nas.nasa.gov/Software/NPB
18. Storz, O., Boddupalli, P.V., Davies, N., et al.: Leveraging the Grid to Provide a Global Platform for Ubiquitous Computing Research. Lancaster University Technical Report CSEG/2/03 (April 2003)
19. Perkins, C.E.: Mobile Networking Through Mobile IP. IEEE Internet Computing 2(1), 58–69 (1998)
20. Phan, T., Huang, L., Dulan, C.: Challenge: Integrating Mobile Wireless Devices into the Computational Grid. In: Proceeds of MOBICOM 2002, pp. 271–278 (September 2002)
21. Stajano, F.: Security for Ubiquitous Computing. John Wiley & Sons, New York (2002)
22. Weiser, M.: The Computer for the 21st Century, pp. 94–104. Scientific American (September 1991)

Reducing Inter-piconet Delay for Large-Scale Bluetooth Scatternets

Chang Wu Yu, Kun-Ming Yu, and Shu Ling Lin

Department of Computer Science and Information Engineering
Chung Hua University, Hsin-Chu, Taiwan, R.O.C.
{cwyu,yu}@chu.edu.tw

Abstract. When more than seven devices to be connected in a Bluetooth scatternet, bridge devices are used to connect two piconets into a scatternet. To deal with possible data transmissions between different piconets, the bridge device must switch to different masters frequently. Suppose that a bridge is serving a piconet and the master in another piconet is calling it at the same time, the calling master has to wait until the bridge completes the previous service. Such transmission delay may accumulate in a long period and the performance of the whole Bluetooth network will degrade significantly. This work tries to smooth the kind of transmission delay in Bluetooth network. This work proposes two new scheduling protocols: the static schedule and the hybrid schedule. This static schedule deals with this kind of coordination among piconets distributedly by applying edge coloring technique. In case of heavy traffic load, the static schedule is expected to perform well. On the other hand, in case of light traffic load, the static schedule may results in long and unavoidable routing delay even there is no transmission between piconets; thus a naive random round-robin schedule in each piconet becomes more appropriate in case of light traffic load. Thus, in the hybrid schedule, each master in its piconet runs round-robin scheme initially; when the traffic load is heavier than a predefined threshold value, it turn to run the static schedule. Also, a new graph model, delay graphs, is proposed to model and estimate the delay time required for the proposed scheduling schemes theoretically. Finally, we conduct simulations by using ns-2 simulator and Bluehoc to demonstrate the efficiency and effectiveness of the proposed scheduling protocols.

Keywords: Bluetooth, Piconet, Scatternet, Scheduling protocol, modeling.

1 Introduction

In Bluetooth architecture, the master is connected with at most seven slaves to form a piconet, and the salves have to synchronize clock information with its masters for future transmission in the same hopping mode. When more than seven devices to be connected, bridge devices are used to connect two different piconets into a scatternet; and the bridge device must deal with any possible data transmissions between them. Suppose that a bridge is serving a piconet and the master in another piconet is calling it at the same time, the calling master has to wait until the bridge completes the

T. Washio et al. (Eds.): PAKDD 2007 Workshops, LNAI 4819, pp. 562–573, 2007.

previous service. Afterward, the calling master connects with the bridge. Such transmission delay may accumulate in a long period and the performance of the whole Bluetooth network will degrade significantly [1, 2].

Inter-piconet scheduling is required to coordinate bridges such that particular slots used in each piconet are reserved for specific master-slave pairs. These slots are called rendezvous points [2]. Generally, in large-scale scatternets, the coordination of the bridges in different piconets is harder than in small-scale scatternets; that is, large-scale scatternets require complex coordination mechanisms to establish rendezvous points in bridge devices [3]. In addition, Misic and Misic's analysis [4, 5] indicates that the crucial factor in minimizing end-to-end packet delay should be in minimizing the end-to-end delay for inter-piconet rather than intra-piconet traffic. The intra-piconet scheduling considers the scheduling problem for a single piconet by tackling three main issues: slave activity, fairness, and efficiency [2]. So far, these two kinds of scheduling problems have been proposed and evaluated extensively [2, 3, 6-21].

This work presents two scheduling protocols: the static schedule and the hybrid schedule to smooth the inter-piconet transmission delay in large-scale Bluetooth networks. The hybrid schedule applies either the static schedule or RR schedule dynamically according to the traffic load of each piconet. Note that some piconets may run the static schedule; however, others may run the hybrid schedule simultaneously in the same scatternet. Theoretically, we also propose a directed graph model, called *delay graph*, to model time delay resulting from inter-piconet and intra-piconet scheduling.

We also conduct simulations by using ns-2 simulator [22] and Bluehoc [23] to demonstrate the efficiency and effectiveness of the presented scheduling protocols. Three scheduling protocols including round robin (RR) scheme, the static schedule, and the hybrid schedule combining with two routing protocols: the routing vector method (RVM) [24] and a new simple routing protocol are used as the basis of comparisons. Three system metrics in evaluating how these protocols perform include throughputs, end to end delays, and packet loss rates. Simulation results demonstrate that the protocols with the static schedule increase throughput than the protocols without.

The remainder of the paper is organized as follows. In the next section, we briefly describe related work, followed with some necessary graph definitions and notations in Section 3. We then introduce our proposed two scheduling protocols in Section 4. Next, we show how delay graphs can be used for modeling the delay time of inter-piconet and intra-piconet scheduling on Bluetooth networks in Section 5. Simulation results will be presented and discussed in Section 6. Finally, Section 7 concludes this paper.

2 Related Work

A comprehensive survey on the issue of Bluetooth scatternet can be found in [2]. In [25], the blue-star build mechanism is applied to construct a scatternet. A device v decides whether it is going to be a master or a slave depending on the decision made by its neighbors with weights. Specifically, device v becomes the slave of the first master among its bigger neighbors that has paged and invited it to join its piconet. In case no bigger neighbors invited v, v itself then becomes a master. Once a device has decided its role, it will communicate to all its neighbors so that they can also make their own decision. In [6], Johansson *et al.* proposed a dynamic window-based scheduling framework for Bluetooth networks. They introduced a new Bluetooth mode, called JUMP mode, which includes a set of communication rules that allow efficient scatternet operation by offering flexibility for a device to adapt its activity in different piconets with respect to different traffic conditions.

In demand-based Bluetooth schedule [7], R. Rao *et al.* used 7 slots for a period so as to arrange piconet schedule. If a piconet consists of 3 slaves, each slave could be at least takes two turns in a period. If one piconet comprises of four slaves, one of these slaves is a bridge device. The time slots are distributed to the remaining slave devices. In [8], Racz *et al.* proposed an approach concerning bridge devices assigning meeting points with their peers so that the sequence of meeting points are generated by using a pseudo-random sequence. Their method removes the need for explicit information exchange between peer devices. Baatz *et al.* [9, 10] utilized the concept presence points, which are some adaptive defined points where a master and its slaves may meet. The length of a specific communication period is not pre-defined, but bases on current link utilization and the number of packets ready to send.

In [18], Capone *et al.* designed and evaluated different simple polling schemes for Bluetooth networks. In [11], Zhang and Cao devised a flexible scatternet-wide scheduling algorithm for Bluetooth networks, which tried to improve the network performance by arranging the polling order based on wasted pools and traffic load. Their algorithm uses a switch table concept which directs bridge devices to switch between the multiple piconets to which it belongs. In [12], Johansson *et al.* proposed a periodic rendezvous approach, called the maximum distance rendezvous point (MDRP) algorithm, whose basis is that rendezvous points are as far away from each other as possible. Latter, Kazantzidis and Gerla [13] extended their rendezvous point approach by not only selecting non-conflicting rendezvous points, but also seeking optimal choices of these points. In [14], Tan and Guttah devised a locally coordinated scheduling that dynamically adjusts the schedule based on workload conditions and the concept of scheduled meetings. In [15, 16], Son *et al.* proposed two inter-piconet scheduling approaches to satisfy measures for quality of service. In [3], Har-Shai *et al.* showed that the coordination of the bridges in different piconets is easier than in large-scale scatternets which require complex coordination mechanisms to establish rendezvous points. They presented and evaluated a load adaptive scheduling algorithm for small scale scatternets. To improve Bluetooth network performance through a time-slot leasing approach, the authors [26, 27] showed that establishing temporary piconets can enhance the speed of transmission. They also suggested that let one of slave play a role of master to poll the other slaves. It constructs a small piconet and sent data directly. After finishing the communication, the master device changes to be a slave device again. In [24], Bhagwat and Segall proposed a routing vector method (RVM), which is based on the concept of route vector, for routing in Bluetooth scatternets. To reduce the overhead of carrying information, piconets in RVM are represented by local identification numbers (LocID) selected locally by relays and the sequence of LocIDS (as opposed to sequence of MAC addresses) are saved in the packet header.

3 Definitions and Notations

A *graph* $G=(V, E)$ consists of a finite nonempty vertex set V and edge set E of unordered pairs of distinct vertices of V. A graph is *simple* if it has no loops and no two of its links join the same pair of vertices. The ends of an edge are said to be *incident* with the edge. Two vertices which are incident with a common edge are *adjacent*. An edge with identical ends is called a *loop*, and an edge with distinct ends a *link*. The *degree* $d_{G(v)}$ of a vertex v in G is the number of edges of G incident with v. We denote the *maximum degree* of vertices of G by $\Delta(G)$. A *walk* in $G=(V, E)$ is a finite non-null sequence $W=v_0 e_1 v_1 e_2 \dots e_k v_k$, where $v_i \in V$ and $e_j \in E$ for $0 \le i \le k$ and $1 \le j \le k$. The integer k is the *length* of the walk. When $v_0, v_1, \dots, v_k$ are distinct, W is called a *path*. A path is *a cycle* if its origin and terminus are the same. A directed graph is a graph with directed edges. A *bipartite graph* $G =(S, T, E)$ is a graph whose vertex set can be partitioned into two subsets S and T such that each of the edges has one end in S and the other end in T. A *k-edge coloring* of a loopless graph G is an assignment of k colors to the edges of G. The coloring is *proper* if no two adjacent edges have the same color. The *edge chromatic number* $\chi'(G)$, of a loopless graph G, is the minimum k for which G is k-edge-colorable. A subset M of E is called a *matching* in $G=(V, E)$ if its elements are links and no two are adjacent in G. The definitions and notations can be found in [28].

4 Designing Scheduling Protocols by Employing Edge Coloring Technique

We present two scheduling protocols: the static schedule and the hybrid schedule as follows. The intuitive idea of the static scheduling protocol originates from the following client-server environment [29]. Suppose that there are several clients request services from a number of servers. The problem here is to schedule those communications between services and clients so that conflicts can be avoided.

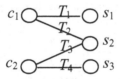

Fig. 1. The Client-Server environment

Suppose that we have two clients $\{C_1, C_2\}$ and three servers $\{S_1, S_2, S_3\}$, C_1 requests service T_1 from S_1 and T_2 from S_2, and C_2 requests service T_3 from S_2 and T_4 from S_3 (Fig. 1). If T_1 and T_4 are scheduled in the first time slot, then T_2 and T_3 should be executed in different time slots to avoid conflicts. As a result, three time slots are required to complete the desired communication (Fig. 2(a)). Alternatively, if T_1 and T_3 are executed in the first time slot, then T_2 and T_4 can scheduled in the second time slot simultaneously (Fig. 2(b)). The number of overall required time slots is two (one less than the previous schedule). Moreover, possible conflicts are also prevented. If we view Fig. 2 as a bipartite graph G, sets $\{T_1, T_3\}$ and $\{T_2, T_4\}$ are two matchings in G, which form an edge coloring by using two colors (i.e., time slots). Fig. 3 presents a schedule with edge coloring technique.

Similarly, the above edge coloring technique can be used for the scheduling problem in the Bluetooth scatternet network. A partial schedule for masters and bridges (with three time slots) in Fig. 4 is presented.

In Fig. 4, bridge device B_1 requires at least three time slots for $\{T_1, T_2, T_3\}$. B_2 requires at least two time slots for $\{T_1, T_2\}$, and B_3 requires the processing time of T_1 and T_2. The Bluetooth network requires additional two time slots for scheduling the remaining slave devices. Determining the least number of required time slots is equivalent to determining the edge chromatic number $\chi'(G)$ of the graph G representing the desired Bluetooth network. An important theorem due to Vizing [28]

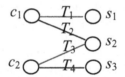

Fig. 2. An effective schedule reduces the overall communication time

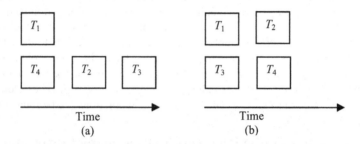

Fig. 3. A schedule with its edge coloring

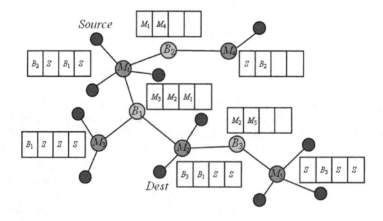

Fig. 4. A schedule with occupied time slots

states that $\chi'(G)=\Delta(G)$ or $\Delta(G)+1$, when G is a simple graph. Evidently, the corresponding graphs of Bluetooth network are simple (i.e., without loops and multiple edges). If every communication between two devices occupies a single time slots, the minimum number of time slots required in the whole network is the maximum degree of the representing graph plus one. Moreover, when G is bipartite, we have $\chi'(G)=\Delta(G)$ [28]. That explains why the network in Fig. 4 requires only five (not six) time slots (colors) for schedule because the maximum degree of the underlining graph is five and the graph is a tree, which is bipartite.

The number of time slots T selected in the Bluetooth network directly affects system throughput. In the static schedule, we determine the value of T after completing the scatternet formation. Actually, we set the value of T to be the maximum degree of the underlining graph plus one at the beginning.

To obtain the desired degree information, every bridge sends its degree value to its masters. When receiving degree information from bridge devices, masters preserve the largest degree value among them. As master and bridge device receive a bigger degree value, they will broadcast it to other masters and bridges; otherwise stop propagating this degree information. Eventually, every master and bridge device obtains the largest degree T of associated network and prepares a schedule by using T time slots in a round.

The static schedule is described as below:

Step 1: Collect the degree information to determine the number of time slots used in a round within each piconet.

Step 2: Schedule master-bridge communications by applying an algorithm modified from Dannie Durand *et al.*'s distributed edge coloring algorithm [29].

Step 3: Schedule master-slave communications by selecting them in appropriate remaining time slots in each piconet.

Step 2 can be implemented by modifying Dannie Durand *et al.*'s distributed edge coloring algorithm [29], which is designed originally for bipartite graphs.

The modified algorithm:

Input: A non-bipartite graph $G=(M, B, E)$ which represents a given Bluetooth network, where M and B are vertex sets representing masters and bridges, respectively and E is the edge set representing all communication links between masters and bridges.

Output: An edge coloring (equivalently a time slot assignment) on G.

While $(G=(M, B, E)$ is not empty)

{

Get *Ncolors* new colors.

For $i=1$ to *Npasses*;

      {

      Bridge selection step: for all bridges in B, assign *Ncolors* to untried edges(s) chosen uniformly at random.

      Master resolving step: for all masters in M, resolve conflicts by selecting uniformly at random.

      }

Delete colored edges and vertices of zero degree from G.

}

Note that a node may be belonging to both M and B because its corresponding device is not only a master and but also a bridge connecting to another master. Above randomized algorithm can be easily applied to select non-conflict time slots for even non-bipartite scatternet. For the reason that every edge connecting a master and a bridge will definitely be selected and assigned a time slot in bridge selection step; its conflicts may occur and resolved in the subsequent master resolving step. For illustration, a possible schedule after Step 2 with occupied time slots is shown in Fig. 4.

In case of heavy traffic load, the static schedule is expected to perform well; in case of light traffic load, however, the static schedule may results in long and unavoidable routing delay even there is no transmission between piconets; thus a naive random round-robin (RR) schedule in each piconet becomes more appropriate in case of light traffic load. In Fig. 4, for example, if Source node tries to send packets to Dest node, the packets have to pass through M_1, B_1, and M_2. Suppose that Source node is scheduled in the fourth time slot to communicate with M_1, M_1 has to wait for the third time slot of the next round to relay the data to B_1. Such delay could be unjustified especially when the traffic load of network is light. Moreover, collecting degree information in the static schedule introduces extra overhead. Therefore, in the hybrid schedule, we try to find a compromise between the static schedule and the naive RR schedule. In the hybrid schedule, each master in its piconet runs RR schedule initially; when the traffic load of a piconet is heavier (determined and decided by the master) than a predefined threshold value, it turn to run the static schedule instead. On the contrary, the hybrid schedule will run RR schedule again whenever the traffic load becomes light again.

Note that the hybrid schedule assume that each master independently detects whether the switching condition is satisfied or not and determines one of RR and the static schedule to be adopted in its own piconet distributedly. Since traffic loads within different piconet may be unequal at any one time, different schedules may be adopted at the same time by different piconets' masters in the same Bluetooth network. In other words, some may run RR schedule while the other may run the static schedule at a time; after a period of time, the adopted schedules may be changed again due to varying work loads. Moreover, neighboring piconets which newly adopt the static schedule and thus forms a connected sub-scatternet require rescheduling for selecting appropriate new time slots.

5 A Graph Model for Inter-piconet and Intra-piconet Delay

Some papers discuss the performance of bridging strategies. Misic and Misic [4, 5] proposed Queuing methods to model access delay, and consider the probability distribution of end-to-end delay time for both local and non-local traffic. To the best of our knowledge, no graph model has been proposed for this kind of delay problem.

In this section, we propose a discrete structure, called *delay graphs*, for modeling the inter-piconet and intra-piconet delay time of the scheduling problem in Bluetooth networks. Given a Bluetooth network, we first represent it as a simple graph $G=(V, E)$, where a vertex (depicted as a circle) in V represents a Bluetooth device and an edge (depicted as a straight line) in E indicates a possible communication between two devices. Moreover, when the device acts a role as a master or bridge, we represent it in the graph with a heavy-line circle, as shown in Fig. 5.

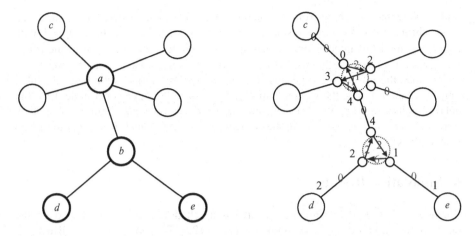

Fig. 5. A Bluetooth network with its delay graph

For a Bluetooth network with its representing graph $G=(V, E)$, we then construct a delay graph $G^*=(V^*, E^*)$ from it as follows:

(1) For each $v \in V$ and its representing device is a slave node; that is, it is neither a master nor a bridge, we add v into V^*.

(2) For each $v \in V$ and its representing device is either a master or a bridge, we add a set of $d_G(v)$ vertices called *super node* $S_v = \{s_{v, 1}, s_{v, 2}, ..., s_{v, d_G(v)}\}$ into V^*. In a word, we use a set of vertices to replace a non-slave vertex.

(3) For each edge $e=(u, v) \in E$, we construct the edge set of G^* as follows. If one of u and v is neither a master nor a bridge, we add $e=(u, v)$ into E^*. Otherwise, if u is a slave vertex and v is a non-slave vertex, we add $e=(u, s_{v, i})$ into E^*, where u is the ith adjacent vertex of v. If both u and v are non-slave vertices, we add $e=(s_{u,j}, s_{v,i})$ into E^*, where u is the ith adjacent vertex of v and v is the jth adjacent vertex of u.

(4) Add a directed edge $[v_i, v_j]$ into E^* if device v_j is scheduled just after device v_i in intra-piconet schedule for all $v_i, v_j \in V^*$.

(5) For each vertex $v \in V^*$, assign an integer $w_v = i$ to v if the device v is scheduled in i^{th} time slot.

(6) Associate each directed edge $[u, v]$ with a nonnegative weight w, where $w=(w_v-w_u)$ mod k if $w >= \alpha$; $w=((w_v-w_u)$ mod $k)+k$ if $w < \alpha$ and a piconet switching occurs. Here $0 \le w \le k-1$ and k is the required time slots in a piconet and α is the required time slots for switching between piconets.

An example of delay graph with $k=5$ and $\alpha=2$ is given in Fig. 5, where node a, e and d are masters but b is a bridge. Note that in the delay graph we represent the intra-piconet schedule of master or bridge nodes by using super nodes with associated directed edges. Specially, the induced subgraph of these vertices in a super node is a directed cycle which represents a cyclic scheduling order of slaves in a piconet. Since the weight of each vertex in the delay graph represents the current time slot assigned to the corresponding device, the weight of each directed edge $[u, v]$ in the super node then indicates the least required delay time for relaying messages through a master or bridge. Note that switching from one piconet to another incurs a delay due to the need to synchronize clocks. As a result, the weight of the edge can be assigned with $w=(w_v-w_u)$ mod k if inter-piconet switching time α is less than or equal to the offset time (w_v-w_u). Otherwise, $w=((w_v-w_u)$ mod $k)+k$ if inter-piconet switching time α is greater than the offset time w because the switching device misses the first scheduled time slot. Accordingly, the weight sum of a directed path from a super node A to B in delay graphs estimates the least accumulated inter-piconet and intra-piconet delay time for packets transmitted from master/slave A to master/slave B through the given Bluetooth network.

6   Simulation Results

Bluehoc [23] is used to simulate the proposed scheduling protocols. Based on the open source simulator network simulator (ns-2) [22], Bluehoc provides a Bluetooth extension to ns. In addition, Bluehoc provides a standard platform for performance comparisons between various routing and service discovery protocols over ad-hoc networks.

In the following simulations, we specify a 100 by 100 meters rectangular region with 100 to 500 nodes. Each of these devices is placed randomly inside the region. We also use BlueSstar formation protocol [25], which is also supported by ns-2, to construct a scatternet. Three *scheduling* protocols RR, the static schedule, and the hybrid schedule combining with two routing protocols: RVM [24] and a simple routing protocol are used as the basis of comparisons. Three system metrics in evaluating how these protocols perform include throughputs, end to end delays, and packet loss rates

Note that the simple routing protocol constructs a shorter routing path for pairs of sources and destinations; however, shorter paths may result in larger power consumption. Moreover, more piconets constructed in the same scatternet may also introduce additional switching time required for those devices included in those newly added piconets. Simulation results with respect to the hybrid schedule and the simple routing protocol in this work are omitted intentionally due to page limit.

In the following simulations, we use constant bit rate (CBR) as traffic model and 20% of 100~500 devices are chosen for generating traffic data. Diverse traffic loads ranged from 50 to 500 packets / second are also considered.

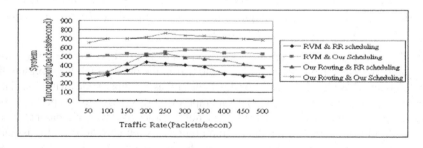

Fig. 6. System throughput

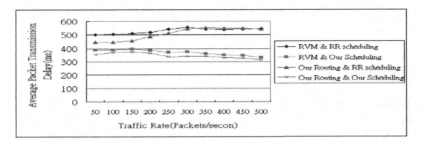

Fig. 7. The packet transmission delay

As the traffic rate becomes heavier, system throughputs of four combinations (including either RVM or the simple routing protocol with either RR or the static scheduling) rise gradually, and then reach peaks, and finally decline slowly. Particularly, Fig. 6 reveals that these protocols combined with the static schedule increase throughput than the protocols without. When comparing the highest throughput, RR combined with RVM routing reaches its peak rate at 420 packets/second. However, the static schedule combined with our routing protocol reaches its peak rate at 780 packets/second, increasing about 85 percent.

Fig. 7 depicts the packet transmission delays. Evidently, the static schedule combined with the simple routing results in shortest packet delay than other combinations do. When combining with RR schedule, both our routing protocol and RVM result in long waiting delay due to unexpected meeting point nature of RR schedule. We also note that, in these scenarios, inter-piconet scheduling deeply affects average packet delay time than routing protocol does. For the same reason, as the traffic rate is increasing, these gaps between protocols combined with RR and that with the static schedule become wider gradually.

Another important metric in Bluetooth network is packet loss rate. In the following simulations, the size of buffer prepared for every master device is 500 packets. Fig. 8 depicts the packet loss rates. The combination of RVM with RR schedule achieves the highest packet loss rate, which is within 0.4% ~ 0.3%. The merit of the static schedule can be easily seen by noting that protocols combined with RR schedule have higher packet loss rates than protocols combined with the static schedule. Particularly, the static schedule combined with the simple routing protocol has the lowest packet loss rate between 0.02% and 0.1%. Moreover, the combination achieves a lower packet

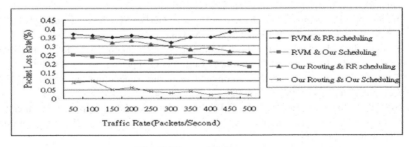

Fig. 8. The packet loss rate

loss rate even as the traffic load become heavier due to conflict-free schedule and shorter routing paths.

7 Conclusions

In this work, we present two scheduling protocols: the static schedule and the hybrid schedule in Bluetooth networks. The static schedule increases system throughput by shortening packet delay time. The hybrid schedule, a mixture of the static schedule and RR schedule, further increase system throughput and shorten the required delay time. We also define the delay graph, which can be used to estimate required delay time in Bluetooth networks.

References

1. Specification of the Bluetooth System Ver. 1.1, http://www.bluetooth.com
2. Whitaker, R.M., Hodge, L., Chlamtac, I.: Bluetooth scatternet formation: a survey. Ad Hoc Networks 3, 403–450 (2005)
3. Har-Shai, L., Kofman, R., Segall, A., Zussman, G.: Load-adaptive inter-piconet scheduling in small-scale Bluetooth scatternets. IEEE Communication Magazine, 136–142 (2004)
4. Misic, V.B., Misic, J.: Modelling Bluetooth piconet performance. IEEE Communications Letters 7, 18–20 (2003)
5. Misic, V.B., Misic, J.: Performance of Bluetooth bridges in scatternets with exhaustive service scheduling. In: Hawaii International Conference on System Sciences, pp. 311–320 (2003)
6. Johansson, N., Alriksson, F., Jonsson, U.: JUMP mode-a dynamic window-based scheduling framework for Bluetooth scatternets. In: MobiHoc, pp. 204–211 (2001)
7. Rao, R., Baux, O., Kesidis, G.: Demand-based bluetooth scheduling. In: IEEE Wireless LAN Conference, pp. 573–579 (2001)
8. Rácz, A., Miklós, G., Kubinszky, F., Valkó, A.G.: A pseudo random coordinated scheduling algorithm for Bluetooth scatternets. In: MobiHoc, pp. 193–203 (2001)
9. Baatz, S., Frank, M., Kjhl, C., Martini, P., Scholz, C.: Adaptive scatternet support for bluetooth using sniff mode. In: IEEE Conference on Local Computer Networks, pp. 112–120 (2001)
10. Baatz, S., Frank, M., Kjhl, C., Martini, P., Scholz, C.: Bluetooth scatternets: An enchanced adaptive scheduling scheme. In: IEEE INFOCOM, pp. 782–790 (2002)
11. Zhang, W., Cao, G.: A flexible scatternet-wide scheduling algorithm for Bluetooth networks. In: IEEE IPCCC, pp. 291–298 (2002)
12. Johansson, P., Kapoor, R., Kazantzidis, M., Gerla, M.: Rendezvous scheduling in bluetooth scatternets. In: IEEE ICC, pp. 318–324 (2002)
13. Kazantzidis, M., Gerla, M.: On the impact of inter-piconet scheduling in Bluetooth scatternets. In: International Conference of Internet Computing, pp. 37–43 (2002)
14. Tan, G., Guttah, J.: A locally coordinated scatternet scheduling algorithm. IEEE Local Computer Networks, 293–303 (2002)
15. Son, L.T., Schioler, H., Madsen, O.B.: Predictive scheduling approach in inter-piconet communications. In: International Symposium on Wireless Personal Multimedia Communications (2001)

16. Son, L.T., Schioler, H., Madsen, O.B.: Hybrid distributed iterative capacity allocation over Bluetooth network. In: IEEE International Conference on Communication Systems, pp. 583–588 (2002)
17. Chawla, S., Saran, H., Singh, M.: QoS based scheduling for incorporating variable rate coded voice in Bluetooth. In: IEEE ICC, pp. 1232–1237 (2001)
18. Capone, A., Gerla, M., Kapoor, R.: Efficient polling schemes for Bluetooth picocells. In: IEEE ICC, pp. 1990–1994 (2001)
19. Das, A., Ghose, A., Razdan, A., Saran, H., Shorey, R.: Enhancing performance of asynchronous data traffic over the Bluetooth wireless ad hoc network. IEEE INFOCOM 1, 591–600 (2001)
20. Zussman, G., Segall, A.: Capacity assignment in Bluetooth scatternets-optimal and heuristic algorithms. ACM/Kluwer MONET 9, 49–61 (2004)
21. Zussman, G., Yechiali, U., Segall, A.: Exact probabilistic analysis of the limited scheduling algorithm for symmetrical Bluetooth piconets. In: Conti, M., Gregori, E. (eds.) PWC 2003. LNCS, vol. 2775, Springer, Heidelberg (2003)
22. Fall, K., Varadhan, K. (eds.): ns notes and documentation, The VINT Project, UC Berkeley, LBL, USC/ISI, and Xerox PARC (1999)
23. Bluehoc simulation, a Bluetooth extension to NS, http://www-124.ibm.com/developerworks/opensource
24. Bhagwat, P., Segall, A.: A routing vector method (RVM) for routing in bluetooth scatternets. In: IEEE International Workshop on Mobile Multimedia Communications, pp. 375–379 (1999)
25. Petrioli, C., Basagni, S., Chlamtac, I.: Configuring BlueStars: Multihop scatternet formation for bluetooth networks. IEEE Transactions on Computers 52, 779–790 (2003)
26. Lin, C.R.: Admission control in time-slotted multihop mobile networks. IEEE Journal on Selected Areas in Communications 19, 1974–1983 (2001)
27. Zhang, W., Zhu, H., Cao, G.: Improving bluetooth network performance through a time-slot leasing approach. In: IEEE WCNC, pp. 151–157 (2002)
28. Bondy, J.A., Murty, U.S.R.: Graph Theory with Applications. Macmillan, London (1976)
29. Durand, D., Jain, R., Tseytlin, D.: Applying randomized edge coloring algorithms to distributed communication: An experimental study. In: ACM Symposium on Parallel Algorithms and Architectures, pp. 264–274 (1995)

Security Analysis and Enhancement of One-Way Hash Based Low-Cost Authentication Protocol (OHLCAP)*

JeaCheol Ha[1], SangJae Moon[2], Juan Manuel Gonzalez Nieto[3], and Colin Boyd[3]

[1] Dept. of Information Security, Hoseo Univ., 336-795, Korea
jcha@hoseo.edu
[2] School of Electrical Eng. and Computer Science, Kyungpook National Univ.,
702-701, Korea
sjmoon@ee.knu.ac.kr
[3] Information Security Institute, Queensland Univ. of Technology, GPO Box 2434,
Brisbane, QLD, 4001, Australia
{juamma,boyd}@isrc.qut.edu.au

Abstract. Choi *et al.* recently proposed an efficient RFID authentication protocol for a ubiquitous computing environment, OHLCAP (One-Way Hash based Low-Cost Authentication Protocol). However, this paper reveals that the protocol has several security weaknesses : 1) traceability based on the leakage of counter information, 2) vulnerability to an impersonation attack by maliciously updating a random number, and 3) traceability based on a physically-attacked tag. Finally, a security enhanced group-based authentication protocol is presented.

Keywords: RFID system, group-based authentication, indistinguishability, traceability.

1 Introduction

Radio Frequency Identification(RFID) systems, consisting of RFID tags, an RFID reader, and back-end database, are expected to replace optical bar codes due to several advantages, such as their low cost, small size, quick identification, and embedded implementation into objects. However, communication using the RF signal between a tag and a reader can create new threats to the security and privacy of a RFID tag, including the leakage of privacy, location tracing, and tag or reader impersonation.

Various attempts have already been made to protect the privacy of a tag using physical technology, such as the 'Kill command' [12], 'Active jamming' [5], and 'Blocker tag' [5] approaches. However, none have been successful. As a cryptographic solution, Weis *et al.* [10,11,12] proposed a hash-lock protocol

---

* This research was supported by the MIC of Korea, under the ITRC support program supervised by the IITA(IITA-2006-C1090-0603-0026).

T. Washio et al. (Eds.): PAKDD 2007 Workshops, LNAI 4819, pp. 574–583, 2007.

and randomized hash-lock protocol. Yet, with the hash-lock protocol, since the *metaID* is unique for each tag, location privacy is compromised due to the fixed *metaID*. Meanwhile, with the randomized hash-lock protocol, the identity of a tag, ID_k is transmitted from the reader to the tag, making the system vulnerable to a replay attack, spoofing attack, and location tracing. Henrici and Müller [4] proposed an ID variation protocol, that is secure against a replay attack, yet location privacy is compromised as the tag's response remains constant until the next authentication session when desynchronization occurs [8]. Ohkubo *et al.* [7] proposed a hash chain-based authentication protocol in which the reader sends a query using two different hash functions, however this scheme is still vulnerable to a replay attack and spoofing attack. In 2005, Lee *et al.* [6] proposed a low-cost RFID authentication scheme in which a tag and the back-end database only perform two one-way hash operations, yet this scheme is still vulnerable to a spoofing attack and location-tracing attack when desynchronization occurs. More recently, Choi *et al.* [1] proposed an efficient RFID authentication protocol for a ubiquitous computing environment, where the tag's ID is static. In [1], the authors claim that their protocol guarantees location privacy due to the use of fresh values in every session, plus an adversary cannot trace the target tag using a physical attack, even when certain secret values are obtained.

However, this paper shows that the protocol developed by Choi *et al.* still has security weaknesses. First, an adversary can trace a tag using leaked counter information. Second, an adversary can impersonate a reader by maliciously updating the random number obtained from the previous session. Finally, in the case of a physically attacked tag, an adversary can easily trace a target tag. Therefore, a low-cost authentication protocol that enhances OHLCAP is proposed to protect against the above attacks.

2   Security Threats to RFID System

An RFID system usually consists of three parts: RFID tags(transponders), the RFID reader(transceiver), and back-end database(Back-end server). An RFID tag includes a microchip for computing and a coupling element, such as an antenna, for communication with the RFID reader. The RFID reader interrogates the tags using an RF signal, then transmits the collected data to the back-end database. However, the channel between the reader and a tag is insecure, as it is based on wireless communication. After the back-end database receives the data from the reader, it transmits certain information to a authenticated tag. The channel between the reader and the database is considered as secure. In this paper, it is assumed that an adversary has the following capabilities:

- **Eavesdropping:** An attacker has a capacity to eavesdrop messages between the reader and the tags due to an insecure channel, then uses the intermediate information or useful responses to try certain enhanced attacks, such as location tracing or a spoofing attack. Therefore, an RFID system should at least protect against information leakage in an insecure channel.

- **Transmitting a malicious message or replaying:** It is assumed that an adversary has the capability to transmit certain malicious messages to the tag or the reader. By transmitting these messages, the attacker can perform a spoofing attack or replay attack.
- **Interrupting a message:** The communication messages between the tags and the reader can be blocked by an attacker. As a result, a message interrupt attack can bring into desynchronization state between the tag and the reader, due to an abnormal closing of a session, malicious blocking of messages, or different updating of ID between the tag and the database. Furthermore, several successive message interrupts can be used by an attacker in location tracing a target tag.

Since the communication between the reader and the tag is performed using an wireless RF interface, the communicated data can easily be tapped by an attacker. The various security threats that can occur with an insecure channel are categorized as follows:

- **Information leakage:** One RFID privacy problem is information leakage about a user's belongings. For example, a user may not want certain information known by others, such as ownership of expensive products, identification of personal medicine, and so on.
- **Impersonation attack:** After an adversary sends a malicious query to a target tag, they collect the responses emitted by the tag. The attacker can then impersonate the reader using the messages collected from the tag. Conversely, an adversary can replay the reader's query to impersonate the target tag. An attacker can also impersonate a legal tag or reader by replaying certain useful messages.
- **Desynchronization attack:** If the current ID for a tag is different to the one in the database, this is referred to as a state of desynchronization. An adversary can block certain transmitted messages between the tag and the reader, creating a dysynchronization state. This state can occur in an ID-renewable RFID system. If the ID of a tag is desynchronized, the tag can be easily traced, as one of emitted values from the tag will be constant, thereby compromising the location privacy.
- **Location tracing attack:** Here, the adversary can seek some useful information on a tag's location trace. This attack is essentially applied to a rigid RFID system in which certain communication messages between the tag and the database are identical to those used in the previous session.

3   Review of OHLCAP

This section briefly reviews Choi *et al.*'s One-way Hash based Low-Cost Authentication Protocol(OHLCAP).

Database	Reader	Tag
Database field		Tag field
$[GI_i][K][ID_{ij}][S_{ij}]$		$[ID][GI][K][S][c]$

	$\xrightarrow{Query,\ r}$	$A^1 = K \oplus c$
		$A^2 = ID + (GI \oplus r \oplus c)$
		$B = H(ID\|(S \oplus GI)\|(r \oplus c))$
Computes c	$\xleftarrow{A^1,A^2,B_R,r}$	$\xleftarrow{A^1,A^2,B_R}$ $B = B_L\|B_R$
Computes IDs for every GI_i		
Finds ID and checks GI		
$B = H(ID\|(S \oplus GI)\|(r \oplus c))$		
Checks B_R	$\xrightarrow{B_L}$	$\xrightarrow{B_L}$ Checks B_L

Fig. 1. OHLCAP: One-way Hash-based Low-Cost Authentication Protocol

3.1 Notations

$H()$: one-way hash function, $H : \{0,1\}^* \to \{0,1\}^l$
ID_{ij} : identity of jth tag in ith group, l bits
S_{ij} : secret key for jth tag in ith group, l bits
GI_i : ith group index, l bits
K : common secret key in DB and all tags, l bits
r : random number generated by reader,
t : random number generated by tag, l bits
c : counter stored in tag, l bits
$Query$: request generated by reader
B_R : right half of message B
B_L : left half of message B
x_p : value of x in previous session
x_c : value of x in current session
$+$: modular addition by $mod\ (2^l - 1)$
$\oplus$: exclusive-or(xor) operation
$\|$: concatenation of two inputs

3.2 Description of OHLCAP

OHLCAP consists of a set-up and mutual authentication phase, as described in Fig. 1.

1) Set-up phase:
– Back-end database: Divides all the tags into n groups, which include m tag identities. The data field of the back-end database is $GI_i\|K\|ID_{ij}\|S_{ij}$.
– Tag: A tag is initialized by a data field, including $ID\|GI\|K\|S$ and a counter c, where K is the same in all tags and GI is the same within a group.

2) Authentication phase:

- Step 1. The reader sends a *Query* and r to a tag.
- Step 2. The tag computes A^1, A^2 and B, then sends them to the reader. The tag increases the counter c whenever it receives a query from the reader.
- Step 3. Upon receiving A^1, A^2, and B_R from the tag, the reader forwards them with r to the back-end database.
- Step 4. The back-end database computes $c' = A^1 \oplus K$ and $ID'_i = A^2 - (GI'_i \oplus r \oplus c')$ using all the group indices $GI'_i, i \in \{1, \ldots n\}$. If one of the computed ID'_i matches one of the stored IDs, the back-end database checks if one of the computed ID'_is matches one of the stored IDs, the back-end database then checks whether the GI'_i contains the ID'_i matching that for the true GI_i group. The back-end database authenticates the tag by checking that the computed B_R equals the received one, then sends the B_L to the reader.
- Step 5. The reader forwards the B_L to the tag.
- Step 6. The tag authenticates the reader by checking the B_L.

4   Security Analysis of OHLCAP

This section analyzes the security weaknesses of OHLCAP and provides attack details.

4.1   Traceability Using Counter Information

With OHLCAP, when responding to a query from the reader, the tag computes $A^1 = K \oplus c$ using a counter c. At this point, an adversary can trace a tag if the tag's messages are caught in two successive sessions. The following explains how OHLCAP is vulnerable to location tracing.

- Assumption: It is supposed that an adversary knows certain tag's responses from two successive sessions, $A^1{}_p = K \oplus c_p$ and $A^1{}_c = K \oplus c_c$. Here, the relationship between the two counters is $c_c = c_p + 1$.
- Attack: The adversary computes $A = A^1{}_p \oplus A^1{}_c = c_p \oplus c_c$. As a result, the secret key K is removed from the equation. The value A always has a distinguishable sign, a 1's-run value from the LSB. Now, the adversary can trace a tag by observing successive 1-runs from the LSB of A.

For example, if the previous counter value is $c_p = 1011010111$ and the current one is $c_c = 1011011000$, then $A = c_p \oplus c_c = 0000001111$, which has four 1-runs. As such, it is easy to determine that A is always one when the LSB of the first counter c_p is zero, *i.e.* if $c_p = 1011010110$ and the current $c_c = 1011010111$, then $A = 0000000001$. Thus, an adversary can trace a tag by observing two successive responses, A^1_p and A^1_c. Furthermore, if the counter is a l-bit string, then the possibility that the target tag and a random tag(with a random counter) cannot be distinguished is $l/2^l$, which is negligible as a function of l.

4.2   Impersonation by Maliciously Updating Random Number

Choi *et al.* claim that reader impersonation is impossible, due to the authentication process between the reader and a tag, making it impossible for an adversary to send a correct last B_L message to the tag. However, an adversary can impersonate a legal reader using a random number from a previous session as follows:

- Assumption: It is supposed that an adversary catches two messages r_p and B_L from the previous session. Also, for the sake of simplicity, it is assumed that the LSB of r_p is zero.
- Attack: The adversary generates a malicious random number r_c, such as $r_c = r_p + 1$, in the attack session, *i.e.*, the LSB of r_p is just changed to one. After sending a *Query* and the malicious r_c to the tag and receiving some responses from the tag, the adversary then sends the same B_L as used in the previous session to the tag as the last message. Since the value B is computed by $H(ID\|(S \oplus GI)\|(r \oplus c))$, if $r_p \oplus c_p$ equals $r_c \oplus c_c$, then B_c is equal to B_p. In this case, the impersonation attack as a legal reader is successful. When $r_p \oplus c_p$ is not $r_c \oplus c_c$, the attack fails.

The following provides more detail on the above impersonation attack. The tag computes A^1, A^2, and B in response to the query from the adversary. If $r_p \oplus c_p$ is equal to $r_c \oplus c_c$, $A^2{}_c$ and B_c will be the same as $A^2{}_p$ and B_p, respectively, computed in the previous session. From the relationship of $c_c = c_p + 1$ and $r_c = r_p + 1$, if the LSB of r_p is zero, the LSB of c_p is also zero, then this attack will be absolutely successful, as $r_p \oplus c_p$ is equal to $r_c \oplus c_c$. However, if the LSB of c_p is one, such an attack is impossible, as $r_1 \oplus c_1$ is not equal to $r_2 \oplus c_c$. Therefore, an adversary can impersonate the reader by sending the random number updated by one and the last B_L message used in the previous session. The possibility of success is $1/2$ when the adversary chooses a previous random number, r_p, where the LSB is a zero bit. Table 1 outlines the possibility of an impersonation attack based on maliciously updating the LSB of a random number.

Table 1. Possibility of impersonation attack by updating LSB of random number unknown bit that can differ from

LSB of r_p (Known bit)	LSB of r_c (Update)	LSB of c_p (Guessing bit)	LSB of c_c (Current session)	$r_p \oplus c_p$	$r_c \oplus c_c$	Success $B_c = B_p$
0	1	0	1	$bb..bbb0$	$bb..bbb0$	O
0	1	1	0 with carry	$bb..bbb1$	$xx..xxx1$	X

b : unknown bit x : unknown bit, it can be different with b

4.3   Physical Attack on Tag

When considering the case of an adversary obtaining the secret key K and group key GI_i by physically attacking a tag, Choi *et al.* claim that the adversary cannot trace the target tag, as the secret value S is unknown, however, tracing is possible without considering S as follows:

- Assumption: It is assumed that an adversary can eavesdrop on A^1, A^2, B_L and B_R between the reader and the target tag. Furthermore, the secret key K and group key GI_i are known through a physical attack.
- Attack 1: The adversary extracts counter c from A^1 using the value K.
  - $c = A^1 \oplus K$

  Then, even though the secret value S is unknown, the tag counter can be compared with the previous one. Thus, all tags can be traced by checking the counter increment by one. Furthermore, if the adversary knows the counter number for the previous session, a malicious random number r_c can be computed, such as $r_c = (r_p \oplus c_p) \oplus c_c$ and $c_c = c_p + 1$. Since the A^2 and B in the attack session are the same as the previous session values, the adversary can easily impersonate the reader, as described in section 3.2.
- Attack 2: Since an adversary can compute the counter c, as shown above, the ID can also be extracted from the eavesdropped messages A^2 and r.
  - $ID = A^2 - (GI_i \oplus r \oplus c)$

  Then, even though the adversary does not know the tag's secret key S, the ID can be extracted from every session related to the ith group. Thus, the adversary can trace a tag by computing the ID for a group that includes a tag corrupted by a physical attack.

The above attack means that the RFID system is compromised with regard to traceability, so the RFID tag can no longer be used.

5   Security Enhancement of OHLCAP

To prevent traceability in the case of using counter information, it is recommended that a random number be used in a tag instead of a counter. However, this requires a random number generator in a tag. Alternatively, a hashed value of a stored number can be used that is changed in each session. If a random number or hashed value is used instead of a counter, the first tracing attack and impersonation attack by maliciously updating the number r_c become impossible. However, if the secret key K and GI_i are compromised, then OHLCAP cannot prevent a tracing attack, as an adversary can compute the counter value from A^1 and fixed ID for a group corrupted by a physical attack.

5.1   Group-Based Low-Cost Authentication Protocol

Accordingly, a new authentication protocol is proposed that is based on a group key. In contrast to OHLCAP, the proposed protocol removes the data fields for the secret key S_{ij} and counter c due to their uselessness. To protect the RFID system from the case of K and GI_i being compromised, the proposed protocol computes three messages: $A^1 = K \oplus t$, $A^2 = GI + (r \oplus t)$, and $B = H(ID\|GI\|r\|t))$. As a result, even though K and GI_i are compromised, the attacker can not extract a tag's ID from B due to the one-way property of the hash function. Therefore, the proposed group-based protocol is at least secure from above three attacks. Fig. 2 shows the process of the proposed group-based

Database	Reader	Tag
Database field		Tag field
$[GI_i][K][ID_{ij}]$		$[ID][GI][K]$

	$\xrightarrow{\quad Query,\ r \quad}$	random number t
		$A^1 = K \oplus t$
		$A^2 = GI + (r \oplus t)$
		$B = H(ID\|GI\|r\|t))$
Computes $t = A^1 \oplus K$	$\xleftarrow{\quad A^1,A^2,B_R,r \quad}$ $\xleftarrow{\quad A^1,A^2,B_R \quad}$	$B = B_L\|B_R$
Computes $GI = A^2 - (r \oplus t)$		
Finds ID in GI by checking B_R		
$B = H(ID\|GI\|r\|t))$	$\xrightarrow{\quad B_L \quad}$ $\xrightarrow{\quad B_L \quad}$	Checks B_L

Fig. 2. Proposed Group based Authentication Protocol

authentication protocol, and the following gives a detailed description of each step.

1) Set-up phase:
- Back-end database: Divides all the tags into n groups. The data field is $GI_i\|K\|ID_{ij}$.
- Tag: A tag is initialized by a data field, including $ID\|GI\|K$.

2) Authentication phase:
- Step 1. The reader sends a *Query* and r to a tag.
- Step 2. The tag generates a random number t and computes $A^1 = K \oplus t$, $A^2 = GI + (r \oplus t)$, and $B = H(ID\|GI\|r\|t))$, then sends A^1, A^2, and B_R to the reader.
- Step 3. The reader forwards A^1, A^2, and B_R with r to the back-end database.
- Step 4. The back-end database computes $t = A^1 \oplus K$ and $GI = A^2 - (r \oplus t)$, then finds the ID in the GI by checking the B_R. The back-end database authenticates the tag by checking that the computed B_R equals the received one, then sends the B_L to the reader.
- Step 5. The reader forwards the B_L to the tag.
- Step 6. The tag authenticates the reader by checking whether the received B_L equals the one computed in Step 2.

5.2 Security and Efficiency Analysis

The security of the proposed protocol was evaluated against the threats described in Section 2: 1) information leakage, 2)impersonation attack, 3) desynchronization attack, and 4) location tracing attack. To obtain secret information from a tag, an adversary must be able to guess the ID. However, an adversary cannot compute the ID from the A^1, A^2, B, and r, due to the security property of a one-way hash function.

Even when an adversary collects a tag's responses, then tries to impersonate a legitimate tag, they cannot compute the hashed messages A^1, A^2, and B without knowing the K, GI, and ID values. Meanwhile, to impersonate the reader, an adversary must send the correct B_L. This is also impossible, as it cannot be computed without knowing the ID value.

In a desynchronization attack, assuming that an adversary blocks the response messages transmitted from a tag, *i.e.*, step 2 in Fig. 2, even though the tag receives the same random number r as in the previous session, the tag sends A^1, A^2, and B in the next session as a response to a query. Therefore, the proposed protocol can protect against a desynchronization attack, as the tag does not emit any useful messages for enhanced attacks, such as location tracing.

In the case of location tracing, the proposed protocol guarantees location privacy by sending different random messages for each session. After the authentication is finished in the previous session, the tag sends A^1, A^2, and B in response to a query in the current session, that is, the same response is not emitted by the tag in the subsequent session. Thus, location privacy is satisfied as A^1, A^2, and B are already refreshed in each session using two random numbers.

When evaluating the storage costs and computational load for the DB and tag, the proposed protocol makes an improvement in the storage costs for the DB as removing the secret key S_{ij} and the counter for each tag. With the proposed protocol, the storage size of the DB is $3l \cdot m$, where l is the length of an ID_{ij}, K, or group index GI_i and m is the number of IDs. Plus, a tag requires $3l$ bits of memory to store an ID, K, and the GI value. The total length of the messages transmitted from a tag to the reader is $2.5l$, while that from the reader to a tag is $1.5l$, except for a *Query*.

The computational cost in the tag can be slightly reduced compared to the original OHLCAP. The main processing in a tag is hash operation like SHA-1[9] which is the most widely used secure hash function. By high design techniques, SHA-1 needs only 405 clock cycles to compute the hash of 512 bits of data in the work of Kaps *et al.* [2], and SHA-256 requires 1,128 cycles in [3]. Therefore, the proposed protocol is also suitable for a lightweight RFID system with limited memory space and low computational power.

6   Conclusion

This paper revealed several security weaknesses of OHLCAP. Thus, to guarantee security against various threats, it is recommended that a random number be used in a tag instead of a counter. If a random number is generated in a tag, a tracing attack and impersonation attack then become impossible. Furthermore, if the secret key K and GI_i are compromised by a physical attack, OHLCAP cannot prevent a tracing attack. Thus, a group-based low-cost authentication protocol is proposed as a more secure version of OHLCAP. The proposed protocol is robust to most threats, such as information leakage, an impersonation attack, desynchronization attack, and location tracing attack.

References

1. Choi, E., Lee, S., Lee, D.: Efficient RFID Authentication Protocol for Ubiquitous Computing Environment. In: Enokido, T., Yan, L., Xiao, B., Kim, D., Dai, Y., Yang, L.T. (eds.) Embedded and Ubiquitous Computing – EUC 2005 Workshops. LNCS, vol. 3823, pp. 945–954. Springer, Heidelberg (2005)
2. Kaps, J.P., Sunar, B.: Energy Comparison of AES and SHA-1 for Ubiquitous Computing. In: Zhou, X., Sokolsky, O., Yan, L., Jung, E.-S., Shao, Z., Mu, Y., Lee, D.C., Kim, D., Jeong, Y.-S., Xu, C.-Z. (eds.) Emerging Directions in Embedded and Ubiquitous Computing. LNCS, vol. 4097, pp. 372–381. Springer, Heidelberg (2006)
3. Feldhofer, M., Rechberger, C.: A Case Against Currently Used Hash Functions in RFID Protocols. In: Meersman, R., Tari, Z., Herrero, P. (eds.) OTM 2006. LNCS, vol. 4277, pp. 372–381. Springer, Heidelberg (2006)
4. Henrici, D., Müller, P.: Hash-based Enhancement of Location Privacy for Radio Frequency Identification Devices using Varying Identifiers. In: Proceeding of the Second IEEE Annual Conference on Pervasive Computing and Communications Workshops, pp. 149–162. IEEE, Los Alamitos (2004)
5. Juels, A., Rivest, R.L., Szydlo, M.: The Blocker Tag: Selective Blocking of RFID Tags for consumer Privacy. In: Proceeding of 10th ACM Conference on Computer and Communications Security 2003, pp. 103–111 (2003)
6. Lee, S., Hwang, Y., Lee, D., Lim, J.: Efficient Authentication for Low-cost RFID Systems. In: Gervasi, O., Gavrilova, M., Kumar, V., Laganà, A., Lee, H.P., Mun, Y., Taniar, D., Tan, C.J.K. (eds.) ICCSA 2005. LNCS, vol. 3480, pp. 619–627. Springer, Heidelberg (2005)
7. Ohkubo, M., Suzuki, K., Kinoshita, S.: Hash-Chain Based Forward-Secure Privacy Protection Scheme for Low-Cost RFID. In: Proceedings of the SCIS 2004, pp. 719–724 (2004)
8. Rhee, K., Kwak, J., Kim, S., Won, D.: Challenge-Response Based on RFID Authentication Protocol for Distributed Database Environment. In: Hutter, D., Ullmann, M. (eds.) SPC 2005. LNCS, vol. 3450, Springer, Heidelberg (2005)
9. National Institute of Standards and Technilogy(NIST) FIPS-180-2: Secure Hash Standard(SHS) (2002)
10. Sarma, S.E., Weis, S.A., Engels, D.W.: Radio-Frequency Identification: Security Risks and Challenges. RSA Laboratories 6(1) (Spring 2003)
11. Weis, S.A.: Security and Privacy in Radio-Frequency Identification Devices. MS Thesis, MIT (2003)
12. Weis, S.A., Sarma, S.E., Rivest, R.L., Engles, D.W.: Security and Privacy Aspects of Low-Cost Radio Frequency Identification Systems. In: Hutter, D., Müller, G., Stephan, W., Ullmann, M. (eds.) Security in Pervasive Computing. LNCS, vol. 2802, Springer, Heidelberg (2004)

An Effective Design of an Active RFID Reader Using a Cache of Tag Memory Data*

Seok-Young Jang, Sang-Hwa Chung, Won-Ju Yoon, and Seong-Joon Lee

Department of Computer Engineering, Pusan National University
Busan, 609-735, Korea
{myuyung,shchung,anospirit,lachesis}@pusan.ac.kr

Abstract. RFID (Radio Frequency Identification) is technology that a reader automatically identifies the data in a tag with a built-in microchip via a radio frequency. The relevant standards for 433 MHz active RFID are ISO/IEC 15961, 15962, and 18000-7. Communication between a reader and a tag in the management of active RFID tag memory data as defined by the standards is not efficient. In this paper, to address efficiency concerns, we present an active RFID reader software design to improve time and transmission rating of the ISO/IEC 18000-7 commands using memory data cached within the reader. We also present a method in which cached memory data are validated before use, to address the cache coherence issue when using multiple readers. For the identification method, we designed two models; one adheres to the RF-interface standard and the other modifies some fields of the RF-interface standard. We designed an active RFID reader and analyzed its efficiency through experiments based on the two models. The experimental results show that the performance on processing time and transmission rating of RF-interface improved by average 60% over the noncache design.

Keywords: Tag Memory Data, Logical Memory, Logical Cache Memory.

1 Introduction

RFID (Radio Frequency Identification) is technology that a reader automatically identifies data in a tag with a built-in microchip via a radio frequency. The 433 MHz active RFID system is suitable for harbor logistics systems because tags in this system are equipped with their own batteries and so have a longer communication range and sensing modules can be added easily [1,2].

There are two ways for RFID readers to access the tag memory: implementing reader commands following ISO/IEC 15961 and 15962 or implementing RF-interface commands based on ISO/IEC 18000-7. With the RF-interface commands, the user will access the tag memory data without any processing. The user can efficiently manage the tag memory data if the middleware system is

---

* "This work was supported by the Korea Research Foundation Grant funded by the Korean Government(MOEHRD)" (The Regional Research Universities Program/Research Center for Logistics Information Technology).

T. Washio et al. (Eds.): PAKDD 2007 Workshops, LNAI 4819, pp. 584–595, 2007.

efficiently built. On this account, ISO/IEC 15961 and 15962 define a management method for the tag memory data using the RF-interface. In the method, tag memory data can be managed using the memory of the active RFID reader to hold a working copy of all the data of the tag memory. This memory in the active RFID reader is named the Logical Memory in the ISO/IEC 15961 and 15962 standards [3,4,5].

Companies developing RFID readers and tags include SAVI, Hi-G-Tek, and KPC. There are a variety of methods for building an RFID system with components from different companies. SAVI supports ISO/IEC 18000-7 commands as a user interface, but does not use ISO/IEC 15961 commands, and tag memory data are managed using their own middleware [6]. KPC supports ISO/IEC 15961 and 18000-7 commands as a user interface [7]. So far there has been no attempt to apply cache memory technique to RFID systems.

In this paper, we raise the question of access efficiency of the ISO/IEC 18000-7 RF-interface when an active RFID reader is implemented using the ISO/IEC 15961 and 15962 standards. In an active RFID reader design based on the standards, frequent communication between a reader and a tag is necessary, which is increasing the chance of data transmission errors and shortening the battery life of active RFID tags. In passive RFID systems, battery life is not a problem, but frequent communication raises the transmission error problem as for active RFID systems. To address efficiency concerns, we present an active RFID reader design to improve time and transmission rating of the ISO/IEC 18000-7 commands using the cached memory data within the active RFID reader. In this paper, this cached memory is named the Logical Cache Memory, because the logical memory of ISO/IEC 15961 and 15962 is cached. We also use a cache data validation method to address the cache coherence issue when using multiple readers. For data validation, we designed two models. The first adheres to the RF-interface standard and uses part of the tag data memory to validate cached data. In the second, some fields of the RF-interface standard are modified and validation information is managed in a system memory space where a tag-ID is stored. We implemented the two models and analyzed their efficiencies through experiments. While we implemented and analyzed an active RFID system, the logical cache memory can also be applied in passive RFID systems that exchange large data volumes over the RF-interface.

This paper is organized as follows. In Section 2, we explain an active RFID reader based on the standards and raise the question of access efficiency. We also present an active RFID reader applying the logical cache memory. In Section 3, experimental results are presented and analyzed. Finally, we present conclusions and future work in Section 4.

2   Design and Implementation of an Active RFID Reader Based on a Cache of Tag Memory Data

First, we describe our development environment. We then describe our implementation of a 433 MHz active RFID reader based on ISO/IEC 15961, 15962, and 18000-7, and analyze the access efficiency of the ISO/IEC 18000-7 commands. To resolve the questions raised, we present an active RFID reader design applying logical cache memory. Finally, we explain the cache coherence issue when using multiple readers, and describe our solution.

2.1   Development Environment

To develop an active RFID system, we used the 433 MHz active reader, which were implemented in our previous work [8]. This reader is equipped with an ARM9 processor, a Chipcon RF transceiver (CC1020), a Xilinx SPARTAN-XC3S400 FPGA, a 64 MB SDRAM module, and other components. This reader also provides an Ethernet interface, an RS232 interface, and interfaces for programming and debugging the FPGA.

2.2   An Active RFID Reader Based on the Standards

Figure 1 shows the structure of the software for a 433 MHz active RFID reader based on ISO/IEC 15961, 15962, and 18000-7. We designed each module based on the context defined by the standards.

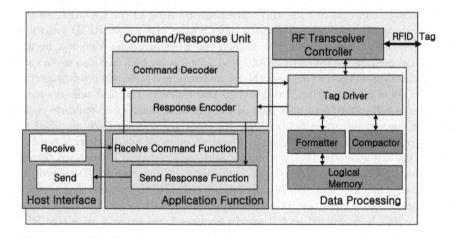

Fig. 1. Structure of software for a 433 MHz active RFID reader

The command/response unit receives a parameter as an octet stream from the application function and parses the parameter. It calls the tag driver function after parsing, then encodes the parameter received from the tag driver in the Type, Length, Value (TLV) form defined by the standards and sends the TLV value to the application function. The tag driver communicates with a tag following the ISO/IEC 18000-7 protocol. The formatter and compactor encode or decode object ID and object following ISO/IEC 15962.

2.3 Efficiency Problem for Command Processing Based on the Standards

Figure 2 shows the ISO/IEC 15691 command processing sequence. When a reader receives a user command, the reader brings tag memory data into the logical memory through the memory read command defined in the ISO/IEC 18000-7 standard, and processes the user command using the logical memory within the reader. There are several user commands in the ISO/IEC 15961, but in most cases, all the memory data in a tag must be brought into the logical memory. In this process, several ISO/IEC 18000-7 commands are repetitively executed in one ISO/IEC 15961 command. Thus, frequent communication between a reader and a tag is necessary, increasing the chance of data transmission errors and shortening the battery life of active RFID tags. In passive RFID systems, battery life is not a problem, but frequent communication raises the transmission error problem as for active RFID systems.

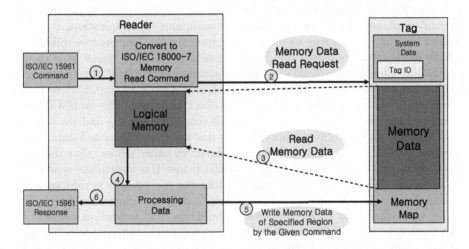

Fig. 2. ISO/IEC 15961 command processing sequence

2.4 An Active RFID Reader Based on a Cache of Tag Memory Data

To avoid frequent communication, when a reader reads tag memory data, the reader maintains a cached copy of the data within its memory. In this paper, the cached copy of the data is called the logical cache memory. However, this raises the problem of maintaining cache coherence, as shown in Figure 3. In a multiple-reader system for managing resources over a wide area, there is an overlap region between readers where tag A is located. If reader A performs a command that updates memory data of tag A and the logical cache memory of tag A within reader A, and thereafter reader B performs a command that relies

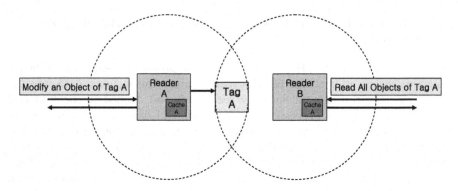

Fig. 3. Data consistency in multiple readers

on the logical cache memory of tag A within reader B, the data used by reader B is no longer valid.

To address cache coherence in a multiple-reader system, readers must receive a valid-response token from the tag using minimum communication, even if a reader has the logical cache memory of the specified tag. For process validations of the cache data, we define the data for validation. In this paper, it is named validation information. Validation information are the ID of the reader that performed the last write operation and a Cache Check Byte (CCB) recorded when updating the tag memory data. To check validity of the cached data, a reader compares two reader IDs first: one is read from the tag and the other is stored in the logical cache memory. If they don't match, the cached data has to be invalidated because data is already updated by other reader, thus there are no need to check the CCB. Otherwise, if the reader IDs match, the reader has to check the CCB so that makes sure whether the cached data is valid or not.

We describe how we designed two models to process validations of the cache data. One adheres to ISO/IEC 18000-7 standards, and validation information are located in part of the tag data memory and command processes use only standard commands. The other method modifies the option/status fields of the ISO/IEC 18000-7 standard, and validation information are located in a system memory space where a tag-ID is stored. When a reader sends a command packet, a reader adds validation information to command packet.

The first method adheres to the standards. The tag source code is not modified and the source code relavant to the logical cache memory is added in the command processing of the reader. Figure 4 shows the processing sequence. When a reader receives a user command, it checks the logical cache memory of the specified tag. If the logical cache memory of the specified tag exists, the reader requests the reader ID and CCB of the tag memory map using the "Read Tag Memory" command of ISO/IEC 18000-7. The reader then compares the reader ID and CCB with those stored in the logical cache memory of the specified tag, which is confirmed valid or not. If it is valid, the reader does not request the tag memory data and processes the ISO/IEC 15961 command using the logical cache memory. If the modified data must be updated following ISO/IEC 15961

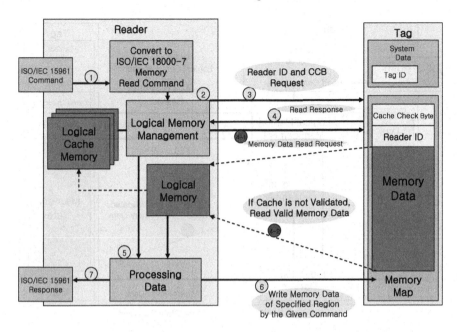

Fig. 4. Improved processing sequence adhering to the standards

commands, the logical cache memory of the specified tag is updated, and the reader updates the data in the tag using the "Write Tag Memory" command in ISO/IEC 18000-7, including validation information (reader ID and CCB).

Figure 5 shows the second processing sequence. The tag source code is modified following the modification facilities of ISO/IEC 18000-7, and the source code relavant to the logical cache memory is added to the reader's command process. When a reader receives a user command, it checks the logical cache memory of the specified tag. If it exists, the reader sets the request validate bit in an option field of the "Read Tag Memory" command of ISO/IEC 18000-7 and adds validation information (reader ID and CCB) to the command packet. When the tag receives this command, it compares validation information of the command packet with its own data. If they match, the tag sets the valid bit of the Modified-Tag-Status, indicating that the Logical Cache Memory is valid. The tag sends a "valid" response, without any memory data. When the reader receives the "valid" response, it uses the logical cache memory of the specified tag. If the modified data must be updated following ISO/IEC 15961 commands, the logical cache memory of the specified tag is updated. The reader sets validation information write bit in an option field of the "Write Tag Memory" command of ISO/IEC 18000-7 and adds validation information to the command packet. The reader updates memory data and validation information in a tag using the "Write Tag Memory" command in ISO/IEC 18000-7.

Both models have the advantage of reducing the RF communication overhead of ISO/IEC 18000-7 by using the logical cache memory, but each model has its

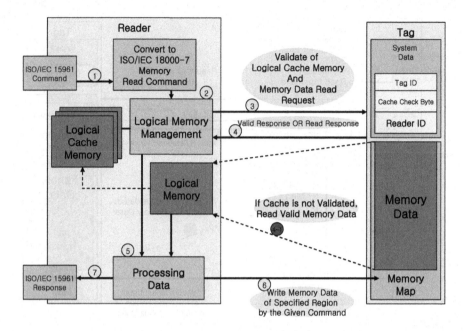

Fig. 5. Improved processing sequence modifying the standards

strength and weakness caused by modifying standards. The first model has an advantage that modifying the tag source code is not necessary because of adhering to the ISO/IEC 18000-7 standard. However, while part of the tag memory map is used as a validation information space, an additional update process is required for validation information. Further, if the cache is stale, one or two additional command processes are necessary than for the standard model without the cache. The second model has the advantage of a greater reduction in RF communication overhead by modifying the standards. It has the same number of RF communications even when the cache is stale. However, it does not adhere to the standard and requires modified tag source code.

3   Experiments and Analyses

Figure 6 shows the environment for our experiments with the reader with built-in software suggested in this paper. The reader is connected to the TCP/IP-based Internet to receive commands from a remote server. To monitor the operation of the reader in real time, each reader is connected to a separate terminal program through its RS-232 interface. The tags used, which were implemented in our previous work [8] and these are equipped with ATMEL microcontrollers (ATmega128L) and Chipcon RF transceivers (CC1020), and they are implemented in accordance with the relevant standards.

Our experiments compared the system in which a reader does not use a logical cache memory with those in which a logical cache memory is validated from a tag

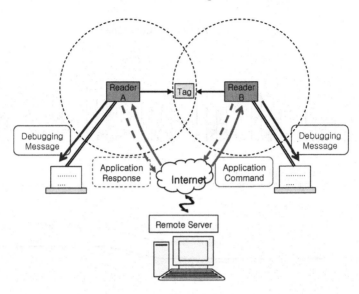

Fig. 6. Experimental environment for the active RFID reader

Table 1. Experimental data

Object ID	Object
CONTAINER	12345
WGT	3870
SHIPDATE	77
ITEM	ROCKET
QUANTITY	500
NUM_OF_PALLET	10

using our two models. We also measured the cache miss penalty when the cache was not valid in adhering to the standard. We measured command operation time, data transmission rate, and number of RF communications in all cases. We also confirmed the coherence protocol in a multiple-reader environment with two readers and one tag.

We executed 11 ISO/IEC 15961 commands, which required all tag memory data and used the object data shown in Table 1. The object data of Table 1 is given in the JDTAV format 1.01 [9].

3.1 Analyses of Processing Time and Transmission Rate

We analyzed "Read All Objects" and "Add Single Object" commands among the command set. The "Read All Objects" command instructs a reader to read all objects, their object IDs and associated objects from the tag. Basically, it is necessary to read out all tag memory data to execute the "Read All Objects." The "Add Single Object" command instructs a reader to write an object and its object ID into

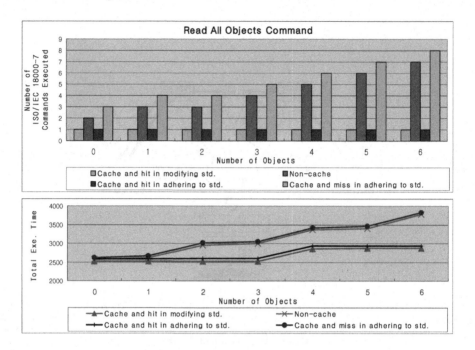

Fig. 7. Experimental results for "Read All Objects"

the tag. A reader needs to verify, before adding the object, that no object with the same object ID already exists. Therefore, a reader reads all objects like the "Read All Ojbects" command, and then a reader checks object IDs of all objects before adding a new object. The experimental results for two commands are shown in Figures 7 and 8. The bar graph shows the number of ISO/IEC 18000-7 commands executed in one ISO/IEC 15961 command and the line graph shows the total execution time of the ISO/IEC 15961 command. The number of RF messages is regularly the minimum number when using a valid cache in "Read All Objects." And "Add Single Object." However, if caches are not used, then the number of RF communications increases as a function of the number of objects stored on the tag.

First, for "Read All Objects," the number of RF communication is always one when using a cache. In this case, ISO/IEC 15961 commands are executed without requiring the execution of multiple ISO/IEC 18000-7 commands. The processing time is about 2.5 s, so that the processing time is almost the same as the wake-up time. When not using a cache, the number of RF communications increases as a function of the number of objects stored on the tag. If the cache is not validated, the unmodified-standard model always had one more RF communication as the cache miss penalty.

Second, for "Add Single Object," the number of RF communications is regularly two or three when using caches and the modified RF-interface standard. The processing time is 2.7 s, regardless of the object count. Given that the wake-up time for tags is 2.5 s, the command processing time is relatively unaffected by

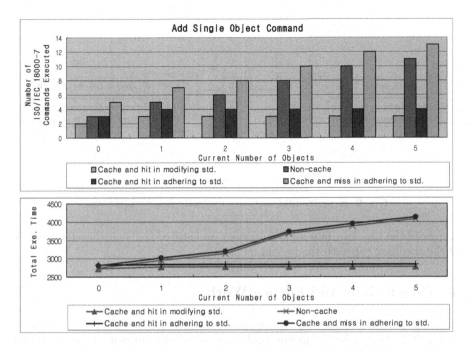

Fig. 8. Experimental results for "Add Single Object"

object count when using caches. On the other hand, when using caches without modifying the RF-interface standard, one more execution was required than with the modified standard. On average, more than 40 bytes of extra communication was required than for the modified standard. When the cache was not used, the number of RF communications increases as a function of the number of objects stored on the tag. In this case, if the cache is not validated, the modified-standard model was comparable to the original model. However, the unmodified-standard model always had two extra RF communications as the cache miss penalty.

The experimental results show that the performance improved average 60% over the noncache design. This value is obtained by averaging the execution time diffrences of the cache design and the noncache design as shown in Figures 7 and 8. The experimental results also show that a reader's RF interface is efficiently managed by using the logical cache memory if a reader need to execute many data exchanged in a long time.

3.2 Cache Coherence Experiments

We confirmed the following sequence in testing the coherence protocol. Reader A and reader B both performed the "Read All Objects" command on tag A, as shown in Figure 9, so readers A and B had the same logical cache memory for tag A. Reader A modified the memory data of tag A through the "Modify Object" command. When reader B performed the "Read All Objects" command on tag A for a second time, reader B received the "invalid" response from tag

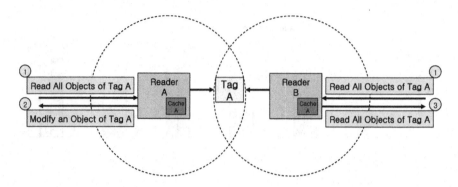

Fig. 9. Experimental sequence for testing cache coherence

A, reread the tag data into logical memory and stored the logical cache memory of the specified tag with the logical memory.

4   Conclusions and Future Work

In this paper, we consider the management of tag memory using ISO/IEC 15961 and 15962. As a solution to the problems raised, we present an active RFID reader that the performance on execution time and transmission rating improved by average 60% through caching the tag memory data in the memory of an active RFID reader. We validate the logical cache memory before using it to ensure cache consistency when using multiple readers. We present more efficient models using the modified or unmodified ISO/IEC 18000-7 standard. In this paper, we study our management method in the basic environment of the tag memory map. Some performance gain is achieved using the normal structure of the tag memory map. However, we also show the many limitations in the structure of command processing in ISO/IEC 18000-7. Recently, methods for tag memory management have been modified in ISO/IEC 18000-7 Version 1 [10]. ISO/IEC 18000-7 Version 1 will have added database commands for management of tag memory data that appear to allow improvement of time and transmission requirements. The logical cache memory model suggested in this paper can be applied to both passive and active RFID systems where many data values are exchanged in the RF interface. It will be necessary to study how to improve access efficiency using caching of tag memory data, modifying structure of the tag memory map, and applying database commands of ISO/IEC 18000-7 Version 1 in the future.

References

1. Finkenzeller, K.: RFID Handbook: Fundamentals and Applications in Contactless Smart Card Identification, 2nd edn. John Wiley & Sons, Chichester (2003)
2. Michael, K., McCathie, L.: The Pros and Cons of RFID in Supply Chain Management. In: Proceedings of the International Conference on Mobile Business, pp. 623–629 (2005)

3. ISO/IEC 15961: Information technology - Radio frequency identification (RFID) for item management - Data protocol: application interface (2004)
4. ISO/IEC 15962: Information technology - Radio frequency identification (RFID) for item management - Data protocol: data encoding rules and logical memory functions (2004)
5. ISO/IEC 18000-7:2004: Information technology - Radio frequency identification (RFID) for item management - Part 7: Parameters for active air interface communications at 433 MHz (2004)
6. Savi Technology, Inc.: Savi SmartChain Site Manager Datasheet (2004), http:// www.savi.com
7. KPC, Inc.: KRS900 (KPC RFID System 900) S/W Specifications (2006), http:// www.kpcnet.com
8. Yoon, W.-J., Chung, S.-H., Kim, H.-P., Lee, S.-J.: Implementation of a 433 MHz Active RFID System for U-Port. In: Proceedings of the 9th International Conference on Advanced Communication Technology, pp. 106–109 (2007)
9. RF-Tag Data Format Specification, Version 2.0 (2002), http://www.acq.osd.mil/ log/rfid/Policy/RF-Tag Data Format Specification, Version 2.0.pdf
10. ISO/IEC 18000-7:2007: Information technology - Radio frequency identification (RFID) for item management - Part 7: Parameters for active air interface communications at 433 MHz (2007)

Privacy Protection Scheme of RFID Using Random Number*

Soo-Young Kang and Im-Yeong Lee

Division of Computer, Soonchunhyang University, #646, Eupnae-ri, Shinchang-myun,
Asan-si, chungchungnam-Do, 336-745, Korea
{bbang814,imylee}@sch.ac.kr

Abstract. With the development in IT technology and with the growing demands of users, a ubiquitous environment is being made. Ubiquitous stands for a convenient environment in which users can get information and services through networks with devices that can communicate. This ubiquitous environment requires device technology, and Radio Frequency IDentification(RFID) is a good example. RFID, a technology that identifies radio frequency, was developed to substitute bar-codes and has various strengths, such as high recognition rates, quick recognition speed, and large storage. Because of these strengths, it is being used in various fields such as the public, distribution, and financial sectors; but it has a weakness in privacy, as readers operating within the same frequency bandwidth can be easily tagged. By acquiring information from the tag, unauthorized people can assess into the personal information of users. The current study, therefore, provides a secure and effective protection method, which can guarantee anonymity towards users with tags.

Keywords: RFID, Privacy Protection, Random Number.

1 Introduction

With the development in IT technology and with growing demands of users, a ubiquitous environment is being made. Ubiquitous comes from a Latin origin, which means 'anywhere, anytime', a convenient environment in which users can get information and services through networks with devices that can communicate. In this ubiquitous environment, demand for communications devices is growing, and Radio Frequency Identification (RFID) is gaining attention as the core of next generation ubiquitous computing. RFID differs from previous identification technology in that it can identify individual tags and has memory to store data. Also, rapid identification speed, high identification rates, and semipermanent duration make it ideal to be used in the distribution, financial, and public sectors. The RFID, which is being used in various application fields, require many tags; and therefore, low-cost passive tags, which get power from the reader, are more popular than high-cost active tags, which have their own power source. Passive tags get power from standard readers and

---

* "This work was supported by the Korea Research Foundation Grant funded by the Korean Government(MOEHRD)" (KRF-2006-311-D00851).

T. Washio et al. (Eds.): PAKDD 2007 Workshops, LNAI 4819, pp. 596–607, 2007.

provide information stored in the tag. In managing warehouses or luggage, immediate reader response may be more effective in aspect of time and cost, but if regular users have objects with tags on them, the objects' information may be easily exposed to a third party. Because of this, characteristics and tastes of users are being exposed, causing privacy encroachment problems. To solve this privacy issue, tags should not respond to unauthorized readers with malicious purposes; and, at the same time, provide accurate information to authorized readers. Also when the tag sends response data, it should be variable to provide anonymity to the user. Also, solutions to prevent position tracking should need to be developed.

Therefore, the current study provides a technique to provide variable tag value for anonymity and protect the user's location from being tracked by a malicious third party. The most general environment was used as basis. Since passive tags, which get power from the reader, were used, a random number generator (RNG) is operated while being attached to the reader.

The new procedure is introduced based on the existing procedures. Chapter 2 deals with security risks and requirements related with RFID systems. Chapter 3 studies previous RFID security, and chapter 4 illustrates the design of the proposal. Chapter 5 analyzes solutions for requirements discussed in chapter 2, and chapter 6 provides a conclusion and the direction of the future research.

2 Security Threats and Requirements

The RFID system consists of the tag, reader, and database. User information or item information is sent to the database when the tag receives power from the reader. The database compares the tag information and stored information, and sends authorization data to the tag after authorization. For this procedure to be processed properly, fabrication of data transfer must be prevented. However, the communication channel between the tag and reader is wireless, thus exposed to third parties. Therefore, this chapter deals with security risks associated with the RFID system and security requirements to solve these problems.

2.1 Security Threats

RFID is vulnerable to various forms of attack because it uses radio frequencies. Therefore, we must look at various forms of attack in RFID usage and solutions to counter these problems. The following describes various forms of security threat associated with the RFID system.

- Eavesdropping: Vulnerability exists in the communication channel that relays data between the tag and reader. Illegal users may access this. Eavesdropping, which is a problem to not only frequency communication but also to radio communication, can be solved by producing variable values so that a third party cannot access to essential values even if he acquires data.
- Traffic Analysis: This is a process in which data, acquired from eavesdropping, is compiled in guessing the next session's value or acquiring essential values. The traffic of data transfer of each session must not be uniform, and random variable

values should be used to prevent third party users from analyzing values used in communication.

- Replay Attack: This form of attack re-sends data acquired through eavesdropping to induce or acquire essential values. When identical or essential values are exposed from the tag during the re-sending process, tracking problems may arise and privacy may be encroached upon. Therefore, variable values should be used in communication to prevent replay attack.
- Tracking Attack: One of the most serious privacy problems of the RFID is that if a set value is exposed during each session, the privacy of the user may be encroached upon even if it is not possible to locate his exact location. To counter this form of attack, variable random numbers or a time stamp should be used.

2.2 Requirements

To solve the aforementioned security threats, the following requirements should be provided.

- Authentication: All components of the system should go through an authentication process. The RFID is comprised of a tag, reader, and database. Each part should provide authentication to each other. The tag should send secret values, which have been previously agreed upon, to each component to become authorized. The database can reply with identification values or secret values to become authenticated by the tag.
- Anonymity: Even if data is acquired from a tag, it should not be trackable to a tag. If identification values are set, anonymity cannot be guaranteed. Therefore, essential values should not be exposed and variable values should be used to send.
- Integrity: Since data is sent through a radio channel, illegal users can copy and modify data. To solve this problem, data should have encryption to have integrity. Passive tags have little power and calculation skills. This enables them to use hash formulas to prevent data copying and modification.
- Confidentiality: Values used in security protocol should not be exposed and only authorized users may share them. All components should share a secret value to authenticate each other. In order to provide confidentiality, a third party should not be able to access the secret value; and the security protocol should be protected based on this confidentiality.
- Efficiency: Although efficiency is not included in the security requirements, passive tags require hash formulas or XOR calculation. Passive tags require applicable security measures, and therefore efficiency should be provided, too.

Security threats noted above require solutions. These solutions will provide an adequate privacy protection technique.

3 Related Works

Numerous security protocols have been developed to protect privacy during RFID usage. In the early periods, MIT developed a Hash-Lock technique in 2003, and further research has ensued. The hash-lock technique is very effective, but has problems concerning location tracking, caused by set identification values, and anonymity problems. Therefore, this chapter will analyze existing techniques, which protect user's privacy by producing random values, and their vulnerabilities to be solved.

3.1 RHLP (Random Hash-Lock Protocol)

This method was developed at MIT to enable low price tags to prevent user location tracking and privacy encroachment problems, using meta ID with the hash-lock method[5]. In the Hash-Lock method, the meta ID that hashes the key value is set. The tag produces a number in response to the reader's question and sends the hashed value, which is calculated with the tag's identification value, to provide security. However, illegal tags may eavesdrop on authenticated tag response values and replay this to the reader. Therefore IDs are easily exposed and random number producers must be attached to the passive tag, which makes its efficiency drop dramatically.

3.2 HIDV (Hash-based ID Variation Protocol)

This method uses various identification values of tags and uses variable ID to protect the privacy of users[1]. This method, if the last session was completed abnormally, H(ID), the identification value, which enables location tracking and hinders anonymity, may be exposed. Also, in updating ID, when communication is abnormally concluded, synchronization problems may occur.

3.3 LCAP (Low-Cost Authentication Protocol)

This method is an improvement to the hash-based ID variation method[9]. This method is much simpler than the Hash-based ID Variation method, and differs from previous methods in that it employs half-half authentication procedures. However, in the event of abnormal conclusion of the last session, R values can be acquired through replay attacks, which enables location tracking and anonymity to be lost.

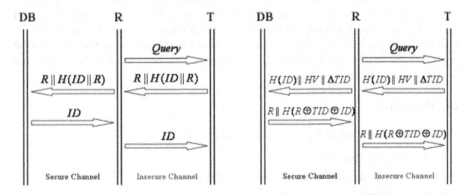

Fig. 1. Right Protocol is RHLP and left Protocol is HIDV

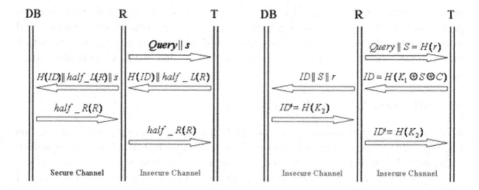

Fig. 2. Right Protocol is LCAP and left Protocol is MAP

3.4 MAP (Mutual Authentication Protocol)

This method was proposed in a different environment from that of previous methods[2]. This method is based on a wireless communication between the reader and database. Unlike previous methods, there is a need to secure the connection between the reader and database. This calls for addition in key updating and ID value updating, which causes lower efficiency. However, each component authenticates each other, giving it advantage in the wireless network environment. If the last section is concluded abnormally because of network instability, location tracking problem may occur since the key value and ID value is not updated, and the privacy problem may occur since it is not able to provide anonymity. Also, updates for key values and IDs, which must be carried out every session, may be difficult to synchronize in the wireless communication environment.

4 Proposed Protocols

Pervious methods have problems in preventing location tracking and not being able to secure anonymity. This study proposes to use random numbers in the reader to protect the privacy of users. Random numbers are protected with secret values, and the use of hash functions prevents copying and forgery. Also, the time frame provides a solution to prevent the replay attacks, and the production of variable values makes it possible to solve synchronization problems.

4.1 Parameters

The following is parameters used on proposed protocols.

- r : Random number
- Δr : Random number and timestamp subtraction
- R_r : Random number right half
- L_r : Random number left half

- TS : Timestamp
- R_TS : Timestamp right half
- L_TS : Timestamp left half
- S_V_n : n-th secret value(n=1, 2, 3...)
- ID : Tag's unique ID
- $metaID$: Hashing ID H(ID)
- $H(\ )$: Secure one-way hash function
- $\oplus$: Exclusive OR function

4.2 RFID Privacy Protect Protocols 1

The proposed method consists of 8 stages and provides security through random numbers. Also the privacy of users can be protected by providing anonymity.

The proposed method is based on the insecure communication channel between the tag and reader. The method provides mutual authentication between the tag and database, and therefore provides anonymity to the tag user and secures him from location tracking.

Step 1. The reader produces a random number r and timestamp TS and conducts an XOR calculation using the secret value S_V, which the components share, to produce $r \oplus S\_V$ and $TS \oplus S\_V$. Also, to prevent r and TS from being copied and forged during transfer, the two values are combined to form a hashed value $H(r \| TS)$. The reader sends the $r \oplus S\_V$ and $TS \oplus S\_V$, $H(r \| TS)$, which it produced, to the tag.

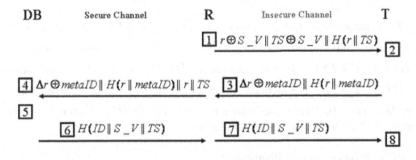

Fig. 3. Proposed Protocol 1

Step 2. This step authenticates values transferred from the reader. The tag conducts XOR calculation by inputting S_V into $r \oplus S\_V$ and $TS \oplus S\_V$ to acquire r and TS . To authenticate r and TS, the tag hashes the two values to produce $H(r \| TS)'$ and compares it with $H(r \| TS)$, the hashed value transferred from the reader. When the two values are identical, r and TS are authenticated, and the values to be sent to the database are produced.

Step 3. To get authenticated by the database, the tag produces r, the difference between r and TS, and conducts an XOR calculation on $metaID$ to produce $\Delta r \oplus metaID$. The hashed value $H(r \parallel metaID)$ is produces and combined with $\Delta r \oplus metaID$ to form $\Delta r \oplus metaID \parallel H(r \parallel metaID)$, which is to be sent to the reader for authentication.

Step 4. The reader receives $\Delta r \oplus metaID \parallel H(r \parallel metaID)$ from the tag and combines r and TS to send $\Delta r \oplus metaID \parallel H(r \parallel metaID) \parallel r \parallel TS$ to the database so that it can produce $metaID$, the anonymous identification value of the tag.

Step 5. The database receives $\Delta r \oplus metaID \parallel H(r \parallel metaID) \parallel r \parallel TS$ from the reader and produces Δr, the difference between r and TS. Afterwards, the $metaID$ is acquired by conducting XOR calculation on the $\Delta r \oplus metaID$. To authenticate the acquired $metaID$, the transferred r is combined to form hash value $H(r \parallel metaID)'$. If the transferred $H(r \parallel metaID)$ is identical with the produced $H(r \parallel metaID)'$, the $metaID$ is authenticated along with the tag with the $metaID$.

Step 6. To acknowledge the tag authentication, the database combines the S_V and variable TS to form a hash value of $H(ID \parallel S\_V \parallel TS)$ to be sent to the reader.

Step 7. To authenticate the authentication of the database, the reader sends the hash value $H(ID \parallel S\_V \parallel TS)$, transferred from the database, to the tag.

Step 8. To authenticate the hash value $H(ID \parallel S\_V \parallel TS)$, transferred from the reader, the tag combines S_V and TS to produce hash value $H(ID \parallel S\_V \parallel TS)'$. If the transferred $H(ID \parallel S\_V \parallel TS)$ and produced $H(ID \parallel S\_V \parallel TS)'$ are identical, the database and tag mutually authenticate each other. Also, after the hash value is authenticated, the tag ID, which is included as a component of the hash value, is authenticated.

4.3 RFID Privacy Protect Protocols 2

The proposed method differs from previous ones in that it provides solutions that can be used in the mobile environment. In the mobile environment, the reader and database can be applied in a wireless environment, and therefore can be perceived as insecure channels. The proposed method is based on the insecure communication channel between the tag and reader. The method provides mutual authentication between the tag and database, and therefore provides anonymity to the tag user and secures him from being tracked of location.

Step 1. The reader produces a random number r and timestamp TS and conducts an XOR calculation using the secret value S_V_1, which the components share, to produce $S\_V_1 \oplus r$ and $S\_V_1 \oplus TS$. Also, to prevent r and TS from being copied and forged during transfer, the two values are combined to form a hashed value $H(r \parallel TS)$. The reader sends the $S\_V_1 \oplus r$ and $S\_V_1 \oplus TS$, $H(r \parallel TS)$, which it produced, to the tag.

Step 2. The tag acquires r and TS by conducting an XOR calculation with the two secret values S_V_1 on $S\_V_1 \oplus r \parallel S\_V_1 \oplus TS \parallel H(r \parallel TS)$. To authenticate the two values, the tag combines and hashes r and TS to produce $H(r \parallel TS)'$. If $H(r \parallel TS)$, the hashed value transferred, and the produced hash value $H(r \parallel TS)'$ are identical r and TS are authenticated.

Step 3. The tag produces Δr, the difference between random number r and timestamp TS, and conducts an XOR calculation with the $metaID$. Also to prevent copying and forgery of the $metaID$, Δr and $metaID$ are combined to produce the hash value $H(\Delta r \oplus metaID)$. Afterwards, $\Delta r \oplus metaID \parallel H(\Delta r \parallel metaID)$ is transferred to the reader.

Step 4. The reader must send a value so that the database can acknowledge the r and TS. To prevent location tracking caused by exposing the same value, 1 will be added to r and TS respectively and produce $S\_V_2 \oplus (r+1)$ and $S\_V_2 \oplus (TS+1)$ by calculating S_V_2 with the XOR calculation. Also, to prevent copying and forgery of $1+r$ and $1+TS$, the two values are hashed to produce $H((r+1) \parallel (TS+1))$. Since $S\_V_2 \oplus (r+1)$, $S\_V_2 \oplus (TS+1)$, $H((r+1) \parallel (TS+1))$ are all values produces to security transfer r and TS, XOR calculation with $metaID$ on $\Delta r+1$ is produced $(\Delta r+1) \oplus metaID$ to securely send the $metaID$. Hash value $(\Delta r+1) \oplus metaID$ is produced to prevent copying and forgery of $\Delta r+1$ and $metaID$, and is transferred to the database.

Step 5. The database conducts XOR calculations with S_V_2 on $S\_V_2 \oplus (r+1)$ and $S\_V_2 \oplus (TS+1)$, which it received from the reader. After subtracting 1 from $1+r$ and $1+TS$, r and TS are acquired. Also $1+r$ and $1+TS$ are hashed to form $H((r+1) \parallel (TS+1))'$. If the transferred hash value $H((r+1) \parallel (TS+1))$ and produced hash value $H((r+1) \parallel (TS+1))'$ are identical, r and TS are authenticated. Then, the database finds the difference between the two values Δr and adds 1 to carry out an XOR calculation, $(\Delta r+1) \oplus metaID$, to acquire $metaID$. This is then searched to see if it exists in the database. If an identical $metaID$ exists, the tag is authenticated and a hash value $H((\Delta r+1) \| \, metaID)$ is produced to check for an identical value for authentication.

Step 6. The database adds 2 to Δr and produces $ID \oplus (\Delta r+2)$, by conducting an XOR calculation with the ID. To prevent the ID from being copied or forged during transfer, a hash value c is produced to transfer $ID \oplus (\Delta r+2) \| H(ID \| (\Delta r+2))$ to the reader.

Step 7. The reader acquires the ID by conducting an XOR calculation with $\Delta r+2$ on the first value received from the database. For ID authentication, $\Delta r+2$ and ID are combined and hashed to produce $H(ID \| (\Delta r+2))'$. $H(ID \| (\Delta r+2))$, transferred from the database and $H(ID \| (\Delta r+2))'$, produced by the reader, are compared and authenticated if they are identical. The reader, to let the tag know of the authentication, sends $H(ID \| (\Delta r+3))$, a hash value produced with the ID and $\Delta r+3$.

Step 8. The tag receives $H(ID \| (\Delta r+3))$ and adds 3 to Δr to produce a hash value of $H(ID \| (\Delta r+3))'$ with the ID. If the transferred hash value $H(ID \| (\Delta r+3))$ and produced hash value $H(ID \| (\Delta r+3))'$ are identical, the database is authenticated, and the tag acknowledges the fact that the database authenticated it. The ID is not updated to prevent problems of non-synchronization, and the same ID is used.

5 Analysis

The proposed method is analyzed according to the RFID security threats and requirements mentioned in chapter 2. This method is capable in overcoming all security threats and can satisfy requirements. The existing research on the Randomized-Hash Lock method, Hash-based ID Variation method, Low-Cost method, and Mutual Authentication method were analyzed and compared with the proposed new method.

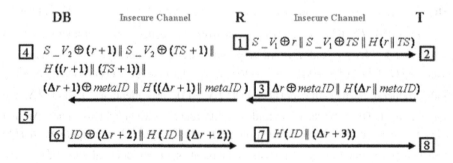

Fig. 4. Proposed Protocol 2

5.1 Security Threat Measures

The security of the proposed method can be verified through the security threat related problems mentioned in chapter 2. There are 4 kinds of security threats, eavesdropping, traffic analysis, replay attack, and tracking attack. The security of the proposed method can be evaluated through the level of dealing these problems.

- Eavesdropping: Vulnerability exists in the communication channel, which is a wireless channel, between the tag and reader, within the RFID privacy security method. Therefore, eavesdropping is possible. However, when the secret value and XOR calculation is used, critical values are not exposed, and the tapped data cannot be used to acquire new values.
- Traffic Analysis: This is a process in which data, acquired from eavesdropping, is compiled to guess the next session's value or acquiring essential values. The proposed method utilizes secret values, so that a malicious third party cannot acquire values even if he analyzes the data traffic.
- Replay Attack: In the RFID privacy security method, the time stamp is reproduced every session, which makes it impossible for the attacker to acquire values even if he retransfers acquired data. To make it even safer, the difference between the random number and time stamp is used.
- Tracking Attack: One of the most serious privacy problems of the RFID is that if a set value is exposed during each session, the privacy of the user may be encroached upon even if it is not possible to locate his exact location. However, this method uses random numbers to produce basic values, and communication starting times are produced with previous data to make different values for every session.

5.2 Requirements Measures

This chapter analyzes and provides solutions to security requirements mentioned in chapter 2. These solutions must satisfy security requirements. The current chapter analyzes how the solutions can be satisfied. Also, the calculation tools of the components and frequency of usage of these tools are analyzed to see if they are efficient.

- Authentication: The tag sends the hashed ID value to the database, and the database checks whether the value exists; if it exists, it is authenticated. The tag produces a hash value received from the database; if the same value is produced, the database is authenticated. The database authenticates the tag. The reader sends the secret and calculated values to the database for it to have authenticated components with secret values to secure each other.
- Anonymity: The tag conducts XOR calculations to ensure user anonymity. Also, acquired hashed values cannot be used to find which tag it has come from. The tag uses secret values and variable time stamps with XOR calculation to make it impossible to exploit. And by using the hashed values, a third party cannot find which tag it has come from.
- Integrity: The method uses a unilateral hash function to safely send secret values of the tag. By checking the transfer of general data and hashed data, we could confirm that copying and forgery could not be carried out during the method.

- Confidentiality: Only authenticated components share secret values, and the authentication process is carried out with XOR calculation of the secret values. Also, confidentiality is guaranteed as only authorized components share the tag ID value and essential value.
- Efficiency: The method uses unilateral hash functions rather than complicated encryption algorithms, considering the limited calculation and storage capabilities of the passive tag. These hash functions must have a small number of calculation frequency, and random numbers are produced by the R.N.G, which is attached to the reader.

Table 1. Analysis according to security threats and requirements

	RHLP	HIDV	LCAP	MAP	**Protocol 1**	**Protocol 2**
Eavesdropping	W	W	W	W	W	W
Traffic Analysis	U	U	U	U	S	S
Replay Attack	U	U	U	U	S	S
Tracking Attack	U	U	U	U	S	S
Authentication	S	S	S	S	S	S
Anonymity	U	U	U	U	S	S
Integrity	S	S	S	S	S	S
Confidentiality	S	S	S	S	S	S
Efficiency	W	S	S	S	U	U
Desynchronization	W	W	W	W	S	S

[Secure in security threats or offer requirements : S]
[Usually secure in security threats or requirements : U]
[Weak in security threats or not offer requirements : W]

6 Conclusion

RFID is gaining its attention as a ubiquitous computing technology in the ubiquitous environment. RFID, with its high identification speed, high recognition rate, and large storage capacity, is being used in various fields such as distribution, finance, and the public sector. However, the reader can be activated by applying power to easily expose its tag information and user information. To solve this problem, a random number is used in the reader to produce variable values to ensure the privacy of the RFID user. The new method solves the non-synchronization problem existing in previous methods, which makes it possible to secure RFID user privacy in the wireless environment. However, since variable values are produced without an update process, calculation frequency increases, and thus efficiency drops.

The research on the light weight should be followed with the other researches to get more competitive in RFID cost. In a ubiquitous environment, the environment in which users can use low-cost devices to access many kinds of services should be built as soon as possible. Also, the more researches must be carried out not only for the convenience of such tools but also to protect the privacy of users.

References

1. Henrici, D., Müller, P.: Hash-based enhancement of location privacy for radio-frequency identification devices using varying identifiers. In: IEEE International Workshop on Pervasive Computing and Communication Security PerSec, pp. 149–153 (2004)
2. Vajda, I., Butty'an, L.: Lightweight authentication protocols for low-cost RFID tags. In: 2nd Workshop on Security in Ubiquitous Computing (2003)
3. Yang, J., Park, J., Lee, H., Ren, K., Kim, K.: Mutual authentication protocol for low-cost RFID. In: Ecrypt Workshop (2005)
4. Feldhofer, M., Dominikus, S., Wolkerstorfer, J.: Strong Authentication for RFID Systems Using the AES Algorithm. In: Joye, M., Quisquater, J.-J. (eds.) CHES 2004. LNCS, vol. 3156, Springer, Heidelberg (2004)
5. Ohkubo, M., Suzuki, K., Kinoshita, S.: Cryptographic approach to 'privacy-friendly' tags. In: RFID Privacy Workshop (2003)
6. Peris-Lopez, P., Hernandez-Castro, J., Estevez-Tapiador, J., Ribagorda, A.: EMAP: An Efficient Mutual Authentication Protocol for Low-cost RFID Tags. In: OTM Federated Conferences and Workshop: IS Workshop (2006)
7. Peris-Lopez, P., Hernandez-Castro, J., Estevez-Tapiador, J., Ribagorda, A.: LMAP: A Real Lightweight Mutual Authentication Protocol for Low-cost RFID tags. In: Workshop on RFID Security (2006)
8. Peris-Lopez, P., Hernandez-Castro, J., Estevez-Tapiador, J., Ribagorda, A.: M2AP: A Minimalist Mutual-Authentication Protocol for Low-cost RFID Tags. In: Ma, J., Jin, H., Yang, L.T., Tsai, J.J.-P. (eds.) UIC 2006. LNCS, vol. 4159, Springer, Heidelberg (2006)
9. Sarma, S., Weis, S., Engels, D.: Radio-frequency identification: security risks and challenges, Cryptobytes, RSA Laboratories, pp. 2–9 (2003)
10. Weis, S., Sarma, S., Rivest, R., Engels, D.: Security and privacy aspects of low-cost radio frequency identification systems. In: International Conference on Security in Pervasive Computing SPC, pp. 454–469 (2003)
11. Weis, S.: Security and privacy in radio-frequency identification devices, Masters Thesis, MIT (2003)
12. Lee, S.-M., Hwang, Y.J., Lee, D.H., Lim, J.I.: Efficient authentication for low-cost RFID systems. In: International Conference on Computational Science and its Applications - ICCSA, pp. 619–627 (2005)
13. Dimitriou, T.: A lightweight RFID protocol to protect against traceability and cloning attacks. In: Conference on Security and Privacy for Emerging Areas in Communication Networks SecureComm (2005)
14. Weis, S.: Security Parallels Between People and Pervasive Devices. In: Workshop on Pervasive Computing and Communications Security - PerSec (2005)

A Hierarchical Composition of LU Matrix-Based Key Distribution Scheme for Sensor Networks*

Mi Wen, Yanfei Zheng, Hui Li, and Kefei Chen

Department of Computer Science and Engineering Shanghai Jiao Tong University
200240, Shanghai, China, P.R.C.
{superwm,zheng-yf,li_hui,kfchen}@sjtu.edu.cn

Abstract. Key pre-distribution techniques for sensing data security provision of wireless sensor networks (WSNs) have attracted much more attention and been studied extensively. But most of these schemes are not scalable due to their linearly increased communication and key storage overhead. Furthermore, existing protocols cannot provide sufficient security when the number of compromised nodes exceeds a critical value. To address these limitations, we propose a hierarchical composition of LU matrix-based key distribution scheme for sensor networks. Our scheme guarantees that two communicating parties can establish a unique pairwise key between them and allows mutual authentication. Compared with existing protocols, our scheme has better performance in terms of network resilience, associated overhead and scalability.

Keywords: Wireless Sensor Network, Key Management, LU Matrix, Security.

1 Introduction

Wireless sensor networks(WSNs) increasingly become viable solutions to many challenging problems for both military and civilian applications, including target tracking, battlefield surveillance, intruder detection and scientific exploration. However, deploying sensors without security in mind has often proved to be dangerous in hostile environments. In wireless communication environments an adversary not only can eavesdrop the radio traffic in a network, but also can intercept the exchanged data. To prevent the malicious node impersonating good nodes for spreading misleading data intentionally, secret keys should be used to achieve data confidentiality, integrity and authentication between communicating parties. Additionally, wireless sensors are not tamper resistant due to their low cost. Thus, the adversary may physically capture some sensors to compromise their stored sensitive data and communication keys. This serious attack is known as node capture attack, which makes the node's operation become under the control of the adversary. Hence key protection should be paid more attention in sensor networks.

* This work is supported by SRFDP of China under grant 20050248043.

T. Washio et al. (Eds.): PAKDD 2007 Workshops, LNAI 4819, pp. 608–620, 2007.
© Springer-Verlag Berlin Heidelberg 2007

Whereas, distribution of a secret key for every possible communication link is non-trivial due to large number of sensors and the resource constraints on sensor nodes. Consider a scenario where N number of sensor nodes are dropped from an airplane in the battlefield. Thus the geographical positioning of the nodes cannot be decided a priori. However, nodes in radio frequency range can be organized into clusters and any two of them can be expected to be able to communicate securely. One option is to maintain different secret keys for each of the pairs. Then each of the nodes needs to store $N-1$ keys. But, this method is not suitable for wireless sensor network due to the limited memory and energy power of the sensor nodes. On the other hand, on-line key exchange is not very popular till date since implementation of public key framework demands processing power at the higher end. Very recently implementations of ECC and RSA on 8-bit CPUs have been proposed [7]. Still a closer analysis of [7] reveals that the algorithms execute in seconds (0.43s or so); whereas the nodes need to calculate the inverse of an integer modulo a prime number. Recent research shows that pre-loading symmetric keys into sensors before they are deployed is a practical method, which is bound to be much faster than the former.

Recently, a basic probabilistic key pre-distribution scheme is proposed in [1]. Here each sensor node is assigned a random subset of keys from a key pool before deployment. After the deployment, if two neighboring nodes have some common keys, they can setup a secure link by the shared keys. This scheme is further extended in [2] by requiring two sensor nodes share at least q-composite pre-distributed keys to establish a pairwise key. Furthermore, two similar random key pre-distribution techniques [3][4] based on polynomial pool to drastically improve network resilience and memory usage efficiency are developed independently. Also, combinatorial set [5][6] is used to distribute keys to sensors prior to deployment. Such a deterministic combinatorial set technique allows each key in the key pool to be assigned to a constant number of sensor nodes. Therefore, the number of nodes with which each sensor shares a pairwise key is fixed. Due to the randomness of the key selection process in key pre-distribution, none of the above key management schemes can guarantee that a pairwise key will be found between two nodes wanting to communicate, and when the number of compromised nodes exceeds a certain threshold their security decreases drastically.

To address this issue, Park and Choil[8] proposes a new key pre-distribution scheme with LU Matrix, which guarantees that any pair of nodes can find a pairwise key between themselves. This is achieved by using the secret information assigned from the lower and upper triangular matrixes decomposed from a symmetric matrix of a pool of keys. Because all the established pairwise keys are distinct to each other, any sensor's compromise cannot affect the secure communication between non-compromised nodes. Although Park et al.'s scheme can provide better network performance and security than previous schemes; it incurs linearly increasing costs (n) in either communications per node or memory per node, so it is not suitable for large-scale sensor networks. And the node-to-node mutual authentication risks the exposure of the secret keys.

All the above schemes assume that wireless sensor networks are distributed flat structures and attempt to establish pairwise keys between any two sensor nodes. As we know, wireless sensors only have short transmission range. Therefore, it is not necessary to setup a pairwise key for any two nodes in a network, only neighboring nodes need to secure their communications. Furthermore, if a wireless sensor network has a hierarchical architecture, only the cluster head and its cluster members need to establish a pairwise key between them, which will significantly improve the performance in the network initialization phase. Gaurav et al.[10] propose a low-energy key management protocol for hierarchical wireless sensor networks, but the drawback is that any cluster head's compromise in the network initialization phase would compromise a large number of nodes' secret keys.

To address the limitations of current key distribution schemes, we propose a hierarchical composition of LU Matrix-based key predistribution scheme. Our scheme also enables any pair of communicating nodes to establish a unique pairwise key between them and provide sufficient security for large-scale sensor networks against node capture attack. This paper is organized as follows: Section 2 gives an overview of the key pre-distribution technique with LU matrix. Section 3 presents the network model and our scheme, and section 4 provides detailed analysis. Section 5 concludes this paper.

2   Background of LU Matrix-Based Key Pre-distribution

In this section, we briefly review the matrix-based key pre-distribution protocols in [8], which is the basis of our new scheme. For simplicity, here we only review the key pre-distribution process in [8]. First of all, we will introduce some definitions.

Definition 1. If the multiplication result of a $n * n$ lower triangular matrix L and a $n * n$ upper triangular matrix U equals to a symmetric matrix K, namely $K = LU$. We say K is the "LU composition" of triangular matrices L and U.

The composition process is omitted here and the details could reference [8] or section 3 of this paper. Suppose we are able to compute a symmetric key matrix K from two triangular matrices L and U as follows.

$$\begin{bmatrix} l_{11} & 0 & 0 \\ l_{21} & l_{22} & 0 \\ l_{31} & l_{32} & l_{33} \end{bmatrix} \cdot \begin{bmatrix} u_{11} & u_{12} & u_{13} \\ 0 & u_{22} & u_{23} \\ 0 & 0 & u_{33} \end{bmatrix} = \begin{bmatrix} k_{11} & k_{12} & k_{13} \\ k_{12} & k_{22} & k_{23} \\ k_{13} & k_{23} & k_{33} \end{bmatrix} \tag{1}$$

For each sensor i in the network, it will be assigned one row from the L matrix and one column from the U matrix respectively. For convenience, the same row and column position are assigned to each node, i.e. L_{ri} (ith row of L) and U_{ci} (ith column of U) are assigned to node i.

In this approach, if two nodes, assume node i (with L_{ri} and U_{ci}) and node j (with L_{rj} and U_{cj}), need to establish a pairwise key between them, they should first exchange their columns, and then compute a product as following:

$$node_i : L_{ri} * U_{cj} = k_{ij}$$
$$node_j : L_{rj} * U_{ci} = k_{ji}$$

Note that K is a symmetric key matrix, $k_{ij} = k_{ji}$. Thus, k_{ij} (or k_{ji}) will be used as the pairwise key between $node_i$ and $node_j$.

It is theoretically possible to use the basic LU matrix-based scheme to guarantee that there is a pairwise key between any two communicating nodes in WSNs. However, each sensor node's storage cost for a row and a column is linearly increasing in terms of the network size if the whole network only share one symmetric key matrix. In this paper, we will highlight this problem.

3 The Proposed Scheme

3.1 The Network Model

Based on the architectural consideration, wireless sensor networks may be broadly classified into two categories viz. (i) Hierarchical Wireless Sensor Networks(HWSN) and (ii) Distributed Wireless Sensor Networks (DWSN). A Distributed WSN is easier for deployment, and there is no fixed infrastructure, and the network topology is not known prior to deployment. Sensor nodes are usually randomly scattered all over the target area. While a hierarchical WSN provides simpler network management, and can help further reduce transmissions. There is a pre-defined hierarchy among the participating nodes and three types of nodes in the descending order of capabilities: (a) base stations, (b) cluster heads, and (c) sensor nodes. Base stations are many orders of magnitude more powerful than sensor nodes and cluster heads. Nodes with better resources, named as cluster heads (CH), may be used to collect and merge local data from sensor nodes and send it to base station. Sensor nodes are deployed around the neighborhood of the cluster heads.

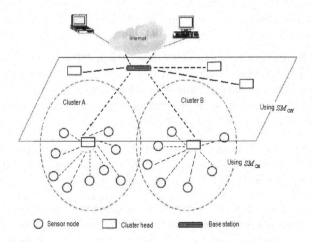

Fig. 1. Hierarchical wireless sensor network architecture

In this paper, we focus on three-tier hierarchical architecture, which is similar to [10], as shown in Figure 1. Where sensor nodes are deployed, the clusters can be formed based on various criteria such as capabilities, location, signal strength etc. Each cluster has a cluster head and a set of sensor nodes. For ease of presentation, we assume there are at most n sensor nodes and m cluster heads in the WSN, each cluster has a cluster head and $\lceil n/m \rceil$ sensor nodes inside. We assume that sensor nodes are static once they are deployed. To reduce the energy consumption and the redundant traffic loads in the network, usually sensor nodes only communicate with its neighbors in the same cluster.

Table 1. The notations used in our scheme

F: a hash function used to calculate hash values
CH_i: cluster head i
C_i: the ith cluster,$1 \leq i \leq m$
S_{i_j}: the ID of sensor node j in cluster C_i,$1 \leq j \leq \lceil n/m \rceil$
SM_{CH}: the keys symmetric matrix for cluster heads
SM_{Ci}: symmetric matrix for sensor nodes in cluster C_i
K_{CH_ij}: the key at ith row and jth column of matrix SM_{CH}
K_{Ci_ij}: the key at ith row and jth column of matrix SM_{Ci}
L_{CH_ri}: the ith row of SM_{CH}'s lower triangular matrix CH_L
U_{CH_ci}: the ith column of SM_{CH}'s upper triangular matrix CH_U
L_{Cj_ri}: the ith row of SM_{Cj}'s lower triangular matrix Cj_L
U_{Cj_ci}: the ith column of SM_{Cj}'s upper triangular matrix Cj_U

3.2 The Hierarchical Composition of LU Matrix-Based Key Distribution Scheme

Our key pre-distribution technique can be divided into three phases: symmetric matrixes generation phase, key pre-distribution phase, pairwise key establishment phase. The notations used in this paper are listed in Table 1.

Phase 1: Symmetric Matrixes Generation Phase. Two different types of symmetric key matrixes are used in our scheme, one is SM_{CH}, shared by the cluster heads and the base station. The other is SM_{Ci}, shared by sensor nodes in C_i and their cluster head. Thus, the size of SM_{CH} and SM_{Ci} are $(m+1)*(m+1)$ and $(\lceil n/m \rceil + 1) * (\lceil n/m \rceil + 1)$ respectively.

Step 1: The base station needs to generate large enough pool of keys in this step (e.g. 2^{20} or more). The keys are selected from a finite field $GF(q)$ to create a symmetric matrix(SM). Where q is the smallest prime larger than the key size.

Step 2: The base station computes m symmetric key matrices $SM_{Ci}(i = 1, ..., m)$, for m clusters and one matrix SM_{CH} for the cluster heads, using the method in [8]. And in order to guarantee that the secret keys between any pair of the communication entities are unique, the elements in all the symmetric key matrices must be different from each other. For clarity, we demonstrate the "LU composition" process of the symmetric key matrix by computing a $3*3$ symmetric key matrix A as follows.

Firstly, a $3*3$ L matrix is formed by selecting (-1,1,2,-3) from the pool of keys generated in the first step. Then a U matrix is calculated such that the multiplication of L and U is a symmetric key matrix A. That's like:

$$\begin{bmatrix} 1 & 0 & 0 \\ 2 & 1 & 0 \\ -1 & -3 & 1 \end{bmatrix} \cdot \begin{bmatrix} u_{11} & u_{12} & u_{13} \\ 0 & u_{22} & u_{23} \\ 0 & 0 & u_{33} \end{bmatrix} = \begin{bmatrix} k_{11} & k_{12} & k_{13} \\ k_{12} & k_{22} & k_{23} \\ k_{13} & k_{23} & k_{33} \end{bmatrix} \tag{2}$$

Now, we get three equations from the left side of (2) in terms of the properties of the symmetric matrix.

$$l_{11}.u_{12} = l_{21}.u_{11} \Rightarrow u_{12} = (l_{21}/l_{11}).u_{11} \Rightarrow u_{12} = 2u_{11}$$
$$l_{11}.u_{13} = l_{31}.u_{11} \Rightarrow u_{13} = (l_{31}/l_{11}).u_{11} \Rightarrow u_{13} = -u_{11}$$
$$l_{21}.u_{13} + l_{22}.u_{23} = l_{31}.u_{12} + l_{32}.u_{22} \Rightarrow u_{23} = -u_{12} - 3u_{22} - 2u_{13}$$

Where l_{ij} is the element at the ith row and jth column of matrix L. We can see that these three equations have five unknown variables ($u_{11}, u_{12}, u_{13}, u_{22}, u_{23}$), and u_{33} is an arbitrary variable. Therefore, there exist many solutions for this matrix. If we randomly set the value of three variables (e.g. u_{11}, u_{22}, u_{33}), the remaining elements can be decided. Assume that $u_{11} = 1, u_{22} = 2, u_{33} = 3$, then we get the following matrices:

$$\begin{bmatrix} 1 & 0 & 0 \\ 2 & 1 & 0 \\ -1 & -3 & 1 \end{bmatrix} \cdot \begin{bmatrix} 1 & 2 & -1 \\ 0 & 2 & -6 \\ 0 & 0 & 3 \end{bmatrix} = \begin{bmatrix} 1 & 2 & -1 \\ 2 & 6 & -8 \\ -1 & -8 & 22 \end{bmatrix} = A \tag{3}$$

Similarly, the symmetric matrices $SM_{Ci}, (i = 1, ..., m)$ and SM_{CH} can be computed. One condition here is that only the rows and columns in CH_L and CH_U or Ci_L and Ci_U are needed to be pre-loaded to the sensor nodes, and when they need to communicate with each other, the keys they established are the elements in SM_{CH} or SM_{Ci}.

Phase 2: Key Pre-distribution Phase. In this phase every node is randomly assigned one row from its corresponding L matrix and one column from its corresponding U. In our three-tier hierarchical structure different secret information will be pre-loaded into different level of nodes.

Base station: To authenticate and secure the communication between the base station and other nodes, base station needs to store a pair of vectors $(L_{CH\_r(m+1)}, U_{CH\_c(m+1)})$ and all the redundant key elements of the key matrices in its memory. The pair of vectors $(L_{CH\_r(m+1)}, U_{CH\_c(m+1)})$ is used to establish a pairwise key with the cluster heads. The redundant key elements are used for the new nodes or replacement of the cluster head when the old one fails or is compromised.

Cluster head CH_i: Each cluster head not only needs to share a symmetric key matrix SM_{CH} with its neighboring cluster heads and base station, but also needs to share a symmetric key matrix SM_{Ci} with its cluster members. That means, it needs to store $(L_{CH\_ri}, U_{CH\_ci}, L_{Ci\_r(\lceil n/m \rceil+1)}$, and $U_{Ci\_c(\lceil n/m \rceil+1)})$ in its memory.

Sensor node S_{i_j}: To reduce the storage overhead of sensors, each sensor node in Ci needs to store two vectors (L_{Ci_rj}, U_{Ci_cj}). As we introduced before, (L_{Ci_rj}, U_{Ci_cj}) will be used to establish a pairwise key with its cluster neighbors.

After the key pre-distribution, each resource-limited sensor node in the network stores only two vectors in its memory. Each vector has $\lceil n/m \rceil + 1$ elements instead of n in [8]. Which greatly reduces the key storage overhead in large scale WSNs.

Phase 3: Pairwise Key Establishment Phase. After key deployment, each cluster member needs to establish a pairwise key with its neighbors and the cluster head to secure the intra-cluster communication. To establish a pairwise key between node S_{i_k} and node S_{i_j}, the following things are needed to be done:

1. Node S_{i_k} sends its column $\{U_{Ci\_ck}\}$ to node S_{i_j}.
2. Upon receiving U_{Ci_ck} from node S_{i_k}, node S_{i_j} computes K_{Ci_jk} and replies with $\{U_{Ci\_cj}, F(K_{Ci\_jk})\}$. Here, $F(K_{Ci\_jk})$ is the hash value of K_{Ci_jk}. When node S_{i_k} receives this value, it can authenticate if their calculated keys are the same one. (While in [7], K_{Ci_jk} is sent in plaintext, which is at the risk of eavesdropping by the adversary.)
3. Upon receiving U_{Ci_cj}, node S_{i_k} computes K_{Ci_kj} and checks if $F(K_{Ci\_kj})$ = $F(K_{Ci\_jk})$.
4. If node S_{i_k} verifies $F(K_{Ci\_kj}) = F(K_{Ci\_jk})$, that is to say $K_{Ci\_kj} = K_{Ci\_jk}$. Node S_{i_k} can use K_{Ci_kj} as its pairwise key with node S_{i_j} and responds with a message $\{ok, F(K_{Ci\_kj})\}$. Else, node S_{i_k} will broadcast a error message $\{err, S_{i\_j}\}$ to its neighbors and its cluster head. The error message will be authenticated by the receiver using source authentication methods, such as TESLA [13] etc.

If the number of error messages with S_{i_j} is more than a certain threshold, the cluster members in C_i will exclude the messages from node S_{i_j}. When base station wants to send messages to cluster heads, they must firstly establish a pairwise key. The process is similar to the process in phase 3, because they can be seen as one high level cluster, where the base station acts as a cluster head and $CH_i, (i = 1, ..., m)$ act as cluster members.

3.3   Remediations

We assume that the compromise or the failure of the cluster head (CH) and the nodes will be detected by the method in [11][12]. When a cluster head CH_i fails, the cluster C_i will get disconnected from the network, because sensor nodes in C_i communicate with the BS through their cluster head CH_i. How do we solve this problem? We cannot select a node from cluster C_i to act as CH_i, simply because the following two reasons: first, it doesn't hold the necessary key information $(L_{CH\_ri}, U_{CH\_ci}, L_{Ci\_r(\lceil n/m \rceil+1)}, U_{Ci\_c(\lceil n/m \rceil+1)})$; Second, nodes are supposed less powerful than CH, the communication range of the nodes may not cover the other cluster heads or the BS. Therefore, the secure solution is to deploy a replacement, that means to remove manually the defective cluster head and load a new CH_i', which has compatible capabilities in terms of computational resources and transmission range with the previous, then the new CH_i' can securely communicate with the nodes of its cluster or the other clusters heads.

When a cluster head CH_i is compromised, the key information pre-loaded to the new CH_i' should be different from the previous. The BS should choose two vectors from its redundant key elements, one is from SM_{CH} and the other is from SM_{Ci}. The pairwise key should be reestablished soon. Note that, the rows in CH_L or Cj_L are linearly independent, and the columns in CH_U or Cj_U are linearly independent either, no matter how many rows (or columns) of CH_L (or CH_U) are known by the adversary, the other unknown rows can't be worked out. Thus the compromise of the cluster heads can not effect the secrecy of the pairwise key between the innocent nodes and innocent CHs. We need not to worry about the exhaust of the redundant key elements, because we also assume that the key elements in each node or CH will be refreshed by the BS periodically, which unicasts the new key elements to the node or CH secretly with the help of previous innocent pairwise keys.

When nodes are failed or compromised, the solutions are similar to those of the cluster heads.

4   Security Analysis and Evaluations

In this section, we evaluate the security property and performance of our proposed scheme by simulating on Crossbow's MICAz platform.

4.1   Security Analysis

Node compromise attack is the main threat for key establishment in WSNs. According to the key pre-distribution method, if there is no compromised node, it is guaranteed that any two sensor nodes can establish a pairwise key.

Now, we study the resilience property of the proposed scheme against node compromise by calculating the fraction of communications links in the rest of the network that are compromised due to key revealing. Figure 2 compares the resilience against sensor node compromise attack for different schemes. For the basic probabilistic [1] and the q-composite key pre-distribution schemes [2], as the number of compromised nodes increases, the fraction of affected communication links in the rest of the network increases quickly. The reason is that in random key pre-distribution schemes [1,2], the same key may be used by several different pairs of sensors, some sensor nodes' compromise may affect the communication links in the rest of the network. As a result, a small number of nodes compromise may affect a large fraction of the rest communication links. For our proposed scheme, as the number of compromised nodes increases, the fraction of affected communication links remains zero. That's because each pairwise key

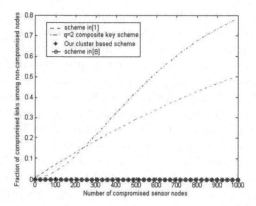

Fig. 2. Number of compromised links between non-compromised sensor nodes vs. Number of compromised sensor nodes

is different from others in our proposed scheme, any sensor node's compromise cannot affect the secure communication between non-compromised nodes. Although [8] can also prevent the key compromising for non-compromised sensor nodes, benefiting from the unique pairwise key between each pair of two communicating parties. But it has a dangerous risk of the pairwise key exposure during the authentication process. While our proposed scheme prevents this risk by using a hash function in that process.

Figure 3 shows the network resilience against cluster head node compromise attack. Suppose there are 20000 sensors and 100 cluster heads in a network. In [10], each cluster head stores 200 sensors' secret keys in its memory. Therefore, any single cluster head's compromise could affect the 200 sensors' secret keys. When the number of compromised cluster heads increases, the number of compromised sensors increases dramatically. In our proposed scheme, only two vectors are stored in each cluster head during the network initialization phase, cluster heads have no idea about the sensors secret keys. Even all the 100 cluster heads are compromised, none of the pairwise keys in the non-compromised sensor nodes could be compromised.

Therefore, the resilience property against the node or cluster head compromise has been greatly improved in our scheme.

4.2 Performance Evaluation

Our scheme has reasonable overheads in terms of storage requirements, computational and communication cost. Furthermore, our scheme has a significant scalability in the large-scale hierarchical sensor networks.

Overhead. Assume that key elements are of the same length of the keys, for simplicity, we assume each cluster has k members inside with the network size of n. Therefore, one row and one column of the secret information stored in each node equals to $2k$ keys storage overhead. Note that, half of the elements in the

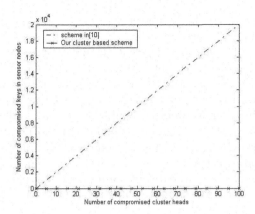

Fig. 3. Number of compromised keys in sensor nodes vs. Number of compromised cluster heads

matrices CH_L (Cj_L) or CH_U (Cj_U) are zero, we may change the storing methods of the rows or columns to reduce the storage overheads as the form of (nonzero portions and one value specifying the number of following zeros in the zero-element part). Thus, the average storage overhead of a row or column is $k/2 + 1$ and the totally storage overhead of a sensor node equals $k + 2$ keys. In the same way, the storage overhead of each cluster head equals to $(k + 2 + (\lceil n/k \rceil + 1) + 2)$ keys. In addition, if the number of nodes in each cluster remains the same, the storage overhead of each node is still $k + 2$ no matter how large the network size increases (the fresh nodes will be organized into a new cluster). If multi-tier network architecture is used, the storage overhead of the cluster head also will be a certain value. This is a great improvement compared with the schemes in [1][2] and [8].

In terms of communication overhead, during the pairwise key establishment phase, each pair of entities only needs to exchange a column and a hash value to establish their pairwise key. Due to the smaller size of cluster in our scheme, all the exchanged messages are smaller than those in [8], and the communication overhead in this phase is at most $(k/2+2)$, which can be done efficiently in resource limited sensor networks. During the Remediation phase, if a CH is compromised, two rows and two columns will be pre-loaded in the new CH's memory, the communication overhead in the pairwise key re-establishment is at most $(k-1)*(k/2+2)+(\lceil n/k \rceil /2+2)$. Since the CH compromise seldom happens, this overhead is affordable to the WSNs. If a sensor node is compromised, the communication overhead in the pairwise key re-establishment is same to the normal key establishment; it is $(k/2+2)$ at the most.

As to the computational overhead, the pairwise key establishment in intra-cluster only needs a vector multiplication and a hash computation. Although it needs one more hash computation compared with [8], it enables to protect the pairwise key during the authentication process. Thus, it is desirable to add a hash computation in our scheme. During the Remediation phase, if a CH is compromised, $(k-1)$ vector multiplications and $(k-1)$ hash computations are needed to re-establish the pairwise keys. If a sensor node, one vector multiplication and one hash computation are needed.

Compared with [3], which based on the Blom's matrix, our proposed scheme has less computation overhead, but $k/2$ more transmission overhead. While [3] has the less transmission overhead, because during key establishment phase only column seeds instead of the columns are transmitted (a node can easily and locally compute the column from the column seed), but [3]involves considerable more computation overhead (n modular exponentiation). Additionally, Bloms scheme is not perfectly resilient against node capture. Instead it has the following λ-secure property: as long as an adversary compromises less than or equal to λ nodes, uncompromised nodes are perfectly secure; when an adversary compromises more than λ nodes, all pairwise keys of the entire network are compromised.

Maximum Supported Network Size. In large-scale wireless sensor networks, the key distribution scheme should be scalable when the number of sensor nodes increases. Here, we will take the Maximum Supported Network Size as the metric to evaluate the scalability of our scheme. When the network size linearly increases, to achieve the required network connectivity in [1][2], the number of keys stored in each sensor also need to increase linearly. However, the constrained physical memory size of sensor nodes limits the number of keys each sensor stores. Thus, the maximum supported network size is limited. [8] also incur linearly increasing cost with the increasing network size.

Given a fixed memory constraint (assume 64-bit keys and less than 4KB of data memory [4]) and guarantee of perfect security against node compromises, Figure 4 shows the maximum supported network size as a function of the number of tiers in our proposed scheme. This figure shows that, the maximum supported network size increases dramatically when we have more tiers within the range shown in the figure (the maximum supported network size in our three-tier architecture is 65026). Indeed, when the number of tiers is bigger, our scheme can support a larger network by adding more tiers without increasing the storage

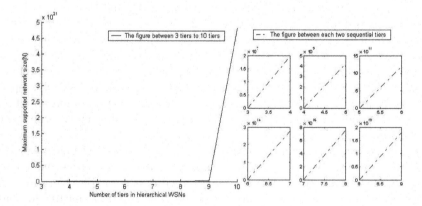

Fig. 4. Maximum supported network size for different number of tiers

overhead or sacrificing the security performance. Therefore, given certain storage constraint, our cluster-based key pre-distribution scheme allows the network to grow, while the schemes [1][2] and [8] have an upper limit on the network size.

5 Conclusion

In this paper, we developed a hierarchical composition of LU matrix-based key distribution scheme for large-scale sensor networks. We also present an in-depth analysis of our scheme in terms of network resilience, associated overhead and scalability. The results show that our scheme can achieve a good level of security, outperform most existing schemes, significantly improve the effectiveness of filtering false data. In addition, our scheme is suitable for large scale sensor networks due to its hierarchical network model and hierarchical composition of the key distribution scheme.

References

1. Eschenauer, L., Gligor, V.D.: A key-management scheme for distributed sensor networks. In: Proceeding of the 9th ACM Conference on Computer and Communication security, pp. 41–47 (2002)
2. Chan, H., Perrig, A., Song, D.: Random key pre-distribution schemes for sensor networks. In: IEEE Symposium on Security and Privacy, pp. 197–213 (2003)
3. Du, W., Deng, J., Han, Y.S., Varshney, P.K., Katz, J., Khalili, A.: A pairwise key pre-distribution scheme for wireless sensor networks. ACM Transation on Information and System Security, 228–258 (2005)
4. Liu, D., Ning, P.: Establishing pairwise keys in distributed sensor networks. ACM Transactions on Information and System Security 8(1), 41–77 (2005)
5. Camtepe, S.A., Yener, B.: Combinatorial design of key distribution mechanisms for wireless sensor networks. In: Samarati, P., Ryan, P.Y.A., Gollmann, D., Molva, R. (eds.) ESORICS 2004. LNCS, vol. 3193, pp. 293–308. Springer, Heidelberg (2004)
6. Lee, J., Stinson, D.R.: A combinatorial approach to key predistribution for distributed sensor networks. In: Proc. of IEEE Wireless Communications and Networking Conference (WCNC 2005), New Orleans, LA, USA (2005)
7. Gura, N., Patel, A., Wander, A., Eberle, H., Shantz, S.C.: Comparing Elliptic Curve Cryptography and RSA on 8-bit CPUs. In: Joye, M., Quisquater, J.-J. (eds.) CHES 2004. LNCS, vol. 3156, pp. 119–132. Springer, Heidelberg (2004)
8. Park, C.W., Choi, S.J., Youn, H.Y.: A Noble Key Pre-distribution Scheme with LU Matrix for Secure Wireless Sensor Networks. In: Hao, Y., Liu, J., Wang, Y.-P., Cheung, Y.-m., Yin, H., Jiao, L., Ma, J., Jiao, Y.-C. (eds.) CIS 2005. LNCS (LNAI), vol. 3802, pp. 494–499. Springer, Heidelberg (2005)
9. Choi, S.J., Youn, H.Y.: A Novel Data Encryption and Distribution Approach for High Security and Availability Using LU Decomposition. In: Laganà, A., Gavrilova, M., Kumar, V., Mun, Y., Tan, C.J.K., Gervasi, O. (eds.) ICCSA 2004. LNCS, vol. 3046, pp. 637–646. Springer, Heidelberg (2004)
10. Jolly, G., Kuscu, M.C., Kokate, P., Yuonis, M.: A low-energy management protocol for wireless sensor networks. In: Proceedings of the Eighth IEEE International Symposium on Computers and Communication (ISCC 2003), Kemer-Antalya, Turkey (June 30-July 3 2003)

11. Shaneck, M., Mahadevan, K., Kher, V., Kim, Y.: Software-Based Attestation for Wireless Sensors. In: Molva, R., Tsudik, G., Westhoff, D. (eds.) ESAS 2005. LNCS, vol. 3813, pp. 27–41. Springer, Heidelberg (2005)
12. Wang, G., Zhang, W., Cao, G., La Porta, T.: On Supporting Distributed Collaboration in Sensor networks. In: IEEE Military Communications Conference (MILCOM) (October 2003)
13. Perrig, A., Canetti, R., Tygar, J.D., Song, D.: The Tesla Broadcast Authentication Protocol. RSA CryptoBytes 5 (2002)

Security Framework for Home Network: Authentication, Authorization, and Security Policy

Geon Woo Kim[1], Deok Gyu Lee[1], Jong Wook Han[1], Sang Choon Kim[2], and Sang Wook Kim[3]

[1] Electronics and Telecommunications Research Institute
161 Gajeong-dong , Yuseong-gu, Daejeon, 305-350, Korea
{kimgw,deokgyulee,hanjw}@etri.re.kr
[2] Kangwon National University
21 Joongang-ro Samcheok, Gangwon-do, Korea
kimsc@kangwon.ac.kr
[3] Kyungpook National University
1370, Sankyuk-dong, Buk-gu, Daegu 702-701, Korea
swkim@cs.knu.ac.kr

Abstract. As a number of home network services are available and home network is expanding into ubiquitous computing environment, we need to protect home network system from illegal accesses and a variety of threats. Home network is exposed to various cyber attacks of Internet, involves hacking, malicious codes, worms, viruses, DoS attacks, and eavesdropping since it is connected to Internet. So in this paper, we propose a home network security framework for guaranteeing reliability and availability including authentication, authorization and security policy system.

Keywords: Home Network, Security, Authentication, Authorization, Access Control, Security Policy.

1 Introduction

Home network is a new IT technology environment for making an offer of convenient, safe, pleasant, and blessed lives to people, making it possible to be provided with a variety of home network services by constructing home network infrastructure regardless of deices, time, and places. This can be done by connecting home devices based on various kinds of communicating networks, such as mobile communication, Internet, and sensor network [1]. With the home network, we can easily control home devices, make use of a number of services such as a VOD service, a remote health care service, a T-commerce service and etc.

However, without guaranteeing reliability and availability, we may be confronted with unexpected results as well as revelation of private information and illegal accesses. Especially, as home network consists of heterogeneous network protocols and a variety of service models, it is likely to be exposed to various cyber attacks of Internet, involves hacking, malicious codes, worms, viruses, DoS attacks, and eavesdropping since it is connected to Internet [2].

T. Washio et al. (Eds.): PAKDD 2007 Workshops, LNAI 4819, pp. 621–628, 2007.

So in this paper, we propose a security framework for home network in order to safeguard against wide range of threats and the possible loss, which basically involves a user authentication mechanism, an authorization mechanism for controlling access, and a security policy mechanism especially designed for home network.

2 Security Framework for Home Network

The Security framework for home network contains a few essential components, such as an authentication component, an authorization component, and a security policy component. They work at each home gateway and often cooperate with a home portal server established by a home service provider in the Internet.

Since the home gateway is installed at the border of each home and every home network packet must pass through it, it is supposed to be a core component and suitable in providing security functions described in security framework for home network. Whenever a new access to home network is found, it should be able to authenticate and authorize it and enforce security policy based on security rules set by the corresponding home security administrator [3].

Figure 1 depicts the overall architecture of secure home network.

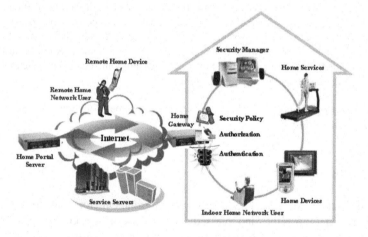

Fig. 1. Architecture of secure home network

Home network consists of several modules such indoor/remote home network user, home devices, home network services, a home portal server, service servers, especially a home gateway and optionally a security manager. The home gateway is supposed to do a central role in home network, work as a bridge between closed home network and Internet, and is responsible for security of each home. Every home network user should be reliable and be able to use home network services in a safe way.

The indoor home network user controls home devices, uses home services via the home gateway and the remote home network user accesses the home network via the home portal server and/or the home gateway, where the home gateway enforces security policy and does authentication and authorization.

Figure 2 illustrates the security framework for home network.

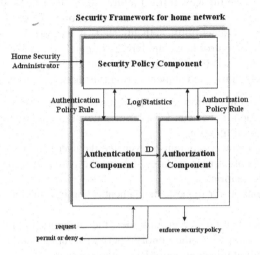

Fig. 2. Security Framework for home network

3 Authentication

Home network needs to authenticate entities that are accessing home network in order to provide home service to only registered members. The entity may be a home network user or a device.

The home network is able to support a wide range of ages of home network users and reflect their own characteristics and needs. As a result, we can selectively use our favorite authentication method among existing authentication methods. The authentication mechanisms are sure to be safe and reliable. Our proposed authentication mechanism satisfies these requirements.

The preconditions of our authentication mechanism are as follows.

- Home network entity must pre-register authentication information to authentication server to use.
- Authentication server is secure from outdoor/indoor attacks.

The authentication server of authentication mechanism means a computing device, which has a database storing authentication information (such as, ID-password, certificate, biometric information, etc.) and loads authentication programs. It can be a home gateway or a home portal server connected to the home gateway at home network. A entity can select a favorite authentication method. That is, the authentication server supplies authentication mechanism using ID-password, certificate and biometric information, etc [6].

The authentication mechanism supplying a variety of authentication methods uses EEAP (Encrypted Extensible Authentication Protocol) [7]. EEAP packet is similar to EAP (Extensible Authentication Protocol) [8] packet. And EEAP is similar to EAP-TTLS (EAP Tunneled TLS) [9] or PEAP (Protected EAP) [10].

The authentication server is different from CA server in that it supports authentication using certificate as well as ID-password, biometric information and stores authentication information registered in the various contents server securely. CA server in PKI (Public Key Infrastructure) does the role issuing and managing certificates and takes a neutral attitude in authentication process.

Once an entity is successfully authenticated, it can use his (or her) authorized service. If it wants to control home appliances remotely, it can control them. And if it wants to use contents by ISPs, it connects the wanted service server.

Figure 3 shows the proposed authentication information in home network

- An entity is authenticated using preferred authentication information, where authentication server can be a home gateway or a home portal server using authentication function.
- The entity connects contents server that the user want to use.
- If the contents server requires log-on, the entity informs the authentication server information.
- The contents server requests the authentication information to the authentication server.
- The authentication server provides the authentication information to the contents server if the entity permits authentication server to give authentication information to the contents server.
- The contents server authenticates the entity using received authentication information.

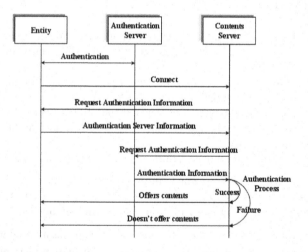

Fig. 3. Authentication Mechanism using Indoor Authentication Server

4 Authorization

The purpose of authorization is controlling access of entity even though it has been successfully authenticated and restricting a privilege and access right. Also, it can minimize the loss when home network system is penetrated and attacked by malicious accesses or unauthorized uses.

We can use an ACL (Access Control List) or a RBAC (Role-based Access Control). The ACL directly established relationships between subjects and resources, where a subject means an entity that is accessing, and a resource an entity that the subject is accessing. The ACL is simple so useful for relatively small-scale network. On the other hand, The RBAC adapts an intermediate component called a role between a subject and a resource, so indirectly sets up relationships between them. The RBAC seems to be adequate for relatively large-scale network.

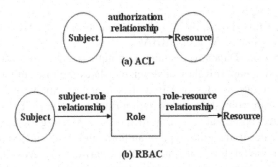

Fig. 4. ACL vs. RBAC

Since home network includes a variety of network protocols and is expected to support many service models such as a client-server model, a peer-to-peer communication model, and hybrid model, it is difficult to definitely decide which mechanism is suitable for home network. Actually we had better use a different authorization model according to the specific home network service. As a result, we need an integrated authorization framework for home network.

Figure 5 shows an integrated authorization framework for home network.

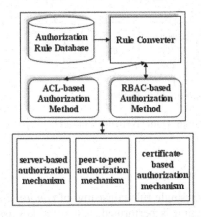

Fig. 5. Integrated authorization framework for home network

Existing authorization mechanisms can be categorized into three fields: server-based authorization mechanism, peer-to-peer authorization mechanism and certificate-based authorization mechanism. A server based authorization mechanism works on client-server model and the server generates and maintains authorization rules, enforces it. This method is relatively simple and easy to apply. A peer-to-peer authorization mechanism is for p2p communication service model. Peer can manage authorization rules by itself or require help of designated authorization server. This model is relatively complicated to implement and there are a few constraints considering database maintenance and H/W specifications of peer device and etc,. A certificate-based authorization mechanism is generally used in open network and conforms to the PKI.

These mechanisms define their own schema to specify the authorization rules, which may be of either an ACL or a RBAC.

Authorization Rule Database contains raw authorization rules and the details are not described in this paper. Rule Converter translates the raw authorization rules into ACL-based authorization rules or RBAC-based authorization rules and the reverse. Also, it can maintain consistent authorization rules between ACL-based authorization method and RBAC-based authorization method by reflecting the changes of one type of authorization rules to the other type of authorization rules immediately.

Authorization model for home network comprises an access control definition module, an access control enforcement module, an information collection module, an access control database, and a log database.

Figure 6 shows the authorization model for home network

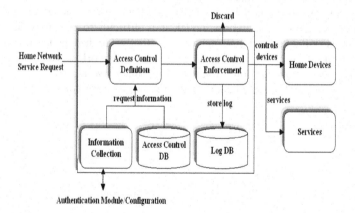

Fig. 6. Home Network Authorization Model

5 Security Policy

Security policy is a set of single rule consisting of *condition* and *action*. Whenever a condition is satisfied, action is performed, where the key issue is how to construct condition. Elements to be contained in condition are as follows

- Time(date, day, duration)
- Event(sensor, user-triggering, state)
- Log(statistics)

Also, we define relationships (interaction, union) among above elements and support recursive structure, which makes it possible to build complex conditions.

Time and event are the basic elements of condition and can be generally used. On the other hand, log-based condition controls access by statistics information. For example, there is a pre-condition that the security policy manager set the policy that children could not use the game service more than 30-hour in a month. Whenever the children access the game service, their usage information may be store at log database. If the above condition is satisfied, connection would be rejected and they can't access it during the month.

Action contains controlling device and providing home services and etc. In order to control device, it should cooperate with corresponding middleware used by controlled device such as UPnP, LnCP, zigbee, UWB, etc.

Figure 7 shows the conceptual architecture and operations on security policy enforcement for home network.

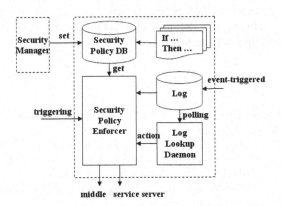

Fig. 7. Security policy system for home network

Considering the features of home network, it must be enough easy for non IT-familiar user to use. In our system, we use a Drag-and-Drop mechanism to establish the security policy, so anyone can handle it if he has been authenticated successfully.

Figure 9 illustrates the functions that the security policy manager provides.

6 Conclusion

Since home network consists of heterogeneous network protocols and contains existing security threats and holes of Internet such as hacking, malicious codes, worms, viruses, DoS attacks and eavesdropping, due to connections to the open network, we need a security framework to safeguard against them, efficiently guarantee reliability and availability.

So in this paper, we propose home network security framework including authentication, authorization, and security policy.

An authentication mechanism authenticates entity that is access home network. Also we can select our favorite authentication method such as ID-password-based authentication method, certificate-based authentication method, and biometric-information-based authentication method.

An authorization mechanism controls access by an entity even though it has been successfully authenticated already and restricts a privilege and access right. When some authenticated user wants to use home network service, the authorization mechanism receives identity-related information from corresponding authentication mechanism and looks for an adequate authorization rule in security policy database of security policy mechanism. Based on the found rule, the access control enforcer does authorization and informs the result to the entity. It may use both ACL and RBAC simultaneously. We propose an integrated authorization framework, where a variety of authorization methods can work collaboratively, and we do not have to care about the specific authorization method.

Security policy specifies the strategy for home network and provides basic rules for other security mechanisms such as authentication mechanism, authorization mechanism, and enforces security policy. In order to efficiently describe the security policy for home network, we define a new language called an xHDL (eXensible Home security Description Language).

References

1. Han, J.-W.: Revitalization Policy of Home Network Industry, 22nd edn. Korea Information Science Society, vol. 9, 09 (2009)
2. Han, J.-W., Kim, D.-W., Joo, H.-I.: Considerations for Home Network Security Framework, 22nd edn. Korea Information Science Security, vol. 9, (2004.09)
3. Kim, G.-W., Kim, D.-W., Lee, J.-H., Hwang, J.-G., Han, J.-W.: Considerations on Security Model of Home Network. In: Proceedings of The 8th International Conference on Advanced Communication and Technology
4. Galtatu, M., Lioy, A., Mazzicchi, D.: Security Policy system: status and perspective. In: ICON 2000. Proceedings IEEE Internet Conference on Network 2000, Networking Trends and Challenges in the New Millennium, pp. 278–284. IEEE Computer, Soc, Los Alamitos (2000)
5. Sanchez, L.A., Condell, M.N.: Security Policy System, Internet Draft, Network Working Group, IETF (November 18, 1998)
6. Lee, Y.-k., Ju, H.-i., Park, J.-h., Han, J.-w.: User Authentication Mechanism Using Authentication Server in Home Network. In: Proceedings of the 8th International Conference on Advanced Communication and Technology
7. Lee, H.-k., Lee, Y.-k., Ju, H.-i., Han, J.-w.: User Authentication Mechanisms for Home Network using Home Server, TTAS.KO-12.0030
8. Abobo, B., et al.: Extensible Authentication Protocol (EAP), IETF RFC3748 (June 2004)
9. Fund, P.: EAP Tunneled TLS Authentication Protocol. IETF draft-funk-eap-ttls-v1-00 (February 2005)
10. Palekar, A.S., Salowey, D., Zhou, J., Zorn, H.: Protected EAP Protocol (PEAP) version 2. IETF draft-josefsson-pppext-eap-tls-eap-10 (2004)

Bogus Data Filtering in Sensor Networks*

Yong Ho Kim, Jong Hwan Park, Dong Hoon Lee, and Jongin Lim

Center for Information Security Technologies (CIST),
Korea University, Seoul, Korea
{optim,decartian,donghlee,jilim}@korea.ac.kr

Abstract. Recently, Zhang *et al.* proposed a location-based threshold-endorsement scheme (LTE) to thwart bogus data injection attacks. This scheme exhibits much greater filtering power than earlier symmetric schemes and results in enhanced energy savings. In this paper, we show that LTE has a significant vulnerability. We also propose an improved scheme that mitigates this weakness and thereby achieves the original claims of Zhang *et al.* without lessening LTE's remarkable filtering power.

Keywords: Security, bogus data injection attack, wireless sensor networks.

1 Introduction

Wireless sensor networks (WSNs) are well recognized as a new paradigm for future communication. Sensor networks consist of a large number of low-cost sensor nodes. Sensor nodes are deployed randomly in hostile areas where they can be exposed to the risk of physical attacks. For instance, adversaries can capture a sensor node to obtain secret information stored within its memory.

In this paper, we focus on bogus data injection attacks such that a compromised node injects bogus data into the network. This attack can not only cause bogus alarms but also the depletion of the already-constrained resources of intermediate nodes on any forwarding path to the sink. Therefore, it is important to filter bogus data reports as early as possible before they reach the sink. To lessen the propagation of bogus data, a few mechanisms [3,5] have been proposed, which rely on the collective decisions of multiple sensors for bogus data detection. However, these schemes have two disadvantages. First, to increase filtering power, each node must store more keys. Second, the mechanisms cannot filter bogus data when t or more sensor nodes are compromised, for a pre-defined threshold value t.

Recently, Zhang *et al.* proposed a location-based threshold-endorsement scheme [4] that can detect and drop injected bogus data reports, where a legitimate data report should be co-signed by t authenticated nodes. Their scheme

---

* "This research was supported by the MIC(Ministry of Information and Communication), Korea, under the ITRC(Information Technology Research Center) support program supervised by the IITA(Institute of Information Technology Advancement)" (IITA-2006-(C1090-0603-0025)).

T. Washio et al. (Eds.): PAKDD 2007 Workshops, LNAI 4819, pp. 629–635, 2007.

is a simplified adaptation of the ID-based threshold signature scheme [2]. When compared with the previous schemes [3,5], LTE features remarkable filtering power which can effectively save the cost of transmitting bogus data reports to the distant sink.

In this paper, we identify a weakness in LTE that seriously threaten its viability. To increase filtering power, LTE applies an asymmetric threshold signature technique where each node is assigned a share of a cell signature key. If an event occurs in any cell, a data aggregation point (AP) elected among the detecting nodes generates an authenticated threshold signature of the data report. However, after launching one normal run of the scheme, the AP will learn both a portion of the shares and the cell key. This must not be intended in the original design of LTE since the cell key must be secure even when the AP is compromised. In this paper, we propose an improved filtering method that mitigates the vulnerability while still inheriting LTE's remarkable filtering power.

2   Preliminaries

2.1   Bilinear Groups and Pairing

Let p, q be two large primes and $\mathbb{G}_1, \mathbb{G}_2$ denote two groups of prime order q, where $\mathbb{G}_1$, with an additive notation, denotes the group of points on an elliptic curve $\mathbb{E}/\mathbb{F}_p$; and $\mathbb{G}_2$, with a multiplicative notation, denotes a subgroup of the multiplicative group of a finite field $\mathbb{F}_{p^\alpha}^*$, for a suitable α. A pairing is a map $\hat{e} : \mathbb{G}_1 \times \mathbb{G}_1 \longrightarrow \mathbb{G}_2$ with the following properties.

1. Bilinearity: for all $P, Q \in \mathbb{G}_1$ and $a, b \in \mathbb{Z}_q^*$, we have $\hat{e}(aP, bQ) = \hat{e}(P, Q)^{ab}$.
2. Non-degenerancy: if P is a generator of $\mathbb{G}_1$, then $\hat{e}(P, P)$ is a generator of $\mathbb{G}_2$.
3. Computability: for all $P, Q \in \mathbb{G}_1$, there is an efficient algorithm to compute $\hat{e}(P, Q) \in \mathbb{G}_2$.

2.2   Location-Based Key Management Scheme

We briefly summarize a location-based key (LBK) management scheme which is a preliminary phase in both our scheme and LTE. First, before network deployment, a trusted authority (TA) performs the following operations.

1. TA constructs two groups $\mathbb{G}_1, \mathbb{G}_2$ and a map $\hat{e}$ as described above.
2. TA chooses two hash functions H and h, mapping arbitrary inputs to nonzero elements in $\mathbb{G}_1$ and to fixed-length outputs, respectively.
3. TA picks a random integer $\kappa \in \mathbb{Z}_q^*$ as the network master secret and sets $W_{pub} = \kappa W$, where W is a random generator of $\mathbb{G}_1$.
4. For each node A with identification information ID_A, TA calculates an ID-based key, $IK_A = \kappa H(ID_A)$.

Next, each node is preloaded with the public system parameters $(p, q, \mathbb{G}_1, \mathbb{G}_2, \hat{e}, H, h, W, W_{pub})$ and its ID-based key IK_A. Once sensor nodes are deployed, each node is furnished with its geographic location. Zhang *et al.* consider two sensor localization techniques using mobile robots or anchors [4]. After localization, each node A possesses its location l_A and LBK $LK_A = \kappa H(ID_A \| l_A)$.

Finally, two sensor nodes A and B exchange their location and identification, and derive a shared key:

$$K_{A,B} = \hat{e}(LK_A, H(ID_B \| l_B)) = \hat{e}(LK_B, H(ID_A \| l_A)) = K_{B,A}.$$

3   Brief Review of LTE

We briefly describe LTE of Zhang *et al.*. Refer to paper [4] for details.

3.1   Generation and Distribution of Cell Keys

A sensor field is divided into $M \times N$ square cells of equal side length r, and each cell is labeled by a pair of integers $\langle m, n \rangle$ for $1 \le m \le M$ and $1 \le n \le N$. Let $\mathcal{K}_{m,n}$ be the cell key of cell $\langle m, n \rangle$ which is computed as $\mathcal{K}_{m,n} = \kappa H(m \| n)$. The basic idea of LTE is to assign a share of $\mathcal{K}_{m,n}$ to each node in cell $\langle m, n \rangle$ by applying a secret-sharing method [4]. Let $ID_{m,n}^i$ denote the ith node with location $l_{m,n}^i$ in cell $\langle m, n \rangle$. The following two approaches are used to distribute cell key shares.

Range-Based Cell Key Distribution. In this approach, a mobile robot plays the important role of distributing shares of $\mathcal{K}_{m,n}$. It is assumed that the mobile robot is equipped with a $(t-1)$-degree polynomial $\mathcal{F}(x) = \sum_{j=1}^{t-1} F_j x^j$, with coefficients F_j randomly selected from $\mathbb{G}_1^*$. First, the mobile robot computes $\mathcal{K}_{m,n} = \kappa H(m \| n)$ and a set of authenticators $V_{m,n}^{\rightarrow} = \{v_{m,n}^{(j)} \mid 0 \le j \le t-1\}$, where $v_{m,n}^{(j)} = \hat{e}(F_j, W)$ and $F_0 = \mathcal{K}_{m,n}$. Next, the mobile robot calculates $\mathcal{K}_{m,n}^i = \mathcal{F}(ID_{m,n}^i \| l_{m,n}^i) + \mathcal{K}_{m,n} \in \mathbb{G}_1$, which is node $ID_{m,n}^i$'s share of $\mathcal{K}_{m,n}$. Finally, $\mathcal{K}_{m,n}^i$ and $V_{m,n}^{\rightarrow}$ are securely sent to node $ID_{m,n}^i$ using the shared key IK_A.

$\mathcal{K}_{m,n}$ can be reconstructed from any t shares, but is not reconstructible from any $(t-1)$ or fewer shares. Let Ω be a t-order subset of all the nodes in cell $\langle m, n \rangle$. We can check the computation as follows:

$$\mathcal{K}_{m,n} = \sum_{i \in \Omega} \lambda_i \mathcal{K}_{m,n}^i$$

where $\lambda_i = \Pi_{j \in \Omega \setminus \{i\}} (ID_{m,n}^j \| l_{m,n}^j) / ((ID_{m,n}^j \| l_{m,n}^j - ID_{m,n}^i \| l_{m,n}^i)$. Note that there is a trade-off between resilience to node compromise and node density with respect to the choice of t.

Range-Free Cell Key Distribution. In this approach, each node is preloaded with the polynomial $\mathcal{F}(x)$ as well as the master secret κ. Once generating its LBK, node $ID_{m,n}^i$ uses κ to first derive $\mathcal{K}_{m,n}$ and then its share $\mathcal{K}_{m,n}^i$. It also computes the authenticator vector $V_{m,n}^{\rightarrow}$. Upon finishing all these operations, it should securely erase κ, $\mathcal{F}(x)$, and all the complete cell keys from its memory.

3.2 Report Generation and En-Route Filtering

Assume that a special event that needs to be reported occurs in cell $\langle m, n \rangle$ and is detected by $s \geq t$ nodes. By local interactions, the detecting nodes can reach an assent on a final report, denoted by Λ.

The nodes first elect a data aggregation point (AP) among themselves. To obtain a threshold-endorsement of Λ, the AP chooses a random $\alpha \in \mathbb{Z}_q^*$, computes $\theta = \hat{e}(W, W)^{\alpha}$, and broadcasts θ to the other nodes. On receipt of θ, each node $ID_{m,n}^i$ detecting the event endorses the report Λ by computing $U_{m,n}^i = h(\Lambda||\theta)\mathcal{K}_{m,n}^i$. It then sends $U_{m,n}^i$ to the AP using the pairwise key shared with the AP. After receiving t or more such endorsements, the AP randomly selects t among the endorsers, denoted by a set notation Ω which may include itself. Then, the AP calculates $U_{m,n} = \sum_{i \in \Omega} \lambda_i U_{m,n}^i = h(\Lambda||\theta)\mathcal{K}_{m,n}$ and $\Upsilon_{m,n} = U_{m,n} + \alpha W$. The threshold-endorsement of Λ is $(\Upsilon_{m,n}, h(\Lambda||\theta))$. The final report $\langle \Lambda, \Upsilon_{m,n}, h(\Lambda||\theta) \rangle$ is sent out by the AP to the sink.

When an intermediate node receives the report, it is verified by some system-wide parameter, p_s, also called the *sampling probability* [4].

3.3 Security Analysis

To tolerate compromise by bogus data injection attacks, Zhang *et al.* [4] applied the secret-sharing method to filtering bogus data. However, after launching one normal run of LTE, a malicious AP (i.e., insider attack) can learn the cell key as well as a few shares. In performing the threshold-endorsement of data reports, the AP obtains $U_{m,n}^i = h(\Lambda||\theta)\mathcal{K}_{m,n}^i$ of endorsers and calculates $U_{m,n} = \sum_{i \in \Omega} \lambda_i U_{m,n}^i = h(\Lambda||\theta)\mathcal{K}_{m,n}$. Since $h(\Lambda||\theta)$ and its inverse $h(\Lambda||\theta)^{-1} \in \mathbb{Z}_q^*$ are easily computed, the AP can calculate the cell key and t shares as $\mathcal{K}_{m,n} = h(\Lambda||\theta)^{-1}U_{m,n}$ and $\mathcal{K}_{m,n}^i = h(\Lambda||\theta)^{-1}U_{m,n}^i$, respectively. Next, the adversary can compute all coefficients of the $(t - 1)$-degree polynomial $\mathcal{F}(x)$ by solving t-variable linear equations. Additionally, if the adversary compromises a single node in another cell $\langle m', n' \rangle$, using only an additional share $\mathcal{K}_{m',n'}^i$, the adversary can calculate out the new cell key as $\mathcal{K}_{m',n'} = \mathcal{K}_{m',n'}^i - \mathcal{F}(ID_{m',n'}^i||l_{m',n'}^i)$. This must not be intended in the original design of LTE since the scheme must achieves an independent level of security for each cell.

4 Improvement

In LTE, the sensor field is divided into square cells of equal size to achieve greater compromise-tolerance. However, if the AP is compromised in one cell C and a single node is compromised in any other cell C', an adversary can fake data reports from C'. Since LTE uses only one polynomial $\mathcal{F}(x) = \sum_{j=1}^{t-1} F_j x^j$, the compromise of a cell directly affects the security of other cells. To deal with this weakness, first of all, distinct cell-polynomials should be used, to achieve an independent level of security for each cell. Also, report generation phase should be fixed, to lessen the propagation of the compromised AP.

4.1   Generation and Distribution of Cell Keys

For each cell, we define cell-polynomials $\mathcal{F}_{m,n}(x) = \sum_{j=1}^{t-1} H(F_j||m||n)x^j$ and cell-authenticator vectors $\overrightarrow{V_{m,n}} = \{v_{m,n}^{(j)}| \ 0 \leq j \leq t-1\}$, where $v_{m,n}^{(j)} = \hat{e}(H(F_j||m||n), W)$ and $v_{m,n}^{(0)} = \hat{e}(\mathcal{K}_{m,n}, W)$. After determining each node's cell $\langle m, n \rangle$, each cell-polynomial $\mathcal{F}_{m,n}(x)$ is induced from a polynomial $F(x)$. The generation and distribution of cell keys is the same as that of LTE (subsection 3.1) except cell-polynomials $\mathcal{F}_{m,n}(x)$ and cell-authenticator vectors $\overrightarrow{V_{m,n}}$ are different for each cell.

4.2   Report Generation and En-Route Filtering

Like LTE, when detecting an event, they elect one node among themselves to be a data aggregation point (AP). To obtain a threshold-endorsement of Λ, each detecting node $ID_{m,n}^i$ chooses a random $\alpha_i \in \mathbb{Z}_q^*$, computes $\theta_i = \hat{e}(W, W)^{\alpha_i}$, and sends it to the AP. After receiving t or more such θ_i's, the AP randomly selects t among the endorsers, denoted by a set notation Ω which may include itself. Next, the AP computes $\theta = \prod_{i \in \Omega} \theta_i^{\lambda_i}$ and broadcasts θ to the other nodes. On receipt of θ, each node $ID_{m,n}^i$ endorses the report Λ by computing $\Upsilon_{m,n}^i = h(\Lambda||\theta)\mathcal{K}_{m,n}^i + \alpha_i W$. It then sends $\Upsilon_{m,n}^i$ to the AP using the pairwise key shared with the AP. Next, the AP calculates $\Upsilon_{m,n} = \sum_{i \in \Omega} \lambda_i \Upsilon_{m,n}^i$. The threshold-endorsement of Λ is $(\Upsilon_{m,n}, h(\Lambda||\theta))$. The final report $\langle \Lambda, \Upsilon_{m,n}, h(\Lambda||\theta) \rangle$ is sent out by the AP to the sink.

Since some endorsers may be compromised, once deriving $\Upsilon_{m,n}$, the AP verifies its authenticity by checking if

$$\hat{e}(\Upsilon_{m,n}, W) = (v_{m,n}^{(0)})^{h(\Lambda||\theta)} \cdot \theta.$$

The verification works since

$$\hat{e}(\Upsilon_{m,n}, W) = \hat{e}(\sum_{i \in \Omega} \lambda_i \Upsilon_{m,n}^i, W) = \hat{e}(\sum_{i \in \Omega} (\lambda_i \mathcal{K}_{m,n}^i h(\Lambda||\theta) + \lambda_i \alpha_i W), W)$$

$$= \hat{e}(h(\Lambda||\theta)\mathcal{K}_{m,n} + \sum_{i \in \Omega} \lambda_i \alpha_i W, W)$$

$$= \hat{e}(h(\Lambda||\theta)\mathcal{K}_{m,n}, W) \cdot \hat{e}(\sum_{i \in \Omega} \lambda_i \alpha_i W, W)$$

$$= \hat{e}(\mathcal{K}_{m,n}, W)^{h(\Lambda||\theta)} \cdot \prod_{i \in \Omega} \hat{e}(W, W)^{\alpha_i \lambda_i}$$

$$= (v_{m,n}^{(0)})^{h(\Lambda||\theta)} \cdot \prod_{i \in \Omega} \theta_i^{\lambda_i} = (v_{m,n}^{(0)})^{h(\Lambda||\theta)} \cdot \theta.$$

If the checks succeed, the AP believes that the final report is valid and sends it to the sink. Otherwise, to find compromised node, the AP proceeds to verify each received $\Upsilon_{m,n}^i$ by checking if

$$\hat{e}(\Upsilon_{m,n}^i, W) = \prod_{j=0}^{t-1} (v_{m,n}^{(j)})^{(ID_{m,n}^i||l_{m,n}^i)^j h(\Lambda||\theta)} \cdot \theta_i.$$

When an intermediate node receives the final report $\langle \Lambda, \Upsilon_{m,n}, h(\Lambda||\theta) \rangle$, the node should verify the report with probability p_s as follows:

The node first computes $h(\Lambda||\theta')$, where

$$\theta' = \hat{e}(\Upsilon_{m,n}, W) \cdot \hat{e}(H(m||n), -W_{pub})^{h(\Lambda||\theta)}.$$

If the report is valid, we will have

$$\theta' = \hat{e}(\Upsilon_{m,n}, W) \cdot \hat{e}(H(m||n), W_{pub})^{-h(\Lambda||\theta)}$$
$$= \hat{e}(\mathcal{K}_{m,n}, W)^{h(\Lambda||\theta)} \cdot \prod_{i \in \Omega} \hat{e}(W, W)^{\alpha_i \lambda_i} \cdot \hat{e}(H(m||n), \kappa W)^{-h(\Lambda||\theta)}$$
$$= \hat{e}(\mathcal{K}_{m,n}, W)^{h(\Lambda||\theta)} \cdot \prod_{i \in \Omega} \theta_i^{\lambda_i} \cdot \hat{e}(\mathcal{K}_{m,n}, W)^{-h(\Lambda||\theta)} = \theta.$$

If $h(\Lambda||\theta) = h(\Lambda||\theta')$, the node passes the report to the next hop. Otherwise, the report is dropped.

4.3   Security Analysis

The impact of a malicious AP must be negligible. However, after launching one normal run of LTE, the adversary can obtain the cell key and at least t shares, i.e, *cell compromise* [4]. This is because this report generation phase of LTE is insecurity. Note that we fix that of the proposed scheme. Since each detecting node $ID_{m,n}^i$ chooses a random α_i which must be secure, the AP cannot know the α_i's of other participating nodes. Although the AP know $\Upsilon_{m,n}^i = h(\Lambda||\theta)\mathcal{K}_{m,n}^i + \alpha_i W$ and $\Upsilon_{m,n} = h(\Lambda||\theta)\mathcal{K}_{m,n} + \sum_{i \in \Omega} \lambda_i \alpha_i W$, it cannot obtain any knowledge of $\mathcal{K}_{m,n}^i$ or $\mathcal{K}_{m,n}$ without α_i.

In LTE, if an adversary compromises t shares of a cell, the adversary can compute all cell keys and all coefficients of $\mathcal{F}(x)$ by solving t-variable linear equations. However, in our scheme, the adversary cannot compute any cell-polynomial of other cell since it is difficult to induce $H(F_j||m'||n')$ from $H(F_j||m||n)$. In other words, a compromised cell-polynomial of a cell doesn't affect the security of another cell.

4.4   Overhead Analysis

Our scheme and LTE have similar computation and communication costs. In general, since a point in $\mathbb{G}_1$ is 170 bits [1], the size of an aggregate signature is 340 bits. Also, if $t = 5$ and the size of a node ID is 16 bits, the size of Ω will be 80 bits. Therefore, the total communication overhead will be 420 bits, less than half the size of an RSA signature. To check the validation of a report, an intermediate node needs to compute two pairings and one pairing exponentiation like LTE. In our scheme, a legitimate data report is verified with t authenticated nodes' signature keys to provide our scheme with the similar performance to the t-threshold signature scheme of LTE. When compared to LTE, much stronger security and resilience compensates for a slight additional overhead.

5   Conclusion

In this paper, we have shown that Zhang *et al.*'s bogus data filtering scheme has a significant vulnerability. LTE applies an asymmetric threshold signature technique to increase filtering power, however, the AP will obtain the cell key and the polynomial $\mathcal{F}(x)$. Both of them must be secure even when the AP is compromised. In this paper, we have presented an improved scheme which fixes the security flaws of LTE without lessening LTE's remarkable filtering power.

References

1. Boneh, D., Lynn, B., Shacham, H.: Short signatures from the Weil pairing. In: Boyd, C. (ed.) ASIACRYPT 2001. LNCS, vol. 2248, pp. 514–532. Springer, Heidelberg (2001)
2. Baek, J., Zheng, Y.: Identity-based threshold signature from the bilinear pairings. In: Proc. Int. Conf. Inf. Tech.: Coding Comput., pp. 124–128 (April 2004)
3. Ye, F., Luo, H., Lu, S., Zhang, L.: Statistical en-route filtering of injected false data in sensor networks. IEEE JSAC, Special Issue on Self-Orgazing Distributed Collaborative Sensor Networks 23(4), 839–850 (2005)
4. Zhang, Y., Liu, W., Lou, W., Fang, Y.: Location-based compromise-tolerant security mechanisms for wireless sensor networks. IEEE JSAC, Special Issue on Security in Wireless Ad Hoc Networks 24(2), 247–260 (2006)
5. Zhu, S., Setia, S., Jajodia, S., Ning, P.: An interleaved hop-by-hop authentication scheme for filtering of injected false data in sensor networks. In: Proc. IEEE Symp. Security Privacy, pp. 259–271 (May 2004)

Streaming Media Securely over Multipath Multihop Wireless Network[*]

Binod Vaidya[1], SangDuck Lee[2], Eung-Kon Kim[3], and SeungJo Han[2,**]

[1] Dept. of Electronics & Computer Eng., Tribhuvan Univ., Nepal
bnvaidya@gmail.com
[2] Dept. of Information & Communication Eng., Chosun Univ., Korea
dandylsd@hanmail.net, sjbhan@chosun.ac.kr
[3] Dept. of Computer Science, Sunchon National Univ., Korea
kek@sunchon.ac.kr

Abstract. With the rapid growth and popularity of wireless LANs (WLANs), the development of ubiquitous services is in demand. Mobile Ad hoc Networks (MANETs) are very attractive for many environments, such as home networking, emergency response scenarios etc. However, real-time voice communication is a critical application for many of these network scenarios. In this paper, we put forward a framework for secure audio streaming over multihop wireless network. And we propose an efficient multipath routing for MANET, while using scalable speech coding and selective encryption in secure RTP. With the simulation results, the performance of the proposed scheme is evaluated.

1 Introduction

With the rapid growth and popularity of wireless LANs (WLANs), the development of ubiquitous services is in demand. So nowadays there is a growing demand for support of media streaming over WLANs.

A mobile ad hoc network (MANET) is a system of mobile nodes that dynamically self-organize in arbitrary and temporary network topologies allowing nodes to communicate each other without any pre-existing communication infrastructure. MANETs are very attractive for many environments, such as home networking, conferencing and emergency response scenarios. However, a real-time voice communication is a critical application for many of these network scenarios. Media streaming over MANET is quite challenging because of dynamic network topology, limited wireless bandwidth, and high bit error rate of wireless links. This paper presents effective use of multipath routing and selective encryption in secure RTP for secure media streaming over wireless mobile ad hoc network.

---

[*] This work was partially supported by University Research Program of Ministry of Information & Communication in Republic of Korea.
[**] Corresponding author.

T. Washio et al. (Eds.): PAKDD 2007 Workshops, LNAI 4819, pp. 636–643, 2007.

2 Related Works

We briefly present related works to multipath routing [1]. Some multipath routing protocols [2, 3, 4] have been proposed for wireless ad hoc networks. Multipath routing protocols based on DSR [5] are MSR [4] and SMR [6]. Whereas Well-known multipath routing protocols based on AODV [7] are NDMR [2], AODVM [3], AODV-BR [8], and AOMDV [9]. SMORT [17] uses the idea of fail-safe alternate path to determine multiple paths.

We also address the techniques for streaming through multiple paths. Basically we can stream the whole flow through a single path as in RMPSR [10] or divide the content into multiple minor flows and stream them through the available paths as in MDSR [11].

Several studies of selective encryption for video and image compression have been performed [12, 13], and some on selective encryption of coded speech have been presented [14].

3 Framework for Secure Multipath Media Streaming over MANET

Voice communication in MANET is challenging due to multihop routing and high bit error rates. Communication might be lost if packet loss rate is high or a communication link is considered broken. One of the methods to improve reliability of transmission over MANET is to use path diversity, i.e. send data simultaneously through multiple paths. As the probability of all the paths breaking down simultaneously is low, the probability of packet loss is reduced. Fig 1 shows possible multiple path connections between source node S and destination node D. Another issue is the need for privacy and security in wireless transmission.

In this paper, we present a framework for secure audio streaming over multipath multihop wireless network. For this purpose, we propose a multipath routing protocol for MANET, which can be efficiently used for media streaming. In order to address quality of service (QoS) requirements in MANETs, we have considered scalable audio coding technique [15] since this scheme shows promise for adaptive and robust real-time media streaming over lossy networks. And to avoid high processing requirements in handheld devices such as mobile phones, selective encryption technique [16] in secure RTP is considered.

3.1 Multipath Routing Protocol

The proposed multipath routing protocol for MANET is a modification of single path AODV protocol. It is basically intended for highly dynamic ad hoc networks in which communication failures occur frequently and designed to compute node-disjoint paths as well as fail-safe paths. Fail-safe [17] is a path between source and destination if it bypasses at least one intermediate node on the primary path. Fail-safe paths are different from node-disjoint and link-disjoint paths, in the sense that fail-safe route paths can have both nodes and links in common.

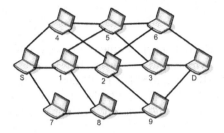

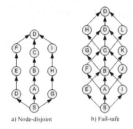

Fig. 1. Multipath MANET

Fig. 2. Type of multipaths

Node-disjoint paths and fail-safe paths are shown in Fig 2. It can be seen that combination of node-disjoint and fail-safe paths allows computation of more alternative paths than in node-disjoint or link-disjoint multipath routings.

As in AODV, there are three types of control messages - Route Request (RREQ), Route Reply (RREP) and Route Error (RERR). In this scheme, a routing table entry has information of path accumulation list and path type. The proposed protocol has two basic phases –route discovery, and route maintenance.

A source node S initiates route discovery process, when it wants to communicate to a destination D, for which it does not have a valid route. The source node S inserts last known destination sequence number, address of the destination, RREQ ID, its own address and sequence number into a RREQ packet and broadcasts it. RREQ packet structure of the proposed scheme is same as that of AODV, except presence of path accumulation list of the route path. The source node appends own address to the route path in RREQ message. Further, when RREQ is forward by intermediate nodes in the network, each node appends its address to it.

When a node receives a RREQ, it checks to determine whether it has received a RREQ with the same source and RREQ ID. If a node receives a RREQ for the first time, it searches for a reverse route to the source. If no reverse route, then create one. An intermediate node receiving the RREQ, replies by sending a RREP if it has a route to the destination. In the proposed scheme, intermediate nodes forward duplicate RREQs that came from at most two different neighbors. This is essential to discover a number of alternate route paths.

In this scheme, the destination is responsible for selecting multiple route paths. When receiving the first RREQ, the destination records the route paths of RREQ and the destination will set the path type to one. Then after copying route paths of RREQ to a RREP packet, the destination node sends it to source node via the route paths of it. Hence the intermediate nodes can forward this packet using the route paths of RREP.

If only source node and destination node are the same between them, the path is node-disjoint with the primary path the destination will set the path type to three. If at least one of intermediate nodes in the route paths in the routing table is different from all of nodes in the route paths of the RREQ, the route is a partially disjoint path, which is defined as fail-safe path and the destination will set the path type to two. If any alternate path is received, the destination

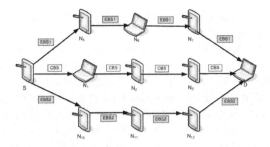

Fig. 3. Traffic distribution over multipath MANET

sends the RREP to the source along the route path of RREP. RREPs follow the reverse paths to reach the source node. After storing the routing information, intermediate nodes relay RREP packet to the next hop according to route path field toward the source.

In route maintenance, normally route links in wireless ad hoc networks are broken frequently due to the mobility of nodes, congestion and packet collisions. The proposed scheme is capable of recovering broken routes immediately. When a node fails to deliver the data packets to the next hop of the route by receiving a link layer feedback from link layer or receives a route error (RERR) packet, it removes entries in its route table that uses the broken link and looks up its routing table if there is another entry for the destination. If it has another entry for the destination, data packets therefore can be delivered through the alternate route and are not dropped when route breaks occur. If it has no another entry for the destination, it sends RERR packet to the upstream node. When the source has no entry for the destination and the session is still active, it would initiate a new route discovery.

In our scheme, scalable audio coding is used for media streaming over multipath MANET. Intially, the source node begins to send the core bit stream (CBS) on the primary path and the enhancement bit streams (EBSs) on the node-disjoint paths, which is depicted in Fig 3. Since the primary path and the node-disjoint path are not correlated, source node uses node-disjoint paths to provide load balancing. When forwarding paths break, nodes receiving core bit stream or enhancement bit stream may use different paths in the routing table to forward packets. Generally, a multihop path is up or down for random periods of time, leading to bursty packet loss. A core layer packet loss is likely to be experiencing a packet loss burst. So the proposed scheme finds an alternate fail-safe path for each node on the primary path as it has higher packet delivery rate.

3.2 Scalable Audio Coding

Scalable audio coding consists of a minimum rate bit stream that provides acceptable coded audio quality, along with one or more enhancement bit streams, which when combined with a lower rate coded bit stream, provide improved speech quality. The standards for scalable audio coding are G.727 [18], and the MPEG-4 audio coding [19].

MPEG-4 Natural Audio Coding Tool Set [19] provides a generic coding framework for a wide range of applications with speech signals. Its bitrate coverage spans from as low as 2 kbit/s to 23.4 kbit/s. It contains two algorithms: HVXC (Harmonic vector excitation coding) and CELP (Code excited linear predictive coding). HVXC is used at a low bitrate of 2 or 4 kbit/s. Higher bitrates than 4 kbit/s in addition to 3.85 kbit/s are covered by CELP. The bit rate scalability of any of these core layer bit rates is possible in increments of 2 kbps, with up to three enhancement layers.

In this framework, we have considered the lowest MPEG-4 CELP rate, 3.85 kbps, because this rate offers the greatest possible efficiency in terms of reduced bandwidth utilization and transmitted power conservation. The case of a 3.85 Kbps core layer and two 2 Kbps enhancement layers, for a total of 7.85 Kbps, has been evaluated.

3.3 Selective Encryption in SRTP

The authors [16] have suggested some enhancements to Secure RTP (SRTP) protocol to enable SRTP to support selective encryption. Basic requirement in selective encryption is to encrypt some media packets and not encrypt others.

SRTP uses MKI (Master Key Index) to indicate index of the master key in use. And current definition of SRTP defines security context at both the ends of the communication and also creates master key, which is used to create session keys for encryption and integrity. But it does not include encryption (cipher) algorithm to be part of MKI, so basically while establishing security context at both ends of the communication, cipher algorithm is negotiated and multiple master keys are established and all the keys uses same cipher suite. SRTP also allows NULL encryption to be supported as valid cipher algorithm.

If SRTP definition is changed by linking encryption algorithm to the master key and each master key can hold its own cipher algorithms, SRTP can support selective encryption without changing protocol syntax. In order to support selective encryption between two endpoints, security context establishment shall establish at least two master keys and one of the master key carries a cipher algorithm and other one uses NULL Cipher. During RTP packet processing by SRTP stack, if encryption for that packet is needed, MKI value will be set to the one that has cipher algorithm attached and if encryption is not needed, MKI value will be set to one that has NULL Cipher.

4   Performance Evaluations

In order to evaluate the performance of the proposed framework for secure media streaming over multipath MANET, we have designed experimental model and simulated using OPNET Modeler [20].

4.1 Simulation Environment

In the simulations, the MANET consists of sixteen mobile nodes which are located inside a 600m x 600m region. Each node is randomly placed in the region initially. We consider a popular random waypoint mobility model. We have used a pause time of 1.0s for all the experiments. The speed of the nodes varies from 1m/s to 10m/s. We use the IEEE 802.11 protocol in the MAC layer working in the DCF mode. The channel has a bandwidth of 1Mb/s. The transmission range is 250 m. UDP is used as transport protocol.

Among these nodes, one is randomly chosen as the streaming source with MPEG-4 speech codec and another node is chosen as the destination. Five UDP traffic flows are introduced as background traffics. Each of these flows has the traffic rate of four packets per second. The size of data payload was 512 bytes. The some, destination and the duration of these background flows are set random. Each of nodes has a queue size of 10 packets.

For the experimental purpose, three scenarios have been considered –a secure audio streaming framework using single-path AODV routing, node-disjoint (ND) multipath routing and proposed multipath routing. In the simulation, two performance metrics were considered: packet loss rate which is the fraction of transmitted packets from the source that are not received at the destination; and end-to-end packet delay which is the difference between the packet transmission time at the source and its reception time at the destination.

4.2 Simulation Results

In order to analyze the simulation results for the framework for securely streaming audio over multipath MANET, we compare performance of the proposed multipath routing with ND multipath routing and AODV in terms of packet loss rate and end-to-end delay with respect to max speed.

Fig 4 shows the average loss rate for three scenarios at different mobility. As the velocity of nodes ncreases, the probability of link failure increases and

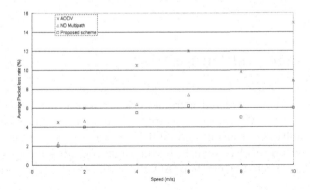

Fig. 4. Packet loss rate

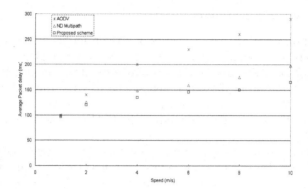

Fig. 5. End-to-end Packet Delay

hence the number of packet drops also increases. It can be seen that the mean packet loss rate in the both multipath routing schemes are reduced than in case of AODV. The proposed scheme has slightly lower packet loss rate than ND multipath scheme whereas much lower loss rate than AODV. With speed of 6m/s, the average loss rates for the proposed scheme and ND multipath scheme are about 6.1%, and 7.5% respectively while that of AODV is about 12%.

Fig 5 illustrates the average packet delay as a function of node speed. For all the scenarios, it can be seen that there is increase in average delay with the increase of node speed. Both multipath routing schemes provide smaller packet delay than AODV. This is because multipath routing protocols have alternate paths and need smaller route discovery time. However, when comparing two multipath routing schemes, it can be seen that the average packet delay in the proposed scheme is smaller than that of ND multipath scheme. This is because in the proposed scheme, frequency of route discovery process is lesser that that of ND multipath routing. The delays of both multipath schemes are gradually increased after speed reaches 4m/s, while delay in AODV increases quickly as velocity increases.

5   Conclusions and Future Work

In this paper we have depicted a framework for secure media streaming over multipath MANET. We have simulated above framework using proposed scheme, ND multipath routing and AODV. It can be seen that performance of the proposed protocol is slightly better than node-disjoint multipath routing scheme and much better than AODV in terms of packet loss rate and end-to-end delay.

In the future work, we will investigate multimedia delivery over multipath hybrid MANET using scalable audio coding technique.

References

1. Mueller, S., et al.: Multipath Routing in Mobile Ad Hoc Networks: Issues and Challenges. In: Calzarossa, M.C., Gelenbe, E. (eds.) Performance Tools and Applications to Networked Systems. LNCS, vol. 2965, pp. 209–234. Springer, Heidelberg (2004)
2. Li, X., Cuthbert, L.: Stable node-disjoint multipath routing with low overhead in mobile ad hoc networks. In: IEEE MASCOTS 2004, Netherlands, pp. 184–191 (2004)
3. Ye, Z., et al.: A routing framework for providing robustness to node failures in mobile ad hoc networks. Elsevier Ad Hoc Networks Journal 2(1), 87–107 (2004)
4. Wang, L., et al.: Multipath source routing in wireless ad hoc networks. In: CCECE 2000, vol. 1, pp. 479–483 (2000)
5. Johnson, D., Maltz, D.: Dynamic source routing in ad hoc wireless networks. Mobile Computing 353, 153–181 (1996)
6. Lee, S.J., Gerla, M.: Split multipath routing with maximally disjoint paths in ad hoc networks. IEEE ICC 2001 3, 867–871 (2001)
7. Perkins, C.E., et al.: Ad hoc on-demand distance vector (AODV) routing. In: IETF RFC, vol. 3561 (2003)
8. Lee, S.J., Gerla, M.: AODV-BR: Backup routing in ad hoc networks. IEEE WCNC 2000 3, 1311–1316 (2000)
9. Marina, M.K., Das, S.R.: Ad hoc on-demand multipath distance vector routing. Wiley Wireless Communications and Mobile Computing 6(7), 969–988 (2006)
10. Wei, W., Zakhor, A.: Robust Multipath Source Routing Protocol (RMPSR) for Video Communication over Wireless Ad Hoc Networks. In: ICME 2004, Taiwan, pp. 1379–1382 (2004)
11. Mao, S., et al.: Video transport over ad hoc networks: multistream coding with multipath transport. IEEE Journal on Selected Areas in Communications 21(10), 1721–1737 (2003)
12. Alattar, A.M., Al-Regib, G.I.: Evaluation of selective encryption techniques for secure transmission of MPEG-compressed bit-streams. IEEE ISCAS 1999, 340–343 (1999)
13. Lookabaugh, T., et al.: Selective encryption and MPEG-2. ACM Multimedia (2003)
14. Gibson, J.D., et al.: Selective Encryption and Scalable Speech Coding for Voice Communications over Multi-Hop Wireless Links. IEEE MILCOM 2004 2, 792–798 (2004)
15. Dong, H., et al.: SNR and bandwidth scalable speech coding. IEEE ISCAS 2002 2, 859–862 (2002)
16. Pallapothu, S., Mahajan, S.: Selective Encryption Support in SRTP. draft-smahajan-srtp-selective-encryption-00.txt, work in progress (2006)
17. Reddy, L.R., Raghavan, S.V.: SMORT: Scalable multipath on-demand routing for mobile ad hoc networks. Elsevier Ad hoc Networks 5(2), 162–188 (2007)
18. ITU-T, 5-, 4-, 3-, and 2-bit/sample embedded adaptive differential pulse code modulation (ADPCM) (December 1990)
19. ISO/IEC JTC1 SC29/WG11, ISO/IEC FDIS 14496-3, Subparts 1, 2, 3, Coding of Audio-Visual Objects Part 3: Audio, ISO/IEC JTC1 SC29/WG11 N2503 (October 1998)
20. OPNET Modeler Simulation Software, http://www.opnet.com

Flexible Selection of Wavelet Coefficients Based on the Estimation Error of Predefined Queries

Jaehoon Kim and Seog Park

Department of Computer Science, Sogang University
1-1 Shinsu-Dong Mapo-Gu Seoul Korea 121-742
{chris3,spark}@sogang.ac.kr

Abstract. In this paper, we introduce a data stream reduction method using lossy wavelets compression. The lossy compression means that compressed data carry as much information about the original data stream as possible while the original data size remarkably reduced. We think that wavelets technique should be an efficient method for such lossy compression. Especially we consider storing a plenty of past data stream into stable storage (flash memory or micro HDD) rather than keeping only recent streaming data allowable in memory, because data stream mining and tracking of past data stream are often required. In the general method using wavelets, a specific amount of streaming data from a sensor is periodically compressed into fixed size and the fixed amount of compressed data is stored into stable storage. However, differently from the general method, our method flexibly adjusts the compressing size based on a heuristic criterion. Experimental results with some real stream data show that wavelets technique is useful in data stream reduction and our flexible approach has lower estimation error than the general fixed approach.

1 Introduction

Recently a great deal of attention has been driven toward processing data stream in mobile computing, ubiquitous computing, and sensor network. For example, mobile healthcare is to use mobile device equipped with biosensors and advanced wireless communication technology (3G/4G) to analyze the chronic conditions of certain disease and detect health emergencies [9]. A traffic control system with smart sensors (or called motes [10]) at major crossroads enable us to monitor and analyze traffic data in real time. These systems include the following queries: "Return the abnormal traffic (or heartbeat), that is, those above a threshold value", "Find the time when inputted data pattern is similar to a special pattern", or "Every minute retrieve the time, the total traffic of S1, and the total traffic of S2 over the last ten minutes if the total traffic from a sensor S1 is larger than the total traffic from a sensor S2". These queries focus on recent streaming data allowable in memory.

The infinite extent of streaming data makes it necessary to periodically store the past data stream in stable storage, and queries on this past data are also important.

T. Washio et al. (Eds.): PAKDD 2007 Workshops, LNAI 4819, pp. 644–655, 2007.

These queries include database queries over the past data stream, together with analysis and data mining: "Return the name and traffic of the crossroad which had the largest traffic for the past two months", "Return the current heartbeat if it has very different pattern against the past heartbeat", or "Generate strong association rules from the crossroads having heavy traffic".

However, it is impractical to store all the data because stable storage still has restricted and low capacity in mobile device and motes. Therefore, the data reduction method such as wavelets [5, 7], histograms, and sampling can be considered to store much more data. That is, streaming data from a sensor can be stored within a restricted storage space after being compressed as compactly as possible, and can be queried after being decompressed. However, the characteristics of streaming data, which are continuous, endless, and flows quickly, remind us of the following problem. As the end of streaming data to be summarized is not defined, it is necessary to periodically summarize the local data up to a specific point. However, if we assume that a single global summarization of all the data input exists, the sequence of local summarizations should be less accurate than the global summarization. For example, consider a continuous 100 megabyte streaming data, although streaming data are endless. It is more accurate to summarize the total 100 megabyte data into a global 20 megabyte histogram than to summarize each 10 megabyte data section into a local 2 megabyte histogram. Thus, it is necessary to efficiently perform the periodic local summarization to make it close to the global summarization.

In this paper, we propose a framework to flexibly adjust storage allocation (or *storage limit m* called by us) for each local time section, especially using wavelets summarization. This approach minimizes the overall estimation error by adjusting the storage limit m flexibly depending on the local estimation error of each time section. Some experiments have shown that our flexible approach has lower overall estimation error than the existing fixed approach.

This paper is organized as follows. The concept of periodic summarization of past data stream, and the problem associated with this, are introduced in Section 2. Section 3 describes our method that uses the local estimation error to solve the discussed problem. Section 4 presents some experimental results showing the correctness of our method, and Section 5 reviews other research related to this paper. Finally the conclusions of this research are presented in Section 6.

2 Periodic Data Stream Reduction Using Wavelets

2.1 Wavelets Summarization

Wavelets [5, 7] can be a good tool for summarizing the discrete changes of streaming data over time. This subsection discusses how wavelets are applied to streaming data summarization. Although linear wavelets have better estimation accuracy, we will focus on the Harr wavelets, which are the simplest and easiest for implementation, for the purposes of exposition in this paper.

Suppose the data distribution from a sensor over time: {(0, 3), (2, 6), (3, 2)}, where it is assumed that there is a time interval between streaming data points being generated from the sensor. Assuming this time interval is one second, the values 3, 6, and 2 are each sensed at the times of 0, 2, and 3 second. Additionally, if missing time samples are represented by the preceding data value, we can easily get the extended values over the entire discrete time domain: {(0, 3), (1, 3), (2, 6), (3, 2)}. Such data vacancies can arise from partial data loss due to transmission error, dropped data due to a high streaming data rate [8], or the non-forwarding of sensed data that are identical to preceding samples (used to reduce network traffic).

First, we perform a wavelet transform (also called a wavelet decomposition) on the one-dimensional "signal" of the extended values: $S = [3, 3, 6, 2]$. The Harr wavelet decomposition is as follows. We first average the value pairs to obtain the new lower resolution signal with values [3, 4], i.e., $(3 + 3) / 2 = 3$ and $(6 + 2) / 2 = 4$. To recover the original signal from the two averaged values, we must also store some detail coefficients (the pair-wise differences of the original values), i.e., $3 - 3 = 0$ and $2 - 6 = -4$. By repeating this process recursively on the averages, we obtain the full decomposition. It is easy to see that the original values can be recovered from the average and differences as in Figure 1.

Resolution	Averages	Detail Coefficients
4	[3, 3, 6, 2]	
2	[3, 4]	[0, -4]
1	[3.5]	[1]

$\hat{S} = [3.5, 1, 0, -4]$
$S(0) = \hat{S}(0) - 1/2\, \hat{S}(1) - 1/2\, \hat{S}(2)$
$S(1) = \hat{S}(0) - 1/2\, \hat{S}(1) + 1/2\, \hat{S}(2)$
$S(2) = \hat{S}(0) + 1/2\, \hat{S}(1) \qquad\quad - 1/2\, \hat{S}(3)$
$S(3) = \hat{S}(0) + 1/2\, \hat{S}(1) \qquad\quad + 1/2\, \hat{S}(3)$

Fig. 1. Harr wavelets summarization

Next, we perform thresholding on the detail coefficients. One advantage of the wavelet transformation is that, in many cases, a large number of the detail coefficients turn out to be very small in magnitude, so truncating these small coefficients from the representation only introduces small errors in the reconstructed signal. We can approximate the original data distribution effectively by retaining only the most significant m coefficients from all N coefficients using the thresholding method.

Matias, Vitter, & Wang [5] discussed some thresholding methods. The first step in the thresholding is to weight the coefficients in a certain way. In particular, for the Harr basis, normalization is performed by dividing the wavelet coefficients $\hat{S}(2^j)$, ..., $\hat{S}(2^{j+1} - 1)$ by $\sqrt{2^j}$, for $0 \leq j \leq log(N - 1)$, e.g., [3.5, 1, 0/$\sqrt{2}$, -4/$\sqrt{2}$]. This weights the coefficients at the lower resolutions more heavily than the coefficients at the higher resolutions. A thresholding method is then heuristically processed.

Method 1. Simply choose the m largest (in absolute value) wavelet coefficients for the quick summarization.

Method 2. Greedily choose m wavelet coefficients for more accuracy. For example, after performing Method 1, and then repeat the following two steps m times:

- Among the $(N - m)$ remaining wavelet coefficients, select the wavelet coefficient whose inclusion leads to the largest reduction in error.
- Among the $(m + 1)$ selected wavelet coefficients, discard the wavelet coefficient whose deletion leads to the smallest increase in error.

2.2 The Problem

Since streaming data are continuous and endless, the limits of data to be summarized are ambiguous. Therefore, we must calculate local summaries (= the m wavelet coefficients) periodically for the fixed amounts of data (= the N original local data), and then store them independently into stable storage. Let us assume that the data limits are an arbitrarily long period (e.g., a month, a year, etc.) and mass stable storage can store the summarized data within the period. Then, the long period can be divided into multiple time sections for periodic summarization and the size of one local summary can be decided. We name the size *storage limit m*. Figure 2 shows this environment. Independently summarized data (equal in size) are stored in each time section of $[t_0, t_1), [t_1, t_2), ..., [t_{i-1}, t_i)$. All intervals close on the left and open on the right.

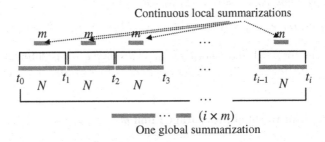

Fig. 2. The periodic summarization

   Let us consider the following problem when using the wavelets technique for this periodic summarization: the local summarization should be more effective so that the sum of the local estimation errors by the local wavelet coefficients can be close to the global estimation error by the global wavelet coefficients from a single global summarization. That is, if $[t_{i-1}, t_i)$ is the last time section and $e_{[t0, ti)}$ is the global estimation error of the global time section $[t_0, t_i)$, it is desirable to satisfy the following: $e_{[t0, ti)} \approx \bigcup_{0 \leq i \leq I} e_{[ti-1, ti)}$. Therefore, in this paper, we introduce a flexible method to gain much lower *overall estimation error* which is the sum of the local estimation errors of all local time sections (= $\bigcup_{0 \leq i \leq I} e_{[ti-1, ti)}$).

3 Our Flexible Approach

3.1 Data Based and Query Based Estimation Error

Before introducing the concept of flexible storage allocation, consider the methods for measuring the estimation error by approximating the original data. We classified them into two groups: data based method and query based method. For the data based method, periodic summarization compresses the original data with a lossy-compression scheme such as the wavelet approach, and calculates the absolute difference between the decompressed value and the original data value. Let D_k be the original kth data value in a specific time section $[t_{i-1}, t_i)$ and D'_k be the decompressed data value. The following error measures are defined:

– Absolute error: $abs\_e_{[ti-1,\ ti)} = \Sigma_k |D_k - D'_k|$

– Relative error: $rel\_e_{[ti-1,\ ti)} = \Sigma_k (|D_k - D'_k| / |D_k|),\ D_k \neq 0.$

The overall estimation error for all the local time sections can also be defined as absolute type (abs_e) or relative type (rel_e), e.g., $abs\_e = \Sigma_i\ abs\_e_{[ti-1,\ ti)}$.

For the query based method, the estimation error is defined as dependent on the result size of a query. Let R_k be the actual size of a query q_k in a specific time section $[t_{i-1},\ t_i)$ and let R'_k be the estimated size of the query. The absolute and relative errors in a specific time section can be defined for the given queries as above. In particular, the p-norm average error has been defined as the estimation error for the given Q queries in the reference [5], and we use it here. For $p > 0$:

– Absolute error: $abs\_e_{[ti-1,\ ti)} = ((\Sigma_{1 \leq k \leq Q} |R_k - R'_k|^p) / Q)^{1/p}$

– Relative error: $rel\_e_{[ti-1,\ ti)} = ((\Sigma_{1 \leq k \leq Q} (|R_k - R'_k| / R_k)^p) / Q)^{1/p},\ R_k > 0.$

For example, for $p = 1$, the *1-norm average absolute error* is defined as $(\Sigma_{1 \leq k \leq Q} |R_k - R'_k|) / Q$, and for $p = 2$, the *2-norm average absolute error* is defined as $\sqrt{(\Sigma_{1 \leq k \leq Q} |R_k - R'_k|^2) / Q}$.

3.2 Flexible Adjustment of the Storage Limit m

Now, we explain a heuristic method based on the above estimation error measurements for the flexible storage allocation. First, we define the following objective estimation error as a criterion for the flexible adjustment.

Definition 1 (objective estimation error (OEE))
This error is the average of all LEEs up to the current time section. Here, LEE symbolizes the *local estimation error* in each local time section.

Table 1. A sample streaming data and obtaining the OEE

Streaming data	Coeffi(fixed)	LEE(fixed)	OEE
$D_{[0,\ 4)} = \{(0, 3), (2, 6), (3, 2)\}$	3.5, –4	2	2
$D_{[4,\ 8)} = \{(4, 2), (5, 3), (6, 5), (7, 1)\}$	3, –4	2	2
$D_{[8,\ 12)} = \{(8, 3), (9, 4), (10, 3)\}$	3.25, –4	1	1.67
$D_{[12,\ 16)} = \{(12, 3), (13, 9), (14, 6), (15, 2)\}$	5, 6	6	2.75
:::	:::	:::	:::

For example, consider a sample streaming data in Table 1. In each local time section, four (N) wavelet coefficients are lossy-compressed into two (m) wavelet coefficients. For the expanded signal $S_{[0,\ 4)} = [3, 3, 6, 2]$, two wavelet coefficients 3.5 and –4 are selected by the Method 1 in Subsection 2.1. The decompressed signal $S'_{[0,\ 4)}$ is [3.5, 3.5, 5.5, 1.5], and the LEE calculated by the absolute data based error measurement is two ($= |3 - 3.5| + |3 - 3.5| + |6 - 5.5| + |2 - 1.5|$). The rest can also be calculated easily, and the OEE becomes two ($= (2 + 2) / 2$) for $D_{[0, 8)}$, 1.67 ($= (2 + 2 + 1) / 3$) for $D_{[0, 12)}$, and 2.75 ($= (2 + 2 + 1 + 6) / 4$) for $D_{[0, 16)}$.

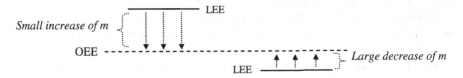

Small increase of m

OEE

LEE

LEE

Large decrease of m

Fig. 3. Flexible adjustment of the storage limit m based on OEE and LEE

Based on the OEE, the following flexible storage allocation is performed. In Figure 3, if LEE > OEE, we can make the LEE close to the OEE by increasing the number of the local wavelet coefficients (m). If LEE ≤ OEE, the m is decreased also to be close to the OEE. However, this adjustment cannot reduce the overall estimation error, because finally the adjusted estimation error becomes close to the OEE which is defined as the average of all LEEs. Our method does provide a benefit through following two important heuristic assumptions:

Assumption 1. When LEE > OEE, let the local wavelet coefficient that causes the biggest estimation error be included first. The effect is that if the current data distribution allows the dramatic decrease of the LEE by adding a few local wavelet coefficients, the LEE can be more reduced with a small amount of the added storage.

Assumption 2. When LEE ≤ OEE, let the local wavelet coefficient that causes the lowest estimation error be excluded first. The effect is that if the current data distribution allows very little increase in the LEE by discarding multiple local wavelet coefficients, the LEE can be a little reduced with a large amount of released storage. The released storage can be assigned to the local time sections identified under Assumption 1. Therefore, unexpectedly, the overall estimation error can be improved.

Table 2. The flexible adjustment for the sample streaming data

Time Section	$Coeffi_{(fixed)}$	$LEE_{(fixed)}$	OEE	$Coeffi_{(flexible)}$	$LEE_{(flexible)}$
$D_{[0, 4)}$	3.5, –4	2	2	3.5, –4	2
$D_{[4, 8)}$	3, –4	2	2	3, –4	2
$D_{[8, 12)}$	3.25, –4	1	1.67	3.25	1.5
$D_{[12, 16)}$	5, 6	6	2.75	5, 6, –4	4
:::	:::	:::	:::	:::	:::

For example, in Table 2, since LEE < OEE in the local time section $D_{[8, 12)}$ and the increase of the LEE by the exclusion of the wavelet coefficient –4 is lowest ($= 0.5$), the wavelet coefficient is removed. Next, since LEE > OEE in $D_{[12, 16)}$ and the decrease of the LEE by the inclusion of the wavelet coefficient –4 is largest (= 2), the wavelet coefficient is added. By this adjustment, the overall estimation error is reduced from 11 (= 2 + 2 + 1 + 6) to 9.5 (= 2 + 2 + 1.5 + 4). Although this shows a simple case, we can expect that the overall estimation error should be improved according to the flexible storage allocation of Assumptions 1 and 2.

Algorithm 1 expands Method 2 in Subsection 2.1 for this flexible storage allocation; steps 4 and 5 were added. In steps 1, 2 and 3, the largest m wavelet coefficients are selected from all N wavelet coefficients of the current local time section ($m \ll N$), and then m wavelet coefficients are greedily reselected. In step 4, to calculate the OEE, the average local estimation error is recalculated including the current local estimation error. In step 5, if LEE > OEE, the wavelet coefficient causing the largest error is first selected among the excluded ($N – m$) wavelet coefficients. Otherwise, the wavelet coefficient causing the smallest error is first excluded from the selected m wavelet coefficients.

It was observed in experiments that if the increased error by the exclusion of Assumption 2 is too large, then the overall estimation error could become worse. The reason is that generally, the change of estimation error by the decrease of m is larger than by the increase. That is, when 1000 (N) wavelet coefficients are compressed to 100 (m) wavelet coefficients, decreasing m into 50 is most likely to make a larger change than increasing m into 150. Therefore, in our algorithm, we limit the reduction of m within the possible proportion of (*objective_estimation_error* - *temp_error*) as in the shaded code. The variable *temp_error* keeps the current local estimation error. For the parameter *ratio*, we used 0.25. In Assumption 1, if the reduced estimation error by the increase of m is very small, it can cause to unnecessarily add many wavelet coefficients, and eventually the waste of storage space. Therefore, we also limit the increase of m within the possible proportion of the storage space released by Assumption 2 as in the shaded code. The variable *num_of_available_m* keeps the size of the released storage space.

For the initialization of the variable *objective_estimation_error*, the algorithm can be performed except step 5 during the some early time sections. That is, the flexible storage allocation of step 5 is not executed during the some early time sections in order to get the more reasonable objective estimation error.

The time complexity of the suggested algorithm is $O(N(logN)logm)$. Although steps 4 and 5 are added, Algorithm 1 still has the original time complexity in the reference [5]. This is because step 5 can be considered as a repetition of step 3.

4 Experimental Results

4.1 Experimental Setup

To evaluate the correctness of our method, we have compared the overall estimation error of our flexible approach with that of the existing fixed approach.

Algorithm 1. Expansion of Method 2 for the flexible storage allocation

```
/* this function summarize one time section. */
function WaveletCompress
input:
    stream_data[]: Stream data of one time section
    N: Number of elements in one time section (or one time section size)
    m: Default compressing size of one time section
    objective_estimation_error: (See Definition 1)
    maximum_local_estimation_error: The largest local estimation error
    up to now
    ratio: A threshold for decreasing coefficients in Assumption 2
output: An optimal wavelet coefficients
begin
    local_estimation_error = 0, temp_m = 0;   /* initialize */
    temp_error = 0, num_of_available_m = 0;

1. Calculate N wavelet coefficients from stream_data[];
2. Select the m largest (in absolute value) wavelet coefficients
among the N wavelet coefficients;
3. for ( i = 0; i < m; i++ ) {
    Among (N - m) wavelet coefficients, select the coefficient whose
    inclusion leads to the largest reduction in error;
    Among (m + 1) wavelet coefficients, discard the coefficient
    whose exclusion leads to the smallest increase in error;
    }

4. Calculate local_estimation_error and objective_estimation_error;
if (local_estimation_error > maximum_local_estimation_error)
    maximum_local_estimation_error = local_estimation_error;
5. temp_m = m;
temp_error = local_estimation_error;
if (local_estimation_error > objective_estimation_error
    && num_of_available_m != 0) {
    while (local_estimation_error > objective_estimation_error ) {
        Among (N - m) wavelet coefficients, select the wavelet coef-
        ficient whose inclusion leads to the largest reduction in
        error;
        Recalculate the current local estimation error;
        m = m + 1;
        if ((m - temp_m) >= (temp_error/maximum_local_estimation_error)
            * num_of_available_m || (m - temp_m) >= num_of_available_m)
            break;
    }
    num_of_available_m -= (m - temp_m);
} else {
    while (local_estimation_error < objective_estimation_error ) {
        Among m wavelet coefficients, discard the wavelet coeffi-
        cient whose exclusion leads to the smallest increase in er-
        ror;
        Recalculate the current local estimation error;
        m = m - 1;
        if ( (local_estimation_error - temp_error) > ratio
            * (objective_estimation_error - temp_error)) break;
    }
    num_of_available_m += (temp_m - m);
    }

6. return the m wavelet coefficients;
end.
```

Some real data streams provided at the website [11] were used in our experiments: stock data, EEG measurements of an albin rat, and light measurements from motes. For the experiments discussed below, the following parameter is constant with the number of elements in one time section N = 4,096, the default compressing size of one time section m = 512, and the reduction threshold *ratio* = 0.25 (See Algorithm 1 for details). We think that one time section size over the same data does not become a significant factor only for comparing two approaches. Also, the compressing size value and the reduction threshold value are reasonable for the comparison. All experiments are performed over 300 consecutive time sections. For the wavelets technique, we used the Harr wavelets that are the simplest and easiest for implementation. For the estimation error measurement (Subsection 3.1), we performed both the absolute data based measurement and the *1-norm* absolute query based measurement. The following query sets were used for the query based measurement:

$$\{X_i \mid X_i = a\}, \{X_i \mid a \leq X_i \leq b\}, \{X_i \mid (X_i - X_{i-1}) \geq a \parallel (X_{i+1} - X_i) \geq a\},$$

the variable X_i defines an ith data element from a sensor and the constants a and b are a real number.

Our experiments were performed on a HP Pavilion with Pentium(R) dual-core processor, 2.8 GHz CPU, 1GB RAM, and all codes were written in Java.

4.2 Flexible *vs* Fixed Storage Allocation

Table 3 shows the improvement ratio by our flexible approach against the existing fixed approach, and the surplus storage space of our approach. The surplus storage space of the fixed approach is surely zero because of the fixed compressing size for all the local time sections, but our flexible compression can have a surplus space (See the first shaded code in Algorithm 1). This surplus space can be used for the later time sections.

The improvement ratio under the query based error measurement shows that the correctness of the fixed approach is each 8.8 %, 17.1 %, 15.7 % less than that of our flexible approach. However, the very low improvement ratio under the data based error measurement shows that our flexible approach is more effective when the estimation error depends on given queries. The query based error measurement is more advantageous to *predefined* queries than *ad hoc* queries, because it keeps more wavelet coefficients significant to given queries (but, note that the selected coefficients can also be relevant to ad hoc queries). The predefined query is one issued before any relevant data has arrived, on the other hand the ad hoc query is one issued after [3].

The graph in Figure 4 shows local estimation errors and storage allocation over 80 consecutive time sections in the experiment, which gained the improvement ratio 15.7 % with light test data in Table 3. The graph compares our flexible approach with the fixed approach. The line graph is for the measurement of local estimation error and the bar graph is for the measurement of compressing size (i.e. local storage allocation). The compressing size of the fixed approach is uniform as 512. From the line graph, we can see that the flexible approach has lower estimation error in many time sections. This is because as in Assumption 2, such time sections having lower estimation error further use the released storage from the other time sections. The bar graph shows this flexible storage allocation.

Table 3. Improvement ratio by our flexible Approach against the exising fixed appraoch

Data Set	Query Based Error Measurement		Data Based Error Measurement	
	Improvement (%)	Surplus Space (byte)	Improvement (%)	Surplus Space (byte)
Stock	8.8	22,604	−3.1	7,580
EEG	17.1	70,808	0	48
Light	15.7	3,148	0.1	6,112

Improvement = {(the overall estimation error of the fixed approach) − (the overall estimation error of the flexible approach)} / (the overall estimation error of thefixed approach) × 100 %.

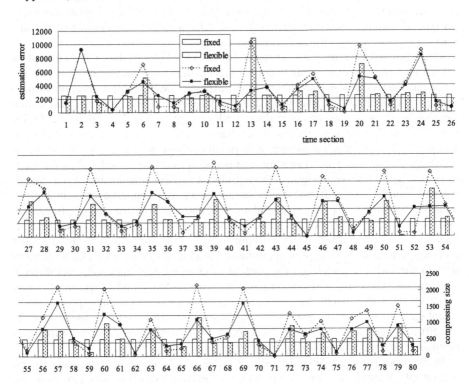

Fig. 4. Trace of estimation error and storage allocation over consecutive time sections: our flexible approach *vs.*the existing fixed approach

5 Related Work

For the data summarization using wavelets technique, Matias, Vitter, & Wang [4] introduced a new approach based upon probabilistic counting and sampling to maintain wavelet coefficients over continuous data stream. Karras and Mamoulis [2] also introduced a new approach based upon maximum error metrics. However, these studies focused mainly on one-pass summarization and fast small-space algorithms within a limited memory space. Moreover, Matias et al's approach is not related with our ordered data summarization over time.

Charu C. Aggarwal et al. [1] introduced the concept of online micro-clustering and offline macro-clustering. Online micro-clustering maintains statistics at a sufficiently high level of (temporal and spatial) granularity so that they can be effectively used by the offline macro-clustering. It satisfies the one-pass criterion for endless fast streaming data, and such micro-clusters represent the current snapshot of clusters that change over the course of the stream, as new points arrive. The offline macro-clustering uses the compactly stored summary statistics of the micro-clusters to interactively compute clusters over user-defined time periods. Therefore, macro-clustering is not constrained by one-pass requirements. Although this paper uses another summarization technique such as clustering, it is related to our concept in that it deals with the periodic summarization into stable storage.

The STanford stREam datA Manager (STREAM) [3] introduced the concept of evaluating a query over a sliding window of recent data from the streams (not over the entire past history of the data streams), and expanded SQL to offer an optional time window specification for streaming data with timestamps. In addition, due to the memory overflow of the streaming data, it allocates system memory dynamically for synopses and data queues. That is, it summarizes older data and stores that data as synopses to hold as much information as possible about the streaming data in memory. Histograms, wavelets, clustering, and decision trees are considered as the summary structures. The Aurora system [8] introduced real time monitoring of the streaming data from a sensor. It executes a stream query as a sequence of eight built-in primitive operations as in the relational algebra, and it introduced query optimization through reordering and resource allocation of the primitive operations. As the continuous and fast characteristic of streaming data cause problems similar to packet dropping in a router, this approach introduced load shedding based on QoS information. In particular, it introduced the concept of dropping less important data through semantic load shedding, which maintains data synopsis based on histograms of output query results. These two leading projects focused on the real time query processing of recent data stream and synopses held only in memory. However, our interest lies in the efficient periodic data stream reduction held in stable storage to store as much streaming data as possible for data tracking and analysis.

In the paper [6], we already introduced the necessity of summarizing data stream periodically, using histograms and concept hierarchy besides wavelets. However, the experiments did not show the significant improvement because they used the data based estimation error measurement. In this paper, we explain our method more in detail especially using wavelets, revise the before Algorithm, and present the improvement and meaning under the query based estimation error measurement.

6   Conclusions

In this paper, we have proposed a periodic data stream summarization method for storing as much information about data as possible with lowering the overall estimation error. The proposed method is to adjust the compressing size of each local time section flexibly based on Assumptions 1 and 2. That is, the basic idea is to exclude the wavelet coefficients less significant for any time section and utilize the saved storage space for the other time sections. Some experimental results have shown that our flexible approach has lower estimation error than the existing fixed approach, especially in the case of using the query based estimation error.

Acknowledgements. This study is supported in part by the Second Stage of BK21. In addition, this work was supported by the Korea Science and Engineering Foundation(KOSEF) grant funded by the Korea government(MOST) (No. R01-2006-000-10609-0).

References

1. Aggarwal, C.C., Han, J., Wang, J., Yu, P.S.: A Framework for Clustering Evolving Data Streams. In: Proc. 29th International Conf. on VLDB, Berlin, Germany, pp. 81–92 (2003)
2. Karras, P., Mamoulis, N.: One-Pass Wavelet Synopses for Maxium-Error Metrics. In: Proc. 31th International Conf. on VLDB, Trondheim, Norway, pp. 421–432 (2005)
3. Babcock, B., Babu, S., Datar, M., Motwani, R., Widom, J.: Models and Issues in Data Stream Systems. In: Proc. the 21th ACM SIGACT-SIGMOD-SIGART Symposium on Principles of Database Systems, Madison, USA, pp. 1–16 (2002)
4. Matias, Y., Vitter, J.S., Wang, M.: Dynamic Maintenance of Wavelet-Based Histograms. In: Proc. 26th International Conf. on VLDB, Egypt, pp. 101–110 (2000)
5. Matias, Y., Vitter, J.S., Wang, M.: Wavelet-Based Histograms for Selectivity Estimation. In: Proc. the ACM SIGMOD International Conf. on Management of Data, Seattle, USA, pp. 448–459 (1998)
6. Kim, J., Park, S.: Periodic Streaming Data Reduction Using Flexible Adjustment of Time Section Size. International Journal of Data Warehousing & Mining 1(1), 37–56 (2005)
7. Stollnitz, E.J., Derose, T.D., Salesin, D.H.: Wavelets for Computer Graphics. Morgan Kaufmann, San Francisco (1996)
8. Tatbul, N., Çetintemel, U., Zdonik, S.B., Cherniack, M., Stonebraker, M.: Load Shedding in a Data Stream Manager. In: Proc. 29th International Conf. on VLDB, Berlin, Germany, pp. 309–320 (2003)
9. Istepanian, R.S., Jovanov, E., Zhang, Y.T.: Introduction to the special section on M-Health: beyond seamless mobility and global wireless health-care connectivity, Guest Editorial. IEEE Transactions on Information Technology in Biomedicine 8(4), 405–413 (2004)
10. Deshpande, A., Guestrin, C., Madden, S.R., Hellerstein, J.M., Hong, W.: Model-Driven Data Acquisition in Sensor Networks. In: Proc. 30th International Conf. on VLDB, Toronto, Canada, pp. 588–599 (2004)
11. Time Series Data Mining Archive, http://www.cs.ucr.edu/ĕamonn/TSDMA/index.html

Secured Web Services
Based on Extended Usage Control

Woochul Shin and Sang Bong Yoo

School of Computer Science, Inha University
Incheon, Korea
woochul@dbsi.inha.ac.kr, syoo@inha.ac.kr

Abstract. With the worldwide dissemination of Internet, Web Service has become a promising paradigm of ubiquitous computing. However, one major stumbling block of using Web Service in ubiquitous environment is the lack of security enforcement. In this paper, we introduce a systematic approach to enhance Web Service with the access control, user authentication, and session management. Delegation of access right among distributed security manager has been added to the Usage Control model and formal description of access control model is defined. A set of Web service operations is devised for procedural implementation of the security model. Using the proposed operations we show how SSO (Single Sign On) can be realized in ubiquitous computing environments.

Keywords: Secured Web Service, Ubiquitous Computing, Access Control, Security Management, User Authentication, Single Sign On.

1 Introduction

Web services have many promising features because they consist of open technologies and share many S/W and H/W components with Internet technologies. However, the same characteristics of Web services also limit the widespread deployment of Web services by organizations. Because Web services basically exchange messages in plain XML texts via open networks, security risks and the types or attack common for Web sites are the same for Web services. Types of security attacks within a Web service environment include malicious attacks, denial of service attacks, dictionary attacks, and internal threats [2].

To secure Web services, a range of XML-based security mechanisms are needed to solve problems related to authentication, role-based access control, distributed security policy enforcement, message layer security that accommodate the presence of intermediaries [14]. A set of specifications are recommended or proposed by world wide organizations such as W3C and OASIS. A partial list of the specifications includes WS-Security [11], XML digital signature [16], XML encryption [15], XKMS [15], SAML [9], XACML [10], and ebXML Message Service [5]. The set of specifications complement each other to build a security framework for Web service. However the practices of security enforcement of each Web service will vary with a wide range of selections of security policies.

T. Washio et al. (Eds.): PAKDD 2007 Workshops, LNAI 4819, pp. 656–663, 2007.

Ubiquitous computing environments can be characterized by the diversity of components and by the dynamic changes of relationship among objects and computing agents. In order to be effective solution for such environments, the security paradigm should satisfy the two requirements:

a) The range of protection should be end-to-end rather than point-to-point because the potential for security breaches is increased.

b) The definition of security policy should cover the dynamic handovers the access privileges among distributed computing agents.

In this paper, delegation of access right among distributed security managers has been added to the Usage Control model [12, 13] and formal descriptions of access decisions are defined. A set of Web service operations is devised for procedural implementation of the security model. By combining access control and session management we can reduce the number of hand shaking that usually required for processes to interact with computing agents. As an application of integrated support of access control and session management, the scenario of supporting SSO (Single Sign On) [3] is presented.

2 Related Work

Providers of services, resources, and digital content need to selectively determine who can access these valuable objects; this is the central objective of access control. As basic modeling methods of access control, there are such traditional approaches such as mandatory access control (MAC), discretionary access control (DAC), and role-based access control (BRAC) [17]. Recently usage control model (UCON) is a systematic access control method that satisfies the requirements of the security and privacy related to business and information systems [12, 13]. The concept of usage represents the right for digital objects. This model includes access control, trust management, and digital right management like such existing methods as MAC, DAC, and RBAC.

More specifically for Internet and Web service applications, quite a few approaches have been introduced. In order to overcome the limitations of SSL and XML security schemas, ESM integrates some popular XML security schemas and constructs a layered model based on SOAP message transport mechanism [1, 4, 8, 18]. ESM model include 6 layers, i.e., network layer, message layer, security layer, transfer layer, validation layer, and application layer. Each layer of ESM model incorporates related protocols (e.g., HTTP, SOAP, SAML, and XKMS) and forms hierarchical structure of an integrated model of Web service security.

SRP(Secure Remote Password) is used for Web service authentication [6]. SRP is one of the most used password-based authentication protocol. Instead of storing the password directly, a verifier obtained from the password through a one-way hash function is stored. One characteristic of the authentication scheme is that the password is never sent across the network, thus avoiding that an intruder spoofs the network and retrieve the password or some information that could make possible a password reconstruction. A session authentication protocol based on standard Web services technologies such as SOAP, XML Signature/Encryption, and SOAP-DSIG is designed [7].

In this paper, we first extend the UCON model in order to handle the delegation of rights among distributed computing agents. It is required especially for ubiquitous computing environments because we should handle dynamic interactions of heterogeneous objects. A variation of hybrid cipher method is used for user authentication and all interfaces are implemented using Web service operations. Design principle of the secured Web service is to provide end-to-end security and minimize the number of handshaking for user authentication and session management. As a working scenario, we present how SSO (Single Sign On) can be achieved using the Web service operations.

3 Extended Usage Control Model (UCON$_{ABCD}$)

UCON model consists of the three core components (i.e. Subjects, Objects, and Rights) and three additional components (i.e. Authorizations, Conditions, and Obligations). The Extended Usage Control Model (UCON$_{ABCD}$) adds one more component Delegation for modeling delegating rights in ubiquitous computing environments. The relationship among components of UCON$_{ABCD}$ is depicted in Fig 1.

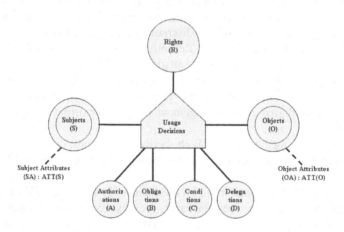

Fig. 1. Components of UCON$_{ABCD}$

The Usage Decisions of UCON$_{ABCD}$ include Pre-Decision of initial authorization process and Ongoing-Decision of dynamic authorization process. Because the right of users will change depending on the context of environments in ubiquitous computing, we must have dynamic authorization process. The dynamic authorization process occurs according to the usage of user's right, time changes, or change of object's status after the initial authorization process. The dynamic authorization process includes the step for investigating the right that the Subject currently uses, the step for investigating the Condition of Object, and the step for confirming if the Subject has completed its Obligations. Application of the authorization processes will be explained for single-sign on process later in this paper.

4 Secured Web Service in Ubiquitous Computing Environments

4.1 Security Manager

The secured Web service environment presented in this paper has SM (Security Manager) at the center of its functional architecture. Authentication using the hybrid method will be discussed in detail next section. Session management employs a mutual authentication method based on RSA method. In the beginning of a session, a session key is delivered in order to share the status of the session. Services participating a session maintains a session secret (i.e., the hash value which digests the session shared information by the session key) in order to authenticate messages from the users. Operations that are provided for session management include StartSession, JoinSession, EndSession, NotifySession, ForwardSession, and QuerySession.

Because the operations for user authentication and session management can be implemented by Web service operations, the security feature can be added to existing Web services by employing the operations. Separate operation sets are required for clients, services, security manager, authentication information DB, and Key/Policy DB.

4.2 User Authentication Using Hybrid Method

Conventional Password and its encryption are not adequate for the user authentication of Web service. Because Web service interactions encoded in SOAP message share ports with standard HTTP traffic, the password information is more vulnerable to any malicious attacks. In this paper, we propose a user authentication using a hybrid cipher method.

Two major cryptographic algorithms are symmetric ciphers and asymmetric ciphers. Symmetric ciphers such as DES and AES use the same key for both encryption and decryption. Asymmetric ciphers such as RSA use different keys for encryption and decryption. Symmetric ciphers are usually computationally faster than asymmetric ciphers, but they have difficulties in key distribution. In this section, both symmetric and asymmetric ciphers are used for user authentication.

4.3 SSO Using Delegation of Access Rights

One example of using the authenticated session management is SSO (Single Sign On). SSO is a specialized form of authentication that enables a user to authenticate once and gain access to the resources of multiple systems. Because there are lots of inter-references among resources stored in distributed computing agents in ubiquitous environment, multiple authentication processes can be avoided by using SSO. Using the extended component (i.e., Delegation) of $UCON_{ABCD}$, we can effectively model the access control of SSO. The access right for each service is described based on the $UCON_{ABCD}$ as follows.

$$S = \{user, IS, ONS, SM, SM'\};$$
$$O = \{assertioninfo, userinfo, service\_object\};$$
$$R = \{access, sign\};$$
$$ATT(user) = \{userid, password, SMmembership\};$$
$$ATT(IS) = \{serviceid, servicegroup\};$$
$$ATT(ONS) = \{serviceid, servicegroup\};$$ \hfill (1)
$$ATT(SM) = \{managerid, urlinfo\};$$
$$ATT(assertioninfo) = \{issuer, subjectinfo, authnstatement\};$$
$$ATT(userinfo) = \{userid, password, group, membership\};$$
$$ATT(service\_object) = \{objectid, objecttype, ownership\};$$

A user who belongs to SM' can be delegated the right from SM (preA + preD).

$$DS = \{SM\}; DO = \{assertioninfo\}; DR = \{authenticate\};$$
$$getPreDRL(user, service\_object, access)$$
$$= \{(SM', userinfo, authenticate), where\ ATT(user) \in ATT(SM'),$$ \hfill (2)
$$assertioninfo = ATT(user);$$
$$allowed(user, service\_object, access) \Rightarrow preA(SM, assertioninfo, sign)$$
$$\cap preDelegated(getPreDRL(user, service\_object, access)$$

A user can access Web Service only after authorized by an SM (preA + preB).

$$OBS = \{SM\}; OBO = \{userinfo\}; OB = \{authenticate\};$$
$$getPreOBL(user, service\_object, access)$$
$$= \{(s', userinfo, authenticate), where\ s' \in OBS, userinfo = ATT(user);$$ \hfill (3)
$$allowed(user, service\_object, access) \Rightarrow preA(SM, userinfo, authenticate)$$
$$\cap preFulFilled(getPreOBL(user, service\_object, access)$$

If a user has right to access a service_object, SM creates assertioninfo and send it to a Web Service (preA + preB + preC).

$$allowed(user, service\_object, access) \Rightarrow preA(user, service\_object, access)$$
$$\cap preA(SM, assertioninfo, create)$$ \hfill (4)
$$\cap preB(SM, assertioninfo, sent)$$
$$\cap preC(user, service\_object, signon)$$

The access control procedure of SSO among the user, LSM (local security manager), and RSM (remote security manager) will be explained in the remaining of this section. A user first access a local Web service (LWS) as in Fig. 2. Then the LWS requests the user authorization to a local Security Manager (LSM). Using the hybrid authorization process described in the previous section, the LSM verifies the user information and send assertion information to the LWS.

Assuming that the user consequently accesses a remote Web service (RWS), the RWS requests authorization to a remote security manager (RSM) as in Fig. 3. Instead of having another hybrid authorization process, the RSM request an assertion to the LSM that already have verified the user's identity. The LSM delegates the access right to the RSM by sending an assertion.

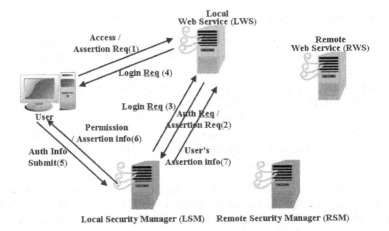

Fig. 2. Procedure of initial access to a local Web service

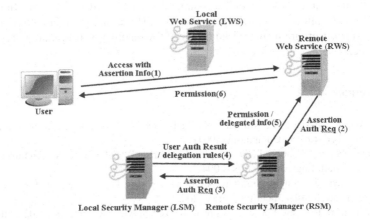

Fig. 3. Procedure of SSO to a remote Web service

5 Conclusion

Ubiquitous computing environments consist of heterogeneous S/W and H/W connected via networks. Many of such components do not have enough computing power to provide effective security measure for valuable resources. Therefore it is essential for components to properly cooperate in order to achieve maximal protection. In this paper, the usage control model (UCON) has been extended by adding one more component (i.e., Delegation) for effective modeling of delegating rights in ubiquitous computing. Based on the formal definition of the extended UCON model, access control operations of Web services have been proposed.

Contributions of this paper can be summarized as follows.

a)   Formal description of usage decision is given for both pre-decision of initial authorization process and ongoing-decision of dynamic authorization process based on the extended usage control model (UCON$_{ABCD}$).
b)   In order to provide end-to-end protection in ubiquitous computing environments, we combine the symmetric and asymmetric ciphers and propose a hybrid user authentication procedure.
c)   Using the modeling methods proposed in this paper, we show how SSO can be realized. Because the user authentication information is reused, the number of interactions can be minimized.

The current set of operations can be augmented for more general ubiquitous computing. One example is the access control method for general sensor networks. Because sensors like RFID (Radio Frequency Identification) may move around world, the proposed UCON$_{ABCD}$ model can be effectively applied. Formal proofing of the correctness of proposed model is also future work.

Acknowledgments. This research was supported by the MIC(Ministry of Information and Communication), Korea, under the ITRC(Information Technology Research Center) support program supervised by the IITA(Institute of Information Technology Assessment).

References

1. Berket, K., Essiari, A., Muratas, A.: PKI-based security for peer-to-peer information sharing. Peer-to-Peer Computing (2004)
2. Chatterjee, S., Webber, J.: Developing Enterprise Web Services An Architect's Guide. Prentice Hall, Englewood Cliffs (2004)
3. De Clercq, J.: Single Sign-On Architectures. In: Proceedings of the International Conference on Infrastructure Security (2002)
4. Djajadinata, R.: Yes, You can secure your Web Services documents (2002), http://www.javaworld.com
5. ebXML: Message Service Specification (ebXML Transport, Routing & Packaging) Version 1.0 (2001), http://www.ebXML.org
6. Flavio, O., Joao, A.A., Pedro, F.: A Web Service Authentication Control System Based on SRP and SAML. In: Proceedings of the IEEE International Conference on Web Service (2005)
7. Handa, S., Maruyama, H.: Session Authentication Protocol for Web Services. In: Proceedings of the 2002 Symposium on Application and the Internet (2002)
8. Mohan, S., Klinginsmith, J., Sengupta, A., Wu, Y.: ACXESS – Access Control for XML with Enhanced Security Specifications. In: Proceedings of 22nd International Conference on Data Engineering, Atlanta, Georgia, USA (2006)
9. OASIS: Assertions and Protocol for the OASIS Security Assertion Markup Language(SAML) (2002), http://www.oasis-open.org
10. OASIS: eXtensible Access Control Markup Language Version 1.0. (2003), http://oasis-open.org

11. OASIS: Web Service Security: SOAP Message Security (2003), http://www.oasis-open.org/comittees/wss
12. Park, J., Sandhu, R.: Towards Usage Control Models: Beyond Traditional Access Control. In: Proceedings of the Symposium on Access Control Models and Technologies (2002)
13. Park, J., Sandhu, R.: The UCONABC Usage Control Model. ACM Transactions on Information and System Security (2004)
14. W3C: Web Service Architecture (2004), http://www.w3.org/TR/ws-arch/
15. W3C: XML Key Management Specification(XKMS) (2001), http://www.w3c.org
16. W3C: XML Signature Syntax and Processing (2002), http://www.w3c.org
17. Yamada, S., Kamioka, E.: Access Control for Security and Privacy in Ubiquitous Computing Environments. The Institute of Electronics, Information and Communication Engineers (2005)
18. Zhang, M., Cheng, Z., Ma, Z., Zang, B.: A Security Model Design in Web Service Environment. In: Proceedings of the 2005 the Fifth International Conference on Computer and Information Technology (2005)

A Digital Rights Management Architecture for Multimedia in P2P

Cheng Yang[1], Jianbo Liu[1], Aina Sui[2], and Yongbin Wang[2]

[1] Information Engineering School, Communication University of China
[2] Computer & Software School, Communication University of China,
Beijing, China
cafeeyang@163.com

Abstract. P2P content sharing is often blamed for copyright infringement, making the establishment of DRM technologies an urgent need. A PDRM (P2P-based Digital Rights Management) system is proposed with the support of Next Generation Internet Project. The system is based on a trust model that focuses on content security, rights management and access control. Encryption, digital watermarking, and packaging technologies are adopted to protect the confidentiality and integrity of contents, and support copyrights verifying and piracy tracing. The structure of rights management integrates the distributed and centralized modes, which not only reduces the burdens of networks and rights server, but also provides controllability. The contents downloaded on the P2P networks can be played only with rights control. To realize access control, the password and identity authentication are used. The PDRM system is implemented to prove that it can provide a more robust Intellectual Property protection solution for P2P content delivery.

Keywords: Peer-to-Peer (P2P), Digital Rights Management (DRM), Public Key Infrastructure (PKI), Identity Authentication, Digital Watermarking.

1 Introduction

P2P networks have grown rapidly in information sharing to havens for trafficking in unauthorized copies of Intellectual Property. P2P file sharing systems, such as Napster, Gnutella and KaZaA, allow contents to be shared between distributed peers. Most P2P networks do not have any content protection (CP) or access control. P2P networks are often blamed for illegally sharing copyrighted materials [1][2].

There are many traditional CP solutions, such as the Microsoft's Windows Media Rights Manager (WMRM), IBM's Electronic Media Management System (EMMS), InterTrust's Rights|System, and RealNetworks's RealSystems Media Commerce Suite (RMCS) [2][3]. However most of them are applicable only to conventional client/server based content delivery. A few CP products are applicable to P2P content delivery. Guofei Gu etc. propose a PLI (Public License Infrastructure)-based CP system to provide digital rights management for users of Peer-to-Peer (P2P) networks. The system is the first distributed CP license service system, which is especially useful for small content providers such as peers in a P2P network. Other researchers

T. Washio et al. (Eds.): PAKDD 2007 Workshops, LNAI 4819, pp. 664–672, 2007.

also have pay attention to integrating CP with P2P networks, such as Tetsuya Iwata etc. of NTT Corporation studying on a CP system suitable for P2P content delivery[4], Bill Rosenblatt of GiantSteps Media Technology Strategies studying on integrating CP with P2P networks[5], Paul Judge and Mostafa Ammar of Georgia Institute of Technology studying on the benefits and challenges of providing content protection in Peer-to-Peer Systems[6].

A novel PCP (P2P-based Content Protection) system is proposed in this paper that will allow content providers to safely delivery their digital media, such as films and television programs.

In this paper, Section 2 introduces the PCP system architecture. From aspects of content packaging, right management and content playing, the section 3.1 discusses the method and format of content packaging, which integrates encryption and digital watermarking. Then we present right agent that is the core of the system in section 3.2. Section 3.3 addresses the playing process supporting PCP.

2 The Architecture of PDRM System

2.1 The Characters of PDRM System

The requirement of P2P content protection is from users and commercial application, mainly focusing on content security, the rights control and copyright protection. In order to create a environment to protect the copyright for the programming, delivery and consuming of digital content, the characters of PCP system as follows:

1) Security: From aspects of content, user and right, the content is packaged with special format through encryption and watermarking, and is granted special right. Users must register to login the system with certificate, identity and credit. The content could be decrypted and the right could be parsed, and played by media player.

2) Controllability: This characteristic includes access control, usage control, transmission control, post control. In access control, the contents should be access according to a right license. In usage control, the special media player must be used to play the content. In transmission control, the encryption and SSL or IPsec are adopted. In post control, the extracting watermark and verifying copyright are used to monitor piracy behaviors.

3) Scalability: PCP can be treated as middle ware and be realized from many aspects such as the functions, modularization, interfaces and rights expression language.

4) Behavior monitor and piracy tracing: The right certificate can be used to monitor and verify the users' behavior. And the piracy tracing can be implemented through watermark extracting and verification.

2.2 The Framework of PDRM System

In PCP system, the contents with copyright desire will be packaged firstly. Then the content packages are delivered to peer-user through P2P networks. Before the content is played, the relevant rights of contents must be purchased. The figure 1 addresses the framework of PCP system.

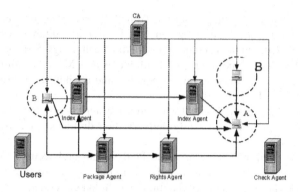

Fig. 1. The Framework of PDRM System

1) CA (Certificate Authority) presides over distributing identity certificates to the entities in P2P network, such as the peer users, index agent, package agent (PA), rights agent (RA), etc.

2) The content owner B calls the API of PA to packages his digital productions using the integrated scheme of cipher and watermark, binds and stores specified rights, informs index agent to create the content indexes for P2P downloading. The index agent also updates indexes using traditional P2P delivery scheme after the B backup obtain the same content.

3) Within the P2P network, the peer user A finds out content through index agent, he starts download from B and B backup. And he could select to purchase one or more rights from RA.

4) In order to obtain rights, RA and A need to bidirectional identity authentication using SSL or IPSec, RA must queries the users manager to verify user's identity information and user's credit information. After authentication, RA generates rights certificate using XrML with stored rights information, distributes to peer user A.

3   The Modules in PDRM System

According to the analysis above, PDRM system includes five modules: Content Packaging, Content Delivery, Right Management, Content Play and Copyright Verifying. The use case digraph is showed in figure 2.

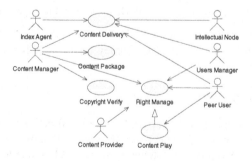

Fig. 2. The Use Case Digraph of PDRM System

3.1 Content Packaging

3.1.1 The Pattern of Content Packaging

The self-determined packaging is adopted. The content manager will call the API of PA directly. Its implementation is simple and flexible. The process of self-determined packaging in PCP system is presented in figure 3.

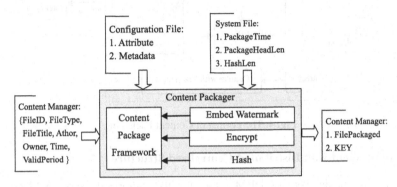

Fig. 3. The Process of Self-determined Packaging in PDRM System

Firstly, By the content manager system, the content owner transfers information about the copyright identifier of his content productions M to PA. Secondly, according to the specified encryption algorithm identifer and watermarking algorithm identifer, PA calls the relevant algorithm modules to create new productions packaged with P-DCF format criterion. Finally, the new media files and the protection keys for content encryption are returned to content manager.

The digital watermarking algorithm is able to embed copyright information W to and extract from the content, such as the video streamings or images. So, M_W, the new content with watermark is created:

$$M_W = F(M, W). \tag{1}$$

The encryption adopts symmetrical AES that is used to encrypt M_W to ensure the confidentiality of content. The Hash function is realized by SHA-1 to abstract digest from the encrypted M_W, which can verify the integrity of content. The watermark will be extracted to verify copyright and trace piracy behavior in the future.

3.1.2 The Format of Content Packaging

Before the content is packaged, other relevant information should be packaged together, such as file name, copyright, as well as some additional methods for the CIA of content. In PDRM system, the content will be packaged is appended a head file including file head, attribute, metadata and Hash digest:

The definition of attribute and metadata will conduce to the scalability of PDRM system. The name of information is presented in attribute list. However, the metadata could be defined concretely according to practical applications. And the position of metadata can be computed from the original address and offset in attribute list.

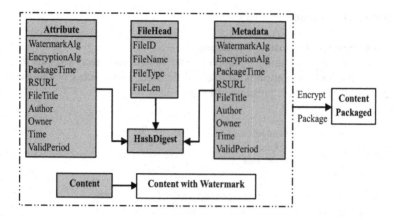

Fig. 4. The Format of Content Packaging

3.2 Right Agent with Distributed-Centralized Structure

Right control is the core of DRM. Only legally authorized users could consume digital content correctly. In PDRM system, RA binds the rights to digital contents, and controls the content access according to different rights.

3.2.1 The Structure of Right Agent

After content owner, such as the content provider, uploads content resource to content maneger, the content and its users must be authorized, which can be done through the interface provided by RA, and then the rights will be saved to database with the given data structure.

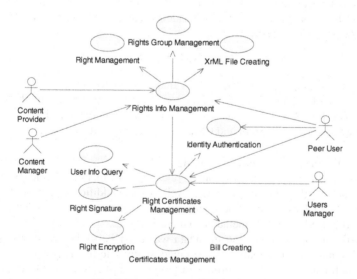

Fig. 5. The Structure of Right Agent

While peer-user wants to browse the usable rights, RA can get the rights information from database and send to peer-user on the network. If peer-user decides to purchase one or more of them, RA can create right certificate to peer-user when the bidirectional identity authentication between RA and peer-user is successful.

RA includes three modules: Rights Info Management, Right Certificates Management and Identity Authentication.

The PDRM system is designed for film and television industry. In order to make every content provider, for example TV station, control his own media resources and usage rights easily and securely, the structure of RA adopts the mix-mode combined distributed and centralized modes. Every TV station can build and control his own RA server that can deal with the right requests from peer-users located at same region, which could not only reduce the burden of networks bandwidth and server computing, but also provide the controllability. Moreover, right certificates and media files are stored and delivered separately, which will bring more flexibility to PDRM system.

3.2.2 XrML-Based Rights Info Management

There are three main functions in Rights Info Management: Right Group Management, Rights Management and XrML File Creating.

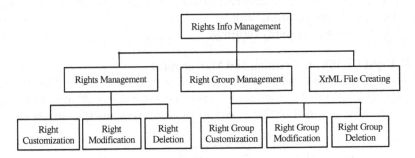

Fig. 6. The Modules of Right Info Management

This module is designed as B/S structure, through which Content Provider can login the RA server by a browser easily. In contrast to C/S structure, in B/S structure, user needs not install special client software, and server software can be upgraded and maintained conveniently.

To make the right info management flexible and scalable, three configuration files are defined, which can be assembled and configured according to practical applications.

1) RightGroupInfo = {RightGroupID, RightGroupRegisterTime,
        RightGroupRegisterUser, RightGroupName,RightGroupType,
        RightGroupProperty,RightGroupPrice,RightGroupModifyTime,
        RightGroupModifyUser }
2) RightMedaData = { Play, Copy, Rework, Print }
3) RightsInfo = { URI, FileName, RightsGroupID,
        RightMedadata, UserGroup,RightsRegisterTime,
        RightsRegisterUser, RightsModifyTime,RightsModifyUser,
        FileType, FileTimeStamp }

URI is the abbreviation of Unified Resource Identifier.

PDRM system adopts XrML (Extensible Rights Makeup Language) to express and standardize rights, which includes four important elements: principle, right, resource and condition. On the basis of them, this module builds a template that contains the license metadata and semantic. The XrML template has a standard structure and syntax that is conduced to data interchange among systems with different structures. By the XrML template, XrML file can be created automatically.

3.2.3 PKI-Based Bidirectional Identity Authentication

To secure the rights purchase process, the bidirectional identity authentication based on PKI is needed between peer-user and RA. The public key of CA is 2048 bits, and the RSA public key of peer user and RA is 1024 bits. Hash algorithm is SHA-1, and signature algorithm is RSA. The interaction between peer-user and RA is:

1) Connecting with Ipsec or SSL: negotiating about the parameters of authentication, such as the pattern of authentication, and session key (Ks).

2) User send RA the $E_{Ks}(C_u, UA_{name}, URI, R_{id})$, where the C_u is identity certificate of user agent, R_{id} is ID of right selected by user.

3) RA verifies peer user's identity certificate, and send user the $E_{Ks}(C_r, RA_{name})$, where the C_r is identity certificate of RA.

4) User verifies RA's identity certificate.

3.2.4 Right Certificate Creating and Management

When the bidirectional identity authentication is successful, RA will query Users Manager about whether the peer-user has enough balance and credit to purchase right. If yes, RA will create right certificate and distribute it to peer-user through IPsec.

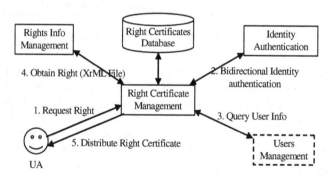

Fig. 7. Right Certificate Creating and Management

The XrML right file is obtained from Rights Info Management module, after that, right certificate will be created:

1) Signing XrML right file to ensure the integrity of right certificate:

$$SR = Sign_{Kr}^{-1}(Hash\ (R)).\tag{2}$$

2) Encrypting SR, XrML file and KEY to ensure the confidentiality of right certificate:

$$ER = E_{Ku}(R, SR, KEY).\qquad(3)$$

After creating is successful, the right certificate is saved into database with the pattern {CerSN, UAname, IssueDate, Rid, URI}.

3.3 Content Player Supporting DRM

In system, peer user could play downloaded contents by a player supporting PCP. When peer-user received the right certificate from RA, the validity of right certificate needs to be verified by judging whether (R, SR) = DKu(ER) and DKr(SR) is equal to Hash(R). When peer user begins to play contents, Right certificate will be checked. If right certificate does not exist, peer-user will be inducted to purchase it. Otherwise, the player opens it and verifies details. And the player will decrypt the content by AES algorithm to play. During the playing, the right certificate needs to be updated after the content is played.

4 Conclusion

In this paper, we propose and implement a PDRM system, a novel DRM system based on peer-to-peer networks for content providers, such as the film and television industry. This system focuses on the content security, rights management and access control, which forms the whole trust model of PDRM system. The content security is carried out through AES encryption, RSA signature, packaging and SSL/IPsec. The rights management with the distributed-centralized structure is more controllable and can reduce the burden of networks bandwidth and server computing. And the downloaded content can be played only with the rights control. The access control demands every user in the PDRM system to register, and needs identity authentication during important interactions such as right purchasing. The implemented PDRM system shows that it can provide a more robust Intellectual Property protection solution for P2P content delivery.

References

1. Qiong, L., Reihaneh, S., Nicholas, P.S.: Digital Rights Management for Content Distribution. In: Proceedings of the Australasian Information Security Workshop Conference on ACSW Frontiers 2003, vol. 21, pp. 49–58. Australian Computer Society, Australia (2003)
2. Guofei, G., Shipeng, L., Shiyong, Z.: PLI: A New Framework to Protect Digital Content for P2P Networks. In: Zhou, J., Yung, M., Han, Y. (eds.) ACNS 2003. LNCS, vol. 2846, pp. 206–216. Springer, Heidelberg (2003)
3. Yinyan, Y., Tang, Z.: A Survey of the Research on Digital Rights Management. Chinese Journal of Computers 28, 1221–1225 (2005)

4. Tetsuya, I., Takehito, A., Kiyoshi, U., Hiroshi, S.: A DRM system suitable for P2P content delivery and the study on its implementation. In: Proceedings of the 9th Asia-Pacific Conference on Communications, Penang, Malaysia, pp. 806–811 (2003)
5. Rosenblatt, B.: Integrating DRM with P2P Networks (2003), http://www.drmwatch.com/resources/whitepapers/article.php/3112631
6. Paul, J., Mostafa, A.: The Benefits and Challenges of Providing Content Protection in Peer-to-Peer Systems. In: Proceedings of the International Workshop for Technology, Economy, Social and Legal Aspects of Business Models for Virtual Goods 2003, pp. 22–24. Nova Science Publishers, New York (2003)

Author Index

Lecture Notes in Artificial Intelligence (LNAI)

Vol. 4651: F. Azevedo, P. Barahona, F. Fages, F. Rossi (Eds.), Recent Advances in Constraints. VIII, 185 pages. 2007.

Vol. 4648: F. Almeida e Costa, L.M. Rocha, E. Costa, I. Harvey, A. Coutinho (Eds.), Advances in Artificial Life. XVIII, 1215 pages. 2007.

Vol. 4635: B. Kokinov, D.C. Richardson, T.R. Roth-Berghofer, L. Vieu (Eds.), Modeling and Using Context. XIV, 574 pages. 2007.

Vol. 4632: R. Alhajj, H. Gao, X. Li, J. Li, O.R. Zaïane (Eds.), Advanced Data Mining and Applications. XV, 634 pages. 2007.

Vol. 4629: V. Matoušek, P. Mautner (Eds.), Text, Speech and Dialogue. XVII, 663 pages. 2007.

Vol. 4626: R.O. Weber, M.M. Richter (Eds.), Case-Based Reasoning Research and Development. XIII, 534 pages. 2007.

Vol. 4617: V. Torra, Y. Narukawa, Y. Yoshida (Eds.), Modeling Decisions for Artificial Intelligence. XII, 502 pages. 2007.

Vol. 4612: I. Miguel, W. Ruml (Eds.), Abstraction, Reformulation, and Approximation. XI, 418 pages. 2007.

Vol. 4604: U. Priss, S. Polovina, R. Hill (Eds.), Conceptual Structures: Knowledge Architectures for Smart Applications. XII, 514 pages. 2007.

Vol. 4603: F. Pfenning (Ed.), Automated Deduction – CADE-21. XII, 522 pages. 2007.

Vol. 4597: P. Perner (Ed.), Advances in Data Mining. XI, 353 pages. 2007.

Vol. 4594: R. Bellazzi, A. Abu-Hanna, J. Hunter (Eds.), Artificial Intelligence in Medicine. XVI, 509 pages. 2007.

Vol. 4585: M. Kryszkiewicz, J.F. Peters, H. Rybinski, A. Skowron (Eds.), Rough Sets and Intelligent Systems Paradigms. XIX, 836 pages. 2007.

Vol. 4578: F. Masulli, S. Mitra, G. Pasi (Eds.), Applications of Fuzzy Sets Theory. XVIII, 693 pages. 2007.

Vol. 4573: M. Kauers, M. Kerber, R. Miner, W. Windsteiger (Eds.), Towards Mechanized Mathematical Assistants. XIII, 407 pages. 2007.

Vol. 4571: P. Perner (Ed.), Machine Learning and Data Mining in Pattern Recognition. XIV, 913 pages. 2007.

Vol. 4570: H.G. Okuno, M. Ali (Eds.), New Trends in Applied Artificial Intelligence. XXI, 1194 pages. 2007.

Vol. 4565: D.D. Schmorrow, L.M. Reeves (Eds.), Foundations of Augmented Cognition. XIX, 450 pages. 2007.

Vol. 4562: D. Harris (Ed.), Engineering Psychology and Cognitive Ergonomics. XXIII, 879 pages. 2007.

Vol. 4548: N. Olivetti (Ed.), Automated Reasoning with Analytic Tableaux and Related Methods. X, 245 pages. 2007.

Vol. 4539: N.H. Bshouty, C. Gentile (Eds.), Learning Theory. XII, 634 pages. 2007.

Vol. 4529: P. Melin, O. Castillo, L.T. Aguilar, J. Kacprzyk, W. Pedrycz (Eds.), Foundations of Fuzzy Logic and Soft Computing. XIX, 830 pages. 2007.

Vol. 4520: M.V. Butz, O. Sigaud, G. Pezzulo, G. Baldassarre (Eds.), Anticipatory Behavior in Adaptive Learning Systems. X, 379 pages. 2007.

Vol. 4511: C. Conati, K. McCoy, G. Paliouras (Eds.), User Modeling 2007. XVI, 487 pages. 2007.

Vol. 4509: Z. Kobti, D. Wu (Eds.), Advances in Artificial Intelligence. XII, 552 pages. 2007.

Vol. 4496: N.T. Nguyen, A. Grzech, R.J. Howlett, L.C. Jain (Eds.), Agent and Multi-Agent Systems: Technologies and Applications. XXI, 1046 pages. 2007.

Vol. 4483: C. Baral, G. Brewka, J. Schlipf (Eds.), Logic Programming and Nonmonotonic Reasoning. IX, 327 pages. 2007.

Vol. 4482: A. An, J. Stefanowski, S. Ramanna, C.J. Butz, W. Pedrycz, G. Wang (Eds.), Rough Sets, Fuzzy Sets, Data Mining and Granular Computing. XIV, 585 pages. 2007.

Vol. 4481: J. Yao, P. Lingras, W.-Z. Wu, M. Szczuka, N.J. Cercone, D. Ślęzak (Eds.), Rough Sets and Knowledge Technology. XIV, 576 pages. 2007.

Vol. 4476: V. Gorodetsky, C. Zhang, V.A. Skormin, L. Cao (Eds.), Autonomous Intelligent Systems: Multi-Agents and Data Mining. XIII, 323 pages. 2007.

Vol. 4460: S. Aguzzoli, A. Ciabattoni, B. Gerla, C. Manara, V. Marra (Eds.), Algebraic and Proof-theoretic Aspects of Non-classical Logics. VIII, 309 pages. 2007.

Vol. 4457: G.M.P. O'Hare, A. Ricci, M.J. O'Grady, O. Dikenelli (Eds.), Engineering Societies in the Agents World VII. XI, 401 pages. 2007.

Vol. 4456: Y. Wang, Y.-m. Cheung, H. Liu (Eds.), Computational Intelligence and Security. XXIII, 1118 pages. 2007.

Vol. 4455: S. Muggleton, R. Otero, A. Tamaddoni-Nezhad (Eds.), Inductive Logic Programming. XII, 456 pages. 2007.

Vol. 4452: M. Fasli, O. Shehory (Eds.), Agent-Mediated Electronic Commerce. VIII, 249 pages. 2007.

Vol. 4451: T.S. Huang, A. Nijholt, M. Pantic, A. Pentland (Eds.), Artifical Intelligence for Human Computing. XVI, 359 pages. 2007.

Vol. 4442: L. Antunes, K. Takadama (Eds.), Multi-Agent-Based Simulation VII. X, 189 pages. 2007.

Vol. 4441: C. Müller (Ed.), Speaker Classification II. X, 309 pages. 2007.

Vol. 4438: L. Maicher, A. Sigel, L.M. Garshol (Eds.), Leveraging the Semantics of Topic Maps. X, 257 pages. 2007.

Vol. 4434: G. Lakemeyer, E. Sklar, D.G. Sorrenti, T. Takahashi (Eds.), RoboCup 2006: Robot Soccer World Cup X. XIII, 566 pages. 2007.

Vol. 4429: R. Lu, J.H. Siekmann, C. Ullrich (Eds.), Cognitive Systems. X, 161 pages. 2007.

Vol. 4428: S. Edelkamp, A. Lomuscio (Eds.), Model Checking and Artificial Intelligence. IX, 185 pages. 2007.

Vol. 4426: Z.-H. Zhou, H. Li, Q. Yang (Eds.), Advances in Knowledge Discovery and Data Mining. XXV, 1161 pages. 2007.